T3-BVC-448

The Ecology of Fishes on Coral Reefs

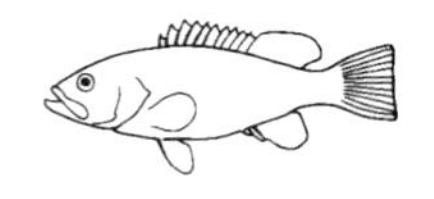

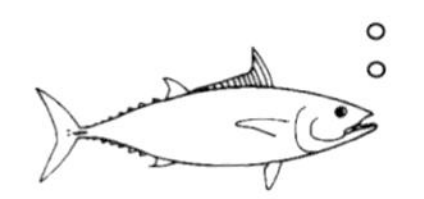

The Ecology of Fishes on Coral Reefs

Edited by
Peter F. Sale
Department of Zoology
University of New Hampshire
Durham, New Hampshire

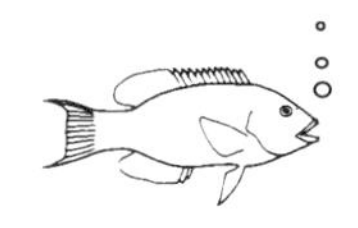

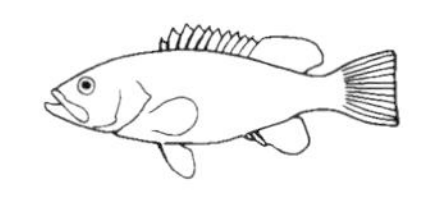

ACADEMIC PRESS, INC.
Harcourt Brace Jovanovich, Publishers
San Diego New York Boston London Sydney Tokyo Toronto

Cover photograph: A school of masked butterflyfish, *Chaetodon lunula*, before a Hawaiian coral reef. Photograph by Edmund Hobson.

This book is printed on acid-free paper. ♾

Academic Press, Inc.
San Diego, California 92101

United Kingdom Edition published by
Academic Press Limited
24–28 Oval Road, London NW1 7DX

Library of Congress Cataloging-in-Publication Data

The Ecology of fishes on coral reefs / [edited by] Peter F. Sale.
p. cm.
Includes index.
ISBN 0-12-615180-6
1. Marine fishes--Ecology. 2. Coral reef ecology. I. Sale, Peter F.
QL620.E26 1991
597'.0526367--dc20 91-12105
CIP

PRINTED IN THE UNITED STATES OF AMERICA

91 92 93 94 9 8 7 6 5 4 3 2 1

Contents

CHAPTER 16. Patterns and Processes in the Distribution of Coral Reef Fishes

David McB. Williams

CHAPTER 17. Predation as a Process Structuring Coral Reef Fish Communities

Mark A. Hixon

CHAPTER 18. Tropical and Temperate Reef Fishes: Comparison of Community Structures

Alfred W. Ebeling and Mark A. Hixon

CHAPTER 19. Reef Fish Communities: Open Nonequilibrial Systems

Peter F. Sale

PART VI FISHERIES AND MANAGEMENT

CHAPTER 20. Coral Reef Fisheries: Effects and Yields

Garry R. Russ

CONTRIBUTORS

Numbers in parentheses indicate the pages on which the authors' contributions begin.

D. R. Bellwood (39), Department of Marine Biology, James Cook University, Townsville, Queensland 4811, Australia

J. H. Choat (39, 120), Department of Marine Biology, James Cook University, Townsville, Queensland 4811, Australia

Peter J. Doherty (261), Australian Institute of Marine Science, Townsville, Queensland 4810, Australia

Alfred W. Ebeling (509), Department of Biological Sciences and Marine Science Institute, University of California at Santa Barbara, California 93106

D. J. Ferrell (156), Fisheries Research Institute, Cronulla, New South Wales 2230, Australia

Mark E. Hay (96), University of North Carolina at Chapel Hill, Institute of Marine Sciences, Morehead City, North Carolina 28557

Mark A. Hixon (475, 509), Department of Zoology and College of Oceanography, Oregon State University, Corvallis, Oregon 97331

Edmund S. Hobson (69), Southwest Fisheries Science Center, Tiburon Laboratory, Tiburon, California 94920

G. P. Jones (156, 294), Department of Zoology, University of Auckland, Auckland, New Zealand

Jeffrey M. Leis (183), Division of Vertebrate Zoology, The Australian Museum, Sydney, New South Wales 2000, Australia

William N. McFarland (16), Catalina Marine Science Center, University of Southern California, Avalon, California 90704

D. Ross Robertson (356), Smithsonian Tropical Research Institute, Balboa, Republic of Panama (APO Miami, 34002-0011, U.S.A.)

Garry R. Russ (601), Department of Marine Biology, James Cook University, Townsville, Queensland 4811, Australia

Peter F. Sale (3, 156, 564), Department of Zoology, University of New Hampshire, Durham, New Hampshire 03824

Douglas Y. Shapiro (331), Department of Marine Sciences, University of Puerto Rico, Mayagüez, Puerto Rico 00709

R. E. Thresher (401), CSIRO Division of Fisheries, Hobart, Tasmania 7001, Australia

Benjamin C. Victor (231), Department of Biological Sciences, University of California at Santa Barbara, Santa Barbara, California 93106

Robert R. Warner (387), Department of Biological Sciences and Marine Science Institute, University of California at Santa Barbara, Santa Barbara, California 93106

David McB. Williams (437), Australian Institute of Marine Science, Townsville, Queensland 4810, Australia

FOREWORD

Some of the most delightful hours of my scientific career have been spent studying reef fishes. To a biologist, scuba diving over a coral reef is roughly like being able to fly through a tropical rain forest. Indeed, in some ways it is better than that, since many groups of fishes carry out their daily activities in plain sight—while the most interesting organisms in the forest are usually concealed or move so rapidly that they are difficult to observe. And, of course, most reef fishes are very beautiful, a bonus for those who investigate their behavior and ecology. In recent decades, numerous biologists have been attracted by these and other benefits to the study of reef fish ecology, and, as this volume demonstrates, they have made a great deal of progress. Scientifically, the understanding of the relationships of reef fishes to each other and to their environments has grown by leaps, and the reef fish system promises to become a standard system for testing ideas in ecology. Such an aquatic system is badly needed because of the bias toward terrestrial systems that characterizes much modern work in ecology.

Sadly, though, all is not bright in the world of reef fish ecology. Like the tropical forests, these ecosystems are under threat. An ever-growing human population, consuming increasing amounts of Earth's bounty per person and often using ecologically malign technologies to supply that consumption, now threatens to destroy these intricate and important ecosystems. Some threats are local, such as the destruction of coral heads, siltation of reefs by soil erosion caused by poor land-use practices ashore, or direct injury caused by the anchors of recreational boats and the fins of passing snorkelers in areas of high tourist activity. Other effects tend to be regional, such as overfishing of reefs by local populations for food or aquarium specimens. Still others have potentially global causes. These include the possible impacts of climatic change, ozone depletion, or the general toxification of the planet with pesticides and other synthetic poisons. The well-publicized problems of coral bleaching in the Caribbean, possibly associated with a warming of the water, may be a harbinger.

With the human population "scheduled" to double in the next half century, and with politicians speaking of a five- or tenfold increase in economic activity, all these threats to the integrity of reef ecosystems (and thus of reef fish communities) seem certain to intensify.

Understanding the ecology of reef fishes is therefore all the more important. Increased knowledge may permit the fishes and the reefs with which they are associated to serve as an early-warning mechanism for deleterious environmental changes in marine systems. Increased knowledge will also make the reefs all the more attractive to both biologists and the growing numbers of enthusiastic amateur reef fish watchers. The more attention paid to the fishes, the greater is the chance of saving them and their (and our) environment.

So this is a particularly auspicious time for this volume to appear. *The Ecology of Fishes on Coral Reefs* can provide a foundation on which to build future research. It also can be seen as a toolbox of arguments for convincing decision makers and the public that such fascinating creatures must be preserved. Not only does the fish reef system have great scientific value, but, if it is preserved, our grandchildren and great-grandchildren can derive great delight from understanding its operation.

Paul R. Ehrlich
Bing Professor of Population Studies
Stanford University

PREFACE

The idea behind this book was first articulated in a conversation I had with Howard Choat early in 1987. I believed then, and still do now, that scientists studying the ecology of fishes on coral reefs had reached that stage in understanding that would justify a book to summarize what was known, and make it readily available to a wider audience. The fact is that significant new ecological insights have been gained by workers in this area in the past few years—ones that are of general interest to ecologists working with other groups. Yet the press of an ever-increasing literature makes it more and more difficult for any scientist to read much beyond his or her primary focus. What I intended was a book about the ecology of reef fish, but written for the wider ecological audience.

Ecological studies of coral reef fishes have made immense strides in the past 20 years and have had several significant impacts on the thinking of ecologists working in other systems and with other organisms. Among the most significant impacts have been: the development of nonequilibrial models of community organization, new emphasis on the role of recruitment variability in structuring local assemblages, development and empirical testing of evolutionary models of social organization and reproductive biology, and new evolutionary, behavioral, ecological, and physiological insights into predator–prey, and plant–herbivore interactions. I believe strongly that the approaches which reef fish ecologists have used in their empirical studies have been particularly powerful, and that this methodology, as well as the results achieved, warrant a wider audience.

Early in June 1987 I wrote to about a dozen prospective contributors, found they shared my belief in the need for such a book, and began to put the framework of this book together. My decision to then move half-way around the world, from the University of Sydney to the University of New Hampshire inevitably intruded and slowed progress! Not until mid-1988 was I able to present a detailed plan to Academic Press. By early December I was able to write to all intending contributors, giving instructions and deadlines, and two years later the manuscript was at the publishers.

The resulting book is comprehensive, but not uniform in coverage, because its content has been dictated by the shape of the current research effort, and by a decision to limit coverage of reef fish behavior. (Study of the behavioral biology of coral reef fishes has been about as extensive as that of their ecology!) The book is divided into six sections, the first of which provides much of the background information needed by readers not familiar with fish or coral reefs. Chapter 1, Introduction, includes a brief outline of content of each section. Subsequent sections deal with feeding ecology, early life history and recruitment, social and reproductive ecology, community organization, and reef fish management. A single reference section completes the book. I would like here to thank the contributors for their enthusiastic and usually prompt responses to my demands, and especially, Donna Sale, my wife, for her considerable help in many phases of the proofing and editing.

This book provides a comprehensive, and up-to-date review of the ecology of coral reef fishes. The contributors, about half each from North America and Australia, provide a healthy diversity of perspectives on a field which has not been without its differences of opinion. They include many of the most active participants in this area of research, and they bring a wealth of direct experience in their writings. Each chapter aims to review one aspect of reef fish ecology and is written for an audience of professional ecologists and graduate students. The emphasis is on current understanding of the ecology of this system, the methods used in gaining that understanding, and the wider significance for the science of ecology. Authors have been encouraged to be brave in pointing toward future developments in the areas covered as a way of stimulating new research by others. My hope is that this book will find its way onto the bookshelves of ecologists who do not study reef fish, into the reading lists of graduate seminar courses, and above all, into the hands of graduate students setting out to take our understanding of reef fish ecology further.

Peter F. Sale

Part I

Basics

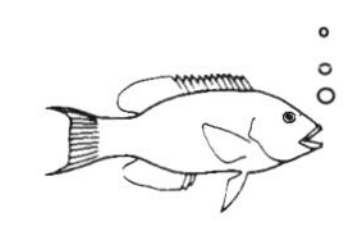

CHAPTER 1

Introduction

Peter F. Sale
Department of Zoology
University of New Hampshire
Durham, New Hampshire

I. Ecology of Coral Reef Fishes

A. A Very Short History

Study of the ecology of coral reef fishes began a relatively short time ago. While a few individuals had attempted observational work much earlier, it was the development of SCUBA and the access which that gave to the reef environment which opened up the possibilities for significant ecological study of this system. Thus the earliest papers date from the mid- and late-1950s, and interest exploded during the 1960s and 1970s. Attempts to review the topic first appeared in the 1970s, and four significant reviews have appeared since (Ehrlich, 1975; Goldman and Talbot, 1976; Sale, 1980a; Doherty and Williams, 1988a).

With such a short history of endeavor, why devote a book to the topic? Why indeed, except that the work which has been done has been particularly fruitful and seems worthy of exposure to a wider audience. As I will summarize briefly, reef fish offer the ecologist the opportunity to explore his or her questions in a reasonably complex system, but one which is particularly suitable for experimental manipulation, and one which permits operating at spatial and temporal scales which are comfortable for the scientist and the granting agency, while still being ecologically useful. In addition, while it would be a mistake to present reef fish ecology as narrowly focused on one or two topics, community-level studies have been preeminent, during a time when community ecology has been undergoing a major paradigm shift from the MacArthurian focus on interspecific competition as the primary organizing mechanism toward a much more flexible view of how assemblages of species are put together and maintained (see references in Strong *et al.*, 1984a;

Diamond and Case, 1986; Roughgarden *et al.*, 1989). Work on coral reef fishes has played a role in this change of perspective, as has some other coral reef ecology (Sale, 1988b). This book will serve as a statement of the present condition in this field.

B. Fascinating Richness and Complexity

The first impression gained by the biologist who dons fins and facemask to look at a coral reef is one of fantastic movement and color. A large part of this effect is due to the abundance and variety of small fishes which appear to swim about, every which way, just centimeters from reach. In time one recognizes that these fishes belong to a large, but still finite set of species, with predictable morphology and behavior, and that far from swimming randomly about, their distribution and their behavior are predictable from day to day, or from place to place. It is also true that they remain a bewilderingly rich set of types to anyone accustomed to fish faunas in other kinds of places.

The hundreds, or sometimes thousands, of species of fish present on a coral reef make these as rich or richer than any other environment for fish on earth. The wide taxonomic range represented—mostly higher perciform families, but including some lower teleosts as well as the chondrichthian sharks, skates, and rays—contrasts these very broadly diverse assemblages to the strangely rich (hundreds of species of fish) African rift lake faunas which are dominated by species of a single family, the Cichlidae. In addition, the broad distribution of many species, most genera, and all families means that reef fish faunas, while they vary in richness and in species composition from place to place, have a uniformity around the world which suggests that their study may lead to principles of some generality for a circumtropical fauna. By contrast, the African rift lakes are each separate pockets of speciation, and ecological patterns and processes can be different from one to another.

Admittedly, the richness of any reef fish assemblages is trivial when compared to that of most insect faunas (Ehrlich, 1975), or to that of the molluscs or polychaete annelids of the same coral reefs. Yet the fish are, in many ways, more amenable to study than are these other reef organisms. Nor is their richness just a taxonomic richness. There is a diversity of form, of habits, of relationships which opens up possibilities for study at the same time that it amazes with the possibilities for existence as a fish on a coral reef.

C. An Eminently Manipulable System

While the diversity of coral reef fish has attracted attention, I believe that other aspects of the nature of coral reefs have made their fishes so attractive as subjects for ecological study. Ecologists have been less prone than other life scientists to stress the values of particular "systems" for doing their research. Yet it is undeniable that ecological progress has been easier in some systems than in others. Our understanding of the populational and community processes in freshwater phyto- and zooplankton assemblages is far ahead of our understanding of these processes in their marine counterparts, perhaps only because of the larger scale, the cost, and the difficulty of staying with a group of oceanic plankton while they "do" their ecology. Our understanding of the ecology of the rocky intertidal is perhaps ahead of our understanding of soft sediment shores only because the residents of the rocky shore live on its surface and are readily accessible to the ecologist, while those of the mud flat live out of sight in a shifting, semifluid medium. Our understanding of the ecology of insects is well ahead of our understanding of the ecology of the trees on which some of them feed, partly because insects live out their lives on spatial and temporal scales more compatible with the usual funding patterns for ecological research. In each of these respects—mobility, accessibility, and temporal and spatial scale of processes—coral reef fishes are an excellent system with which to work.

1. Mobility and Scale

Coral reef fishes are mainly small and sedentary. The great majority of species are "aquarium" species less than 30 cm in maximum length, and size frequency distributions of local assemblages are dominated by the smaller individuals and species. Indeed, in some dominant reef fish families, such as the Pomacentridae (damselfishes) or the Labridae (wrasses), the larger species are preponderantly temperate zone outliers. While these small fishes vary in the degree to which they move about, it is generally true that reef fishes are more sedentary than are other comparably sized vertebrates (Sale, 1978b). One contributing reason for this is, undoubtedly, that they live in a highly structured environment produced by the complex architecture of the corals, and an environment which differs markedly in structure from place to place on scales of meters. Whatever the cause, the result is that reef fish provide rich assemblages of small species which operate on scales easily encompassed by the mobility of the average diver.

In presenting this picture of the small, sedentary reef fish, I have omitted mentioning that virtually all species have a dispersive pelagic larval phase lasting several days or weeks, during which they may travel considerable distances (Sale, 1980a). This phase of the life history, with its considerably larger spatial scale was ignored by reef fish ecologists until the mid-1970s, but as several ensuing chapters make clear, it is now being given the attention it requires. Nonetheless, reef fish are relatively sedentary through most of their lives.

2. *Accessibility*

Reef fish are obviously accessible to the diver, and diving is a simpler process in the shallow, warm, clear waters of a coral reef than in most other aquatic environments. With modern equipment and a little experience, the diving ecologist has much the same freedom, ease of movement, and comfort over a reef that he or she might have on a forest floor. Like birds in a forest, reef fishes are active animals which continue their activities in the presence of a diver. Individuals can be watched and followed. Indeed, it is probably easier to follow a reef fish about than it is to follow a bird through a forest, because it is easier to float over a reef than to creep through the bushes beneath the trees! A number of reef species, particularly of the families Holocentridae (squirrelfishes), Apogonidae (cardinalfishes), and Haemulidae (grunts), are primarily nocturnally active, and many of the smaller species, especially among the Gobiidae (gobies) and Eleotridae (gudgeons), and some larger ones among families like the Muraenidae (moray eels), are sufficiently cryptic, moving about within the reef's numerous caverns and crevices, that their activities are not easily accessible to divers. Less has been learned about these species, but they constitute a minority of reef fish. In addition, some species are limited to deeper reef environments in which the time of a diver using conventional SCUBA is quite limited. These also have received less attention.

My own prejudice has always been to focus on ecology by first watching what individuals do. One can learn more, and more quickly, about interspecific interactions, dietary habits, habitat preferences, and special adaptive traits by watching individual animals going about their business. This approach is well rewarded on the coral reef. Indeed, the biologists who first began in the 1950s and 1960s to use SCUBA to study reef fishes were interested in their behavior. Modern reef fish ecology has grown from this behavioral beginning, rather than from the earlier ecological and fisheries management studies which used traditional trapping and tagging approaches to get indirect information. It is also true that this "fish watching" origin of reef fish ecology has helped maintain its separation from fisheries biology, with separate approaches, separate questions and hypotheses, and separate literatures that are only weakly cross-referenced. Nevertheless, it is abundantly demonstrated in the following chapters that this "fish watching" approach to ecology has been beneficial in generating new insights at the individual, populational, and community scales of ecological investigation. Even our understanding of nutrient cycling in reef environments has been enhanced by this "fish watching" approach (Robertson, 1982; Meyer *et al.*, 1983)!

As well as being visually accessible, reef fish have scales and rates of movement which leave them physically accessible to the diving ecologist. True, some are harder to capture than others, but diving ecologists carry a long list of tricks which, used appropriately, can capture any species alive (though not always any individual) or provide a sample of preserved specimens.

In addition, the reef environment is one which facilitates access to the fishes and their ecology. Many studies reported in subsequent chapters will make use of particular coral reef features in their design. In particular, reef fish ecologists have used small patch reefs, either naturally occurring or artificial ones constructed from either concrete blocks or natural reef materials, as separate replicates in their designs. Patch reefs are a common formation in protected shallow waters and, naturally separated by expanses of open water offering little cover, they provide the ecologist with essentially discrete groups of fish living their lives on a single patch reef to which they recruited after settlement from the plankton.

The hard substrata of coral reefs are among the softest rocks in the world, making life easy for the ecologist who wishes to manipulate the environment, whether to mount a camera to record activity (Smith and Tyler, 1973b), fasten down a cage to exclude predators (Doherty and Sale, 1986), or put a patch of living coral where none existed before in order to attract settling fish (Williams and Sale, 1981; Sale, 1985). There are few rocky habitats in which an ecologist would contemplate removing half the living space to test an idea, in the way Robertson (1984) did when he carted away half a patch reef to explore whether the pomacentrids living there were limited by the availability of space in which to establish their territories! The nature of this environment has combined with the accessibility of the fish to foster a strongly manipulative approach to ecology.

3. *An Appropriate Scale*

I have already touched on the spatial scales used by reef fish. They live, as juveniles and adults in shallow seas, close to a solid substratum, and usually also close to land. Suitable study sites are often accessible to the diver from shore, and certainly from a small boat. Only questions directly concerning the larval biology require being at sea, and much of this work can be done with a small boat or by "island-hopping" to sample the coastal waters at separate locations in an archipelago or along a coast.

Temporal scales are also appropriate. Reef fish with annual life cycles are known, and many live only a few years. Still, a surprising number of species are turning out to have longevities of a decade or more—a fact which must be taken into account when evaluating the results of manipulative studies only a couple of years in duration. Despite this, these temporal scales are far closer to our own than are those of trees and corals, on one hand, or microorganisms, on the other. We are likely to be able to obtain useful findings from studies with the durations commonly dictated by funding and graduate education patterns.

II. A Crash Course for the Nonspecialist

A. A Book for Ecologists Who Work in Other Systems

My intention is that this book be accessible to ecologists who work in other systems, and that each chapter be comprehensible on its own. For this reason, there is some overlap in ideas presented, although usually with different perspectives in each case. Some use of specialized terminology is unavoidable, including use of numerous Latin names for species. The fact is that there are no widely accepted common English names for most coral reef fishes. The following brief comments are intended to help the reader get started.

B. Reef Fish Taxonomy

The fish of coral reefs, and certainly the ones which have received most attention from ecologists, are advanced perciform teleosts, the product of a dramatic radiation during the early Tertiary. Their radiation was apparently in conjunction with the radiation of modern scleractinian corals which took place at the start of the Tertiary. The most characteristic groups, in the sense of being most completely associated with coral reef environments, are:

1. three labroid families: the Labridae or wrasses, the Scaridae or parrotfishes, and the Pomacentridae or damselfishes;
2. three acanthuroid families, the Acanthuridae or surgeonfishes, the Siganidae or rabbitfishes, and the Zanclidae or moorish idols containing in its single genus, *Zanclus,* perhaps the archetypical coral reef fish;
3. two chaetodontoid families, the Chaetodontidae or butterflyfishes and the Pomacanthidae or angelfishes. Species of these families may vie with the moorish idol for title of Archetype of the Reef!

Numerous other families are important on coral reefs, and while a great amount of ecological work has been done on members of the 8 families mentioned, there have been studies of members of many other families besides. Table 1 of Chapter 8 lists all 100 families known to have coral reef representatives, classifying them in terms of reproductive mode. The most important fishes, in the sense that they are both seen on reefs and studied by ecologists, are the following:

1. the speciose and abundant Blenniidae (blennies) and Gobiidae (gobies), characteristically strongly demersal and site-attached fishes;
2. the highly abundant, nocturnally active, small predators on invertebrates and smaller fishes represented by the predominantly Indo–west Pacific

Apogonidae or cardinalfishes, and the predominantly Caribbean Haemulidae or grunts;
3. the bizarre Ostraciidae (boxfishes), Tetraodontidae (puffers), and Balistidae (triggerfishes), which are never numerous, always conspicuous because of bold shapes and colors, and represent some of the most highly evolved extant teleosts;
4. abundant and important larger predators and piscivores belonging to the Holocentridae (squirrelfishes), the Serranidae (rock cods and groupers), the Lutjanidae (snappers), and the Lethrinidae (emperors).

Of these, only the Holocentridae is an older beryciform family in an otherwise overwhelmingly perciform fauna.

Despite the taxonomic diversity, it remains true that most reef fish ecology has been investigated using a very few species. These are among the commonest species on coral reefs, but not necessarily a randomly drawn sample of them.

C. Basic Life Cycles

The statement that reef fish generally have a bipartite life cycle with a dispersive larval phase will be made in many chapters of this book. Several chapters in Part 3 provide detailed ecological information about this important dispersive phase. Here I want to briefly stress the diversity of life cycles and reproductive modes found within this general plan.

Reef fish vary greatly in where, when, and how they reproduce. While the majority shed sperm and eggs in midwater (usually in pairwise spawnings), some scatter them over the substratum, some prepare and defend demersal nests, and some carry the fertilized eggs in the mouth until hatching. One remarkable pomacentrid (*Acanthochromis polyacanthus*) forms stable pairs which jointly defend first a demersal nest and then the clutch of hatchlings, foregoing the pelagic phase entirely (Robertson, 1973b; Thresher, 1985a). The Syngnathidae (pipefishes and sea horses) brood their eggs in a pouch borne by the male parent, and one species at least has an elaborate monogamous social structure, perhaps made possible by this reproductive mode (Gronell, 1984).

While coral reef fishes tend to be highly fecund iteroparous species, they produce clutches of eggs on daily, weekly, fortnightly, monthly or less frequent schedules, their clutches vary greatly in size from tens to hundred's to thousands of eggs at a time, and their juveniles, settling after the pelagic phase, have high or quite low mortality in the first couple of weeks in the reef environment. My point is that there is considerable variation on the general theme, a variation which ecologists are beginning to use in understanding the adaptiveness of reef fish life histories.

Reef fishes are also flexible in how they determine sex. In addition to conventionally (to us?) gonochoristic species in which the sex of individuals is fixed early in development, there are numerous hermaphroditic species. While some of these are simultaneous hermaphrodites, in which pair members may take turns in performing as male or female on a single day (Fischer and Petersen, 1987), most are sequential hermaphrodites. Most of the latter are protogynous (female first, then male), but some are protandrous (male first). Chapters 12 and 14 deal specifically with this feature, but the subject crops up in many other places throughout the book.

D. Coral Reefs—Special Environments

Finally, a few comments about the coral reef environment. For fishes, coral reefs are predominantly a hard substratum, but one which is topologically far more complex than any other. Reefs offer a diversity of shapes and sizes of shelter—shelter which is used by the fishes. Most coral reef fishes are diurnally active, while the remainder are active primarily at night. Virtually all retreat to shelter during the part of the day when they are inactive, and this factor alone may be responsible for the strong association between these active but demersal fishes and the structures of their environment.

Coral reefs are also a very patchy environment. At scales of hundreds of kilometers, reefs are scattered across the tropical seas. At smaller scales they each offer a series of habitat zones distinctive in physical and other features, and within each zone, at scales of meters, there exists a patchwork of physically distinctive sites due to the differing morphologies of corals of different species and to the intermixing of coral colonies with rubble, sand, and limestone platform substrata. Figure 1 diagrams the habitat zones which can be found on most reefs as a terminological aid. However, reefs can differ markedly and the fore reef slope of a reef growing in relatively protected waters is a very different kind of place than that of a reef facing the full force of the open South Pacific. Similarly, while many reef fish ecologists have made use of patch reefs, the patch reefs they have used vary greatly in size, in complexity, and in the hydrographic conditions in which they sit.

Coral reefs are also changeable, even ephemeral on geological time scales. They are the result of a balance between active growth and calcification in shallow warm waters, continued mortality and erosion caused by physical and biological agents (including some parrotfishes), and the tectonic actions of lifting, sinking, and drift of continents.

The modern Great Barrier Reef, the most expansive development of reef in the world at present, has not been around for much longer than the oldest Egyptian pyramids. Reefs began to grow in what is now the Great Barrier Reef as rising sea levels inundated an alluvial plain some 9000 years ago.

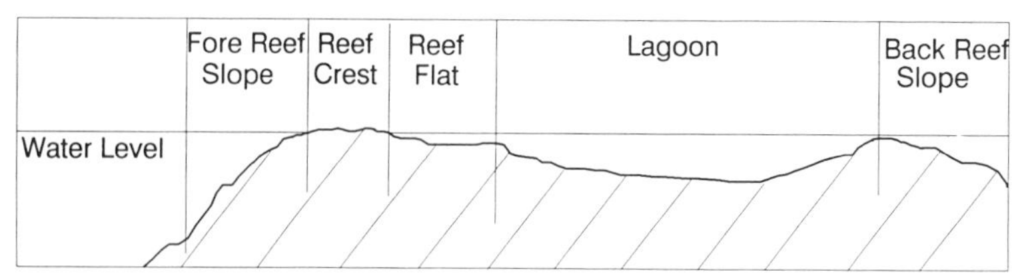

Figure 1 A cross section of a "typical" coral reef with the various physiographic zones named. Although different authors will use the same names for analogous parts of reefs in different regions, the extent and physical nature of these zones vary greatly depending on tidal range, exposure, depth of the underlying platform, and geological history.

Growing vertically at first, reefs caught up with the rising sea surface and began to develop their modern range of shallow habitats 4000 to 6000 years ago. No reefs existed along that coast of Australia for 100,000 years before that time (Davies *et al.,* 1985). It is worth keeping this changeability of coral reef habitats, as well as their structural complexity, firmly in mind when seeking to understand the ecology of the fishes which live there.

III. Structure of This Book

A. Part I: Basics

This book is organized into six sections. The first, of three chapters, including this one, is introductory. In Chapter 2, McFarland discusses the visual ability of reef fishes, and the nature of light in the world they inhabit. Vision is the primary sense used by reef fishes in their daily activities. It is worth understanding their visual sensory capabilities and the differences between their eyes and ours, as well as the special features of the light environment they occupy.

We know somewhat less about reef fish capabilities with other senses, and these are not dealt with specifically. The acoustic sense is clearly important, particularly in social interactions in some species, and useful introductions can be found in the work by Tavolga *et al.* (1981) and Hawkins and Myrberg (1983). Olfactory responses may also be more important than has been assumed. Sweatman (1988) has demonstrated experimentally a chemically mediated selection of habitat at settlement for one genus of damselfish.

In Chapter 3, Choat and Bellwood review what is known about the origins of reef fish faunas in the early Tertiary and provide an excellent summary of the phylogenetic interrelationships among this diverse fauna. They also provide a picture of what a reef fish is and how it spends its life.

B. Part II: Trophic Ecology

Part II comprises four chapters dealing with aspects of the trophic ecology of reef fishes. In Chapter 4, Hobson reviews planktivory by focusing on the morphological adaptations which have made the "plankton-picking" mode of feeding practical and on the distinctions among the diurnally active planktivores and those which are nocturnally active. The latter tend to feed on larger organisms which are not present in the water column during the daytime. The remarkably ordered day–night transition which occurs at dusk and dawn as fish either retire to or emerge from shelter is also examined.

In Chapter 5, Hay examines herbivory from the perspective of marine plants. Herbivory on coral reefs, largely by fish, is sufficiently intense that reef algae are maintained primarily as low-biomass, but highly productive turfs. Hay reviews an interesting tale of chemical defenses used by plants to minimize predation.

Choat considers herbivory from the opposite perspective in Chapter 6. He examines the morphological and other specializations which permit feeding on reef algae and gives special attention to the poorly understood role of endosymbionts in digesting this material. He also poses the question of why the most successful families of coral reef herbivores have not extended into the temperate zone.

In the final chapter in this section, Chapter 7, Jones *et al.* review the feeding ecology of those numerous reef fishes which feed upon other benthic animals. Their goal is to assess the role of predation by fishes in the ecology of the numerous benthic invertebrate groups found on coral reefs and in the evolutionary development of these groups.

C. Part III: Larval and Juvenile Ecology

Part III includes four chapters which look in different ways at the early life history of reef fishes. In Chapter 8, Leis provides a thorough review of the larval ecology of coral reef fishes. He considers their morphological and behavioral adaptations for pelagic life, their distribution horizontally and vertically while at sea, and their activities during this part of the life cycle. He also explores what is known about the processes of settlement and transition to the juvenile stage. He closes with a brief consideration of the biogeographic implications of the pelagic phase in reef fishes.

Victor focuses on the process of settlement in Chapter 9. He deals with the methods used for study of settlement timing and extent, particularly those using microstructural information in fish otoliths (small structures in the inner ear). He reviews timing, size, and age at settlement, and he uses information on age at settlement to explore biogeographic patterns in distribution of reef fishes.

Chapter 10, by Doherty, deals with variable recruitment to reef fish populations. He examines the extent of recruitment variation on several spatial and temporal scales and, like most reef fish ecologists, defines recruitment as occurring at or soon after settlement to the reef environment. His concern is for the extent to which variations in rates of recruitment determine subsequent local patterns of abundance of reef fish species. He considers causes of this variation and approaches to monitor it.

Victor and Doherty raise the possibility that it is the vagaries in the process of settlement of reef fishes which determine much of subsequent ecology in coral reefs, because these vagaries set the sizes of local cohorts. In Chapter 11, Jones carefully reviews recent ideas about the determinants of sizes of local reef fish populations, distinguishing between recruitment and various postrecruitment processes which might play a role. His chapter completes this section by taking us from the biology of the bipartite life cycle to the ecological implications of resulting recruitment variation, and then to the need for a more pluralistic view of the determinants of population size and structure. In doing this, Chapter 11 also lays the groundwork for Part V.

D. Part IV: Reproductive and Life History Patterns

Part IV completes the biological review of reef fishes with three chapters considering the reproductive and life history patterns in this group. In Chapter 12, Shapiro draws attention to the variability in social organization which exists in some reef fish species. He attributes this variation to differing mixes of individuals set up by a variable pattern of recruitment and to the differential expression of behavioral traits which this might lead to.

Robertson explores the considerable variation among species in patterns of spawning activity in Chapter 13. He develops the argument that these variations in reproductive behavior are a response to selective forces acting on adult fishes. Most extant explanations of the evolution of patterns of spawning behavior have been based on the adaptiveness of particular patterns for the resulting larvae.

In Chapter 14, Warner takes the discussion of variability in behavioral traits a step further. He returns to the intraspecific variation in social behavior, as considered by Shapiro, and makes a convincing argument for using this variation in explicit tests of evolutionary ecological theory. Taken together, his chapter and the two preceding it make a powerful case for using the richness apparent in behavioral traits and in social organization of reef fish species to test general hypotheses in evolutionary ecology.

E. Part V: Community Organization

Part V includes five chapters concerned with the organization of reef fish assemblages. In Chapter 15, Thresher attempts a comprehensive biogeographic evaluation of community, population, and behavioral ecology of reef fishes. His chapter provides an overview of the global-scale patterns in distribution of species and higher taxa, as well as an examination of some biogeographic patterns of behavioral traits—particularly traits associated with reproduction and larval life. He provides a bridge between the chapters of Parts III and IV and the rest of Part V.

In Chapter 16, Williams tackles a more limited task by describing the taxonomic patterns apparent in local assemblages at different locations within the Great Barrier Reef Province. These patterns, of some complexity, were the first broad-scale spatial patterns to be documented by adequate ecological sampling, and the same concern for methods used will be needed if the kinds of comparisons which Thresher discusses are to be taken further. Williams also considers causes of the patterns detected, and in so doing he demonstrates how scanty our knowledge is when we try to deal with broad spatial patterns in this system.

Despite the evidence of high rates of mortality among young reef fishes (Doherty and Sale, 1986), the role of predators feeding upon reef fishes has been largely neglected as a subject of ecological study. In Chapter 17, Hixon reviews the role of piscivory in determining the structure of local assemblages. At this smaller spatial scale, it is possible to use experimental approaches, and conclusions can be somewhat more firmly drawn. His chapter has important links to Chapters 10 and 11, by Doherty and Jones, respectively, as well as to those which follow it.

Chapter 18, by Ebeling and Hixon, is a thorough review of the similarities in community organization between coral reef and temperate rocky reef fishes. The message is clear: coral reef fishes do not possess an ecology which is different in kind to that of their less speciose temperate kin. In both kinds of system, a number of processes interact to create community organization, and the relative importance of processes varies in space and time.

In Chapter 19, Sale sets forth the view of coral reef fishes as inhabiting open, nonequilibrial ecological systems. His view of community organization, and in particular the plurality of organizing processes, is shared with several earlier chapters, particularly those of Jones, and Ebeling and Hixon.

F. Part VI: Fisheries and Management

Part VI includes a single chapter reviewing the role of reef fish communities in supporting fisheries. In Chapter 20, Russ introduces fisheries management concepts and approaches, considers the importance of reef fisheries, and discusses the impact of fishing on reef fish communities and populations. Despite the local importance of coral reef fisheries in many regions, their management has not received the attention it might, and Russ's closing plea for experimental testing of management schemes is both timely and an apt finish to this book. In the end, new understanding of any ecological phenomenon must find its way into resource management before it will be of real value to humanity.

CHAPTER 2

The Visual World of Coral Reef Fishes

William N. McFarland
Catalina Marine Science Center
University of Southern California
Avalon, California

I. INTRODUCTION

The activity patterns of coral reef fishes tend to be either diurnal, nocturnal, or crepuscular. Although varied senses are utilized by coral fishes, the behaviors of most species are largely dependent on vision. In this chapter emphasis is placed on the relations between the photic environment, the visual system, and the general behaviors of reef fishes. Readers interested in other senses of fishes are referred to some general works (Atema *et al.*, 1988; Davis and Northcutt, 1983; Hodgson and Mathewson, 1978) and *The Handbook of Sensory Physiology* (Vols. I–IX, 1971–1981).

II. LIGHT IN TROPICAL SEAS: ATTENUATION AND HUE

The penetration of light into tropical seas has been the subject of numerous investigations (for general reviews see Jerlov, 1968, 1976; Preisendorfer, 1976; Kirk, 1983; for studies related to vision see McFarland and Munz, 1975a; Loew and McFarland, 1990). In water the degree of light transmission along an optical path is determined by scattering and absorption by the water molecules and by dissolved and particulate matter. In tropical seas, light transmission is more the result of its interactions with water than with inclusions, because dissolved substances and particulate matter are usually present at low concentrations. As a result, the light wavelengths maximally transmitted occur between 450 and 470 nm. To a diver the ocean appears bluish along all lines of sight, except when looking upward near the surface (Fig. 1).

In shallow environments, bottom reflections affect the spectrum and light intensity along different lines of sight (Fig. 1). The presence of coral sand in

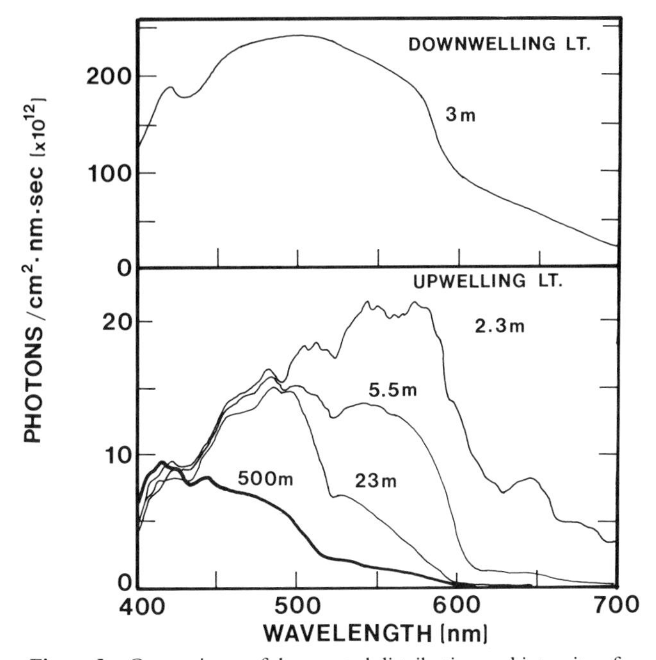

Figure 1 Comparisons of the spectral distribution and intensity of downwelling and upwelling light in the open sea and over a coral sand bottom. The spectral distribution of downwelling light 3 m beneath the sea surface is essentially independent of the bottom depth, unless particulate inclusions vary with depth (e.g., sand particles or plankton). The single curve displayed represents downwelling light at 3 m in a clear tropical sea. The upwelling irradiance spectra were obtained with the spectroradiometer detector positioned 1 m beneath the sea surface at the indicated depths. Note that maximum intensity for upwelling light in the open ocean (500 m) occurs below 450 nm and shifts toward longer wavelengths as depth to the bottom decreases. Horizontal radiance, as for downwelling light, is essentially independent of total depth and contains mostly photons below 500 nm (see Fig. 5B).

the water (produced by currents or turbulence) reduces clarity and broadens the spectrum (McFarland and Munz, 1975a). In addition, over the course of a day the excretions of reef fishes, and especially of herbivores like parrotfishes, have a similar effect. In the morning, water clarity can be exceptional and bluish when viewed horizontally, but by midafternoon the excretions can seriously reduce clarity and the hue becomes greenish (unpublished observations at Enewetak, New Guinea, and the Caribbean).

III. THE VISION OF FISHES

The eyes of all vertebrates share a similar generalized morphology (Walls, 1967). Most fishes, for example, possess an external cornea and a lens for focusing light rays on the retina. The retina is composed of photoreceptor cells that absorb and transduce light into neural signals. These signals are transmitted to the neural retina for initial processing (bipolar, horizontal, ganglion cells, etc.) and then via the optic nerve to the brain for further analysis. Several monographs detail the anatomical, physiological, and behavioral characteristics of fish vision (Nicol, 1988; Douglas and Djamgoz, 1990; Heuter and Cohen, 1991).

The retina of vertebrates contains two distinctive types of photoreceptor cells—rods and cones, which are defined by cell morphology and distinctive responses to light intensity. Rods function best under dim light, as during nighttime—this is termed scotopic vision (scoto = darkness). Cones operate best at higher light levels, as during daytime—this is referred to as photopic vision (photo = light). At intermediate intensities, as during twilight, neither the rods nor the cones function optimally—a condition termed mesopic vision (meso = middle). All rods and cones, however, contain a photoabsorbing visual pigment molecule composed of a protein (the opsin) and a prosthetic group (the chromaphore) that is the aldehyde of either vitamin A_1 or A_2. A_1-based rod pigments are referred to as rhodopsins, and A_2 pigments are called porphyropsins.

All visual pigments have similar broad absorbance spectra that differ primarily in the wavelength at which they maximally absorb light (defined as the λ_{max}, in Fig. 2). By measuring the absorbance spectra of a fish's visual pigment, the spectral band to which the fish should be most photosensitive can be established. If only a single visual pigment is present in a fish's photoreceptor cells, different hues cannot be discriminated. This is termed monochromatic vision. When multiple pigments with different λ_{max}'s are present, the possibility for hue discrimination exists. In most instances multiple pigments are present only in cone cells and, therefore, color discrimination is associated almost exclusively with photopic conditions. Aquatic environments tend to be broadly monochromatic, that is, the prevalent backlight contains mostly photons at short (blue), middle (green to yellow), or longer wavelengths (orange to red). The absorbance spectrum of a visual pigment therefore will approximate the light spectrum characteristic of a habitat, or it will not. For example, a pigment with a λ_{max} at 470 nm is "matched" to the general "bluish" spectrum of a tropical sea, whereas a pigment with λ_{max} at 540 nm is not. In visual terms the backlighting should appear "bright" to a fish with a matched visual pigment. If the pigment is "offset" from the spectrum of the backlighting then the backlight should appear to be "dim." These two condi-

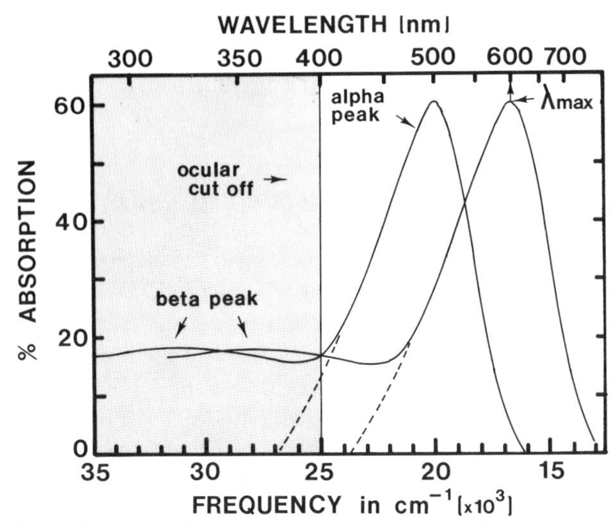

Figure 2 Comparison of the spectral absorptions of two visual pigments. The curves are plotted on a frequency scale in reciprocal centimeters (lower abscissa) and against wavelength (upper abscissa). Note that the curves are identical in shape and differ only in their spectral locations. Because of this similarity, visual pigments can be identified by the wavelength (or frequency) at which they maximally absorb light, a wavelength defined as the λ_{max}. If plotted on a linear wavelength scale the absorption curves are not identical and appear broader at higher λ_{max} values. The λ_{max} coincides with the broad primary alpha absorption band. Visual pigments can also absorb light over the shorter wavelengths associated with the very broad secondary beta absorption band. Beta-band absorptions are usually dismissed in considerations of photon capture (Govardovskii, 1976), and the pigment's absorption is assumed to follow the indicated dashed lines. In most vertebrates, however, the ocular preretinal media act as shortwave cutoff filters, as indicated for this example by the shaded area below 400 nm. Shortwave cutoff filtering tends to diminish the importance of beta-band absorptions, unless the cutoff occurs below 400 nm and/or the visual pigment is shifted toward longer wavelengths (e.g., the indicated pigment with a λ_{max} at 600 nm).

tions, as pointed out by Lythgoe (1968), can affect visual detection by enhancing the contrast of dark and bright targets. This is important in aquatic animals because absorption and scattering of light rapidly reduce the visibility of objects over short optical paths (Lythgoe, 1979). The spectral location of a fish's visual pigment(s) therefore is potentially critical to its survival.

IV. Correlations of Scotopic Vision and Light

A. Visual Pigments

1. *Variation Among Species*

Perhaps the most extensive studies of visual ecology of coral reef fishes are attempts to explain the distribution of visual pigments in terms of behavior and incident submarine light (Lythgoe, 1966, 1968, for several Mediterranean species; Munz and McFarland, 1973, 1975, 1977, and McFarland and Munz, 1975a,b, for 167 species of Pacific coral reef fishes). In these studies rod pigments were extracted with digitonin from the retinae of dark-adapted fishes and their absorbance spectra were measured.

The λ_{max} for the rod visual pigments (rhodopsins only) from the Pacific coral reef fishes ranged from 480 to 502 [mean = 493 nm ± 4.5 (SD)]. This seems narrow compared to the reported range for all freshwater and marine fishes (i.e., 467 to 551 nm). The sample of seven Mediterranean species studied by Lythgoe (1966) yielded similar values (490 to 503 nm). We concluded that the clumping of visual pigment absorption maxima at about 493 nm in coral reef fishes was adaptive and probably resulted from strong selective pressures (Munz and McFarland, 1973). Before examining this possibility, I emphasize that rod pigments serve vision mostly at low light levels, as during twilight and night, and at depths where light is dim during daytime. Knowledge of behaviors and photic conditions during these periods of the day is relevant when attempting to explain the tight clustering of the rod pigments about 493 nm.

2. *Reasons for Uniformity of Rhodopsin Characteristics*

We might expect some variation in rhodopsin features depending on whether fish are diurnally or nocturnally active. However, classifying 167 coral reef fishes by diel feeding habits reveals no correlations with spectral position of rod pigment (Table 1).

Table 1 Diel Feeding Habits and the Spectral Location of the Rod Visual Pigment for 167 Pacific Coral Reef Fishes and 12 Tropical Pelagic Species.[a]

	Most common feeding time			
	D	N	D-N	C
No. species	102	40	11	16
Percent sample	60	24	7	9
Range λ_{max}	480–506	481–502	484–499	490–501

[a] Spectral location is λ_{max} in nm. Feeding habits are: D = diurnal, N = nocturnal, D-N = feeding both day and night, and C = mostly crepuscular.

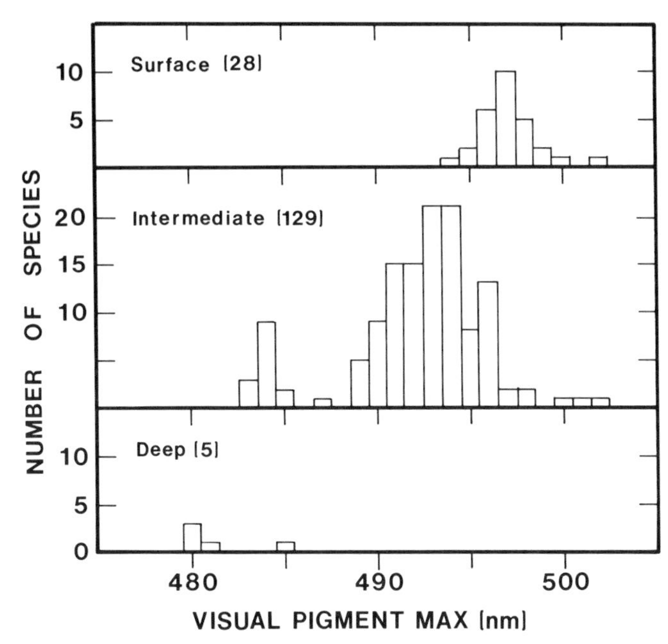

Figure 3 Frequency distribution of the rod visual pigments of Pacific coral reef fishes as a function of depth. Surface species are considered to be most common from 0 to 5 m; intermediate species from 2 to 20 m; and deeper species from 20 to 200 m. Considerable overlap exists, but the trend toward visual pigments that are spectrally positioned at shorter wavelengths is clear.

In temperate and deep-sea fishes the rod pigment is usually blue-shifted in deeper-living species. The usual interpretation, termed the Sensitivity Hypothesis, is that by matching the pigment absorbance to the prevailing bluish light, increased photoabsorption is achieved (Wald *et al.,* 1957; Denton and Warren, 1956, 1957; Munz, 1957, 1958, 1964). A similar relation with depth is present in coral reef fishes (Fig. 3; for details see Munz and McFarland, 1973).

Given the clarity of water about coral reefs it is unlikely that rod pigments serve any reef species during daytime except perhaps at depths where light intensity might hover around and below cone thresholds. If this threshold approximates 0.7 μW per cm^2, the average value to initiate light–dark adaptation in fishes (Blaxter, 1988), then scotopic vision could function efficiently only below 280 m during midday (threshold depth calculated for Type IB water; Jerlov, 1968). Most of the reef fishes investigated occur commonly above 100 m, although a few species, such as *Chromis verater, Holocanthus arcuatus,* and *Bodianus bilunulatus,* are most abundant in Hawaii at 100 m or more (Brock and Chamberlain, 1968). Earlier and later in the day the threshold for cone vision would occur at shallower depths. But calculations reveal that the threshold depth does not rise to 100 m until a few minutes before sunset and after sunrise. I infer therefore that the rod pigments of reef fishes function mostly near and during twilight and at night (exceptions may involve fishes that inhabit dark reef interstices during the day and fishes, such as wrasses, that seek cover in the reef close to sunset).

To optimize photoabsorption of downwelling light at 30 m depth during the night (also daytime), a rod pigment with a λ_{max} anywhere between 450 and 550 nm would do (Fig. 4D and F); a 550-nm pigment would provide even more photoabsorption above 30 m. However, during twilight, photons are most abundant between 450 and 500 nm, and the λP_{50} shifts only from 490 nm near the surface to 481 nm at 30 m (see Fig. 4 for explanation of P_{50}). The difference results from reductions of "green" to "orange" photons in the solar spectrum during dusk and dawn (Fig. 4; for details see Munz and

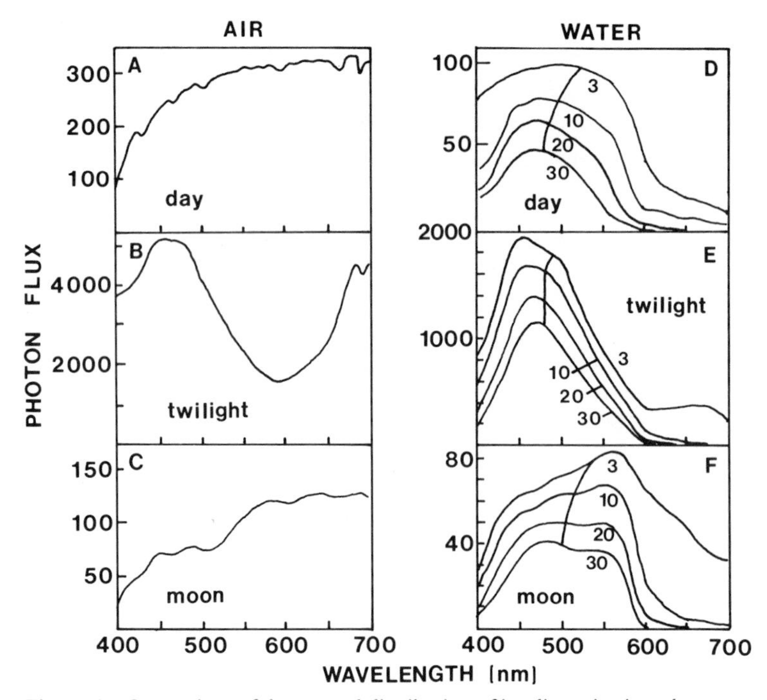

Figure 4 Comparison of the spectral distribution of irradiance in air at the sea surface and at various depths beneath the sea surface during midday, twilight, and moonlight. All irradiance values are for downwelling light, expressed as photons per cm^2 per nm per sec; scaling is photons $\times 10^{12}$ for day, $\times 10^7$ for twilight, and $\times 10^6$ for moonlight. The λP_{50} values for the underwater spectra represent the wavelengths at which the total number of photons between 400 and 700 nm are equally distributed.

McFarland, 1973, 1977, and McFarland and Munz, 1975a). To optimize photoabsorption during dusk and dawn, visual pigments with λ_{max} values between 480 and 495 nm are best—exactly the range of most reef fishes (Fig. 3). But what selective force might act? Hobson concluded in 1972 that predation before and during twilight has been and is the dominant selective force in shaping the general behaviors of coral reef fishes.

A general explanation for the clustering of the rod pigments of coral reef fishes invokes interactions between ambient light, visual constraints, behavior, and the selective power of predation. This answer endorses the Sensitivity Hypothesis. As emphasized by Lythgoe (1968), the Sensitivity Hypothesis implies that targets appear as dark silhouettes unless they are "brighter" than the background (i.e., the target radiates more photons than the background in the wave bands that include the background). The latter circumstance likely applies only for reflective targets viewed near the surface (such as the silvery flanks of a fish) and over very short ranges. Because reflective targets possess less inherent contrast than dark targets when viewed against a bright background, they are more difficult to detect with a matched pigment. Pigments with λ_{max}'s near 493 nm match the underwater spectrum at dusk and dawn and, therefore, improve a fish's ability to detect targets as dark silhouettes against a background that is dim but, nevertheless, "brighter" than the target. In depths shallower than 20 m, a pigment with λ_{max} at 530 nm would improve photoabsorption of downwelling light at night and diminish it during twilight (Fig. 4E and F). At greater depths during the night and at all depths during twilight, however, the actual λ_{max} range for the rod pigments of reef fishes is optimal.

B. Is the Spectral Position of a Rod Pigment Significant?

To compare the photon catch of visual pigments with different λ_{max} values requires summation of the light absorbed by each pigment under similar photic conditions over its total absorption spectrum (here considered to be 400 to 700 nm). Using this computational method and our 1973 study of coral reef fishes, Govardovskii (1976) calculated that rod pigments with a λ_{max} near 493 nm would improve photoabsorption by only 5 percent as compared to a pigment at 540 nm. Because these absorptive differences are small, he concluded that the clumping of rod pigments in coral fishes had little to do with maximizing photosensitivity. The foundation for this conclusion is that most analyses have ignored the possible increase in photoabsorption of a visual pigment at short wavelengths (see difference between beta-peak and alpha-peak absorptions in Fig. 2).

I have confirmed Govardovskii's computations for downwelling twilight irradiance at 3 m. For the photic conditions specified, his interpretation is probably correct. However, the calculations utilized only underwater irradi-

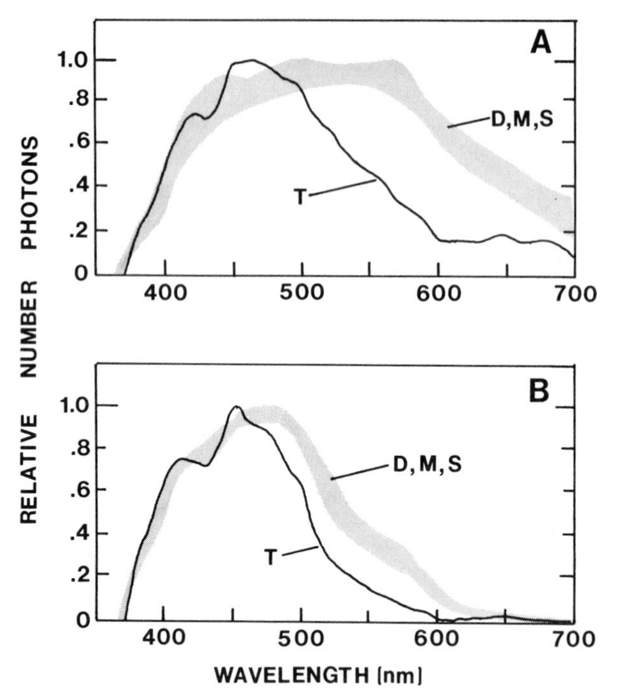

Figure 5 Comparison of downwelling irradiance (A) and horizontal radiance (B) 3 m beneath the surface of Enewetak lagoon for midday, twilight, moonlight, and starlight. D = midday; T = twilight at 28 min postsunset; M = moonlight; S = starlight. The total number of photons per cm^2 per sec between 400 and 700 nm is recorded in Table 3. Shaded area encompasses the spectra for midday, moonlight, and starlight.

ance spectra of downwelling light (Figs. 4E and 5A). Scotopic visual tasks take place against radiant backlighting (Loew and McFarland, 1990) that is usually spectrally different from the downwelling irradiance spectra used to "represent" the twilight or night sky (McFarland and Munz, 1975c). The spectral shift of downwelling irradiance toward shorter wavelengths close to the surface from midday into twilight, for example, is pronounced when compared to changes in horizontal spectral radiance (Fig. 5). In addition, horizontal radiance is spectrally similar throughout a day (Fig. 5B) and over a wide range of depths (1 to 10 m during midday by our measures). One would infer that if the rod pigments of coral reef fishes are matched to twilight then it must apply only near the surface and involve only downwelling light, for the rod pigments are approximately matched to the horizontal radiant background of the water column at all times of the day (Fig. 5B).

What happens to underwater spectra at night after twilight has passed? Like daylight and moonlight, the night sky contains more photons at long than at short wavelengths (Fig. 4). In fact at 3 m depth the spectral distributions for downwelling daylight, moonlight, and starlight irradiance (as indexed by λP_{50}) are between 520 and 540 nm (Fig. 5A). Given the similarity in the day and night irradiance spectra, the horizontal spectral radiances should also be similar, which measurement and calculations support (Fig. 5B).

How might these varied photic conditions relate to the clumping of the rod pigments of coral reef fishes about 493 nm (actually from 480 to 500 nm)? The Twilight Hypothesis, which is a special corollary of the Sensitivity Hypothesis, was based on the apparent match between the rod pigment λ_{max} of coral fishes and the wavelength at which the mean number of photons in downwelling light occurred during twilight (i.e., the λP_{50}). That prey become vulnerable targets during twilight when viewed from below is dramatized by the evacuation of coral reef fishes from the water column toward the substrate

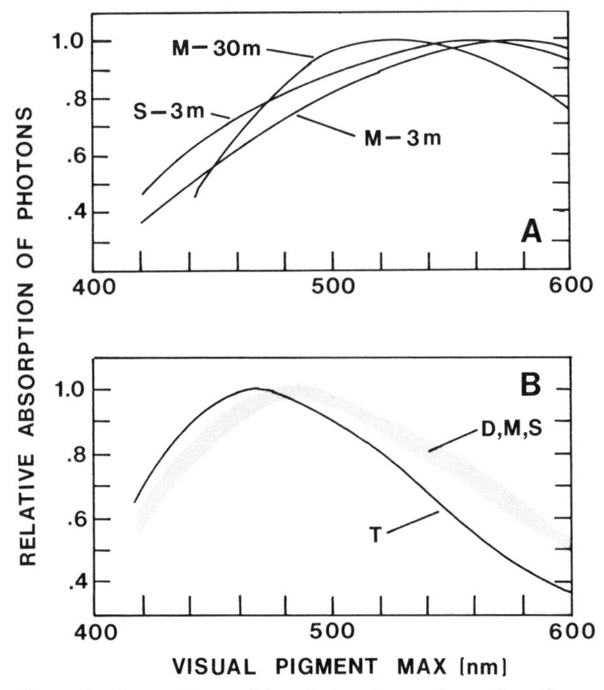

Figure 6 Comparisons of the relative photon absorption of visual pigments with different maximum absorption wavelengths (λ_{max}) toward downwelling irradiance (A) and horizontal radiance (B) at different periods of the day. D, T, M and S are daylight; twilight at 28 min postsunset, moonlight, and starlight. All values were computed following Govardovskii (1976) and refer to 3 m depth, with the exception of the curve at 30 m for moonlight.

during dusk (Hobson, 1965, 1972), and while undergoing dark adaptation (Munz and McFarland, 1973). Twilight therefore is a dangerous time to remain in the water column (Hobson, 1972, 1979; Major, 1977). Although the precise photic conditions when a reef predator strikes at its prey may vary, the position of prey near the substrate during twilight suggests that a target is more likely to be detected along other lines of sight than upward. For the horizontal line of sight computations reveal that during twilight, different rod pigments will differ considerably in total photon capture of the backlighting (Fig. 6B). Compared to a rod pigment with a λ_{max} at 493 nm, a pigment at 520 nm would absorb only 80 percent of the horizontal radiance at twilight, and a 540-nm pigment only 65 percent as much light. It is difficult to dismiss the probability that the rod pigments of coral reef fishes optimize photon capture along all lines of sight during twilight.

What about nighttime when light is even dimmer than during the waning of twilight? The nighttime spectra of downwelling irradiance at 3 m (Fig. 5A) allow computation of the total photon capture for different rod pigments and indicate that pigments positioned above 540 nm are best (Fig. 6A). During moonlight the maximum number of photons in downwelling light achieve a broad peak between 450 and 560 nm at 30 m (Fig. 4F) and similar conditions theoretically apply for starlight. Why then are there no known rhodopsins located above 502 nm in reef fishes? It is important to reemphasize that the downwelling spectral irradiance of daylight, moonlight, and starlight is similar and that of twilight is unique (Table 2 and Figs. 4 and 5A), whereas their horizontal radiance distributions do not differ greatly (Fig. 5B). The computed relative absorptions of different rod pigments to horizontal radiance at all times of the day therefore are similar (Fig. 6B). Maximal absorption along the horizontal plane at 3 m would be achieved by pigments with λ_{max}'s at daylight = 490 nm, moonlight and starlight = 480 nm, and twilight = 470 nm.

What inferences can be drawn from what is currently known? It seems clear that the spectral location of the rod visual pigment of most reef fishes optimizes photon capture for all lines of sight during twilight. This holds also for most lines of sight at night except when looking upward near the surface

Table 2 Comparison of the Mean Wavelength of Photon Distribution (λP_{50}) at 3 m for Different Times of Day

	λP_{50} (nm)			
Measurement condition	Midday	Twilight	Moonlight	Starlight
Downwelling irradiance	521	480	539	520
Horizontal radiance	468	452	461	470

(Fig. 6A). The close spectral match between the rod pigments and the spectral distribution of underwater light for most lines of sight during twilight and at night speaks forcefully in support of the Sensitivity Hypothesis—evolutionary forces have selected the rod pigments of coral reef fishes to optimize photon capture of the available light when they are most vulnerable and/or active during twilight and at night. The optimal solution for coral reef fishes is to possess blue-sensitive rod pigments.

An alternative hypothesis suggests that rhodopsins located at long wavelengths are unstable and, therefore, they are positioned at short wavelengths (Stiles, 1948; Barlow, 1957; Lythgoe, 1984, 1988). Rhodopsins, however, range from 467 nm in deep-sea fishes (Dartnall and Lythgoe, 1965) to 526 nm in another deep-sea fish (O'Day and Fernandez, 1974). A pigment located at 525 nm, which presumably would be stable, would represent a clear compromise for nighttime vision and improve photon capture of downwelling radiance by about 10 percent (Fig. 6A). But none of the nocturnal reef fishes studied so far possesses a rod pigment with λ_{max} above 502 nm (Table 1).

It has been suggested that thermal events might agitate a rhodopsin molecule sufficiently to allow an infrared photon to be absorbed and thus produce spurious visual excitation (see Lythgoe, 1984, for a review). The idea is questionable, however, and particularly so because others doubt that dark isomerizations lead to visual excitation (Hubbard, 1958; Crescitelli, 1991). Lythgoe (1988) emphasizes that at levels of absolute visual thresholds, reducing physiological noise in the photoreceptor is critical. But this circumstance prevails only at extremely low light intensities, specified as a condition in which a rod receives only one photon every 40 minutes! It is unlikely that such low intensities are ever encountered by reef fishes unless they are ensconced deep in the reef at night. Nevertheless, to be useful behaviorally, visual information must be processed over very short time intervals. It is important to ask, therefore, whether the photon fluxes that might enter a fish's eye during twilight, or at night, are sufficiently above threshold intensities to excite a useful response in the retina.

C. Vision at Night

Fishes that emerge from the reef at night to feed include holocentrids, pempherids, apogonids, sciaenids, and priacanthids. Their feeding behavior(s), as compared to those of diurnal fishes, tends more toward stealth, a behavior we observed for nocturnal temperate fishes (Hobson *et al.,* 1981; see also Hobson, Chapter 4). Visual detection and strikes at prey likely depend more on maximizing sensitivity and on motion detection than on high visual resolution. To maximize sensitivity, the rods of fishes, as in humans, are connected

Table 3 Total Number of Photons at Different Times of Day at 3 m beneath the Surface of Enewetak Lagoon[a]

Time of day	Downwelling irradiance	Horizontal radiance
Midday	44×10^{15}	67×10^{7}
Twilight	84×10^{9}	17×10^{6}
Moonlight	18×10^{9}	25×10^{6}
Starlight	84×10^{7}	24×10^{5}

[a] Horizontal radiance is reported for a 6 degree cone. Values are photons per cm^2 per sec per 400 to 700 nm.

into unified receptor fields so that multiple absorptions by different rods can sum to produce visual excitation in the CNS (Walls, 1967). Although a single rod can respond to the absorption of one photon, the absolute threshold for human scotopic vision requires a corneal flux of 50 to 100 quanta per second. This corresponds to the capture of 4 or more quanta during the integration time of a receptor summation field. Objects seen at threshold level by humans, however, are perceived as vague impressions of reality (Pirenne, 1962b). Biologically meaningful vision in humans requires higher light intensities (Barlow, 1972) and, I assume, the same relation holds for fishes.

The radiances measured or computed for a depth of 3 m indicate that close to one million or more photons per second bombard the cornea of a fish with a pupillary aperture of 1 cm^2 (Table 3). Using these flux values, it is possible to compute the photon absorption per rod for various optical densities. Unfortunately data on rod densities and eye dimensions that are necessary for these calculations are unavailable for any coral reef species. However, data exist for several temperate marine reef fish that have tropical counterparts (Pankhurst, 1989). In these fishes, rod densities vary from 10^5 to 10^7 rods per mm^2, with diurnal species having lower and nocturnal species higher densities. The potential rate of photon capture by an individual rod or group of rods will depend on the radiant flux, the preretinal absorption, the area of the retina illuminated by the radiant field, the size and number of rods in the illuminated field, and the optical density (O.D.) of the rod pigment.

In computing the values presented in Table 4, a preretinal absorption of 15 percent was assigned and Pankhurst's rod density data were assumed typical for a diurnal coral reef fish. An O.D. of 0.1 is close to the peripheral retinal absorption of a human but low for a teleost. Optical densities between 0.25 and 0.4 are more realistic. A diurnal reef fish when viewing a target during late twilight against the horizontal radiant backlight would absorb 200 or more photons per rod each second (Table 4). However, under starlight skies, photon capture would decline to approximately 30 to 40 photons per rod per second, and under overcast skies to 3 to 4 photons. Such low photon capture rates probably are close to the threshold values of useful vision. For example, if

Table 4 Comparison of the Photon Absorption of a Fish with Different Visual Pigment Densities under Various Radiant Conditions[a]

	Photon absorption per rod per sec			
Radiant condition	O.D. 0.10	O.D. 0.25	O.D. 0.40	O.D. 1.0
Sunset + 8 min	10,000	38,700	53,300	78,600
Sunset + 28 min	100	200	280	420
Moonlight	150	300	420	620
Starlight	13	30	36	60
Overcast	2	3	4	7

[a] All values are based on horizontal radiance of a 6 degree cone at 3 m depth. Overcast night intensity levels were computed on the basis of a 90 percent reduction in horizontal radiance of starlight. Values are rounded to the nearest hundreds, tens, or integer.

a receptor field sums 100 rods and has an integration period of 10 msec, then the absorption by the receptor field under overcast conditions will be 3 to 4 photons per response interval, a value close to the absolute threshold of vision in humans (Pirenne, 1962a). Increasing the summation of rods and the integration time would improve target detection, but at the expense of resolution and motion analysis (Lythgoe, 1988). The computational results imply that a diurnal reef fish's scotopic vision becomes light limited during twilight, a probability that is reinforced by their behavioral inactivity during this period of the day and at night. If the rod vision of diurnal reef fishes is constrained physiologically at twilight intensity levels, then how do nocturnal fishes manage to "see" at night?

D. Visual Adaptations in Nocturnal Reef Fishes

1. Retinal Illumination

As the eye enlarges with growth, the spherical lens in teleosts enlarges in proportion to the retina. As a result, Mathiesson's ratio (focal length lens/lens radius) remains constant. Typically, this ratio is 2.55 for teleosts. This implies that retinal illumination for a given level of intensity is constant among ages and across most species of fish.

In a given radiant source, retinal illumination can be increased by decreasing the distance of the retina from the lens. While in many deep-sea teleosts Mathiesson's ratio is close to 2.55, in species with tubular eyes a secondary retina is placed close to the lens and, although target images may be somewhat defocused, the illumination of a unit area of this retina would be increased. In *Cyclothone,* a migrating mesopelagic fish, Mathiesson's ratio is much less than 2.55, resulting in increased illumination of the entire retina (Locket, 1977). In at least two nocturnal coral reef fishes the retina is closer to the lens than the average reported by Mathiesson in 1882 (see Fernald, 1988). In the squirrelfish, *Holocentrus rufus,* the ratio is closer to 2.2, and in the barred soldierfish, *Myripristis jacobus,* the ratio is as low as 2.1. This would increase retinal illumination by 20 to 22 percent compared with teleosts where Mathiesson's ratio holds. These differences, however, require confirmation by comparing several diurnal reef species with more nocturnal reef species. A recent tabulation indicates that Mathiesson's ratio varies between species and that the average is closer to 2.4 than to 2.55 (Fernald, 1988). By using 2.4 rather than 2.55, the increase in retinal illumination in the squirrelfish and the barred soldierfish would amount to only 10 percent. Whether such a small increase in absorption would yield a significant improvement in vision is moot, but it may be critical when light is limiting. The possibility invites more exacting studies of optical dimensions and the refractive power of the lenses of diurnal and nocturnal reef fishes.

2. *Photoreceptor Modifications*

At present the rod densities for coral reef fishes have not been quantified. However, the area of the retina devoted to rods in a nocturnal fish exceeds that of a diurnal reef fish (Fig. 9 in Munz and McFarland, 1973) and increases the chance of photon absorption. Nocturnal temperate species have rod densities of 10^6 to 10^7 rods per mm^2, which is at least 10-fold greater than in diurnal species (Pankhurst, 1989). In addition, the thickness of the photoreceptor layer that contains the rod outer segments tends to be greater in nocturnal fishes (Table 5).

Because light passes through the length of a rod outer segment, increasing the outer segment length increases the amount of visual pigment through which light can penetrate and, therefore, the capacity of the rod to absorb light. In many deep-sea fishes several rod outer segments are stacked on top of each other and this, presumably, increases photon capture. Stacked layers of rods are also present in the Holocentridae (8 species; unpublished observa-

Table 5 Dimensions of the Visual Pigment Containing Outer Segments of the Photoreceptor Cells of Two Nocturnal Coral Reef Fishes[a]

Species	Rods	Cones	
		Singles	Twins
Myripristis jacobus	1.8 × 14.4	1.5 × 7.0 × 1.0	3.6 × 8.1 × 2.5
Apogon bimaculatus	1.0 × 48.0	2.7 × 7.3 × 1.8	3.6 × 6.3 × 3.0

[a] All dimensions are in micrometers and read in the order of outer segment basal cell width × cell length × tip width. In rods the basal and tip width are the same. Note that the rods in *Myripristis* are stacked into three to four layers.

tions). Other nocturnal fishes achieve an increase in photon capture by having very long rod outer segments (Table 5). It is likely that the optical density in these nocturnal coral reef fishes approaches 1.0, as it does in some deep-sea fishes (Locket, 1977), but this needs to be measured. It has been suggested that only the photoreceptors near the focal plane (= vitread cells) may be functional (Shapley and Gordon, 1980), but I find this hard to accept for nocturnal species where light is often limiting (Lythgoe, 1979, 1988). In any case, increased length of the visual pigment containing outer segments and the number of rods per retinal area must increase photon capture and therefore be adaptive to nocturnal reef fishes.

3. *Tapeta*

The presence of a reflective tapetal layer behind the photoreceptors is common in nocturnal species, and reef fishes are no exception. The eyeshine from reef croakers, pempherids, and especially priacanthids, as created by reflections from a diver's light, provides a vivid display of the importance of tapeta in enhancing photon catch. By almost doubling the chance of absorption, the presence of a tapetum could, for example, increase the photon catch during starlight and, perhaps, make vision useful even during overcast nights.

It is not clear whether nocturnal planktivores silhouette their prey against the "brighter" downwelling light from the surface (see Hobson, Chapter 4, for a discussion). Such a tactic would considerably increase photon capture and the chance of target detection. Another aspect of nocturnal vision may be the use of bioluminescence. Rapid movements under water at night produce turbulence and, in the presence of bioluminescent plankton, can brightly outline a target. Exactly how important this may be in nocturnal prey detection in reef fishes is unclear. But our observations in California kelp forests led us to infer that it is critical to survival and that stealthy movements of prey and predator at night minimize the chance of detection (Hobson *et al.*, 1981).

V. Correlations of Photopic Vision and Light

A. Do Fish See Color?

Coral reef fishes are for the most part vividly colored. Although yellow, brown, white, and black body colors predominate, vivid blues, greens, and reds are common (Hailman, 1977). It is likely therefore that many reef fishes possess a well-developed sense of color. Many species possess at least two or more visual pigments in their cone cells (Lythgoe, 1988) and distinct neurophysiological responses to different wavelengths of light (e.g., color opponency) are demonstrable (Svaetichin, 1956). The demonstration of multiple cone visual pigments does not in itself prove that they function to distinguish color. Multiple visual pigments may, for example, merely act to optimize the contrast of reflective (or nonreflective) objects when viewed against a monochromatic background (Lythgoe, 1968; McFarland and Munz, 1975b). Clear examples of the ability to distinguish various hues by reef fishes remain to be documented by quantitative behavioral experiments.

B. Cone Cells and Pigments

Cone cells of reef fishes, as in other teleosts, are ordered into a definable mosaic pattern. The arrangement usually consists of a row or square pattern consisting of double and single cones (Munz and McFarland, 1973). In many species the double cones are identical twin cones, whereas the singles consist of one or two types. The exact function of this ordered mosaic remains unclear, although the arrangement may serve to enhance motion detection.

The two members of double cones often contain the same visual pigment but may contain spectrally different pigments (Loew and Lythgoe, 1978; Levine and MacNichol, 1979). For the Pacific tropical marine fishes whose visual pigments we characterized by extraction methods, 63 percent (112 species) possessed pigment(s) in addition to the rod pigment. For 17 percent (31 species) the secondary pigment constituted more than 10 percent of the total extractable pigment (Munz and McFarland, 1973). Using a computer-assisted iterative procedure, we evaluated the spectral position of the additional pigments (Munz and McFarland, 1975) and concluded that they represented cone pigments. This was confirmed in collaboration with Dr. Paul Liebman by direct microspectrophotometry (MSP) of the rod and cone pigments of the mahi mahi, *Coryphaena hippurus* (Munz and McFarland, 1977). In general the cone pigments of these reef fishes tend to be blue and green sensitive.

Subsequent analysis of cone pigments in reef fishes using MSP confirms the general presence of a blue- to blue-green-sensitive cone pigment (λ_{max} ca. 450 to 500 nm) and of a green-sensitive one (λ_{max} ca. 520 to 530 nm). For example, in an extensive study of a variety of fishes, Levine and MacNichol (1979) report λ_{max} values of 495 and 522 nm for the double cones and 495 nm for the single cones of *Chaetodon* spp. In *Gramna lateo* the double cones also contained two pigments at 489 and 521 nm and a single cone had a pigment at 440 nm. In the nocturnal soldierfish, *Adioryx* spp., there were two pigments, the single cones containing one at 440 nm and the identical twins one at 520 nm. A similar distribution of cone pigments occurs in several Caribbean species (Table 6). Determination of the cone pigments from a diverse group of reef fishes, however, remains a desideratum. The lack of a longwave-sensitive cone pigment, as so commonly found in freshwater and temperate marine teleosts (see Loew and Lythgoe, 1978; Levine and MacNichol, 1979; Lythgoe, 1988), is not precluded, at least in those species where bold color patterns

Table 6 Maximum Absorption Wavelength for the Visual Pigments of the Rods and Cones from Several Caribbean Coral Reef Fishes[a]

Species	Rods	Single cones	Twin cones
Squirrelfish			
Holocentrus ascensionis	500	440	515–520
Blackbar soldierfish			
Myripristis jacobus	490–495	440–450	520
Barred cardinalfish			
Apogon binotatus	495	450	525
Copper sweeper (post larvae)			
Pempheris schombergi	495–500	500	525
Lizardfish			
Synodus intermedius	—	—	520
White grunt			
Haemulon plumerii	500	500	525–530
French grunt			
Haemulon flavolineatum	—	500–510	530

[a] Measurements were made by microspectrophotometry and are reported in nanometers (W. N. McFarland and E. R. Loew, unpublished observations).

predominate (e.g., labrids). Attempts to define the cone pigments of labrids using MSP have been frustrated by the small size of their cone cells and the presence of large amounts of melanin screening pigment in retinal preparations.

The blue- to blue-green-absorbing cone pigments of coral reef fishes (440 to 500 nm) likely serve to match the generally bluish backlighting typical of tropical waters and, therefore, optimize the visibility of nonreflective targets (McFarland and Munz, 1975b). In contrast, the green cone pigments (ca. 515 to 540 nm) probably optimize the detection of reflective targets over reasonably short ranges (to perhaps 20 m) because they are less sensitive to the bluish water background (Lythgoe, 1968; McFarland and Munz, 1975b). This general hypothesis is supported by the MSP-based investigations of Levine and MacNichol (1979, 1982). Indeed, a sensitivity to blue and to greenish-yellow for fishes that inhabit bluish water and a sensitivity shifted to green and to reddish light for fishes that inhabit greenish water were suggested in 1973 by Lythgoe and Northmore in their review of colors under water. The predominance of matched and offset cone pigments in reef fishes led us to conclude that dichromaticity evolved to enhance the visibility of 'dim' and 'bright' targets under water and, furthermore, provided the foundation of a color vision system (McFarland and Munz, 1975b).

In addition to the contrast enhancement of targets, color vision is likely critical to reef fishes in many other ways. Although multiple cone pigments have been demonstrated in coral reef fishes, quantitative behavioral studies to evaluate the abilities of reef fishes to discriminate colors may be more critical for our understanding of the visual environment of reef fishes than studies of the physiological basis of their color vision.

C. Visual Acuity

The ability of vertebrates to resolve detail is limited by many factors, for example, neural integration, photoreceptor densities (i.e., graininess of the retina), and, ultimately, the optical properties of the cornea and lens (Lythgoe, 1979). Because the refractive index of the cornea is similar to that of water, the cornea adds little to resolving power. As a result, the focusing of images on a fish's retina is accomplished by a highly refractive and nondeformable spherical lens (Walls, 1967). Accommodation results from active movement of the lens toward (for distance vision) and away (for close vision) from the retina (Sivak, 1973).

In fishes the ability to resolve detail is considered to be limited by the "graininess" of the retina—as indexed by the density of cone photoreceptors. For example, for an eye of given size, the higher the cone density the greater the ability of a fish to resolve detail. The theoretical visual acuity—as indexed by the "morphological" minimum separable angle (MSA), which is calculated from data on lens diameter and cone densities and assumptions concerning Mathiesson's ratio (see Tamura and Wisby, 1963, for details)—has been measured for numerous species of fishes (see examples in Tamura and Wisby, 1963; Fernald, 1988; Pankhurst, 1989). Few examples exist for coral fishes, although visual acuity is most likely higher in diurnal fishes than in crepuscular and nocturnal species (Munz and McFarland, 1973). Visual acuity improves as a fish grows (Hairston *et al.,* 1982), particularly in the juvenile phase.

In only a few cases has the MSA index been compared with behavioral data concerning the size of particles that a fish actually eats (Hairston *et al.,* 1982). An example for reef fishes is provided by the bluehead wrasse, *Thalassoma bifasciatum,* for which we measured the MSA and the size of planktonic food actually consumed by different-sized individuals (Fig. 7). One would expect wrasses to have high visual resolution, which they do, but their generally small eyes optically limit resolution. To compensate for this optical constraint they achieve high cone densities (>100,000 cones per mm^2 of retina). Blueheads are largely planktivorous when young, but even the largest males feed partly on plankton. All sizes of bluehead wrasses collected in the field fed on copepods as small as 0.3 to 0.4 mm in length. Because the visual acuity of the larger individuals is greatest, they can detect objects at greater distances. The net effect is to increase the search volume approximately 13-fold for a large adult as compared to a small juvenile (Fig. 7). Because of their discrete feeding

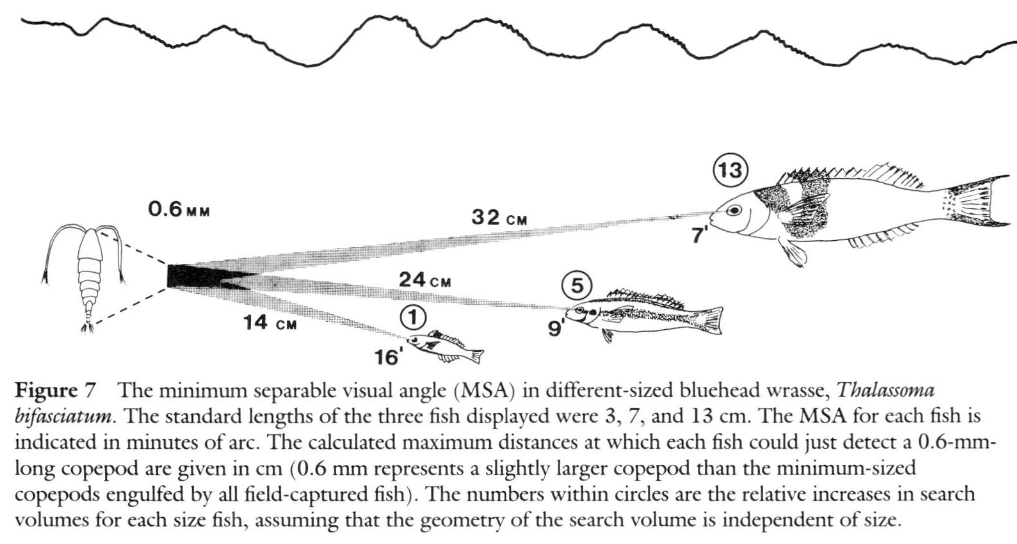

Figure 7 The minimum separable visual angle (MSA) in different-sized bluehead wrasse, *Thalassoma bifasciatum*. The standard lengths of the three fish displayed were 3, 7, and 13 cm. The MSA for each fish is indicated in minutes of arc. The calculated maximum distances at which each fish could just detect a 0.6-mm-long copepod are given in cm (0.6 mm represents a slightly larger copepod than the minimum-sized copepods engulfed by all field-captured fish). The numbers within circles are the relative increases in search volumes for each size fish, assuming that the geometry of the search volume is independent of size.

habits, planktivorous reef fishes provide a model system for further comparisons of actual feeding performance with visual abilities.

The cone densities from different parts of the retina reveal that many reef fishes possess higher acuity along some lines of sight than along others. For example, in *Thalassoma duperrey,* the dorsal retina achieves densities of 113,000 cones per mm^2, twice that of the ventral retina (Munz and McFarland, 1973). Looking downward at the substrate to detect tiny food objects against the heterogeneous reef substrate is certainly an important behavior for this omnivorous picker (Hobson, 1972). Similar differences exist in many other diurnal reef species. In general, however, high cone densities and therefore potentially high visual acuity are associated with most visual axes (up, down, forward, backward, and laterally), a circumstance that must endow diurnal coral reef fishes with excellent surveillance of their surroundings. Threatening movements by potential predators would be readily detected and, indeed, diurnal reef fishes often remain in the water column until dusk, seemingly oblivious to the nearby presence of predators (Hobson, 1968; Munz and McFarland, 1973).

Recent investigations of several species of reef fishes further refine these earlier results (Collins and Pettigrew, 1988a,b). By examining the density of retinal ganglion cells, which presumably reflects the underlying density of photoreceptor cells, they found that many species lack visual streaks (i.e., retinal areas of potentially higher visual acuity), while other species possess streaks, which are often oriented horizontally or nearly so. Collins and Pettigrew correlate the latter condition with fishes that associate with sand bottoms and require enhanced surveillance along the horizontal visual axis. In addition, many reef species are capable of considerable eye movements to allow them to fixate areas of the retina with highest visual acuity on objects of interest, parrotfishes being exemplary models. Certainly further studies of cone distributions along with ganglion cell distributions if coupled to behavioral experiments will lead to a better understanding of how visual acuity serves the diverse kinds of fishes that inhabit reef ecosystems.

VI. Topics for Further Study

A more exacting analysis is needed of photic conditions in and about coral reefs throughout all periods of the day. In several instances I had to calculate the radiance field against which a fish might detect a target. This was necessary because the sensitivity of the spectroradiometers used was exceeded during starlight and on overcast nights. In addition, radiant measurements should be extended whenever possible from the surface to depths of 60 or more feet and along varied lines of sight. Data should be expressed in quantal units

(=photons) rather than the typical energy values, so the data can be effectively applied to visual absorption processes (Arnold, 1975). Innovations in the light sensitivity of spectroradiometers should allow this type of radiant data to be compiled over the next few years. Spectral radiant measures should include the UV-A (ca. 300 to 400 nm) as well as the visible wavelengths (400 to 750 nm), because recent studies reveal that several freshwater fishes can "see" in the near-UV (Loew and McFarland, 1990). It is likely that UV vision is present in some coral reef fishes, given the fact that UV-A light is abundant in clear seas (McFarland, 1986), but this remains to be demonstrated.

Although data exist on the rod pigments of a great number of coral reef fishes, the cone pigments of too few species have been measured for us to develop generalizations. Over the last decade, microspectrophotometers have become available to more investigators and provide the only exacting way of defining the cone pigments of a large number of species. For example, knowing whether a fish has only one, two, three, or more cone pigments and the wavelengths over which they are most absorptive would provide the basis of evaluating its ability to distinguish color. This type of information would be useful for designing behavioral protocols to test a fish's resolution to discriminate hues.

A better grasp of the retinal anatomy of coral reef fishes is required to allow computations of photon capture under the varied light conditions when coral reef fishes are active. This requires careful measures of rod and cone dimensions, their densities, their types and positions, their summation with the neural retina, transmission characteristics of the cornea, lens, and ocular fluids, and measures of visual pigment optical densities (e.g., Fernald, 1988; Pankhurst, 1989). I have purposely said little about direct physiological measure of vision in reef fishes in this chapter. However, electroretinograms and spike responses in the optic tectum can provide insights that are difficult to obtain by other methods. These methods, however, require sophisticated laboratory apparatus and a kind of experience that most fish ecologists do not have time to accumulate.

Although anatomical and physiological studies of vision in coral reef fishes can provide a wealth of insightful information, the most meaningful data to fish ecologists are more likely to be derived from behavioral studies of vision. How well a species can distinguish color, or at what light intensity threshold it can detect a target, is best answered with a behavioral experiment. That such experiments are fraught with difficulty in execution and interpretation should be appreciated by the unwary investigator. Critical to such behavioral studies is the ability to accurately measure the light fields and stimuli that are presented to the fish. But even when all factors are controlled rigorously, different protocols testing the same question often give different answers (Muntz and Northmore, 1973). Nevertheless, a direct behavioral test provides the closest tie between experiment and a coral fish's natural behaviors. Even though modern instruments will make the task easier, it will be very time-consuming to quantify how vision serves coral reef fishes. I emphasize and recommend, therefore, that investigators attempt to utilize species about which the natural behaviors are well established. To that end, the behaviors of the various coral reef fishes described in this volume provide a basis for progress.

Acknowledgments

The view of fish vision presented in this chapter mostly reflects my field experiences with coral reef fishes. Many scientists and students have shared these experiences and, although all of them have influenced my impressions of fish vision, four individuals stand out—Ted Hobson for teaching me to look critically at what fishes do; Ellis Loew for showing me how optics and electronics can make possible the investigation of a seemingly impossible visual ecological task; Fred Crescitelli for transferring his exuberance about visual pigments and their relationship to animal behavior to me; and Fred Munz for the exciting years of exploration we spent together in the field, in the laboratory, and in virtually endless discussions about science. I owe all of them my thanks and friendship. Three scientists have critically read this manuscript—John Ebersole, Robert Warner, and Ned Pankhurst. I thank each of them for their helpful comments. This is contribution 144 of the Catalina Marine Science Center.

CHAPTER 3

Reef Fishes: Their History and Evolution

J. H. Choat and D. R. Bellwood
Department of Marine Biology
James Cook University
Townsville, Queensland, Australia

I. INTRODUCTION

Reef fishes exist today as a circumtropical assemblage clearly linked to coral reefs (Sale, 1980a). We recognize similar taxonomic and ecological categories from a wide range of tropical reef environments. A fauna with this widespread distribution and association with such a distinctive habitat prompts a number of questions. Where did the present-day reef fish fauna come from? Did the ancestral groupings lie within the present teleost fauna? Did they evolve on reefs or did migrations from other habitats play a role? Did they accumulate slowly on reefs or did the fauna become established rapidly in this habitat? How has the evolution of the reef fish fauna contributed to the present-day assemblage? Recent work in phylogenetics, paleontology, and marine geology has provided opportunities to resolve these questions (Sale, 1988b).

We identify three main requirements in the development of an evolutionary and historical perspective on reef fish biology. First, it demands a working definition of what is meant by a reef fish. To do this we need more comprehensive descriptions of the fishes and their habitats. A majority of studies have focused on the behaviors and life histories of particular groups of fishes such as pomacentrids, labrids, and scarids. We tend to use their characteristics when thinking about reef fish assemblages as a whole. There is a need to expand the information base to cover a greater variety of species.

Second, there is the question of a biogeographic and paleontological perspective. What mechanisms are responsible for present-day patterns? The recent history of reef fish studies has been dominated by debates concerning the generality of ecological patterns (Gladfelter *et al.*, 1980; Sale, 1980a). It is not surprising that the nature of these generalizations has been influenced by the particular study locations. What is surprising is that we do not have a more informed perspective on the processes responsible for the present-day distribution and structure of reef fish faunas.

Third, it is necessary to develop a phylogenetic perspective. For many life history features, especially those associated with reproduction and larval dispersal, current thinking emphasizes the adaptive significance of certain traits (Doherty *et al.*, 1985). There has been relatively little work on the evolutionary history and phylogeny of reef fishes and the structural constraints that may have influenced the expression of reproductive and life history features.

Our approach must be preliminary for three reasons. (1) The material on which to base comprehensive distribution lists has only recently become available. (2) The fossil history of reef fishes is poorly known. (3) Phylogenetic studies of key groups of reef fishes are only just emerging.

The organization of this chapter is as follows. Reef fish are defined in ecological, structural, and distributional terms. The defining characteristics are then considered in the context of the coral reef habitat and fish/habitat interactions. This provides the framework for a review of the evolutionary history of reef fishes. The final section considers reef fishes in terms of their history and the possible influence of structural constraints on their biology.

Although reef fish studies have been conducted over the full geographic range of coral reef habitats, two major centers of activity and two schools of study have dominated the field for the last ten years. These are represented by the work of P. F. Sale and colleagues on the Great Barrier Reef of Australia and that of R. R. Warner in the Caribbean. Both groups have used labroid fishes as their primary study organisms. In addition to these groups, the literature on reef fish has been enriched by the studies of D. R. Robertson, who has addressed numerous topics relating to the biology of labroid and acanthuroid fishes. From these studies there has emerged a picture of reef fish populations as open systems in the terminology of Warner and Hughes (1988), in which the link between parental reproduction and recruitment back to the reef environment may be tenuous at the scale of readily definable populations (Doherty and Williams, 1988a). In this context the importance of settlement and recruitment processes in structuring local populations of fishes is the focus of vigorous debate.

The taxonomic basis for all such studies has been provided primarily by the works of J. E. Randall, with major contributions from V. G. Springer and more lately G. R. Allen. Not only have these works provided a basis for establishing biological identities in complex assemblages of reef fishes but they have also laid the foundations for credible distributional and biogeographic studies (Springer, 1982). These studies have provided the authors with critical information for the development of this chapter.

II. What Are Reef Fishes?

Reef fishes have a number of defining characteristics. These are (1) group features that characterize assemblages of reef fishes as a whole, (2) ecological characteristics, (3) habitat associations, (4) distributional patterns, and (5) taxonomic characteristics and (6) structural features.

A. Group Characteristics

The most striking feature of reef fish is their diversity, in terms of both species numbers and the range of morphologies. For example, an estimated 4000 species of fishes live on coral reefs and associated habitats of the Indo-Pacific (Springer, 1982), 18% of all living fishes. As generic and family revisions of reef fishes are completed, this number is likely to increase as many supposedly widespread taxa are being redefined as complexes of closely related species. Although the fact of high diversity is well established, it is still unclear if the absolute abundances or biomasses of coral reef fishes are greater than those in other reef environments. Obtaining robust estimates of reef fish abundances is a major priority.

Morphological diversity also occurs in many forms, from highly specialized feeding structures to variability in fish sizes. An example of this is the Labridae, a widespread family that reaches its greatest diversity in the reefs of the Indo-Pacific. Here the smallest and largest members of the family co-occur—*Minilabrus striatus,* which reaches only 30 mm SL (standard length), and the largest labrid *Chelinus undulatus* at 2290 mm SL. Reefs are also dominated by groups of very small species (P. J. Miller, 1979). In addition, reef fishes display great diversity within particular feeding guilds and among different feeding categories. For example, the range of structural modifications associated with feeding on benthic invertebrates has its greatest expression in reef fishes. Reefs also harbor categories of fish that are rare or absent from most other habitats. Grazing herbivores are an example.

B. Ecological Characteristics

With one exception, reef fishes have life cycles in which relatively sedentary juvenile and adult phases alternate with dispersive larvae and juveniles. The adults display complex reproductive schedules in which numerous small eggs develop within the paired ovaries. In many taxa, protogyny is the normal mode of sexual ontogeny. Fertilization is external. A significant minority of species have some form of egg care but most teleosts display broadcast spawning. Eggs are shed directly into the water column and proceed through an obligate development period leading to juvenile fishes that are competent to settle on a reef substratum. In many groups of reef fishes the presettlement open-water phase comprises fully developed juveniles. The overwhelming frequency of the bipartite life cycle may have diverted attention from the great diversity of reproductive and developmental programs of reef fishes. Many reproductive behaviors and life history features are represented. As size at age and growth rate data slowly accumulate, it is becoming apparent that even small reef fishes tend to be long-lived.

Reef fish assemblages are made up of complexes of ecologically similar species co-occurring within localized areas. Particular guilds of species feeding on the sessile benthic biota and smaller mobile invertebrates of reefs are the best examples. These generate much of the diversity just identified. The interactions between species within these complexes have been the focal point in many studies. Reef fishes rely heavily upon vision and invariably occur in clear waters. This environment allows complex behavior patterns to diversify, behaviors that are usually associated with the development of distinctive color patterns. Reproductive and ecological interactions in such fishes are associated with complex color patterns and dominated by vision.

C. Habitat Associations

Reefs constitute highly fragmented habitats that are patchy at several spatial scales. Reef fishes occupy the reef proper but the definition must be more specific than this. In certain groups the majority of species recruit directly onto reefs and remain within this habitat for their entire lives. The planktonic stage terminates in settlement on reef substrata. These fish include scarids, acanthurids, siganids, chaetodontids, pomacanthids, and many species of labrids and pomacentrids. Members of these groups not only remain associated with reefs but also display highly circumscribed patterns of movement. Many are associated with particular structural and biotic features of the reef.

D. Distributional Features

The habitat associations described in the foregoing can be used to define reef fishes by distributional characteristics. Most species have broad geographic distributions relative to temperate water reef species, which reflects the widespread distribution of coral reefs. Certain taxa are almost always associated with reefs and achieve their greatest abundance in these habitats over their full geographic range. This association may be so characteristic as to define biogeographic boundaries. Scarids and acanthurids are examples. Others have some representatives that are very characteristic of reefs and also species that extend into different habitats and higher latitudes, well beyond the boundaries of reefs. Serranids, scorpaenids, and some types of plectognaths and labrids are good examples.

E. Taxonomic Characteristics

Tertiary fish faunas have been and still are dominated by perciform teleosts. Of the 445 families of fishes listed by Nelson (1984), 150 are perciforms and these contain the most speciose families. This dominance has its greatest expression in reef fish faunas so that their great diversity is developed within relatively few taxa. For example, Myers (1989), in cataloging the fishes of the extensive reef habitats of Micronesia identified 103 families of reef fishes of which 51 were perciforms. Perciforms also made up 86% of the 20 most speciose families on Micronesian reefs. In programs that have sampled reef habitat fish faunas for abundances of individuals (Williams and Hatcher, 1983), perciforms are overwhelmingly dominant. Other groups, such as elasmobranchs, anguillids, scorpaenids, holocentrids, and sygnathids, all have reef representatives. However, the total species number and the abundances in such groups form only a small proportion of the reef fish community.

F. Structural Features

Fishes that are consistently associated with coral reef habitats have characteristic structure and morphologies. These are lateral compression and higher body planes, increase in the proportion of locomotory and feeding musculature as a function of entire body weight, development of the physoclistous swim bladder to adjust buoyancy and the elaboration of the alimentary tract, modification of the jaws and pharyngeal apparatus to provide increased power and a more efficient feeding mechanism, and migration of pelvic fins to the thoracic position and the development of the pectorals for locomotion and orientation. Development of these characteristics may have resulted in an increase in the musculature and skeletal structures associated with limb girdles anterior to the visceral cavity. Ratios of body depth to length are also relatively high, resulting in some compression of the visceral cavity.

III. Reefs as Habitats

Present-day reefs are structurally complex formations with an external framework provided by skeletal elements of large, rapidly growing colonial organisms of high diversity (Fagerstrom, 1987). They are the result of constructive processes that produce calcium carbonate and erosive processes that reduce skeletal elements to sediment (Hubbard *et al.*, 1990). The resulting structures usually have high topographic relief that is continually modified by the active growth of corals. Reefs profoundly affect the local environment. They modify the hydrological environment in such a way that their influence extends a considerable distance into the water column (Wolanski and Hamner, 1988). The sediment produced by physical and biological destruction of calcium carbonate accumulates as aprons around the reef base or on lagoonal floors and, by filtering into the reef framework, contributes to its structure (Hubbard *et al.*, 1990). The term reef may also be used to describe rock substrata especially in temperate waters. In this chapter the term is restricted to tropical shallow-water structures of biological origin.

Although hard substrata and biologically produced structures occur at most depths in the ocean, the consideration of reef fish faunas is usually restricted to shallow areas that occur within the photosynthetically active depth range. Scleractinian coral reefs are the dominant form in tropical waters. Their growth, driven by the photosynthesis of their symbionts, produces a calcareous matrix exposed to direct tropical sunlight covered with a complex biota of sessile organisms. The composition of reefs and the responses of corals to depth and water movement are similar across wide areas. This results in reefs with a high degree of similarity. They are in effect habitat units that are replicated many times in tropical seas.

Reefs have their greatest development in shallow, clear waters with impoverished nutrient content (Hubbard, 1988), but there are also abundant fringing reefs on high-rainfall coasts. "Modern" tropical reefs made their appearance in the late Cretaceous (Rosen, 1988) and are characterized by symbiotic relationships between sessile invertebrates and unicellular algae. A significant amount of reef primary production results from symbioses involving zooxanthellae, corals, clams, and ascidians. Most coral reefs also have a component of calcareous algae (Littler and Littler, 1984).

Many species of scleractinian corals have high growth and turnover rates. They produce by death or fragmentation substantial surface areas of calcareous matrix that is colonized by highly productive turfing, filamentous, and encrusting algae (Steneck, 1988). A feature of algal turfs is that they can trap organic detritus and provide sites for bacterial growth. These highly productive complexes are a major source of accessible primary productivity for grazing benthic herbivores and detritivores (Steneck, 1988).

There are numerous interactions between grazing and boring organisms, live scleractinian corals, and the calcareous matrix of the reef. Both internal (invertebrate) and external (fishes) bioerosion agents convert biologically produced calcareous structures to sediment. At reef bases and lagoons, this sediment provides a distinctive habitat for invertebrates and a medium for meiofaunal and bacterial growth. These are exploited by benthic-feeding fishes and larger invertebrates.

Reefs interact with tidal- and wind-generated current systems to produce hydrological features that both concentrate planktonic organisms and retain nutrients and particulate food materials in areas adjacent to reefs (Wolanski and Hamner, 1988). Broader-scale patterns of reef distribution can modify current systems and entrain dispersal patterns of planktonic and pelagic organisms along consistent tracks among reefs (Dight *et al.*, 1988).

These features focus attention on the importance of reefs as fragmented habitats with a great deal of intrinsic patchiness. Patchiness is an important characteristic of temperate marine (almost exclusively coastal) habitats, tropical (biogenic) reefs, and the rocky shore habitats of lacustrine environments. However, tropical coral reefs have a characteristic degree and scale of patchiness that is quantitatively and qualitatively different from the other examples. This reflects the nature of reefs and the rapid responses of reefs to changes in sea level (Hubbard, 1988). As biogenic structures, reefs can maintain themselves as habitats in the face of changes in sea level and foundation subsidence (Rosen, 1984).

The degree and pattern of reef patchiness depends on their origin and relationship to continental masses. On a broad spatial scale, reefs offer clues to the past distributions of major geological features. In present terms they are habitat units with a high degree of similarity and replication that serve as links in a distributional chain connected by dispersive life history stages over areas of otherwise open ocean. The considerable geographic ranges of reef fish faunas and individual species reflect the role of reefs as biogeographic staging posts (Rosen, 1984). Staging posts are important in maintaining patterns of dispersion and gene flow within faunas and in defining biogeographic units.

Coral reefs are subject to different patterns of physical disturbance when compared to higher-latitude shallow-water habitats. They do not experience the same magnitude and consistency of wave-generated turbulence as hard substratum environments of higher latitudes. This reflects prevailing global weather patterns and also the physical structure of coral reefs. Although the structure of reefs of the open Pacific reflects the swell-dominated ocean regime (Hubbard, 1988) there are always major areas of reef environments in shallow waters that are sheltered from consistent and predictable influences of strong wave forces.

The lack of such physical disturbance can result in long-term persistence of complex small-scale structures that provide shelter sites for fishes. However, tropical reefs are subject to major weather-generated disturbances in the form of cyclones, although even this is not a feature of all coral reef habitats, for example, in the Red Sea. Although cyclones occur regularly in many reef habitats, their local effects are unpredictable on a scale of hundreds of kilometers and tens of years.

Reefs are subject to seasonal influences that may be considerable. These take the form of enhanced runoff following seasonal rains and changes in prevailing wind direction. Such seasonal influences will be enhanced in reefs with continental affinities and result in major incursions of turbid coastal waters and their associated nutrients at intervals determined by rainfall patterns. Among-year disturbances occur on a scale of tens of years and are attributable to localized effects of cyclones and medium-term meteorological changes (e.g., El Niño).

Coral reefs are also subject to varying degrees of biological disturbance chiefly from fluctuations in the numbers of echinoderms. The asteroid *Acanthaster planci* causes coral mortalities over wide areas within reefs (Moran, 1986). It seems likely that there will be flow-on effects for some fish assemblages although there is only limited evidence for this at present (Williams, 1986a). Widespread and rapid mortalities of the echinoid *Diadema* in the Caribbean have been associated with major changes in abundances and proportional cover by algae, which in turn has major implications for recruitment patterns in other sessile organisms (Hughes *et al.,* 1985).

Despite the obvious relationships between biological distribution patterns and the physical properties of reef structure there have been few attempts to examine the processes that modify reef structure on biologically relevant time scales. Two observations are changing this view. First, it is now apparent that a number of reef organisms have extended life spans. Second, environmental processes that affect reef structure and morphology may occur on comparable time scales. Coral growth and erosion and the accumulation of by-products may significantly modify patterns of habitat structure and primary production within the life spans of some reef fishes. Changes in sea level may have effects over a relatively few generations (Potts, 1984; Woodroffe and McLean, 1990). The interaction between biological and geological processes in determining fish distributions is an important aspect of reef fish ecology and requires greater attention.

In summary, coral reefs comprise habitats with striking structural and taxonomic similarities distributed across wide areas. This reflects their biogenic origins in which structural pattern is determined by the growth of modular organisms. Coral reefs are patchy on several spatial scales. Within reefs there is a great deal of habitat diversification associated with the responses of scleractinian corals to changing depths and exposure. Within each general habitat, important structural features such as lagoonal patch reefs and reef front gutters are replicated many times. These represent habitats for many sedentary species. Reefs themselves are usually aggregated at different scales, from clusters within local areas to the great complexes of reefs across the Pacific plate (Springer, 1982).

Reefs are subject to a disturbance regime that is qualitatively different from that of higher-latitude reef environments. Major structural and biological disturbances occur unpredictably in space with a periodicity of roughly 10 to

15 years. Although seasonal effects have some influences on reefs, especially those in coastal environments, it is among-year variation that results in the significant disturbance patterns of groups of reefs. Within these time periods the complex fine-scale structures that characterize reef habitats are likely to remain undisturbed.

IV. Key Features of the Reef Fish–habitat Relationship

A. Reef Fishes: Taxonomic and Ecological Features Evaluated

The key features of the reef fish–habitat interaction can be clarified by a more explicit definition of reef fishes. They constitute three major taxa:

1. the chaetodontid fishes, comprising the families Chaetodontidae and Pomacanthidae;
2. the acanthuroids, comprising the Acanthuridae, Siganidae, and Zanclidae;
3. the labroids, comprising the Scaridae, Pomacentridae, and Labridae.

These perciform fishes represent the main groupings of reef species. Most, with the exception of some of the labroids, have distribution patterns that correspond to those of coral reefs. Moreover they are associated with the reef habitat for the entire postsettlement life cycle.

Chaetodontids, acanthuroids, and labroids share two ecological characteristics. First, each grouping contains extensive guilds of ecologically similar species. These guilds of co-occurring species are the most distinctive elements of reef fish faunas. Within each of these groups there are also species that show divergence of feeding habits, especially in the direction of planktivory. Details of the diversity and differentiation within the acanthuroids and herbivorous labroids are considered in Chapter 6. The general characteristics for each of the groups are as follows.

The chaetodontid fishes comprise two major circumtropical reef-associated families, the Chaetodontidae and the Pomacanthidae. The Chaetodontidae comprise 114 species in 10 genera of which 78% are in the genus *Chaetodon* (Allen, 1981). Ninety percent of the species occur in the Indo-Pacific. Approximately 50% of the Chaetodontidae feed on coral and many others feed on cryptic reef invertebrates. Algae is also a significant component of the diet of many species. A minority of species, such as *Chaetodon marleyi,* occur in nonreef environments.

The Pomacanthidae are made up of 74 species within 7 genera with most species occurring on the reefs of the western Pacific. The genus *Centropyge* is numerically dominant with 28 species in total. Pomacanthids display three major feeding behaviors (Allen, 1981)—sessile invertebrates, especially sponges; herbivory; and planktonic invertebrates. The dominant food sources are sponges, tunicates, and algal turfs. Members of the genus *Geniacanthus* are planktivores. A review of chaetodontid biology is provided by Allen (1981).

Structurally the chaetodontid fishes are readily recognizable by their extreme degree of lateral compression and small terminal jaws with bristlelike dentition. There are a number of families of chaetodontoid like fishes with similar structural characteristics. These include the Ephippidae, Scorpidae, Scatophagidae, and Enoplosidae. Although members of these (especially the Ephippidae) occur on reefs, most extend into other tropical habitats and beyond the latitudinal limits of reefs. Recent evidence suggests that the Ephippidae, Scatophagidae, and Acanthuroidei form a monophyletic assemblage (Tyler *et al.*, 1989). Their association with the chaetodontid fishes is unclear. It is not known whether this larger grouping of chaetodontoid fishes is monophyletic.

Acanthuroids are dealt with in Chapter 6, and with the exception of *Prionurus* they are explicitly reef species with herbivory and detritus feeding being the main trophic mode. Planktivory is strongly developed in a number of species of acanthurids in the genera *Naso* and *Acanthurus*.

Labroids are made up of three main groups—scarids, pomacentrids, and labrids. Scarids and the herbivorous groups of pomacentrids are dealt with in Chapter 6. Planktivory and feeding on benthic crustacea are strongly developed in the pomacentrids and all trophic groups of this family extend into temperate waters.

Labrids are a very large group containing approximately 50 genera and 500 species. Most are carnivores feeding on a variety of invertebrates. Small and intermediate sized species feed mainly on crustaceans; larger species concentrate on molluscs (Sano *et al.*, 1984b). Although labrids are highly characteristic of coral reefs, many species have exclusively temperate-water distributions and may extend to the southern and northern limits of perciform fish distributions. Both acanthuroids and labroids represent coherent phylogenetic groupings.

The dominant feature of chaetodontoids, acanthuroids, and labroids is their exploitation of the sessile biota that defines the reef habitat. This is accomplished in two ways. First, they feed directly on the tissues and metabolic by-products of corals and associated symbionts, a feeding pattern exhibited by chaetodontids and some labroids, including the labrid genus *Labrichthys* and the scarid *Bolbometopon muricatum*. Second, acanthuroids and many labroids feed on the algal complexes growing on exposed calcareous matrix.

The complex structures of coral reefs provide the physical habitats and shelter sites that accommodate many size classes and especially small individuals of invertebrates. Many species of fishes exploit the invertebrate fauna characteristic of coral colonies, rubble drifts, and algal turfs. This feeding mode is best developed in labrids, fishes that display an extraordinary range of morphologies associated with crustacean feeding in reef environments. While this mode of feeding is not confined to coral reefs, it finds its most characteristic expression in tropical labrids.

The common features of these groups may be interpreted in terms of their feeding biology and reproductive behaviors. Their structural and anatomical features allow for precise movements and orientation over a complex substratum. The jaw and suspensorium provide the ability to feed with precision and continuity from a sessile benthic biota in which the individual food items are small. These features have their greatest expression in the herbivores, which require continuous harvesting and processing of algal turfs. Reproduction and social behavior in these groups are usually associated with complex interactions dominated by visual signals. Complex species and sex-specific color patterns are most enhanced in these groups. All the foregoing groups are exclusively diurnal foragers.

There are numerous additional groups of teleost fishes that are associated with reefs. They differ from the preceding taxa in terms of their habitat associations and distribution patterns. Reef fishes as defined here are exclusively perciforms. Three other abundant groupings of teleosts—scorpaenids, plectognaths, and pleuronectids—may be common on reefs but exploit nonreef tropical habitats and achieve their greatest diversity and abundance in subtropical and temperate waters. They will not be considered further. In addition, two abundant taxa of small perciforms, Gobiidae and Blenniidae, are characteristic of reefs but also occur in a wide range of tropical and temperate habitats.

Other groups are also abundant in both reef and nonreef tropical habitats. These are largely predators on mobile invertebrates and other fishes or planktivores. They exploit features of the reef habitat that occur in other environments and are reef associated rather than reef fishes.

One characteristic grouping includes predators on mobile invertebrates and fishes (Muraenidae, Holocentridae, Apogonidae, Haemulidae, Lethrinidae, Lutjanidae, Mullidae, and Serranidae) and planktivores and pelagic piscivores (Caesionidae, Carangidae). The former group may be further subdivivded into species that use structural features of the reef as shelter or aggregation sites and show active nocturnal foraging (muraenids to lutjanids) and those that forage diurnally (mullids and many serranids). Numerous other groups, including ambush predators (cirrhitids, scorpaenids, synodontids), are also common on reefs (Norris and Parrish, 1988), but they are abundant in other habitats.

Some groups are more closely associated with reefs than others. Lutjanids and serranids, although abundant through most reef environments (eastern Pacific reefs are an exception), also extend into deeper waters, inshore environments, and subtropical regions. Haemulids (the commuters of Parrish, 1989) and lethrinids are more obviously reef associated, and a minority of predatory taxa, for example, aulostomids, are restricted to reef environments.

Planktivores, usually dominated by caesionids, exploit the hydrological features generated by reefs that retain and concentrate plankton. These in turn serve as focal areas for the foraging of open-water piscivorous predators such as carangids, sphyraenids, and scombrids. Members of these groups are capable of feeding in open water but attain high numbers adjacent to reefs. They also serve as an important link in the returning of energy and nutrients derived from open-water feeding to the reef via defecation (see the following discussion).

The key feature of reef-associated species and planktivores is that they are linked to habitat features that occur in a wide variety of shallow-water marine environments but are most strongly developed in tropical reef habitats. This is in contrast to the chaetodontids, acanthuroids, and labroids listed earlier that have obligate associations with the coral reef biota.

Fishes on coral reefs display a wide range of sizes from gobiids maturing at 15 mm SL to epinephaline serranids that reach 2600 mm SL (Fig. 1). The largest species, members of the families Serranidae and Labridae and the complexes of lutjanidlike fishes, are carnivores that feed on other fishes and larger mobile invertebrates. The smallest species, primarily gobiids, feed on minute mobile invertebrates, primarily crustaceans, the food source also of most newly settled reef fishes. Carnivores provide the smallest and largest examples. A number of very small species are omnivores. The majority of acanthurids, siganids, scarids, and the herbivorous pomacentrids lie in the intermediate size range, a feature that appears to be associated with their mode of feeding. In two primarily herbivorous groups, acanthurids and scarids, the upper size range is extended by the presence of large planktivorous or coral-feeding species (Chapter 6).

Similar ecologically related size distributions can be detected in certain groups of cichlid fishes (Fig. 2). Lacustrine cichlids, especially those of the African rift lakes, are freshwater analogs of coral reef fish faunas having the same pattern of trophic organization, including a major component of grazing herbivores (Fryer, 1959; Fryer and Iles, 1972). The complex of species feeding on "aufwuchs" (benthic algal assemblages that are comparable to algal turfs) has a size range at the lower end of the scale similar to that seen in coral reef herbivores. Predatory and planktivorous species are larger. Both groups of advanced perciforms, the marine chaetodontoids, acanthuroids and labroids on the one hand, and the lacustrine cichlids on the other, may have the same pattern of ecological and phylogenetic constraints.

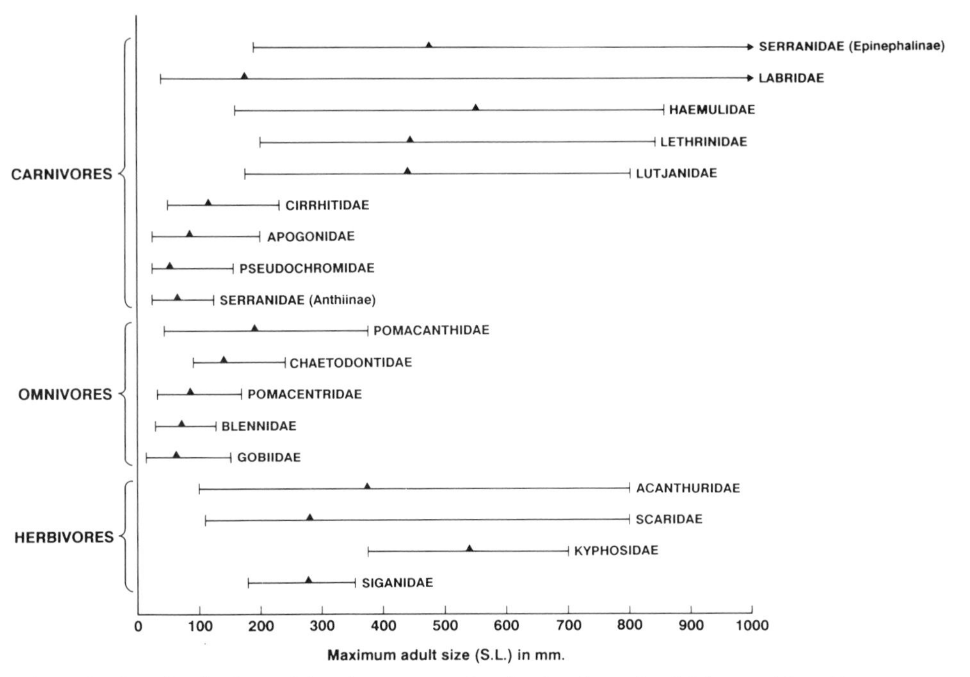

Figure 1 Size distributions of abundant groups of reef and reef-associated fishes partitioned by taxon and feeding behavior. The size ranges and modes of the standard lengths are shown. Size ranges of acanthurids and scarids are extended by planktivorous and coral-feeding species in the 600- to 800-mm size range (see Chapter 6). Size estimates were obtained from Masuda *et al.* (1984) and Myers (1989).

B. Reef Fish–Habitat Interactions

Coral reefs have particular structural and biological features that make them unique as marine habitats. Not only do reefs contain complex structures, but the scale of surface heterogeneity is very great and is repeated many times within small areas. Relatively small areas of habitat can accommodate a large range of sizes of fishes. Because the substratum is biogenic it will grow and be modified on short time scales. A basal structure of calcium carbonate means rapid growth but also rapid erosion, which will in turn produce other habitat features. Biological relationships play an important role in these processes. They may be straightforward physical habitat–animal relationships or complex reciprocal interactions between the reef-forming invertebrates and the fishes.

The structures of scleractinian corals provide physical habitats and shelter sites that accommodate fish sizes of several orders of magnitude. There are two

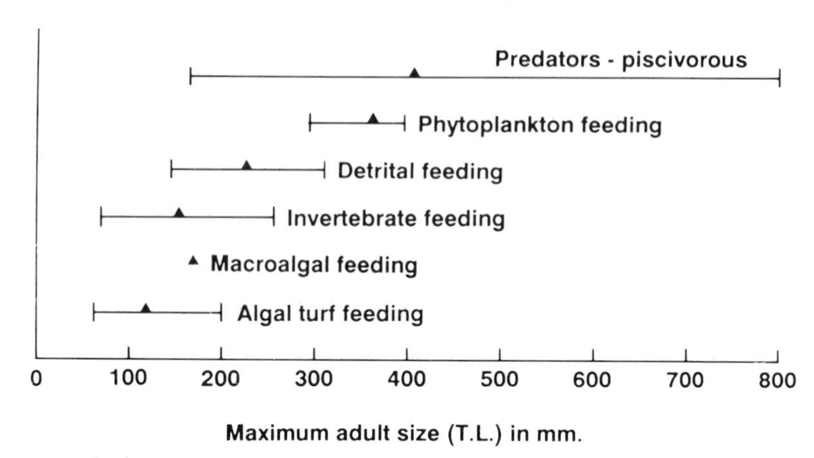

Figure 2 Size distributions of the main feeding groups of African rift lake cichlids. The size ranges and modes of the standard lengths are shown. Size estimates were obtained from Fryer (1959) and Fryer and Iles (1972).

elements to this relationship. First, many species of fish reach sexual maturity at a small size (<100 mm), and they are abundant on coral reefs (P. J. Miller, 1979) and use corals as permanent shelter sites. Second, the reef is colonized by very large numbers of settling fishes, most of which are cryptic or require access to shelter during the critical periods of establishment. The abundance of very small fishes on coral reefs may be partially explained in physical terms.

Fishes exploit corals in a number of ways. These may involve direct interactions such as feeding on their tissues and associated symbionts and on their by-products such as mucus. A number of species in the groups listed earlier show very specific modifications in this respect. A more complex and indirect set of interactions results from the feeding activities of grazing and browsing herbivores. Fishes may be detrimental to coral growth and distribution on a local scale through feeding activities that promote algal growth. This occurs with larger species of herbivorous pomacentrids (Potts, 1977; Risk and Sammarco, 1982). Grazing fishes substantially modify the standing crop of algae on reefs by intensive feeding, which has both negative and positive effects on corals. Newly recruited corals may be removed by grazers, but more importantly, grazing reduces algae that might otherwise overgrow corals (Horn, 1989).

Grazing by fishes, especially scarids, erodes calcareous structures and produces sediment as a by-product of this feeding (Chapter 6). This results in the development of sedimentary aprons at reef bases and in lagoons, which provide feeding sites and habitats for large numbers of deposit- and suspension-feeding invertebrates. These animals in turn constitute the major food source for diurnally foraging mullids and nocturnal lethrinids, lutjanids, and haemulids. There is also evidence of reciprocal interactions between the latter groups of fishes and coral growth, in that fishes may facilitate the redistribution of nutrients that is important to corals via site-specific defecation (Meyer *et al.*, 1983).

Fish feeding may retain nutrients and energy within reef systems via defecation, and this occurs in a number of trophic contexts. The best examples are provided by the feeding activities of planktivorous fishes. Buoyant biological particles and zooplankters may be accumulated near reef systems by topographically generated fronts (Wolanski and Hamner, 1988) and serve as a food source for numerous species of planktivores (Hobson and Chess, 1978). A substantial portion of the energy intake of these species is provided from reef-derived sources, especially reproductive products of invertebrates and fishes. Moreover, much of this energy destined for dispersal away from the source reef by currents (Johannes, 1978a) is retained by planktivores and returned to the reef via defecation. Defecation by planktivores can provide additional sources of nutrients for other species of fishes, especially for herbivores (Robertson, 1982).

Additional examples of rapid and localized nutrient recycling to the reef biota are provided by herbivores and benthic carnivores. Polunin (1988) provided examples of a rapid return of territorial herbivores' fecal products to the reef biota. Benthic carnivores (especially haemulids and lethrinids) may forage into nonreef habitats and concentrate nutrient returns to specific areas of reef (Meyer *et al.*, 1983; Parrish, 1989). In the cases of carnivores, which produce energy and nutrient-rich feces (Bailey and Robertson, 1982), there is a relationship between the reef physical structure, fish feeding, and nutrient cycling.

In summary, fish–habitat interactions on reefs take three general forms. First, there is the direct relationship between reef structure and shelter, being most obvious in terms of small fishes. Second, there is a feeding interaction involving reef fishes proper and the sessile biota, including algae. This has a number of important secondary effects, including mediation of the interaction between algae and corals and the generation of sediment-based habitats. Third, there is the role of reef structure and the feeding patterns of planktivores and reef-associated carnivores. These groups exploit prey groups associated with reef-generated habitats, including hydrological fronts and sediment fields. Feeding in these habitats results in a link between fish activities and the recycling of nutrients from adjacent habitats through the reef system.

V. The Origin of Reef Fishes

A. Taxonomic Affinities and Fossil History

Given that one can identify or characterize reef fish communities as distinct assemblages in terms of both morphology and familial composition, we are in a position to begin to assess the evolutionary history of reef fishes. This can be approached through the following questions.

When did reef fishes first appear, and where? The exact evolutionary history of reef fishes is not clear. The vast majority of reef fishes are in the Series Percomorpha, Order Perciformes. The Perciformes is the largest vertebrate order with about 7800 species in 150 families. Unfortunately the Perciformes cannot be rigorously defined cladistically. It is a nonmonophyletic assemblage and appears to be both polyphyletic, that is, having arisen from several distinct acanthopterygian lineages, and paraphyletic in that it appears to have given rise to several more specialized percomorph orders. The interrelationships between the major groups within the Perciformes likewise cannot be clearly defined (Lauder and Liem, 1983).

However, it is generally accepted that the perciform fishes arose from a beryciformlike acanthopterygian stock during the late Cretaceous over 65 million years ago (Carroll, 1987). The Beryciformes are the earliest and most primitive of the percomorphs and are represented today on reefs by the Holocentridae (squirrelfishes, e.g., *Sargocentron,* and soldierfishes, e.g., *Myripristis*). The Perciformes and Beryciformes appear to share an immediate common ancestor. Both groups were present in the late Cretaceous. During this period and in the early Tertiary (to 50 million years before present, or Myr BP) the perciform fishes underwent a period of extremely rapid anatomical evolution (Carroll, 1987).

Of the 20 perciform suborders living today, 6 have no fossil record (although these include only 14 of the 1367 living genera). Of the remaining 14 suborders 11 are represented by living genera that date to the Paleocene (54–65 Myr BP) or Eocene (54–30 Myr BP). The remainder are represented by living genera that date to the Oligocene (38–26 Myr BP) or Miocene (26–5 Myr BP) (Carroll, 1987). Thus within 20 Myr of the first appearance of perciform fishes they achieved a level of morphological complexity and diversity that is almost indistinguishable from that of living fishes. From paleontological records, therefore, it appears that reef fishes arose at the beginning of the great radiation of perciform fishes, which is a dominant feature of the Tertiary period and within the first few million years of the Tertiary we see the sudden appearance of an almost complete reef fish community.

This appearance coincides with major changes in the nature of coral reefs (Rosen, 1988). Although reefs and reef-building corals have had a long tenure in the fossil record, the first appearance of the coral taxa that dominate reefs today, the scleractinian corals, was not until the late Cretaceous/early Tertiary. The early Tertiary was a major period of reorganization and expansion of coral reefs after a period of decline and extinctions of many forms at the end of the Cretaceous (Fagerstrom, 1987). In the Tertiary, coral reefs took on a modern appearance, in that for the first time they were dominated by scleractinian corals. Throughout the Tertiary, coral reefs were a prominent feature of tropical shallow seas in many regions. The first records of the scleractinian genera that dominate reefs today (*Acropora, Porites,* and *Pocillopora*) are from the Eocene. The Paleocene and Eocene span a transition from the Mesozoic to modern coral forms (Rosen, 1988). With these corals the modern reef fish community appeared.

The key to understanding the evolution of reef fishes lies in the events that occurred during the late Cretaceous to early Tertiary (70–50 Myr BP). It is fortunate that the best fossil deposits of coral reef fishes date from this critical period. Most of what we know of the evolution of reef and reef-associated fish assemblages comes from the fossil records of southern Europe. This region was inundated by a shallow tropical sea that was part of the northern margin of the Tethys Sea during the late Cretaceous and early Tertiary periods. This region had areas of coral reef throughout most of the Tertiary, up to at least the mid-Miocene (Rosen, 1988).

The ancient fish communities of southern Europe are known from several fossil deposits. Late Cretaceous fossil deposits in southern Italy (about 75 Myr BP), which appear to have been associated with reef structures, contain beryciform and perciform fishes, but no reef fish families. Collections from the Eocene, on the other hand, have a rich beryciform component (Sorbini, 1979) and numerous reef fish families.

These deposits are unique in terms of the number and quality of the specimens and are of paramount importance in the study of the evolution of reef fishes, due to (1) the nature and age of the material, (2) the community diversity, and (3) the quality of individual specimens. Almost all of what is known of the evolution of reef fishes is derived from these deposits. They represent the oldest fossil record of most reef fish families.

B. The Fossil Beds of Monte Bolca: A Snapshot of Reef History

The Eocene deposits of Monte Bolca in northern Italy are the most important fossil teleost assemblage and are exceptional in both quantity and quality. The specimens have come from two sites, Monte Postale and Pesciara, located several hundred meters apart on the slopes of Val Cherpa in the foothills of the

Italian Alps. Over 227 species of fish in 177 genera have been recorded from these deposits, representing 80 families in 17 orders (Blot, 1980).

The Eocene fishes of Monte Bolca are clearly a reef fish community. Among the 80 families recorded by Blot (1980), the following families are represented: Ephippidae (e.g., *Eoplatax*); Chaetodontidae (e.g., *Pygaeus, Parapygaeus*); Pomacentridae (e.g., *Odonteus*); Labridae (e.g., *Phyllopharyngodon*) (Bellwood, 1991); Acanthuridae (e.g., *Acanthonemus, Acanthurus, Tylerichthys, Naseus*); and Zanclidae (e.g., *Eozanclus*) (Blot and Voruz, 1975). The fossil beds also contain a genus of acanthuroid fishes, *Ruffoichthys,* with strong similarities to present-day siganids (Sorbini, 1983a). These families now occur almost exclusively on reefs and as noted earlier may be regarded as the characteristic taxa that define modern reef fish assemblages.

The Bolca collections are remarkable in that they provide a complete community picture of reef fish fauna at this time. Entire fish communities appear to have died in mass mortality events, with no evidence of selectivity in mortalities or subsequent preservation. The collections also appear to provide an unbiased picture of the community structure. These mass mortality events killed large numbers of fish of every ecological form, from benthic bothids and rays, to reef acanthurids, to open-water scombrids and clupeids. Subsequent preservation is excellent, with 1-m-wide batoid rays and 2-cm-long gobies preserved with equal clarity. The large fins of the batfishes *Eoplatax* are preserved in expanded position usually with no evidence of physical damage (Sorbini, 1983a).

The most widely accepted explanation for these mass mortality events is that they were the result of algal blooms. Comparable phenomena may be seen today in shallow bays and lagoons. Periodic algal blooms in the ancient shallow lagoonal waters are believed to account for the large numbers of fish in the deposits, including diverse ecological forms. The deposits include crabs and other invertebrates that presumably died along with the fishes, which may account for the fact that the fishes show no signs of damage due to scavengers. The surrounding sediments suggest that the dead fish settled in a sheltered area with evidence of low oxygen levels and high sedimentation rates. As a result, the Bolca specimens are extremely well preserved.

The evidence available from the fossil record has been largely neglected in the study of the evolution and biogeography of reef fishes. Theoretical problems notwithstanding, some possible reasons are quite clear: (1) Most fossil specimens are represented by fragmentary remains, which limits the value of the specimen for cladistic analyses as the number of available characters is limited and often makes identifications difficult. (2) Many identifications are based on phenetic criteria, that is, overall similarity. Such identifications, especially if based on fragmentary remains, are highly subjective and are difficult to place in classifications based on cladistic criteria. Because of the number of questionable identifications that have arisen over the years, the fossil record as it stands for many reef fish taxa is of limited value in evolutionary or biogeographic studies.

These problems are exemplified by the fossil record of the Scaridae. Of the seven extinct species and six fossil fragments placed in the Scaridae, six were found to be *incertae cedis*, five could be placed in other families, one identification was tentatively accepted (material not located), and one identification was accepted. The earliest known date for the family was changed from the Eocene (50 Myr BP) to the Miocene with an age of no more than 26 Myr BP and probably 15 Myr BP (Bellwood and Schultz, 1991).

Despite these problems, fossils have a great deal to offer, especially those from Monte Bolca. The Bolca fossils are almost invariably whole fish and preservation quality is so high that details of all hard internal structures are preserved, which enables detailed comparisons to be made with living species. The preservation of bones is particularly valuable as the osteology of recent fishes represents the basic structural feature which is used in systematic studies and is widely used in cladistic analyses. In some specimens the preservation is so good that squamation and pigmentation patterns are still visible.

The two most important features of the Bolca fish fossils are (1) that individual species are readily identifiable and bear a striking resemblance to living forms and (2) that an almost complete community is preserved, with the most highly specialized forms characteristic of reefs today being represented. The similarity between the Bolca specimens and living forms is remarkable. Species can be readily placed in recent families and, although placed in different genera, the differences between modern and fossil forms are often minimal. Many of the component species are almost identical morphologically to modern representatives. *Eoplatax* shows great similarities to the modern batfishes *Platax*. The similarities are even more significant in the case of the Zanclidae, a family of distinctive and highly modified acanthuroids. *Eozanclus* from Eocene deposits is virtually indistinguishable from the Recent moorish idol *Zanclus*.

By the early Eocene, reef fish communities not only included most of the families characteristic of reefs today, but these fishes had already achieved a level of morphological development equal to that observed in present-day families. Liem and Sanderson (1986) describe the pharyngeal apparatus of labrids as the most highly integrated and specialized pharyngeal jaw apparatus among acanthopterygian fishes, yet labrids with these complex structures were already present in the Eocene and the major labrid lineages had already differentiated (Bellwood, 1991). This points to rapid evolution followed by an exceptionally long period of relative stasis. Reef fishes evolved rapidly within a 20-Myr period, and thereafter remained virtually unchanged for the duration of the Cenozoic Epoch, a period of over 50 Myr. What can this tell us about present-day reef fish assemblages?

The Eocene deposits of Monte Bolca represent the first record of a "modern" coral reef community. Indeed, it is the first record of many reef fish families. While having strong overall similarities to present-day reefs, the community differed in two important respects, which give an indication of the unique significance of this assemblage in the evolution of reef fish communities.

First, the Eocene Bolca community has distinct links with older assemblages from the Mesozoic (248–65 Myr BP). It is particularly rich in beryciform and generalized perciform fishes, a reflection of the dominant fauna prior to the Eocene. The presence of pycnodonts provides one of the strongest links with older assemblages of fishes. Blot (1980) records three species from the family Pycnodontidae from Monte Bolca. These species represent the remnants of a diverse order of fishes (the Pycnodontiformes) that were dominant in shallow waters throughout most of the Mesozoic (Carroll, 1987). The pycnodonts apparently died out in the Eocene.

Second, despite the similarities between the ancient (Eocene) and modern fish communities, some important differences are apparent. For example, on modern reefs, the herbivorous community is dominated by the Acanthuridae and Scaridae (Bouchon-Navaro and Harmelin-Vivien, 1981). Today, these two groups have comparable species diversities. Their latitudinal and longitudinal distributions are similar, and within individual reefs their distributions broadly overlap. In the Monte Bolca deposits, the Acanthuridae are well represented with numerous species. In comparison, the Scaridae have not been recorded from these deposits to date. The first record of the Scaridae is not until the Miocene (approximately 15 Myr BP). Contrary to the published record (Steneck, 1983; Carroll, 1987), scarids as the main piscine excavating herbivores of the reef environment (Kiene, 1988) do not appear to have been an important element of the pre-Miocene fauna. This suggests that some previous scenarios concerning the response of sessile organisms to excavating herbivores in pre-Miocene reef habitats (Steneck, 1983) may need reevaluation.

C. The Major Features of Reef Fish History

The morphological stability of reef fishes, especially acanthurids, zanclids, labrids, and platacids, stands in marked contrast to that of other vertebrate groups. Mammals, for example, have a much more labile history, having experienced a period of radiation in the late Cretaceous/early Tertiary followed by the extinction of many archaic forms. The subsequent Tertiary history was characterized by a gradual change of morphology in many forms and the appearance of many modern groups within the last 10 million years (Carroll, 1987).

When we look at reef fish communities today, the antiquity of the taxa must be addressed. To put this into perspective, most higher taxa of reef fishes had evolved long before the first corals settled on the Great Barrier Reef, which is a relatively modern structure. When we consider proximal factors that may appear to determine feeding modes, distributions, reproductive patterns, etc., these must be viewed in the light of the antiquity of many perciform groups.

The fossil record suggests that factors influencing morphological form occurred early in the evolution of reef fishes. Alternative histories of the habitat/morphology interaction in reef fishes are suggested. Either (a) conditions in reef areas have changed little during the Tertiary and present-day fish/habitat interactions are an accurate mirror of reef fish history or (b) fishes have maintained similar structures and morphologies in the face of fluctuating habitat features. This suggests that once a particular morphological plan is adopted, subsequent change may be very difficult. Whichever alternative is correct, the morphology of present-day reef fishes appears to have been established coincident with the origin of modern reefs.

Where did reef fishes evolve? The evidence to date suggests that this occurred in the equivalent of present-day reefs. The few deposits from which we can obtain reef fish fossils are in sites that indicate that they were laid down in shallow tropical coastal marine environments in the vicinity of coral reefs (Sorbini, 1983a,b). The pantropical distribution of many reef fish families and the strong association today between reefs and reef fish families suggest that most reef fish families were widespread at least prior to the closing of the Middle East land bridge (15–20 Myr BP).

Bolca, the best fossil data base, is in many ways unique, but in analyses of past distributions it represents a sample of one. Bolca can tell us minimum ages of taxa, their morphology, and that taxa were present together in the region, but little about past distributions. The best indications of past distributions of reef fishes may be through the use of coral historical biogeography. Rosen (1984, 1988) and Rosen and Smith (1988) have discussed the historical biogeography of scleractinian corals in the context of differentiation of the Atlantic and Indo-Pacific regions. They have suggested a widespread tropical distribution for Eocene scleractinians and biotic differentiation of the Atlantic and Indo-Pacific faunas predating important vicariance events such as the closing of the Tethyan seaway.

Eocene scleractinian coral distributions were widespread but the faunas were not cosmopolitan (Rosen and Smith, 1988). There is increasing evidence of regional divergences in the coral biotas before major vicariance events in the Miocene. For example, although the modern Indo-Pacific and tropical Atlantic regions came into being as physical entities during the Miocene, the coral biotas on either side of the Middle East land bridge were differentiated well before this (Rosen and Smith, 1988). A long history of Atlantic and Indian Ocean faunal differentiation has been recognized. Similar arguments for a Middle and East Pacific isolation and divergence of coral faunas prior to the Miocene have been made (Rosen and Smith, 1988).

For corals, the appearance of an Indo-Pacific fauna distinct from the existing Indian Ocean and Pacific faunas was associated with vicariance events in the Miocene. These included the northward movement of Australia and the associated uplift of landmasses in the West Pacific. Much of the evidence relating to the fossil history of scleractinian corals suggests sequences of fragmentation and divergence since the Cretaceous and not the development of the present fauna from a pantropical cosmopolitan assemblage (Rosen and Smith, 1988).

Fish fossil assemblages do not begin to match those of scleractinian corals. We have assumed that true reef fishes have had a significant association with scleractinian corals from their first appearance. Sorbini (1983b) argues that the Indo-Pacific affinities and associations with pantropical groups in the Bolca reef fish assemblage indicate an ancient widespread distribution and connection with a generalized Tethyian tract. However, the lack of any specific Bolca–Atlantic connections makes it difficult to separate past widespread distributions from recent dispersal events. The former can only be resolved with corroborating fossil data as found in *Mene* as described by Sorbini (1983b). Better hypotheses of past distributions can come from two sources: cladistic analyses and fossil data. There is an urgent need for a greater geographic spread of fossil sites and identifications in a cladistic framework.

In summary, we suggest that reef fishes evolved in shallow tropical seas on, or in the vicinity of, coral reefs. Fossil evidence suggests that this evolution was relatively rapid, occurring within 20 Myr. The Monte Bolca fossils give us a minimum age of many groups of 50 Myr BP. By this time the reef fish community closely resembled that of modern reefs, with most families represented. Individual families and genera had already reached a level of morphological complexity comparable to that of modern forms and, having reached this level, remained virtually unchanged for the next 50 million years. The antiquity of many reef fish groups is of critical importance in the study of the present. The notion of a widespread Eocene distribution is a plausible starting point. If the notion of correspondence with corals is pursued, then it suggests that divergences among reef fish faunas occurred prior to the Miocene. Their history is uncertain. One thing, however, is clear: the appearance of reef fishes predates many significant biogeographic events. Reef fishes with all their morphological specializations have maintained their integrity in the face of a sequence of major vicariance events that produced biogeographic regions such as the Indo-Pacific.

VI. Morphology, Phylogeny, and History: Reef Fishes Revisited

Research on reef fishes to date has emphasized their ecological and life history features with a marked bias toward particular groups of labroids. The focus has been on ecological interactions and their demographic consequences. Evolutionary aspects of the arguments are often implied rather than spelled out explicitly. Nevertheless, the concept of adaptation and the role of natural selection have played an important part in developing our current perspectives in reef fish biology.

In this chapter we have emphasized the distributional and structural features of reef fishes as a basis for defining this assemblage and for reviewing their relationship with the reef habitat. We have also reviewed what is known of the fossil history and past distributions of these groups. Two salient features emerge from this. First, the fishes have distinctive morphologies and structure associated chiefly with feeding on a coral reef sessile biota. Second, they are an ancient group. The first records of reef fishes reveal an Eocene fauna with all the distinctive features of advanced perciforms already in place.

The structural and therefore anatomical features of reef fishes have been retained through a long and often turbulent history of climatic and geological change of the shallow-water tropical environment (Fagerstrom, 1987; Rosen and Smith, 1988). Clearly these features have served reef fishes well, as they have emerged from this history with the integrity of the reef fish fauna, so clearly defined in the Bolca deposits, intact. However, the structural features of reef fishes which may be related to their interaction with sessile biotas may have constrained other aspects of their biology.

The dominant features of reef fishes are lateral compression and a substantial investment in skeletal structure. This occurs not only as a well-developed axial skeleton but also as numerous ancillary structures. The enhanced development of fin spines in siganids and chaetodontids, caudal knives in acanthurids, and pharyngeal apparatus in scarids are cases in point. Ratios of body length to depth are also modified, resulting in high-bodied, laterally compressed fish with a narrow shortened visceral cavity. As a result of the major investment in skeletal structures, the visceral cavity must accommodate a substantial swim bladder. Feeding on the sessile reef biota is associated with the development of an elongate and anatomically complex alimentary tract and a large liver. For groups such as the Acanthuridae, additional structures like fat bodies must be accommodated in posterior extensions to the visceral cavity (Fishelson *et al.*, 1985a).

If we regard the visceral cavity as the site of a number of biological processing units then there are obvious structural constraints on the amount of space available for the different organ systems. These include those associated with the processing of food, storage and elaboration of the resultant materials, maintenance of position in the water column, and the development of reproductive products. All are essential functions that must be carried out in a limited volume and so some degree of compromise and constraint is to be expected.

Does this have an effect on the organization of the reproductive system and subsequent life history features of reef fishes? Most perciforms have high fecundities with single litters measured in thousands, with either eggs or larvae dispersed into the water column. A significant minority of species show some form of egg care. What mechanisms have driven these life history patterns? Why do not perciforms commonly produce small numbers of large offspring as do elasmobranchs? The latter group has demonstrable evolutionary success (Schaeffer and Williams, 1977), and its species are consistent if not abundant members of the reef fish fauna, yet they lack the twin features of high fecundity and a dispersive larval stage. There are obviously no long-term evolutionary disadvantages to the elasmobranch life history pattern.

Provision of live bearing is not restricted to elasmobranchs. Many marine teleost groups, most notably the scorpaeniform genus *Sebastes* and the perciform embiotocids, are live-bearers. Other groups, such as apogonids, are mouth brooders. There are no examples of coral reef perciform live-bearers. Some taxa of brotulid ophidiiform fishes are rare, but they are clearly documented examples of live-bearing coral reef teleosts (P. J. Miller, 1979). Live-bearing does occur with considerable frequency within teleosts, especially in some freshwater groups, and many, such as siluriforms, may produce small numbers of large yolky eggs. However, given the diversity of coral reef perciforms, the uniformity of high fecundity and dispersive spawning is noteworthy.

The anatomy of chaetodontids and acanthuroids must constrain the size of offspring. The internal organization of the viscera and size of the visceral cavity may not permit the simultaneous development of even modest numbers of large offspring. The taxa of reef fishes reviewed here would be obvious examples of this type of constraint, for all have a relatively large and complex visceral mass. The ability to feed on and process reef sessile biota may predetermine the form of other activities (i.e., reproductive output).

Large size at birth requires a relatively long developmental period within the body. The internal configuration of reef perciforms would allow for the development of only very small numbers of offspring under conditions of internal development. If there is a numerical threshold required for consistent survival to maturity in fishes in reef habitats, then a minimum number of propagules must be produced. The production of more numerous offspring requires a reduction in mean offspring size at birth, and this can be accomplished in two ways. For teleosts, which achieve a large adult size and have an internal configuration that allows development of a relatively large ovarian mass, a few major spawning episodes will produce large numbers of offspring. For small teleosts or those with little free space in the visceral cavity, this must be accomplished by continual spawning and rapid production of eggs.

The frequency and periodicity of spawning by reef fishes vary but few comprehensive records that estimate the true output are available. Accurate estimates of reproductive output will be difficult to obtain as low-frequency sampling with gravimetric and histological analysis will underestimate true egg production. Recent information suggests that the degree of underestimation will be considerable (Thresher, 1984). Individual labrid and acanthurid fishes undergo daily spawning episodes that must result in a major throughput of high-energy reproductive products over very short time periods (Fishelson *et al.*, 1987; Hoffman and Grau, 1989).

The significance of reef fish life histories and the dispersive larval phase is usually approached from an adaptive viewpoint. How might natural selection generate, maintain, and fine-tune life history parameters? While natural selection undoubtedly plays some role in this process, general features of the life cycle such as small egg size, high fecundities, and an ontogeny that spans several orders of magnitude of size may reflect more fundamental features of teleost biology, that is, basic constraints on the reproductive function imposed by adult size, body proportions, and internal structure. The size of individual reproductive products may be constrained by the space available in the visceral cavity. This would also influence the frequency of spawning if high production of eggs is to be maintained in the face of a reduced ovarian standing crop. There may be an interesting relationship between body structure, fecundity, and frequency of spawning.

This represents an alternative viewpoint to discussions of life histories and reproductive biology in reef fishes. Not only may visceral architecture play a central role in determining the reproductive patterns in chaetodontid and acanthuroid fishes, but the spawning modes displayed may be plesiomorphic and representative of the reproductive patterns of the earlier biotas. Pre-Cretaceous radiations of actinopterygian fishes provide evidence of similar patterns of structural development. The most obvious candidates would be the Paleozoic paleonisciform fishes of the suborder Platysomoidei (Moy-Thomas and Miles, 1971) and the pycnodontid neopterygians of the Mesozoic. Both have representatives with high body planes and lateral compression, which would result in a reduced visceral cavity (Fig. 3). The same types of constraints could have influenced the reproductive patterns of these fishes.

Some of these species were associated with reef habitats (Sorbini, 1983a). From the evidence of jaw architecture it is doubtful that these groups were herbivorous (Fig. 3). The tendencies in body form in paleoniscoid and neopterygian fishes were not associated with the development of a small terminal mouth. Only in the perciform teleosts did the characteristic combination of

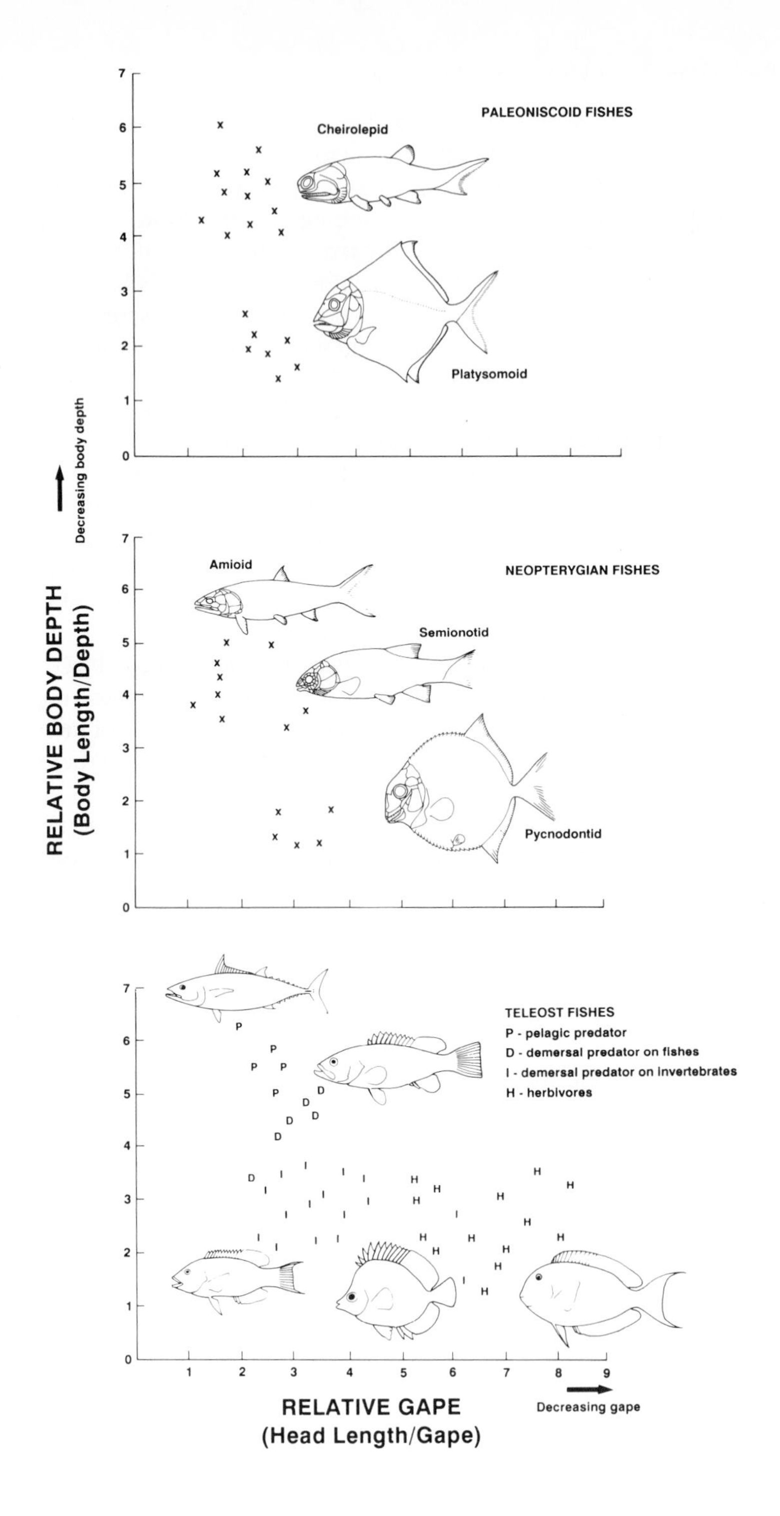

PALEONISCOID FISHES
Cheirolepid
Platysomoid
NEOPTERYGIAN FISHES
Amioid
Semionotid
Pycnodontid
TELEOST FISHES
P - pelagic predator
D - demersal predator on fishes
I - demersal predator on invertebrates
H - herbivores
RELATIVE BODY DEPTH
(Body Length/Depth)
Decreasing body depth
RELATIVE GAPE
(Head Length/Gape)
Decreasing gape

body form and jaw structure associated with feeding on sessile reef biotas appear. Earlier groups of actinopterygian fishes may have been subject to the same pattern of anatomical constraints as perciform reef fishes. However, the ability to exploit reef algae and small colonial invertebrates was apparently restricted to post-Cretaceous perciforms (Fig. 3).

The relationship between rapid turnover of reproductive products and frequent spawning adjacent to reef habitats may have more general implications for reef biological systems. The abundances of labroid and acanthuroid fishes in reef systems and their frequent spawning means that a significant proportion of the productivity of reef fishes is exported via reproductive propagules. However the abundances of planktivorous fishes (Williams and Hatcher, 1983; Hobson and Chess, 1978) and the rate of feeding on their fecal products by herbivores (Robertson, 1982) mean that much of this productivity may be returned directly to the reef system at the point of potential departure. How much productivity is expressed in terms of somatic growth of fishes and how much via reproductive products is an important issue for reef dynamics.

Over the last ten years much of the emphasis in reef fish biology has tended to focus on patterns of settlement and early recruitment stages. The information available on postsettlement phases of reef fish assemblages tends to focus on larger carnivores of commercial importance (Munro and Williams, 1985). An understanding of the unique features of reef fish biology and their evolutionary significance requires more comprehensive work on the metabolism, growth rates, and reproductive schedules of postsettlement chaetodontid, acanthuroid, and labroid fishes.

Our summary emphasizes dual aspects of reef fish biology: their structural features and evolutionary history. Even though the fossil record is sparse, the material it contains is of crucial importance. Two cases illustrate this. First, the

Figure 3 Body form and structure of actinopterygian fishes. The graphs are ratios of body length/depth plotted on head length/gape to display trends in the body form and feeding behavior of three groups of actinopterygian fishes. Paleozoic paleoniscoid fishes are represented by cheirolepids and platysomoids; Mesozoic neopterygian fishes are represented by amiids, semionotids, and pycnodontids; and present-day perciform teleosts are partitioned by feeding group. Fusiform large-mouthed fishes (predators on nektonic organisms) are represented in the upper left quadrant of the lower graph; laterally compressed large-mouthed fishes (predators on mobile invertebrates) in the lower left quadrant; and laterally compressed small-mouthed fishes (herbivores and colonial invertebrate feeders) in the lower right quadrant. The lower right quadrant is occupied only by post-Cretaceous perciform herbivores and benthic invertebrate feeders. Body length is the standard length; body depth is the greatest body depth; head length is the snout to posterior opercular margin; and gape is based on the length of the upper jaw. Data are from Moy-Thomas and Miles (1971), Carroll (1987), and Masuda *et al.* (1984).

ancient nature of reef fish assemblages and the rapidity with which complex structural configurations such as the labrid pharyngeal apparatus became established prompt a reexamination of ideas concerning reef fish–habitat relationships and their evolution. Second, the unexpectedly late appearance of scarids in the fossil record, despite the superb beds at Bolca, requires some rethinking of the evolution of reef biota likely to be influenced by excavating herbivores. The material available is unequivocal enough to allow interpretation of past events in terms of present reef fish biology with some confidence.

The hypotheses generated are not readily testable in the light of present information. Three investigative programs are required. First, comparative studies of related groups such as lacustrine cichlids and temperate reef labrids and embiotocids will help clarify ideas concerning structural patterns and reproductive tactics. The fact that a single pomacentrid (*Acanthochromis polyacanthus*) has a body structure and reproductive mode that are almost indistinguishable from those of cichlids (Robertson, 1973b; Thresher, 1985a) adds to this interest. Second, more cladistic analyses of labroid, acanthuroid, and chaetodontid fishes are needed. These studies might be usefully associated with molecular and genetic techniques that access patterns of relatedness through sequencing of mitochondrial DNA. Third, careful scrutiny of the fossil representatives of these groups and the search for additional deposits are an urgent priority.

Acknowledgments

The authors are grateful to the following for discussion of the material for this chapter: J. M. Leis, M. Gomon, J. Paxton, A. M. Ayling, K. D. Clements, G. R. Russ, J. E. Randall, B. C. Russell, B. D. Mapstone, and L. Sorbini. Support for studies of present-day reef fishes was provided by the Australian Research Council. Assistance with travel to the Museo Civico di Storia Naturale di Verona and the Naturhistorisches Museum Wien, was provided by the Australian Research Council and the James Cook University Research Funding Panel.

Part II

Trophic Ecology

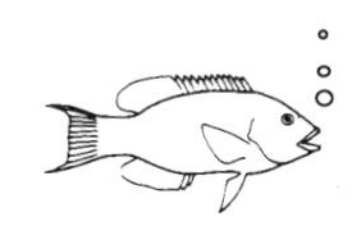

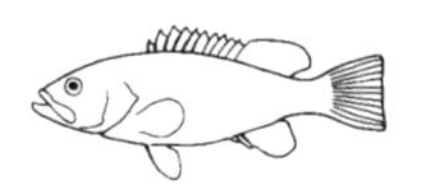

CHAPTER 4

Trophic Relationships of Fishes Specialized to Feed on Zooplankters above Coral Reefs

Edmund S. Hobson
Southwest Fisheries Science Center
Tiburon Laboratory
Tiburon, California

I. INTRODUCTION

Fishes specialized to feed on zooplankters are major components of coral reef communities. They are distinctive animals, with many features that relate to their way of feeding. Although most fishes are planktivorous as larvae and early juveniles (Durbin, 1979), generally only species with appropriate adaptations remain obligate planktivores as adults (Davis and Birdsong, 1973; Hobson and Chess, 1976, 1978). This chapter examines how these species have accommodated the special needs of planktivory while retaining close ties to reef structures, generally as places to shelter when at rest or threatened.

Virtually all coral reef planktivores are among the highly diverse acanthopterygians (spiny-finned teleosts), which are the most highly evolved fishes (Gosline, 1971). Acanthopterygians dominate in coral reef communities (Smith and Tyler, 1972, Schaeffer and Rosen, 1961); for example, 98% of the species noted during a census of Hawaiian reefs are members of this group (Hobson, 1974).

Despite the great diversity of forms among coral reef acanthopterygians, however, virtually every major family includes species that are specialized as planktivores. For example, there are diurnal planktivores among the Serranidae, Chaetodontidae, Pomacentridae, and Balistidae, and there are nocturnal planktivores among the Holocentridae, Priacanthidae, and Apogonidae (Starck and Davis, 1966; Randall, 1967; Hobson, 1974). Most planktivorous acanthopterygians, including all species considered in this chapter, feed with visually oriented strikes at individual prey (Zaret, 1972; Confer and Blades, 1975; Durbin, 1979; unpublished observations). Furthermore, the major threats to planktivores feeding above coral reefs (at least during the day) come from visually oriented attacks of large piscivorous fishes (Hobson, 1968, 1972). With vision so important, the trophic relations of coral reef planktivores—both as predators and as prey—are profoundly influenced by variations in levels of incident light (Hobson, 1972; Stevenson, 1972; Collette and Talbot, 1972). Many are adapted to specific photic conditions (Hobson, 1972; Munz and McFarland, 1973) and the vast majority feed strictly by day or by night (Hobson, 1974, 1975).

A. Need for Innovative Methods of Study

Studies of trophic dynamics among fishes and zooplankters above coral reefs have required innovative sampling procedures. The traditional methods used to sample zooplankton, for example, were developed for work in open water offshore and are ineffective in confined spaces among reefs. So studies that would sample coral reef plankton have used nets pushed by divers (Emery, 1968; Hobson and Chess, 1978) and tethered to anchors on the reef (Johannes *et al.,* 1970; Hobson and Chess, 1978). And traps have been set on the seabed to sample the many otherwise benthic organisms that periodically join the plankton (Porter and Porter, 1977; Alldredge and King, 1977, 1980; Hobson and Chess, 1979, 1986). Similarly, assessments of trophic relations among fishes based on specimens collected from above water are limited by problems of relating gut contents to occurrences of organisms in the environment. To better define prey selection by planktivorous fishes, hand-held spears have been used to select specimens known to have been feeding at the specific time and place where the plankton was sampled (Hobson, 1968; Hobson and Chess, 1976, 1978).

Perhaps the greatest advantage these methods have over traditional shipboard operations for study of trophic interactions, however, is that they put the investigator at the site of the interactions. From there one can directly observe vital spatial and temporal relationships and behaviors that can only be inferred from above water. These methods and advantages, with emphasis on events throughout day and night, have been refined by my studies with James R. Chess of interactions among reef fishes and zooplankters at widespread locations in the Atlantic and Pacific oceans. This chapter incorporates data, observations, and impressions gained from these studies in a synthesis of present knowledge.

II. Diurnal Planktivores

Planktivorous reef fishes that feed by day typically form aggregations in the water column (Fig. 1), and from a distance it is often difficult to distinguish one species from another. Despite differing limitations on adaptive change

Figure 1 Feeding aggregation of *Chromis atrilobata* (Pomacentridae), a diurnal planktivore in the Gulf of California, Mexico.

related to their differing phylogenies, these fishes have acquired similar features in response to problems they have in common. This is especially evident in adaptations to the size of their prey and to feeding in exposed positions above the reefs. As a result of the remarkable degree of morphological and behavioral convergence that has developed, they tend to resemble one another more than they do other members of their own families that feed on the benthos (Davis and Birdsong, 1973).

A. Adaptations to a Diurnal Planktivorous Diet

Most diurnal reef planktivores feed primarily on swimming crustacea, particularly calanoid and cyclopoid copepods, but larvaceans or fish eggs are favored by some (Hiatt and Strasburg, 1960; Hobson and Chess, 1978; Sano *et al.*, 1984b). These prey vary in form and behavior, but those accessible during the day are largely transparent and visible mainly through their pigmented parts, such as eyes, or even gut contents (Zaret, 1980). So while planktivorous fishes generally take the larger of zooplankters available to them (Ivlev, 1961; Coates, 1980), they have been known to take the smaller of two similar forms if this one is more heavily pigmented (e.g., Zaret, 1972; Zaret and Kerfoot, 1975).

Whatever problems there may be with pigmentation, however, clearly it is an advantage to be smaller (Hobson and Chess, 1976; Obrien, 1979). And so zooplankters above coral reefs during the day tend toward not only transparency, but also reduced size. Judging from my studies with Chess, virtually all zooplankters taken by planktivorous reef fishes during the day are less than 3 mm in their greatest dimension.

Because planktivorous reef fishes have evolved from early ancestors that were adapted to feed on relatively large prey (Schaeffer and Rosen, 1961; Gosline, 1971), successful feeding depends on the performance of features modified for tasks very different from their original purpose. Particularly important have been modifications of head and jaws, including dentition, that permit even relatively large individuals to consume tiny organisms in open water. In contrast to the large, generalized mouth of the ancestral form (Gosline, 1971), most diurnal planktivores have a small mouth that in many is sharply upturned and with highly protrusible, often toothless jaws (Davis and Birdsong, 1973). Thus, whereas the primitive mouth functioned to grasp large prey, the modern planktivore mouth functions to engulf small prey.

Most of the evolution from the primitive condition, however, occurred in nonplanktivorous progenitors of modern planktivores. Protrusible jaws, for example, characterized the early acanthopterygians (Alexander, 1967; Gosline, 1971) and are now widespread among fishes (see Motta, 1984, for a review). As Gosline (1981, p. 11) stated: "The acanthopteran (acanthopterygian) system of premaxillary protrusion . . . appears to form part of the inheritance of all higher teleosts." It was to a large extent the potential of this feature that led Schaeffer and Rosen (1961, pp. 198–199) to state: "It is primarily the acanthopterygian mouth that has given rise to the enormous variety of specialized . . . feeding mechanisms for which teleosts are so well known." Among these specialized feeding mechanisms are those of diurnal planktivores, which have been refined by selection pressures specific to the planktivorous habit.

So even though some of the most protrusible jaws occur in such nonplanktivorous forms as the piscivorous *Luciocephalus pulcher* (Lauder and Liem, 1981) and the benthivorous *Gerres* spp. (Schaeffer and Rosen, 1961), the feature has nevertheless proven especially adaptive in planktivory. For example, when pomacentrids of the planktivorous genus *Chromis* are within a few centimeters of a target, their jaws shoot forward and engulf the prey (Fig. 2). According to Davis and Birdsong (1973, p. 299), jaw protrusion in planktivores "creates suction which draws prey into the cavity." This view is consistent with Lauder and Liem's (1981, p. 266) assessment that "an underlying assumption of most current research on advanced teleost feeding mechanisms is that protrusion is correlated with increased suction efficiency." But Lauder and Liem went on to show that the extremely protrusible jaws of *Luciocephalus pulcher* (among the most protrusible in teleosts) produce very little suction.

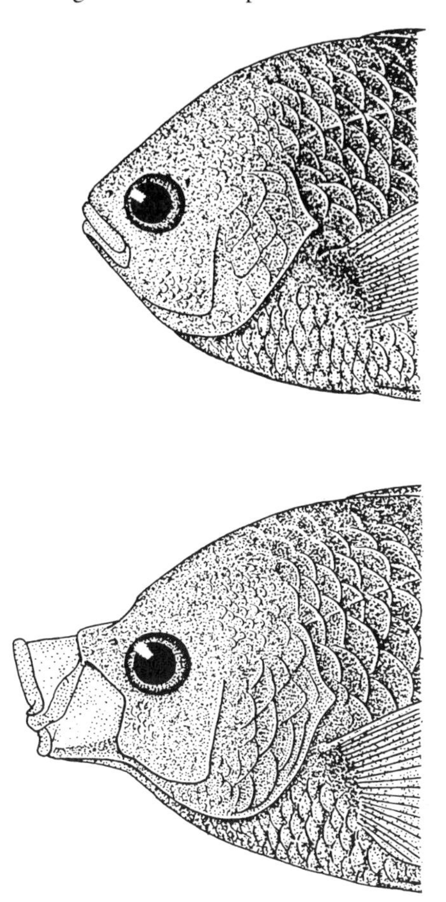

Figure 2 The protrusible upper jaw of *Chromis punctipinnis*, a typical feature of diurnal planktivores, permits projecting the mouth to engulf prey without swimming forward.

The assumption that jaw protrusion functions mainly to produce suction was earlier questioned by Nyberg's (1971) study of feeding mechanics in *Micropterus salmoides* (Centrarchidae). That species, Nyberg concluded, uses protrusion less to create suction than to bring the jaws more quickly to the prey. The advantage is greatest, he observed, in smaller bass moving slowly during the attack, and he related this to a trend in acanthopterygian evolution in which smaller, more protrusible mouths are associated with slower approaches to prey. Certainly planktivorous fishes above tropical reefs represent a culmination of this trend. *Chromis* spp., for example, generally move little, if at all, toward their prey; instead, they depend largely on their extremely protrusible jaws to get their mouths to the target. This is the "ram-jaw" feeding mode described by Coughlin and Strickler (1990), who used high-speed cinemaphotography and video to analyze feeding by *Chromis viridis*. They concluded that this species uses "ram-jaw," low-suction feeding to capture evasive prey, but decreases jaw protrusion and increases suction when prey are less evasive.

Although the small mouth is another planktivore feature widespread among nonplanktivores, the oblique orientation of this mouth, so common among planktivores, seems to be a feature especially adaptive in fishes that would capture tiny, motile organisms in open water. Probably the advantage of this arrangement comes not from the orientation of the mouth, but rather as a consequence of the shortened snout, which places both eyes in position to train simultaneously on small targets immediately ahead (W. A. Starck, cited in Rosenblatt, 1967a).

Gill rakers, a general feature of acanthopterygians that prevents ingested prey from escaping through the gill openings, tend in planktivores to be long and especially numerous (thus closely spaced). This increases their effectiveness in fishes that would feed on small prey (Davis and Birdsong, 1973).

Despite these adaptive tendencies, however, trophic features vary among planktivores—even among close relatives. The mechanisms and structures involved in jaw protrusion, for example, vary among species of *Chromis* (Emery, 1973). So while similarities related to their common purpose unite the group, each species has been to some varying extent customized by distinctive elements of its evolutionary history.

B. Adaptations to Diurnal Threats from Predators

Interspecific convergence among diurnal planktivorous fishes in features that increase mobility are even more striking than the convergence in features that enhance feeding. Particularly evident are tendencies toward streamlined bodies and deeply forked caudal fins—features that increase swimming speeds (Norman, 1947). Significantly, these features generally are more developed in fishes that range farther from the reef (Fig. 3), which indicates an increasing need for swimming efficiency in open water. Consistent with this is a tendency to be larger, which might be expected considering the direct relation between body length and swimming speed.

Planktivores that forage farther above the reef would find that streamlining helps them maintain station in the stronger currents that flow there. They would also benefit from a speedy flight to the reef, because the survival of small fishes exposed to predators in sunlit open water often depends on how fast they get to shelter when threatened (Hartline *et al.,* 1972; Hobson and Chess, 1978).

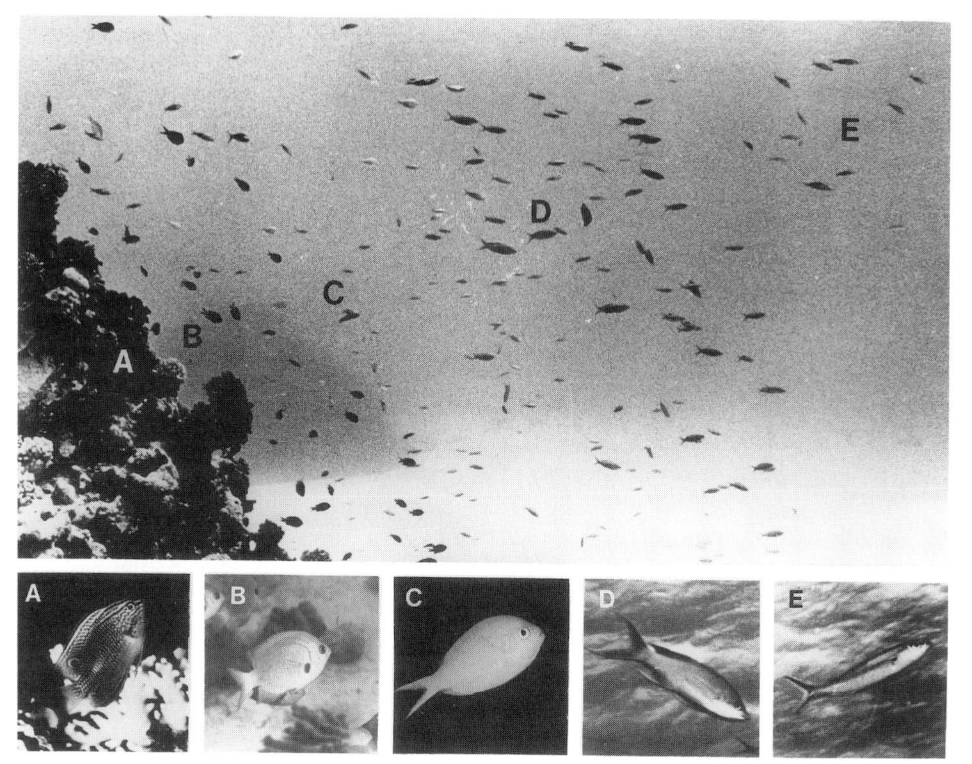

Figure 3 Planktivorous fishes foraging above a reef in the lagoon of Enewetak Atoll. Note that those farther from the reef have features that increase their swimming speed and ability to maintain station in currents. (A) *Pomacentrus vaiuli,* (B) *Chromis agilis,* (C) *Chromis viridis,* (D) *Anthias pascallus,* (E) *Pterocaesio tile.*

Diurnal planktivorous fishes in open water also tend to aggregate, a widely recognized defensive behavior in fishes (Eibl-Eibesfeldt, 1962; Hobson, 1968). That aggregating and a quick retreat to cover are adaptive in defense becomes evident when certain open-water predators appear, notably large *Caranx* spp. (Carangidae). In response to this threat, the planktivores assembled to feed above the reefs abruptly close ranks, and then often dive to the reef below.

Despite the convergence among so many unrelated species in features that increase swimming speed, and the obvious adaptive value of these features, certain other planktivorous fishes have accommodated the same threats by taking virtually the opposite evolutionary course. Instead of being more streamlined than their benthic relatives, species of the planktivorous pomacentrid genera *Dascyllus* and *Amblyglyphidodon* are actually deeper-bodied (and have longer fin spines) than benthivorous pomacentrids (Hobson and Chess, 1978). These features retard rather than promote swimming speeds, so if maintaining station in currents and accelerated flight are important, how do these fishes compensate for their reduced swimming efficiency?

They can, of course, avoid strong currents simply by feeding elsewhere, but what about the problem with predators? I suggest that threats from predators are reduced by their greater body depth and spine development, because these features increase the chance of becoming lodged in the mouth or pharynx of predators that would attempt to swallow them. An effective combination of deep bodies and strong fin-spines probably protects benthivorous as well as planktivorous chaetodontids and pomacanthids (Hobson and Chave, 1972). It is significant that the planktivorous species of these two families are, as a group, indistinguishable from their benthivorous relatives on the basis of external morphologies. Many, in fact, are congeners, for example, the planktivorous and benthivorous species of *Chaetodon* (Hobson, 1974).

Clearly, the exceptionally deep bodies characteristic of these predominantly benthivorous families are suited to activities in open water, although perhaps not where strong currents flow. It would seem that while the streamlined bodies and deeply forked caudal fins of many diurnal planktivorous fishes promote eluding predators, the exceptionally deep bodies and long fin-spines of certain others tend to discourage predators. Both combinations may promote access to zooplankton in open water during the day by reducing threats from predators (Fig. 4).

Planktivorous fishes in open water above reefs become increasingly vulnerable with decreasing light, and they respond by moving closer to shelter. Thus, they forage lower in the water column when the sky is overcast than when it is clear, and they descend briefly when clouds pass in front of the sun on otherwise clear days (Hobson, 1972; Stevenson, 1972).

C. Distribution and Movement

Diurnal planktivores are most numerous along reef edges adjacent to deeper water, probably because their major prey—holoplankters from open water (Hobson and Chess, 1978, 1986)—are most accessible there. Consider, for example, their distribution (excluding juveniles) among reef habitats off the island of Hawaii, as determined by visual assessments in belt transects (Hobson, 1974). Diurnal planktivores were of 12 species and constituted 45% of the fishes counted along the drop-off into deeper water, but were only of 6 species constituting 14.6% of the fishes counted over the inner reef.

The abundance of diurnal planktivores near the reef edge comes not only from the many that reside there, but also from others that migrate each morning from nocturnal shelter sites elsewhere on the reef (Hobson, 1972,

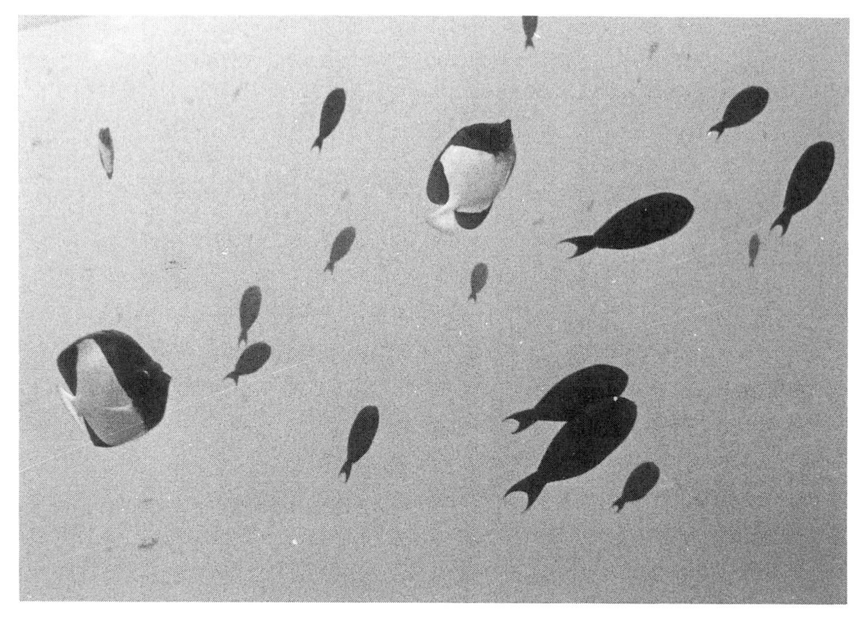

Figure 4 The planktivorous chaetodontid *Hemitaurichthys polylepis* (left) is as deep-bodied as the benthivorous members of its family, whereas the planktivorous acanthurid *Acanthurus thompsoni* (dark fish) is far more streamlined than the benthivorous members of its family.

1973, 1974). Although reef edges tend to be irregular structures rich in shelters used by diurnal planktivores, apparently these shelters cannot accommodate all the planktivores that feed there. That the migrators are among the larger diurnal planktivores (Hobson, 1974) suggests either that there is a shortage of the more spacious shelters at the feeding site or that the smaller species lack the capacity to migrate.

Diurnal planktivores concentrate along the reef edge and feed mainly on transient zooplankters from open water, even though resident zooplankters of comparable type and size are widespread and accessible on many reefs. I refer to the resident zooplankters that occur in dense, generally monospecific swarms close to reef structures, including various copepods (e.g., *Acartia* spp. and *Oithona* spp.) as well as mysids (e.g., *Anisomysis* sp. and *Mysidium* spp.) (Emery, 1968; Hobson and Chess, 1978; Hamner and Carleton, 1979).

These diurnal swarms can be immense. For example, Hamner and Carleton (1979) described a swarm of the copepod *Oithona oculata* encircling a Palauan bay that was more than 100 m long and estimated to contain 75 million individuals. Similarly, Emery (1968) observed among reef formations in the Florida keys swarms of the copepod *Acartia spinata* that ranged in size from just a few to over 60 m^3, with densities of about 110,000 individuals per m^3.

The zooplankters in these swarms, however, seem generally unavailable to predatory fishes. Probably this is because predators confronted with multiple targets have difficulty distinguishing individuals (Welty, 1934; Eibl-Eibesfeldt, 1962; Hobson and Chess, 1978). In concluding that swarms protect their members from predators, Hamner and Carleton (1979, p. 11) noted that planktivores "swim through the swarms as if they were invisible."

Certainly there is some predation on these swarms, just as there is predation on fish schools. For example, Emery (1968) saw predators take copepods from swarms in the tropical Atlantic, and Hobson and Chess (1978) found mysids in the gut contents of diurnal planktivores taken near mysid swarms in the central Pacific. But considering how many individuals there are in these swarms, this predation is probably insignificant. Swarming, therefore, would seem to be an adaptive behavior that reduces the vulnerability of reef zooplankters to reef planktivores.

Because diurnal planktivores feed mainly on transients, they depend on water currents to supply them with food. It is well known that currents are important to feeding planktivores (Stevenson, 1972; Thresher, 1983a), and even casual observations note that there are currents where planktivores are abundant. In fact, it is because their food is transported by currents that planktivorous fishes are able to feed in the stationary aggregations so characteristic of them. As Stevenson (1972) pointed out, the aggregations dissolve when currents slacken and the planktivores are forced to swim about in search of prey. Despite their dependence on currents as transporters of food, however, planktivorous fishes find it increasingly difficult to maintain station when currents exceed optimal velocities (which vary with the species), and ultimately they are forced to shelter (Hobson and Chess, 1976, 1978).

Planktivorous fishes also are distributed relative to their size. Feeding at reef edges, for example, is most characteristic of larger individuals. Thus, species concentrated close to the drop-off (depth about 25 m) into deeper water off the island of Hawaii (Hobson, 1974) are, at sizes exceeding 10 cm SL, the larger of the planktivorous reef fishes there (e.g., the chaetodontids *Chaetodon miliaris* and *Hemitaurichthys polylepis,* the pomacentrid *Chromis verater,* the acanthurids *Acanthurus thompsoni* and *Naso hexacanthus,* and the balistid *Xanthichthys auromarginatus*). The smaller species are more widely distributed. Examples include the pomacentrids *Chromis agilis* and *C. hanui* (combined as *C. leucurus* in that report), which have maximum sizes of about 7 cm SL. Although these two are numerous along the drop-off, they also are abundant above the inner reef, where the larger species are comparatively scarce.

A similar difference in distribution with size exists above the shelf of sand and isolated patch reefs that rims the windward side of the lagoon at Enewetak atoll (Hobson and Chess, 1978, 1986). In that setting, zooplanktivorous fishes of all sizes abound above patch reefs along the shelf's edge (depth about

10 m), but only the smaller subadults and very small species (less than about 6 cm) are numerous above patch reefs on the inner shelf (depth < 5m). [Relatively large herbivorous planktivores are abundant above patch reefs on the inner shelf, where they feed mainly on drifting plant fragments (Hobson and Chess, 1978), but herbivorous planktivores are not considered here because many of their trophic relations are unlike those of zooplanktivorous species.]

Differences in distribution with size can be related to differences in trophic relationships. As noted earlier, the concentration of larger zooplanktivores at the shelf edge is consistent with diets dominated by transient organisms from open water (Hobson and Chess, 1978). The major planktivore above patch reefs along the shelf edge at Enewetak is the adult of *Chromis viridis*, which is larger than most other planktivores at this site even though its maximum size is only about 7 cm SL. [This species was previously known, and identified in that report, as *C. caerulea* (see Randall *et al.,* 1985a).] Tony Chess and I collected 21 of these fish (6.0–7.3, $\bar{X}$ = 6.5 cm SL) from above the patch reefs during the day, and of the 1168 zooplankters in their guts, 1047 (89.6%) were 2.5 to 3.0-mm individuals of *Undinula vulgaris*. This calanoid copepod is a major component of the plankton in the Enewetak lagoon, but during the day it is largely absent above the shelf (Gerber and Marshall, 1974; Hobson and Chess, 1986). In fact, the relatively shallow water above shelf patch reefs during daytime is virtually without zooplankters of any kind larger than 1.5 mm (Hobson and Chess, 1986), which probably is why there are so few larger zooplanktivorous fishes there in daylight.

On the other hand, the smaller zooplanktivorous fishes so abundant above patch reefs on the inner shelf at Enewetak during the day find their food readily available. Chess and I collected subadults of three species whose large plankton-feeding aggregations were prominent above the inner-shelf patch reefs: *Chromis viridis* (*n* = 13; 9–40, $\bar{X}$ = 25.5 mm SL), *Dascyllus aruanus* (*n* = 10; 16–35, $\bar{X}$ = 27.5 mm SL), and *Rhabdamia gracilis* (*n* = 20; 20–36, $\bar{X}$ = 30.9 mm SL). None of the varied zooplankters in their guts—mostly crustacea, but also including larvaceans and fish eggs—were larger than 1.0 mm, and plankton collections taken at the time found these smaller forms to be abundant in the water column.

That the larger planktivorous fishes feed mainly on transient zooplankters at the reef's perimeter is particularly evident above reefs swept by currents. Here these fishes concentrate above the upcurrent edge of the reef and are relatively few above the reef downcurrent. Above a southern Californian reef swept by currents that reverse with the tide, for example, Bray (1981) found that larger (>15 cm) individuals of *Chromis punctipinnis* (but not the smaller ones) concentrate to feed on cladocerans, copepods, and larvaceans at whichever end of the reef faces the current, changing ends with each reversal of current.

Furthermore, plankton samples that Bray took both upcurrent and downcurrent of the feeding *C. punctipinnis* showed that these aggregations remove from the inflowing current a significant proportion of the organisms suitable as prey.

A greater availability of suitable prey above upcurrent parts of a reef explains diel movements that I observed in the planktivorous *Clepticus parri* (Labridae) above a reef off St. Croix, U.S. Virgin Islands. From early to mid morning each day, larger individuals (>15 cm) of this species migrated toward the upcurrent end of the reef, and then from mid to late afternoon they returned downcurrent. The similarity of this pattern to the diel migrations that some Hawaiian planktivores make between inner-reef shelter sites and outer-reef feeding grounds (Hobson, 1972, 1973) suggests that the larger *C. parri* found their prey more available upcurrent.

But if prey of the larger planktivores become increasingly scarce in currents that flow over a reef, how is it that prey of the smaller planktivores apparently remain abundant? Part of the answer is that during the day the smaller zooplankters (less than about 1 mm) are vastly more numerous in currents flowing over the reefs. Perhaps a more important reason, however, is that during the day these smaller zooplankters may be relatively safe from reef predators when more than a few meters above the substrate. This is because the predators that threaten them most are themselves relatively small animals that become increasingly vulnerable with distance from shelter in daylight. Smaller planktivorous fishes that feed by day, for example, generally stay within a meter of the reef. And the various mysids and other predaceous crustacea that range into the water column to feed on still smaller zooplankters do so only at night (Hobson and Chess, 1976, 1978). Thus, during the day, smaller zooplankters may enjoy what is in essence a refuge from reef predators in all but the lower levels of currents flowing over a reef. And as the current flows along, individuals from this refuge would be there to replace those of their kind consumed at the lower levels.

D. Trophic Links with Open-Water Communities

It is widely recognized that planktivorous fishes are a major trophic link between coral reef and open-water communities (Emery, 1968; Davis and Birdsong, 1973). Food webs that have been constructed to represent coral reef trophic systems, however, assume that the primary flow of energy from planktivores to other elements of the community is through piscivorous predators (e.g., Polovina, 1984). But while undoubtedly some diurnal reef planktivores are consumed by predators (e.g., Choat, 1968; Hartline *et al.*, 1972), the numbers taken are relatively few owing to effective defenses, as described earlier.

Probably the major path of energy from diurnal reef planktivores to other components of the reef trophic system is through feces. Feeding planktivores produce prodigious amounts of feces that rain down upon the reef throughout the day, and much if not most of this material is consumed by other fishes (Robertson, 1982). Chess and I have found that when zooplankters are especially abundant in the water column, those consumed by planktivores pass through the guts so rapidly that they appear in the feces with little sign of digestion. In this situation the secondary consumers probably gain more energy from the zooplankton than do the planktivores that had consumed them first. Although planktivore feces are most readily ingested by fishes before settling on the seafloor, significant amounts are taken from the bottom by herbivores and detritivores (Robertson, 1982). Most of this coprophagy occurs below the feeding aggregations, but some material still in planktivore guts at day's end is vented at nocturnal shelter sites elsewhere on the reef, and according to Bray *et al.* (1981) this represents a significant importation of energy to the reef benthos.

III. Crepuscular Changeover

Like virtually all members of the coral reef community, planktivorous reef fishes and their prey find twilight to be a time of transition between distinctive diurnal and nocturnal modes. The changeover process is an orderly complex of responses to specific levels of diminished daylight, with the morning and evening sequences being essentially mirror images of one another (Hobson, 1972; Collette and Talbot, 1972; Helfman, 1986b). Here I describe events during the evening changeover that involve planktivorous fishes and zooplankters.

A. Transition from Day to Night: Fishes

When the diurnal planktivores descend toward the seafloor late in the afternoon they provide the first clear indication that the reef community is shifting toward its nocturnal mode. Although the beginnings of this process cannot be distinguished from the highly variable diurnal condition described earlier by 30 min before sunset it is clear that many diurnal planktivores are descending toward shelter on the reef.

Just as smaller individuals stay closer to the reef while active during the day, they also are the first to shelter in the evening. Consider, for example, the species of *Chromis* in Hawaii (Hobson, 1972): The smallest, *C. vanderbilti* (maximum size about 5 cm SL), stays within a meter or so of the reef on even the brightest days and is dispersed under cover on the reef by 15 min before sunset. The next smallest, *C. agilis* (as *C. leucurus* in that paper; maximum size about 7 cm SL), ranges 1–2 m above the reef during the day and is under cover within a few minutes after sunset. Finally, two larger species of similar size, *C. ovalis* and *C. verater* (maximum size about 14 cm SL), range throughout the water column during the day, and the last of them take cover about 15 min after sunset. The final descent of the larger planktivores marks the beginning of the "quiet period," a 15- to 20-min interval when the waters above the reef appear relatively empty of fishes (Hobson, 1972). It is believed that reef fishes tend to avoid exposed positions at this time because of sharply increased threats from predators. Actually, as the diurnal planktivores demonstrate in descending to cover in order of increasing size, most diurnal fishes have left the water column before the quiet period begins. Apparently the smaller fishes are threatened first—the smallest well before sunset—and after that the danger continues to grow in the fading light, threatening progressively larger fishes, until by the start of the quiet period virtually all diurnal fishes that had been in the water column have been affected. Thus, the timing of the quiet period and other changeover events appears to be determined by levels of light. These events occur earlier under cloudy skies (Collette and Talbot, 1972) and are extended in the longer twilight at higher latitudes (Helfman, 1981, 1986b; Hobson, 1986).

The nocturnal planktivores enter the waters above the reef at nightfall in the same order by size that the diurnal species leave—smaller ones first. The smallest appear while some of the diurnal species still hover low in the water column, well before the quiet period has begun. The first of them are less than 3 cm in size and not readily visible (many are highly transparent). Among the first to rise above the reef are juvenile *Apogon* spp. (Apogonidae), which are prominent members of tropical reef communities worldwide. A previous discussion of the early appearance of largely transparent juvenile apogonids (Hobson *et al.*, 1981, p. 23) stated:

> This entry into exposed locations when many piscivorous predators hunt most effectively might seem in conflict with the quiet period concept. But in the dim light we are not surprised that these inconspicuous little fish seem to go unseen by the visual hunters that so seriously threaten the more visible adults. Certainly these juveniles go unseen by human eyes at this time, except under close inspection with a diving light, and so fail to detract from the aura of inactivity that characterizes the quiet period.

On many tropical reefs the quiet period comes to an abrupt end about 30 min after sunset, when hordes of *Myripristis* spp. (Holocentridae) surge into the water column from their daytime shelters in the reef (Hobson, 1972) (Fig. 5A and B). Other planktivorous fishes follow in short order, including *Priacanthus* spp. (Priacanthidae). As is the case among their diurnal counter-

Figure 5 A reef in the lagoon at Enewetak Atoll (A) 30 min after sunset and (B) 2 min later, as numerous *Myripristis* spp. emerge from shelter.

parts, while most nocturnal planktivores forage above their diurnal refuges, some migrate to more distant feeding grounds (Hobson, 1973). Among the migrators are *Myripristis murdjan*, *M. amaena*, and *Priacanthus cruentatus*, which head seaward in groups after emerging from Hawaiian reefs about 35 min after sunset (Gosline, 1965; Hobson, 1972). Other migrators include *Pempheris schomburgki* in the tropical Atlantic. After emerging from shelters along the inside of a Virgin Island reef at nightfall, members of this species swim in schools close to the coral as they cross the reef to nocturnal feeding grounds in the water column outside (Gladfelter, 1979) (Fig. 6).

B. Transition from Day to Night: Zooplankters

Transient zooplankters of types consumed by diurnal planktivores greatly increase in size and number above shallow-water reefs after dark. At Enewetak Atoll, for example, the major prey of diurnal planktivores—calanoids of 1–3 mm—increased more than fivefold in numbers above reefs on the lagoon shelf during the two hours immediately after sunset. During the day these calanoids were abundant as prey of reef planktivores only at the shelf's edge and

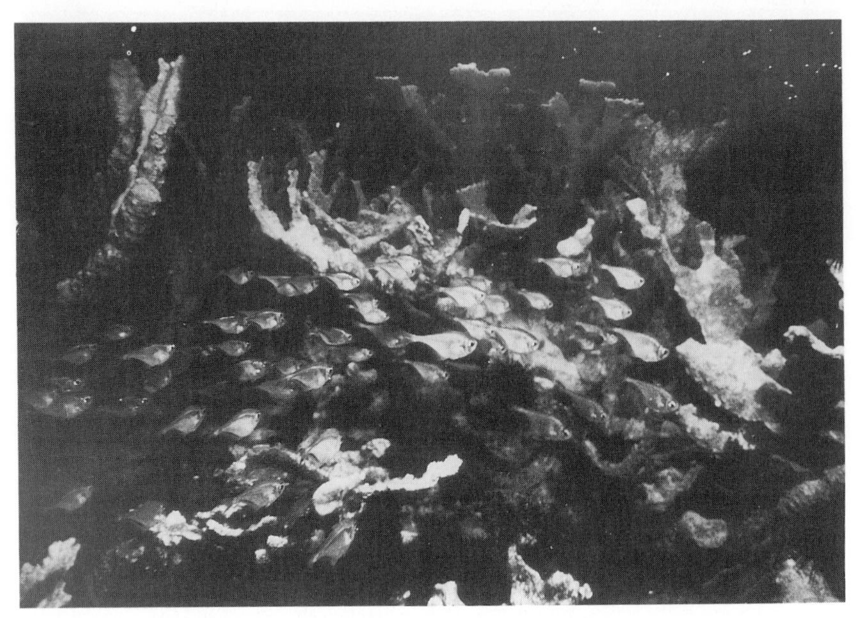

Figure 6 A school of *Pempheris schomburgki* (Pempheridae) following its crepuscular migration route to feeding grounds along the seaward side of a reef at St. Croix, U.S. Virgin Islands. Silvery specks at upper right are atherinids that have dispersed from diurnal schools among the coral for nocturnal feeding above the reef.

in water that had flowed through the deeper passes from the open sea (except for individuals trapped in shallow water at sunrise, as detailed in the following). But beginning during late twilight, calanoids of this size were among the many forms that became increasingly numerous in the waters above the shelf. It was obvious that they had arrived in currents that flowed over the interisland reefs from the open sea outside the atoll—currents that carried relatively few such forms during the day. Considering that Enewetak is in the path of the trade winds and the oceanic Equatorial Current, it seems probable that the increased numbers of zooplankters in water flowing over the interisland reef at night include organisms that shortly before had risen to the surface from ocean depths windward and upcurrent of the atoll (Hobson and Chess, 1986).

A rich assortment of reef residents join the open-water transients above reefs late during the changeover. They include the holoplankters that had been in swarms close to benthic substrata during the day, mostly copepods and mysids (Emery, 1968; Hobson and Chess, 1978; Hamner and Carleton, 1979), but that at nightfall disperse throughout the water column. Even more prominent, however, are a variety of semipelagic organisms that enter the water column during the night from positions in or on the seafloor. Many of these are primarily benthic organisms that spend only a relatively short period of time in the water column. They include various polychaetes, ostracods, copepods, mysids, isopods, amphipods, and crustacean larvae (Alldredge and King, 1977, 1980; Robichaux *et al.*, 1981; McWilliam *et al.*, 1981). The residents vary widely in size, from well under 1 mm to more than 10 mm, and generally include the largest of the zooplankters above reefs at night (Hobson and Chess, 1978, 1986).

IV. Nocturnal Planktivores

The fishes that forage on zooplankters above tropical reefs at night have been strongly influenced by the difficulty of visually locating prey in dim light, as evidenced by the exceptionally large eyes of *Myripristis* spp. (Fig. 7). The differences in morphology and behavior between nocturnal planktivores and their diurnal counterparts shows that selection pressures affecting planktivore form and function differ greatly between day and night.

A. Adaptations to a Nocturnal Planktivorous Diet

The major prey of most nocturnal reef planktivores are among the relatively large (>2 mm), semipelagic residents of the local habitat that rise into the water column at some time during the night (Hobson, 1974; Hobson and Chess, 1978; Gladfelter, 1979). Other important prey are larger holoplankto-

Figure 7 An aggregation of the nocturnal planktivore *Myripristis amaenus* (Holocentridae) in a Hawaiian reef cave during the day.

nic residents, such as mysids, and still others are larger transients from open water, like euphausids (Hobson and Chess, 1978). Conspicuously absent from this diet, however, are the relatively small transient holoplankters that are major prey of the diurnal planktivores, even though these organisms are more abundant at night than during the day (Hobson and Chess, 1978, 1986).

One might take these findings as evidence that nocturnal planktivores make a choice based on preference for larger prey, citing the widespread belief that planktivorous fishes choose the largest organisms available to them (Zaret, 1980). A corollary to this belief is that incentive to take the smaller zooplankters is lost in the presence of larger alternatives, which is consistent with the observation that natural selection for abilities to consume smaller prey is much reduced at night. (Consider, for example, that nocturnal planktivores generally lack the highly evolved modifications of head and jaws that enable even the larger diurnal planktivores to consume tiny prey in open water.) But despite compelling arguments that support this reasoning, I believe larger zooplankters are taken at night, not because they are preferred or that it is more efficient to do so, but rather because they simply are more vulnerable.

That nocturnal planktivores feed mainly on the larger components of the plankton may have less to do with preference for larger prey than with inability to see the smaller ones. Studies that have shown preference for larger prey have been done in daylight (e.g., Brooks and Dotson, 1965; Coates, 1980) and so fail to consider how lack of light would influence what is available to planktivorous fishes at night. Most nearshore fishes that use vision to feed at night have sacrificed visual acuity for visual sensitivity, which limits their ability to see smaller objects (Walls, 1967; Munz and McFarland, 1973). The greatly enlarged eyes of *Myripristis* spp. and *Priacanthus* spp., for example, permit these fishes to sense visual cues in the nocturnal water column, but apparently these cues do not include the smaller zooplankters. This limitation readily accounts for the observed lack of selection for features that permit a diet of smaller prey, noted earlier, because obviously predators that feed by sight would have little use for trophic mechanisms that accommodate prey too small for them to see.

An inability to see the smaller zooplankters would also explain why my studies with Chess have found that generally even the smallest nocturnal reef planktivores are limited to zooplankters larger than about 1 mm, even where comparably sized diurnal planktivores feed primarily on organisms smaller than this. One might attribute this to the smaller mouths generally found in "comparably sized diurnal planktivores," but our studies have shown that nocturnal planktivores feed on larger prey even when they have feeding structures similar to those of their diurnal counterparts.

Consider, for example, the juveniles of two closely related apogonid species, *Apogon cyanosoma* and *Rhabdamia gracilis* [considered congeners by Lachner (1953), with *A. cyanosoma* as *A. novae-guineae*]. Despite similar feeding mechanisms (both have the generalized trophic features of basal percoids), the juveniles of *A. cyanosoma* feed at night and consistently take larger zooplankters, whereas the juveniles of *R. gracilis* feed by day and consistently take smaller zooplankters. This finding is based on data from specimens that Chess and I collected above a patch reef on the lagoon shelf at Enewetak during one night and the following day. Here, the 45 calanoid copepods in the guts of five nocturnal *A. cyanosoma* (20–27, $\overline{X}$ = 23.6, mm SL) were 2 to 3 mm in size, whereas the 70 calanoids in the guts of five diurnal *R. gracilis* (17–37, $\overline{X}$ = 27, mm SL) were 0.3 to 2 mm. Clearly this difference in size of prey, day compared to night, was determined by something other than size of mouth in these morphologically similar predators.

The absence of larger copepods in the diet of the diurnal *R. gracilis* is readily explained by the scarcity of these prey in the daytime water column above the shelf, as noted earlier. But similar reasoning cannot account for the absence of smaller copepods in the diet of the nocturnal *A. cyanosoma* because these are so abundant at night—far more numerous than the larger ones, in fact. Although it is possible that the smaller zooplankters were simply ignored when the larger ones became available at night, I would expect the juvenile *A. cyanosoma* to have taken at least some of them if they could, and the fact that they did not seems best explained by visual limitations.

I know of one nocturnal planktivore that feeds regularly on zooplankters smaller than 1 mm, but this exception to the generalization can be explained. *Praneseus pinguis* (Atherinidae), which is peripherally associated with reefs in the lagoons of western Pacific atolls, feeds heavily on organisms as small as 0.2 mm during nighttime excursions from nearshore schooling sites to offshore feeding grounds. But this fish feeds right at the water's surface, where submarine starlight and moonlight are strongest (Hobson and Chess, 1973), so its visual capabilities likely differ from those of species adapted to dimmer light at greater depths.

Presumably factors other than size contribute to the relative vulnerability of various zooplankters at night. Many of the semipelagic residents, in particular, are more opaque than the more numerous holoplankters among which they occur after dark, and these are appropriate targets in dim light. Others are awkward swimmers, for example, caprellid amphipods, and therefore are less elusive and also more likely to create the turbulence and resulting bioluminescence thought to direct many nocturnal attacks (Hobson *et al.*, 1981).

That nocturnal planktivores have generalized feeding mechanisms does not mean they lack specialized trophic features. The mouths of some resemble the mouths of diurnal planktivores in being sharply upturned (Fig. 7), so obviously the advantage of this arrangement, noted earlier for diurnal species, has adaptive value among nocturnal planktivores as well. I have mentioned the exceptionally large eyes of many, which certainly are highly specialized means to orient visually in dim light and presumably to locate prey. Significantly, eyes of species active at middle depths tend to be larger than eyes of species active at the base of the water column, probably because the light available close to the seafloor is increased by reflected moonlight and starlight (Hobson *et al.*, 1981).

B. Adaptations to Nocturnal Threats from Predators

There is morphological and behavioral evidence that threats from predators are significantly reduced at night. For example, the streamlined bodies and deeply forked caudal fins so widespread among diurnal planktivores are much less developed among their nocturnal counterparts. So if these features are adaptive in providing the speed needed to evade attacks, as suggested earlier, their general absence among the nocturnal species could mean there is less threat of attack after dark (Hobson, 1973, 1979).

Similarly, the tendency to aggregate while feeding in the water column, which is thought to be a defense against predators (Eibl-Eibesfeldt, 1962) and is so strong by day, is much weaker at night. Some, including various *Myripristis* spp., are active in loose aggregations, especially in moonlight, but many others, including various *Apogon* spp., are solitary (Hobson, 1968, 1974). Some individuals of *Priacanthus cruentatus* school high in the water column as they migrate seaward shortly after nightfall, but other individuals of this species are solitary as they remain above the reef during the night (Hobson, 1974).

Even those nocturnal planktivores commonly described as "schooling fishes" based on their daytime assemblages, for example, various clupeids and atherinids (Randall, 1967), generally disperse at night in favor of independent activity and small, loosely associated feeding groups (Starck and Davis, 1966; Hobson, 1968). So if aggregations are adaptive by reducing the threat of attack, the nocturnal condition would indicate relative freedom from such threats.

Certainly the visually directed attacks that threaten smaller fishes during the day would be limited by the sharp decrease in ambient light at night. Nevertheless, it would seem that at least some of the reduced nocturnal threat comes from conventional attackers. Examples include *Caranx marginatus* and *Selar crumenophthalmus* (Carangidae), which hunt smaller fishes at night in the Gulf of California (Hobson, 1968). These two species have morphologies typical of aggressive, open-water predators, but with large eyes that may allow them to attack in the ordinary way using the much lower levels of ambient light produced by moon and stars. Support for this possibility comes not only from the tendency of nocturnal planktivores like *Myripristis* spp. to aggregate under moonlight, but also from the countershaded nocturnal coloration of *Myripristis* spp., that is, dark above and light below (Hobson, 1968). If Longley (1917) and others are correct that such countershading effectively conceals midwater fishes from predators, this feature in nocturnal planktivores would suggest that they are threatened by predators that use light from above—presumably moonlight or starlight—to direct their nocturnal attacks.

The major visual cues above coral reefs on darker nights or in deeper water, however, probably are the emissions of luminescent organisms, which must have profound effects on nocturnal interactions between predators and prey (Hobson, 1966; Hobson *et al.*, 1981). Certainly the straightforward charge typical of predators that threaten diurnal planktivores would be less effective amid luminescent organisms on dark nights. Because luminescence among the plankton is greatly increased around moving objects, an aggressive charge by a large predator would be advertised as soon as launched, giving prey time for evasive maneuvers. Furthermore, while a charging predator could target on luminescence around its prey, I question whether this tactic would be successful if the attack is launched from some distance away. This is because the luminescent organisms that elicited the attack would be left behind as the fish darts away, thus leading the attacker to the wake of the fleeing fish (Hobson *et al.*, 1981). On the other hand, nocturnal piscivores may attack the leading edge of the luminescent trail.

But even if nocturnal planktivores are relatively free of attacks like those that threaten diurnal species, they may be threatened by some other type of predator in situations where defenses effective in daylight are irrelevant (Hobson, 1973). A predator that is effective amid luminescent organisms at night might be one that hovers motionless, waiting for prey to betray themselves by movement that produces turbulence and resulting luminescence. This tactic would be in essence an ambush, and effective only at short range. Although the predator could be led by luminescence to prey some distance away, probably its approach must involve great stealth to avoid turbulence and resulting bioluminescence (Hobson *et al.,* 1981). In this regard, one might consider the carangid *Selene brevoorti*, a predator that feeds on smaller fishes at night in the Gulf of California (Hobson, 1968). The extraordinary morphology of this species suggests unusual feeding behavior; in fact, its deep body, laterally compressed to an extreme, might well prove adaptive in maneuvering with minimum turbulence to approach unseen close to prey in the dark.

C. Distribution and Movement

Nocturnal planktivores generally are more widespread throughout the reef area than are their diurnal counterparts. Presumably this distinction has developed because the nocturnal species feed mainly on reef residents that are themselves widespread over the reef, whereas the diurnal species take mainly open-water transients that are most available at reef edges. In addition, a decreased threat of predation at night is evident in the tendency of the nocturnal species to disperse, at least during moonless nights, while the diurnal species tend to cluster.

Although most nocturnal planktivores feed near their shelter sites, many migrate to feeding grounds elsewhere—much as do certain diurnal planktivores. Some of the migrators go to other parts of their home reef, where they nevertheless continue to feed mainly on semipelagic reef residents, apparently they find them more abundant in these other places. An example is *Pempheris schomburgki,* which migrates during twilight from shelters along the inside of a Virgin Island reef to feeding grounds along the outside (Gladfelter, 1979). Other migrators that apparently go to the edge of the reef or beyond and feed on open-water transients. For example, while some individuals of *Priacanthus cruentatus* remain to feed in the local area after emerging from Hawaiian reefs, many others migrate seaward; and when they return, about 40 min before sunrise, their stomachs are full of pelagic organisms, including cephalopods and crab megalopae (Gosline, 1965; Hobson, 1972, 1974).

Nocturnal and diurnal planktivores tend to differ in their distributions relative to the effects of water currents on availabilities of their major prey—resident and transient zooplankters, respectively. The nocturnal planktivores are less numerous in stronger currents, perhaps because their resident prey tend to avoid currents that would carry them from their home grounds. The diurnal planktivores, on the other hand, concentrate in currents that carry their transient prey from deeper water (Hobson and Chess, 1978, 1986). This is not to say that nocturnal planktivores always avoid currents, or that diurnal planktivores always are attracted to currents. To the contrary, nocturnal planktivores often feed to advantage in currents that can be tolerated by their resident prey (Thresher, 1983a), and diurnal planktivores avoid currents that exceed velocities in which they themselves can maintain station (Hobson and Chess, 1978).

Furthermore, stronger currents from deeper water should benefit nocturnal planktivores that prey heavily on larger transient zooplankters that appear above reefs after dark, as does *Myripristis pralina* at Enewetak (Hobson and Chess, 1978). Similarly, if the seaward twilight migrations of *Myripristis murdjan, M. amaenus,* and *P. cruentatus* in Hawaii are headed for feeding grounds above the outer drop-off, as surmised (Hobson, 1972, 1974), then they too would benefit from currents strong enough to supply them with zooplankters from the open sea.

Many nocturnal planktivores have distinctive defensive needs that affect their distributions during diurnal periods of rest. Unlike the diurnal planktivores, which generally are solitary while at rest during the night, many nocturnal planktivores are highly gregarious in their daytime resting mode. This is particularly evident in the clupeids, engraulids, and atherinids that shelter in dense, inactive schools close to reefs during the day. Examples from the tropical Atlantic include *Jenkinsia lamprotaenia* (Clupeidae) and *Allanetta harringtonensis* (Atherinidae), which school among reef structures by day and disperse to feed above the reef at night (Starck and Davis, 1966). Other nocturnal planktivores, more secretive than the schoolers, aggregate by day deep in the shadows of reef caves and crevices. These include *Myripristis* spp. in Hawaii (Hobson, 1972, 1974) and *Pempheris schomburgki* in the Virgin Islands (Gladfelter, 1979).

D. Trophic Links with Open-Water Communities

Nocturnal planktivores that feed mainly on resident zooplankters do not represent direct trophic connections with open-water communities, as do most diurnal planktivores. They are, of course, intermediate links in the transfer of energy between the two realms because so many of their resident prey consume oceanic materials. Their importance in this role, however, is diluted by the many other predators that consume these same prey close to the reef. On the other hand, those nocturnal planktivores that feed on transient zooplankters from the open sea do represent direct trophic connections with the oceanic realm.

The subsequent flow of energy from nocturnal planktivores to other elements of coral reef communities goes primarily through piscivorous predators, which, as noted earlier, probably is not true of diurnal planktivores. In contrast to the diurnal species, which seem relatively secure from predators, many of the nocturnal planktivores are primary prey—particularly the schooling clupeids and atherinids (Randall, 1967; Hobson, 1968). These fishes are most vulnerable to predators not while active in the water column at night, or even while at rest in schools during the day, but rather during the crepuscular transition between these conditions (Hobson, 1968).

It seems unlikely that feces of nocturnal planktivores can compare with feces of diurnal planktivores as transporters of energy to other elements of the community. Two problems limit the effectiveness of nocturnal coprophagy among fishes: first, the feces are more difficult to see in the dim light of night and, second, the more dispersed nocturnal planktivores do not offer the focal points of activity that greatly facilitate coprophagy by day. Furthermore, because the nocturnal planktivores take larger and fewer prey, digestion is slower and evacuation less frequent.

Thus, the continuous rain of feces beneath aggregations of diurnal planktivores that makes coprophagy such a viable feeding mode during the day is unlikely to be duplicated at night. Still, feces that settle on the reef beneath nocturnal planktivores, at either feeding sites or resting places, are likely to enter the coral reef food chain through vertebrate or invertebrate detritivores. One study found exceptional growth among corals at the daytime resting places of juvenile grunts (Haemulidae), which are nocturnal planktivores, and attributed this to nourishment from the excretory and fecal products of these fishes (Meyer *et al.*, 1983).

V. Vulnerability of Zooplankters above Reefs

It is clear that zooplankters of more than about 1 mm in size are highly vulnerable to planktivorous fishes when in the water column above coral reefs during the day. Most that experience this vulnerability are of open-water species and have encountered the reef habitat by chance. There are few reef residents among them because most of these shelter by day in swarms or under benthic cover. That this sheltering behavior is adaptive as a defense is evident in the intensity of predation suffered by transients that come within reach of reef planktivores.

Zooplankters are less vulnerable above coral reefs at night, as evidenced by the great numbers of them there during that period. The nocturnal increase involves reef residents that had abandoned their daytime shelters and also open-water transients that had arrived in the reeftop flow after upcurrent populations made their nightly ascent into the surface waters. Those taken by nocturnal planktivores tend to be relatively large, opaque forms that are unavailable by day; the far more abundant smaller forms so heavily exploited by diurnal planktivores are relatively unimportant as prey of nocturnal planktivores despite their greatly increased numbers above the reef after dark. It would appear that vulnerability to nocturnal planktivorous fishes involves features visible in dim light.

That zooplankters above reefs are more generally vulnerable in daylight is reaffirmed at dawn. At this time, reef residents that had been dispersed in the water column during the night rejoin swarms or return to benthic shelter, while open-water transients are left exposed to the developing threat from diurnal planktivores. Although the number of transients arriving in the reef-top flow had dropped sharply after upcurrent populations made their regular predawn descent into the depths, many that had been carried above the reef during the night are still there at daybreak and these become vulnerable to diurnal planktivores (Hobson and Chess, 1986). The problem of transients from the open sea trapped in relatively shallow water at daybreak is one widely experienced by zooplankters that make diel vertical migrations in deep water close to reefs, shelves or banks. Reports have cited occurrences above continental shelves (e.g., Issacs and Schwartzlose, 1965; Pereyra *et al.*, 1969) and at the edge of submarine canyons (Chess *et al.*, 1988).

The vulnerability of open-water zooplankters trapped by a relatively shallow sea bed at daybreak exaggerates the more widespread vulnerability of transients that experience incidental contact with reef habitats during the day. These are problems of organisms lacking an effective defense in situations outside their evolutionary experience. Clearly they are a rich and vulnerable source of food for reef fishes, as demonstrated by the range of specialized features these fishes have acquired to exploit them.

I suggest, therefore, that it is the vulnerability of organisms in unfamiliar surroundings that accounts for most planktivory among fishes above coral reefs, at least during the day. It should be expected that zooplankters adapted to the pelagic environment would be vulnerable to predators in reef habitats because they are unlikely to have defenses specific to that setting. Their encounters with reefs must be exceptional events and features that would adapt them for this experience are likely to be maladaptive in their normal open-water habitat.

VI. Topics for Further Study

Certain key features of topics considered in this chapter call for further study, for example:

1. We need more data on the distribution of planktivorous fishes and zooplankters relative to size. I have described the larger planktivorous

fishes concentrated at reef edges, with smaller species and subadults more widely distributed over inner reefs, and attributed this to the pattern of availability of zooplankters of different sizes. This assessment is based on widespread observations, sampling, and impressions, but more data are needed. Confirmation calls for extensive censusing of the planktivorous fishes with concurrent sampling of the plankton, both with emphasis on distribution by size.

2. It is clear that currents are important in the trophic dynamics between fishes and zooplankters, but we need to refine our knowledge of the relationships involved. I have noted that diurnal planktivores tend to concentrate in currents because these carry their transient prey, whereas nocturnal planktivores tend to concentrate away from stronger currents because these are avoided by their resident prey. But there are many exceptions to these generalizations that should be thoroughly analyzed because they provide insight into the relationships involved. For example, one might consider circumstances where nocturnal planktivores feed on transient zooplankters in currents, or on resident zooplankters in relatively weak or intermittent currents.
3. Judgments on adaptive significance generally are based heavily on inference, and they call for further study. For example, I have suggested that streamlining is adaptive both in accelerating flight to shelter and in maintaining station in current. With one complementing the other, the two could be an exceptionally powerful evolutionary force. Whether either or both are active in a specific instance may be evident in the extent of streamlining in planktivores exposed to the prevailing current, especially if there is a concurrent measure of relative danger from predators. The same analysis would provide some basis for accepting or rejecting my suggestion that an exceptionally deep body and strong fin spines allow some fishes to feed in the water column with an acceptable degree of safety. These species would not of course be expected where there is need to maintain station in a strong current.
4. The role of trophic interactions among reef fishes and zooplankters in transferring energy from open water to the reef community is a rich area for study. This chapter identifies major routes involving predation and coprophagy, but the energetics at each link remain to be quantified.
5. What is the nature of nocturnal threats to planktivorous fishes? Nocturnal planktivores lack certain morphological and behavioral features thought to protect their diurnal counterparts from predators, and I have noted this as evidence that threats from predators are reduced at night. It is possible, however, that nocturnal planktivores simply have no use for these defenses because they face different kinds of threats.

Acknowledgments

For their comments and suggestions on drafts of the manuscript I thank Alfred Ebeling, William Gosline, Gene Helfman, Richard Rosenblatt, and Ronald Thresher. I also benefitted from the artistic talents of Ken Raymond, who drew Figure 2.

CHAPTER 5

Fish–Seaweed Interactions on Coral Reefs: Effects of Herbivorous Fishes and Adaptations of Their Prey

Mark E. Hay
University of North Carolina at Chapel Hill
Institute of Marine Sciences
Morehead City, North Carolina

I. INTRODUCTION

Coral-reef seaweeds come in a tremendous variety of forms, ranging from small, rapidly growing filaments only a few centimeters tall, to larger, canopy-forming macrophytes more than a meter tall, to rock-hard corallines that grow as thin crusts adhering tightly to the substrate and that may be as much as 90% calcium carbonate by dry weight. These different types of algae often differ dramatically in their growth rates, competitive abilities, herbivore deterrent characteristics, distributions on reefs, and susceptibility to grazing by coral-reef fishes (Littler and Littler, 1980; Hay, 1981a–c, 1984a, 1985; Gaines and Lubchenco, 1982; Steneck and Watling, 1982; Littler *et al.,* 1983a,b; Carpenter, 1986; Lewis, 1986; Steneck, 1986, 1988).

In a similar way, herbivorous fishes on coral reefs differ dramatically in the morphology and power of their mouthparts, their digestive physiologies, and their foraging ranges and behaviors. As an example, the fused teeth and pharyngeal mill of parrotfishes allow them to feed on tough or heavily calcified algae that could not be effectively used by surgeonfishes, damselfishes, or blennies. Thus, effects of fishes on seaweeds will depend on interactions between specific types of fishes and the specific morphological, structural, or chemical characteristics of the seaweeds they are attacking (Lewis, 1985; Hay *et al.,* 1988a; Steneck, 1988). The different types of herbivorous fishes and their variable effects on coral reef communities are discussed more thoroughly by Horn (1989) and Choat (Chapter 6, this volume).

On shallow fore reefs, herbivores commonly remove most algal biomass leaving only encrusting corallines (which are relatively resistant to damage from grazing), the basal portions of small filamentous algae (which recover very rapidly following grazing because of their rapid growth rates), and chemically or structurally defended macrophytes (which are low-preference foods relative to the filaments or undefended macrophytes) (Hay, 1981a, 1984a; Carpenter, 1986; Lewis, 1986; Morrison, 1988; Steneck, 1988). Despite the low biomass of seaweeds on most tropical reefs, the extremely high growth rates of these algae make shallow reefs among the most productive habitats on earth; most of this tremendous productivity is consumed by local herbivores (Hatcher and Larkum, 1983; Carpenter, 1986) and is thus immediately incorporated into the reef community. Because of the high productivity of coral reefs and the tight coupling between production and consumption on many portions of reefs, plant–herbivore interactions on reefs may be among the most important interactions affecting community structure as a whole. With the exception of a very few seminal papers in the 1960s and 1970s (Randall, 1961b, 1965; Ogden *et al.*, 1973), the most rigorous investigations of seaweed–herbivore interactions on reefs have been published within the last decade. Despite this short period of study, tremendous advances have been made in our understanding of the role of herbivory (especially by fishes) in coral-reef communities and of how seaweeds have adapted to attack by herbivores.

Herbivorous fishes have a profound impact on the distribution, abundance, and evolution of reef seaweeds (Hay, 1981a, 1985, 1991; Lubchenco and Gaines, 1981; Gaines and Lubchenco, 1982; Steneck, 1983, 1986, 1988; Carpenter, 1986; Lewis, 1986; Hay and Fenical, 1988). On shallow fore reefs, fishes are often estimated to consume from 50 to 100% of total algal production (Hatcher, 1981; Hatcher and Larkum, 1983; Klumpp and Polunin, 1989; Russ, 1987; Carpenter, 1986, 1988) and to take as many as 40,000 to 156,000 bites/m^2/day (Hatcher, 1981; Carpenter, 1986). Grazing on these portions of coral reefs equals or exceeds grazing rates for any other habitat, either terrestrial or marine. Because herbivores, including fishes, appear to be limited by nitrogen rather than carbon (Mattson, 1980; Horn, 1989), and because seaweeds are low in nitrogen relative to fish tissues, fishes probably need to consume many times their minimal carbon requirements to acquire sufficient nitrogen. As one example, Hatcher (1981) estimated that herbivorous fishes on the Great Barrier Reef consumed up to 10 times more carbon than was needed to meet their basic metabolic requirements. The great abundance of herbivorous fishes, coupled with their high metabolic rate (relative to many invertebrate herbivores) and their need to process large amounts of plant material to acquire adequate nutrition, mandates that fishes have a large impact on reef seaweeds.

Numerous experimental studies have documented the profound effect that reef fishes have on present-day seaweeds (Lubchenco and Gaines, 1981; Hay,

1985; Carpenter, 1986; Lewis, 1986; Morrison, 1988) and persuasive arguments have been made that the evolution of herbivorous reef fishes during the Eocene may have dramatically changed the structure and character of reef communities in general (Steneck, 1983). In this chapter I will focus on the ways that reef seaweeds avoid, deter, or tolerate herbivorous fishes and therefore persist on modern coral reefs.

II. Refuges in Space or Time

Seaweeds may diminish the impact of herbivorous fishes by occupying habitats where fishes rarely feed or by occurring in habitats so favorable for growth that production exceeds herbivory even if absolute rates of grazing are high. Numerous studies have documented dramatically different rates of grazing on different portions of reefs and have noted the importance of these between-habitat differences in generating spatial refuges for seaweeds that are highly susceptible to grazing fishes (Hay, 1981a, 1984a, 1985; Hay *et al.*, 1983; Lewis, 1986; Morrison, 1988). Fewer studies have noted that the critical measurement is not how much is eaten but rather how much is eaten relative to the rate of production (Hatcher, 1981; Hay, 1981c; Hatcher and Larkum, 1983; Carpenter, 1986). The degree to which seaweeds "escape" grazing fishes thus depends as much on their habitat-specific growth rates as on the rates at which fishes graze the plants in different habitats. As an example, areas defended by territorial pomacentrids have higher standing stocks of palatable seaweeds than undefended areas and have thus been widely interpreted as one type of spatial refuge for palatable seaweeds. However, it is no longer clear whether this refuge is generated by reduced herbivory or by increased potential for growth. Measurements by Klumpp *et al.* (1987) indicate that both the biomass-specific growth rate and area-specific growth rate of algae in territories exceed the growth rate of algae outside of territories by 1.5 to 3.4 times, and Russ (1987) found no significant difference in the proportion of algal production consumed inside versus outside of the territories he studied. The refuge value of these territories for palatable algae may thus result from pomacentrids somehow increasing the growth rate of algae in their territories rather than decreasing the absolute rate at which algae are lost to herbivores. Several recent studies have assessed herbivory as a proportion of algal production (Hatcher, 1981; Hatcher and Larkum, 1983; Carpenter, 1986, 1988; Russ, 1987); this is a clear improvement over earlier methods and should be encouraged.

Spatial changes in the effectiveness of foraging (percentage of production that is consumed) by herbivorous fishes create mosaics of habitats that differ in the degree to which they favor poorly defended but competitively superior seaweeds versus well-defended (or herbivore-tolerant) but competitively inferior ones (Hay, 1985; Lewis, 1986; Morrison, 1988). If herbivory is experimentally reduced, the mosaic nature of the community is decreased and species typical of heavily grazed communities are often excluded, apparently as a result of competition (Lewis, 1986; Morrison, 1988). This pattern occurs across a large range of spatial scales and can explain differences in seaweed communities on a scale of kilometers, meters, centimeters, or millimeters (Hay, 1985). Seaweed spatial refuges from herbivores in general have been reviewed by Lubchenco and Gaines (1981) and Gaines and Lubchenco (1982). Spatial refuges from reef fishes in particular have been reviewed by Hay (1984a, 1985). Since these more lengthy treatments are available, I will summarize only the major points.

A. Among-Habitat Refuges

Herbivory by fishes is highest on shallow fore reefs and significantly lower on deeper fore reefs, shallow unstructured reef flats, sandy lagoons, and deep sandy plains (Randall, 1965; Hay, 1981a,c, 1984a,b; Hay *et al.,* 1983; Hatcher and Larkum, 1983; Lewis, 1985, 1986; Morrison, 1988). Effects of fish grazing may also decrease as one moves from outer-shelf to midshelf to inshore reefs (Russ, 1984a,b; Scott and Russ, 1987). Habitats where rates of algal growth clearly exceed rates of fish grazing serve as spatial refuges from herbivorous fishes and are usually dominated by poorly defended macrophytes that are rapidly eaten by fishes when transplanted into nearby habitats with higher abundances of grazing fishes (Hay, 1981a,c, 1984a; Hay *et al.,* 1983; Lewis, 1985, 1986). In some instances, effective refuges may be located only a few meters, or even centimeters, away from areas of high herbivore effectiveness (Hay, 1985, 1986; Littler *et al.,* 1986). Spatial mosaics produced by differences in fish grazing therefore create habitats with differing selective regimes and support differing communities of seaweeds within what would otherwise be a more homogeneous habitat; this enhances between-habitat diversity and the maintenance of species richness on reefs (Hay, 1981a, 1985).

As one example, when Lewis (1986) experimentally excluded larger grazing fishes from the shallow portions of a Caribbean back reef, macroalgal cover increased from 2 to 30% within 10 weeks, while control areas did not change. In the areas protected from fishes, small filamentous algae, which were tolerant of herbivory, and certain macroalgae and corals, which were resistant to herbivory, declined significantly. Field observations suggested that the more herbivore susceptible macroalgae overgrew the less susceptible species when herbivory was reduced. When herbivore exclosures were removed, almost all macroalgae were eaten by grazing fishes within 48 hr. Field experiments conducted by Carpenter (1986) and Morrison (1988) also showed that herbivore-susceptible macroalgae tended to dominate when herbivory was reduced. These findings lend credence to Steneck's (1983) hypothesis that macroalgae are competitively superior to many organisms typical of shallow modern reefs and that reefs probably could not have existed in their present form until after the evolution of herbivorous fishes that removed most macroalgae from these habitats.

B. Within-Habitat Refuges (Spatial and Temporal)

Spatial mosaics of fish grazing also occur on smaller spatial scales within habitats that are otherwise relatively uniform. Numerous investigations have focused on the importance of territorial herbivorous fishes in generating spatial mosaics of herbivore impact (Hixon, 1983, 1986; Hixon and Brostoff, 1983; Klumpp *et al.*, 1988). Through aggressive defense of their algal mats, territorial pomacentrids create patches of intermediate grazing intensity and affect algal biomass, species richness, evenness, and diversity (Hixon, 1983, 1986; Hixon and Brostoff, 1983), as well as rates of nutrient turnover, community productivity, and reef erosion (see Klumpp *et al.*, 1988, for references). These territorial fishes thus create patches of differing community structure on an otherwise more homogeneous background; this raises the species richness of the community as a whole (Hixon and Brostoff, 1983).

Interactions between fishes and the physical environment are less thoroughly studied but also appear to be important in generating within-habitat refuges for seaweeds on reefs. As one possible example, several studies have shown that grazing rates of herbivorous reef fishes are highly correlated with water temperature (Hatcher, 1981; Carpenter, 1986; Klumpp and Polunin, 1989). If lowered temperature depresses feeding rates more than rates of algal production, then palatable seaweeds will have a greater probability of escaping herbivorous fishes during cooler periods of the year. In a similar way, numerous studies have shown that both the density of herbivorous fishes (Bouchon-Navaro and Harmelin-Vivien, 1981; Williams and Hatcher, 1983; Russ, 1984b; Lewis and Wainwright, 1985) and the rate at which they graze seaweeds (Vine, 1974; Hay, 1981c, 1984b; Hay *et al.*, 1983; Hay and Goertemiller, 1983; Morrison, 1988) decrease with depth on most coral reefs. If grazing along depth gradients decreases faster than the decrease in algal production caused by diminishing light, then deeper areas could serve as spatial escapes because production would exceed consumption. Greater algal biomass on deeper sections of some reefs has been noted (Van den Hoek *et al.*, 1978), but the importance of herbivory in generating this pattern has not been adequately evaluated.

Herbivorous fishes might use deeper habitats less because their prey are less productive there, because the decreased light levels make them more vulnerable to their own predators (Hobson, 1972; Cerri, 1983; McFarland, Chapter 2, this volume), or due to some combination of these or other factors. The importance of predators is suggested by the increased use of deeper areas on heavily fished reefs where larger predators have presumably been depleted (Hay, 1984b) and by the avoidance of shallow unstructured areas where productivity is obviously high but where grazing fishes appear to be more exposed to attack (Hay *et al.,* 1983; Hay, 1985; Lewis and Wainwright, 1985; Lewis, 1986). Several studies suggest that susceptibility to predation strongly affects foraging by herbivorous reef fishes (Hixon, Chapter 17, this volume), and numerous correlative studies suggest that areas heavily used by predators may provide spatial refuges for palatable seaweeds (see Hay, 1985, for possible examples). However, there are no studies on reef fishes that unambiguously document the direct effects of predator avoidance.

Recent studies suggest that competition with herbivorous sea urchins may also be important in explaining why herbivorous fishes increase their use of deeper habitats on reefs that are heavily fished. On heavily exploited reefs in the Caribbean and in eastern Africa, urchin predators are rare and urchin densities are high, suggesting that fishing removes urchin predators and allows large increases in urchin densities (Hay, 1984b; McClanahan and Muthiga, 1988). Since urchins exploit algal resources to lower levels than fishes (Carpenter, 1986), herbivorous fishes may be forced to use deeper areas where urchins are less abundant (Morrison, 1988). When high densities of urchins (7–10/m^2) were removed from experimental sections of reefs in the U.S. Virgin Islands, algal standing stock, density of parrotfishes and surgeonfishes, and grazing rate of fishes all increased significantly relative to adjacent control areas (Hay and Taylor, 1985). On a much larger spatial scale, when the abundant urchin *Diadema antillarum* experienced a Caribbean-wide epidemic that killed greater than 93% of all *Diadema* (Lessios, 1988), fish grazing on shallow reefs in Jamaica and St. Croix increased by three- to fourfold, and in Jamaica, herbivorous fishes that had been found only on the deep reef began to occur in shallow reef habitats (Carpenter, 1988; Morrison, 1988). One to two years after the *Diadema* die-off in St. Croix, the total number of parrotfishes and surgeonfishes in four reef habitats monitored by Carpenter (1990) increased a significant two- to fourfold and remained at elevated levels when the study ended three years after the *Diadema* die-off. These studies all suggest that high densities of grazing urchins may affect both the density and foraging behavior of herbivorous fishes on some reefs. Since urchins and fishes differ in their foraging behavior and their resistance to various algal defenses (Littler *et al.,* 1983b; Carpenter, 1986; Hay *et al.,* 1987a,b; Morrison, 1988), urchin–fish interactions could also affect the distribution of reef seaweeds.

C. Microhabitat Refuges

Benthic organisms that are unpalatable to grazing fishes can produce microhabitat refuges for more palatable prey, and these refuges can occur on a scale of only centimeters. In North Carolina, palatable red and green seaweeds gain significant protection from herbivorous fishes by growing epiphytically on the unpalatable brown alga *Sargassum filipendula;* if the palatable algae are moved only 5 cm away from *Sargassum,* they are rapidly eaten (Hay, 1986). In the absence of herbivorous fishes, palatable species grew better alone than when on *Sargassum.* In the presence of herbivorous fishes, palatable species growing alone were eaten to extinction and only those growing on unpalatable species persisted. *Sargassum* was therefore a competitor that suppressed the growth of palatable species but it also provided microsites of reduced herbivory that prevented fishes from driving the palatable species to local extinction. For those species most susceptible to grazing fishes, the cost (reduced growth) of being associated with an unpalatable competitor was much less than the cost of increased consumption (death) in the absence of that competitor. Under these conditions, one competitor can have a strong positive effect on another and these associational plant refuges can provide an important mechanism for maintaining species richness within communities that are dominated by a few unpalatable species (Hay, 1986).

Although these refuges were initially interpreted as arising from simple visual crypsis (Hay, 1986), more recent studies suggest that chemistry of the unpalatable species may play a significant role (Littler *et al.,* 1986; Pfister and Hay, 1988). As an example, on Caribbean reefs the chemically defended (Hay, *et al.,* 1987a) alga *Stypopodium zonale* produces an associational refuge for more palatable species; numerous species palatable to fishes are more common near the base of *Stypopodium* than several centimeters away (Littler *et al.,* 1986). If *Stypopodium* plants are removed, palatable seaweeds that are nearby experience greater losses to grazing fishes than do palatable seaweeds near *Stypopodium* plants that are not removed. Plastic mimics of *Stypopodium* provide some protection to palatable seaweeds but not as much protection as real plants (Littler *et al.,* 1986). Thus, a portion of the associational refuge may be generated by the mere presence of a nonfood plant, but the chemical repugnance of the unpalatable plant also appears to play a role. Grazing on palatable seaweeds is also reduced if they grow in close proximity to chemically defended (Targett *et al.,* 1986; Hay *et al.,* 1987a, 1988c; Paul and Van Alstyne, 1988a) seaweeds in the genera *Dictyota* and *Halimeda* (Hay, 1985) or to the unpalatable sea fan *Gorgonia ventalina* or the fire coral *Millepora alcicornis* (Littler *et al.,* 1987).

In the tropical Pacific, the unpalatable green alga *Chlorodesmis fastigiata* produces a cytotoxic compound that deters grazing by reef fishes (Paul, 1987;

Paul and Fenical, 1987; Wylie and Paul, 1988) but stimulates grazing by a specialist herbivorous crab that is protected from predators by its physical association with the alga (Hay *et al.,* 1989). Thus, seaweeds that deter fish grazing may serve as spatial refuges for both plants and animals that are less well defended from attack (for more examples, see the review by Hay, 1991).

The topography of the substrate can also produce refuges from herbivorous fishes. Brock (1979) used microcosms to study the effects of herbivory and microhabitat complexity on species richness over a 43-day period. Herbivory was altered by manipulating the density of juvenile parrotfish and microhabitat complexity was altered by covering the bottom of the microcosms with mesh of different sizes. On smooth substrate, both species richness and the biomass of benthic organisms decreased with increasing fish densities. When microhabitat complexity was added, increasing fish density had minimal effect on the biomass and species richness of benthic prey. Hixon and Brostoff (1985) found similar patterns in field experiments. In their study, natural coral substrates maintained a higher diversity of algae under intense fish grazing than did smooth plates of the same coral. Under low rates of grazing, these surfaces differed very little in algal species richness. Thus, the roughened surface appeared to support more species because it provided refuges from herbivorous fishes, not because it provided microhabitats for resource partitioning. Field experiments in intertidal habitats on the Pacific coast of Panama show similar patterns (Menge *et al.,* 1985). Holes and crevices serve as important refuges from grazing fishes and other consumers, and the heterogeneity of consumer effects produced by these refuges is important in maintaining local species richness.

III. Seaweed Characteristics That Deter Feeding by Reef Fishes

Many species of tropical seaweeds live on shallow reef-slope habitats where grazing is intense and where they are constantly accessible to foraging fishes. A few, such as *Halimeda,* may even become extremely abundant under these conditions. Because of the high rates of grazing that typify these habitats, coral-reef seaweeds offer one of the best opportunities for investigating plant defenses against herbivores. Most studies of seaweed defenses against reef fishes have concentrated on chemical or morphological deterrents. Several recent reviews address a broad spectrum of topics regarding seaweed chemical defenses and the chemical ecology of marine organisms in general (Paul and Fenical, 1987; Hay and Fenical, 1988; Van Alstyne and Paul, 1988; Duffy and Hay, 1990; Hay, 1991; Paul, 1991). The general hypotheses and data supporting or contesting the importance of morphological defenses against herbivores are also available in several recent papers (Littler and Littler, 1980; Hay, 1981b, 1984a; Steneck and Watling, 1982; Littler *et al.,* 1983b; Steneck, 1983, 1985, 1986, 1988; Lewis, 1985; Padilla, 1985, 1989; Lewis *et al.,* 1987). In this paper, I will focus on algal characteristics that affect feeding by fishes; for effects against other herbivores, see the papers just cited.

A. Structural and Morphological Deterrents

Seaweed morphology and structure vary tremendously, from small filamentous species, to 60-m-long kelps, to crustose corallines that are highly calcified and tightly adherent to the substrate. A generalized theory of how the probability of being eaten changes as a function of seaweed morphology has been proposed by Littler, Steneck, and co-workers (Littler and Littler, 1980; Littler *et al.,* 1983a,b; Steneck and Watling, 1982; Steneck, 1983, 1986, 1988). They predict that the difficulty of grazing a particular algal form increases in the following order (see Fig. 1 in Steneck and Watling, 1982, or Table 1 in Littler *et al.,* 1983a): microalgae, filamentous algae, sheetlike algae, coarsely branched algae, leathery or rubbery algae, jointed calcareous algae, and crustose coralline algae. This hierarchy is based in large part on increasing algal toughness and decreasing food value due to a greater portion of the thallus being composed of indigestible structural materials. Patterns of algal susceptibility that roughly support the predicted pattern have been shown for herbivorous molluscs (Steneck and Watling, 1982) and sea urchins (Littler and Littler, 1980; Littler *et al.,* 1983a,b). Tests with reef fishes have been variable. Littler *et al.,* (1983b) found patterns supporting the model. Hay (1984a) used a larger data set and found that extreme ends of the spectrum differed significantly but that the model was of very limited value in the center portion of the morphology–susceptibility spectrum (see also Lewis, 1985) and that many of the tougher, calcified seaweeds also possessed chemical defenses, making it difficult to separate the effects of chemical from morphological or structural deterrents in these types of correlative assays (see also Paul and Hay, 1986).

A few studies have directly addressed the effects of algal morphology and phenotypic plasticity on susceptibility to herbivorous reef fishes. Steneck and Adey (1976) studied the coralline alga *Lithophyllum congestum,* which is common on Caribbean reefs. On shallow reef slopes where fish grazing is intense, it grows as a smooth crust; on reef flats where fish grazing is diminished, it occurs as a crust that produces upright branches. The production of upright branches increased its growth rate and the rate at which it could produce spores, but parrotfish grazing precluded this more productive form from many areas of the reef. Geographic surveys of temperate and tropical communities show that temperate communities have an abundance of branched corallines in shallow subtidal habitats while tropical reefs tend to support corallines with upright branches only on shallow reef flats or on the deeper portions of reefs (Steneck, 1986) where herbivory by fishes is reduced (see Section II). In addition to the geographic contrast, it also appears that branched corallines became much less important in the tropics following the evolution of parrotfishes in the Eocene (Steneck, 1985). Thus, where herbivorous reef fishes are abundant, they appear to select against upright forms and for crustose forms.

Like other sessile organisms, seaweeds may adjust their morphology to prevailing environmental conditions, including local grazing by fishes. As an example, clonal seaweeds can form either loose aggregations that grow rapidly but are susceptible to attack by fishes, or tightly packed and highly branched turfs that grow slowly but are relatively resistant to attack by fishes (Hay, 1981b). Subtidally, the tightly packed form occurs primarily in areas subject to high fish grazing. Additionally, several reef seaweeds have two morphologically distinct forms that differ in growth rate and susceptibility to herbivory; transition between the forms is mediated by changes in fish grazing (Lewis *et al.*, 1987). The best-studied example is the brown alga *Padina*, which occurs as a prostrate thin-branching form in areas of high grazing by herbivorous fishes (this form is relatively resistant to grazers). However, within 96 hr of excluding herbivorous fishes, the plant dramatically changes its growth form to a rapidly growing upright blade that is very susceptible to fish grazing but is more competitive and may even overgrow and kill corals (Lewis, 1986; Lewis *et al.*, 1987).

Paleontological assessments of changes in seaweed structure and morphology that coincide with the evolution of different herbivore types are impossible for most seaweeds because they rarely fossilize. However, the fossil record for calcareous algal crusts is substantial and grazing scars attributable to different types of herbivores are often preserved on their surface. Using fossilized corallines and available information on present-day interactions between corallines and various herbivores, Steneck (1983) argues that the evolution of herbivorous fishes (especially parrotfishes) significantly increased herbivory on coral reefs and dramatically changed selective regimes affecting reef seaweeds and reef communities in general. Coincident with the evolution of these fishes, coralline crusts with characteristics that reduced the effects of reef fishes radiated, while their assumed parent taxon (the solenopores, which lacked these characteristics) diminished in importance and eventually went extinct. Because crustose seaweeds are generally less damaged by herbivores than are upright species, the effects on upright seaweeds should have been even more dramatic.

B. Chemical Deterrents

Over 600 secondary metabolites have been isolated from seaweeds, with the majority of compounds being produced by tropical genera (Faulkner, 1984, 1986, 1987). Most seaweed natural products are terpenoids, with many being sesquiterpenes and diterpenes. Other types of compounds include acetogenins, unusual fatty acids, amino acid-derived substances, phlorotannins (algal polyphenolics), and metabolites of mixed biosynthesis that often have terpenoid and aromatic portions. Seaweed metabolites differ markedly from those of terrestrial plants in that they are commonly halogenated and they almost never contain nitrogen (a limiting resource for most seaweeds). The only "seaweeds" that often produce nitrogen-containing metabolites are the nitrogen-fixing blue-green algae (really cyanobacteria). Although many seaweed metabolites have long been known to be very bioactive in pharmacological assays (see Faulkner, 1984, 1986, 1987; Paul and Fenical, 1987), their ecological functions under natural conditions have been addressed only recently (see Hay and Fenical, 1988; Paul, 1991). Many have been shown to significantly deter feeding by herbivores, and the literature lists about 40 compounds that have been tested as feeding deterrents against reef fishes (Table 1); 23 of these significantly depressed feeding.

Larger seaweeds that grow on structurally complex reef slopes where herbivorous fishes are common are almost always chemically rich. A few of the best-studied examples among the green, red, and brown seaweeds are discussed in the following sections.

1. Green Seaweeds (Chlorophyta)

About 70 secondary metabolites have been isolated from the Chlorophyta, with most being sesquiterpenoid and diterpenoid compounds produced by seaweeds in the tropical order Caulerpales (Paul and Fenical, 1987). Concentrations of these compounds range from trace amounts to about 2% of whole-plant dry mass (Paul and Fenical, 1987). However, compounds are not equally distributed throughout the plant and some compounds have been reported to constitute as much as 2–6% of the blotted wet mass of young plant portions (Hay *et al.,* 1988c).

Calcified green algae in the genus *Halimeda* provide one of the most thoroughly studied examples of seaweed chemical defenses against fishes. *Halimeda* is spectacularly successful in the tropics and is one of the very few upright macroalgae that is often abundant on reefs with high densities of herbivorous fishes. *Halimeda* species typically produce several diterpenoids that deter feeding by reef fishes (Table 1) and that are structurally related to the powerful insect antifeedant warburganol. These compounds are often strong cytotoxins and are active against bacteria, fungi, and the sperm, fertil-

Table 1 Seaweed Secondary Metabolites and Their Effects on Feeding by Various Reef Fishes[a,b]

Seaweed species	Compound	Effect	Tested against	Reference
I. Green Seaweeds (Chlorophyta)				
Avrainvillea spp.	Avrainvilleol	0	*Zebrasoma flavescens*	16
Caulerpa spp.	Caulerpenyne	*	*Sparisoma radians*	15
		0	*Siganus spinus*	10
		0	*Zebrasoma flavescens*	16
	Caulerpin	0	*Zebrasoma flavescens*	16
Caulerpa brownii	*Caulerpa* "diacetate"	0	*Zebrasoma flavescens*	14
		0	*Siganus argenteus*	14
		0	Reef fishes (Guam)	14
	Trifarin and dihydrotrifarin	0	*Zebrasoma flavescens*	14
		0	Reef fishes (Guam)	14
Chlorodesmis fastigiata	Chlorodesmin	*	*Siganus spinus*	10
		0	*Siganus doliatus*	3
		*	*Zebrasoma flavescens*	16
		*	Reef fishes (Guam)	9
		0	Reef fishes (Australia)	7
Cymopolia barbata	Cymopol	0	*Siganus doliatus*	3
		*	Reef fishes (Caribbean)	1
Halimeda spp.	Halimedatrial	*	Caribbean parrotfishes	5
		*	Caribbean surgeonfishes	5
		*	Reef fishes (Guam)	11
Halimeda spp.	Halimedatetracetate	*	*Sparisoma radians*	15
		0	*Siganus spinus*	10
		*/0	Reef fishes (Guam)	9, 11
Pseudochlorodesmis furcellata	An "epoxylactone"	*	*Siganus spinus*	13
		*	Reef fishes (Guam)	9
Tydemania expeditionis	A "diterpenoid"	*	*Siganus spinus*	10

(*continued*)

Table 1 *(Continued)*

Seaweed species	Compound	Effect	Tested against	Reference
Udotea spp.	Udoteal	*	*Zebrasoma flavescens*	14
		0	*Siganus argenteus*	14
		*/0	Reef fishes (Guam)	9, 14
	Flexilin	*	*Zebrasoma flavescens*	14
		*	*Siganus argenteus*	14
		0	Reef fishes (Guam)	14
II. Red Seaweeds (Rhodophyta)				
Laurencia spp.	Elatol	*	*Siganus doliatus*	3
		0	*Zebrasoma flavescens*	14
		*	*Siganus argenteus*	14
		*	Reef fishes (Caribbean)	1
		*	Reef fishes (Guam)	14
	Isolaurinterol	*	Reef fishes (Caribbean)	1
	Debromolaurinterol	*	*Siganus doliatus*	3
	Pacifenol	0	*Siganus doliatus*	3
	Prepacifenol	0	*Siganus doliatus*	3
	A "chamigrene"	0	*Siganus doliatus*	3
	(?) Aplysin	0	*Siganus doliatus*	3
		0	Reef fishes (Caribbean)	1
Laurencia snyderae	Chlorofucin	*	*Siganus doliatus*	3
Laurencia cf. *palisada*	Palisadin-A	0	*Siganus doliatus*	3
		+/0	*Zebrasoma flavescens*	14
		0	*Siganus argenteus*	14
		*/0	Reef fishes (Guam)	14
	5-Acetoxy-palisadin-B	0	*Zebrasoma flavescens*	14
		0	Reef Fishes (Guam)	14
	(?) Aplysistatin	0	*Zebrasoma flavescens*	14
		0	*Siganus argenteus*	14
		0	Reef fishes (Guam)	14

Ochtodes secundiramea	Ochtodene	*	*Zebrasoma flavescens*	16
		*	Reef fishes (Caribbean)	12
		*	Reef fishes (Guam)	12
	Chondricole-C	0	Reef fishes (Caribbean)	12
		0	Reef fishes (Guam)	12
Sphaerococcus coronopifolius	Sphaerococcenol-A	0	*Siganus doliatus*	3
Vidalia obtusaloba	Vidalol-A	0	*Siganus doliatus*	3
III. Brown Seaweeds (Phaeophyta)				
Dictyota spp.	Pachydictyol-A	*	*Zebrasoma flavescens*	16
		*	*Siganus doliatus*	3
		*	*Lagodon rhomboides*	6
		*	*Diplodus holbrooki*	2
		*	Reef fishes (Caribbean)	1
	Dictyol-E	0	*Siganus doliatus*	3
		*	*Lagodon rhomboides*	6
		*	*Diplodus holbrooki*	2
	Dictyol-B	*	*Siganus doliatus*	3
	Dictyol-B acetate	*	Reef fishes (Caribbean)	8
	Dictyol-H	*	*Siganus doliatus*	3
	A "spatane diterpene"	0	*Siganus doliatus*	3
	A "dolastane class diterpene"	0	*Siganus doliatus*	3
Dictyopteris spp.	Dictyopterenes-A and B	*	*Siganus doliatus*	3
		*	Reef fishes (Caribbean)	4
	Zonarol	0	*Siganus doliatus*	3
Stypopodium zonale	Stypoldione	*	*Siganus doliatus*	3
	Stypotriol	0	*Siganus doliatus*	3
		*	Reef fishes (Caribbean)	1

(*continued*)

Table 1 *(Continued)*

Seaweed species	Compound	Effect	Tested against	Reference
	Geranylgeranylmethyl quinone	0	*Siganus doliatus*	3
IV. Blue-Green Seaweeds (Cyanobacteria)				
Microcoleus lyngbyaceus	Malyngamide-A	*	*Zebrasoma flavescens*	16

[a] * = Compounds that significantly deterred feeding; 0 = compounds that had no significant effect on feeding; + = compounds that significantly increased feeding. Multiple symbols separated by a slash indicate differing results at different times or locations. A (?) before a compound name indicates that the compound might be a degradation product instead of a true algal metabolite. Compound names in quotation marks do not have official common names. See the cited reference for the compound structure. Tests that list reef fishes, *Siganus doliatus*, or Caribbean parrotfishes or surgeonfishes as the grazers were conducted in the field; others were conducted in the lab. Most assays listed here tested compounds at concentrations of 1–2% of plant dry mass or less; these appear to be ecologically relevant concentrations for these compounds on reef (Hay and Fenical, 1988). Structures of the compounds and lists of the specific species producing them can be found in the cited references or in the work by Faulkner (1984, 1986, 1987). When only a genus is listed, multiple species, but not all species, within that genus produce the compound(s).

[b] References: (1) Hay *et al.*, 1987a; (2) Hay *et al.*, 1987b; (3) Hay *et al.*, 1988a; (4) Hay *et al.*, 1988b; (5) Hay *et al.*, 1988c; (6) Hay *et al.*, 1988d; (7) Hay *et al.*, 1989; (8) Hay *et al.*, 1990; (9) Paul, 1987; (10) Paul and Fenical, 1987; (11) Paul and Van Alstyne, 1988a; (12) Paul *et al.*, 1987; (13) Paul *et al.*, 1988a; (14) Paul *et al.*, 1988b; (15) Targett *et al.*, 1986; (16) Wylie and Paul, 1988.

ized eggs, and larvae of sea urchins (Paul and Fenical, 1986). The major metabolites produced by most species of *Halimeda* are the diterpenoid trialdehyde halimedatrial and the diterpenoid tetracetate halimedatetracetate (Paul and Fenical, 1986). When palatable seaweeds are treated with these compounds (see Hay and Fenical, 1988, for methods) and placed on herbivore-diverse areas of coral reefs in both the Caribbean (Hay *et al.,* 1988c) and Pacific (Paul, 1987; Paul and Van Alstyne, 1988a), they experience significantly less loss to grazing fishes than do adjacent controls. Halimedatrial is a stronger and more consistent deterrent than halimedatetracetate (Paul, 1987; Paul and Van Alstyne, 1988a) (Table 1).

Seaweeds may also allocate their defenses in space and time so that they are maximally effective against reef fishes. As an example, extracts of *Halimeda* from areas with low densities of herbivorous fishes may be less powerful feeding deterrents than extracts of the same species from areas where herbivorous fishes are abundant (Paul and Van Alstyne, 1988a). In an even more unusual example, *Halimeda* has somehow decoupled photosynthesis and growth and is thus able to produce its new growth almost exclusively at night when herbivorous reef fishes are inactive (Hay *et al.,* 1988c). These youngest plant portions have 3–4.5 times the food value of older portions but are only moderately more susceptible (Hay *et al.,* 1988c) or no more susceptible (Targett *et al.,* 1986; Paul and Van Alstyne, 1988a) to herbivorous fishes, as a result of their high concentrations of chemical defenses. Young portions contain very high concentrations of the most potent feeding deterrent (halimedatrial) (Hay *et al.,* 1988c) and extracts of these young white tips are significantly more deterrent to reef fishes than are extracts of mature green tips (Paul and Van Alstyne, 1988a). Additionally, the newly produced portions of *Halimeda* remain unpigmented until just before sunrise because the valuable, nitrogen-containing molecules associated with photosynthesis are not placed in the new growth until light is available and they can start producing income for the plant (Hay *et al.,* 1988c).

In the first 2 days after new portions are produced, calcification (a structural defense) increases, the tissues become much less valuable as a food, and chemical defenses decrease; concentrations of chemical defenses in older plant portions are less than 10% of those in the youngest portions (Hay *et al.,* 1988c; Paul and Van Alstyne, 1988a). In the older portions, chemical defenses consist almost exclusively of the less deterrent compound halimedatetracetate, rather than the more deterrent compound, halimedatrial, which is concentrated in the newest portions. However, recent investigations by V. J. Paul (personal communication) indicate that the less deterrent compound is immediately converted (apparently enzymatically) to the more deterrent compound when *Halimeda* tissue is macerated as would happen in the pharyngeal mill of parrotfishes.

The chemistry of the tropical green alga *Caulerpa* has also been extensively studied and this genus produces several chemically interesting terpenoids (Paul and Fenical, 1987). However, although it is commonly assumed to be chemically defended (Norris and Fenical, 1982; Paul and Fenical, 1986; Targett *et al.*, 1986), numerous tests of its metabolites against grazing fishes in the Caribbean and Pacific have almost uniformly shown these compounds to have little, if any, effect (V. J. Paul, personal communication) (Table 1). Additionally, studies have indicated that some species of *Caulerpa* are often consumed at moderate to high rates when exposed to reef fishes (Hay, 1984a; Lewis, 1985; Paul and Hay, 1986). The ecological role of *Caulerpa's* unique metabolites are, at present, unclear.

2. *Red Seaweeds (Rhodophyta)*

Rhodophyta produce biologically active compounds ranging in structure from simple brominated phenols and aliphatic halo-ketones to more complex monoterpenes, sesquiterpenes, and diterpenes (Fenical, 1975, 1982; Erickson, 1983; Faulkner, 1984, 1986, 1987). Concentrations range from trace amounts to as much as 3–5% of plant dry mass. Most of the well-studied genera produce several related compounds. The extreme example is *Laurencia,* which is reported to produce over 400 different compounds (most are terpenoids) representing 26 different structural classes (Fenical, 1975, 1982; Erickson, 1983). One of the better-known *Laurencia* metabolites is the chamigrene-class sesquiterpenoid elatol. It is cytotoxic, ichthyotoxic, insecticidal, and deters feeding by a range of reef fishes (see Table 1). Some other *Laurencia* metabolites show similar activities; however, there are also several that are structurally very similar to elatol but have no effect on fish feeding (Hay *et al.,* 1988a) (Table 1).

3. *Brown Seaweeds (Phaeophyta)*

Seaweed polyphenolics, which are hypothesized to function like terrestrial tannins (Ragan and Glombitza, 1986; Steinberg, 1985, 1988; Van Alstyne, 1988), are produced only by brown algae. However, in contrast to the complex biochemical origins of phenolics in terrestrial plants, algal polyphenolics are all derived from the simple C_6 precursor phloroglucinol and are thus more appropriately termed phlorotannins to distinguish them from the chemically different terrestrial tannins (Ragan and Glombitza, 1986). Phlorotannins have been demonstrated to deter feeding by herbivorous invertebrates (Steinberg, 1988) but their effects on herbivorous fishes are largely unknown. Their geographic distribution suggests they do not serve as important defenses against coral-reef fishes. Unlike other algal secondary metabolites, which are most often produced in tropical species (Faulkner, 1984, 1986, 1987), phlorotannins occur in high concentrations in Fucales from temperate habitats, but in low concentrations in tropical Fucales (Steinberg, 1986; Steinberg and Paul, 1990). Extracts of phlorotannin-rich temperate seaweeds deterred feeding by herbivorous fishes on tropical reefs in Guam (Van Alstyne and Paul, 1990), so the near absence of these compounds in tropical seaweeds does not appear to result from their ineffectiveness against reef fishes.

The most common secondary metabolites produced by brown seaweeds in tropical and warm-temperate seas are terpenoids, acetogenins, and terpenoid-aromatic compounds of mixed biosynthetic origin (McEnroe *et al.*, 1977; Faulkner, 1984). Chemically rich genera include *Dictyota, Dictyopteris, Zonaria, Stypopodium, Pachydictyon,* and *Glossophora,* all in the order Dictyotales. The bicyclic diterpene alcohol pachydictyol-A (Fig. 1) is a well-studied example and is representative of a family of compounds (the dictyols) that have been extensively tested as defenses against reef fishes (see Table 1). Although pachydictyol-A lacks significant biological activity in pharmacological assays in the lab (see Hay *et al.*, 1987a, for references), it significantly deters feeding by reef fishes in Guam, Australia, the Caribbean, and the temperate Atlantic (see Table 1). With one exception, related dictyols (dictyol-E, dictyol-H, dictyol-B, and dictyol-B acetate) have also been consistently effective feeding deterrents (Table 1).

4. *Blue-Green Seaweeds (Cyanobacteria)*

Unlike other seaweeds, some cyanobacteria produce nitrogen-containing secondary metabolites; these compounds are often strongly bioactive against fungi, bacteria, and cancer cells (Moore, 1977, 1981; Mynderse *et al.*, 1977). Many of the metabolites are halogenated; this is especially common in the family Oscillatoriaceae. The only blue-green metabolite that has been tested against a reef fish was malyngamide-A from *Microcoleus lyngbyaceus*. It significantly reduced grazing by *Zebrasoma flavescens* (Wylie and Paul, 1988). Given the inflammatory and carcinogenic effects of compounds from cyanobacteria (Moore, 1977; Cardellina *et al.*, 1979), the chemical ecology of this group deserves more attention than it has received to date.

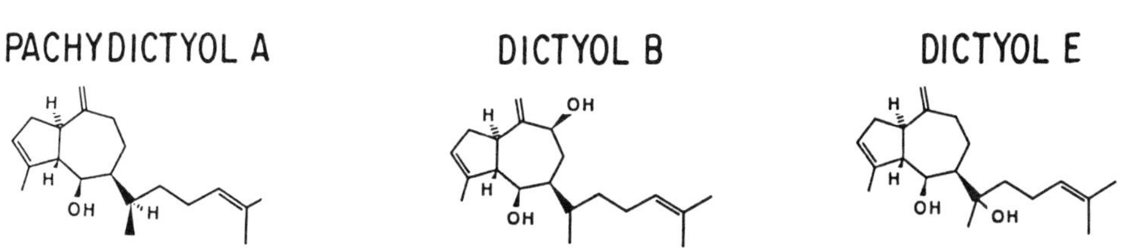

PACHYDICTYOL A

DICTYOL B

DICTYOL E

Figure 1 The structures of three dictyols.

5. *Physiological Effects on Herbivorous Fishes*

Many seaweed metabolites have been extensively investigated in pharmacology assays (Faulkner, 1984, 1986, 1987; Paul and Fenical, 1987), however, very little is known of their natural effects when consumed by reef herbivores. When the omnivorous temperate fish *Diplodus holbrooki* consumed a diet containing 1% pachydictyol-A from the brown alga *Dictyota,* the fish's growth was slowed by about 50% compared to paired control fish that were held in the same aquaria and fed the same amount of food without pachydictyol-A (Hay *et al.,* 1987b). A similar experiment using *Diplodus holbrooki* and adding the *Laurencia* metabolite elatol as 0.5% of the diet indicated that elatol affected survivorship rather than growth. After 2 weeks of feeding, 55% ($N = 20$) of treatment fish and only 10% of control fish had died (M. E. Hay, work in progress). These are the only data assessing even the gross physiological effects of algal metabolites on reef fishes. Given the different feeding modes and digestive physiologies of herbivorous fishes (reviewed by Horn, 1989), this is an area that deserves more study. As one potential example of how physiological or digestive variance among fish species might have important consequences for seaweeds, kyphosids selectively consume brown seaweeds that are relatively unused by other fishes (Horn, 1989). The guts of these fishes may be packed with seaweeds such as *Dictyota* that produce dictyols, which deter feeding by most fishes against which they have been tested (Table 1). Since kyphosids are the only fishes known to contain hindgut microflora that appear to fermentatively digest seaweed material (Rimmer and Wiebe, 1987), it is possible that this digestive process makes them less susceptible to certain algal defenses.

6. *Structure–Function Relationships*

In general, ecologists do very poor chemistry, chemists do very poor ecology, and the methodologies and social customs (chemists often wear ties) of each discipline are largely nonoverlapping. Given this reciprocal ineptness, it is easy to see how false generalities can be erected and defended by both camps in a desperate attempt to avoid dealing with the complexities and subtleties of the other discipline. One of the most often cited generalities of chemically mediated plant–herbivore interactions is that tannins (phlorotannins for seaweeds) are nontoxic digestibility reducers and that alkaloids, terpenes, etc., are toxins. These generalities allow ecologists to erect broad dichotomies about the ecology and evolution of digestibility reducers versus toxins (see Feeney, 1976; Coley *et al.,* 1985) and to deal with classes of compounds as a group (i.e. total phenolics, total terpenes, etc.) rather than having to do the difficult (and at times impossible) work of isolating, identifying, and quantifying each separate compound. As with many generalities that are unusually productive, this one now appears false for marine (Hay and Fenical, 1988) as well as terrestrial (Karowe, 1989; Bernays *et al.,* 1989) systems. Effects of compounds on herbivores are a unique product of the particular compound and the particular herbivore (Zucker, 1983; Hay and Fenical, 1988; Hay *et al.,* 1988a; Paul *et al.,* 1988b; Hay, 1991; Bernays *et al.,* 1989). Although some compounds such as pachydictyol-A from *Dictyota* and elatol from *Laurencia* appear to be broadly effective deterrents against most fishes, and some families of compounds, such as the dictyols, appear to be uniformly deterrent (Table 1), these apparent patterns obscure our inability to predict biological function from the structure of the compound being considered. In the following I outline a few of several available examples.

a. Variance Among Compounds With Similar Structures Hay *et al.* (1988a) assessed the effects of numerous closely related metabolites from brown and red seaweeds in the genera *Dictyota* and *Laurencia* on feeding by the rabbitfish *Siganus doliatus.* Although most of the compounds tested were terpenoids, they differed in their carbon skeletons and chemical functionalities. Such studies provide an opportunity to look for general correlations between the structural features of compounds and their effects on feeding by reef fishes. These authors concluded that neither the general structure of compounds nor their activity in pharmacological assays were useful predictors of antifeedant properties. As examples, (1) pachydictyol-A and dictyol-E differ by only one hydroxyl substitution (see Fig. 1) and neither possesses significant biological activity in pharmacological assays, however, pachydictyol-A strongly decreased grazing and dictyol-E had no effect; and (2) dictyol-B and dictyol-E differ only in the *position* of one hydroxyl group, yet dictyol-B significantly depressed feeding and dictyol-E did not. *Laurencia* metabolites showed similar patterns. Elatol, pacifenol, prepacifenol, and the unnamed chamigrene have very similar carbon skeletons and levels of halogenation (for chemical structures see Hay *et al.,* 1988a). However, elatol was a strong deterrent while the others had no effect. Additional studies indicating that similar compounds differ markedly in their effect on feeding by reef fishes are available (Hay *et al.,* 1987a,b, 1988d; Paul *et al.,* 1988b).

b. Variance Among Herbivores Just as structurally similar compounds may vary in their effects on feeding by a single species of fish, a single compound can vary dramatically in its effects on different fish species. Table 1 shows several cases where herbivorous reef fishes differ in their response to a given compound. Fishes may also differ dramatically from other types of herbivores in their response to algal metabolites. The diterpene alcohols pachydictyol-A and dictyol-E produced by various species of *Dictyota* provide the most extensively studied contrast in this regard. In the temperate Atlantic, dictyol-E strongly deters feeding by the herbivorous fishes *Diplodus holbrooki* and *Lago-*

don rhomboides and the sea urchin *Arbacia punctulata;* pachydictyol-A is less deterrent for the fishes and completely ineffective against the urchin (Hay *et al.,* 1987b, 1988d). In contrast, pachydictyol-A is a strong deterrent of numerous tropical reef fishes (see Table 1) and of the tropical sea urchin *Diadema antillarum* (Hay *et al.,* 1987a), while dictyol-E did not deter the Pacific reef fish *Siganus doliatus* (Hay *et al.,* 1988a) and even stimulated feeding in field assays against reef fishes in the Caribbean (M. E. Hay, work in progress).

In addition to the patterns discussed, there may be general differences in the effectiveness of seaweed metabolites against large mobile consumers like fishes versus small sedentary herbivores (mesograzers) like certain amphipods, polychaetes, crabs, and ascoglossans. The hypothesis has been advanced that small herbivores that live on the plants they consume should evolve to live on seaweeds chemically defended from fishes since these seaweeds might provide sites of reduced predation because they are less often visited by fishes. Investigations of this process have been conducted in the temperate Atlantic (Hay *et al.,* 1987b, 1988d), the Caribbean (Hay *et al.,* 1988b, 1990), and the tropical Pacific (Paul and Van Alstyne, 1988b; Hay *et al.,* 1989); these studies all support the contention that sedentary mesograzers living in close association with seaweeds that are chemically repellent to fishes experience dramatically reduced rates of predation (see the review by Hay, 1991).

It is now clear that compounds with similar structures can differ dramatically in their effects on herbivore feeding, and that compounds that deter one herbivore can have no effect or may even stimulate feeding in another. Often cited generalizations regarding the functions and mechanisms of action of various classes of secondary metabolites are of limited, if any, value (see Hay and Fenical, 1988; Bernays *et al.,* 1989) because compounds differing by only one hydroxyl group or in the position of that group can show great variance in their activities, and because modes of action almost certainly change with the identity of the herbivore consuming the plant (Karowe, 1989; Bernays *et al.,* 1989; Hay, 1991). For progress to be made in our understanding of the chemically mediated interactions between reef fishes and seaweeds, ecologists will need to increase their understanding of the chemical complexities involved and avoid lumping compounds and their activities into convenient, but ecologically unsubstantiated, groups.

7. How Fishes Perceive Seaweed Chemical Defenses

No rigorous data are available on fish perception of chemical defenses, and the absence of such data seriously limits our ability to understand and interpret seaweed–fish interactions. When palatable seaweeds were treated with halimedatrial from *Halimeda* and placed on reefs where they could be grazed by either parrotfishes or surgeonfishes, parrotfishes tended to bite both treatment and control blades but to bite treatment blades only once and to take several bites from control blades (Hay *et al.,* 1988c). Surgeonfishes showed a significant ability to avoid treatment blades without biting them. These observations suggest that parrotfishes may need to taste an unfamiliar alga before rejecting it, while surgeonfishes may be able to sense some aspects of algal chemistry without biting the seaweed. Numerous studies of the effects of specific compounds on feeding by reef fishes have been conducted in the field by treating palatable algae with seaweed metabolites, putting these along with appropriate controls at various locations on natural reefs, and determining how much of each treatment is eaten over a period of a few minutes to hours (see Hay and Fenical, 1988, for references). The finding that reef fishes often show strong avoidance of the treated seaweeds relative to the identical-looking controls means that fishes are rapidly able to assess differences in "quality" among morphologically identical patches of seaweeds. Since herbivorous reef fishes are abundant, highly mobile, can consume almost all local algal production, and can rapidly assess and utilize algal resources in accordance with their value, it seems reasonable to suspect that they have been one of the major factors selecting for chemical (and other) defenses in seaweeds.

C. Nutritional Deterrents

The potential deterrent effects of low nutritional quality have been mentioned (Lubchenco and Gaines, 1981) but never addressed experimentally. If critically low availability of a necessary nutrient (e.g., nitrogen) were to evolve as a defense against herbivorous fishes, this strategy should work best if alternate foods of good quality were rarely available as supplementary sources of nutrients (i.e., ice cream and beer are both poor foods but are consumed anyway because adequate nutrition can be provided from other sources). Therefore, this strategy should be most effective for species that grow in monocultures; seagrass beds would be a good habitat in which to study this possibility.

IV. Tolerance of Herbivory

The dominant plants in areas most affected by herbivory represent the extremes of susceptibility to herbivores—small filamentous forms that are very susceptible (Carpenter, 1986; Lewis, 1986) and heavily calcified, crustose corallines that are very resistant (Steneck, 1986). Both of these forms may be dependent on herbivorous fishes to prevent their exclusion by larger macroalgae (Lewis, 1986). Relying on herbivores to remove competitors may entail loss of the plant's own tissues. By growing very rapidly, and by having basal portions that escape herbivores because of the topographic complexity of the substrate, small filamentous algae persist on reefs despite large losses to grazers

(Carpenter, 1986; Lewis, 1986). This strategy of "tolerating" rather than deterring herbivory allows inconspicuous filamentous algae to make up much of the plant biomass on grazed areas of coral reefs. These plants are in fact more productive on a mass-specific basis when they are grazed because cropping prevents self-shading, and herbivore excretions may increase available nutrients (Carpenter, 1986, 1988).

In some cases, losses to herbivores may be minimized if ingested propagules or vegetative portions remain viable and are dispersed by the herbivore—similar to seed dispersal by birds and mammals. This process has been studied for several types of herbivores (Santelices and Ugarte, 1987), including a temperate fish (Paya and Santelices, 1989). Numerous species of filamentous algae not only survive gut passage, but gut passage may significantly increase the production of motile spores and the growth rate of sporelings relative to uningested controls. This process has not been studied in tropical reef fishes but clearly deserves investigation.

V. Summary

Herbivory on coral reefs is intense and herbivorous fishes account for a large portion of the plant mass removed. Both present-day ecological studies and paleontological considerations suggest that reef fishes are a major force affecting the organization of reef communities and selecting for traits that allow seaweeds to avoid, deter, or tolerate feeding by fishes. The seaweed traits discussed here were considered separately; however, in the field, morphological, structural, chemical, and nutritional characteristics act in concert to affect seaweed susceptibility to fishes, and these interactions may also be coordinated with temporal and spatial patterns affecting refuges from fishes. Such an integration of defenses should be particularly important in tropical reef environments where selection for antiherbivore traits is intense and the diversity of herbivore types is high, thus decreasing the probability of any single defensive trait being effective against the generalized guild of herbivores. Most previous studies focus on how one particular trait affects susceptibility to fishes. Few investigations have assessed the potential importance of covariation of deterrent traits within a species and how interactions among these traits might affect susceptibility to reef fishes. Such studies have been conducted on a limited number of seaweeds in which both chemical and structural defenses are relatively well studied (e.g., *Halimeda*). These studies suggest that seaweeds use combinations of morphological, structural, chemical, and nutritional defenses and that allocation of these various types of deterrents may be changed dramatically over even very short periods of time (hours) in accordance with diel or spatial changes in the foraging activity of herbivorous fishes

(Hay *et al.*, 1988c; Paul and Van Alstyne, 1988a). These complex defensive strategies are probably common among reef seaweeds and deserve greater attention.

ACKNOWLEDGMENTS

Preparation of this paper was supported by NSF Grants OCE 89-00131 and 89-11872. The manuscript was improved by suggestions from Howard Choat, J. Emmett Duffy, Sara Lewis, and Robert Steneck. As always, William Fenical patiently suffered my questions regarding the chemistry of seaweed secondary metabolites.

CHAPTER 6

The Biology of Herbivorous Fishes on Coral Reefs

J. H. Choat
Department of Marine Biology
James Cook University
Townsville, Queensland, Australia

I. INTRODUCTION

Herbivores attract special attention. As consumers of primary production they channel food materials and energy to the remaining members of the food chain. Although plant material is abundant and readily harvested, the extraction of energy and food material is a major challenge to the digestive systems of animals (Mattson, 1980). Because plants are sessile, procuring food presents few problems. The major constraints lie in the processing of material containing substantial amounts of cellulose. The innovative aspects of herbivory are to be found in the mechanisms by which plant material is processed and assimilated (Bowen, 1984; Horn, 1989).

Marine herbivores generally, and fishes specifically, interact with a vegetation which is comprehensively different from that encountered in terrestrial habitats (Littler and Littler, 1984). Moreover the herbivorous fishes of coral reefs are among the most abundant and widespread groups of vertebrate herbivores. Their tenure in the fossil record suggests that they have maintained these properties for most of the Tertiary period (Carroll, 1987). A major part of the fascination with these fishes is their ability to sustain large populations and high growth rates (Russ, 1984a; Russ and St. John, 1988) on food sources which have a small standing crop (Hatcher, 1983; Russ, 1987; Steneck, 1988) and which are low in protein (Bowen, 1979).

Unlike many fishes found on coral reefs, representatives of the most characteristic herbivore groups do not extend beyond the environmental boundaries set by reefs (Horn, 1989). What factors restrict these groups to coral reef environments? This question goes to the heart of understanding the unique properties of reef fish faunas. The paleontological evidence argues for an extended evolutionary history associated with reef environments and ample opportunity for such fishes or their derivatives to extend beyond these habitats (Carroll, 1987). Yet they have not and their present association with coral reefs is explicit enough to define biogeographic boundaries (Gaines and Lubchenco, 1982).

Why are these fishes able to process plant material efficiently enough to sustain net gains in somatic and reproductive growth in reef environments and yet lose this capability in subtropical and temperate waters? Posing the question in this fashion argues that we should focus on the population and biogeographic consequences of feeding behavior and the ability to process and digest plant material.

This chapter will deal explicitly with fishes rather than with the more general question of plant–herbivore interactions. There will be two themes. First, studies on digestive physiology of tropical marine fishes are in a preliminary stage. This is in contrast to terrestrial studies, where research on digestive physiology has provided the foundation for the study of herbivory (Stevens, 1988). In marine environments the focus has been on trophodynamics and energy flow, plant–herbivore interactions, and fish behavior and demography. The studies by Bowen (1979, 1984) on freshwater fishes and those by Horn (1989) on temperate marine species have provided a framework for future research.

Second, herbivorous reef fishes are a diverse group. There has been a tendency to place them into broad functional categories, which provides a basis for developing general models of plant–herbivore interactions (Steneck, 1988). This also provides a great deal of analytical convenience for the estimation of trophic fluxes (Hatcher, 1983; Hatcher and Larkum, 1983; Carpenter, 1986). However, herbivorous fishes are not as uniform as these approaches imply. The variation that they exhibit is finely scaled and occurs in novel forms but exists nevertheless. It must be described and incorporated into any general explanations of behavior and distribution patterns.

This chapter will not attempt a comprehensive review of herbivorous reef fishes. This has been more than adequately accomplished by Horn (1989), who has also made a plea for more studies on digestive processes. A series of topics highlighted in the recent literature will be reviewed in the context of the themes outlined here. These focus on two areas: (a) within-habitat studies of herbivores, including the role of endosymbionts, the ontogeny of herbivory, and among-species differences in feeding patterns, and (b) broad-scale studies on the differences in herbivore abundance, diversity, and feeding behavior across environmental gradients. A context for this review material will be provided by a description of previous research priorities and the general features of tropical herbivores.

II. Significance of Previous Studies

Herbivorous fishes have been implicated in three important processes on reefs. The first concerns trophodynamics. Herbivores provide the link for the flow of energy to the remaining consumers in the reef ecosystem. Second, they may have a profound effect on the distribution and composition of plant assemblages in reef environments. They have the potential to influence not only patterns of distribution and sizes of plants but also their rates of production and internal composition. Third, the interactions among herbivorous fishes, especially territorial species, have been used as a basis for developing demographic and behavioral models for reef fishes in general (Sale, 1980a; Doherty and Williams, 1988a).

These topics incorporate a great deal of the current research on reef fishes. They have provided functional schemes for classifying fishes, algal assemblages, and habitat variables and a valuable conceptual underpinning for the subject. They have also influenced our perspectives in a number of important ways.

A. Trophodynamics

Studies of the trophodynamics of reef systems have built on the pioneering work of Odum and Odum (1955) and focused on the productivity of turf algae. The emphasis has been on the interaction between fishes and attached turf algae as a major pathway of distribution of energy to other reef consumers. The role of herbivorous fishes has been defined in these terms (Hatcher, 1981, 1983; Klumpp and Polunin, 1989). For the purpose of estimating rates of primary production and its removal by grazing species, it has been convenient to visualize the fishes as comprising ecologically uniform groups (Carpenter, 1986).

As a consequence, trophodynamic studies have tended to view herbivore–plant interactions as a simple system of grazing fishes and algal turfs (Klumpp *et al.*, 1987). A number of fishes, however, ingest detrital material, algal mats, and associated organisms growing on sand surfaces, and probably bacteria. The factors determining rates of accumulation and primary productivity in these assemblages may be different from those influencing algal turfs. It is unclear how much energy passes to herbivorous fishes by these routes, but it may be substantial.

Herbivorous fishes do not constitute an ecologically uniform group. Studies dealing with patterns of energy flow have tended to eclipse the fact that herbivorous fishes constitute a diverse group targeting a wide range of primary producers. It is clear that uniformity in feeding and food-processing mechanisms is not an appropriate benchmark for evaluating the biology of these fishes.

B. Plant–Herbivore Interactions

Feeding by herbivores is considered to have two important consequences for reef plants. First, the composition and structure of coral reef plant assemblages are often explained in the context of herbivore activity (Ogden and Lobel, 1978; Lewis, 1986). Life histories, morphology, and the abundance of algae can depend on the level of herbivory they receive (Steneck, 1988; Duffy and Hay, 1990). These arguments have been extended to fossil assemblages (Steneck, 1983) and are based on the concept of major functional groupings of herbivores (Hatcher, 1983; Carpenter, 1986). As with trophodynamics, placing herbivores in functional groups tends to obscure important within-taxon differences in feeding activities and their consequences.

Second, plant–herbivore studies on coral reefs have discussed potential plant defense mechanisms (Hay, 1984a; Duffy and Hay, 1990) and emphasized the selective feeding on, or avoidance of, particular plant taxa (Ogden and Lobel, 1978). This is an important aspect of plant defense theory. The emphasis on selective feeding is partly a consequence of assessment protocols based on feeding choice experiments (Ogden and Lobel, 1978; Hay, 1981a; Paul *et al.,* 1990) or observations of fish feeding on stands of macroscopic algae (Lewis, 1986). While selection of some taxa certainly occurs, the usual feeding patterns may be nonselective with respect to algal species (Hatcher, 1983). This reflects the relative sizes of the herbivores and plants and must have an important influence on the evolution of plant–herbivore interactions and on the processes of digestion and assimilation of plant material.

Less effort has been devoted to plant–environment interactions and the physiological requirements of tropical reef algae. In many cases, changes in algal assemblages between reef habitats and along environmental gradients can be plausibly attributed to concomitant changes in herbivore activity. However, these are also associated with marked changes in the turbidity, nutrient status, temperature, and turbulence of the aquatic environment. There have been relatively few attempts to distinguish between the role of herbivores and plant–environment interactions as explanations for the structure of tropical algal assemblages. The studies by Hay (1981b), Hatcher and Larkum (1983), and Littler and Littler (1984) provide exceptions.

C. Demography and Behavior

The present focus of demographic and behavioral studies concerns the relationship of herbivorous fishes to their resources. Pomacentrids have been a key group in these studies, but the findings tend to be extrapolated to other taxa. Are such fishes limited by the resources available to postsettlement individuals or are numbers determined by presettlement processes (Doherty and

Williams, 1988a)? The need to provide robust estimates of resource abundance is an important aspect of such studies.

Behavioral studies usually take a more indirect approach. Here the emphasis has been on interactions between territorial and roving herbivores. Interspecific competition has been seen as an important influence on the behavior of territorial and roving species. The limiting resource is assumed to be algal turfs (Robertson and Polunin, 1981). In the case of both demographic and behavioral studies the need to define resources in terms more explicit than simply algal turfs is clear.

III. General Features of Herbivorous Fishes and Their Resources

A. Plant Resources

A logical starting point for a discussion of herbivorous reef fishes is a summary of the food resources available to them. Compared to their terrestrial counterparts, reef herbivores are faced with a relatively simple array of plants. Aquatic plants do not require a major commitment to supporting structures or external coverings to suppress the rate of desiccation, structures that are of limited food value to herbivores. Most plants on tropical reefs are extremely small. It is also far less likely that marine plants will evolve specific defenses against herbivores to the extent observed in terrestrial environments (Hay and Fenical, 1988).

Six distinct classes of plant resources are present on reefs, and it has been traditional to emphasize those types of algae growing on hard substrata (Littler and Littler, 1984; Steneck, 1988). However, there are a number of additional categories which should be considered (Larkum, 1983). These vary in their patterns of abundance and availability to fishes and may be subdivided into (1) algae associated with the reef matrix, including boring filamentous, turfing, crustose, and macroscopic algae; (2) large plants growing on the sedimentary aprons and in lagoons of reefs, including macroscopic algae and angiosperms; (3) mats of filamentous algae, diatoms, and bacteria in or on sediments and rubble beds; (4) planktonic and drift material in the water column; (5) symbiotic algae associated with sessile invertebrates; and (6) detrital material derived from all the foregoing plant sources.

The relative contributions of these primary producers to herbivore food chains have not been documented, although it seems likely that algal turfs and mats and accumulated organic material are the main targets (Russ and St. John, 1988). The reviews by Littler and Littler (1984) and Steneck (1988) focused on three major categories of algae: algal turfs, crustose algae, and macroalgae. All are attached to the reef matrix. The most widespread and ubiquitous assemblage is the algal turfs which comprise complexes of a number of species less than 10 mm in height (Scott and Russ, 1987; Steneck, 1988). Turfs also harbor diatoms and bacteria and accumulate organic detritus. Crustose algae consist of heavily calcified species and are characteristic of exposed substrata and may benefit from increased grazing (Steneck, 1982). Macroalgae (greater than 10 mm in height) are the minority component on coral reefs and usually occur in restricted localities (Lewis, 1986).

There are sources of primary production other than those growing on hard substrata. The contribution from algal mats on sand and rubble is not well documented. The importance of macroscopic algae and angiosperms over soft substrata varies geographically. Parrish (1989) suggested that seagrass beds as a food source are more important in the Caribbean than in the Pacific. The contribution of symbiotic algae to herbivore food chains is likely to be small, although it may be important to particular species. At present the role of organic detritus and bacteria is poorly understood, although this also could be substantial.

Planktonic algae in the form of phytoplankton cells are present in reef waters but will be at low abundances. A more important source of primary productivity in the water column is the material identified as detrital drift (Parrish, 1989) derived from algal turfs and consumed by pomacentrids (Hobson and Chess, 1978). Although the significance of such planktonic herbivory has been recognized by some workers (Williams and Hatcher, 1983), there have been few attempts to assess its contribution to reef herbivory.

The relative simplicity of these aquatic plant resources has not resulted in a clear understanding of the process of herbivory in marine organisms. One of the most important conclusions from the review by Horn (1989) was our state of ignorance concerning the digestive physiology of herbivorous fishes. Similar concerns have been expressed by Bowen (1979). Although a number of estimates of algal primary production, ingestion rates, and assimilation efficiencies are available, the identity of critical resources and their digestive processing are still far from clear.

In summary, most of the plant materials consumed by herbivorous reef fishes are small, structurally simple, and occur in complex assemblages. These can trap sediment and organic detritus and provide a habitat for epiphytic diatoms and bacteria. Most algae ingested are in the form of small fragments of the thalli with low surface area to volume ratios. For many of these fragments, cell structure and contents may not be greatly modified after passage through the gut. Most reef fishes investigated have high consumption rates and rapid gut throughput times (Horn, 1989); they apparently must process a great deal of organic material very quickly.

Under these circumstances and especially where algal fragments are mixed with calcareous material, the return per feeding episode could be relatively small. In this context, additional sources of nitrogen such as animal material would be important. Such a system would foster high feeding rates and rapid passage through the alimentary tract. High rates of primary productivity would be necessary to sustain such feeding patterns.

B. Taxonomic Groupings of Fishes

The primary taxa considered in this chapter are listed in Table 1. These include the families Acanthuridae, Pomacentridae, Scaridae, and Siganidae, the herbivores most characteristic of reef environments. They fall into two major and distinctive groupings of perciform fishes, namely, labroids and acanthuroids.

Another potentially important group is the family Blenniidae, which is usually overlooked on account of their cryptic habits. Although there are now studies dealing with herbivorous blennies (Roberts, 1987), insufficient detail is available to provide for a review of this group. Other taxa contain some herbivorous representatives (Pomacanthidae, Gobiidae) or occur in coral reef environments but achieve their greatest abundances elsewhere (Kyphosidae). There are also a number of species of uncertain trophic status such as mugilids which feed on detrital material in reef environments.

The main groups of herbivores are fairly speciose. The acanthurids contain approximately 76 species (allowing for species in the process of description), the scarids 79, and the siganids 27. Of these three families the Scaridae have the greatest number of genera (nine), but most are contained within the genus *Scarus*. There are four monotypic genera. The majority of scarids graze turf algae from hard substrata, although some Indo-Pacific species of *Scarus* graze mainly over sand. Some members of the subfamily Sparisomatinae feed on marine angiosperms and *Bolbometopon* grazes substantial amounts of live coral.

A similar taxonomic pattern is seen in the Acanthuridae with most species occurring within the genus *Acanthurus;* there are only six genera, one of which, *Paracanthurus,* is monotypic. There is a greater diversity of feeding habits within this family. Members of the genus *Ctenochaetus,* comprising seven species, are primarily sediment and detritus feeders; the single species of *Paracanthurus* is a planktivore as are approximately 70% of the 15 species of nasiid surgeonfishes. Within the genus *Acanthurus* (40 species) there are a number of species which graze over sand while a minority of species are planktivores. Most members of the genera *Acanthurus* and *Zebrasoma* feed over reef substrata and graze turf algae (Jones, 1968; Robertson and Polunin, 1981).

Siganids have fewer species, are more conservative in their feeding habits, and have a more restricted distribution than acanthurids or scarids. Most feed on algal turfs but some species also take larger algae. There is evidence that this

Table 1 Main taxa of Herbivorous Fishes Considered and Their Distribution Patterns

Familes and genera (spp.)	First appearance in fossil record	Distributional data[a]				
		IWP	PP	EP	C	EA
Acanthuroids						
Family Acanthuridae	Eocene; 50 mybp					
Acanthurus (~40)		30	20	4	4	1
Ctenochaetus (7)		6	5	1	0	0
Paracanthurus (1)		1	1	0	0	0
Zebrasoma (7)		6	4	0	0	0
Naso (13)		12	9	0	0	0
Prionurus (6)		3	0	2	0	1
Family Siganidae						
Siganus (27)	Eocene; 50 mybp	27	6(?)	0	0	0
Subgenus *Lo* (5)						
Labroids						
Family Scaridae	Miocene; 13 mybp					
Scarus (58)		45	25	4	6	1
Hipposcarus (2)		2	1	0	0	0
Cetoscarus (1)		1	1	0	0	0
Bolbometopon (1)		1	1	0	0	0
Calotomus (5)		5	2	1	0	0
Leptoscarus (1)		1	1	0	0	0
Nicholsina (2)		0	0	1	1	1
Cryptotomus (1)		0	0	0	1	0
Sparisoma (8)		0	0	0	6	2
Family Pomacentridae[b]	Eocene; 50 mybp					
Omnivores						
Abudefduf (13)		11	6	1	2	1
Chrysiptera (25)		16	6	0	0	0
Pomacentrus (60)		55	4	3	0?	0?
Territorial Herbivores						
Stegastes (30)		10	5	5	8	2
Plectroglyphidodon (10)		8	8	0	0	0
Dischistodus (7)		7	0	0	0	0
Hemiglyphidodon (1)		1	0	0	0	0
Parma (9)[c]		9	0	0	0	0
Microspathodon (3)		0	0	2	1	1
Hypsypops (1)[c]		0	0	1	0	0

[a] IWP, Indo-West Pacific; PP, Pacific Plate; EP, Eastern Pacific; C, Caribbean; EA, Eastern Atlantic.
[b] Pomacentridae are divided into omnivorous and territorial herbivore groups. Distribution data for some pomacentrid genera are provisional.
[c] *Parma* and *Hypsypops* have representatives in southern and northern temperate reef habitats, respectively.

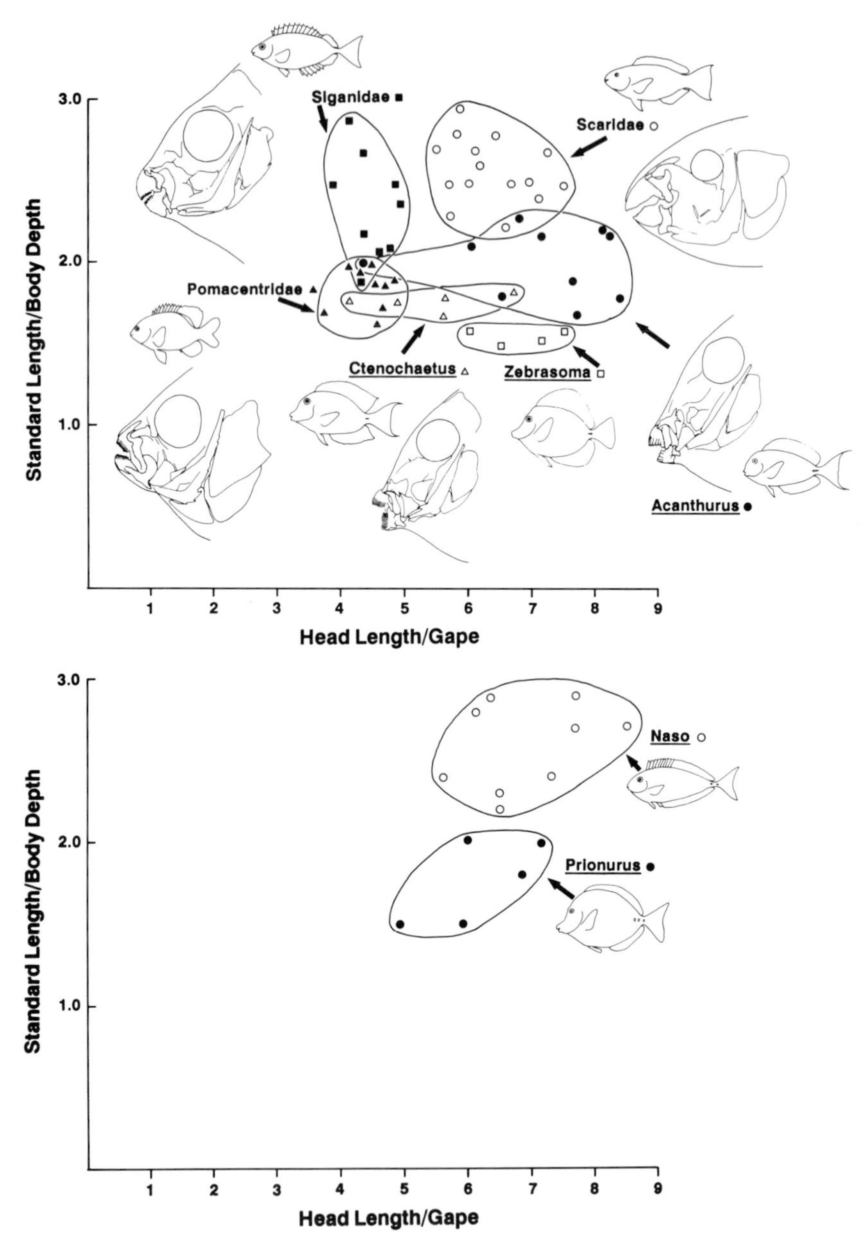

Figure 1 Bivariate plots of structural features of representive Acanthuridae, Pomacentridae, Scaridae, and Siganidae. For the Acanthuridae, separate plots are provided for the genera *Acanthurus, Ctenochaetus, Naso, Prionurus,* and *Zebrasoma*. Each point represents the ratio of standard length to body depth plotted against the ratio of head length to gape (expressed as length of the maxilla) for representative species from each group. The measurements were taken from photographs of fresh material except in the case of *Prionurus*. Increases along the

group is more selective of large algae than the other families (Lundberg and Lipkin, 1979; Paul *et al.*, 1990).

The Pomacentridae have more species (approximately 300) than the other groups. In comparison the pomacentrids are more diverse in their feeding habits although almost all species appear to take some algal material. The most speciose genera are either aggregating planktivores (e.g., *Chromis, Dascyllus*) or omnivores (*Pomacentrus*, *Chrysiptera*). However, within the latter group a number of species have pronounced tendencies toward herbivory and territorial behavior, for example, *Chrysiptera biocellata, Pomacentrus grammorhynchus,* and *P. wardi*. The most explicitly herbivorous members of the Pomacentridae are the large territorial species in the genera *Plectroglyphidodon, Stegastes, Dischistodus,* and *Hemiglyphidodon*.

With the exception of *Stegastes* these are largely Indo-Pacific groups and tend to have relatively localized geographic distributions. *Stegastes* is the exception in that it has an extended distribution with strong representation in the eastern Pacific and the tropical Atlantic. *Microspathodon* is a further example of a genus of large territorial pomacentrids with a restricted distribution. Species of the genus *Parma* are large territorial herbivores restricted to reefs on the southern fringe of coral reef environments in Australia and New Caledonia and to subtropical waters of Australia and New Zealand. They are very similar to *Hypsypops* of the waters of subtropical Mexico and California.

Historically, herbivorous fishes show abrupt discontinuities in the fossil record. Most are perciform teleosts which make a sudden appearance in early Eocene strata. There is little evidence of fishes with the structural and morphological characteristics of herbivores in the earlier radiations of actinopterygian fishes. Although many herbivores are distinct at their first appearance in the post-Cretaceous record, their history has been confused by misidentifications of reef fish material (Chapter 3).

C. Structural Features

Herbivorous reef fishes may be recognized by a series of highly characteristic morphological and structural features (Fig. 1). Almost all tropical marine herbivores are perciform teleosts and in terms of current systematics are considered to be advanced members of this group. The characteristic

head length/gape axis imply a smaller gape relative to head length; increases along the standard length/body depth axis imply increasing elongation of the body. For reasons of clarity, *Naso* and *Prionurus* are shown on separate plots. Structural features of the jaws and dentition for each family plus *Ctenochaetus* are shown.

structural trademarks of tropical herbivores are as follows. Morphologically they are high-bodied fishes with marked lateral compression. The mouth is terminal with a small gape. In most species the occiput is rounded and the snout obtuse, although a minority of acanthuroids have projecting snouts with small terminal jaws. The pectoral limb girdle and associated musculature are well developed and support pectoral fins which play an important role in locomotion and orientation during feeding. Pelvic fins have migrated to the thoracic position.

Grazing and browsing is relatively simple in terms of jaw movements, involving episodes of continuous feeding with numerous rapid bites. Derivation of oral jaws capable of such feeding involves reduction of the maxillary and dentary elements (Fig. 1). The cranial elements and suspensorium serve as a relatively rigid platform for the oral jaws (Clements and Bellwood, 1988). This is in contrast to carnivorous groups in which feeding episodes incorporate the hyomandibular and opercular elements in an integrated pattern of expansion of the oral cavity. In herbivores the articulations between the oral jaws and the structural platform of the anterior cranial elements become critical features of the functional design. In most cases the dental morphology of herbivores is described as simply fused teeth. However, the form of the dentition and jaw architecture varies considerably among groups of taxa (Fig. 1) and provides for differences in feeding behavior on a microscale.

The result is oral jaws of reduced gape appropriate for continuous harvesting of small food items. Innovative development occurs in the teeth and, in the case of scarids and ctenochaetid acanthurids, in secondary articulations in the jaw elements (Fig. 1). The primary requirement of the jaws is collecting, not processing. For labroid fishes the pharyngeal apparatus assumes the primary processing role; in acanthuroids, major processing must occur in the alimentary tract. The combined features of body, cranial, and jaw morphology provide for relatively slow swimming and precise control of body position during episodes of continuous rapid grazing over small areas of the reef substratum, the trademark of herbivorous reef fishes. These structural features are also associated with characteristic patterns in the alimentary tract. The reader is referred to the work by Lobel (1981) and Horn (1989) for details.

The taxon-specific differences in structure of the jaw elements and in body proportions are summarized in Fig. 1. There are clear patterns of structural differentiation among members of the genera *Acanthurus, Scarus,* and *Siganus*. Scarids and some siganids (especially schooling species such as *S. argenteus*) are more fusiform than acanthurids. Most acanthurids have very small jaws. Members of the genus *Naso* have tendencies for increase in size and elongation of the body and overlap with scarids. Pomacentrids, and the surgeonfish genera *Ctenochaetus, Zebrasoma,* and *Prionurus,* all have tendencies to develop high body planes but are differentiated by gape size, dentition, and jaw structure.

D. Diversity and Distribution Patterns

Herbivores occur throughout the world's oceans but their distribution patterns are modified by striking gradients in diversity and abundance. First, herbivores are concentrated in shallow-water habitats within depth zones determined by active photosynthesis; their abundance declines rapidly with depth (Bouchon-Navaro and Harmelin-Vivien, 1981; Steneck, 1988). Second, they display trends in latitudinal diversity with a marked decline in species numbers from tropical to temperate regions (Horn, 1989). These trends correlate with systematic environmental features. At latitudes of up to 28°, herbivorous fishes in the preceding taxa predominate. The diversity and species richness of this fauna decline abruptly beyond the limits of coral reef environments. In temperate and boreal reefs the dominant herbivore faunas are invertebrates comprising echinoids, gastropods, and crustacea.

A number of species of herbivorous fishes are found in cold temperate waters. These fall into three groups:

1. Representatives of typical coral reef fishes which extend into higher latitudes. These are a minority.
2. Species which are subtropical and often abundant at the limits of coral reef environments and which extend into temperate reef habitats. These may be abundant on warm temperate reefs.
3. Species characteristic of cold temperate environments and which do not penetrate into tropical waters. There are few species in this group.

The global pattern of herbivore distribution and especially that of fishes are complicated by the presence of significant longitudinal trends (Table 1). For example, the relative importance of different herbivorous groups in tropical waters is ocean dependent (Sale, 1980a). In the Caribbean, both echinoids and fishes are important (Carpenter, 1986). On Indo-Pacific and Australian reefs especially, fishes are the dominant group. Such variation occurs on a geographic scale, which reflects the action of historical factors such as the evolution and regional biogeography of particular groups which have developed in different ocean basins. Arguments about the role of environmental factors in determining patterns of herbivore distribution on a global scale must acknowledge the influence of historical events at particular localities (Sale, 1980a). Studies on the fossil history and phylogenies of reef fishes are important in this context (Springer, 1982; Sorbini, 1983b). Distribution patterns are reviewed more comprehensively in Section V, F.

IV. Summary

The foregoing observations highlight the important features of tropical herbivorous fishes, their food sources, and the environment. The main features are:

1. The food resources are of small individual size, are structurally simple, and occur in complex mixtures of algal species, detritus, and sediment. Most resources are identified as turf algae but other materials certainly contribute. Algal assemblages also provide habitats for microfauna and meiofauna.
2. The fishes are characterized by modifications to the oral jaws and suspensorial elements which result in a small gape and rapid and continuous feeding episodes. The body morphology and fin structure allow for precise orientation and maintenance of position in space.
3. The groups comprising the four major herbivorous taxa are fairly speciose, have very wide distributions, and, with the exception of scarids, have been established in reef environments since the Eocene. Not all are herbivores, members of the Pomacentridae and the Acanthuridae especially may be planktivorous.
4. For the great majority of species the association with reefs is very tight. Some representatives of these taxa extend into subtropical and temperate waters but the numbers are relatively small.

The major trademark of these fishes is rapid and continuous bouts of feeding from a resource base with a low standing crop and a low nutrient value. Their feeding and digestive regime is linked to fast turnover times and high primary productivities. Feeding in most cases is nonselective. In these circumstances the role of selective digestion and assimilation must be considered. The key to understanding variability in these fishes lies in the processing, not the collection, of food.

These observations focus attention on the nature of resources targeted by the different species. Do species ingesting resources from the same algal turf complexes assimilate the same materials? What differences are seen among members of these taxa which feed on plankton, macroscopic algae, or detritus? What modifications are necessary for feeding in those species which leave reef environments or extend into subtropical and temperate environments? An understanding of these processes in coral reef fishes will provide a perspective on the broader question of herbivorous fish biogeography.

V. Review Topics

A. Role of Endosymbionts in Digestion of Plant Material

The possible sources of energy and nutrients in algal turf complexes are diverse. It is clear that fishes have the capacity to extract material from the algae itself by a variety of means, including acid digestion, mechanical trituration (Lobel, 1981), and probably microbial fermentation (Rimmer and Wiebe, 1987). However, demonstrations of assimilation of algal material are still relatively few (Horn, 1989). The studies by Anderson (1987) have shown, for example, that girellids have the capacity to assimilate components of algal cell walls although the mechanism is unknown.

One of the more promising areas of investigation of digestion and assimilation of plant material concerns those fishes which maintain floras of endosymbiont organisms in the alimentary tract. Most models of digestion and assimilation in herbivores have been developed for herbivores which have slower and more selective modes of feeding and longer gut transit times. What are the prospects of maintaining floras of endosymbionts in species with rapid transit times of gut contents? Although this is a potentially exciting field, it is in a very preliminary phase. For example, a recent review of the bacterial flora of fishes (Cahill, 1990) provided no examples relating to herbivorous fishes.

However, a number of fishes harboring complex floras of gut microorganisms have recently been identified. The primary example is provided by two herbivorous species of kyphosid, *Kyphosus cornelii* and *K. sydneyanus,* in which Rimmer and Wiebe (1987) identified a microflora in the hindgut cecum. This was described as a hindgut fermentation chamber in which the microflora can digest algal material. Rimmer and Wiebe (1987) recorded the presence of volatile fatty acids in the hindguts of both species and argued that microbial fermentation played an important role in the nutrition of these herbivores. Although the presence of volatile fatty acids within the alimentary tract is evidence of fermentative digestion by microorganisms, it does not in itself demonstrate that this is the principal route of nutrition.

Kyphosids conform to a number of criteria suggested for the successful development of a symbiotic relationship between host and gut microfloras (Bjorndal, 1987; Horn, 1989). These include a constant food supply and a slow transit time through the gut to allow for microbial reproduction. Kyphosids have a gut transit time of 21 hours (Rimmer and Wiebe, 1987). Can endosymbiont floras develop in coral reef herbivores such as acanthurids with gut transit times of 2–4 hours?

Fishelson *et al.* (1985b) recorded populations of microorganisms from the gut of *Acanthurus nigrofuscus* which included bacteria, flagellates, and large numbers of a unique protist. They argued, however, that the protists were not involved in primary digestion. Clements *et al.* (1989) examined a wide range of herbivorous reef fish taxa, including 26 members of the family Acanthuridae. The protists were found only in herbivorous and detritivorous members of the Acanthuridae, being absent from planktivorous acanthurids and members of the families Kyphosidae, Pomacentridae, Scaridae, Zanclidae, Siganidae, and Blenniidae. Within the acanthurids the occurrence of these forms was correlated with host feeding ecology. The protists were also associated with a wide range of other microorganisms, including bacteria and flagellated and ciliated protozoans.

When hosts contained very little food material (early morning periods), protists were located in the folds of the intestine, but during the latter part of the day they were evenly distributed across the intestinal lumen. Fishelson *et al.* suggested that protists which were retained among the folds of the intestinal lining when the gut is emptied during the night emerged back into the intestinal lumen at the recommencement of feeding. The rate at which an alimentary microflora becomes established in newly settled juveniles may be rapid, less than one week (K. D. Clements, personal communication). Species-specific coprophagy in small acanthurids has been observed. We would expect behavioral interactions among acanthurids which would promote the reinfection of individuals and the transfer of microbes between generations (Troyer, 1984).

Additional evidence of active gut microfloras associated with herbivory is provided by Sutton and Clements (1988). They detected large numbers of non-*Vibrio* agar-digesting bacteria in the hindgut of the herbivorous acanthurid *Acanthurus nigrofuscus*. This bacterial flora was absent from the stomach and was distinct from the floras recorded from the hindguts of carnivorous, planktivorous, and detritus-feeding fishes. Moreover the bacteria was not detected in algal turfs. This suggests that both a specialized bacterial flora and the protists are present in the hindgut of *A. nigrofuscus*.

Evidence is accumulating that despite rapid gut transit times, some reef herbivores, especially acanthurids, may harbor endosymbionts actively involved in the digestion and assimilation of plant material. However, more biochemically explicit demonstrations of the role of microorganisms in the digestion and assimilation of plant material are required before this relationship can be established. Appropriate examples are provided by the literature on insect digestive mechanisms (Hogan *et al.*, 1985).

B. Nutrients, Growth, and Size Structure

Growth in fishes requires access to sources of nitrogen. However, it is now clear that fishes do not require greater absolute amounts of protein than other vertebrates to achieve maximal growth (Bowen, 1979). Indeed, tropical reef fishes are not only able to maintain high individual growth rates but also to support abundant and widespread populations on diets characteristically low in protein (Russ and St. John, 1988). Whatever the sources of this protein, it appears that such fishes are exceedingly efficient in its assimilation and subsequent use for growth.

There is still doubt concerning the sources of nitrogen used by herbivorous reef fishes. Some can be extracted from plant material, and small invertebrates, especially crustaceans, can also contribute to the nitrogen budget of herbivorous species. Horn (1989) reviewed the evidence for growth by fishes on an entirely herbivorous diet and concluded that less than definitive results have been obtained. Part of the difficulty arises from the fact that the resources consumed by tropical herbivores are a complex mixture of small items which are difficult to manipulate experimentally. A further problem lies in the uncertainty of contributions to nitrogen requirements from a variety of sources.

For example, Robertson (1982) demonstrated that the feces of many fishes are rapidly consumed by herbivores. Depending on the trophic status of the fish, herbivores would be able to derive significant amounts of nitrogen from this source. The feces of planktivorous fishes contained higher proportions of protein than did algal food sources (Bailey and Robertson, 1982) and were invariably consumed by herbivores while in the water column (Robertson, 1982). Large schools of actively feeding planktivorous fishes are characteristic of reef front areas adjacent to high densities of herbivorous fishes (Williams, 1982; Williams and Hatcher, 1983). Defecation may provide an unmeasured source of nitrogen for herbivorous fishes as both groups are abundant and frequently display complementary distributions.

The patterns of growth and schedules of reproductive output in reef fishes emphasize the critical role of nitrogen. Although most species have asymptotic growth patterns with respect to length (Russ and St. John, 1988), increase in body mass occurs throughout life. More importantly, reproductive output is maintained throughout adult life and may reach high levels under conditions of continuous daily spawning (Fishelson *et al.*, 1987; Hoffman and Grau, 1989). Both somatic and reproductive growth would require access to protein and, as with most herbivores, nitrogen may be limiting.

If herbivores are constrained by inadequate supplies of dietary nitrogen they might be expected to have evolved mechanisms which would enhance its harvesting and processing. Mattson (1980) has reviewed possible adaptations for this purpose, although mostly in terrestrial herbivores. To what extent would these adaptations apply to coral reef herbivores?

Herbivorous reef fishes are unlikely, except under a limited set of circumstances to have the capacity to select individual plants or parts of plants relatively rich in nitrogen. There is usually little evidence of selectivity

(Robertson and Gaines, 1986; Montgomery *et al.*, 1989), although recent work with the Pomacentridae (Klumpp and Polunin, 1989) suggests that some members of this group may feed selectively from defended turfs. For the same reasons, reef fishes are unlikely to switch among plant species or parts. Bowen (1984) argues that a species may enhance the quality of its diet by choosing appropriate materials, however, most reef fishes may have difficulty in doing this.

Mattson (1980) suggested that increased body size might be expected in animals restricted to a nitrogen-poor diet. Among the postulated advantages of increased body size were lower mass-energy requirements for larger animals, increased mechanical advantages in foraging, and the ability to support the complex digestive systems of herbivores. Horn (1989) has produced evidence that supports an association of increased body size and herbivory in three groups of fishes. Two of these (Stichaeidae and Odacidae) are temperate water herbivores. The relationship between feeding behavior and body size in herbivorous reef fishes is of general interest.

Body size in reef herbivores appears to be constrained when compared with other trophic groups; there are few very small or very large species. Although some large species of herbivorous fishes occur, the evolution of larger body size is not obviously associated with herbivory. The relationship between size and diet in a number of groups of herbivores is complex, and a number of factors appear to be involved.

Size distributions of the various functional groups of herbivores can be examined in the context of feeding categories (Fig. 2). For species of acanthurids, scarids, and siganids, those species which feed on algal turfs over hard substrata have size range varying from 15 to 550 cm, with modal sizes ranging from 20 to 33 cm. Those species of acanthurids and scarids which feed over sand have a smaller size range. Sand-grazing scarids have a smaller modal size than those which consistently graze over hard substrata; for acanthurids this pattern is reversed. Members of the acanthurid genus *Ctenochaetus* which feed on sediment are smaller with a modal size of 16 cm. Certain groups of scarids associated with seagrass beds have a relatively small size with a modal length of 19 cm. These represent phylogenetically primitive groups which occur in habitats that are peripheral to reefs.

Certain reef herbivores, including the nasiid surgeonfishes and the kyphosids, feed on larger algal turfs and in some cases (*Naso unicornis*) on macroscopic algae. Others, such as the scarid *Bolbometopon muricatum*, take substantial amounts of living coral. These groups are consistently larger than the turf-grazing acanthurids, scarids, and siganids, with *B. muricatum* being the largest piscine herbivore.

Within the acanthurids, scarids, and siganids the majority of species feed on algal turfs growing on the reef surface. For the acanthurids and siganids these

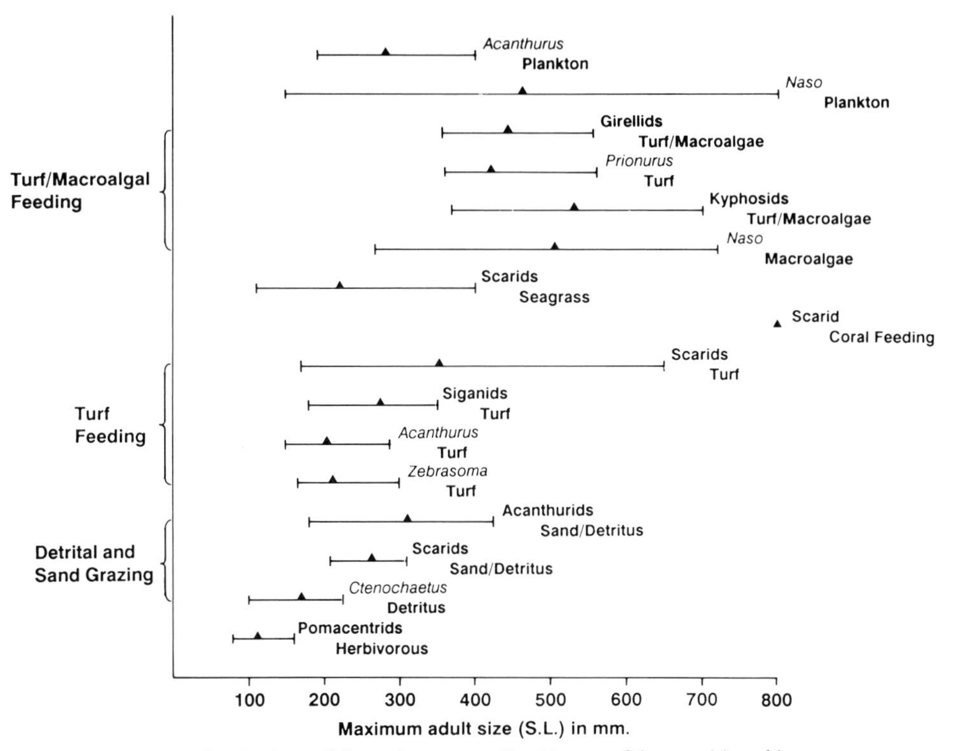

Figure 2 Size distributions of the main groups of herbivorous fishes partitioned by taxon and feeding behavior. The size ranges and modes of the standard lengths are shown. Size estimates were obtained from Allen (1975), Masuda *et al.* (1984), and Myers (1989).

have very similar size ranges and modes. Scarids feeding on turfs may be substantially larger. The largest examples within these trophic categories are the scarids *S. gibbus* and *S. rubroviolaceus,* both of which reach total lengths in excess of 50 cm and which consistently graze over hard substrata. Larger scarids are relatively rare members of the herbivorous fish assemblage.

The species in the upper size ranges of these groups display increasing diversity in feeding patterns. The largest, *B. muricatum,* selects living corals. The next largest category is the nasiid acanthurids. The largest individuals (*Naso lopezi*, *N. hexacanthus*, *N. annulatus*) are open-water feeders concentrating on macroplankton. A second group, which includes species such as *Naso unicornis* and *N. tuberosus,* feeds on larger algal species including fucoids such as *Turbinaria.* A similar size range and feeding pattern is seen in prionurid surgeonfishes and kyphosids, which, although present on reefs, extend into subtropical and in the case of the latter into temperate areas. Moreover, one of the largest species within the genus *Scarus, S. ghobban,* is not strictly a reef species, as more than any other scarid it extends into deeper water and into subtropical areas where it feeds on macrophytes.

Pomacentrid fishes also show complex patterns of size distribution. These are consistently smaller than turf-feeding acanthurids, scarids, and siganids. The largest sizes are achieved by territorial species in the genera *Dischistodus* and *Stegastes*. Other herbivorous species in the genera *Plectroglyphidodon* and *Pomacentrus* are still smaller. In the majority of these species the food resources are algal turfs and mats within defended territories and the associated invertebrate biota. The extent to which territorial pomacentrids rely on an invertebrate component in the diet is still unclear.

Efficient feeding on turfs or detrital mats requires numerous rapid bites to be taken from a thin turf of primary producers. These cover a locally heterogeneous surface. Gape and jaw architecture would seem to be a major constraint in this feeding mode. Once a herbivore develops beyond a certain size threshold then efficient feeding and selection of substrata would be difficult. This may explain why algal turf and detrital feeders have a characteristic size range.

Species beyond this size range target different types of food, including macroscopic algae, scleractinian corals, and larger planktonic organisms. There also appears to be a lower size threshold for algal turf and detritus feeders. However, more information is required on the feeding biology of smaller herbivores such as salariid blennies. The reasons for these patterns are likely to be complex and not explainable in terms of metabolism alone. History and the phylogenetic background for each group are also likely to have an important influence on size.

C. Herbivorous Feeding and Ontogeny

All species must pass through juvenile stages during the course of their ontogeny. Even allowing for rapid growth, the juveniles of many turf-feeding acanthurids, scarids, and siganids spend between 12 and 18 months feeding over hard substrata. What are the food sources of such small individuals? Current opinion suggests that in the case of herbivores generally (White, 1985) and fishes specifically (Horn, 1989), juveniles must pass through a period of carnivorous feeding before developing a herbivorous diet. Periods of rapid growth require access to greater amounts of nitrogen than can be supplied by plant material.

A number of current examples cited by Horn (1989) support this view. Additional evidence is presented by Bellwood (1988b), who investigated the diets of newly settled scarids. Bellwood examined the alimentary tract contents of juvenile scarids ranging from 7.5 to 35 mm SL. Analysis of tract contents showed that the diet of these scarids changed progressively during early ontogeny from carnivory to herbivory. In juveniles up to 11 mm SL the diet was dominated by harpacticoid copepods. Growth was accompanied by an increase in the proportion of algal filaments and fragments within the gut. Estimates of growth over this period suggest that for juvenile scarids the period of carnivory will last for approximately one month. Subtropical herbivores such as *Kyphosus cornelii* also have initial episodes of carnivory (Rimmer, 1986).

Bellwood also cited studies of initial feeding in fishes with herbivorous or omnivorous diets as adults. In two groups, Blenniidae (Labelle and Nursall, 1985) and Siganidae (Bryan and Madraisau, 1977), herbivory was observed in the early postsettlement stages. The latter study involved a rearing program in which the fish were followed from hatching through metamorphosis. Bellwood (1988b) made direct observations on early postsettlement stages of a number of species, including *Acanthurus dussumieri, A. olivaceus, Zebrasoma scopas, Z. veliferum* (Acanthuridae), *Siganus spinus, S. doliatus* (Siganidae), and *Centropyge bicolor* (Pomacanthidae).

Alimentary tracts of all individuals of these species examined were dominated by algal filaments with estimates of the percentage comprising algae ranging from 70 to 100%. The estimate of 70% was obtained from specimens of *A. olivaceus* ranging from 28 to 30 mm. The remaining 30% was made up of sand, which suggests that the sand-grazing habit of this species is developed early in life. Individuals of the acanthurids and siganids contained small numbers of harpacticoid copepods, usually less than one per fish. The highest number and greatest diversity of crustacea was recorded from *C. bicolor*.

Adults of all of these species are herbivores with the exception of *Centropyge*, which may be an omnivore. The sizes of individuals examined ranged from 17.5 to 30 mm but because individuals of all of these species settle at a relatively large size (Leis and Rennis, 1983), the study material probably represented early postsettlement stages. The feeding habits at this stage are similar to those of the adults. It seems probable that for acanthuroid fishes at least, herbivory commences with settlement.

These observations encourage examination of other acanthuroid fishes, especially the planktivorous members of the genus *Naso*. Preliminary work (K. D. Clements, personal communication) suggests that species such as *Naso vlamingi* which feed in open water as adults commence life as herbivores. This is an intriguing reversal of the normal sequence of events in feeding ontogeny and requires more study.

Arguments concerning the development of diet in fishes have been based primarily on metabolic grounds, however, there is likely to be a strong phylogenetic component. Acanthuroid fishes typically settle at relatively large sizes with well-developed sensory, locomotor, and alimentary systems (Leis and Rennis, 1983; Johnson and Washington, 1987). Both siganids (Bryan and Madraisau, 1977) and acanthurids (Randall, 1961a; K. D. Clements, personal communication) undergo an internal metamorphosis which involves lengthening of the alimentary tract at the time of settlement. Presumably presettlement individuals have been feeding as planktivores as they have relatively short alimentary tracts. For acanthuroids the critical aspects of food processing occur within the alimentary tract. When individuals commence settled life they are capable of herbivore-based metabolism following structural changes to the gut. These appear to occur rapidly, during which time the newly settled fishes are quiescent and cryptic.

Labroids generally and scarids in particular settle at smaller sizes (Leis and Rennis, 1983). Direct observation of newly settled scarids on Australian reefs suggests a size at settlement for species of *Scarus* of 6 to 10 mm (Bellwood, 1988b; Bellwood and Choat, 1990). Herbivory in the Scaridae is based on the development of the pharyngeal mill, its associated musculature, and the sacculate alimentary tract. The functioning of the mill, which mechanically reduces algal material, may be size dependent, working efficiently only when the individual has reached a certain size or mass. Scarids may feed on copepods while growing to the appropriate size not because they are metabolically incapable of herbivory but because their particular mode of feeding has a size threshold which is a phylogenetic constraint.

D. Territorial Defense and Feeding Patterns

A number of fish species defend defined areas of reef substratum against a range of other species. Aggressive defense is focused on those species with similar feeding habits, that is, grazing of turf algae and its associated biota. Consistent defense of clearly defined areas is associated with the development of distinctive algal turf assemblages with a greater standing crop (Russ, 1987) and arguably greater productivity (Russ, 1987; Klummp *et al.,* 1987) than adjacent nondefended areas. This has important implications for resource availability. Some authors consider such territorial behavior to be evidence of resource limitation in herbivorous fishes (Robertson and Gaines, 1986).

Defended sites and their algal turfs may cover substantial areas of the reef substratum, in one case approximately 70% on flats adjacent to the reef crest (Klumpp *et al.,* 1987). A majority of the daily algal productivity is ingested by herbivorous fishes, resulting in a substantial yield to grazers from the defended area (Polunin, 1988; Russ, 1987). Resident species may return substantial amounts of nutrients and other organic materials to the defended site by excretion and defecation (Polunin and Koike, 1987). For these reasons such sites are assumed to have an important role in reef trophodynamics (Klumpp and Polunin, 1989).

A useful analogy for algal turfs is that of a doormat, as they can trap and accumulate sediments and organic material from a variety of sources. These in turn provide a culture medium and habitat for a variety of small invertebrates and microorganisms, including bacteria. Finally, the algal turf itself provides a habitat for a variety of epiphytes, including diatoms and small filamentous algae. These components are characterized by both their diversity and their small size.

Access to defended areas involves complex patterns of antagonistic and synergistic interactions among a diverse range of herbivorous fishes. Many species feed from the defended algal turf complexes and may consume substantial amounts of the daily primary production (Polunin, 1988). Some do so in the context of mutualistic associations (Robertson and Polunin, 1981); others circumvent the defenses to feed on the site (Robertson *et al.*, 1976); others are simply ignored (Choat and Bellwood, 1985). It is unclear to what extent different components of the defended turfs are selected by these groups. More importantly we have little idea of the critical resource targets (those which provide the basis for growth and reproduction) of each species. Without this information it is difficult to identify important resources for herbivorous fishes or establish the basis of the interactions within and between species.

1. The Interaction between **Acanthurus lineatus** *and* **Ctenochaetus striatus**

The implications of territorial defense in herbivorous reef fishes will be considered in the context of a specific case history, which examines the feeding biology and interactions among two species of abundant and widespread acanthurid fishes, *Ctenochaetus striatus* and *Acanthurus lineatus*. The latter species is aggressive to a number of herbivores, defends well-defined areas of algal turf, and is usually restricted to the flats adjacent to reef fronts. It has been classified as a herbivore concentrating on a characteristic suite of turf algal species (Robertson and Gaines, 1986). *Ctenochaetus striatus* is more problematical being variously classified as a herbivore or detritivore (Robertson and Gaines, 1986; Nelson and Wilkins, 1988; Montgomery *et al.*, 1989). Members of the genus *Ctenochaetus* are probably the most abundant of the grazing and browsing reef fishes and occupy a wide range of reef habitats (Russ, 1984a).

The literature contains different interpretations of *C. striatus* feeding behavior. Nelson and Wilkins (1988) identified *C. striatus* as a sediment feeder and noted that the nitrogen content of stomach samples (0.4%) was higher than that found in reef sediments (0.02–0.04%), suggesting that the diet is enriched from nonsediment sources. It seems likely that the most important target is organic detritus within the algal mat. These authors also noted that the nitrogen content of *C. striatus* stomachs was an order of magnitude less than that recorded for herbivores with algal diets. They argued that this species would need to spend a large amount of time feeding in order to compensate for the lower nutrient composition of their diet relative to that of browsing acanthurids.

However, Montgomery *et al.* (1989) recorded *C. striatus* as spending less of its time in feeding than the herbivorous acanthurids *Acanthurus nigrofuscus* and *Zebrasoma xanthurum,* which had an exclusively algal diet. These findings contrast with those of Nelson and Wilkins (1988) and suggest that *C. striatus* is in fact highly efficient in terms of harvesting and assimilating nutrients. The widespread distribution of *C. striatus* in reef environments and its status as one of the most abundant reef-associated species support this.

Over much of its range *A. lineatus* shares its territory with *C. striatus,* showing little evidence of interspecific aggression, although it reacts strongly to the presence of other acanthurids (Robertson and Gaines, 1986; Choat and Bellwood, 1985). Choat and Bellwood (1985) demonstrated that at one locality both species feed over precisely the same substratum, showing little differentiation in terms of feeding microhabitats. This poses the question as to why *A. lineatus* accommodates *C. striatus* while acting aggressively toward other acanthurids and scarids. An obvious answer is that each species feeds on different components of the algal turf and they do not overlap in resource requirements. Other observations suggest alternative explanations.

The literature identifies *A. lineatus* explicitly as an algal feeder and *C. striatus* with less certainty as a detritivore. Analysis of *C. striatus* tracts (Robertson and Gaines, 1986) revealed a predominance of unidentified organic material (detritus) and a minor component of algal fragments, whereas *A. lineatus* contained mainly algal turf (Robertson and Gaines, 1986). Finely divided organic material when mixed with algal fragments is difficult to differentiate by light microscopy. It is possible that both species were targeting and assimilating organic detritus but harvesting it in different ways.

If organic material is an important source of nutrients for each species then the relationship may be competitive under some circumstances. Alternatively if it could be demonstrated that sediment removed from the turfs by *C. striatus* enhanced algal growth then the relationship might be synergistic. The description of grazing behavior among species of co-occurring herbivorous catfishes (Power, 1990) has some parallels. *Acanthurrs lineatus* would benefit under these circumstances if the primary resource target for this species was algae.

Figure 3 summarizes the comparative feeding biology of these two species. Structural features of the jaws, dentition, and alimentary tract are consistent with differentiation of feeding habits. Both species feed over similar microhabitats but have different dentitions and jaw structures. Analysis of the stomach contents reveals expected differences in the amount of algae consumed. Harvesting of algae could also result in large amounts of organic detritus being

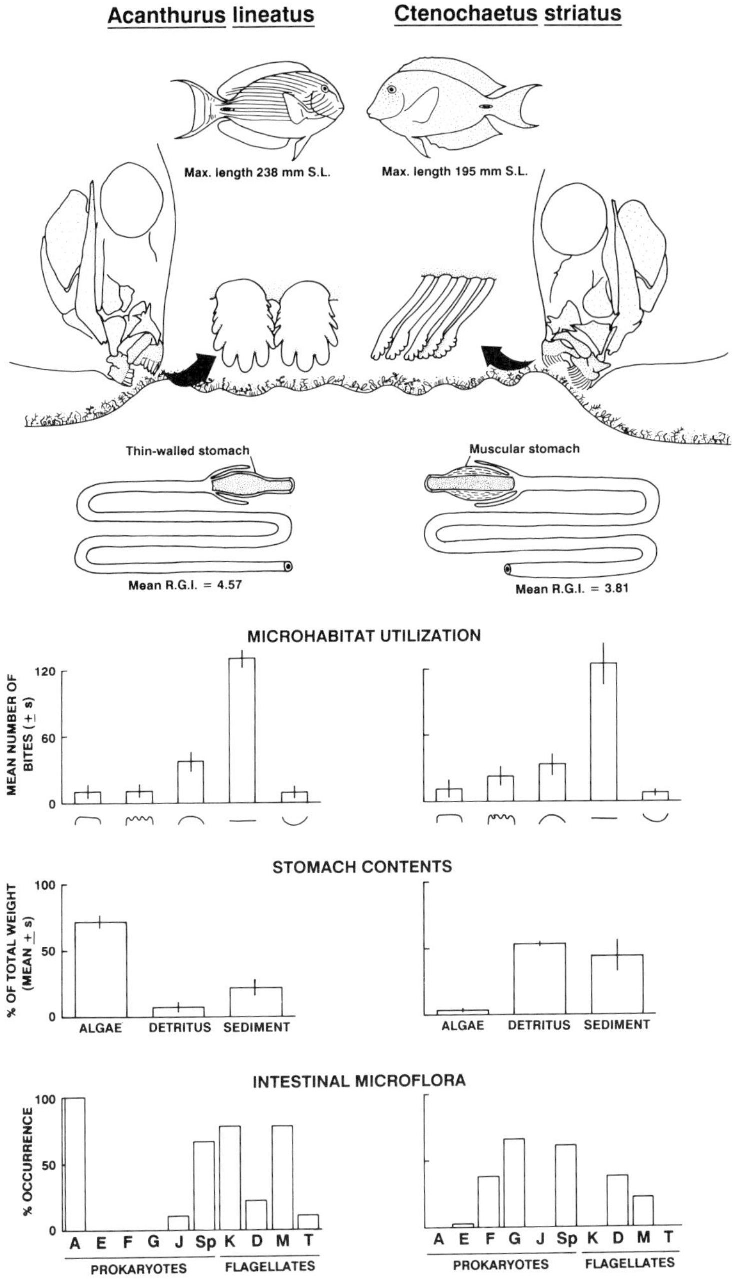

Figure 3 Comparison of the feeding biology of *Acanthurus lineatus* and *Ctenochaetus striatus*. Information for each species includes (1) morphology: jaw architecture, dentition, and alimentary tract (tract as function of body length); (2) feeding microhabitat utilization; (3) stomach contents (expressed as percentage of total weight, $n = 10$); and (4) intestinal microflora. Data sources: microhabitat utilization, Choat and Bellwood (1985); stomach contents, L. Axe (unpublished observations); intestinal microflora (Clements *et al.*, 1989; K. D. Clements, unpublished observations).

taken in. However, careful processing of the algae revealed little detritus in *A. lineatus* stomachs. *Ctenochaetus striatus* contained substantially greater amounts of organic detritus, larger amounts of sediment, and relatively little algae (Fig. 3). It is still possible that the small amounts of detritus in *A. lineatus* are an important resource. The actual resource targets of *C. striatus* have not been identified, although organic material itself and microorganisms associated with sediments are the most plausible candidates. A difference in the assimilation of material by each species is indicated by the different suite of intestinal microorganisms in each species (Fig. 3).

On the evidence obtained, each species is consuming a different set of resources with a focus on turfing algae and organic detritus, respectively. However, until it can be determined what is actually being assimilated and the importance of organic detritus to the growth and reproductive schedules of each species, the status of the interaction between them will be problematical.

On a broader scale the patterns of territorial defense displayed by *A. lineatus* are not always consistent with observations on diet. This species is highly antagonistic to small scarids (Choat and Bellwood, 1985) but ignored larger species capable of removing substantial amounts of material with each bite (Bellwood and Choat, 1990). Moreover the nature of interactions among the participating species vary on both local (Choat and Bellwood, 1985) and geographic (Robertson and Polunin, 1981) scales. The interpretation of territorial defense in such fishes is still an open question.

The phenomenon of territory sharing among potentially competing herbivorous fishes has been discussed by Robertson and co-workers. Robertson and Polunin (1981) have argued that the relationship between territorial acanthurids and pomacentrids is mutualistic as pomacentrids assist in defense. They also demonstrated that the pomacentrid *Stegastes fasciolatus* took large numbers of copepods from the defended turfs. It is possible that the priority of pomacentrid feeding in turfs is access to crustaceans, which are more abundant in defended areas (Zeller, 1988; Klumpp *et al.*, 1988).

E. Bioerosion

The term bioerosion implies the removal of material from the reef matrix by biological processes. One such process is grazing of the reef substratum by herbivorous fishes, especially scarids (Kiene, 1988). In the process of grazing, material is scraped or excavated from the surface of the reef matrix or from living coral and reduced to sediment. Many fishes also recycle sediment by taking it up directly from reef aprons or lagoonal floors or from algal turfs where it can accumulate.

Both Horn (1989) and Hutchings (1986) have identified scarids and acanthurids as major bioeroders and sediment producers on reefs. Many species in these groups pass substantial amounts of sediment through their alimentary tracts and redistribute it over the reef surface. Although these groups and especially some scarids can remove calcareous material from the reef, it is uncertain how much of their feeding actually produces new sediment. Much of the sediment defecated by acanthurids and scarids may represent recycling only.

The oral jaws of scarids are highly modified for a scraping mode of feeding (Bellwood and Choat, 1990). A minority of species display features of the jaw articulation and musculature which result in a powerful bite and significant excavation of the reef matrix. These species are the important piscine bioeroders of reef systems. Other scarids which scrape the surface of the matrix are more likely to recycle sediment rather than produce new material (Bellwood and Choat, 1990). Given the functional differences observed in the Scaridae it seems highly unlikely that acanthurids with their less robust jaw structures would contribute significantly to the process of bioerosion. The presence of calcareous material in the alimentary tracts of many species of grazers may be a reflection of how well algal turfs accumulate sediment.

The most important bioeroding species in the Indo-West Pacific region are members of the *Scarus gibbus* and *Scarus sordidus* complexes. These species are characteristic of reef crests and fronts (Russ, 1984b) and feed preferentially on convex surfaces (Bellwood and Choat, 1990). This suggests that most of the important processes attributable to piscine bioerosion occur on the growing crests of reefs and that the sediment produced contributes to the apron surrounding the reef base. Because this suite of excavating species feeds primarily on dead corals and reef surfaces, the balance between coral recruitment, growth, and death in reef crest areas is of considerable interest.

Some species of grazing fishes appear to feed on living corals and contribute to the sedimentation process by breaking down living coral skeletons. The most important species in this context is the largest scarid, *Bolbometopon muricatum*. There are very few observations on the feeding of this species and it is essential that some estimates of the rate of removal of live coral growth be obtained for comprehensive estimates of bioerosion. As with *S. gibbus* and *S. sordidus,* this species is characteristic of reef crests and fronts.

In summary, piscine bioerosion may be an important agent of structural change on the reef. Estimates of the rate and geological consequences of bioerosion require a number of qualifications. First, bioerosion by fishes is confined to relatively few species of scarid but these may contribute more substantially to the removal of reef matrix than previously thought. Second, such bioerosion would be more significant on reef crests than in other reef habitats. This may have important consequences for reef outward growth.

Third, the form and rate of bioerosion may vary on a geographic scale.

Within the Indo-Pacific, important bioeroding species may be absent from inshore reefs. There may also be biogeographic differences, as most Caribbean scarids are scrapers rather than excavators (Gygi, 1975; Frydl and Stearn, 1978), but in this environment echinoids are an additional important source of new material (Ogden, 1977). The important messages from bioerosion studies reinforce the theme of this chapter, namely, within-family species differences must be evaluated prior to any comprehensive description of the process.

F. Latitudinal Distribution Patterns

Herbivorous fishes display sharply truncated and asymmetrical distributions along latitudinal gradients. Most species of grazing and browsing fishes observed on coral reefs show abrupt declines in species numbers between 24 to 28° latitude. Below these latitudes reef fish faunas show little change with distance from the equator (Sale, 1980a), which reinforces the concept of well-defined fauna which changes abruptly on reaching an environmental threshold. Although herbivorous fish do occur in higher latitudes, there is a decline in species diversity and abundance as one passes from temperate to tropical regions. The pattern of this decline varies in the Northern and Southern Hemispheres and with the form of the continental margins. Associated with the changing patterns of fish abundance are changes in the nature and standing crop of algae and invertebrates.

Horn (1989) reviewed a number of hypotheses which would account for trends in herbivore diversity and abundance. These include:

1. the evolutionary hypothesis that insufficient time has elapsed to allow herbivorous species of tropical origin to invade temperate habitats (Mead, 1970),
2. the effects of temperature on digestive physiology (Gaines and Lubchenco, 1982; Horn, 1989), and
3. environmental influences on algal productivity, standing crop, and composition (Horn, 1989).

The arguments concerning evolutionary rates do not fit with recent information concerning the geological age of herbivorous taxa (Chapter 3 and Table 1). Most taxa have had almost the entire Tertiary period to expand into temperate habitats. The most promising avenues of research appear to be in the interaction of digestive physiology and algal structure and productivity along environmental gradients. The evolutionary history and phylogenies of different groups of fishes are also likely to exert a strong influence on the patterns.

The most appropriate places for examining gradients of herbivore and algal abundances are the margins of continents with major north–south axes. The disposition of continental masses is such that the margins are north–south and their coasts provide natural laboratories for the investigation of environmental gradients. However, continental margins have been subject to historical and biogeographic processes which have resulted in marked differences in the marine faunas (Springer, 1982).

Continuous latitudinal gradients occur on the east and west coasts of the Americas and on the west coast of Africa. However, the biogeographic history of these coasts and the lack of suitable habitat have resulted in very poor herbivorous fish faunas. The west coast of Africa, for example, has little reef environment and few tropical herbivores: there are four scarids, one acanthurid, and no siganids. This is partly due to biogeographic factors but also to the influence of freshwater drainage from major continental river systems. The east and west coasts of the Americas have more diverse herbivore faunas than Africa although they also are depauperate when compared with those of the Indo-West Pacific. Certain areas of the tropical coasts of the eastern Americas are also subject to river runoff.

The geological and biogeographic history of the east and west coasts of Australia and the archipelagoes to the north have provided an ideal set of circumstances for the comparison of herbivorous fishes across latitudinal gradients. Because of the proximity to areas of exceptionally high reef fish diversity, Australia has hosted whole tropical communities for long periods. Southern Australia also supports a suite of subtropical and temperate water herbivores which are unique to the Southern Hemisphere.

Figure 4 provides composite information from a number of sites along the Australian east and west coasts and from the New Zealand east coast spanning a latitudinal gradient from 12 to 46° South. Additional material was obtained from island sites such as the Kermadecs. Four major groups are considered: tropical grazing species (acanthurids, scarids, and siganids), territorial pomacentrids, tropical and subtropical browsing species (acanthurids, girellids, and kyphosids), and temperate browsers (the endemic Southern Hemisphere families Aplodactylidae and Odacidae).

The important distributional features of each group are as follows. Tropical grazing species are largely restricted to reef environments and show an abrupt decline beyond 23°S. A number of species do reach localities beyond 25°S and at least one reef acanthurid, *Acanthurus nigrofuscus,* and one siganid, *Siganus argenteus,* achieve growth to reproductive maturity at 34°S. Southern distributions are partly due to a local factor, a strong southward current which carries larvae down the eastern Australian coastline. These distributions represent range extensions of tropical species.

Territorial pomacentrids, the most abundant reef herbivore in numerical

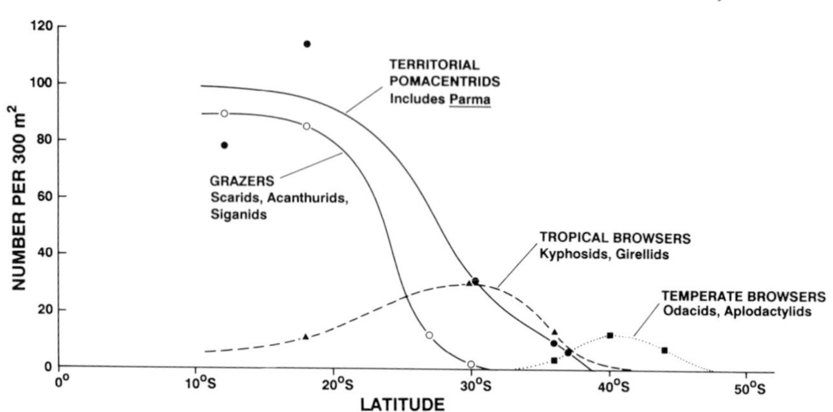

Figure 4 Estimates of abundance of four groups of herbivorous fishes from a latitudinal gradient along the eastern Australian and New Zealand coasts and adjacent islands. The groups represented are (1) tropical grazing herbivores: acanthurids, scarids, and siganids; (2) tropical and subtropical browsing herbivores: girellids and kyphosids; (3) territorial pomacentrids; and (4) temperate browsing herbivores: aplodactylids and odacids. The data points are estimates of abundance from visual transect counts adjusted to mean number per 300 m^2. Details of the sampling sites are as follows: latitude 12°S, Lizard Island GBR (J. H. Choat and R. Birdsey, unpublished observations); latitude 18°S, John Brewer Reef GBR (J. H. Choat and R. Birdsey, unpublished observations); latitude 27°S, Flinders Reef, grazers only (J. H. Choat, unpublished observations); latitude 30°S, Kermadec Islands (Schiel *et al.*, 1986); latitude 36°S, New Zealand (Choat and Ayling, 1987); latitude 37°S, Victoria, Australia (Jones and Norman, 1986); latitude 41°S, New Zealand (Choat and Ayling, 1987); latitude 44°S, New Zealand (D. R. Schiel, unpublished observations); latitude 51°S, Auckland Islands (Kingsford *et al.*, 1989); latitude 27°S, grazers only; latitude 37°S, territorial pomacentrids only.

terms, show a more gradual decline with latitude and extend farther southward to 40°S. They constitute important elements of the temperate water reef fauna (Jones and Andrew, 1990). Unlike the other tropical grazers, territorial pomacentrids include species which are herbivores but restricted largely to temperate waters. The best examples are members of the genus *Parma,* of which there are nine species in Australia (Allen and Hoese, 1975).

Large browsing species such as kyphosids occur in coral reef habitats but achieve their greatest abundances in subtropical environments. Acanthurid fishes of the genus *Prionurus* are one of the few representatives of the tropical families Acanthuridae, Scaridae, or Siganidae which are not based on coral reefs. Members of the genus have a broad, essentially subtropical Pacific distribution and achieve their greatest abundances in marginal reef habitats. Examples are *P. laticlavius* from the Galapagos Islands, *P. punctatus* from the Gulf of California, and *P. maculatus* and *P. microlepidotus* from the New South Wales coast.

Similar patterns have been recorded for kyphosid and girellid fishes, although again few estimates of abundance are available. In the Abrolhos Islands, Western Australia (29°S), the herbivore fauna is comprehensively dominated by two species of large browsing kyphosids, *K. cornelii* and *K. sydneyanus,* which feed on macroscopic algae (Rimmer and Wiebe, 1987). The Kermadec Islands (30°S) are similar in that they represent marginal scleractinian coral habitats at the southern limit of reef formation. At these islands the herbivore fauna was dominated by *Kyphosus fuscus*, *Girella cyanea,* and *G. fimbriatus* (Schiel *et al.,* 1986).

The temperate browsers (aplodactylids and odacids) are mainly cold-water species with the latter reaching 46°S. Aplodactylids but not odacids may extend into subtropical environments and were recorded from the Kermadec Islands (Schiel *et al.,* 1986). Odacids are confined to New Zealand and the southern coasts of Australia and have only three herbivorous species. Aplodactylids contain five species and have a distribution extending from southern Australia to Peru and Chile (Russell, 1987). They are similar to the herbivorous Stichaeidae of the temperate coasts of California, being relatively slow-moving browsers of red and green algae in shallow turbulent waters. Odacids are labroids which feed mainly on the canopies of fucoid and laminarian algae (Clements and Bellwood, 1988).

Temperate habitats also support either seasonally herbivorous groups such as sparids (Hay, 1986) or those which take some plant material in their normal diet. The best examples of the latter are the Hemiramphidae and Monacanthidae, which are often associated with and feed on marine angiosperms (Klumpp and Nichols, 1983; Conacher *et al.,* 1979). Both groups also take large amounts of animal matter during feeding. Members of these groups may be particularly important at different geographic localities. For example, monacanthids are very diverse and abundant in southern Australia (Hutchins and Swainston, 1986).

There are parallels between the west coast of the Americas and the east coast of Australia (territorial pomacentrids, aplodactylids, and stichaeids), and browsing kyphosids and girellids are present in the temperate habitats on each coast. There are no odacids in northern waters but the northern subtropical regions harbor prionurid surgeonfishes. Tropical latitudes of the western Americas have relatively few grazing herbivores, which is consistent with the eastward trend of reduced diversity across the tropical Pacific.

The eastern coast of America is very different from the west coast of America and the east coast of Australia. Kyphosids are present and seasonally herbivorous sparids can reach high densities at 35°N (Hay, 1986). The eastern coast of Africa, with a high diversity of tropical species and a large number of species of potentially herbivorous sparids in temperate and subtropical waters, would provide an interesting comparison with Australia. In each locality, evolutionary and biogeographic events have produced unique features in the herbivore fauna.

The latitudinal distribution of fishes parallels those of major functional groups of algae. The tropics are dominated by diverse, highly productive stands of very small algae. Areas transitional between coral reefs and temperate water reefs are of special interest. They support mixtures of large and small turfing species and macroscopic algae. In some localities, scleractinian corals and laminarian algae co-occur (Hatcher and Rimmer, 1985). Temperate regions support large stands of macroscopic algae dominated by laminarians and fucoids. These occur in shallow turbulent water associated with an understory of large turfing species. Temperate reef habitats have large areas of substratum covered with crustose and articulated coralline algae, turfing red and green algae, and sporelings of macroscopic brown algae (Ayling, 1981; Chapman, 1981; Kennelly, 1987a).

Algal turfs of tropical reefs are among the most productive plant communities in the world (Adey and Steneck, 1985). The study by Hatcher and Rimmer (1985) suggests that productivity of turfs on one subtropical reef may be reduced relative to coral reefs. Productivity rates of encrusting and turfing algae on temperate reefs are substantially lower than those of coral reefs (Chapman, 1981). However, certain subtropical and most temperate reefs support laminarian algae, which are also among the most productive plant communities in the world (Mann, 1982).

What relationships can be established between the structure and productivities of algal assemblages and the herbivorous fish faunas? The primary feature of tropical grazers is the rapid nonselective ingestion of algal turf and organic material. Many species ingest calcareous sediment either to help process algal fragments or possibly as an additional food source. Processing in the alimentary tract is rapid resulting in high turnover rates of the algal turf complexes. Such a feeding regime could be sustained only under conditions of high primary production rates and requires morphological features which permit a large part of the daily activities to be spent in feeding.

Territorial pomacentrids appear to represent a special case of the grazing mode. Such species may be selective in their feeding and the concepts of gardening and differential grazing to enhance productivity have been suggested (Lassuy, 1980; Klumpp and Polunin, 1989; Jones and Andrew, 1990). Rapid recycling of nutrients on algal turfs may increase local productivity (Polunin and Koike, 1987; Polunin, 1988). It is also likely that these species are taking advantage of additional protein sources provided by increased microfaunas in defended algal turfs (Zeller, 1988; Klumpp *et al.,* 1988). The important feature is that these species exert a considerable localized influence on algal assemblages.

The herbivorous faunas in subtropical areas may be dominated by larger selectively browsing fishes with relatively few grazing species (Schiel *et al.,* 1986; Rimmer and Wiebe, 1987), although both here and in temperate water areas large tracts of encrusting and turfing algae are present. Hatcher and

Rimmer (1985) have shown that grazing intensity and impact on algal turfs in the Abrolhos Islands (29°S) are reduced relative to tropical reefs.

Kyphosids and girellids feed on larger elements of the algal flora and in the former group at least demonstrate increased selectivity in feeding and slower passage of algal material through the alimentary tract (Rimmer and Wiebe, 1987). In temperate waters, girellids are also selective, feeding on large turf and macroscopic algae (Russell, 1983a).

The phenomenon of large abundances of prionurid surgeonfishes on subtropical reef habitats is consistent for a number of localities. For example, in the Gulf of California at 28°N (Montgomery *et al.*, 1980), *Prionurus punctatus* was more abundant than scarids by a factor of eight. This species feeds on algal turfs (Montgomery *et al.*, 1980). Observations from the New South Wales coast at a number of localities ranging from 24 to 30°S recorded the greatest abundances of *Prionurus* at the 30° sampling location (K. D. Clements, personal communication). Feeding on large algal turfs was recorded.

In fully temperate waters (below 35° latitude) three types of herbivorous fish are present. Odacids are actively swimming labroids which feed on large fucoid and laminarian algae. It is of interest to note that in Californian waters, *Girella nigricans* and *Medialuna californiensis* consume laminarian algae (Harris *et al.*, 1984), although these groups generally take red or green algae. Aplodactylids and stichaeids are negatively buoyant, slow-moving species which selectively browse large understory red and green algae. Territorial pomacentrids maintain and feed on algal mats with a highly distinctive composition.

The main change in feeding habits for herbivorous fishes across increasing latitudes is a focus on larger elements of the algal flora, greater selectivity, and longer processing times. This is not solely a function of a decrease in algal turf assemblages, for there are substantial areas of temperate reef habitat which support turfing and encrusting algae. However, these are low-productivity equivalents of the turfs of coral reefs and are dominated by invertebrate grazers (Chapman, 1981).

Fishes with a grazing mode similar to acanthurids, scarids, and siganids are unable to maintain populations on temperate reefs. This does not reflect the amount of algae present, but rather that the widespread turfing assemblages have low productivities. Grazing fishes with high metabolic rates requiring a continual turnover of algal and detrital material cannot be sustained on these assemblages. Moreover, much of the seasonal pulses of algal turf production (Ayling, 1981; Kennelly, 1987a) is consumed by echinoids (Ayling, 1981; Jones and Andrew, 1990), which have a metabolic regime which allows them to persist at high densities when algal productivity is low (Andrew, 1989).

Territorial pomacentrids maintain specific algal assemblages which support large numbers of small invertebrates. Pomacentrids may also enhance local productivity by defecation. By maintaining a relatively small size and manipulating the structure of algal turfs, pomacentrids appear to have a greater capacity for feeding choice than most other tropical herbivores. They may also extend this feeding selectivity to increase their intake of animal protein by exploiting the fauna of their turfs. These may be precisely the features that permit territorial pomacentrids to extend well beyond the boundaries of coral reefs and to establish themselves in temperate reef environments, for these species carry their turf assemblages with them. For example, on reefs of the subtropical Kermadec Islands, no members of the families Acanthuridae, Scaridae, or Siganidae were recorded from the sampling sites, although territorial pomacentrids of the genera *Parma* and *Stegastes* were abundant (Schiel *et al.,* 1986).

Reductions in the densities of invertebrate grazers on temperate reefs lead to establishment of stands of laminarian and fucoid algae. However, recruitment of these groups is highly seasonal and their sporelings do not represent a continuous source of herbivore nutrition. Although laminarian algae have high productivity rates, much of this is in a form not readily exploited by herbivores (Mann, 1982). The only piscine herbivores which regularly exploit such algae occur at comparatively low densities (Russell, 1977; Horn, 1989).

Figure 4 encapsulates a number of issues in the debate concerning the latitudinal distribution of herbivorous fishes. There is clearly no intrinsic barrier to the development of herbivory in temperate waters as the stichaeids, aplodactylids, odacids, and pomacentrids demonstrate. Although the diversity and abundances of herbivores decline with increasing latitude, the proportion of herbivores in the fish fauna may not be different from that in tropical regions (Russell, 1977; Jones and Andrew, 1990).

A decline in diversity with latitude is a general feature of many biological groups and is readily observed in reef fishes with broad distributions. The diversities of invertebrate feeders, planktivores, and piscivores all decline with increases in latitude. This is not a special property of herbivores, however, it is presently unclear how the abundances of these groups behave with respect to latitude. Does the total abundance of invertebrate feeders decrease with increasing latitude? What is the mean abundance of invertebrate feeders on temperate as opposed to tropical reefs? Some information is now available for herbivores which strongly suggests that both diversity and abundance decline with increase in latitude. Abundance data are required for other trophic groups.

The continuing debate about distributional gradients in herbivores often confuses two issues. The first is the decline in species diversity and possibly abundance with increased latitude which characterizes most groups. It is a general phenomenon which is not specific to particular taxa or trophic assemblages. The second is the group-specific rate of change in diversity with latitude.

In herbivores the decline in diversity with latitude is steep relative to that of invertebrate feeders. Carnivores extend farther into high-latitude habitats than do herbivores. This does not reflect the absence of plant material as boreal habitats have high standing crops and productivities of algae. Nor does this pattern reflect relative evolutionary age, as herbivorous fishes have been present since the Eocene. There are both metabolic and phylogenetic implications. It appears that there may well be an environmental threshold, perhaps set by temperature which limits the distribution of herbivores (Horn, 1989). However, this relationship is unlikely to be predictable or symmetrical. The uncertainties of evolutionary history are illustrated by the odacid fishes, which will always be a complicating issue in this debate.

VI. Discussion

In terms of external morphology and feeding behavior, herbivorous reef fishes give an impression of uniformity. This is superficial. It is a consequence of the diversity, composition, and size of the plant materials on which they feed. The real nature of the diversity in herbivorous fish biology is best illustrated by a comparison with a carnivorous group of reef fishes such as the Labridae.

Carnivores rely on resources which are frequently mobile and exhibit complex behavior patterns, considerable size ranges, and a great deal of structural diversity. The major challenges lie in the procuring and initial handling of these prey, that is, with food quantity (Bowen, 1984). Once procured the subsequent preparation, digestion, and assimilation are relatively straightforward as animal tissues are rich in essential materials and contain few surprises as unpleasant as the cellulose which surrounds each cell of a plant.

For coral reef herbivores the problems of procuring food are not difficult biological challenges. This requires a relatively stable grazing platform, which is provided by the laterally compressed body, modified fins to provide local orientation, and terminal jaws and teeth capable of rapid harvesting of a fine algal turf or mat. The difficulties are in the variable quality of food materials and in their resistance to processing and digestion.

Variable food quality poses particular difficulties for tropical herbivores, for the opportunities for selection and choice are limited. It is tempting to argue that such reef fishes achieve necessary intakes of energy and food material by a tactic of very rapid processing of plant and detrital substances. Hatcher (1981) estimates that some herbivores consume up to 10 times more carbon than needed to meet their basic metabolic requirements. The implication is that they must process large amounts of plant material to obtain adequate nutrition. Bowen (1984), however, has suggested that the intake of materials needed for growth is not increased by increasing the rate of food consumption. The relationship between rate of food consumption, intake of energy, and the maximization of protein intake in the context of a nonselective feeding regime promises to be one of the more compelling areas of research.

The variety of resources accessed by herbivores and the means of accomplishing this may be as diverse as those exhibited by carnivores. The evolutionary arena within which resources are matched with the consumer is not the external environment of behavior, sensory capabilities, and predator–prey interactions. For herbivores the critical arena is within the alimentary tract and involves biochemical interactions mediated by gut histology and microorganisms. Here resource targets are selected and processed. Although we are ignorant as to the details, it appears likely that different components of algal assemblages are internally targeted by different species.

Starting at the level of morphology one can describe a striking diversity of teeth, pharyngeal apparatus, and alimentary tracts in herbivorous fishes. There are also indications of diverse digestion regimes in the variety of gut pH values and histological structure recorded by Lobel (1981). However, this is merely the tip of an iceberg of physiological and biochemical mechanisms involved in extracting nutrients and energy from the contents and the walls of plant cells.

The feeding biology and inferred digestive physiology of tropical herbivores cannot be readily partitioned on taxonomic or metabolic grounds. Labroids show marked among-family differences. Scarids are essentially concrete mixers taking in a mixture of plant, organic, and calcareous material, reducing it to a slurry with the pharyngeal jaws, and passing it through a sacculate intestine which for obvious reasons has a high pH. Preliminary indications (Clements *et al.,* 1989) are that microorganisms do not play an important role in digestion. On this diet, scarids achieve a fairly rapid growth rate (Russ and St. John, 1988) and achieve a large size (Fig. 2). Although data on age-specific life history features are scarce, scarids may reach maturity in three to five years and have average longevities of eight to twelve years (D. R. Bellwood, personal communication).

The other labroid group, pomacentrids, have a very different feeding pattern and are more successful in accessing plants and other food materials of high quality from algal turfs than most other groups. The combination of small size, territorial defense, and a manipulative feeding strategy allows them a greater degree of behavioral selectivity of algae than other herbivores. This is associated with a life history pattern characterized by a small asymptotic size, relatively slow growth rates, and life spans in the vicinity of 15 to 18 years.

Compared to scarids, acanthurids are more diverse and delicate feeders harvesting a variety of plants and organic materials which are processed in a gut environment characterized by a complex microflora. No such microflora is developed in other acanthuroids. Growth rates are slower than those of scarids, although comparable sizes are reached. Although it appears that siganids may be more selective algal feeders than acanthurids better information on grazing behavior in the field is urgently required. Acanthurids also achieve relatively high growth rates, at least in culture (Russ and St. John, 1988).

Even though growth and demographic data are poor, it is clear that there is no general model for tropical herbivore growth based on metabolic arguments. Pomacentrids, which arguably are able to select high-quality food items, have slow growth rates and achieve small sizes. Scarids, probably the most recent of the reef herbivores in evolutionary terms and with a highly nonselective feeding mechanism, have relatively rapid growth rates, higher than those of some carnivores (Russ and St. John, 1988). Phylogenetic considerations will be important in any comparisons of growth and demographic performance of the major groups.

Reef herbivores may target a variety of resources but are subject to one compelling uniformity. Whatever they are doing, it stops once they leave the reef habitat. Their outstanding evolutionary and demographic success is constrained to a particular habitat setting. Yet subtropical and temperate reefs appear to provide many of the conditions necessary for their herbivorous mode of life. They certainly accommodate other forms of herbivory. The answers to such biogeographic questions will require a comprehensive research effort. However, it is clear that a major part of this effort must involve an investigation of the variety of digestive mechanisms and their interaction with the changing environmental conditions along latitudinal gradients.

ACKNOWLEDGMENTS

The following people discussed ideas and material used in this chapter: J. M. Leis, P. J. Doherty, G. R. Russ, D. R. Bellwood, N. V. C. Polunin, J. E. Randall, K. D. Clements, W. Foley, M. Meekan, B. D. Mapstone, A. J. Robertson, G. Jones, and N. L. Andrew. I am grateful to L. Axe for access to material concerning *Ctenochaetus* and to R. Birdsey for data on the abundances of territorial pomacentrids. D. R. Bellwood produced the figures. Support was provided by the Australian Research Council and the James Cook University Research Funding Panel.

CHAPTER 7

Fish Predation and Its Impact on the Invertebrates of Coral Reefs and Adjacent Sediments

G. P. Jones
Department of Zoology
University of Auckland
Auckland, New Zealand

D. J. Ferrell
Fisheries Research Institute
Cronulla, New South Wales
Australia

P. F. Sale
Department of Zoology
University of New Hampshire
Durham, New Hampshire

I. INTRODUCTION

There has been considerable speculation regarding the impact of predation by carnivorous fishes on the ecology and evolution of benthic invertebrate communities of coral reefs, as well as those inhabiting the sediments enclosed in lagoons or surrounding these reefs (Bakus, 1966, 1969, 1981, 1983; Randall, 1974; Goldman and Talbot, 1976; Parrish and Zimmerman, 1977; Hixon, 1983; Reaka, 1985; Glynn, 1988). A series of papers by Bakus develops the theme that fish strongly influence the distribution, density, and productivity of coral reef invertebrates and are the major factor in maintaining a nonequilibrium state and high species diversity on tropical reefs. Further, Bakus has argued that fish predation is a major selective force leading to protective and defensive mechanisms which have evolved in invertebrates, including morphological, physiological, and behavioral adaptations.

A generalization has emerged, suggesting that the importance of fish predation pressure increases along a gradient from temperate to tropical regions (Bakus and Green, 1974; Green, 1977; Vermeij and Veil, 1978; Bakus, 1981, 1983; Bertness, 1981). In reviewing the data base on predator–prey interactions in coral reef systems, attention must be drawn to the unexpected lack of information upon which this idea is based. Most of the experimental work designed to examine the impact of fish predators has been done in temperate waters (Choat, 1982; Sih *et al.,* 1985). Fish predators have a central place in the ecological models developed for benthic communities in higher latitudes, particularly those of soft sediments and seagrasses (Peterson, 1979; Heck and Orth, 1980). The suggestion that their importance is even greater on or near coral reefs is at present largely without foundation.

Researchers focusing on trophic interactions on coral reefs have been preoccupied with herbivory (see Hay, Chapter 5, and Choat, Chapter 6). Information on the impact of fishes on invertebrate communities has, until very recently, concentrated solely on the effects of coral feeders and herbivores on coral recruitment and distribution (Birkeland, 1977; Kaufman, 1977; Neudecker, 1977, 1979; Potts, 1977; Brock, 1979; Wellington, 1982; Fitz *et al.,* 1983). Fishes have been attributed an important role as consumers in tropical intertidal communities (Bertness, 1981; Bertness *et al.,* 1981; Garrity and Levings, 1981; Lubchenco *et al.,* 1984). Interactions between carnivorous fishes and mobile invertebrates have only recently been examined subtidally, both on hard substrata (Wolf *et al.,* 1983; Reaka, 1985; McClanahan and Muthiga, 1989) and in the surrounding sediments (Alheit, 1981; Alheit and Scheibel, 1982; Jones *et al.,* 1988; St. John *et al.,* 1989).

The effects of predatory fishes on benthic communities may be due not only to consumption, but also to disturbance of the habitat. Disturbance from elasmobranch and teleost feeding has been well documented for temperate sediments (Reidenauer and Thistle, 1981; Thistle, 1981; Van Blaricom, 1982; Sherman *et al.,* 1983; Billheimer and Coull, 1988). Suchanek and Colin (1986) identified representatives from 15 families of coral reef fishes which contribute to bioturbation of the sediments due to burrowing or feeding. Predation and disturbance caused by these fishes are likely to have different effects on different species. Therefore, it is important to separate the two sources of impact, even though this is technically very difficult (Jones *et al.,* 1988).

The attention which has been given to herbivory in no way reflects the trophic composition of coral reef fish communities. Table 1 lists the relative proportions of species in different trophic categories, from seven different studies on the diets of coral reef fishes. Benthic invertebrate predators were the most numerous in five of the seven studies, accounting for between 27 and 56 percent of the species. The range for herbivores is only 7 to 26 percent. Planktivores appear to be most numerous in the very deep reefs (30–300 m) at Enewetak (Thresher and Colin, 1986), but this may not be typical of shallow-water fish assemblages. The unusually high figure for piscivores at One Tree Reef appears to be because it includes many essentially invertebrate feeders, which occasionally consume fish (Goldman and Talbot, 1976).

The different categories of benthic carnivorous fish, including coral-polyp feeders, feeders on other sessile invertebrates, and those consuming mobile invertebrates, may all potentially influence the composition of their prey communities. In addition, omnivores [which appear to be primarily carnivo-

Table 1 Proportion of Fish Species in Different Trophic Categories from Seven Tropical Studies[a]

Trophic category	H & S[b]	Rand[c]	G & T[d]	Hobson	W & H	Sano	T & C
Herbivores	26	13	22	7	15	18	20
Planktivores	4	12	15	18	20	15	38
Benthic invertebrate feeders	49	44	27	56	53	41	33
Coral feeders	6	1		9	5	9	
Sessile animal feeders	8	6		13	3		
Mobile invertebrate feeders	35	37		34	45		
Omnivores	13	7		10	4	19	
Piscivores	10	25	38	7	8	4	8
Others (e.g., cleaners)				2		2	1

[a] Figures given are percentages of the number of species surveyed unless otherwise noted. The category of benthic animal feeders is the sum of the three subcategories below and indented from it. The seven studies were: H & S, Hiatt and Strasburg (1960); Rand. Randall (1967); G & T, Goldman and Talbot (1976); Hobson, Hobson (1974); W & H, Williams and Hatcher (1983); Sano, Sano *et al.* (1984b); and T & C, Thresher and Colin (1986).
[b] Hiatt and Strasburg (1960) classified some species in more than one category.
[c] Randall (1967) indicated that the survey was biased toward important sport and commercial species.
[d] Goldman and Talbot (1976) used only the four categories shown. Herbivores were included by them into a category of grazers, which included coral feeders. Figures given are percent weight of all groups, averaged over three habitat types: fore reef, lagoon, and back reef.

rous (e.g., Sano *et al.*, 1984b)], planktivores consuming larval invertebrates, and nocturnal planktivores consuming demersal zooplankton may also play a role.

The aim of this chapter is to review the evidence necessary to establish the importance (or otherwise) of fish feeding on benthic invertebrate communities. These steps include (a) identifying who the predators are from analysis of diet and feeding selectivity; (b) establishing when, where, and how intense predation pressure (or level of disturbance) occurs from information on spatial and temporal patterns in abundance and foraging; and (c) analyzing controlled manipulations to assess how differing levels of predator pressure alter the distribution, abundance, and structure of prey populations. In addition, we critically examine the methodologies used to assess impacts. We also discuss the relationship between ecological impacts and the proposed importance of fish predators as selective agents in the evolution of prey characteristics. Our goal is to assess the validity of the early speculations about the role of predators in the light of present evidence, and to draw attention to areas of work which may help resolve this issue in the future.

II. Patterns in Carnivore Diets and Feeding Selectivity

An examination of the impact of fish predators on benthic communities should not proceed without first establishing what species, or what categories of individual, are consuming its members (Choat, 1982). However, manipulative experiments have been frequently conducted without this fundamental information (e.g., Bakus, 1964; Day, 1977, 1985; Wolf *et al.*, 1983). Diets are usually determined from direct behavioral observations, stomach content analysis, or both (Neudecker, 1985). There has been a considerable number of general dietary studies on coral reef fish communities throughout the world (Hiatt and Strasburg, 1960; Randall, 1967; Vivien, 1973; Hobson, 1974; Goldman and Talbot, 1976; Sano *et al.*, 1984b; Parrish *et al.*, 1985; Thresher and Colin, 1986; Parrish, 1987b), and such studies enable us to focus on the individual species most likely to have an impact on certain types of prey. However, prey are usually assigned to broad taxonomic or pseudotaxonomic categories, which may not allow the subtleties of predator impacts to be interpreted. Also, such studies pool considerable intraspecific variation in the composition and sizes of prey consumed.

Little comment has been made on the types of prey consumed within broad trophic categories. In terms of species, feeders on mobile invertebrates are far more numerous than those consuming corals or other sessile animals (Table 1). Parrish *et al.*, (1985) found that benthic crustaceans accounted for 76 percent of food in the diets of fish foraging over the reefs of the northern

Hawaiian Islands. This was mainly crabs, with shrimps, stomatopods, and amphipods also present in significant numbers. Other invertebrates such as molluscs were of minor importance by comparison. Williams and Hatcher (1983) also commented on the apparent trophic importance of fish consuming mobile invertebrates such as crustaceans. Studies on carnivores foraging over soft sediments suggest that molluscs are the most important prey category in these systems (G. P. Jones, D. J. Ferrell, and P. F. Sale, unpublished observations).

Hobson (1982) has argued that coral reef fishes are more specialized than their temperate counterparts, but there are insufficient data to support this contention. Dietary items must be distinguished to species level in order categorize fish predators as generalists (which may impact a variety of prey types) or specialists (which may directly influence only one or a few prey species). Neudecker (1985) examined the diets and feeding rates of chaetodontid and pomacanthid fishes from direct feeding observations and stomach contents. Some butterflyfishes were classified as generalized coral feeders, as they were frequently observed to shift onto other prey types. Pomacanthids were classified as sponge specialists.

Many apparently specialized feeders have been shown in reality to have quite flexible diets. For example, the sea urchin *Diadema antillarum* has been reported to be the major prey item consumed by the triggerfish *Balistes vetula* (Randall, 1967). However, following the mass mortality of this urchin, the triggerfish was seen to switch to a diet of crabs and chitons (Reinthal *et al.*, 1984). Similarly, two toadfishes which formerly fed on *Diadema* switched to feeding on a broad range of mobile benthic invertebrates and fish (Robertson, 1987). The pufferfish *Arothron meleagris* is normally a coral feeder (Guzman and Robertson, 1989), however, following the mass mortality of corals, this species has been observed to consume coralline algae. Hence, spatial and temporal changes in the impact of invertebrate feeders may depend on dietary changes, which in turn may be influenced by changes in the relative abundances of different prey types.

Patterns of intraspecific variation in diets, including differences among areas (e.g., habitat zones, local sites), times (e.g., seasons, years), and ontogenetic stages, are not well documented for coral reef fishes [cf. temperate reef fish (Jones, 1988a)]. Alheit and Scheibel (1982) describe the ontogenetic decline in the occurrence of harpacticoids in the guts of six fish species feeding over soft bottoms in Bermuda Sound. The importance of different prey types also changes considerably between samples collected at different times of the year (Alheit, 1981).

Selectivity, or some measure of the consumption of prey relative to their abundance or biomass in the field, requires quantitative sampling of prey within fish stomachs and in the field. Few have attempted to measure the availability of prey for this purpose (but see Alheit, 1981; Alheit and Scheibel, 1982; Parrish *et al.*, 1985; Guzman and Robertson, 1989). Parrish and co-workers compared overall weight of different prey types in guts (for whole community) with the relative weights from field samples. Crustaceans and molluscs appeared to be consumed selectively, while sponges, worms, and echinoderms were discriminated against (Fig. 1). Important patterns also occurred within taxonomic groupings. For example, large crustaceans dominated in the diet (e.g., crabs) whereas small crustaceans dominated in the field samples (e.g., tanaids). Alheit and Scheibel (1982) examined differences in prey selection by three species of haemulids in Harrington Lagoon (subtropical Bermuda). They found significant differences between periods of the year when food was abundant and when it was scarce. Guzman and Robertson (1989) described site-specific differences in the prey selected by *Arothon meleagris*.

We have examined species-specific patterns in prey selection in two large carnivorous fishes feeding over soft sediments within One Tree Lagoon. The slate bream, *Diagramma pictum*, feeds by taking bites of sand and filtering out molluscs and crustaceans using gill rakers. It is clearly a functional "grazer" in the sense that it consumes mollusc species in proportion to their abundance (Fig. 2). In contrast, the spangled emperor, *Lethrinus nebulosus*, appears to select the bivalve *Tellina robusta* and trochid *Umbonium guamensis*, and avoids the cardid *Fragum* sp. (Fig. 2). Individual feeding bites made by this species appeared to result in the consumption of a single prey organism.

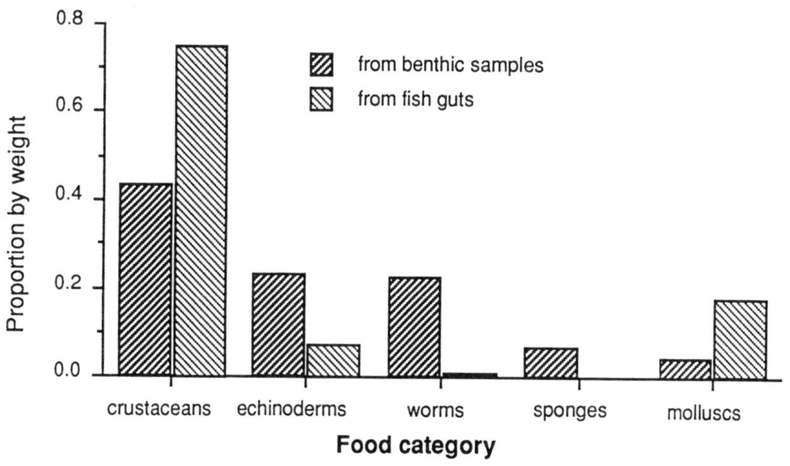

Figure 1 Comparison between weight of prey consumed (from all fish guts) and the composition by weight of prey (from patch reef habitats), for a fish assemblage from the northwestern Hawaiian Islands. Data are from Parrish *et al.* (1985).

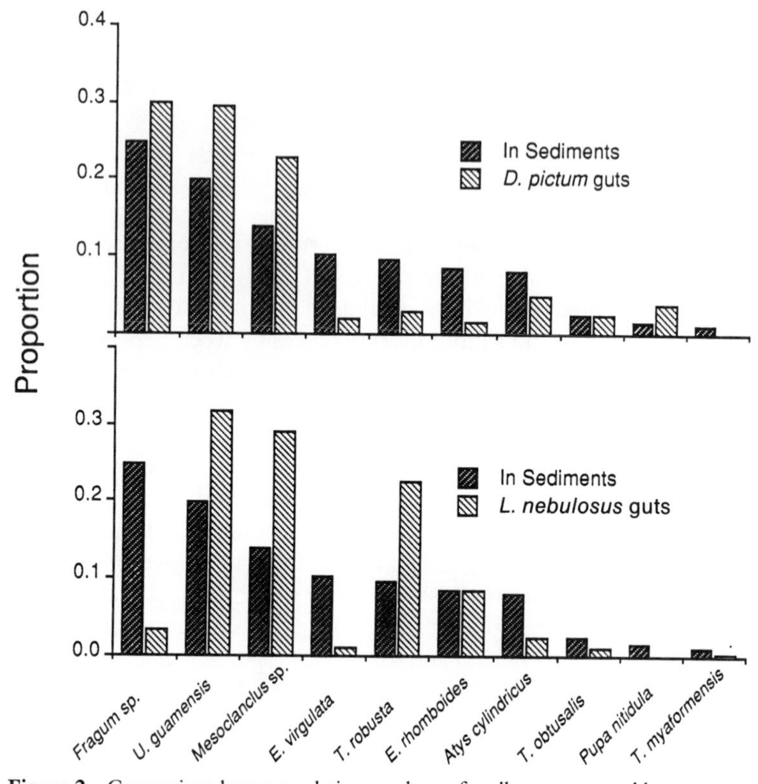

Figure 2 Comparison between relative numbers of molluscs consumed by *Diagramma pictum* and *Lethrinus nebulosus* and numbers present in their soft-sediment feeding grounds in One Tree Lagoon. (Fish and mollusc samples were collected from Shark Alley, December 1987; gut contents were examined from five fish of each species; molluscs were sampled in eight 0.1-m^2 cores.)

Diagramma and *Lethrinus* appear to select individuals of some prey species on the basis of prey size (Fig. 3). Both consume *T. robusta* from the upper end of the size range observed in the field. The mean size for *Lethrinus* (7.9 mm + 0.2 S.E.) is significantly larger than that for *Diagramma* (7.0 mm + 0.3), and both means are significantly larger than the mean size available (5.2 mm + 0.3). The gastropod *Umbonium* has a more restricted size range than *Tellina,* and the size distribution in the field closely corresponds with that based on measurements of prey size from stomach samples of the two fish species.

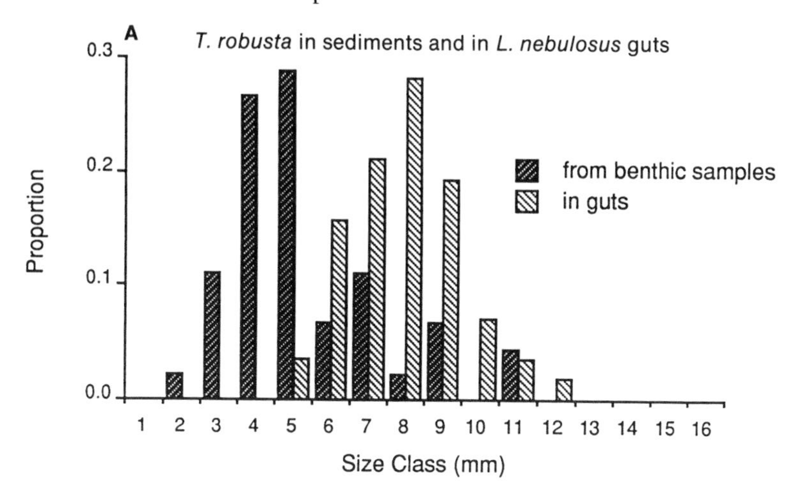

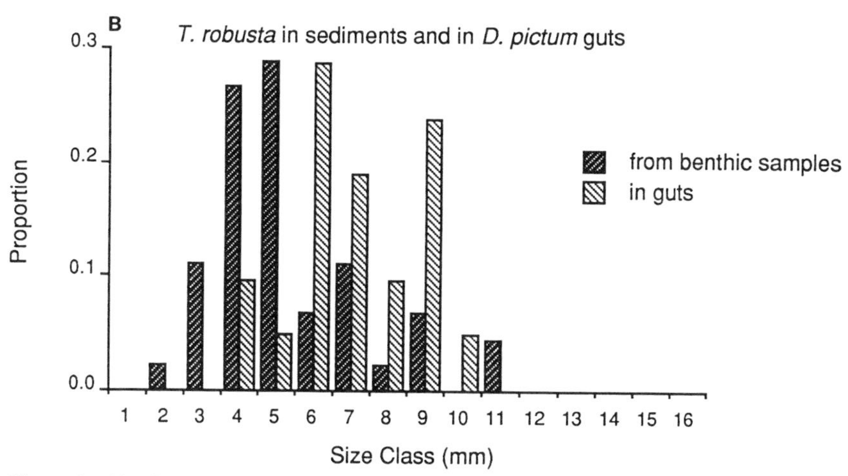

Figure 3 Size frequency distributions of two mollusc species, *Tellina robusta* (A,B) and *Umbonium guamensis* (C,D; on next page), from the stomachs of the fishes *Diagramma pictum* and *Lethrinus nebulosus* and from field samples collected from their soft-sediment feeding grounds in One Tree Lagoon. (See Fig. 2 legend for sampling protocol.)

III. Patterns in Carnivore Abundance and Foraging

Once the predators have been identified, a description of spatial and temporal patterns in their abundance provides a framework for interpreting where and

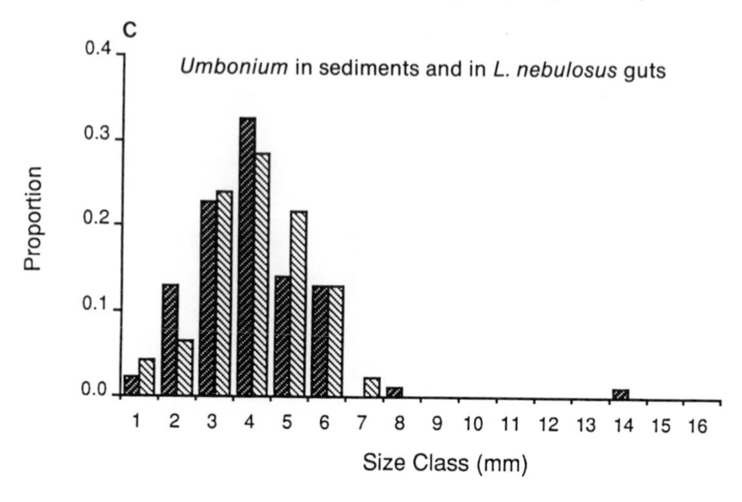

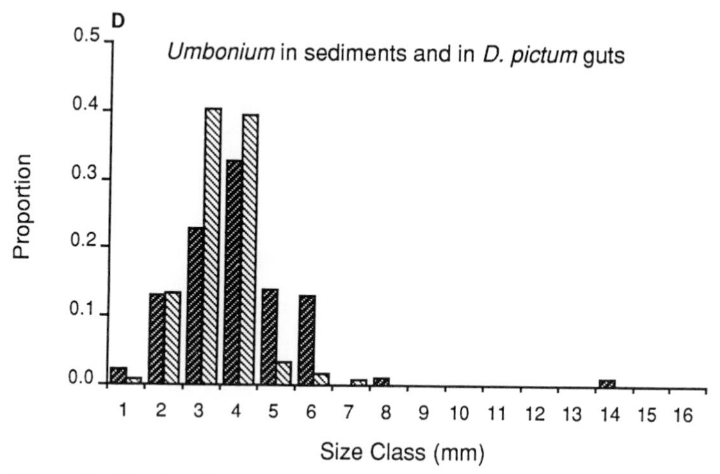

Figure 3 (*Continued*)

when prey populations are likely to be affected. Considerable spatial variability has been detected in the magnitude of the effects of fish predation (e.g., Neudecker, 1979; Keller, 1983; Reaka, 1985). However, without information on changes in the abundance of predators, the reasons for these patterns are not clear. Such data may allow predictions to be made of how often a particular prey organism will be predator-limited.

At present, data on the distribution and abundance of coral reef fishes primarily concern planktivores and herbivores (Doherty and Williams, 1988a;

Jones, Chapter 11). Williams and Hatcher (1983) describe broad-scale patterns in the trophic organization of reef communities across the breadth of the central Great Barrier Reef. Invertebrate feeders appear to be equally important at inshore, midshelf, and outer shelf reefs, although there are considerable changes in the abundance of individual taxa along this gradient (Table 2). The most important family in terms of biomass is the wrasses (Labridae). In New Caledonia, invertebrate feeders increase in abundance from coastal areas along a 50-km transect to the offshore barrier reef (Kulbicki, 1988).

The trophic organization of reef communities changes among zones within individual reef systems, with invertebrate feeders being proportionally most important in lagoons and back reef regions (Goldman and Talbot, 1976). Most species, including carnivorous fishes, exhibit major changes in distribution and abundance among habitat zones (e.g., Nagelkerken, 1977). However, we know little of the demographic factors affecting spatial and/or temporal patterns in the abundance of carnivorous fishes (but see Eckert, 1984, 1987).

Within habitats, local variation in feeding pressure will relate closely to the degree of patchiness in foraging in space and time. Some benthic carnivores are active during the day (e.g., wrasses) and others are active during the night (e.g., many haemulids). Since many benthic invertebrates migrate into the water column at night (Alldredge and King, 1977; Jacoby and Greenwood, 1988), the impacts stemming from these different foraging strategies will differ.

We examined spatial variation in the feeding pressure and/or disturbance caused by rays feeding over soft sediments by quantifying the distribution of their feeding pits across One Tree Lagoon (Fig. 4). The greatest densities of

Table 2 Comparison of the Taxonomic Composition (Percent of Total Wet Weight from Each Zone) of Motile Invertebrate Feeding Fishes of an Inshore, Midshelf, and Outer Shelf Reef of the Great Barrier Reef[a]

Family shelf	Inshore	Midshelf	Outer
Apogonidae	7.6	4.5	0.4
Balistidae	0	3.8	5.0
Chaetodontidae	3.1	2.1	5.8
Holocentridae	4.8	3.4	15.1
Labridae	28.3	34.0	30.9
Lethrinidae	5.2	14.6	6.0
Lutjanidae	19.0	3.8	12.4
Nemipteridae	4.8	10.4	16.9
Pomadasyidae	21.4	20.8	6.0
Others	5.9	2.4	1.6

[a] After Williams and Hatcher (1983).

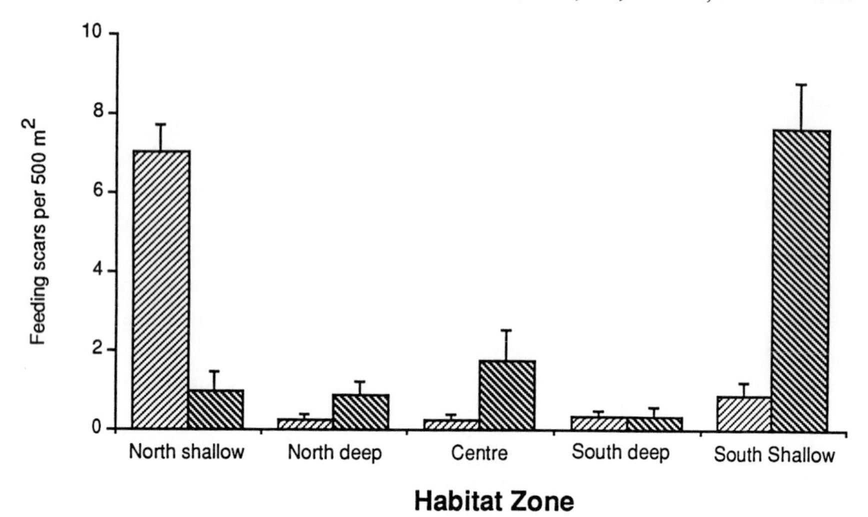

Figure 4 Mean density of ray feeding scars (per 500 m^2 ± S.E.) at two random locations from five soft-sediment lagoonal zones at One Tree Lagoon. Means are from nine 5 × 100-m random transects counted at each location during May 1986.

feeding pits were found at certain locations in the northern shallow and southern shallow perimeters of the lagoon. Although feeding intensities are variable in space, rays appear to prefer the relatively coarse sediments at these sites (see Jones *et al.*, 1990). Accordingly, a ray exclusion experiment was set up in one of these zones.

We also examined spatial and temporal variation in teleost feeding pressure across One Tree Lagoon by quantifying feeding scars or indentations made by teleost fishes (see also Billheimer and Coull, 1988). When carnivores such as *Lethrinus* and *Diagramma* take bites from the sand they leave scars which remain recognizable for periods of up to several days (Fig. 5). The densities of these feeding scars within random 1-m^2 quadrats were monitored at three "locations" within each of four broad lagoonal "zones" at four "time" intervals over a one-year period. Feeding pressure differed significantly among "zones" ($F_{[3,8]} = 4.2, p < 0.05$), a result clearly due to low levels of feeding in the central zone (Fig. 5). This may reflect low densities of fish in this zone and/or the low abundances of molluscs observed in this region (see Jones *et al.*, 1990). Within zones, feeding pressure was extremely patchy and variable, as indicated by a significant interaction between the factors "Location" and "Time" ($F_{[24,1104]} = 5.9, p < 0.05$). It was not possible to identify any locations or times for which teleost feeding was consistently high; this probably reflects the mobile and gregarious foraging activities of these fishes.

If such patchiness and temporal variability in fish feeding pressure is the

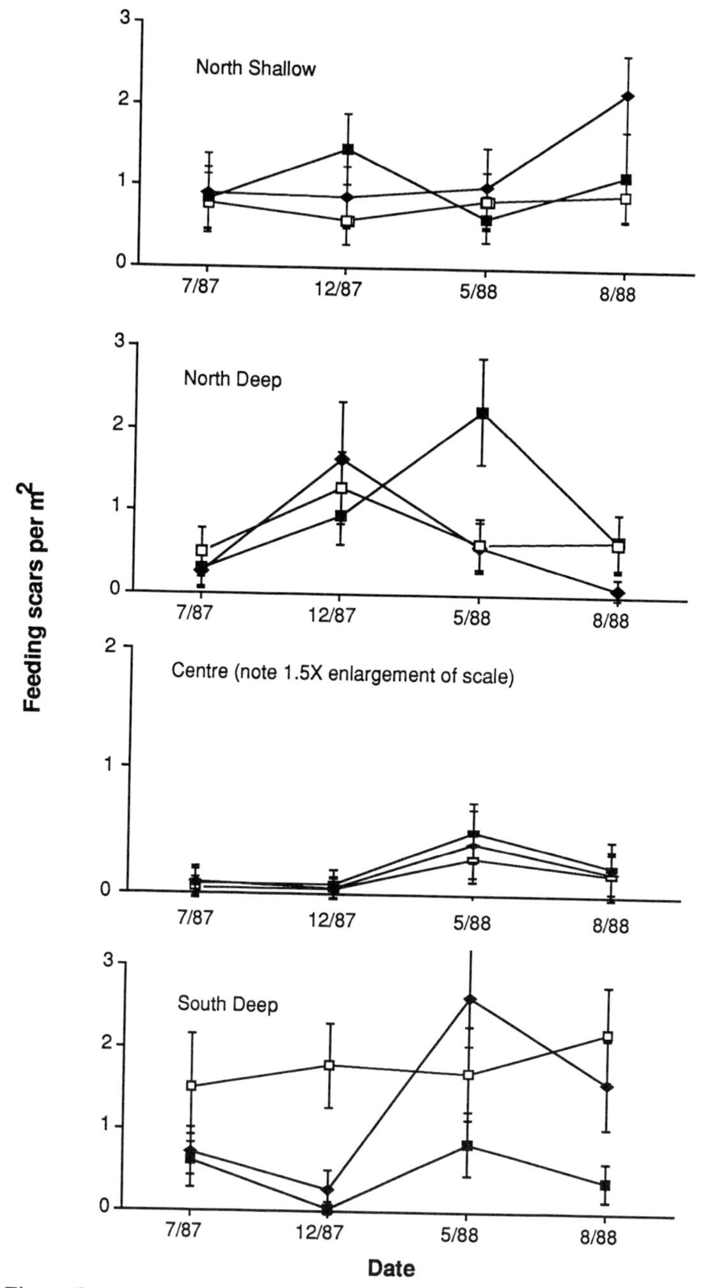

Figure 5 Mean density of teleost feeding scars (per $m^2 \pm$ S.E.) at three random locations within four soft-sediment lagoonal zones at One Tree Lagoon. Each location was sampled four times over a one-year period between August 1987 and August 1988. Means come from ten 1 × 1-m quadrats counted at each location and time.

rule, experiments set up to detect it must be repeated at different locations and at different times to assess the overall importance of this process. At present, while there has been some spatial comparisons (Neudecker, 1979; Keller, 1983; Reaka, 1985), the importance of both spatial and temporal changes in fish abundance and foraging behavior remains undetermined.

IV. Impact on Coral Reef Invertebrate Assemblages

In this section, the effects of carnivorous fish feeding and disturbance on the distribution and abundance of corals, other sessile invertebrates, mobile invertebrates associated with coral substrata, and intertidal animals are examined. Indirect effects due to herbivore grazing or the cultivation of algal mats will also be considered.

A. Corals

The majority of experiments designed to detect effects of fish feeding on coral distribution, abundance, or demography have done just that. Effects can be divided into those which are direct and negative (resulting from fish-induced coral mortality) and those which are indirect (either positive or negative effects resulting from the removal or cultivation of coral competitors such as algae) (Glynn, 1988).

Neudecker (1977) described a direct effect of coral feeders such as the triggerfish *Balistipus undulatus* on the coral *Pocillopora damicornis*. This coral appears to be restricted to reef flat, crest, and lagoons as a result of high grazing pressure on the deeper slope. Individuals transplanted to deeper water showed improved chances of survival inside cages, in comparison to those exposed to fish feeding. While this coral can survive and grow well in deeper habitats, it appears to be excluded from these zones by fish feeding. In later work (Neudecker, 1979), the impact of coral feeders was found to be patchy and reduced transplanted colonies in size in deeper water, but did not cause mortality. Some damselfishes appear to kill corals directly by removing coral polyps, which results in the opening up of the substratum for growth of their algal mats (Kaufman, 1977; Wellington, 1982). The impact of coral-feeding fishes may be enhanced following the mass mortality of corals, to the extent that they reduce the rate at which reefs recover from physical disturbances (Glynn, 1985, 1988; Knowlton *et al.*, 1988; Guzman and Robertson, 1989).

It has been suggested that fish predation during coral spawning may be an important source of larval coral mortality (Westneat and Resing, 1988). However, it has not been established that this translates into an effect on coral abundances.

There may be a range of indirect effects on the distribution and abundance of corals, the most obvious involving herbivorous fish (Glynn, 1988). Birkeland (1977) found that herbivorous fishes enhanced the survival of coral recruits on settlement plates in the Caribbean. In an experimental microcosm, coral recruitment increased in relation to increased grazing pressure by the herbivorous parrotfish *Scarus taeniurus* and was further enhanced in the presence of refuges (Brock, 1979). Fitz *et al.* (1983) described similar rates of coral settlement on artificial surfaces which were fully caged and open and cage controls in deep water. However, more individuals survived to larger sizes on uncaged surfaces, suggesting that herbivores indirectly control the survival of coral recruits by removing algae. Territorial damselfishes appear to indirectly suppress coral recruitment by the cultivation of dense mats of algae (Birkeland, 1977). However, increases in the recruitment of corals within damselfish territories have also been observed (Sammarco and Carleton, 1981). Dense algal turfs frequently overgrow small corals. The algal-sediment turf inside territories of *Dischistodus perspicillatus* induces heavy mortality in transplanted pieces of the coral *Acropora* (Potts, 1977).

Various other indirect effects should be mentioned. Selective predation be corallivores on dominant corals can promote the abundance of less preferred species (Glynn, 1988). For example, selective predation by *Chaetodon* on the coral *Montipora* has a positive effect on the competitively inferior *Porites* (Cox, 1986). Territorial fish may indirectly affect corals by altering patterns in the foraging of grazing invertebrates such as echinoids (Glynn, 1988). Randall (1974) argued that the production of sediment by herbivorous fishes, and the deposition of the crushed remains of hard-shelled prey consumed by mobile invertebrate feeders, may have greater effects on coral reefs than those due to corallivores.

A comprehensive study by Wellington (1982) is an interesting example in which zonation of corals can be explained by the differential effects of a territorial damselfish on the survival of different coral species. Low numbers of the massive coral *Pavona gigantea* in shallow water at a fringing reef in the Gulf of Panama appear to be caused by the omnivorous damselfish *Eupomacentrus acapulcoensis,* which kills coral to establish algal mats. Pocilloporid coral fragments suffer high mortality and substantially reduce growth rates in deep areas as a result of grazing by pufferfish and parrotfish. Damselfish territories are a refuge for *Pocillopora* survival. Hence, the distribution of damselfish, through direct effects on one coral species (*Pavona*) and indirect effects on another (*Pocillopora*), controls the vertical distribution of these dominant coral species (Fig. 6).

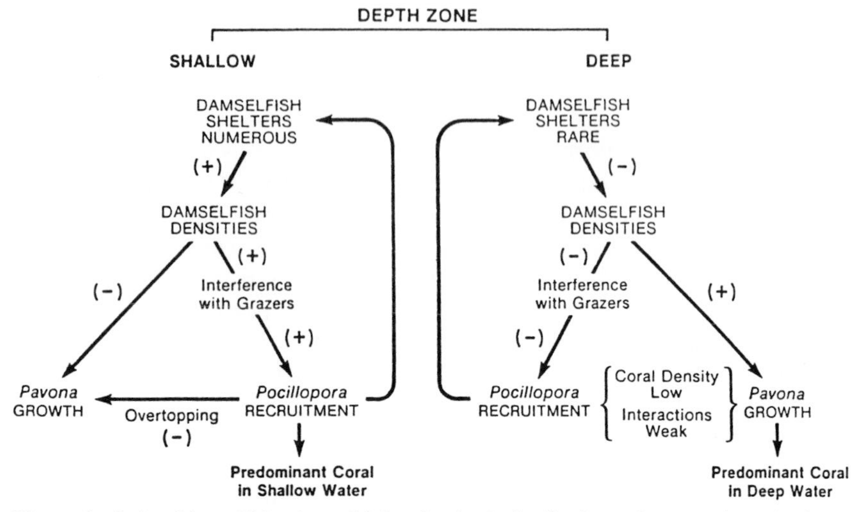

Figure 6 Role of damselfishes in explaining the depth distributions of two coral taxa in the Gulf of Panama. (After Wellington, 1982.)

B. Other Subtidal Sessile Organisms

The distribution and abundance of other sessile species have also been attributed to fish feeding. Bakus (1964) argued that sponge distributions at Fanning Island were controlled by grazers, and sponges transplanted from refuge areas out onto the reef slope were rapidly consumed. Encrusting communities inside caves at Heron Island appear to be influenced by fish predation (Day, 1977, 1985). Diversity was 20 percent greater on grazed surfaces than on those inside cages and protected from grazing. Predation appears to be selective, with the preferred species being ascidians. This allows competitively inferior taxa such as bryozoans to survive, since ascidians are removed by fish when they are quite small. Ascidians appear to persist only in areas which are natural refuges, for example, beneath the stinging hydroid *Lytocarpus*.

The opinion that fish predation is an important source of mortality for sessile organisms is not universal. For example, Jackson and Buss (1975) found intense competitive interactions occurring in sessile, cryptic assemblages, and no evidence that predation ameliorates such interactions.

C. Subtidal Mobile Invertebrates

The impact of fish predation on mobile invertebrates such as crustaceans, molluscs, and echinoderms has been examined using artificial reefs and cage exclusions (Wolf *et al.*, 1983; Reaka, 1985). Wolf *et al.* (1983) made observations on the colonization of three different types of artificial reefs, one providing habitat for both fish and invertebrates, one habitat for fish only, and one habitat more suitable for invertebrates only. They argued that fish predation affected the rate of colonization by stomatopods and the recruitment of polychaetes, but did not affect the remaining more secretive or cryptic taxa. It is not clear how this experiment distinguishes between predation impact and habitat selection by stomatopods and polychaetes.

Less equivocal evidence comes from experiments in which artificial reefs were employed in conjunction with fish exclusion cages and barriers to invertebrate movement (Reaka, 1985). Reaka examined the effects of fish predation at six different reef habitats in St. Croix (West Indies). Effects of predation on stomatopods were only detected in shallow habitats when barriers to immigration and emigration by the invertebrates were put in place. In certain deep habitats, however, the effects of fish predation on stomatopods were measurable in the absence of barriers to migration. The impact seemed to be greater on mobile species such as stomatopods than on the relatively immobile species.

McClanahan and Muthiga (1989) provide evidence that predation by fishes controls the abundance and vertical distribution of the burrowing sea urchin *Echinometra mathaei*. They compared areas which have a history of heavy fishing pressure with those protected from fishing and found that urchin densities were higher on fished reefs. It was argued that finfish account for 90 percent of urchin mortality, but the predatory fish involved were not identified. Increasing urchin mortality rates with depth were also attributed to fish predation, suggesting that this explains patterns in the vertical distribution of *E. mathaei*. Finally, they suggested that spatial variation on the size structure of populations was controlled by fish predators. The fish effect was later attributed to the balistids *Balistaphus undulatus* and *Rhinecanthus aculeatus* (McClanahan and Shafir 1990). This work showed that sea urchin diversity was greatest at intermediate predation intensities.

Since many mobile invertebrates live associated with algae and corals, indirect effects of grazers are also potentially important. Lobel (1980) found significantly higher numbers of small motile invertebrates inside the territories of *Eupomacentrus planifrons* than in other patches of algae. Experimental removals of the damselfishes resulted in an increase in feeding by both herbivores and carnivores, and a consequent decline in the abundance of small invertebrates such as crustaceans, polychaetes, and molluscs. Risk and Sammarco (1982) suggested that algal mats inside damselfish territories provide refuges and food sources for juvenile boring invertebrates, which may accelerate bioerosion of the reef. Indirect effects on motile invertebrates resulting from predation by fishes on live corals have not been investigated.

D. Intertidal Assemblages

In contrast to temperate regions, it has been suggested that fish predation has a major influence on intertidal assemblages on tropical, hard substrata (Lubchenco *et al.,* 1984). Cook (1980) described an instance at Enewetak Atoll in which the pulmonate limpet *Siphonaria* suffered 21 percent population loss by fish predation during a single high spring tide. This indicates that fish can exploit the uppermost reaches of the intertidal zone. Comparisons among fished and protected reef flats in Kenya suggest fish have a major effect on the abundance of both sea urchins and gastropods (McClanahan and Muthiga, 1988 and McClanahan, 1989).

Most of the work done on intertidal predation by fishes has been conducted in Panama. Bertness (1981) found that predation intensity on hermit crabs by tropical fish was significant and decreased with increasing tidal height. Lubchenco *et al.* (1984) list at least ten carnivorous fishes which frequent the intertidal zone in this region. Their exclusion experiments suggested that fish consumers are a major factor in maintaining consistently low covers of sessile invertebrates on these shores. Garrity and Levings (1981) argue that the dispersion of intertidal gastropods is related to refuges from fish predation. Ortega (1986) indicated that fish had little impact on shell losses of gastropods in Costa Rica, although some shell crushing occurred.

V. Impact on Soft-Sediment Assemblages

Much less work has been done on the impact of reef-associated fishes upon assemblages inhabiting the carbonate sediments enclosed in or surrounding coral reefs. A wide variety of small, burrowing fishes and large, roving teleost fishes and elasmobranchs feed over the sand or adjacent seagrass beds, but shelter in the vicinity of reefs by day or night (Hobson, 1973; Ogden and Zieman, 1977; Parrish and Zimmerman, 1977). St. John *et al.* (1989) examined the distribution of the meiofauna and a common burrowing, predatory fish, *Valenciennia longipinnis*. No differences in the abundances of different meiofaunal taxa were found between areas where *V. longipinnis* were present in high numbers and where they were absent. However, the chances of detecting a real impact in this way depend on how mobile the meiofauna are. Since many of these taxa migrate into the water column at night, there is considerable opportunity for horizontal movement (Jacoby and Greenwood, 1988). It is notable that an inclusion experiment designed to measure the effects of gobies on soft-sediment meiofauna in the Caribbean detected an effect within 40 hours (Fitzhugh and Fleeger, 1985), that is, numbers of meiofaunal taxa of prey were reduced in 1-m^2 plots on a very short time scale, over which the possibility for prey movement was much reduced.

The impact of fish predation on the recruitment of two sea urchins, *Lytechinus* and *Tripneustes*, was examined in the seagrass meadows inside Discovery Bay, Jamaica (Keller, 1983). Keller detected a predation effect at an exposed site where urchin recruitment was high, but not at a sheltered site where recruitment was low. The predators responsible for this were not identified.

Jones *et al.* (1988) constructed large emergent fences (enclosing a 13-m^2 area) to examine the impact of rays and other large carnivorous fish on mollusc communities inhabiting a sand flat near the perimeter of One Tree Lagoon. Plots enclosed in complete fences, partial fences, and open controls were monitored over a one-year period. Fish exclusion resulted in a significant increase in the gastropod *Cerithium tenellum,* with two other common molluscs, *Rhinoclavis aspera* and *Tellina robusta,* decreasing relative to controls (Fig. 7). Some differences in population structure were also observed. These effects may have arisen due to direct predation or disturbance by fish feeding. However, changes in sediment composition inside complete fences were observed, raising the possibility that all effects were unrelated to fish feeding.

Large fish exclusion cages (enclosing 25 m^2) were constructed at three locations adjacent to reefs in deeper water within One Tree Lagoon (Jones *et al.*, in press). *Lethrinus nebulosus* and *Diagramma pictum,* key predators on molluscs in these areas, were frequently observed foraging in cage control and open control plots at each location. No major differences among treatments in the abundance of the common prey species *Tellina robusta* emerged over the two years of the experiment (Fig. 8). Abundances differed among places and times, but these patterns were independent of fish feeding pressure.

Alheit (1981) examined fish feeding rates and prey standing crops as a measure of predation pressure on harpacticoid populations in a shallow lagoon in subtropical Bermuda. Alheit showed that pomadasyids and sparids remove 3 percent of the total biomass of food species per day and suggested that fish are responsible for dramatic decline in the biomass of the soft-bottom zoobenthos from April to September. In a later paper, Alheit and Scheibel (1982) calculated that only 0.02 percent of the standing crop of the harpacticoid *Longipedia helgolandica* is present in the stomach of the most important predator *Haemulon aurolineatum.* The conclusion based on this result was that prey consumption by fishes is unimportant. Without information on prey turnover rates and other sources of mortality, it is not clear how one decides whether a particular percentage of biomass removal is ecologically significant.

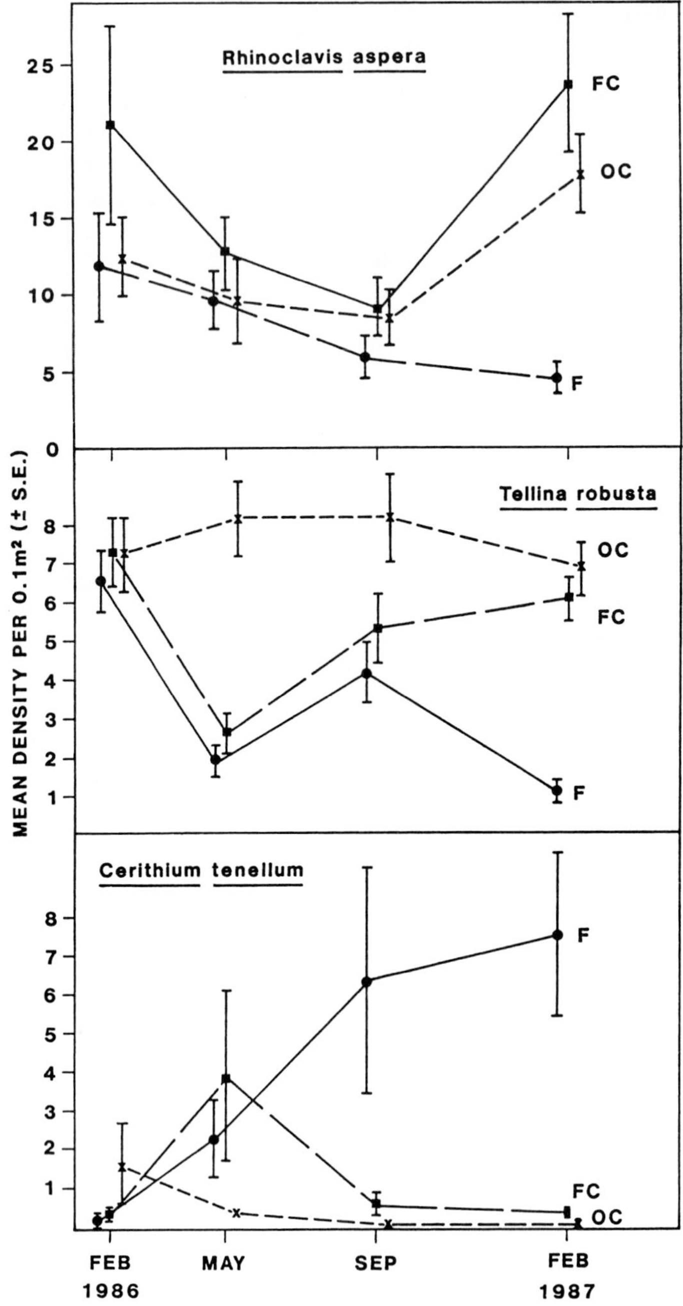

Figure 7 Influence of fish exclusion fences (F), fence controls (FC), and open controls (OC) on patterns of change in the mean densities of three mollusc species in the shallow-water sediments of One Tree Lagoon. (After Jones *et al.,* 1988.)

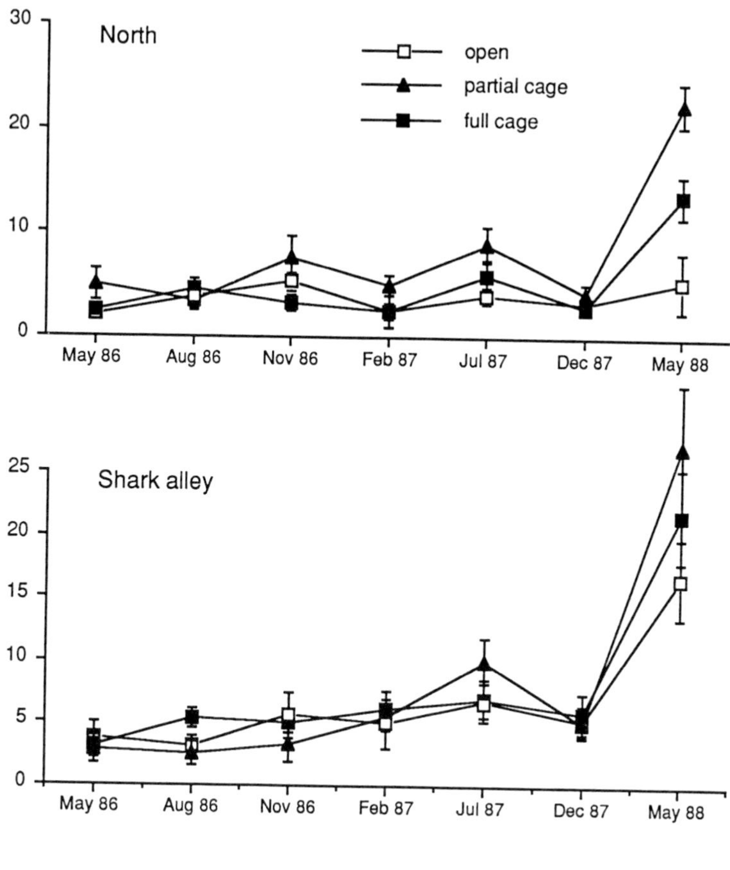

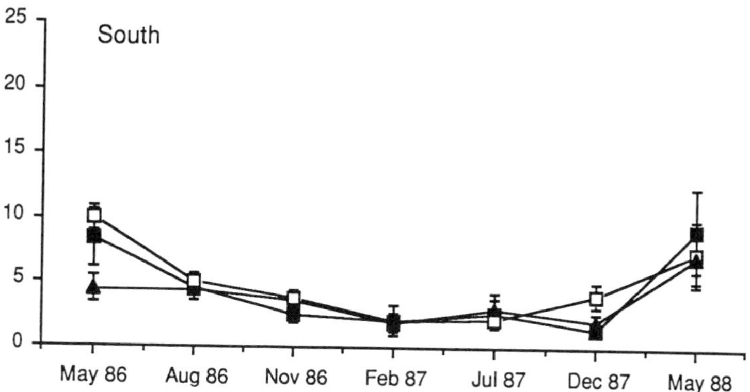

Figure 8 Influence of fish exclusion cages (full cage), cage controls (partial cage), and open controls (open) on the abundance of an important prey mollusc, *Tellina robusta,* at three locations in a deep channel in One Tree Lagoon (previously unpublished information; see Jones *et al.,* 1988, for mollusc sampling procedure). Error bars = S.E.

VI. Critique of Methodologies Employed to Assess Impacts

A wide variety of techniques have been employed to examine predator impacts. If one takes the conclusions from the experimental work reviewed at face value, then the inescapable conclusion would be that fish predation is exceedingly important some of the time. However, the story cannot end without a more critical appraisal of the methodologies used. In terms of background information, a disturbing number of experimental investigations have failed to identify the fish predators involved. How do we know that it is fish and not some other category of predator, and how can we assess whether or not the experiment is adequately testing this factor? Another problem is an almost total absence of data on spatial and temporal variation in predator abundance and feeding pressure to place the experiments in context. An experimental effect detected at a limited number of high fish density sites may be irrelevant to the prey assemblage as a whole. Likewise, failure to detect an effect may result from the chance placement of the experiment in a low fish density habitat.

Some lines of evidence have been based on observational data alone, for example, correlations or lack of correlations between fish abundance and invertebrate densities (St. John *et al.,* 1989). Such comparisons are at most suggestive of fish impacts, and failure to find correlations is fairly weak evidence against their importance.

Assessment of impact based on standing crop estimates of the biomass of prey contained in guts as a percentage of the standing crop in the field is particularly dubious (Alheit, 1981; Alheit and Scheibel, 1982). What rule can be applied to say that 3 percent of standing crop removed per day is significant and 0.02 percent per day is not? It is not standing crop estimates, but rather rates of prey consumption relative to mortality attributable to other causes which are important. Potential impact may be calculated with knowledge of mean turnover rates in stomachs, mean fish abundance, and mean turnover rates of prey in the field. However, all these estimates have error, which must be taken into account in the final calculations. Alheit's estimates of 3 and 0.02 percent may not be significantly different.

The use of artificial reefs as analogs of natural reef systems (e.g., Wolf *et al.,* 1983; Reaka, 1985) must be cautioned against. Real effects of fish on the rates of colonization of such reefs do not indicate that prey numbers are naturally controlled by predators. The response of both predator and prey to novel substrata and shelter placed in their habitat may differ from that exhibited under totally natural circumstances. The same conclusions apply to experimental microcosms (e.g., Brock, 1979).

The use of fish exclusion experiments employing cages or fences has a long history in temperate regions (see Choat, 1982; Jones, 1988a), and they continue to be a major tool in predation studies on coral reefs and adjacent sediments (Day, 1977, 1985; Neudecker, 1977; Fitz *et al.*, 1983; Keller, 1983; Doherty and Sale, 1986; Reaka, 1985; Lubchenco *et al.*, 1984; Jones *et al.*, 1988). Difficulties in the interpretation of caging experiments, problems of cage artifacts, and other uncertainties have been addressed by a number of workers (e.g., Virnstein, 1978; Hurlberg and Oliver, 1980; Doherty and Sale, 1986; Jones *et al.*, 1988). Many of the mistakes made in temperate waters have been repeated by the uncritical application of caging techniques in tropical communities. A number of common problems lay nearly all the caging experiments described open to question. These include failure to assess the effectiveness of complete cages in excluding appropriate fish, failure to use and assess the effectiveness of cage controls (in terms of both predator and prey behavior), and failure to monitor effects of cages on environmental factors such as light and sedimentation regimes.

Fish inclusion experiments are another method by which their effects upon prey communities are frequently measured (e.g., Gilinsky, 1984; Fitzhugh and Fleeger, 1985). These experiments have the advantage of being able to precisely control predator density and detect the threshhold density at which effects become apparent. An unavoidable problem is that one cannot control for changes in the behavior of predators in unnaturally confined conditions.

Transplant experiments have been employed to measure the ability of prey organisms to survive predation outside their normal habitat or distributional range (Bakus, 1964; Neudecker, 1977, 1979). If predators do eat them (when unprotected), this does not mean their distributional limits are set by this process. Underwood and Denley (1984) use the analogy of transplanting polar bears to the desert. If the polar bears die, this does not mean that physical conditions determine the southern distribution of polar bears.

In summary, the collection of papers described gives an impression that fish have an impact on a whole range of benthic invertebrate assemblages—an impression which does not stand up to close scrutiny. The purpose of criticizing the various methodologies was not to show that any one technique was invalid. All methods have their problems and all results when viewed in isolation are equivocal. We believe that progress will only be made in this area if individual workers apply a range of techniques, both observational and experimental, and address the methodological problems inherent in each technique employed.

VII. Fish Predation as Functional Explanation of Prey Characteristics

Structural, physiological, and behavioral characteristics of invertebrate species, which appear to reduce their chances of being eaten by fishes, are frequently cited as evidence for the importance of fish predation in assemblages of such species (Bakus, 1969, 1983; Wolf *et al.,* 1983). Morphological characteristics such as skeletal and tube protection, and cryptic coloration and form, are often considered adaptive responses to fish predation (Bakus, 1969). The buildup of chemicals which are toxic to fishes in a variety of invertebrates is seen as an evolved defensive mechanism (Bakus and Green, 1974; Bakus, 1981). Behavioral patterns such as vertical migration in demersal zooplankton (Alldredge and King, 1977), nocturnal activity (Tertschnig, 1989), boring and burrowing (Bakus, 1969; Reaka, 1980; Reaka and Manning, 1981), and synchronized spawning (Babcock *et al.,* 1986) have all been seen as adaptive evolutionary responses to fish predation on invertebrates. These assertions usually lead on to very speculative coevolutionary arguments to explain many of the predatory characteristics of the fishes themselves.

Whether all of these characteristics did evolve as protective mechanisms or as a response to other selective agents, or other evolutionary mechanisms, is a debate we wish to avoid. Suffice to say that the role of fish predation in the evolution of invertebrates has not been critically examined and may be impossible to determine (see also Glynn, 1988). In ecological time, supposed adaptations are frequently insufficient to protect prey species. For example, Reaka (1985) showed that the armor and apparent defensive behaviors in stomatopods are usually insufficient to deter fish predation. It is possible that in many cases, invertebrates would be immune to fish predation regardless of such things as their color, the depth of their burrow, or their period of active foraging.

Our questions concern the importance of fish predation in limiting or structuring invertebrate populations on present, ecological time scales. The practice of asserting on the basis of prey behavior that fish must currently play an important role is fraught with difficulties. Past ecological impacts may have had important selective influences, but they may not be in operation today (Connell, 1980). For example, fish predation does not appear to be of major significance in affecting the abundance of molluscs in soft sediments of One Tree Lagoon. While it is interesting to speculate whether or not this has always been the case, there is no method available which exposes the history of such ecological relationships.

VIII. Conclusions

We listed three important steps in examining the numerical impact of carnivorous fishes on invertebrate communities, whether dealing with corals, other sessile organisms, or mobile invertebrates. These were: (1) showing what fish predators are involved in consuming invertebrates from the assemblage of interest; (2) measuring variation in the feeding pressure exerted by predators due to patterns in abundance and foraging; and (3) application of manipulative experiments to establish causal links between changes in invertebrate populations and predation pressure. In reviewing studies which address this relatively recent ecological problem, two major deficiencies were apparent. First, there are few comprehensive studies for which all three steps have been carried out. While there are isolated instances in which fish predation seems a likely cause, it is equally possible that it is neither fish nor predation that is involved. The second criticism concerns the rather "uncritical" application of experiments such as caging and transplants, which seem to ignore the long history of debate generated by their use in other systems.

The question frequently asked in the reviewed studies was: Do fishes have an impact on benthic population and assemblage structure? The answer was commonly, although not universally, yes. Predators, without a doubt, consume prey! Therefore if the technique, the experimental design, and the associated statistical test are powerful enough, the answer is guaranteed to be yes. There is something intrinsically wrong with asking a question and then expending enormous amounts of time and energy answering it, when there can only be one answer.

A more appropriate question is: How important is the impact of fish on benthic communities in relation to other processes and how does this importance vary from place to place and time to time? In other words, is predation generally important or universally similar in its effects? Predation studies on coral reefs must follow current trends in examining impacts relative to other ecological factors such as variable recruitment, competition, other forms of disturbance, etc. We strongly question the historical (Bakus, 1964) and currently expressed (Lubchenco *et al.,* 1984) generalization that predation by fishes has the major impact on benthic communities on coral reefs, and that this effect is greater than in temperate regions. We strongly question the adaptationist approach of ascribing sundry structural, physiological, and behavioral characteristics of species to predation as a historical or current selective agent. Predation may play or have played these roles, but a method which could establish this has not yet been applied.

ACKNOWLEDGMENTS

We extend our greatest thanks to all those who assisted us in our work on fish predation at One Tree Reef. This text has benefitted by comments and assistance from U. Kaly, R. Robertson, G. R. Russ, and K. Tricklebank.

Part III

Larval and Juvenile Ecology

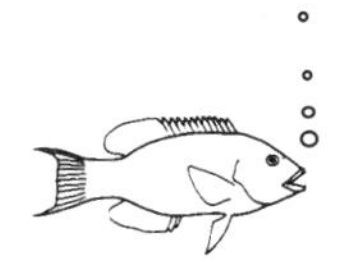

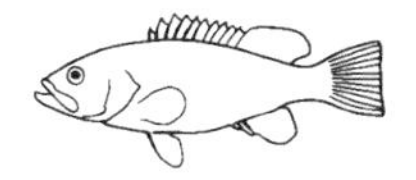

CHAPTER 8

The Pelagic Stage of Reef Fishes: The Larval Biology of Coral Reef Fishes

Jeffrey M. Leis
Division of Vertebrate Zoology
The Australian Museum
Sydney, New South Wales
Australia

I. INTRODUCTION

Most coral reef fishes have a pelagic, usually larval, stage resulting in a life history with two distinct and very different phases. The pelagic and benthic stages differ in almost all characteristics from morphology to size, habitat, food, and behavior. Probably no other single factor distinguishes marine fishes so completely from their terrestrial vertebrate counterparts. Of the approximately 100 families of bony fishes associated with coral reefs, only four plus a single species of a fifth family are known to lack a pelagic early life history stage. This is regardless of the mode of reproduction (Table 1). The dual nature of reef fish life history has profound implications for the biology of these species, and for our attempts to understand that biology.

Almost every bony fish on the reef has passed through a pelagic phase, and a substantial portion of its life, expressed as either size or age, may have taken place in the pelagic environment. Recent otolith-based estimates of the duration of the pelagic phase show that it is species dependent, probably flexible, and ranges from 9 to well over 100 days (Brothers and Thresher, 1985; Victor, 1986a,b, 1987a; Thresher *et al.*, 1989). Size at settlement ranges from about 8 to 200 mm (Table 2). At the upper end, this may represent 80% of size at sexual maturity or 50% of maximum size (Table 2). At the lower end, size at settlement in parrotfishes (Scaridae) is 8–20% of size at sexual maturity and 2–10% of maximum size (Bellwood and Choat, 1989).

Because many adult reef fishes are relatively sedentary (Sale, 1980a) and the pelagic stage may disperse at scales ranging from meters to thousands of kilometers, the pelagic stage is more likely to determine the geographical size

Table 1 Pelagic Stage and Reproductive Characteristics of Bony Coral Reef Fishes[a]

FAMILIES WITH NO PELAGIC STAGE: Plotosidae, Batrachoididae, Sciaenidae?,[b] Pholidichthyidae, *Acanthochromis polyacantha* (Pomacentridae), *Ogilbia?* (Bythitidae). n = 4 + 1 genus? + 1 species

FAMILIES WITH PELAGIC STAGE:

LIVE-BEARERS: Bythitidae (possibly except *Ogilbia*), Clinidae. n = 2

DEMERSAL EGGS: Gobiesocidae, Pseudochromidae, Acanthoclinidae, Pomacentridae (except *A. polyacantha,* see above), Congrogadidae,[c] Tripterygiidae, Chaenopsidae, Blenniidae, Schindleriidae(?), Gobiidae, Microdesmidae, Siganidae, Balistidae, Tetraodontidae. n = 14

BROODED EGGS: Solenostomidae, Syngnathidae, Grammidae, Apogonidae,[d] Opistognathidae, Dactyloscopidae.[c] n = 6

PELAGIC EGGS: Moringuidae, Xenocongridae, Muraenidae, Ophichthidae, Congridae, Synodontidae, Ophidiidae, Carapidae, Lophiidae, Monocentrididae, Anomalopidae, Pegasidae, Aulostomidae, Fistulariidae, Dactylopteridae, Scorpaenidae, Synanceiidae, Platycephalidae, Serranidae, Grammistidae, Priacanthidae, Malacanthidae, Carangidae, Lutjanidae, Caesionidae, Gerreidae, Haemulidae, Inermiidae(?), Sparidae, Lethrinidae, Nemipteridae, Mullidae, Monodactylidae(?), Pempherididae(?), Kyphosidae, Ephippididae, Chaetodontidae, Pomacanthidae, Cirrhitidae, Cheilodactylidae, Cepolidae, Mugilidae, Sphyraenidae, Polynemidae, Labridae, Scaridae, Uranoscopidae, Creedidae, Mugiloididae, Callionymidae, Acanthuridae, Psettodidae(?), Bothidae (including Paralichthyidae), Cynoglossidae, Soleidae, Ostraciidae, Diodontidae. n = 57

SEMI PELAGIC EGGS (eggs with tendrils that may attach them to objects or join them into free-floating clumps): Hemiramphidae, Belonidae, Atherinidae. n = 3

MORE THAN ONE TYPE OF REPRODUCTIVE MODE (different species have different modes): Clupeidae (**P,D**), Antennaridae (**P,B**), Plesiopidae (**B,D**), Labrisomidae (**L,D**)[e] n = 4

UNKNOWN REPRODUCTIVE MODE: Pseudotrichonotidae,[c] Ogocephalidae, Isonidae, Holocentridae, Centriscidae, Caracanthidae, Aploactinidae, Kuhliidae, Trichonotidae, Kraemeriidae. n = 10

[a] Listed are families with at least one member closely associated with coral reefs (families and their sequence are after Nelson, 1984), and only these members are considered for the characteristics noted. **L** (live-bearers), **D** (demersal), **B** (brooded), and **P** (pelagic) refer to reproductive mode. Characteristics are principally after Leis and Rennis (1983), Thresher (1984), Moser *et al.* (1984), and Leis and Trnski (1989), **?**, lack of pelagic stage uncertain. **(?)**, uncertain reproductive mode.

[b] The only larvae of coral reef sciaenids known are apparently not pelagic (Powles and Burgess, 1978), but many nonreef sciaenids have a pelagic larval stage.

[c] Larvae unknown, but presumably pelagic.

[d] McFarland and Ogden (1985) state, without documentation, that some apogonids "have completely eliminated the planktonic larval phase." This seems doubtful.

[e] Only *Starksia* and *Xenomedea* contain live-bearers, and larvae of these genera are unknown.

of population units than the adult stage. In fact, the pelagic stage of most species probably has a broader (albeit nonreproductive) range than does the demersal, adult stage (Leis, 1986b).

Coral reef fishes are highly fecund, particularly by terrestrial vertebrate standards. Most estimates of annual egg production range from 10,000

Table 2 Size at Important Transitions in the Life History of Some Coral Reef Fishes[a]

Species	Size at Settlement (mm)	Size at Maturity (mm)	Maximum Size (mm)	Reference
Gobiidae				
Eviota epiphanes	7.7–9.0[b]	11.6	15.7	Lachner and Karnella (1980)
Serranidae				
Anthias squamipinnis	11–15[c]	17	73	Shapiro (1981), Avise and Shapiro (1986)
Tripterygiidae				
Eanneapterygius atriceps	12–15[b]	<20	30	Jordan and Evermann (1905)
Synodontidae				
Saurida sp. 2 (age in days)	31[c] (22–56)	73–100 (86–110)	252 (?)	Thresher *et al.* (1986)
Chaetodontidae				
Hemitaurichthys polylepis	60	?	127	Burgess (1978)
Diodontidae				
Diodon hystrix	180–191	?	>500	Leis (1978b)

[a] These species have a large size at settlement relative to maximum adult size. Size at settlement is a range derived from largest pelagic individual and smallest settled individual.
[b] Largest pelagic individual is based on unpublished observations.
[c] Based solely on smallest settled individuals.

to over 1,000,000 per female (Sale, 1980a). However, this is coupled with a mortality approaching 100%, the majority of which take place during the pelagic stage. Indeed, the life of the larval fish has been characterized as "precarious" (Doherty, 1981). Adult population size may be determined during the pelagic stage (Doherty, 1983a), but even if it is not, the heavy pelagic-stage mortality has a large influence on adult populations. Indeed, the answers to vital questions in ecology and resource management regarding recruitment variability lie in understanding the pelagic stage (Doherty and Williams, 1988a).

So, there are very good reasons for studying the pelagic stage of coral reef fishes. In spite of this, work on the pelagic stage has lagged far behind that on the adult, benthic stage (see reviews by Richards, 1982; McFarland, 1985; Richards and Lindeman, 1987). The reasons for this are varied. The study of adult reef fishes in a coherent way is itself young, and larval studies can proceed only from a base built on an understanding of adult taxonomy and ecology (Sale, 1980a; Doherty and Williams, 1988a). Studies of larval fish ecology are challenging and expensive even in ideal conditions, and conditions in the tropics are less than ideal. Coral reefs have a highly diverse fauna and are remote from temperate centers of marine research. In addition, larval and pelagic juvenile stages are very difficult to identify, their identities may have not been established, they are extremely dilute by temperate standards, and techniques of laboratory culture for them are virtually nonexistent. Convenient temperate features such as short, sharp spawning seasons are largely absent. Finally, the pelagic stages of many coral reef fishes reach considerable sizes (Table 2), meaning that one must begin to study plankton, but finish by studying nekton.

One might validly view reef fishes as pelagic animals with a benthic reproductive phase. Size and age have been mentioned. Further, more biomass and energy flow may be tied up in the pelagic stage than in the benthic stage (R. E. Johannes, personal communication). An indication of this is the degree to which epipelagic predators such as tunas and dolphinfishes prey on the pelagic young of reef fishes (see Section IV, B, Growth and Mortality). That this extreme view cannot be dismissed emphasizes the importance of the pelagic stage and our poor understanding of it.

This chapter will review what is known about the pelagic life history stage of coral reef fishes. This has been gleaned from a variety of sources. Much of it is anecdotal and, in many cases, is so limited that I have had to include information on nonreef, tropical shorefishes to provide reasonable coverage. I tried to avoid work on clupeiform fishes except for those few that are closely associated with coral reefs (spratellodin clupeids). A definitive review is not yet possible, but I suggest where our meager store of knowledge on the pelagic stage of coral reef fishes should lead in future research. I believe the evidence supports a portrayal of these pelagic organisms as active rather than the helpless, passive animals of convention. To treat fish larvae as a conservative property of sea water such as salinity is to mislead. The pelagic phase of coral reef fish life history has usually been treated as a black box. Current knowledge sheds some light into this black box and reveals complexity that was unsuspected 15 years ago. The pelagic environment is far from unstructured, and the pelagic reef fishes that occupy this environment are anything but uniform in their interactions with it.

II. Larval Fish Morphology and Identification

A. Definitions

The great variety in development and ecology among species means that any terminology for the pelagic stages of reef fishes will have deficiencies. One can only aim to reduce these (Leis and Rennis, 1983; Moser *et al.*, 1984). **Larval** is a morphological/developmental term that does not necessarily imply small size. The end of the larval stage is at an arbitrarily defined developmental point or morphological transition: this definition often varies among authors. The transition from larva to juvenile, often called **metamorphosis,** may be abrupt or gradual. **Pelagic** is an ecological term denoting open water without reference to size of the organism. At the end of the pelagic stage is a transition (**settlement**) from open water to benthic habitat. For most reef fishes, pelagic is equivalent to the term presettlement (Kingsford, 1988), but **presettlement** is inappropriate for reef-associated semipelagic fishes such as carangids or caesionine lutjanids that do not settle. Some reef fishes may settle into nonreef habitats or into cryptic habitats on the reef (see Section V, The End of the Pelagic Stage), so it is not safe to assume that the smallest fishes visible on the reef (**recruits**) are newly settled. **Planktonic** has connotations of both small size and passivity, and the latter assumption is inappropriate for larval fishes. The definition of **postlarva** varies so widely among authors (see Moser *et al.,* 1984) that its usage merely invites confusion.

Therefore, larval and pelagic are not synonomous. In many species, metamorphosis takes place in the pelagic environment (e.g., many tetraodontiform fishes; see Fig. 1C), while in others it is not complete until well after settlement (e.g., chaetodontids). Settlement is not always a synonym of recruitment, and it may be reversible in some cases. Some species may attain sexual maturity prior to settlement (Pillai *et al.,* 1983; Leis and Moyer, 1985), but this is confirmed only for the paedomorphic family Schindleriidae (Watson and Leis, 1974).

The pelagic environment is three dimensional, so distributional studies may consider either **concentration** (individuals per volume of water) or **abundance** (individuals per area of sea surface, integrated through a water column of specified depth). **Young larvae** means those prior to completion of formation of the caudal—or tail—fin (preflexion). In laboratory-reared larvae, flexion is complete about two weeks after hatching (Leis, 1986a). **Old larvae** have a complete caudal fin (postflexion) (Fig. 1).

B. Identification

The pelagic stages of reef fishes are very different morphologically from the benthic adults (Fig. 1). This is due primarily to three factors. First, fish larvae do not hatch fully developed, so the larval stage is a developmental one during which many structures first appear or proceed to a functional state. Second, reef fish eggs are typically small (pelagic eggs larger than 1.5 mm are rare), so newly hatched larvae are small. Third, larvae have morphological features that are presumably adaptations for pelagic existence (Fig. 1). These are undoubtedly disadvantageous on the reef and are lost at about the time of settlement.

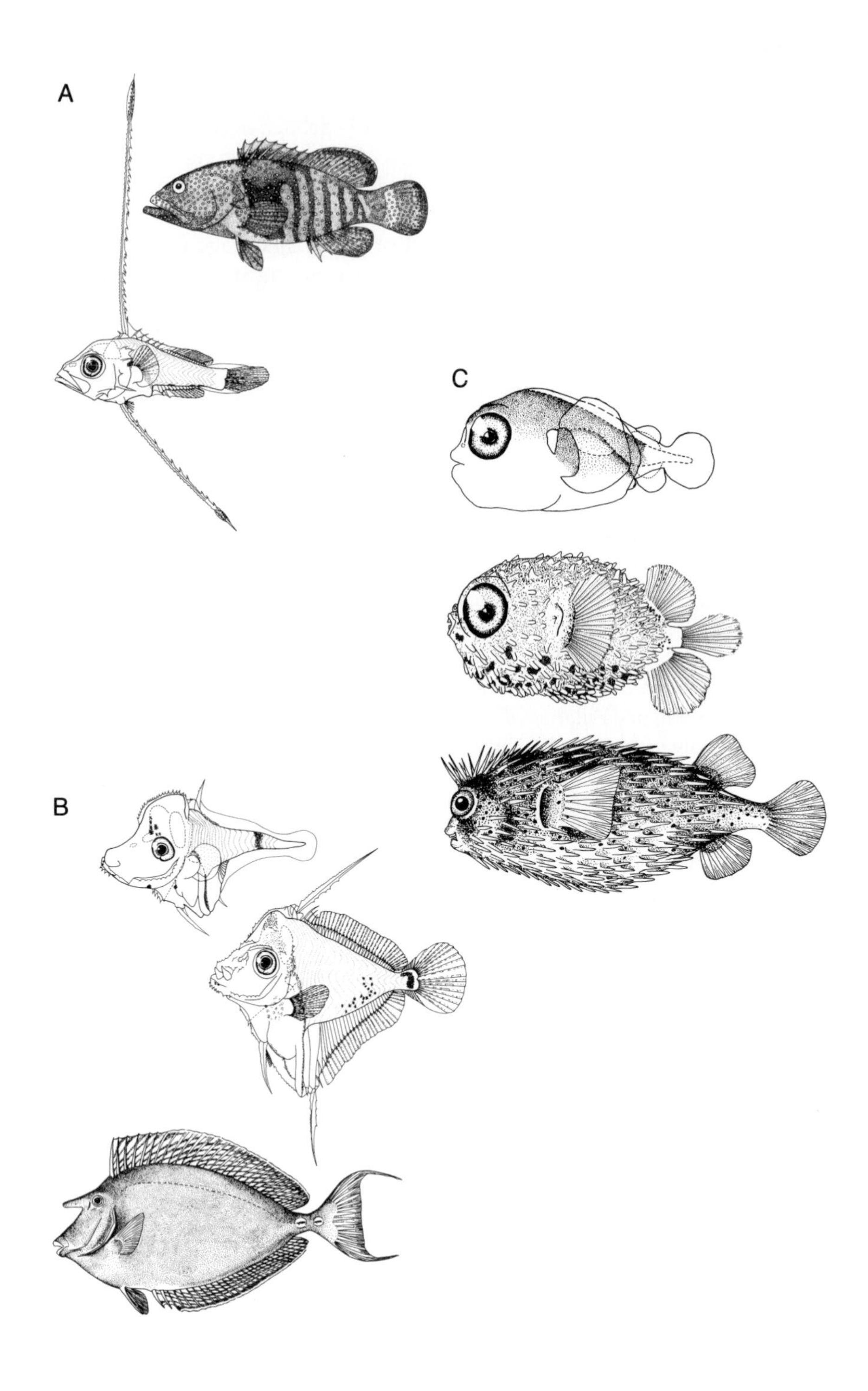

A
B
C

For these reasons, pelagic young are difficult to identify to genus or species. This problem has received attention recently (Leis and Rennis, 1983; Moser *et al.*, 1984; Okiyama, 1988; Leis and Trnski, 1989), and the would-be larval fish researcher now has a starting place for identification.

III. Where Are the Pelagic Stages of Reef Fishes?

A. Sampling Problems

Difficulties in sampling pelagic reef fishes are principally due to five factors. None are unique to the pelagic young of reef fishes, but when taken together, they present formidable problems. These are: markedly patchy distributions, rarity, no single collecting methodology suitable for all ages, spatial patterns that change temporally, and the markedly three-dimensional nature of distributions.

It is often assumed that larval fish are patchily distributed (e.g., Doherty *et al.*, 1985), but there are few formal demonstrations of this for tropical shorefishes. Variance to mean ratios among replicate larval fish samples are often very high, but temporal variation in this is great (unpublished observations). In a shallow, subtropical embayment, Houde and Lovdal (1985) found that patchy distributions occurred at scales of 10–1000 m, but on many days the fish eggs and larvae were randomly or uniformly distributed. Patchiness was not correlated among taxa on collection dates, nor was it correlated with abundance or wind speeds. Patchiness of anchovy eggs and larvae were not correlated, indicating that different biological processes, including behavior of the larvae, were operating to produce the distributions of the two different stages. In contrast, multispecies patches of "at least kilometres" of settlement-

Figure 1 Pelagic and reef-associated stages of three coral reef fishes. Pelagic stages are often characterized by transparency, attenuate and ornamented fin spines, head spination, large heads, and large eyes. These adaptations to pelagic life are lost in the reef-associated stage. Pelagic stages are after Leis and Rennis (1983); reef-associated stages are after Fischer and Bianchi (1984). (A) Grouper or coral cod, genus *Cephalopholis* (Serranidae): 5.8-mm pelagic larva is highly specialized, note fin and head spines (bottom), and 300-mm reef-associated adult (top). (B) Surgeonfish, genus *Naso* (Acanthuridae): 3.0-mm "young" pelagic larva (top), 8.3-mm "old" pelagic larva (center), both highly specialized in body shape and fin and head spines, and 300-mm reef-associated adult (bottom). (C) Porcupinefish, genus *Diodon* (Diodontidae): 2.4-mm "young" pelagic larva (top) is unspecialized, 4.8-mm pelagic juvenile (center) in most respects is a miniature adult, but it will not settle until it reaches 75–125 mm in length, and 200-mm reef-associated adult (bottom). (Adult figures reprinted with permission from W. Fischer and G. Bianchi, 1984. "FAO Species Identification Sheets for Fishery Purposes. Western Indian Ocean." Food and Agriculture Organization, Rome.)

stage larvae were reported by Doherty (1983b) based on anecdotal light collections, and Williams (1986a) reported "a multi-specific patch of larvae (apogonids, gobies and lutjanids) at least 7 km in diameter" in netted samples. From otolith and recruitment data, Victor (1984) estimated that a patch of labrid larvae was 46 km wide. However, the estimate was dubious because it was equivalent to the size of Victor's sampling universe.

No single method is adequate to sample the full range of pelagic stages. Because the pelagic stage of reef fishes is a development and growth stage, different sampling methods are generally required depending on the abundance, age (size), vagility, and general behavior of the larva or juvenile (Choat *et al.,* 1991). When different methods are used, they must be intercalibrated, and this is seldom done (Choat *et al.,* 1991). Sampling method will determine the view one obtains of the pelagic environment. Visual surveys are rarely possible because of the small size, transparency, and rarity of the larvae. A standard plankton net will not quantitatively sample larvae much bigger than 10 mm, nor will it filter enough water to capture larger, rarer individuals of any but the most abundant species. Larger or faster nets may lessen these problems, but they introduce others, including loss of smaller larvae through net mesh. Purse seines may lessen avoidance, but they sample small volumes and are limited to the surface. Aggregation devices such as lights depend on behavior of larvae, water clarity, and current speed in ways that differ among species and developmental state. Aggregation devices capture primarily the settlement stages of a restricted number of taxa. Efficiency of all types of gear varies on a diel basis: both avoidance and diel vertical migration ensure that samples taken during daylight will not be equivalent to those taken at night. All this means that estimates of concentration and abundance are rarely absolute, and care must be taken to ensure that even relative (i.e., within method) comparisons are specific to developmental state or size. In particular, field estimates of growth and mortality must be treated cautiously.

Compared to other zooplankters, reef fish larvae are rare, with concentrations of all taxa combined seldom exceeding 5 per cubic meter, except in specific, localized situations. This is particularly true in oceanic waters. Given the diversity of reef fishes, this means that numbers of individual species per sample are usually very low, even if substantial volumes of water are sampled. In fact, it is not unusual for the larvae of even a common species to be absent in the majority of samples (Leis, 1987, 1989). In this regard, T. A. Clarke (personal communication) argues that some of the conclusions of Hawaiian studies involving absence or rarity of particular taxa inshore could be artifacts of low abundance and low volume sampled. A common means of attempting to deal with this problem is to pool data across taxa, replicates, or treatments. This introduces its own problems. Another characteristic is that data frequently do not meet the assumptions of parametric statistical procedures, even after transformation (Jahn, 1987). All of these characteristics lead to problems in data analysis and have led to a large literature on ways to deal with them (see, e.g., Fasham, 1978; Pennington, 1983; Downing *et al.*, 1987; Jahn, 1987).

A further complicating factor is that distributions of fish larvae are ephemeral both because the animals are only temporarily pelagic and because the water they occupy moves. This calls for extreme care in sample design (Andrew and Mapstone, 1987) and an appreciation that levels of variance measured in a pilot study may have little relevance at other times or at nearby locations (e.g., Houde and Lovdal, 1985).

Finally, distributions of pelagic reef fishes are highly structured in all dimensions. Therefore, choice of a scale appropriate to the question while taking into account biological and logistical constraints is particularly important. This choice will determine sampling methodology and dictate that detection of some structure will be beyond the scope of the investigation while other structure will be integrated and lost. For example, a purse seine is ideal for sampling at small scales (tens of meters), while a towed net cannot be used at very small scales. But the purse seine will, for a given number of samples, result in higher estimates of variance than will a plankton net because the latter integrates much structure along its tow path and samples a larger volume (Choat *et al.*, 1991). The purse seine cannot sample subsurface organisms while the towed net can, so the vertical distribution behavior of the individual taxa will place additional constraints on sampling methodology.

The literature contains many examples of conclusions reached without the benefit of objective analysis (i.e., statistical evaluation) of the data presented. In many cases this is due to the problems noted here: meaningful statistical analysis cannot be performed. This may be due to poor sample design, lower abundance or higher variance than estimated from pilot studies, or use of samples for analyses other than those originally intended. Admittedly, statistical tests are often misapplied, and statistical power is rarely evaluated, but I attempt to indicate which conclusions are not based on statistical testing.

B. Distributions and Physical Oceanography

Distributions of pelagic organisms such as larvae of reef fishes cannot be understood fully without an understanding of the physical oceanography of the study area. Unfortunately, few studies of reef fish larvae have adequately addressed this requirement. This is due to ignorance of physical oceanography by biologists, reluctance by physical oceanographers to work in the very difficult areas near reefs at scales relevant to the biological problems, and difficulties in persuading funding agencies that such integrated, albeit expensive, studies are worthwhile. The physical oceanographic features of relevance to larval fish studies have been well reviewed by Bakun (1986) and Hamner and Wolanski (1988). A promising approach to determining circulation around coral reefs based on numerical hydrodynamic simulations is given by Black and Gay (1987) and Black *et al.* (1990). Power (1984) provides an excellent discussion of the requirements, both biological and physical, for adequate modeling of larval fish distributions. Finally, Hatcher *et al.* (1987) provide an introduction to scaling analysis for reef biologists, including both physical and biological parameters, and an example concerning larval reef fish distribution (see also Richards and Lindeman, 1987). In sum, these papers make it clear that circulation in coral reef areas is very complex; that topography has an overriding effect on the circulation; that a variety of physical mechanisms exist that make retention of larvae near reefs not only possible, but probable; and that some of these mechanisms can operate with passive larvae, but most require some degree of behavioral input from the larvae.

Biologists must obtain a minimal understanding of physical oceanography to be able to ask the right questions and to interpret the answers. A single example will suffice. In coral reef lagoons without a pycnocline, physicists validly do not distinguish surface water from deep water when calculating a flushing time. However, a problem arises if the level at which the water exits from the lagoon is restricted vertically either because of a sill in a deep pass or because flow out is principally over the reef flat. In this case, planktonic animals that avoid surface waters and occur in depths greater than the sill depth (even if they are otherwise passive) will have a much longer residence time in the lagoon than that calculated by a physical oceanographer. Thus, it is vital that the biologist understand both the vertical distribution of the animal and the basis upon which the estimate of flushing time was made. So, when Bakun (1986) states that lagoons "appear to have such short flushing times that retention of significant numbers of larvae through their entire pelagic phase is unlikely," the biologist must understand that the larvae were treated as water particles (passive in three dimensions), and the absence of physical stratification in the lagoon presumes that each water particle is equally likely to exit. If this fits the animal in question, then Bakun (1986) is correct; if not, then he is probably incorrect.

A brief examination of the few papers that have seriously attempted to relate larval reef fish distributions to physical oceanography follows. The purpose is to demonstrate the potential of such integration and to point out some of the pitfalls.

Miller (1974) attempted to combine physical and biological data from a small reef in the Hawaiian Islands to explain the distributions he encountered. Unfortunately, his measurements were all made at the surface and were therefore two dimensional, and he invoked three-dimensional processes (vertical distribution of larvae and upwelling) in his explanation. Miller (1979b) attempted to explain with an upwelling model the distributions of tuna larvae based on surface tows. He had no data on the behavioral portion of the model he proposed, and his physical data were limited to two bathythermograph casts. Miller's ideas are intriguing, but they remain speculative.

Leis (1982a,b) combined measurements and literature reports of currents with measurements of the temporal and spatial distribution of pelagic fish eggs to characterize the flow in a Hawaiian study area. A tidal eddy was inferred from these, and small-scale upwelling was inferred from differences in current vectors between subsurface and surface measurements. These physical data were then used to help explain distributions based on oblique plankton hauls. Leis (1982a) concluded that the distributions of the reef fish larvae and coastal holoplankters could not be explained by passive drift alone, and that (largely unspecified) behavioral input by the larvae and plankters was required. Unfortunately, the physical data were inadequate to do more than infer the proposed physical processes.

On the Great Barrier Reef, Williams *et al.* (1984) assumed that larvae were passive and used depth-integrated current data taken away from reef influence to predict potential movement of larvae. This approach is useful but flawed because of the rarely met assumption of passivity. Few larvae are uniformly distributed vertically, at least during the day (see Section III, D, Vertical Distribution), which invalidates depth-integrated models. Further, recent work (Black and Gay, 1987; Black *et al.*, 1990) indicates that passive retention of larvae around reefs is much greater than previously assumed, and Kingsford *et al.* (1991) shows that tidally induced fronts around reefs could increase the detectable size of the reefs. Williams *et al.* (1984) were careful to restrict application of their model to the area for which they had current measurements. Others may be tempted to apply these results to other areas, but currents in the Great Barrier Reef 500 km from the study area of Williams *et al.* (1984) differ enough to make such application invalid (Frith *et al.*, 1986).

Lobel and Robinson (1983, 1986, 1988) attempted to combine physical and biological data to show that reef fish larvae were retained near an island by a mesoscale eddy. The physical data seem (at least to this biologist) to be adequate, but the biological data are clearly inadequate. Only ten plankton tows were taken, there was no replication, and identities of the fish larvae were not revealed until the 1988 paper (most were of oceanic fishes). That paper reported that fewer than 100 reef fish larvae were captured. In another report on the same data set, Lobel and Robinson (1985) perceptively noted that their data "do not allow statistical trends analysis": they have rightly avoided statistical analysis.

Leis (1986a) combined current measurements with horizontal and vertical distributional data on fish larvae to conclude that circulation at two sites around Lizard Island was favorable for retention of larval stages but was unfavorable at a third. Retention was detected only at one site. Here, favorable circulation combined with appropriate vertical distribution behavior by the larvae was held responsible for retention. Unfortunately, the physical measurements made were not adequate to fully portray the complex circulation at this windward site, although the data were consistent with published descriptions of it at another site. Leis (1986a) speculated that retention was not detected in a small lagoon (a favorable site) for biological reasons. No retention was detected at the unfavorable site.

Black and Gay (1987) and Black *et al.* (1990) modeled circulation around reefs in the Great Barrier Reef. Unfortunately, the models are not fully field tested, are depth-integrated, and assume passive larvae.

All of these studies suffered from a greater or lesser inadequacy in either biological or physical data or both, which points out the need for full collaboration between physicist and biologist. The physicist must be familiar with the complex and difficult oceanography near reefs or the wrong advice may be given or inadequate measurements made. The physicist must describe simply the complex circulation at the appropriate scale, keeping in mind the many processes that Bakun (1986) reviews. The biologist must be fully aware of the logistical problems of sampling reef fish larvae and be skilled at sample design and identification of larvae. Considerable insight into the biology and behavior of the larvae must be gained before the study in order to design properly both biological and physical programs. Otherwise, unrealistic depth-integrated models that assume passive drift will continue to result. It seems clear that circulation around reefs is so topography dependent that applicability of physical studies at one reef to other reefs will be minimal. It is here that numerical models (Black and Gay, 1987; Black *et al.*, 1990) may provide answers, although healthy skepticism is necessary until such models are fully field tested.

The available studies provide some interesting ideas for future testing and have been more successful at raising questions than they have been at explaining the distributions they detected. This should not be surprising: study of the biology of pelagic reef fishes is still in a descriptive, pattern-seeking stage. It is in the description of this pattern and in making clear the importance of larval behavior that these studies have made their greatest contribution.

C. Horizontal Distribution

1. Introduction

Distribution of the pelagic stage of reef fishes has received more attention than any other aspect of their biology. In part, this is because distribution is probably the most easily studied aspect of the pelagic stage—at least for the smaller larvae. But in addition, an understanding of where the larvae spend their time away from the reef is basic to all other study on the pelagic stage.

Larvae and pelagic juveniles of reef fishes can be found from water over a coral reef (see Section III,C,3) to the open ocean many hundreds of kilometers from any reef (Leis, 1983; Victor, 1987a). Although the details of distributions are beginning to emerge, understanding of their causes lags far behind. The factors likely to determine the distribution of fish larvae are: (1) spawning behavior of adults; (2) hydrographic structure at a variety of scales and the interaction of this with reef topography; (3) duration of the pelagic period; (4) behavior of the larvae; and (5) larval mortality and growth and their variations in space and time. We have not achieved good measurements of all five or their relative importance for a single species, let alone a comparison of how they vary temporally, spatially, or among species. Information on spawning, hydrography, and pelagic duration are becoming available; information on behavior can sometimes be inferred from distributional information, but is otherwise in very short supply; and information on growth and mortality are essentially unavailable.

The traditional view—that larvae were essentially passive plankters whose distribution was determined entirely by the currents—has largely been discarded (Leis and Goldman, 1984, 1987; Leis, 1986a; Williams, 1986a). However, in only a few cases do we know much more than this. The distributional information summarized in the following sections provides a complex picture of taxon-specific, repeatable patterns at all scales. Some types of larvae apparently remain within a few hundred meters of their natal reef, others apparently maintain distributions within 5 km of reefs, many are widely distributed over continental shelves, while a number have distributions more like larvae of oceanic fishes than reef fishes. Many types of larvae probably have distributions that vary ontogenetically, although relatively few have been confirmed as such. This emphasizes the importance of including consideration of age or developmental state in distributional studies. All of these larvae originate from essentially the same place—the reef—and their differing interactions with the physical environment produce their differing distributions.

2. *Horizontal Distribution at Large Scales (>50 km)*

Some of the earliest quantitative work on the distribution of reef fish larvae came about as a by-product of surveys of the distribution and abundance of pelagic, commercial fishes on scales of hundreds to thousands of kilometers (Ahlstrom, 1971, 1972; Moser *et al.*, 1974; Richards, 1984; Richards *et al.*, 1984) or from general surveys on research expeditions (Nellen, 1973a,b; Leis, 1983, 1984). While useful in showing the extent to which fish larvae can become dispersed, such very large scale studies reveal little that is relevant to determining either where most larvae spend their pelagic period or how the distributions of larval reef fishes are caused.

A number of Chinese and Russian large-scale surveys with potentially useful information are difficult to evaluate because of language difficulties, dubious identifications, and subjective analyses (Fursa, 1969; Chen and Wei, 1978, 1982; Gordina and Bladimirtsev, 1987; Belyanina, 1987).

More useful are surveys of the distribution of the larvae of commercial groundfishes, primarily over the continental shelf (Dekhnik *et al.*, 1966; Houde *et al.*, 1979, 1986; Young *et al.*, 1986; Janekarn, 1988). Distributional data from such surveys show large differences between the assemblages of larval reef fishes over the shelf and those in the open ocean. In addition, where the scale of sampling has been appropriate, they have shown strong, if temporally varying, cross-shelf patterns of distribution.

Among the relatively few investigations aimed specifically at the larvae of reef fishes are two where the data presented and the nature of the subjective analyses do not seem to support the stated conclusions. Bourret *et al.* (1979) worked up only a small proportion of their samples, and preliminarily concluded that: reef fish larvae were distributed in rings around the atolls of the Tuamotu Islands without any windward/leeward effect; smaller larvae were most abundant close to the atoll; and "older postlarvae" were most abundant away from the atoll. Wyatt (1982) concluded from few samples and generalized identifications that reef fish populations around Jamaica were self-recruiting (i.e., recruits were derived from larvae originating locally).

3. *Horizontal Distribution at Medium (5–50 km) to Small (<5 km) Scales*

The first studies of distribution of reef fish larvae at small to medium scales were carried out in Hawaii, where Miller (1974, 1979a) and Watson and Leis (1974) noted that the inshore larval fish assemblage was not a good reflection of the adult fish assemblage. The larvae of many taxa of reef fishes were missing inshore. Sale (1970) noted that larval acanthurids were abundant in midwater trawls offshore. Leis and Miller (1976) showed that many of the "missing" reef fish larvae could be found 5–12 km offshore, and that the inshore larvae were primarily from nonpelagic eggs and the offshore larvae were primarily from pelagic eggs (Fig. 2). Bourret *et al.* (1979) did not find this sort of distribution in the Tuamotus, and T. A. Clarke (personal communication) argues that its reported presence in Hawaii could be an artifact of rarity of larvae and small volumes of water sampled inshore. Subsequent work in Australia provides a more complex picture (see below), with larvae of many taxa with pelagic eggs being abundant over the continental shelf, and larvae of some taxa from nonpelagic eggs abundant in the open ocean.

Miller (1979a) subjectively identified windward–leeward differences in

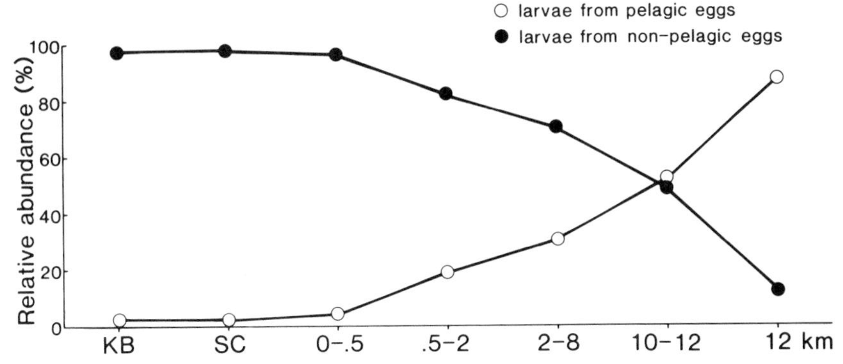

Figure 2 Changes in relative abundance of two types of reef fish larvae with distance from the Hawaiian Island of Oahu. Numbers on horizontal axis are kilometers from shore, KB is a barrier-reef enclosed bay, and SC is a channel through that barrier reef. Data are replotted from Leis and Miller (1976).

larval fish distributions around three Hawaiian islands based on surface tows. He thought the differences might be related to assumed, taxon-specific vertical distributions of larvae, but felt hydrology was the major determinant of larval fish species composition at a given location. Leis (1982a) examined larval fish distributions along a 3-km-long on–offshore transect at a single leeward Hawaiian location and identified strong abundance and concentration gradients (Fig. 3). Ontogenetic changes in distribution were identified in two species of reef fish larvae. T. A. Clarke (personal communication) detected generally similar zonation off windward Oahu. Because the larvae originated from the same inshore reefs, but had taxon-specific patterns of offshore distribution, Leis (1982a) argued that hydrography alone could not explain the observed distributions. The larvae must be participating behaviorally in determining their distributions, unless they are subject to taxon-specific spatial patterns of mortality.

Sale (1970) speculated that acanthurid larvae were kept from expatriation by mesoscale eddies formed in the channels between the Hawaiian Islands. Lobel and Robinson (1983, 1985, 1986, 1988) provided measurements of such an eddy off the island of Hawaii, but their claims that such eddies entrain and retain larvae of reef fishes remain unsupported (see Section III,B, Distributions and Physical Oceanography). The significance of such eddies in structuring the distribution of reef fish larvae near islands remains, therefore, unclear.

The majority of recent work on distribution of larval reef fishes has been carried out in Australian waters. These studies can be summarized by type of habitat investigated.

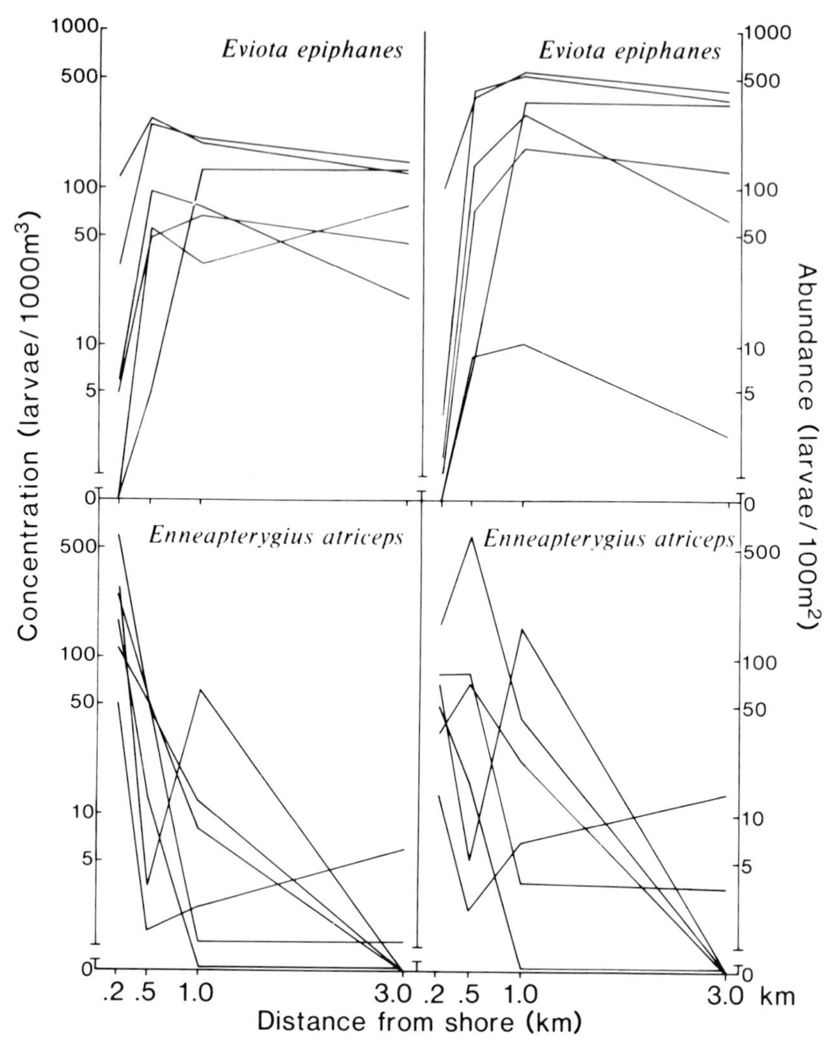

Figure 3 Offshore distributional gradients of larvae of two reef fishes that occupy inshore reefs and spawn demersal eggs as adults, Oahu, Hawaii. Figure on left gives concentration and on the right is abundance (note log scale). Comparison of the two values emphasizes the effect of water depth, which varies from 6 m at 0.2 km to >300 m at 3.0 km. Station at 0.2 km is over coral reefs; rest of stations are over sand bottom. Values are means of two replicates at each station on each of six sampling transects. Data are from Leis (1978a, 1982a). Top: Neritic distribution of the goby *Eviota epiphanes* (Gobiidae). Bottom: Inshore distribution of the triplefin blenny, *Enneapterygius atriceps* (Tripterygiidae).

a. Atoll Lagoons (Leis, 1986b and unpublished observations). Larval fishes are abundant in atoll lagoons, but only a few taxa seem to complete their pelagic stage there. If larvae of a species in a variety of developmental states are found in a lagoon, it is assumed to be completing its pelagic stage there. On this basis a few species from the following families use atoll lagoons throughout the larval stage: Apogonidae, Atherinidae, Blenniidae (Nemophini, Omobranchini), Callionymidae, Clupeidae (*Spratelloides*), Gobiidae, Microdesmidae, Pomacentridae, Pseudochromidae (?), Schindleriidae, and perhaps a few others. These conclusions are supported by preliminary analyses of samples from French Polynesian atolls (Leis *et al.,* in prep.). These few taxa are very abundant within atoll lagoons, but most seem able to complete the pelagic stage in the open ocean as well as in the lagoon. Why more species do not utilize atoll lagoons and the protection they offer from expatriation is not known.

b. Small Reef Lagoons on the Continental Shelf (Leis, 1981, 1986a; Schmitt, 1984a). Young larvae of many types are found in abundance in these lagoons, but larger/older larvae are usually very rare. This indicates that although spawning takes place, the pelagic stage is usually not completed within the lagoon. However, a few taxa of the following families do utilize these lagoons throughout the larval stage: Apogonidae, Atherinidae, Clupeidae (*Spratelloides*) (?), Gobiidae, Pseudochromidae (?), and Schindleriidae (?). Those not yet confirmed are indicated by (?).

c. Continental Shelf (either reef-enclosed or not) (Leis and Goldman, 1984, 1987, and unpublished observations; Leis, 1986a; Young *et al.,* 1986; Milward and Hartwick, 1986; Williams *et al.,* 1988). Large numbers of both young and old larvae of many taxa are found in these shelf waters, indicating that many species complete the pelagic period in these relatively open waters. For many taxa, strong cross-shelf medium-scale (5–50 km) gradients of abundance are found. Clear and consistent cross-shelf patterns of larval fish assemblages are often (Fig. 4) but not always found. However, this merely means "that different taxa do not consistently co-occur at the scales sampled" (Williams *et al.,* 1988) rather than that cross-shelf patterns of individual taxa are absent. Milward and Hartwick (1986) argue that distribution of the larvae in their samples from the inner shelf agrees closely with that of the adults, but this was based on all larvae captured, not just reef fish larvae. There are strong differences in larval fish assemblages and in abundance of many taxa between the shelf and the open ocean, indicating major differences at the edge of the continental shelf (Fig. 4).

d. Immediate Vicinity of Reefs (<5 km) (Leis, 1981, 1986a, and unpublished observations; Leis and Goldman, 1984, 1987; Leis *et al.,* 1987). Strong windward–leeward differences in abundance of individual taxa and in assem-

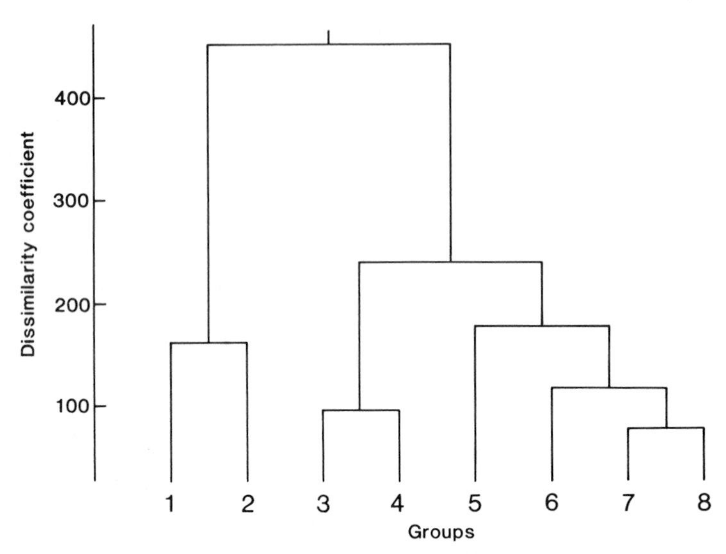

Figure 4 Dissimilarity dendrogram of 273 larval fish samples over the continental shelf off northwest Australia (after Young *et al.*, 1986). Two cross-shelf transects were sampled repeatedly from September 1982 to October 1983. A Maurichi net (A) was used from September to December and an Isaacs–Kidd net (B) was used from January to October. The groups are (1) off- and outer-shelf sites Sep–Dec (net A) and two net B samples from outer shelf in Oct; (2) off- and outer-shelf sites Jan–Oct (net B); (3) inner and midshelf sites Sep–Oct (net A); (4) inner and midshelf sites Nov–Dec (net A); (5) mostly midshelf sites Jan–Apr (net B); (6) mostly midshelf sites Jun–Oct (net B); (7) mostly inner-shelf sites Jan–Feb and Oct (net B); (8) mostly inner-shelf sites Feb–Oct (net B).

blages are generally found (Fig. 5). These may be related to differences in vertical distribution among taxa. Waters within a few hundred meters of windward reef fronts can have very high abundances (3–100 times those of other areas) of both young and old stages of a number of shallow-living larvae (Tripterygiidae, Pomacentridae, Mullidae), indicating that these taxa may complete their pelagic stage there. Tripterygiids seem to be consistent in this, while the other families vary among reefs, perhaps due to differences among reefs in circulation.

Evidence from waters close to leeward reefs is not as clear. The lee of Lizard Island (<500 m from shore) had high numbers of many taxa, but only of young larvae, an observation also suggested by preliminary analysis of data from atoll lagoons of French Polynesia (Leis *et al.*, in prep.). Leis (in prep.) investigated larval distribution of 64 reef fish taxa in the Great Barrier Reef lagoon in the lee of an outer reef. He concluded that seven taxa of five families (apogonids, blenniids, gobiids, microdesmids, and tripterygiids) remained

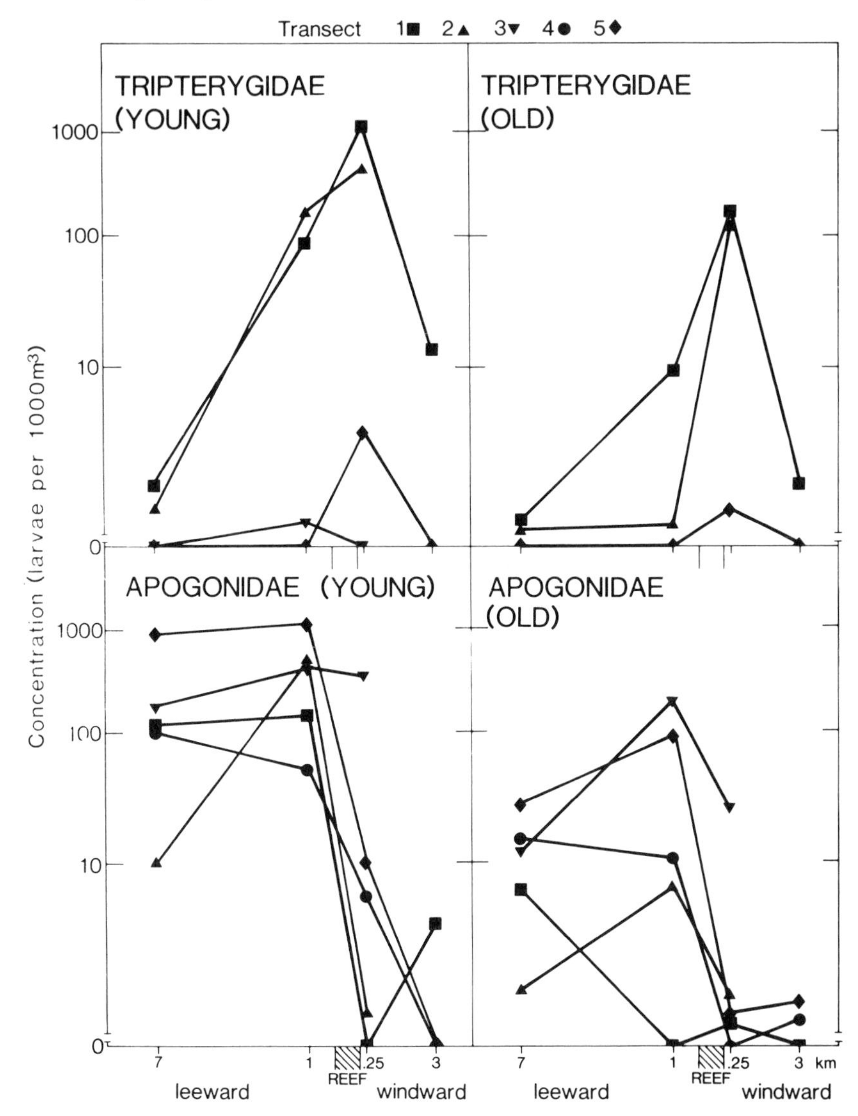

Figure 5 Windward/leeward differences in concentration of young and old stages of two types of reef fish larvae around a ribbon reef in the Great Barrier Reef. Both types of larvae hatch from nonpelagic eggs. Values are means of two replicates at each station on each of five sampling transects from October 1979 to February 1980. Data are replotted from Leis and Goldman (1984). Top: Windward distribution of triplefin blenny larvae (Tripterygiidae). No tripterygiid larvae were present on transect 4, and only a few young larvae were present on transect 3. Bottom: Leeward distribution of cardinalfish larvae (Apogonidae).

near the reef during their pelagic stage. These seven taxa were usually more abundant 2 km from the reef than 1 km from it among its large, detached patch reefs, but were 1.5–12 times more abundant at these locations than they were more than 4 km from the reef. However, most taxa either had no consistent distributional pattern or were most abundant (to 6 times the abundance in other areas) 8 km from the reef.

In only a few cases can the distributions of larvae in the vicinity of reefs be interpreted as simple dilution gradients, that is, high at the source (the reef) and decreasing with distance from the source. A few taxa—the most consistent being tripterygiids—seem to remain near reefs throughout their pelagic stage. See also Kingsford and Choat (1989) for temperate examples of such distributions. It is for these taxa that settlement back to natal reefs is most likely.

e. Open Ocean (Leis, 1986b and unpublished observations). Some larvae occurred in greatest abundance or only in the open ocean, although probably <25 km from reefs. This includes many species of carangids, labrids (Fig. 6), acanthurids, chaetodontids, anthiine serranids, eteline lutjanids, and blenniids (Salariini) and a few species of a number of other families, including epinepheline serranids, caesionine lutjanids and mullids. Some taxa seemed to move only a few hundred meters away from the reefs, others had peak abundance 2–10 km from the reefs, while others were rare near the reefs but had not peaked in abundance 20 km off.

f. Ontogenetically Changing Distributions Some taxa have distributions that change as the larvae get older (Leis and Reader, 1991; J. M. Leis, unpublished observations). Chanid larvae off Lizard Island hatch from pelagic eggs in the Coral Sea, move into the continental shelf lagoon, and across it, at least to Lizard Island, and probably to the mainland coast. Some are retained for a time in the lee of the outer ribbon reefs. Young labrid larvae lack a consistent distributional pattern, but older labrid larvae have an offshore distribution in the Coral Sea (Fig. 6). Conversely, young pomacentrid larvae are most abun-

Figure 6 Distribution of young and old stages of two types of reef fish larvae in the Great Barrier Reef Lagoon and with distance offshore into the Coral Sea off Lizard Island. Values are medians of six samples within the marked blocks (lagoon, A, B, C, E) on each of four sampling transects, November 1984 to February 1985, but only two samples were taken in block A on the second cruise. The median value is plotted in the center of each block, but samples were randomly located within each block (see Leis *et al.,* 1987, for sampling details; data are from Leis, in prep.). Top: Damselfish larvae (Pomacentridae) were distributed relatively near the reef when both young and old, but older larvae had a more restricted distribution. Pomacentrids spawn demersal eggs. Bottom: Wrasse larvae (Labridae) lacked a consistent distributional pattern when young, but were most abundant in the open ocean away from the reef when older. Labrids spawn pelagic eggs.

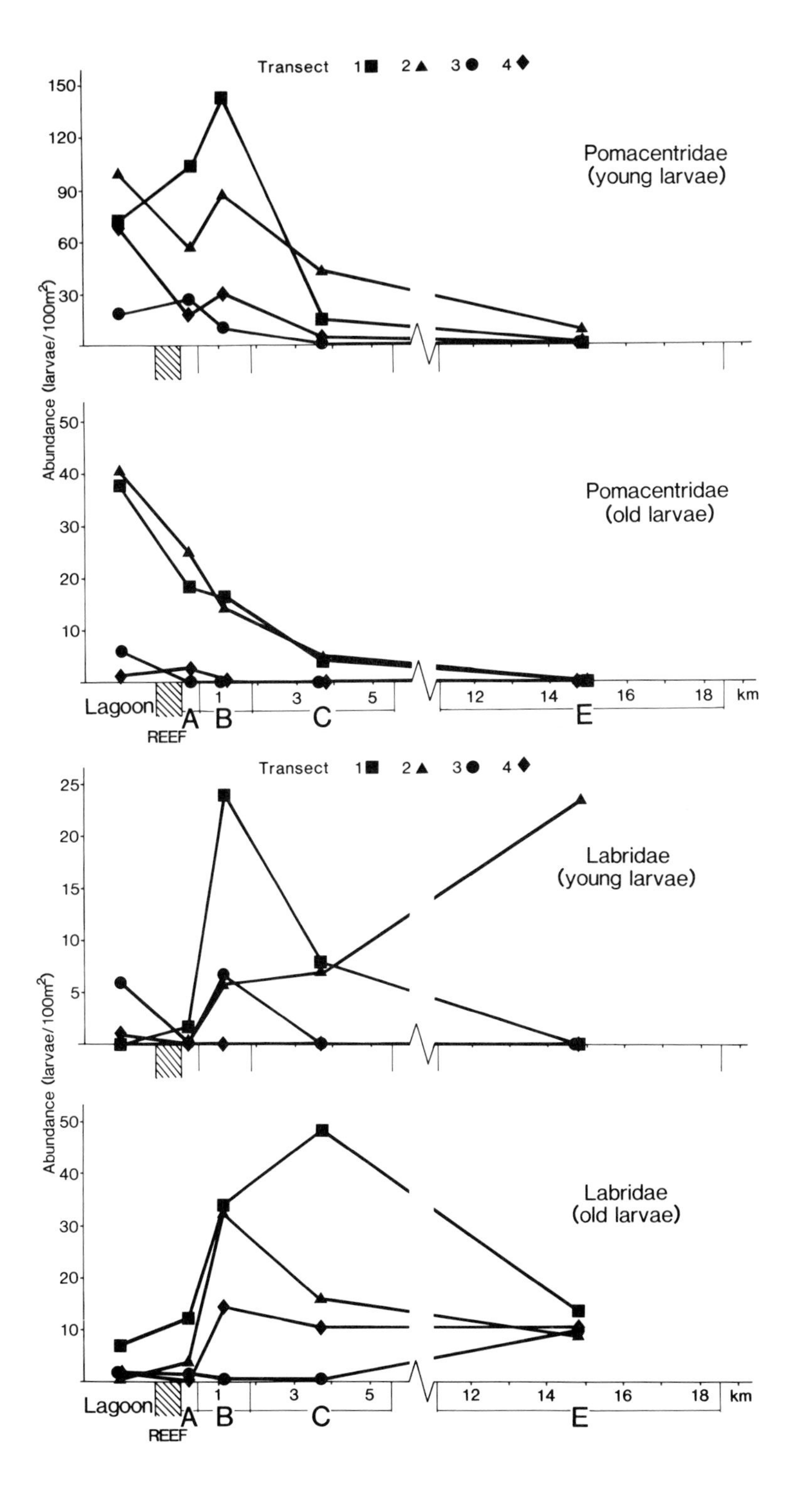

Transect 1 2 3 4
Pomacentridae (young larvae)
Pomacentridae (old larvae)
Abundance (larvae/100m²)
Lagoon
REEF
A B C E
km
Labridae (young larvae)
Labridae (old larvae)

dant within a few kilometers of the outer reefs in the ocean and lagoon while older larvae have highest abundances within the lagoon (Fig. 6). In the same lagoon, young larvae of both blenniids and microdesmids lack a consistent distributional pattern, but old larvae are most abundant 2 km to the lee of the reef (1.8–18 times that of other sites). Ontogenetically changing distributions will probably prove to be common once detailed studies of individual species are made.

g. Hydrographic, Topographic, and Floating Structures Superimposed on these medium-scale patterns are small (0.5–5 km) to very small scale (<0.5 km) distributions intimately connected with hydrographic structure and the interaction of hydrography with reef topography. Pelagic reef fishes can be found in conjunction with fronts and convergences in concentrations that are orders of magnitude higher than those in nearby unstructured water (Kingsford, *et al.*, 1991; J. H. Choat and J. M. Leis, unpublished observations; see Kingsford and Choat, 1986, for temperate examples). The physical structures (reviewed by Wolanski and Hamner, 1988; Hamner and Wolanski, 1988) do not concentrate the animals. Rather they provide the opportunity for concentration given the appropriate behavior by the animal. Most often this involves vertical position maintenance near the surface.

The taxa involved are a small, but diverse subset of reef fishes, including atherinids, hemiramphids, belonids, sphyraenids, mullids, kyphosids, dactylopterids, monacanthids, and a few species of pomacentrids, blenniids, and apogonids, usually older stages. Surface hydrographic features are very important in structuring the distributions of these fishes and their food (M. J. Kingsford, personal communication). Thus not only distribution, but also growth and survival may be affected. Subsurface hydrographic structure may be equally important in organizing the distributions of pelagic reef fishes that occur away from the surface, but this remains to be examined (see Lasker, 1975, for a temperate example).

Floating objects in the tropics attract large numbers of pelagic stages of reef fishes (e.g., Fine, 1970; Gooding and Magnuson, 1967; Hunter and Mitchell, 1967, 1968; see Kingsford and Choat, 1985, for temperate examples). The taxa involved are much the same as those that concentrate in surface hydrographic structures plus carangids. These fishes can accumulate very quickly and may come from below the surface (J. H. Choat, personal communication). Fishes accumulate around floating objects actively and most likely do so to obtain shelter from predators or to feed on other organisms attempting to do so.

These physical features are potentially very important, not only for patchiness (and sample design) but also for position maintenance (Kingsford *et al.*, 1991) and reefward return (see Section V). However, relatively few taxa are known to be associated with these features.

4. Larvae on or in the Reef

Is it possible that larvae of some reef fishes remain in the water directly over the coral reef? Surprisingly, there is relatively little relevant information.

A number of primarily anecdotal reports exist of fish larvae being found in close association with the reef itself (e.g., Emery, 1968; Hobson and Chess, 1978, 1986; Vaissiere and Seguin, 1984). Goldman *et al.* (1983) quantitatively sampled fish larvae in various coral reef habitats at Lizard Island, Great Barrier Reef, with a net mounted on an underwater "scooter." They caught more larvae over the "reef margin" than over the "reef flat," and the "off reef floor" (interreef epibenthos at 25–30 m) had higher concentrations of old—but lower concentrations of young—larvae than did the reef margin. Unfortunately, none of these investigators identified the larvae so it is difficult to evaluate the significance of these reports.

Larvae of reef-associated clupeids of the genera *Jenkinsia* (Atlantic) and *Spratelloides* (Indo-Pacific) are often found very close to the coral substrate (Powles, 1977b; Leis, 1986c). Similarly, the larvae of the sciaenid genus *Pareques* were captured sheltering among echinoid spines and under a ledge in the tropical western Atlantic (Powles and Burgess, 1978) and have not been reported elsewhere. Neustonic atherinid larvae are often found in very shallow water in reef habitats (Schmitt, 1984a; J. M. Leis, unpublished observations). It is not clear to what extent larvae of these taxa require near-reef habitat, or the proportion of the larval population that is reef associated.

In the Virgin Islands, Smith *et al.* (1987) used mostly diver-controlled methods to sample larval fishes over reefs. Their samples were dominated by *Jenkinsia* (50%), blennioids (25%), and gobioids (17%). They compared their qualitative catches with larval fishes taken well offshore and concluded the larval fish assemblages in the two areas were very different. Many of the types of larvae that dominated their over-reef samples were rare or absent offshore. Unfortunately, differences in sampling gear, times, and depths confounded their comparisons. Undoubtedly, larval assemblages very near reefs differ from those very far from reefs (Leis, 1982a; Marliave, 1986; Leis and Goldman, 1987; Kingsford and Choat, 1989), and studies in tropical waters (Section 3) have identified strong gradients and a number of assemblages between the reef and offshore locations. The important question must be, is the assemblage sampled by Smith *et al.* (1987) really an over-reef assemblage, or merely the inshore portion of a more widespread nearshore assemblage? Neither Smith *et al.* (1987) nor the authors cited in the previous two paragraphs used statistical analyses.

The only convincing demonstration that waters over coral reefs constitute an important habitat for the larvae of reef fishes is that of Kobayashi (1989). He verified statistically that older larvae of two gobiid species were 3–50 times more abundant over the reef slope of Hawaiian patch reefs than away from the reef, and that other types of larvae were more abundant away from the reef. Differences involving the gobiids disappeared on dark nights, but not on moonlit nights, and were quickly reestablished in the morning. This strongly indicates that the goby distributions were actively maintained and were vision dependent. Kobayashi (1989) found very few reef fish larvae of any sort in samples taken over the top of the reef.

Off the reef, over soft bottoms at 15–30 m near Lizard Island, Great Barrier Reef, Leis *et al.* (1989) found concentrated in the epibenthos during the day the larvae of a few families that contain reef species. These were Callionymidae, Monacanthidae, Pinguepididae (=Mugiloididae), Platycephalidae, Pseudochromidae, and Schindleriidae. Thus larvae of a few reef fish species may occupy the soft-bottom epibenthos, but the vast majority of larvae of reef fishes were found in the remainder of the water column. This does not preclude the possibility that older, more vagile stages able to avoid the sled net used by Leis *et al.* could occupy the epibenthos, but there is no evidence for this.

Much more work in the water over coral reefs is required to determine if many types of fish larvae utilize these waters, but it currently appears that relatively few do so. It is noteworthy that in the one clear demonstration that larvae of reef fishes were remaining over a coral reef (Kobayashi, 1989), the association broke down at night. Further, young larvae were pelagic in the more open water column and only the older larvae were closely associated with the reef slope. That is, the pelagic stage was not completed in the water over the reef.

D. Vertical Distribution

An understanding of larval fish vertical distribution is essential for progress in larval ecology. Unless vertical distribution and its temporal variations are known, sampling cannot be planned properly, horizontal distributions cannot be interpreted, the relations between larval fish concentration and variables ranging from temperature to food availability cannot be determined, and movement of larvae in currents cannot be understood or predicted.

Vertical distribution of reef fish larvae in coastal waters (i.e., <100 m deep) is better studied than in oceanic waters. Eight studies are available of vertical distribution in coastal waters [Hawaii (Watson, 1974; Leis, 1978a), Great Barrier Reef (Liew, 1983; Leis, 1986a, 1991), Cuba (Dekhnik *et al.*, 1966), Florida (Robison, 1985), and West Africa (Aboussouan, 1965)]. In each, there are taxon-specific patterns of vertical distribution that have little or no spatial or temporal variations, except day/night differences. Few taxa were examined in more than one study, but among studies there does seem to be a general similarity in patterns for related taxa (except in the family Blenniidae, where there are apparently tribe-specific patterns). The quality of the information varies, but vertical distribution has been studied in ten families as well as those listed in Table 3. Aboussouan (1965), Liew (1983), and Dekhnik *et al.* (1966) did not analyze data statistically. The latter paper also suffered from misidentifications of some larvae, and there was no indication as to whether the samples were taken during day or night.

Overall, there is more vertical structure during the day than at night. During the day, more types of larvae are deep, but at any chosen depth stratum (including the neuston), some types of larvae will be found to have peak concentration (Table 3). At night, things tend to be more uniform, resulting

Table 3 Patterns of Vertical Distribution of Reef Fish Larvae in the Lagoon of the Great Barrier Reef in February/March 1983 (see Fig. 7)[a]

Pattern	Day	Night
Neuston	Atherinidae (1)	
Upper	Mullidae	
	Pomacentridae (3)	
Upper–middle	Apogonidae (1)	Microdesmidae (1)
	Holocentridae (1)	
	Pomacentridae (3)	
	Tripteryiidae	
Avoid neuston	Apogonidae (5)	Apogonidae (1)
	Balistidae	Balistidae
	Blenniidae (1)	Callionymidae
	Callionymidae	Syngnathidae
	Microdesmidae (1)	
Increase with depth	Apogonidae (9)	Siganidae
	Bothidae	
	Gobiidae	
	Labridae (3)	
	Lethrinidae	
	Lutjanidae (1)	
	Nemipteridae	
	Scaridae (2)	
Deep	Acanthuridae	
	Labridae (2)	
	Pseudochromidae	
	Serranidae (1)	
	Synodontidae	
	Syngnathidae	
No pattern	Labridae (1)	43 taxa

[a] Pattern named after the depth stratum in which the highest concentration was found (from Leis, 1991). The number of species is given in parentheses. Lack of parentheses indicates analysis at family level.

from spreading of daytime patterns (Table 3 and Fig. 7). In most cases at night, no differences among vertical strata are found (Table 3). Most types of larvae avoid the neuston during the day, but at night many enter it. However, even at night, a few taxa maintain either their daytime peak in concentration or a different, nighttime peak. The "classic" pattern of vertical distribution [i.e., deep during the day and shallow at night (Haney, 1988)] is seldom found.

Watson (1974) hypothesized that most larvae of reef fishes had a taxon-specific, preferred light level that determined daytime depth. The less-structured nocturnal distributions were attributed to loss of orientation at night. This interpretation is supported by the observation that some larvae maintained their daytime vertical distributions, although somewhat shallower, on moonlit nights, but not on moonless nights (Watson, 1974). Most larvae apparently feed only during the day (see Section IV,A, Feeding), and

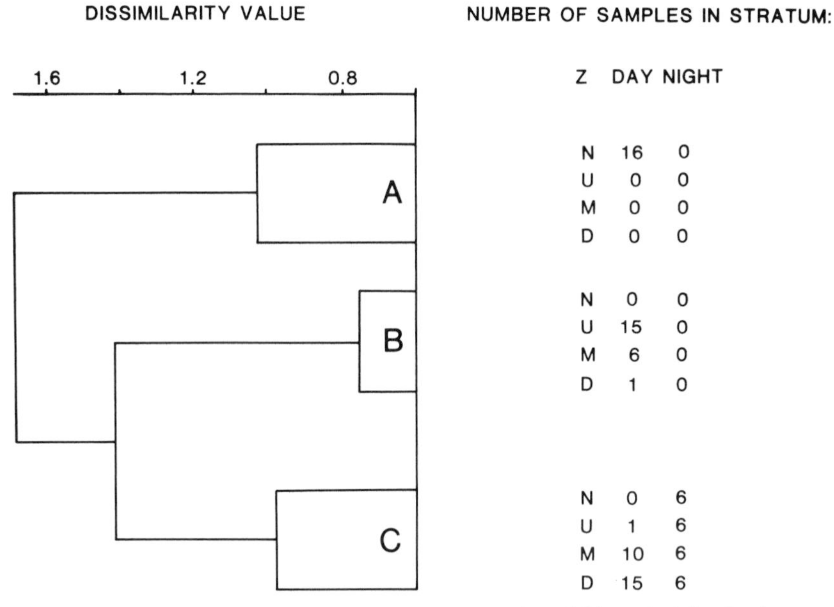

Figure 7 Dissimilarity dendrogram (Canberra-Metric) of larval fish vertical distribution samples at a single location in the Great Barrier Reef Lagoon over eight days in February and March 1983 (see Table 3). Samples were taken day and night in sets of four strata: N, neuston; U, upper (0–6 m); M, middle (6–13 m); and D, deep (13–20 m). Samples clustered into three groups during the day: A, neuston; B, upper-middle; and C, middle-deep. All night samples regardless of stratum clustered with the day middle-deep group (C). Thus, vertical distribution was highly structured during the day, but virtually unstructured at night (from Leis, 1991) as a result of nocturnal upward spreading of the more abundant deep-living larvae and downward spreading of shallow-living larvae. (Reprinted with permission of Springer.)

cessation of depth maintenance at night when feeding does not take place could conserve energy (Watson, 1974; Hunter and Sanchez, 1976). Further, many types of reef fish larvae have inflated gas bladders at night (Leis and Rennis, 1983; Leis and Trnski, 1989), which could also lessen energy expenditure (Hunter and Sanchez, 1976). If larvae decrease or cease swimming and rely on gas-bladder-induced buoyancy at night, spreading of vertical distributions would be expected.

Larvae of some fishes migrate upward in the late afternoon/early evening, move downward in the middle of the night, move up again just before sunrise, and then descend again to the daytime depth (Miller *et al.*, 1963; Parin, 1967). This has not been shown clearly yet for larvae of reef fishes—it has rarely been looked for—but there are suggestions that one apogonid (*Foa brachygramma*) may possess such a pattern (Watson, 1974).

Although very little information exists on vertical distribution of reef fish larvae in oceanic waters, they seem to be confined primarily to the upper 100 m. Burgess (1965) stated that acanthurid larvae off Florida were most abundant above 100 m. Belyanina (1975) concluded that in the Caribbean Sea and Gulf of Mexico, "larvae of coastal and reef fishes: Gobiidae, Labridae, Pomacentridae, Acanthuridae, Chaetodontidae, Scorpaenidae, Callionymidae, Serranidae and others" were taken in the upper 100 m. In the Caribbean, Hoss *et al.* (1986) described the distribution of lutjanids (most abundant above 40 m), serranids (peak abundance at 20–40 m), and scarids (peak abundance at the bottom of the isothermal layer, 40–80 m, depending on site). No diel vertical movement was reported. Off Hawaii, "island (=shorefish) larvae" were virtually all found in the upper 60 m, with peak density between 30 and 50 m during the day (Boehlert, 1986 and personal communication). At night, distributions were perhaps more widespread. Boehlert discussed the distributions of six taxa (Bothidae, Callionymidae, Gobiidae, Lutjanidae, Labridae, and Pomacentridae): distributions were taxon-specific, but all taxa had concentration peaks within the upper 100 m. Some diel movement was reported. Larvae from a Hawaiian study of vertical distribution of mesopelagic fishes "indicated clearly that most, if not all, larvae of inshore taxa occur in the upper 100 m at night in offshore waters" (T. A. Clarke, personal communication). All studies except Clarke's were based on few samples, and none utilized statistical tests.

Much more work is needed in oceanic waters. In particular, the conclusion that shorefish larvae are restricted to the upper 100 m needs to be rigorously examined. We also need to know how this is related to vertical distribution of temperature and other factors. Watson (1974) argued that the use of shallow-water distributions as a scale model of vertical distributions in oceanic water may be reasonably accurate. This hypothesis cannot be meaningfully tested at present.

IV. What Are Reef Fishes Doing while Pelagic?

What a larva is doing (or capable of doing) will depend on its state of development. A newly hatched larva may be capable of doing very little, while a settlement stage larva will have behavioral, sensory, and swimming abilities similar to those of an adult fish. At present, information on ontogeny of feeding, behavior, or sensory capabilities is limited.

A. Feeding

The available literature on feeding in marine fish larvae is based primarily on temperate clupeiform and pleuronectiform species . Most coral reef fishes are perciforms, and available comparisons (see the following) indicate that extrapolation of feeding study results on temperate clupeiform and pleuronectiform larvae to larvae of reef fishes is dubious.

Several field studies of feeding by larvae of tropical shorefishes are available (Liew, 1983; Jenkins *et al.,* 1984; Houde and Lovdal, 1984; Finucane *et al.,* 1990), but only three considered larvae of reef fishes (Randall, 1961a; Watson, 1974; Schmitt, 1986). Most studies examined the gut contents of larvae from plankton tows, identified food items into broad taxonomic categories, and assigned proportions to them. This has shown that while many larvae eat a wide variety of microzooplankton taxa, and that dietary overlap may occur, there is a fair specialization in most species, and extreme feeding specialization in some (Table 4). Diel feeding patterns were examined in nine species: five fed exclusively or almost exclusively during the day, three fed more during the day than at night, and one fed equally both day and night (Randall, 1961a; Watson, 1974; Houde and Schekter, 1978; Liew, 1983). On the basis of laboratory experimentation, Kawamura and Hara (1980) concluded that larval milkfish (Chanidae) could not take food in the dark. Two studies examined food available to the larvae and concluded that they were selective feeders (Houde and Lovdal, 1984; Schmitt, 1986). Ontogenetic shifts in diet were identified in most cases where they were looked for. There is little indication of how stereotyped feeding behavior is in fish larvae. Schmitt (1986) documented shifts in diet with shifts in food availability, and Houde and Schekter (1978, 1981) documented species differences in abilities to utilize increased food levels. Most of the field studies reported relatively high incidence of feeding by larvae.

Houde and Schekter (1978, 1981) compared laboratory feeding characteristics in larvae of one species each of tropical/subtropical clupeiform, pleuronectiform, and perciform fish. They concluded that the perciform was better able to increase growth rate as prey levels increased, and better able to utilize prey efficiently at low prey levels than were larvae of the other orders (i.e., the

Table 4 Diets of Larvae of Tropical Shorefishes

Taxon	Primary food	Reference
Acanthuridae		
Acanthurus triostegus[a]	Various zooplankters	Randall (1961a)
Apogonidae		
Foa brachygramma[a]	Tintinnids	Watson (1974)
Atherinidae		
Hypoatherina tropicalis[a]	Various zooplankters but avoided tintinnids	Schmitt (1986)
Blenniidae		
Omobranchus rotundiceps and unidentified	Copepods, nauplii, bivalve larvae	Watson (1974), Houde and Lovdal (1984)
Bothidae		
Arnoglossus, Asterorthombus, and *Engyprosopon* spp.	Larvaceans	Liew (1983)
Grammatobothus spp.	Copepods	Liew (1983)
Callionymidae		
Callionymus decoratus[a]	Copepods	Watson (1974)
C. pauciradiatus	Copepods, nauplii, bivalve larvae	Houde and Lovdal (1984)
Carangidae		
Atule mate	Copepods	Watson (1974)
Gobiidae		
Psilogobius mainland[a]	Copepods	Watson (1974)
Unidentified	Copepods, nauplii, tintinnids, mollusc larvae	Houde and Lovdal (1984)
Haemulidae		
Orthopristis chrysoptera	Copepods, nauplii	Houde and Lovdal (1984)
Paralichthyidae		
Pseudorhombus elevatus	Copepods	Liew (1983)
P. arsius	Larvaceans	Liew (1983)
P. spinosus and P. diplospinus	Chaetognaths, larvaceans	Liew (1983)
Psettodidae		
Psettodes erumei	Copepods, fish larvae	Liew (1983)
Sciaenidae		
Cynoscion nebulosus	Copepods, nauplii, tintinnids, bivalve larvae	Houde and Lovdal (1984)
Scombridae		
Scomberomorus spp., *S. semifasciatus*	Fish larvae (clupeoids)	Jenkins *et al.* (1984)
S. queenslandicus, S. commerson	Fish larvae (variety), larvaceans	Jenkins *et al.* (1984)
S. cavalla, S. maculatus	Fish larvae (variety)	Finucane *et al.* (1990)
Soleidae		
Achirus lineatus	Copepods, mollusc larvae, rotifers, dinoflagellates	Houde and Lovdal (1984)
Sparidae		
Archosargus rhomboidalis	Copepods, nauplii, bivalve larvae	Houde and Lovdal (1984)

[a] Coral reef fish.

perciform larva may be better able to survive when prey levels are low). Houde and Schekter (1981) showed that the perciform larva (the sparid *Archosargus rhomboidalis*) was able to survive and grow at food concentrations lower than those previously considered necessary for survival (based on laboratory work on temperate clupeiform and pleuronectiform larvae).

The available, limited information on feeding of the larvae of tropical shorefishes indicates strong specificity in diet and ontogenetic shifts in diet in most species. All studied taxa feed on zooplankton, either holoplankton, meroplankton, or both. Most larvae apparently feed primarily during the day. There may be some flexibility in diet, and ability to respond to changed feeding conditions varies among species. The larvae studied thus far are relatively unspecialized in their jaw morphology and dentition, except for the scombrids and psettodids, which have very large jaws and teeth and specialized diets. Larvae of many reef fishes have highly specialized jaws and teeth and would be expected to have specialized diets.

There are no data to support the idea that starvation is a major cause of mortality of reef fish larvae in tropical, supposedly food-sparse, waters (Doherty *et al.,* 1985). On the contrary, the high incidence of feeding reported by most studies (see the foregoing) indicates that the fish larvae are finding adequate food. However, starving larvae may disappear quickly from the water column as a result of increased predation or the increase in metabolic rates expected in warm, tropical waters. It is intriguing that larvae of some tropical shorefishes have specialized in feeding on other fish larvae (Table 4). In tropical waters, fish larvae are very dilute and rare compared to the other food items listed in Table 4. If encountering adequate levels of food were a major problem for fish larvae in tropical waters, specialization on such a rare commodity as other fish larvae is not obviously adaptive (unless this is an example of optimal foraging). A final indication of plentiful food for pelagic stages is the observation that widths of daily rings in otoliths are frequently greater prior to settlement than after (e.g., Fowler, 1989), implying higher growth rates in the pelagic than the benthic environment. Perhaps the levels of food required for survival of tropical fish larvae have been grossly overestimated (Houde and Schekter, 1981), but application of measurements of larval "condition" (e.g., Theilacker, 1986) will be needed to determine this.

B. Growth and Mortality

There is much taxonomic and aquacultural literature on rearing fishes in the laboratory. However, in most cases rearing conditions depart so much from field conditions that they have little, if any, relevance to the real pelagic environment (Blaxter, 1975). Hence, the majority of this literature will be ignored. The growing literature on otolith-based approaches to measuring growth and mortality is reviewed by Victor (Chapter 9). With these sources of information excluded, there is little left.

Randall (1961a) identified a negative correlation between size at settlement and water temperature for the surgeonfish *Acanthurus triostegus*. He was unable to determine if the correlation was due to temperature-related changes in growth rate or duration of the pelagic period.

Houde and Schekter (1981) reared *Archosargus* (Sparidae) and *Achirus* (Soleidae) larvae in the laboratory and found that at all food concentrations, increases in weight of 20–50% per day were attained. Increases in length were 0.1–0.6 mm per day. Both species could consume greater than 100% of their body weight in food per day.

Houde (1987) estimated from laboratory rearings that *Haemulon plumieri* (Haemulidae) egg mortality was 20% (eggs hatched in 1 day), yolk-sac larvae mortality was 20% per day (yolk was exhausted 2 days after hatching), and post-yolk-sac larvae mortality was 16% (this stage lasted 12 days). Growth was estimated to be 34% per day prior to settlement. Estimates of this sort are available for only this one reef fish, but it is unlikely they are representative because they were obtained from laboratory conditions where predators were absent.

While there is general agreement that mortality during the pelagic stage is high, the effect upon recruitment of variation in mortality at different stages is controversial. Many who study temperate fish larvae feel that most presettlement mortality takes place during the egg and yolk-sac larva periods (e.g., Dragesund and Nakken, 1971; de Ciechomski and Sanchez, 1984; Hewitt *et al.*, 1985). If this is the case, it means that larvae of most species retained in a standard 0.5-mm-mesh plankton net would have already passed through their period of greatest risk. However, Houde (1987) states that "modest variability in mortality rates...during egg and yolk-sac stages exercise relatively little control on recruitment level."

Hunter (1984) states that "present evidence indicates predation plays the dominant role in larval mortality," although this may vary depending on species and developmental stage (Theilacker, 1986). The pelagic stages of reef fishes may be consumed by a wide variety of predators, including corals (Barlow, 1981), pelagic cnidarians and ctenophores (Purcell, 1981, 1984, 1985), other larval fishes (see Table 4), and adult fishes, both reef (Hobson and Chess, 1978; Hamner *et al.*, 1988) and coastal pelagic (Colin, 1976). Hunter (1981, 1984) reviewed predation on larval (mostly temperate, pelagic) fishes and mentioned a number of other pelagic, invertebrate predators that probably prey on larvae of reef fishes as well. There are, however, no estimates of the absolute or relative importance or magnitude of these sources of predation upon pelagic reef fishes.

Pelagic stages of reef fishes are preyed upon by adult, oceanic, epipelagic fishes such as tunas and dolphinfishes. In the Pacific, 19% of the total volume of identified fishes in the guts of yellowfin tuna were pelagic reef fishes (Reintjes and King, 1953), while in the Atlantic these made up 9–83% (mean = 34%) of the volume of identified fishes in the guts of tunas, depending on tuna species and area (Dragovich, 1970). A total of 32 families of reef fishes was represented.

Johannes (1978a) argues that high predation pressure over and near coral reefs has evolutionarily forced the young stages of reef fishes off the reef and into the pelagic environment. However, the data on relative predation rates necessary to test this idea are lacking.

C. Behavior

There are very few observations, and no real studies of behavior, of the larvae of coral reef fishes. For example, how the larvae utilize their fantastic morphological specializations (Fig. 1) is unknown. Speculation about behavioral capabilities is rife (Moser, 1981), but data are scarce. The one exception is the study of vertical distribution that is, in essence, a study of larval behavior; this was reviewed earlier (but see Pearre, 1979, on the problems of interpreting vertical distribution data). The preceding distributional data imply that larval behavior is important in determining where larvae spend the pelagic period.

Colin (1982) reared larvae of the labrid *Lachnolaimus maximus* and observed that at 17 days posthatch "over one-half of the larvae formed mucous bubbles around themselves at night while floating free in the water near the surface . . . and broke free of the bubbles within seconds after the lights were turned on." They also tended to stay under floating material at the surface during the day and were colored like *Sargassum*.

Metamorphosing individuals of the gobiid *Elacatinus* were capable of resting on the coral, but spent part of their time swimming (Colin, 1975). These metamorphic fishes did not react to premetamorphic larvae that swam by, but other metamorphic individuals were chased off the coral. In general, the more metamorphosed individuals chased away the less metamorphosed fish.

Powles and Burgess (1978) found 10 *Pareques* (Sciaenidae) larvae in a small cave in the Florida Keys. These 4.4- to 6.3-mm larvae remained in a group and actively maintained their positions in a slight current. A *Pareques* larva of a different species was found among the spines of a sea urchin in Columbia. The 6.3-mm larva maintained its position despite wave surge and had an estimated swimming speed of 20 cm/sec.

Larval *Spratelloides gracilis* (Clupeidae) (9–13 mm) schooled in close proximity to a reef and were adept at avoiding a hand net (Leis, 1986c). The schools were highly coordinated and had a high degree of control over their positions. Settlement-stage *Haemulon flavolineatum* (Haemulidae)

(8.5–11.5 mm) formed mixed "socially interacting" schools with mysids (McFarland and Kotchian, 1982). This association lasted 5 days and re-formed every 14–15 days as newly settled fish arrived.

Leis and Trnski (1989) reported an unidentified tropical soleid larva (10.2 mm) swimming against a slight current at the surface. It was pigmented white and closely resembled a patch of foam.

Sweatman (1983, 1985a,b, 1988) determined that pelagic stages of several reef fishes actively selected or avoided (depending on species) settlement sites occupied by adults of two *Dascyllus* spp. (Pomacentridae). At least two species (which settle at about 10 mm) used olfactory cues to make this selection. Similar, but less definitive, observations have been made on settlement-stage *Amphiprion* (Pomacentridae) and opistognathid larvae (Thresher, 1984). This shows that settlement-stage larvae have considerable sensory, behavioral, and swimming capabilities.

Late larvae of some temperate fishes that are benthic and solitary as adults form schools. Breitburg (1987b, 1989) observed the larvae of an estuarine gobiid, *Gobiosoma bosci,* schooling in the epibenthos. Immediately prior to settlement, groups of up to hundreds of individuals schooled tightly within 0.5 m of the substrate. Late larvae of temperate blennioid and gobiesocid species formed schools in nearshore waters (Marliave, 1977). Both young and old larvae of tripterygiids and gobiesocids schooled loosely to maintain their position in the immediate subtidal over rocky reefs (Kingsford and Choat, 1989). There are as yet no reports of this sort for coral reef fishes, but aggregated settlement, which has been reported for several species, implies presettlement schooling (e.g., Doherty, 1983b; Avise and Shapiro, 1986; Sweatman, 1988).

There is no information on larvae of reef fishes, but burst swimming speeds of fish larvae in general are 10–20 body lengths/sec, and cruising speeds are 1–5 body lengths/sec (Blaxter, 1986). Perciform fishes may have higher cruising speeds than do other types of larvae (Blaxter, 1986), and there are clearly species-specific differences in swimming performance. All of these estimates are based on laboratory studies, usually on laboratory-reared animals. Cobb and Rooney (1987) advised caution in interpretation of such data after determining that wild lobster larvae in the field had swimming speeds up to 80% higher than did laboratory-reared larvae in the laboratory. In addition, anchovy larvae in large tanks swim much faster than those in small tanks (Theilacker and Dorsey, 1980). Therefore, application of laboratory-derived swimming speeds to the real world is open to question (see Blaxter, 1975).

Larvae of some reef fishes are photopositive at certain stages, and this behavior can be utilized to capture them (Doherty, 1987b). Pelagic stages of other reef fishes actively collect around floating objects (see Section III,C,3).

These behaviors may be useful for ecological sampling and to obtain larvae for experimental purposes and behavioral observations.

The small amount of information available indicates that larvae of coral reef fishes are relatively sophisticated behaviorally and not passive plankters. Differences in behavior among species are implied by these observations. At present, there is no indication how stereotyped larval behavior is. Larvae may rely on camouflage rather than transparency more than is currently recognized. Much more work—both observational and experimental—is needed on larval behavior as this holds the key to understanding the extent to and ways in which larvae participate in determining their positions.

D. Sensory Capabilities

There is no information on the sensory capabilities of reef fish larvae, and very little on tropical marine fish larvae of any sort. The following is based primarily on Blaxter's (1986) review with additions as noted. Most larvae have a pure cone retina initially, and rods may not form until metamorphosis (Kawamura *et al.,* 1984). This means night vision is likely to be poor prior to metamorphosis. In support of this, Kawamura and Hara (1980) concluded that larval milkfish (Chanidae) "could not take food in the dark," and that feeding in milkfish larvae "depends totally on vision." Free neuromasts are present at hatching, and the lateral line forms later at a species-specific size: this means there will be species-specific differences in mechanoreception. It may be necessary for the swim bladder to be filled with air before hearing is possible, and the size at which it fills with air varies among species. Olfactory pits and two types of receptor cells are present at hatching (Ishida and Kawamura, 1985), and taste buds are present between 1 and 14 days posthatch.

The only marine, tropical larval fish for which any substantial information is available on sensory capabilities is the milkfish (*Chanos chanos*) (Kawamura, 1984). By 10–15 mm, at the end of the pelagic stage, eyes are well developed, with a few rods present; numerous free neuromasts with well-developed cupulae are present on the head, and a few are present on the trunk, but there is no lateral line; the inner ear can be "considered functional enough for hearing and equilibrium maintenance"; the olfactory pit is not lamellated, but is roofed over in half the larvae; and numerous taste buds are present. Unfortunately, we have no idea what these sense organs are capable of detecting during the larval stage.

Sweatman (1988) shows in field experiments that some coral reef species use olfactory cues when settling upon the reef, and he argues that it is highly likely that other species do likewise.

V. THE END OF THE PELAGIC STAGE

A. When Are Pelagic Stages Ready to Settle?

A lower limit to the duration of the pelagic stage is set physiologically and phylogenetically by maximum rates of growth and development. This will probably be less variable than the upper limit, which is affected by a large suite of physical, physiological, behavioral, phylogenetic, and ecological factors. Therefore, the frequency distribution of durations for any taxon should be asymmetrical with a long upper tail. This may not be obvious unless sample sizes are large, and most estimates of pelagic duration are based on small sample sizes (Thresher and Brothers, 1985; Brothers and Thresher, 1985; Victor, 1986a; Wellington and Victor, 1989; Thresher *et al.*, 1989). The five cited papers give 416 species estimates of pelagic duration based on otolith rings: 8 are based on examination of 20 or more individuals, and only 4 on 30 or more individuals.

Pelagic stage duration not only differs among taxa, but may also differ geographically (Victor, 1986a) and perhaps seasonally (Randall, 1961a).

The degree of flexibility in pelagic stage duration is generally unknown. However, observations that the size at settlement (Breder, 1949; Randall, 1961a) and the number of otolith rings inside a presumed settlement mark (e.g., Victor, 1986a,c; Fowler, 1989) have wide ranges in some taxa indicate that such flexibility does exist in the field as it does in reared fish (Leis, 1987). To what extent this depends on better conditions of growth (e.g., food availability and quality), or lack of reef for settlement, or is under the control of the larva, is unknown. On the other hand, the lack of variation in size at settlement in some species does not necessarily mean that the pelagic duration is inflexible; for example, Victor (1986c) has argued that growth slows substantially once a given size is reached in the pelagic stage of at least one wrasse. Flexibility in duration of the pelagic stage apparently varies among taxa. Labrids and chaetodontids have high flexibility (Victor, 1986a; Leis, 1989; Fowler, 1989), while pomacentrids and pomacanthids apparently have more limited flexibility (Thresher and Brothers, 1985; Thresher *et al.*, 1989; Wellington and Victor, 1989), but low sample sizes for most species estimates make unclear the degree to which these families really differ.

Therefore, larvae may be competent to settle for some time before they leave the pelagic environment. This may extend the duration of the pelagic period by a factor of two or more in some species (Victor, 1986a,c; Fowler, 1989; see Kingsford and Milicich, 1987, for a temperate reef fish). The length of this competency period is generally unknown, as is the actual result of its termination. Do larvae simply become incapable of metamorphosis after a time, or do they metamorphose even if a suitable site has not been found? Or do they remain "larval" and pelagic indefinitely until they perish or settle? Do larvae become increasingly less selective about settlement sites once competent? None of these questions can be answered at present. Kingsford and Milicich (1987) note that it has not been demonstrated that pelagic stages must arrive at a reef at a fixed time to survive, but few species have been examined in the detail necessary to demonstrate this.

Taxa apparently vary not only in mean duration of the pelagic phase, but also in flexibility in extending it, and in whether they continue to grow once competent to settle. R. E. Thresher (personal communication) argues that limited flexibility is the common case and that high flexibility would be expected if returning to the reef was a major problem for pelagic stages. It seems to me that given the current, limited data base, no firm conclusion is possible. In only four families have more than a few species been examined, and numbers of individuals examined per species are usually low. The balance of the slim evidence available seems to point toward flexibility, which at least varies among species but is demonstrably high for some.

B. Movement from the Pelagic Environment to the Reef

When a pelagic reef fish is ready to settle, it may be within meters of a reef or hundreds of kilometers from one. In either case, it must get to a reef if it is to complete its life cycle. In 1970 Sale stated that "shoreward return of pelagic larvae has been generally thought of as a passive process." This view is probably still accepted by most who study adult reef fishes. But Sale went on to propose that the return to shore of acanthurid larvae may not be entirely passive, and today few larval fish biologists would maintain that it is. Most reef fishes at the end of their pelagic stage are relatively large and are capable swimmers (Buri and Kawamura, 1983); many are juveniles. Some might need nothing other than their own swimming abilities to regain the reef. Of course, this begs the question as to how the larvae determine the right direction in which to swim. Since Sale's 1970 paper, a variety of physical mechanisms have been proposed as being involved in reefward return [see the excellent paper by Bakun (1986) for more detail]. Most require some behavioral input from the larvae. It is likely that most or all of these mechanisms act in some situations or for some species. None is a general solution for return to the reef, although some may be utilized by several species. The intent here is to list them and note some of the apparent limitations of each.

Randall (1961a) first proposed that mesoscale current eddies caused by flow around reefs and islands could return larvae to the reef. This idea has been taken up by several investigators (e.g., Sale, 1970; Emery, 1972; Powles, 1977a; Lobel and Robinson, 1983, 1986), but in spite of increased sophistication in physical measurements, no one has demonstrated that these eddies actually trap larvae of reef fishes, let alone transport them from the open ocean to shore. Lobel and Robinson (1983, 1986) show that at least on the lee coast of the island of Hawaii eddies do sometimes sweep over the coastal reefs, thereby providing the potential for reefward transport. However, such eddies exist only on the downcurrent (usually downwind) sides of reefs and islands that generate them; thus they cannot be involved in larval return on more than half of any reef or island—the upcurrent side. Further, such eddies are frequently located some tens of kilometers from shore and move away from their source island at up to 10 km per day. Therefore, if they do trap reef fish larvae, the eddies seem at least as likely to remove the larvae from the island as to return them to shore. However, if an eddy containing larvae moves away from one reef, it may come into contact with another reef, thus transporting larvae between reefs.

Tidal flows are reversing in nature but in the topographically complex environment of a coral reef, tidal flows seldom result in lack of net movement (Leis, 1986a). This provides the opportunity for reefward transport. Tide-induced eddies (usually connected with topographic features such as points or banks) are smaller than the mesoscale eddies just discussed, are more shore connected, and can occur on any side of a reef. A tide-induced eddy off leeward Oahu transported pelagic fish eggs reefward, a role it could also perform for larvae (Leis, 1982a,b). Tanaka (1985) stated that larvae of the temperate sparid *Chrysophrus major* were transported to the mouth of Shijiki Bay by tidal currents and trapped there by a "circular current" (tidal eddy?) until metamorphosis when they actively migrated inshore. Tidal jets through narrow reef passes in the Great Barrier Reef (and probably other reefs) are thought to transport a variety of plankters from the open ocean into the lagoon (Leis and Goldman, 1984; Wolanski and Hamner, 1988). However, this has not been demonstrated for larval reef fishes. These jets drag with them water from below the thermocline, providing a mechanism for return of deep-living larvae through relatively shallow passes. In shallow water, M. K. El Moudni and J. P. Renon (personal communication) captured large numbers of settlement-stage reef fish larvae entering Rangiroa Atoll lagoon (Tuamotu Islands) through a hoa (shallow, reef flat spillway) on incoming tidal flow at night. Similarly, Dufour (1988) captured settlement-stage larval reef fishes on the reef crest at night on Moorea (Society Islands).

Tides interacting with reef topography may induce fronts that extend 1–2 km from the reef. Slicks form along these fronts, the orientation of which changes with the state of tide. Settlement-stage fishes may be able to detect a reef from some distance because of these slicks (Kingsford *et al.*, 1991), thus considerably increasing the detectable size of a reef.

Internal waves propagate along density discontinuities (most frequently the thermocline), and on the continental shelf they transport water shoreward in surface slicks. Several studies (Shanks, 1983, 1985; Kingsford and Choat, 1986; Shanks and Wright, 1987) have identified these surface slicks as a means for shoreward transport of neustonic animals. Internal-wave-produced slicks can transport only neustonic animals, may be disrupted in windy conditions, and are absent in unstratified water, so the times, places, and species upon which they can have a major influence may be limited. However, it may be just these limited periods and places that provide small "windows" for concentrated recruitment events.

On the lee side of an island or reef, the wind pushes surface water away from shore, and this is replaced by upwelling of somewhat deeper, more offshore water (von Arx, 1948). Miller (1979b) proposed that this small-scale upwelling could bring deeper-living offshore larvae to the surface close to shore. Unfortunately, there are no solid data regarding the spatial scales of such a system nor evidence that any larvae are so transported.

At the water surface there is a wind-driven layer a few meters thick that moves faster and in a different direction than deeper water. The upper layer's movement is increasingly downwind nearer the equator (Bakun, 1986). Off the windward sides of reefs, the wind-driven surface layer provides a means for shallow-living larvae to return to the reef (Leis *et al.,* 1987; Leis and Reader, 1991). This effect is not limited to the neuston, but the depth to which it penetrates is not clear and probably varies with wind conditions. Larvae in the upper layer would move faster than those in deeper water in any case and, without regard to direction and given a random distribution of reefs, would have a greater chance of encountering a reef than deeper-living larvae. Thus, moving into the surface layer is potentially a very general means for increasing encounters with reefs. Unfortunately, little is known about the vertical distribution of settlement-stage larvae. Enhancement of this effect may arise for neustonic larvae that aggregate in surface slicks of Langmuir cells, because the slicks have a greater downwind velocity than "rippled" surface water (Bakun, 1986).

Ekman transport refers to movement of the surface "Ekman layer" (<100 m, with most of the transport in the upper 10–30 m according to Bakun, 1986) at 90 degrees to the left or right of the wind (depending on hemisphere). Ekman transport is the net transport integrated through the entire layer over which the effect operates. Flow at each depth varies, becoming more to the left or right (depending on hemisphere) and slower with depth. So, although the entire Ekman layer has a net transport at 90 degrees to the wind, transport in any direction is possible depending on depth within the upper 100 m (e.g., Atkinson *et al.,* 1981). Transport by the entire Ekman layer either toward or away from shore has been considered important to recruitment (e.g., Parrish *et al.,* 1981), but I am unaware of any consideration of the possible effect of vertical distribution within the Ekman layer on direction and speed of transport of larvae.

Bakun (1986) mentioned two other possible physical mechanisms for reefward return. If water column stratification breaks down nearshore (e.g., due to turbulence) but persists offshore, there will be density-driven shoreward flow at the surface and bottom (and offshore transport in the middle) at least offshore of the area where stratification has broken down. Surface wave transport (Stokes transport) is net movement in the direction of wave propagation. Transport is small and results in a primarily surface movement toward shore. In shallow water (depth < wavelength), interaction with the bottom also results in a shoreward flow at the bottom and offshore flow in the middle of the water column. Bakun's (1986) discussion was theoretical, and I am unaware of any application of these two mechanisms to specific situations or animals.

Thus there are several physical mechanisms that could assist larvae in returning to the reef. None has been fully evaluated, and some may prove to be of limited or no involvement in return to the reef. It is likely, however, that there will be strong differences among species and locations as to the importance of each mechanism. These need to be related to differences in behavior among species in order to obtain an understanding of how reefward return is accomplished.

C. Transition

Many reef fishes settle while still clearly larval, while others remain in the pelagic environment for a considerable period (judging by size) after transition to a juvenile stage. Acanthurids and chaetodontids are examples of taxa that settle with larval characters still present (Leis and Rennis, 1983). Examples of fishes with pelagic juvenile stages are tetraodontiform fishes (Leis, 1978b), mullids (M. C. Caldwell, 1962), priacanthids (D. K. Caldwell, 1962), and some kyphosids (unpublished observations). Other fishes become associated as pelagic juveniles with floating objects such as algae before settling onto a reef (Kingsford and Choat, 1985). There are even suggestions that some reef fishes may reach sexual maturity prior to settlement (Pillai *et al.*, 1983; Leis and Moyer, 1985).

The degree of change at settlement varies widely, from little more than a color change to a major morphological metamorphosis. Those that are pelagic in a juvenile phase superficially appear to change little upon settlement, but close examination may reveal major changes to internal organs and chemical composition of body stores (Corbin, 1977).

Some reef fishes may settle into habitats other than coral reefs. Intermediate settlement sites such as tide pools (Breder, 1949; Randall, 1961a), estuaries

(Keener *et al.,* 1988), seagrasses (Bell *et al.,* 1987; Lindeman, 1989), algal beds (Bellwood and Choat, 1989), or mangroves (Starck, 1971; Lindeman, 1989) have been noted. The importance of such sites is controversial and may differ between the Atlantic and Indo-Pacific (Parrish, 1987a). Additional work since Parrish's review indicates that Indo-Pacific mangrove areas are not important habitats for either larval or juvenile coral reef fishes (Janekarn and Boonruang, 1986; Robertson and Duke, 1987; Little *et al.,* 1988), but other habitats have not been examined closely. Although most species studied thus far seem to recruit in or near adult habitat on the reef, the distinction between settlement and recruitment is often blurred in the literature, and actual settlement sites are known for few species.

For a few species of reef fishes, Sweatman (1983, 1985a,b, 1988) shows that diel timing of settlement is not random, that the substratum and resident fishes influence the number and types of settlers, and that olfaction plays a role in these processes. R. E. Thresher's (personal communication) work on *Amphiprion* (Pomacentridae) provides similar results. This implies considerable behavioral and sensory sophistication and swimming ability on the part of the settlement-stage fishes.

Anecdotal observations in aquaria and in the field (unpublished observations; P. J. Doherty, personal communication) indicate that at least some pomacentrids reenter the pelagic environment at night after apparent settlement onto the reef. A temperate reef-dwelling labrisomid (*Paraclinus*) may "presettle" to soft substrate during the day and reenter the water column at night (Stephens *et al.,* 1986). This behavioral flexibility could provide a means of testing reef habitat (Marliave, 1981) or avoiding daytime predation on pelagic stages near the reef.

The transition from pelagic environment to reef environment is seemingly a dangerous one. First, the prospective settler must pass through the "wall of mouths" that surrounds many reefs (Hamner *et al.,* 1988). Second, an animal well adapted both morphologially and behaviorally to the pelagic environment must quickly become well adapted to the reef environment. The animal must be extremely vulnerable during this transition. There are many anecdotal observations of the rapid color changes associated with such a transition, but the morphological changes take time (Randall, 1961a; Victor, 1983b). How most species cope with this seemingly vulnerable period between settlement and recruitment is unknown. However, various behaviors have been reported that seem to be responses to high predation pressure. These include burying in sand (Victor, 1983b), settlement into intermediate habitats, forming schools (Avise and Shapiro, 1986), hiding in holes (R. E. Thresher, personal communication), and settling into corals (Colin, 1975).

From what little is known about the end of the pelagic period of reef fishes it seems there is great variety among taxa in many, if not most aspects.

VI. Why a Pelagic Stage?

Although a few types of larvae complete their pelagic stage in the immediate vicinity of their natal reef, the large majority apparently do not, but move some distance from the reef to more open water. There has been much speculation as to why this should be so. Potential explanations have to contend with the few species that successfully complete their pelagic stage near reefs (Leis, 1981, 1982a, 1986a) and the reef-obligate elasmobranchs, all of which lack a pelagic stage. There are large taxon-specific differences in how far offshore the larvae travel, and any explanation must account for this.

Johannes (1978a) proposed that the low abundance of most types of fish larvae nearshore was due to very high predation pressure in the immediate vicinity of reefs. This could operate because mortality rates nearshore were so high that larvae were forced offshore in an evolutionary sense. Alternatively, there may be no offshore movement, but those larvae near the reef might merely be eaten, leaving only those offshore beyond reach of the reef-based predators. Unfortunately, data on relative rates of predation nearshore and offshore are lacking. For offshore movement to be advantageous, decreased loss to predation has to more than offset increased loss to expatriation. Although the latter should be higher offshore, the relative magnitude of the two losses has not been seriously evaluated.

Barlow (1981) maintained that dispersal during the pelagic stage of reef fishes was adaptive because reef fishes live in a patchy, uncertain environment. Although the idea of risk spreading probably has validity, Barlow detracts from his case by providing a number of factual errors and misinterpretations of published studies, principally in his examples. However, more importantly, Barlow's idea does not explain why so many taxa spend the pelagic stage so far offshore. In addition, the idea is essentially untestable and does not adequately deal with the problems noted in the first paragraph of this section.

Bourret *et al.* (1979) proposed that an offshore pelagic stage is an energy-saving mechanism because "drifting . . . in a rather slow ocean current demands less energy than [position] maintenance swimming in the reefs where tidal currents are always strong" (translation from the French). Drifting may demand less energy than swimming, but for this strategy to be advantageous, energy saved by drifting must outweigh losses from expatriation and any differences in food quality or quantity. The relative effects of these have not been evaluated.

Doherty *et al.* (1985) suggested on the basis of a computer simulation that in an environment of patchy food, dispersal will result in better survival of fish larvae than will remaining near a reef. There is no field evidence to support this conclusion, and unfortunately, several of the assumptions or conditions in their model are either inaccurate, inappropriate, or contradictory, thus rendering the model of very limited heuristic value. Included are those relating to near-reef hydrography (see Black and Gay, 1987; Black *et al.*, 1990); vertical migrations of larvae (see earlier discussion); a model explicitly for a continental shelf reef system, but which used concepts of larval food availability based on tropical oceans [the latter have much lower zooplankton concentrations than do more enclosed, shallow waters such as continental shelves (Gerber, 1981)]; and overestimates of the concentration of food required to sustain fish larvae (Houde and Schekter, 1980, 1981; Hunter, 1981; and Section IV,A, Feeding).

Bakun (1986) nominated pulverization of larvae in the surf zone as a good reason for larvae to avoid the waters immediately around reefs. I know of no data relevant to this suggestion, but if pulverization were a major problem for fish larvae, they should be found closer to the lee side than the windward side of reefs.

A tacit assumption of nearly all work on the biology of reef fish larvae is that the distributions, or behaviors, or patterns that are observed are adaptive (Shapiro *et al.*, 1988). In the case of distributions, for example, we assume that larvae from an area of highest abundance are at least as likely to survive and recruit as larvae from anywhere else. We then search for reasons to explain why such a distribution is adaptive. However, as pointed out by Bakun (1989), "the circumstances affecting a given sample of quite typical larvae may be quite irrelevant to eventual recruitment to a population; i.e., it may well be that only a very small subgroup of quite nontypical larvae that find themselves in very special circumstances are the only ones with any prospects at all of survival." R. E. Johannes has been expressing a similar view for some years, but not, as far as I am aware, in print. In a similar vein, Thresher (1985a) points out that reef fish larvae may settle in large numbers into habitats unsuitable for reproduction. Ecologists are subject to this problem whenever they move from observation of pattern to explanation without experimentation. Experimentation with pelagic stages of reef fishes is very difficult, and we are still in the pattern discovery phase in our investigations of them. Explanations, therefore, must be regarded cautiously.

VII. Use of Larvae in Zoogeographic Analyses

Because the pelagic stage of reef fishes is the presumed dispersal stage, it is particularly relevant to zoogeographic analyses. The use in zoogeographic studies of information from the pelagic stage of reef fishes (and invertebrates) has taken five broad approaches. These are discussed below.

a. Given Enough Time, Anything Can Happen (Heck and McCoy, 1978). This traditional view states that because most reef species have pelagic larvae, oceanic transport is possible (e.g., Kay and Palumbi, 1987). This hypothesis is usually applied on an *ad hoc* basis, is untestable, and stalls further investigation. While it is true that most species do have a pelagic stage, not all with similar distributions do (e.g., elasmobranchs). If most species do have a pelagic stage, why don't all have similar distributions? The usual retort—that the larval stages differ in their dispersal capabilities—is almost invariably applied without any real knowledge of the capabilities.

b. Larval Morphology as a Predictor (Leis, 1983, 1984). This view predicts that the more specialized the larva is for pelagic existence, the more widespread the species will be (e.g., Rosenblatt and Walker, 1963; Barlow, 1981). This approach provides testable predictions. However, "specialized" is seldom defined clearly enough for realistic testing, nor is it really possible to consistently rank taxa along a specialization scale. For example, it is not possible to say a *Coradion* is more (or less) specialized than a *Chaetodon* larva, and the latter are much more widespread (see Leis, 1989). When tested, the predictions of this hypothesis have not been upheld. Leis (1983, 1984) considered larvae that were thought to have dispersed across the East Pacific Barrier and concluded that "morphological specializations are not a reliable indicator of dispersal potential." More recent analysis of larval morphology and adult distributions in the caesionine lutjanids and the serranid *Plectropomus* (unpublished observations) supports this statement (but see Williams, Chapter 16, for an alternate view).

c. Pelagic Duration as a Predictor (Victor, Chapter 9; Thresher, Chapter 15). This approach predicts that within limits the longer the pelagic duration, the more widespread the species [Brothers and Thresher (1985) and Thresher and Brothers (1985) provide tests of this hypothesis]. As both Thresher (Chapter 15) and Victor (Chapter 9) review this approach, I will make only two points. For zoogeographic analyses, it is not the average duration that is important, but the upper extreme. This upper extreme will occur in a vanishingly small proportion of the individuals and will be very difficult to detect, particularly in view of the expected skewed distribution of duration frequencies (see Section V, The End of the Pelagic Stage). Second, this view implicitly assumes that passive drift is the overwhelming factor in larval distribution and that all larvae are alike in this regard. The available evidence (discussed earlier) is counter to both assumptions.

d. Current Patterns (Usually on very Large Scales) as Predictors (Zinsmeister and Emerson, 1979 and Leis, 1986b). This approach uses the current pattern to predict distribution by assuming that larvae drift passively with currents (e.g., Hourigan and Reese, 1987; see Williams *et al.*, 1984, for use at medium scale). The current pattern is usually drawn from a standard source (e.g.,

Sverdrup *et al.*, 1942). The assumption of passive drift is probably not valid, but the main problem is that large-scale currents are much more variable than most biologists realize or than standard sources indicate. The extreme effect of the recent El Niño when portions of the Pacific equatorial current system changed speed and direction is just one example of this variability (Brothers and Thresher, 1985; see also Power, 1984). The variability is so great over moderate time scales (tens of years) that if current were the major determinant of distributions, most species would be very much more widely distributed than they are. Further, because currents vary in velocity vertically, it is essential to choose a particular depth. But vertical distribution behavior of pelagic stages is usually unknown and may vary on a diel and ontogenetic basis, so choice of correct depth and, therefore, correct speed and direction may be impossible.

e. Ecological Requirements of Larvae as a Predictor (Leis, 1986b). Larvae have ecological requirements that vary among species, and unless these are met, the pelagic stage cannot be completed in a given area, and the species cannot persist there (Leis, 1986b). This is, in effect, an ecological deterministic view. Given the variable nature of ocean currents, larval dispersal over extremely wide areas is inevitable. That species are not even more widespread than they are requires explanation. One class of potential explanations involves larval ecological requirements. These can in some cases be inferred from distributional studies at small to medium scales, and then be used to predict the adult distribution at zoogeographic scales (Leis, 1986b). In the few cases where these predictions have been tested, they have been upheld (Leis, 1986b and unpublished observations). Williams (Chapter 16; see also Doherty and Williams, 1988a) takes a similar approach when he argues that patterns of recruitment of reef fishes on the Great Barrier Reef are due to larval availability (i.e., distribution), which is in turn determined either by larval habitat requirements (survival) or by larval behavior. More testing is needed, but this seems to be a useful approach.

A related approach is to search for the larvae of the animal in question in so-called zoogeographic barriers (e.g., Scheltema, 1988). Unfortunately, this information does not provide a clear answer to the questions involved: "presence or absence of larvae . . . alone may be insufficient because although it admits the possibility [of dispersal], such information says nothing about the actual success of larval dispersal and its relationship to species ranges" (Scheltema, 1988). Further, "although the presence of larvae in the plankton provides evidence that dispersal actually happens, their absence cannot prove that it does not occur" (Scheltema, 1988).

While some of these approaches have potential to provide testable zoogeographic hypotheses, they require considerable knowledge of the biology of the animal, and this is seldom available. However, this is not an acceptable reason for persisting in the use of invalid approaches, *ad hoc* explanations, or approaches that provide only untestable hypotheses. In the final analysis, any one approach or an approach that ignores information on vicariance explanations and adult ecological requirements is unlikely to provide realistic answers.

VIII. Conclusions

In the 1970s when reef fish ecologists began to consider the pelagic environment to be an important determinant of the processes and patterns seen on the reef, virtually nothing was known about the pelagic stage of reef fishes (Ehrlich, 1975). This stage was treated as a black box wherein answers were to be found to the complex questions emerging on the reef. Somewhat wishfully, it was thought that the "simple" pelagic environment with its larval bath of pelagic stages, most of which could be treated as equivalent, would be much easier to understand than the demonstrably complex reef. Unfortunately for this view, efforts during the last 15 years to peer into the black box of larval biology have shown the pelagic stage to be behaviorally and ecologically as complex as the adult stage.

The pelagic environment, which seemed physically structureless and uniform, has been revealed as highly structured, complex, and heterogeneous both horizontally and vertically (e.g., Bakun, 1986; Hamner and Wolanski, 1988; Wolanski and Hamner, 1988). The interactions of the pelagic environment with reefs and topography in general are complex and incompletely understood, but present understanding forces upon reef ecologists a new perspective that emphasizes the importance of topography and small-scale processes. Physical oceangraphers used to regard these things as noise to be avoided or filtered out in attempts to understand the circulation. It is now recognized that, far from being noise, this is the music to which the biological systems of the reef, especially its pelagic stages, dance.

Investigation of any aspect of pelagic stage biology reveals taxon-specific pattern, showing that the traditional idea that a larva is a larva and all are equivalent must be discarded. There is no such thing as a general larval distribution or behavior. Each taxon has its own distributions and behaviors, and based on very limited data, there seems to be geographic consistency in these (for similar conclusions regarding temperate reef fishes see Kingsford, 1988; Kingsford and Choat, 1989). Reef fish larvae seem to be active animals; they certainly are not passive, aimlessly drifting seeds.

We know most about the distribution of the pelagic stage, but know relatively little about how these distributions came about. This will require better understanding of physical oceanography and larval behavior, but present understanding does allow some valuable inferences.

The behavior of larvae seems to influence distribution patterns, and although there is little direct information on larval behavior, there are many indirect indications of its importance. From my own work on larval distribution, six factors are strongly implicated as important in determining distributions: (1) adult spawning location and timing, (2) hydrography, (3) topography, (4) vertical distribution of larvae, (5) horizontal swimming by larvae, and (6) larval behavioral capabilities and flexibility. Of these six factors, four are behavioral. It is important to keep in mind that larval input of this sort varies among species, and among sizes (=ages) within the pelagic stage.

The size and development state of the pelagic stage at the time it is put into the pelagic environment also varies among species (Leis and Rennis, 1983; Leis and Trnski, 1989). The more developed a larva, the more likely it will be able to affect its distribution. Therefore, larvae that hatch in a relatively developed state (e.g., from nonpelagic eggs) are more likely to be able to remain near their natal reefs throughout the pelagic stage than are larvae that hatch poorly developed (e.g., from pelagic eggs). Taxa with the former type of larvae are more likely to be self-recruiting. The taxa with larval distributions most indicative of retention near reefs are from nonpelagic eggs. However, there is no simple relationship between spawning type and larval distribution: many larvae from demersal eggs are found far from reefs (e.g., *Chromis,* salariin blenniids, certain apogonid species), while many larvae from pelagic eggs are found in highest abundance in lagoons or in continental shelf seas (e.g., callionymids, haemulids, epinepheline serranids).

Studies of larval distribution have identified two classes of taxa that probably are self-recruiting: those that complete their pelagic phase within lagoons and those that remain near reefs throughout their pelagic stage. Relatively few taxa are involved. Other taxa may be self-recruiting, but for the majority of taxa investigated, distributions give no indication that they are. However, many of these taxa reach relatively large size before settlement and could actively swim to a reef from some distance away. Available evidence could not identify as self-recruiting larvae that moved more than a few kilometers from the natal reef and then returned. For these, approaches other than the type of distributional studies currently available must be developed.

Larval distributions apparently can determine adult distributions over tens of kilometers (Williams, Chapter 16; Doherty and Williams, 1988a) and over zoogeographic scales (Leis, 1986b). At smaller scales, either larval habitat requirements or larval behavior could determine larval distributions, but at larger scales, habitat requirements are more likely to be decisive. This shows clearly the need for more work on both behavior and habitat requirements of the pelagic stage of reef fishes.

Most of the research on the pelagic stage of reef fishes has been done in the central and western Pacific. Thresher (Chapter 15) argues that there are fundamental biological differences between the reef fishes of different tropical regions, including differences in the pelagic stage. Comparative studies are needed, but it could be misleading to extrapolate findings among regions.

Most of the distributional work on pelagic stages of reef fishes has been carried out either in Hawaii or in the northern Great Barrier Reef near Lizard Island. The former is an archipelagic system well out on the Pacific Plate with relatively poorly developed reefs and a relatively depauperate fish fauna. The latter is a continental shelf system close to the Indo-Pacific center of diversity, with reefs and fish fauna to match. The distributional patterns in the two areas are very similar so far as they can be compared at present. The principal difference is a simplification of patterns in relatively depauperate Hawaii. This is particularly impressive considering the differences between the areas.

Future research on the pelagic stage of reef fishes needs to address the following questions, which have not been given any serious attention.

1. Where and when are larvae put into the pelagic system? What are the temporal and spatial variations in these?
2. What is the fine-scale circulation, particularly its vertical structure?
3. What is the vertical distribution of larvae in the open ocean and how is this related to physical structure such as thermoclines?
4. What degree of generality can be found? Is topography such an overriding factor that each situation will be unique? Are there regional differences (e.g., Atlantic versus Pacific)?
5. What are the swimming and sensory capabilities of reef fish larvae? Real data are needed, not extrapolations from temperate taxa such as salmonids, clupeoids, and pleuronectiforms.
6. What are the capabilities, flexibilities, and day/night differences in behavior of reef fish larvae?
7. What are the physical and biological requirements of reef fish larvae, and how do these differ among taxa?
8. Are distributions of larvae adaptive?

This does not mean that other areas of research on the pelagic stage of reef fishes should be neglected. Our store of basic knowledge in nearly all areas is woefully inadequate. Richards (1982) and Boehlert (1986) make proposals for integrated, multidisciplinary studies of reef fish larvae. These are expensive proposals and are of the sort required eventually if understanding of the pelagic stage is to be reached. However, it seems preferable to first examine the questions posed here before proceeding with such studies, because we will then have a much better basis of knowledge from which to proceed, and more efficient studies can be designed.

Acknowledgments

This review could not have been attempted without the generous cooperation of many people who provided access to their work—published, semipublished, and unpublished. M. J. Kingsford, P. F. Sale, and R. E. Thresher critically reviewed the manuscript, offering many constructive suggestions. T. Trnski prepared the figures, and S. E. Reader and S. Bullock provided editorial assistance. My unpublished work cited here was supported by MST Grants 80/2016 and 83/1357, ARC Grant 86/0873, and the Australian Museum. Many thanks to all, and to Peter Sale for challenging me to write this review.

CHAPTER 9

Settlement Strategies and Biogeography of Reef Fishes

Benjamin C. Victor
Department of Biological Sciences
University of California at Santa Barbara
Santa Barbara, California

I. Introduction

Virtually all coral reef fishes undergo a profound transition from life as a larva adrift in the oceanic plankton to a settled existence closely associated with the coral reef structure. The importance of this transition to population dynamics and a variety of other aspects of the ecology of reef fishes is reviewed in other chapters. Despite the widely acknowledged significance of the process of settlement, until very recently there has been a notable dearth of basic information on size, age, and behavior at settlement. This can be explained, perhaps, by the fact that the transition is swift, often occurring overnight, and typically beyond even the most enterprising ecologist's eye. With the development of new techniques, however, a wave of interest has developed in documenting many of the details of the settlement transition.

In this chapter I shall review three seemingly disparate subjects: the ecology of settlement, the biogeography of reef fishes, and the use of daily increments on the otolith for aging. In fact, these subjects are closely linked. Since the larval period is, no doubt, the dispersal phase for reef fishes and coral reefs are some of the most patchy and isolated habitats on earth, it is only logical to assume that the age at settlement (i.e., the duration of the planktonic larval stage) is a major determinant in the geographic distribution of reef fishes. Whether this truism reflects reality is debatable, for it appears that, with the limited information available, little of the complex biogeography of reef fishes can be explained by variation in larval duration.

Finally, the otolith aging technique is inextricably linked with the study of settlement. The technique was first introduced and validated as a welcome tool for studying newly settled recruits. Although the applications of the method extend to all phases of the life history of reef fishes, it is uniquely adapted to the study of very young fishes. Daily otolith increments are particularly wide and clear on the otoliths of larvae and juveniles and, coincidentally, these are the stages of reef fishes that are often the most elusive and difficult to study in any other way.

II. Settlement Strategies of Reef Fishes

A. Introduction

The majority of reef fishes undergo a distinct metamorphosis around the time of settlement, developing opacity and color, acquiring scales, and exhibiting a change in behavior. Some wrasse larvae in the Caribbean bury themselves in the sand for several days before emerging transformed as a juvenile (Victor, 1983b), while damselfish transform in hiding overnight (Robertson, *et al.*, 1988). Some unusual larval forms, such as the large, transparent, leaf-shaped leptocephalus larvae of the eels, bonefish, and tarpon, undergo a profound metamorphosis that takes many days.

A major transformation at settlement is certainly not the rule among reef fishes. In some groups, such as the squirrelfishes (Holocentridae) and angelfishes (Pomacanthidae), competent larvae are well developed and resemble slightly silvery versions of settled juveniles and thus do not change much after settlement. Others, such as the grunts (Haemulidae), drums (Sciaenidae), mojarras (Gerreidae), and sweepers (Pempheridae), remain small and transparent for some time after settlement. A few reef species confound simple definitions entirely and on occasion drift in the ocean for an indefinite period (usually associated with floating debris or algae), developing all the characteristics of juveniles and sometimes attaining large sizes before being delivered inshore. Among the Caribbean fishes following this strategy are the tripletail (*Lobotes surinamensis*), great barracuda (*Sphyraena barracuda*), grey triggerfish (*Balistes capriscus*), and the sergeant major damselfish (*Abudefduf saxatilis*).

The diversity of modes of settlement has resulted in some confusion in the naming of various stages. The traditional taxonomic definition of larvae as the stage up to completion of fin development is ecologically meaningless, since virtually every late-stage reef fish larva in the oceanic plankton would then have to be considered a pelagic juvenile. Furthermore, some relatively undeveloped, yet settled, fishes would have to be considered "benthic larvae." In fact, the reporting of newly settled recruits of the reef drum (*Pareques* sp.) as "larvae" by Powles and Burgess (1978) led Kingsford and Choat (1989) to believe that some reef fishes complete their larval development while resident in the waters over reefs without ever entering the offreef plankton. McFarland *et al.* (1985) also called fish in the first stage after settlement "postlarvae" to distinguish them from the larger "pre-juveniles."

Clearly, the most useful ecological definition of larvae is that they are presettlement fishes that live in the plankton and show some morphological adaptations to that habitat, such as transparency, large melanophores, or a silvery overlay. After settlement, fish are associated to some degree with the reef substratum and should then be considered juveniles.

I use the term settlement strategy to describe the suite of adaptations to the settlement transition, such as the size, age, and state of development of settling larvae, the temporal patterns of settlement, and the behavior of larvae at settlement. I believe these sets of attributes qualify as strategies because certain combinations seem to co-occur and are often characteristic of groups of reef fish species. I certainly do not mean to suggest that reef fishes follow a few basic strategies in settlement. In fact, as the following passages should demonstrate, the diversity of settlement strategies and the number of exceptions to any trend are remarkable. In this aspect of their life history, coral reef fishes fully live up to their deserved reputation (at least among vertebrates) for encompassing as wide a variety of specializations as it seems possible to evolve in any one habitat.

B. Techniques for Assessment

Most projects to date have attempted to approach the moment of settlement by focusing on the smallest juveniles that can be found settled on the reef. This simple method can be effective as long as the study area is clearly delineated and immigration of alien juveniles is uncommon. Obviously, the more frequent the surveys are, the more information can be obtained. The most useful settlement studies have monitored the appearance of new recruits on a daily basis (D. McB. Williams, 1980, 1983a; Victor, 1982, 1983a,b, 1984, 1986b; McFarland *et al.,* 1985; Ochi, 1985; Robertson *et al.,* 1988; Wellington and Victor, 1989). Robertson *et al.* (1988) managed to census new recruits of a variety of reef species at dawn and dusk. Both natural reef areas and artificial patch reefs have been used for surveying settlement. The advantage of using artificial substrates is that new recruits can be detected immediately (because hiding places can be eliminated) and losses to predation can be reduced.

Monitoring juveniles on the reef can only elucidate the process of larval settlement from one perspective. An alternative is to approach settlement by surveying the largest planktonic larvae one can find. Typical plankton surveys usually capture incomplete series of larvae with few mature individuals. Ideally, we should be able to follow competent larvae (defined as those ready to settle) as they move inshore and decide where and how to settle. One approach has been to capture larvae at a light at night over the reef, either by hand (Victor, 1986b; Smith *et al.*, 1987) or by trap (Doherty, 1987b). Another method is to tow plankton nets in waters directly over the reef (Victor, 1986b; Leis and Goldman, 1987; Smith *et al.*, 1987; Kingsford and Choat, 1989). It is also possible, although thus far only anecdotal, to actually follow an individual larva in the water as it settles to the reef (D. McB. Williams, 1980).

The capture of fish larvae at a light at night (nightlighting) is a peculiar collecting technique that is very effective at capturing larvae over the reef just prior to settlement (Fig. 1). Competent fish larvae are strongly attracted to lights at night, a characteristic they share with moths and squid, although no persuasive explanation exists for this behavior. I have made nightly collections of Caribbean fish larvae at a nightlight over several long periods between 1981 and 1983 in the San Blas Islands of Panama. I collected larvae with a dip net under a light over reef waters only about 1 m deep (Victor, 1986b). Smith *et al.* (1987) towed a small fine-meshed plankton net through the beam of a light mounted on a deep underwater structure at St. Croix in the Caribbean. Doherty (1987b) reported the results of captures from a set of light traps placed overnight in shallow waters near Lizard Island on the Great Barrier Reef of Australia. I have found that the capture of fish larvae at a nightlight is significantly correlated with the density of larvae in the plankton determined from plankton tows made in the same area on the same night (Victor, 1986b).

These three methods produced very different sets of data. Doherty (1987b) and I both found that larvae attracted to the nightlight are almost always competent, and in some species I captured they had already started developing juvenile markings. Smith *et al.* (1987) report that they captured early- and intermediate-stage larvae as well, but they do not quantify or identify those larvae and thus it is not clear whether they were mostly newly hatched larvae being released from the reef.

Our results clearly demonstrate that the predominance of competent larvae in nightlight collections is not a product of size selectivity of the technique. In my collections from the Caribbean, for example, the mean size of the 27 species of goby larvae appearing at the nightlight ranged widely from 5.1 to 21.1 mm SL, but the range of size within a type was remarkably narrow (Fig. 2). The larvae of a single type tended to be fully mature, regardless of size, with complete fin development (e.g., Fig. 3). The samples often included individuals with metamorphic melanophore patterns (the tiny surface melanophores that develop on settled juveniles and are distinct from the few and large melanophores characteristic of transparent planktonic larvae). Furthermore, the larvae caught at the nightlight appeared to be similar in size to the smallest individuals seen on the reef. In fact, for the wrasses and parrotfishes (for which I have recorded the size of new recruits on the reef), the size of larvae at the nightlight closely matched the size of new recruits (Fig. 4). Whether this

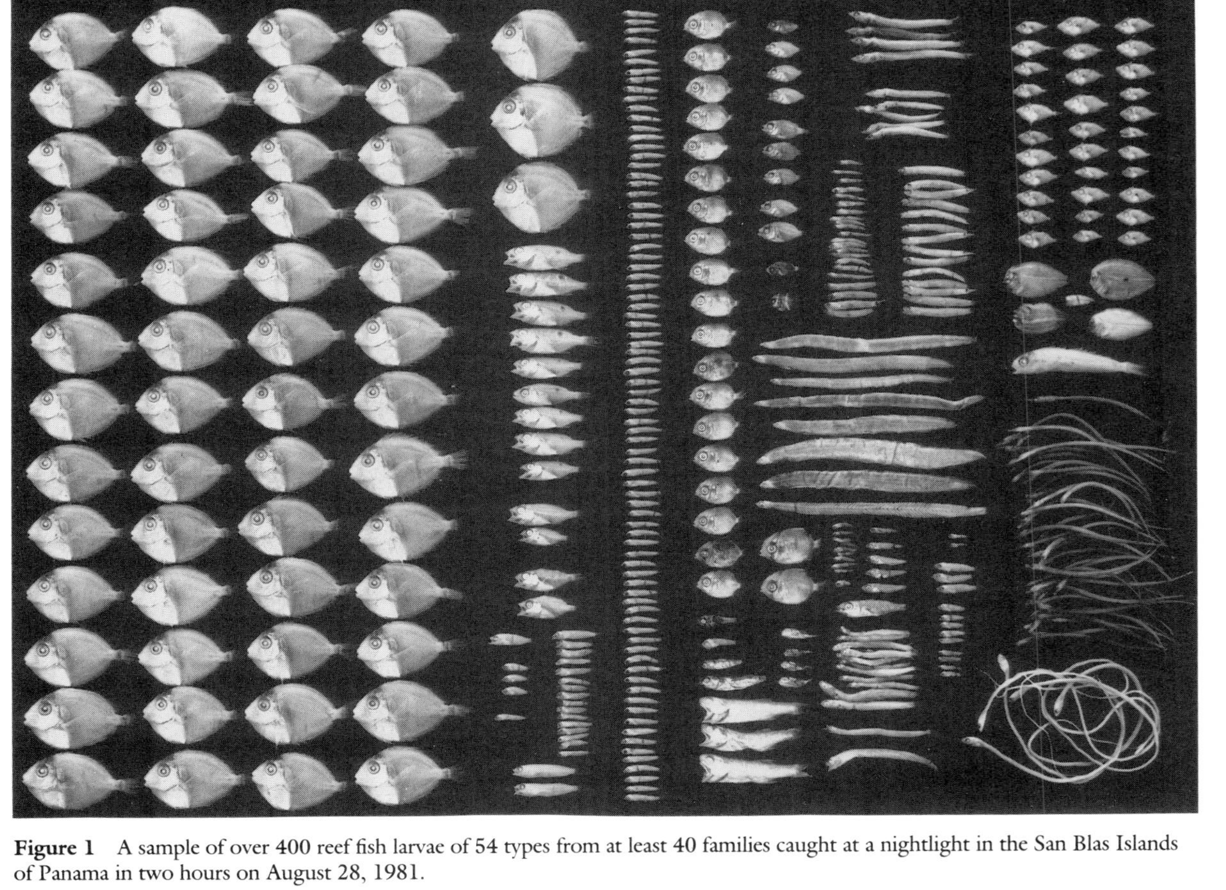

Figure 1 A sample of over 400 reef fish larvae of 54 types from at least 40 families caught at a nightlight in the San Blas Islands of Panama in two hours on August 28, 1981.

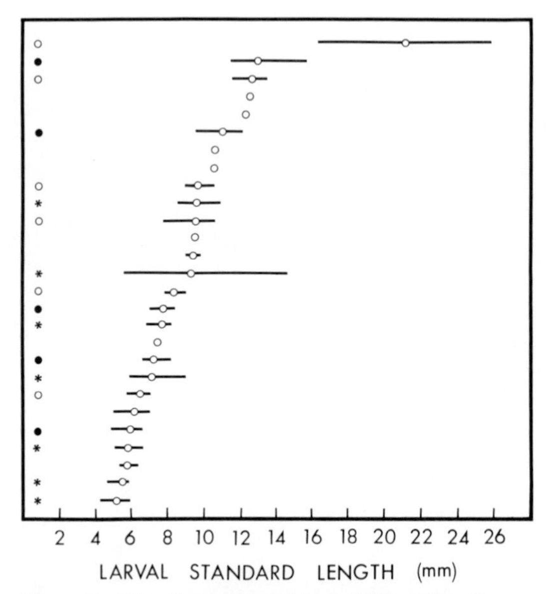

Figure 2 The distribution of sizes of 27 species of unidentified goby larvae caught at the nightlight in Panama. Circles and bars represent mean and range. The symbols at left indicate sample size: no symbol, 0–10; open circle, 10–50; solid circle, 50–100; asterisk, over 100.

reflects the fact that only mature larvae are present in inshore waters over reefs or whether it is only mature larvae that are attracted to light is not known.

The species composition of the three surveys was very different. Doherty's traps captured mostly damselfish larvae, which were uncommon in both my and Smith's surveys, despite their apparent abundance as juveniles on Caribbean reefs. My collections, like Smith's, comprised mostly clupeid, gobiid, and blennioid larvae. I also captured large numbers of labrid and scarid larvae, which Smith found to be rare in his nightlight collections. At present, it is

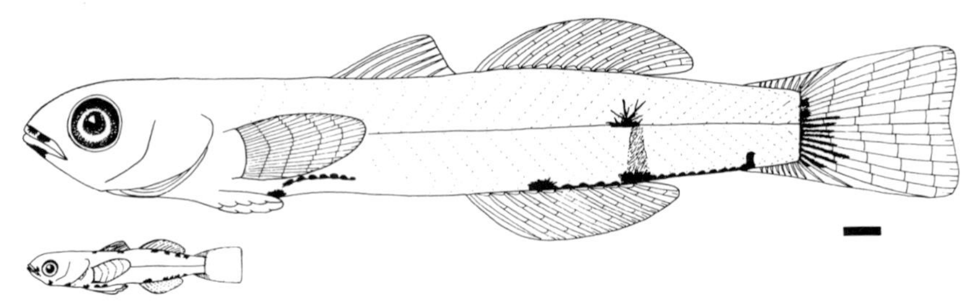

Figure 3 The largest and smallest type of goby larvae caught at the nightlight, drawn to scale. The bar represents 1 mm.

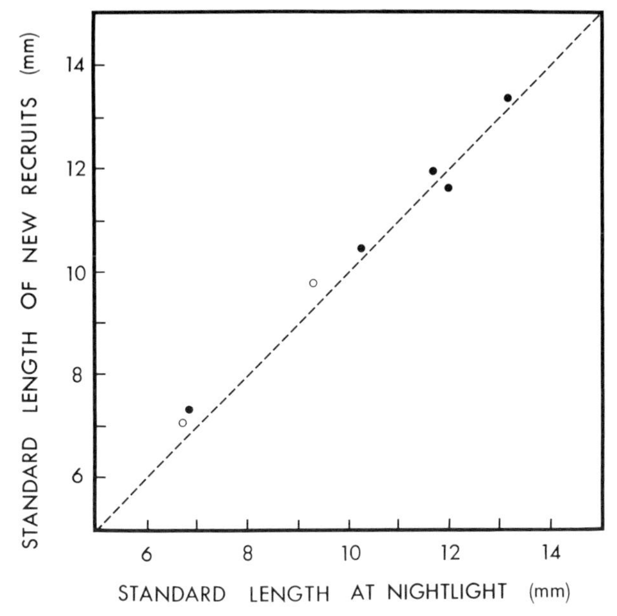

Figure 4 The relationship between the size of new recruits captured on the reef and the size of larvae of the same species caught at a nightlight. Open circles represent means for the parrotfishes *Scarus* spp. and *Sparisoma* spp., while the solid circles represent the means for the wrasses *Doratonotus megalepis, Halichoeres bivittatus, H. poeyi, Thalassoma bifasciatum,* and *Xyrichtys* spp.

impossible to discern whether these differences are a product of varying geographic areas, habitats, depths, times, lighting, nets, or even something else, for it would be difficult to devise three more different nightlighting techniques for reef fishes.

Simple survey techniques have provided data on the size at settlement, the temporal and spatial patterns of settlement, habitat selection, and mortality rates of new recruits. In addition, the use of daily increments on the otoliths of young fishes to determine age has greatly expanded our ability to extract information. This method permits the calculation of planktonic larval duration, age at settlement, and growth rates both in the plankton and on the reef. The application of this technique to settlement studies will be reviewed in the last part of this chapter.

C. Timing of Settlement

Most settlement of reef fish larvae probably occurs at night. Robertson *et al.* (1988) surveyed isolated artificial reefs at dawn and dusk, removing all fish found at each visit. The results indicated that 93% of the new recruits captured (of at least 18 different species) were found at dawn. Considering that there may have been movement of newly settled larvae from the surrounding seagrass beds after dawn, it is likely that virtually all settlement occurred during the night. When visual censuses were done on natural reefs with many inaccessible hiding places, new recruits of the bicolor damselfish, *Stegastes partitus,* tended to be observed more often at dusk. This difference was most probably a result of newly settled larvae avoiding detection by hiding for a number of hours after settlement (Robertson *et al.,* 1988).

Surveys of planktonic larvae are usually undertaken in offshore waters, and, as a result, little is known about the distributions of reef fish larvae in waters over the reef. Night plankton tows I have made over the reef typically yield many more reef fish larvae than daytime tows. One possible explanation is that larvae may be vertically migrating up at night and remaining deep, and thus off the reef, when it is light. Leis (1986a) found that most types of larvae he sampled in nearshore waters bordering the reef preferred deep water during the day and moved upward at night. This fits with the observation that planktonic fish larvae are unusual sights while working on the reef during the day. The adaptive significance of settlement while it is dark is obvious; visual predation would be intense on larvae attempting to settle during the day. For perhaps a similar reason, larvae appeared at the nightlight in much lower numbers on nights when the moon was up. These findings correspond well to observed settlement patterns; many independent surveys report lower settlement rates during the week around the full moon, when the moon is not only brighter but up for most of the night (Victor, 1983a, 1984, 1986b; McFarland *et al.,* 1985; Ochi, 1985; Robertson *et al.,* 1988) [apparently there is no such clear pattern at One Tree Reef in Australia (Williams, 1983a; Sale, 1985)].

The details of the temporal pattern of larval settlement are well documented for some reef fishes and have been reviewed in other chapters. In general, there appears to be a broad spectrum among coral reef fishes in the pattern of occurrences of larvae. While many species of reef fish larvae appear somewhat regularly, in random or periodic short cycles, a few species tend to settle in large numbers on rare occasions. Robertson (1988b) recorded an unusual mass settlement of queen triggerfish in Panama. This species settled in very low numbers over many years of observations and then, over several days in 1985, settled in densities 50 times greater than the totals for any other year. Several types of larvae showed up at the nightlight with a similar pattern. The large species of goby larva illustrated in Fig. 3, for example, appeared on only two occasions over the years of sampling, both in 1981. In one of the episodes, 23 individuals were caught in one hour, comprising about half of the reef fish larvae caught that night.

D. Size at Settlement

Coral reef fish larvae settle over a very broad range of sizes, although there tends to be some similarity in size at settlement within a family. The estimated size at settlement for 45 species of damselfishes, for example, ranged from about 6 to 14 mm SL (Wellington and Victor, 1989; additional data in Thresher and Brothers, 1989). Other families, however, exhibit a great variety of sizes at settlement (e.g., Leis, 1989). Based on these studies and on nightlight data, it appears that the majority of reef species settle between 7 and 12 mm SL.

The smallest size at settlement that I have recorded among the Caribbean reef fishes belongs to the reef cubbyu drum *Pareques acuminatus;* newly settled recruits measure from 3.8 to 4.5 mm SL, with a mean of 4.12 mm (n = 9) (Fig. 5). On the other extreme, I have collected mature larvae of the pearlfish (*Carapus* sp.), a species specialized for living within the body of sea cucumbers, that are on average 143.2 mm SL and reach 174 mm SL. Robertson (1988b) reports that the queen triggerfish settles at a length of about 60 mm SL. A few of the chaetodontid species in the Pacific are reported to appear on the reef at 50–60 mm SL, so these may also be settling very large (Leis, 1989). Some Caribbean fishes that show up at the nightlight at large sizes (30 mm SL or more) are the lizardfishes, squirrelfishes, surgeonfishes, goatfishes, and the trumpetfish. Members of these families in the Pacific must settle at large sizes as well, given the upper size limit of planktonic larvae of these families reported captured in plankton tows from that region (Leis and Rennis, 1983).

There is a set of species that settle very small and retain transparency for some time while associated with the reef. In the Caribbean, the reef drums, as well as the sweepers (Pempheridae), mojarras (Gerreidae), and the grunts (Haemulidae), appear to follow this strategy. The french grunt, *Haemulon flavolineatum,* has been studied in detail by McFarland *et al.* (1985), who found that they settle at about 6.9 mm and then pass through some clearly defined behavioral stages from a relatively undeveloped "postlarva" that roams around in the water column over sandy areas to a juvenile stage that migrates regularly to and from patch reefs at dawn and dusk.

No apparent order seems to be imposed on this variety of settlement sizes, making it difficult to explain in adaptive terms how they have arisen. The only pattern I have discerned is that those species that settle small and transparent tend to be found in back reef habitats such as shallow rubble, sand, or seagrass. Reef drum recruits are usually found in very shallow areas well behind the reef, while grunts settle in the same kind of back reef habitat (Shulman, 1985a). Mojarra recruits also frequent sandy seagrass areas inshore. An exception, however, is the other family of reef fishes that settle relatively undeveloped, the sweepers, which are found on the fore reef. The two types of

Figure 5 A new recruit of the reef drum (*Pareques acumineatus*) captured on the reef in the San Blas Islands of Panama.

larvae caught at the nightlight at sizes almost as small as the reef drum, the emerald clingfish, *Acyrtops beryllina,* and the sole, *Achirus lineatus,* are limited to back reef seagrass and muddy substrates, respectively. Furthermore, the smallest settling member of some large groups of reef fishes is often the one found mostly in shallow back reef habitats, such as the beaugregory damselfish, *Stegastes leucostictus,* and the dwarf wrasse, *Doratonotus megalepis.*

Why these types of fishes tend to settle only in shallow back reef areas is not clear. Shulman (1985a) suggests that predation pressure is lower in these back reef habitats than on the coral reef itself. This would lead more vulnerable small larvae to settle into those habitats. Even if there is lower predation in back reef areas, settling small and undeveloped still has severe costs in mortality. Shulman and Ogden (1987) report that less than one in ten grunt recruits survives a month on the reef and only about one in a thousand survives for a year. In contrast, other reef fishes that settle larger and more developed have two to three orders of magnitude higher survivorship to one year (Doherty, 1982, 1983a; Victor, 1986b; Eckert, 1987; Sale and Ferrell, 1988; Hunte and Cote, 1989). Interestingly, Robertson (1988b) found that queen triggerfish juvenile mortality was extremely high, despite the fact that it is one of the largest settling species. It was noted, however, that the mass settlement probably induced much higher mortality than is typical for the species.

In contrast to the effects on mortality of settling small, there appears to be no cost in growth rate. The growth rates after settlement for drums and grunts in the Caribbean are somewhat higher than for wrasses, which settle at larger sizes (Fig. 6). Even the redlip blenny, one of the largest settling reef fishes (at about 40 mm SL), grows at a similar rate (about 10 mm per month) after settlement (Hunte and Cote, 1989).

Presumably, there should be some major advantage to settling small that outweighs the massive losses to early mortality on the reef. It is likely that species that settle small and undeveloped are also settling after only a short time in the plankton (see the next section). They may be adapted to remaining in inshore waters and thus avoid the losses experienced by fishes that spend a longer time as larvae and are advected far offshore into the oceanic plankton. Theoretically, these strategies differ more in style than substance, since the

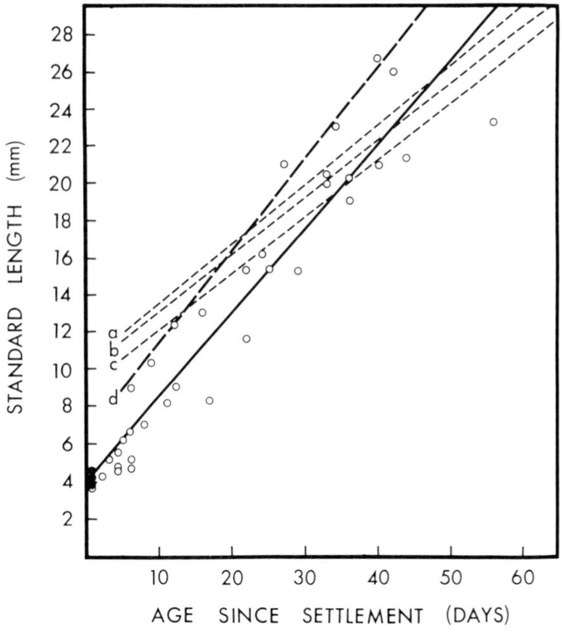

Figure 6 The early growth of reef drum juveniles (circles and solid line) compared to that of some other reef fishes: a, b, and c denote the wrasses *Halichoeres poeyi, Thalassoma bifasciatum,* and *H. bivittatus,* respectively (Victor, 1991), and d denotes the french grunt (Brothers and McFarland, 1981).

high juvenile mortality rates of early-settling species may be analogous to the mortality rate of planktonic larvae of other species. After all, they are of the same age. One could easily accept that the age-specific mortality schedule could be the same, with some species merely moving onto the reef earlier than others. If this were true, looking for a major advantage to either strategy would be missing the point.

In any case, it does appear that the larvae of most species that settle small and undeveloped are not found in offshore waters. Richards (1984) surveyed the larvae captured in oceanic plankton tows in the Caribbean and, although large numbers of other reef fishes such as wrasses and parrotfishes were caught, drums occurred in only one inshore collection and no grunt or mojarra larvae were captured. Interestingly, the dwarf wrasse (the smallest-settling wrasse, which settles in inshore shallow seagrass) was apparently totally absent from the Richards collection, even though hundreds of other wrasse larvae were caught. Apparently, those larvae destined to settle out small (and presumably after a short larval life) are not advected far offshore.

There may also be a fundamental connection between inshore larval distribution and back reef settlement habitat independent of the scenario involving the small size of larvae that I have developed here (Fig. 7). It could be that back reef habitats occur more continuously along coastal regions and thus back-reef-associated species need not be advected away from the coast, while fore reef species are more likely to encounter patchy reef areas if they move far offshore as larvae and then drift in. This hypothesis would be supported if large-settling back reef species were also found only in inshore waters. As this illustration demonstrates, the answers to these questions await much more thorough plankton and settlement studies.

Within species, the size at settlement does not appear to vary much. In many large series of a single species that I caught over several years of nightlight sampling, the entire range of lengths encompassed only 1 or 2 mm (e.g., Fig. 2). With some exceptions, this pattern is consistent enough to permit one to recognize larval types within a family simply by size. In a few other types, a small fraction of individuals were not fully developed and were significantly smaller than the average.

The coefficient of variation (cv) for larval standard length (SL) within a type was generally low, between 0.03 and 0.08 in my collections. Robertson *et al.* (1988) report a cv of 0.062 for the length of new recruits of the bicolor damselfish that is just about identical to the cv for length at settlement I found for three species of wrasses: 0.063, 0.061, and 0.055 (Victor, 1991). In fact, the cv's for length at settlement of a wide variety of species are surprisingly similar. Wellington and Victor (1989) report values between about 0.04 and 0.07 for ten damselfish species, and even the winter flounder from Newfoundland was found to have a cv of 0.051 for size at metamorphosis (Chambers and

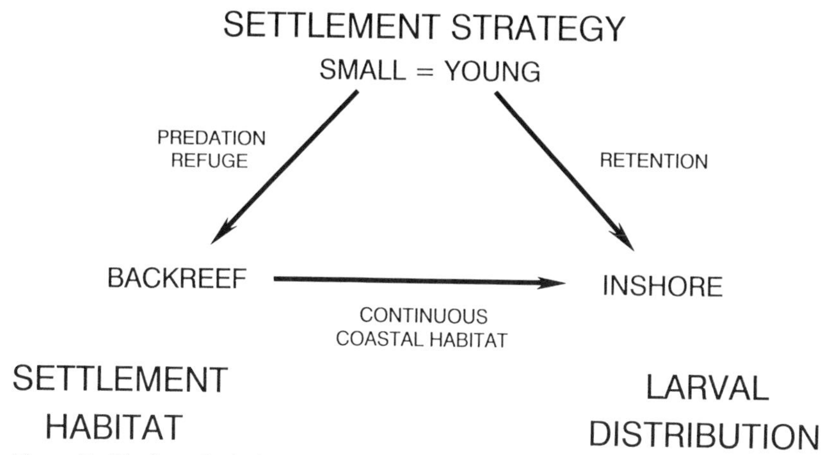

Figure 7 The hypothetical relationship between settlement strategies and settlement habitat and larval distribution. There may also be a causal relationship between settlement habitat and larval distribution independent of size or age at settlement.

Leggett, 1987). The relatively low variability in this aspect of reef fish settlement strategies may well indicate the selective importance of settling at an optimal size.

E. Age at Settlement

The planktonic larval duration (PLD) of reef fishes is more variable than the size at settlement, ranging from no planktonic phase at all to many months. Virtually all of the thousands of coral reef species in the Indo-Pacific do have a planktonic larval phase [with the notable exception of the damselfish, *Acanthochromis polyacanthus* (Robertson, 1973b)]. In the Caribbean, the toadfishes (Batrachoididae) are the only reef-associated fishes known not to have a planktonic larval stage (Hoffman and Robertson, 1983). Given that some reef species can settle at very small sizes, it is curious why more species have not opted for large eggs that hatch out larvae directly onto the reef. Alternatively, giving birth to live young should also be a feasible strategy, given the tiny return on eggs released into the plankton (although live-bearing is generally uncommon among marine fishes and it could be difficult to evolve toward that adaptive peak). I suspect that despite the short-term adaptive advantage of onreef reproduction, there is a long-term necessity for a wide-dispersal phase to avoid progressive local extinctions in the patchy geography of coral reefs.

With the advent of the daily otolith aging technique there has been a burst of information on the planktonic larval duration of reef fishes. There have been broad surveys of a selection of reef species (Brothers *et al.*, 1983; Brothers and Thresher, 1985), as well as intensive surveys within some families of reef fishes, for example, 31 angelfishes (Thresher and Brothers, 1985), 100 wrasses (Victor, 1986a), 67 damselfishes (Thresher *et al.*, 1989), and 100 damselfishes (Wellington and Victor, 1989). The PLDs for a number of other species have also been reported: the french grunt (McFarland *et al.*, 1985), a Hawaiian damselfish and two gobies (Radtke, 1985; Radtke *et al.*, 1988), the queen triggerfish (Robertson, 1988b), the bicolor damselfish (Robertson *et al.*, 1988), and three chaetodontids (Fowler, 1989). I have also estimated the PLD for the reef drum from daily otolith increments on newly settled individuals.

One of the more notable features of the pattern of planktonic larval durations is that the shortest PLDs are consistently about two weeks. A large set of species settle after 14 to 18 days in the plankton, including french grunts and reef drums in the Caribbean, and the earliest-settling representatives of the Indo-Pacific angelfishes, wrasses, and damselfishes. The only exception reported, the Indo-Pacific anemonefishes (*Amphiprion* spp.), are documented to settle after only about one week in the aquarium (Thresher *et al.*, 1989), although it remains to be clarified whether they settle this young in the wild (the immediate availability of settlement sites in the aquarium could induce early settlement). On the other hand, there seems to be no simple limit to how long the larval duration can extend. Some species of wrasses routinely spend more than three months in the plankton (Victor, 1986a, 1987a), while Radtke *et al.* (1988) report that two gobies living in freshwater streams in Hawaii typically spend four and five months in the plankton before migrating inshore. In general, however, the majority of reef fishes appear to have a PLD of 20–30 days. Families that tend to have longer larval durations (wrasses and chaetodontids) appear to exhibit more variability between species than those families with shorter larval durations (damselfishes and angelfishes).

The relative scarcity of large data sets for single species limits the analysis of variability of PLD within species. Overall, species with longer PLDs tend to exhibit greater variance (Wellington and Victor, 1989; Fowler, 1989). The grunts and drums have very short PLDs with a very narrow range; they settle during a window of about three days from about 14–17 days old (Brothers and McFarland, 1981; B. C. Victor, unpublished observations). A large number of species have relatively short PLDs with low variance, including most of the damselfishes and angelfishes (Thresher and Brothers, 1985; Thresher *et al.*, 1989; Wellington and Victor, 1989). Other groups, such as the wrasses and chaetodontids, display a variety of patterns, including short, relatively invariant PLDs as well as longer, more variable ones (Victor, 1986a; Fowler, 1989; Leis, 1989).

Little is known about geographic variation in PLD both overall as well as within a species. One interesting trend is for Hawaiian and tropical eastern Pacific species of both damselfishes and wrasses to have distinctly longer larval lives than their western Pacific and Caribbean congeners (Victor, 1986a; Thresher and Brothers, 1989; Wellington and Victor, 1989). Geographic variation within a species has only rarely been examined. Thresher *et al.* (1989) reported only minor variation between populations of a damselfish in the western Pacific, although Victor (1986a) found that Hawaiian populations of six species of wrasses had significantly longer larval durations than their western Pacific conspecifics. Recent data indicate that Baja California populations of several eastern Pacific wrasses have extremely short PLDs, often half or even less that of more southerly populations (G. M. Wellington and B. C. Victor, unpublished observations). Clearly, the potential for extreme geographic variation exists, although what factors determine these differences remain totally unknown.

F. Relationship between Size and Age at Settlement

A good deal less information is available on the relationship between size and age at settlement, since there are few studies that document both size and age for the same species, especially with a sufficiently large sample size. The interaction of these variables between species has been examined only for a group of damselfish species (Wellington and Victor, 1989) and some wrasse species (Victor, 1991). Among the damselfishes there is only a weak correlation between the mean size and mean age at settlement ($r = 0.49$), driven mainly by the tendency for species with longer PLDs to settle larger than average. The wrasses show no obvious relationship; the blackear wrasse and the bluehead wrasse both settle at about 12 mm SL, but the former has a PLD of 21–28 days and the latter has one of 38–78 days. For reef species in general there is some broad relationship; the small-settling Caribbean species, the grunts and drums, have the shortest PLDs, while the queen triggerfish, a large-settling species, has a PLD of about 75 days (Robertson, 1988b).

These patterns are primarily determined by planktonic larval growth rates, since size at hatching is comparatively very small and therefore should contribute little to subsequent size at settlement. This is especially true for closely related species that are likely to hatch at very similar sizes. It is apparent that larval growth rates can vary greatly, even between closely related species. As an example, one can calculate a crude measure of the average daily larval growth rate by dividing the mean change in size between hatching (assume about 1 mm) and settlement by the mean larval duration for a selection of these species. Despite its short larval duration, the reef drum grows at only about 0.19 mm/day, while the french grunt grows at 0.37 mm/day. In contrast, the queen triggerfish grows at 0.75 mm/day. The extremes for the damselfishes are 0.24 and 0.68 mm/day. The two wrasses that settle at the same size grow at very different rates: 0.24 mm/day for the bluehead wrasse and 0.42 mm/day for the blackear wrasse. Clearly, even closely related species must be doing very different things in the plankton before settlement.

Two studies have examined the relationship between size at settlement and age for individuals within a species (Victor, 1986c; Wellington and Victor, 1989). The latter study reported ambiguous results; only two of ten species showed a significant positive correlation, while two others had a significant negative correlation. Most exhibited no significant interaction. The lack of an interaction could be the result of small sample sizes (about ten per species), or could suggest that these larvae grew at different rates in the plankton and reached settlement size at a variety of ages unrelated to the ultimate size at settlement. More detailed studies are required to confirm these results.

When there is a positive relationship between size and age at settlement it could be caused by either (1) relatively constant growth rates and varying larval durations (the simple explanation) or (2) differing growth rates with slower-growing larvae settling larger (if faster-growing larvae settled larger it would tend to yield a negative correlation). Despite both Occam and his razor, there is no evidence for the simpler explanation. Victor (1986c) examined a large sample of bluehead wrasse recruits and found a very clear positive relationship between size and age at settlement ($r = 0.72$, $p < 0.0001$, $n = 47$) (Fig. 8). Daily growth rates back-calculated from the width of daily otolith increments indicated that the increase in settlement size with age was the result of some individuals delaying metamorphosis after reaching settlement size. Those individuals that delayed metamorphosis had reached the typical settlement size at the same age as average individuals, indicating that their growth rate until then was the same. During the period of delay, however, their growth rate was lowered by two-thirds. In this species, then, it appears that the correlation between size and age at settlement is a product of different larval growth rates (although only when averaged over the entire larval life).

Keeping things complicated, Chambers and Leggett (1987) performed a similar analysis on winter flounder in Newfoundland and also found a positive correlation between size at metamorphosis and age. However, their correlation was caused not by a delay period with a slower growth rate as in the bluehead wrasse, but, curiously, by an unexplained tendency for slower-growing larvae to metamorphose late (to be expected) but also larger than average (i.e., even later than necessary).

The critical data to resolve the questions outlined in the preceding sections have yet to be collected. What is needed is a comprehensive set of data for each species: the pattern of production of eggs, the behavior and distribution of

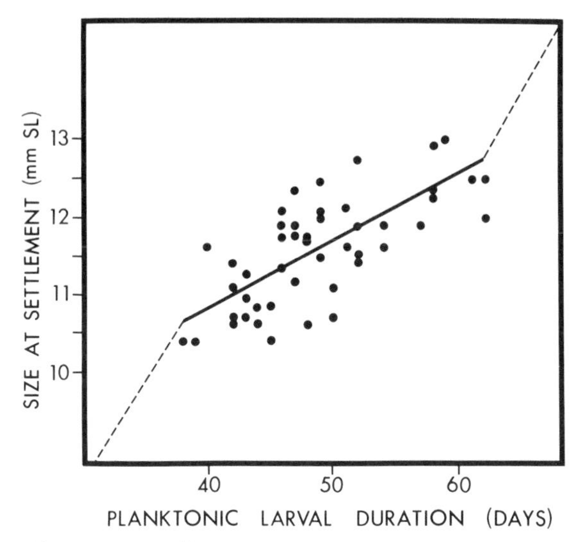

Figure 8 The effect on size at settlement of increasing larval duration for the bluehead wrasse, *Thalassoma bifasciatum*. The dashed lines before and after the solid line represent the estimated precompetent and postsettlement growth trajectories.

larvae in the plankton, the larval duration and the size at settlement for a large sample of recruits, the rate of settlement into various habitats, the mortality schedules of juveniles, and the population dynamics of adults. In addition, experimental studies inducing settlement into different habitats would also be needed to discern the adaptive significance of the existing patterns. It should then be possible to pinpoint which factors are molding the suite of adaptations that comprise the settlement strategy of reef fish species.

III. Planktonic Larval Duration and Biogeography

A. Dispersal and Its Role in Biogeography

It seems intuitively obvious that the degree of dispersal ability should play a role in determining the range of a species, especially for marine species with a planktonic larval stage. Nevertheless, it has proven difficult to detect any major effect of planktonic larval duration on the geographic range of coral reef fishes. Several studies have addressed this relationship (Brothers and Thresher, 1985; Thresher and Brothers, 1985; Victor, 1986a; Thresher *et al.*, 1989; Wellington and Victor, 1989), and all have agreed that little of the observed variation in species range seems to be accounted for by the estimated PLD. Overall, there is some tendency for species with a very short PLD to be restricted in distribution and for species with a very long PLD to be widely distributed. For the vast majority of reef fishes in between, however, the PLD has no predictive value for inferring geographic range.

Three families of reef fishes have been intensively surveyed in an attempt to detect a pattern: the angelfishes (Thresher and Brothers, 1985), wrasses (Victor, 1986a), and damselfishes (Thresher *et al.*, 1989; Wellington and Victor, 1989). The angelfish and damselfish species have relatively short PLDs, all less than 40 days, which are not very variable within species. In contrast, wrasse species display a wide variety of PLDs, from 17 to 103 days, which can vary greatly within species. When the PLDs for angelfishes and damselfishes were plotted against the number of "biogeographic areas" occupied, no significant correlation, or even apparent trend, was evident (Thresher and Brothers, 1985; Thresher *et al.*, 1989). Wellington and Victor (1989) eschewed the problematic method of adding up biogeographic areas and simply grouped species into a widespread group and a restricted group. The mean PLD for the two groups came out virtually identical.

Upon closer examination, however, both studies detected some small effect. Thresher *et al.* (1989) found a significant difference between the PLDs of species found "wholly off the Pacific plate" and those "broadly distributed on both sides of its western margin." The difference between the means was, however, less than five days and the range of PLDs for the two sets of species overlaps almost completely. Whether small differences such as these reflect any real ecological significance is arguable. The only significant relationship found by Wellington and Victor (1989) was that genera with shorter PLDs tend to be restricted to the western Pacific and genera with longer PLDs tend to have species in other regions as well.

The results for wrasses were more clear. All the species caught in Palau (in the western Pacific) with a PLD of less than 20 days have very restricted ranges, and most of the species with PLDs over 35 days have wide ranges that sometimes extend to Hawaii. Furthermore, the three Indo-Pacific wrasses whose ranges extend as far as the Central American coast have particularly long PLDs, about two months or more. There was, however, no clear relationship for the many species with PLDs between 20 and 35 days; some have very restricted ranges and others have extensive ranges within the western Pacific region (although none reaches Hawaii or the eastern Pacific coast).

As soon as the study was broadened to include other regions of the Pacific, the pattern became less clear. Wrasses caught in Hawaii and in Central America were all found to have relatively long PLDs, but some endemics appeared to have longer PLDs than their more cosmopolitan congeners. In fact, the individual with the longest PLD recorded in the study, 121 days, was a Hawaiian endemic. Furthermore, it has recently been found that the wrasse with the smallest range of all, *Halichoeres discolor,* endemic to the tiny island of Cocos near Costa Rica, has a PLD no different to its congeners that range widely up and down the coast of Central America (G. M. Wellington and B. C. Victor, unpublished observations).

B. Methodological Problems and a Protocol for Biogeographic Studies

Given these discordant results, it is imperative that we find some way to sharpen our focus on this question. Several shortcomings in the approach need to be corrected. First, since most of the reef areas of the tropical Indo-Pacific are not very isolated and there are geographic stepping stones connecting them, the absolute area or the number of biogeographic areas occupied is not a real measure of dispersal ability. The degree of isolation needs to be taken into account.

Second, the estimate of the PLD, typically derived from otolith analyses, may not be a good estimate of the true planktonic duration, especially with the small sample sizes characteristic of most of these studies. The maximum PLD of a relatively large sample should better reflect the real dispersal ability of propagules, since it is the occasional long-distance traveler that colonizes new areas, not the typical recruit. This is especially true given the recent discovery of reef fish larvae hundreds of miles from the nearest reefs (Leis, 1983; Victor, 1987a); after all, it is these larvae that are dispersing and not the vast majority that may simply remain near inshore waters.

In addition, some species are able to extend their planktonic existence as juveniles or even become pelagic as adults. This fact could account for some unexpected findings. For example, the only exception in the pattern of damselfish genera with mean PLDs less than about 25 days being restricted to the western Pacific was the genus *Abudefduf,* which is known to metamorphose and drift indefinitely in association with floating algae (Wellington and Victor, 1989). Robertson and Foster (1982) observed that some adults of the wrasse *Epibulus insidiator* in Palau swam up and off the reef with floating debris that passed by. This could explain why, despite a relatively short 30-day PLD, this species ranges all the way from East Africa to Hawaii.

A third problem is geographic variation in larval duration. Thresher *et al.* (1989) reported only slight geographic variation for a damselfish within the western Pacific. Victor (1986a), however, found that Hawaiian populations of wrasses had significantly longer PLDs than their western Pacific conspecifics. The problem with geographic variation is that it can easily lead to mistaken conclusions. For example, if one were to compare the larval durations of a wide-ranging species collected in Hawaii with a species endemic to the western Pacific, one would find that the wide-ranging species had a longer larval period (e.g., Brothers and Thresher, 1985). However, it may well be that the western Pacific populations of the wide-ranging species have a larval duration no different to the endemic, since individuals captured in Hawaii consistently have longer PLDs.

These results dictate a more refined protocol for studies evaluating the role of larval dispersal ability in determining the biogeography of reef fishes. One should first identify the likely source populations for an area and taxon under study and then collect large samples of specimens at that site. Source areas could be determined from prevailing current flows or inferred from diversity clines. Ideally one should also collect planktonic larvae in the oceanic waters between the putative source area and the study area and document PLD and growth rates for comparison. In this manner, without the scattershot approach that compounds all sources of variation into one analysis, we should be able to discern to what degree reef fish distributions are a product of dispersal abilities.

C. What Is Determining the Biogeography of Reef Fishes?

Dispersal abilities do not seem to account for most of the observed patterns of distribution and there does not appear to be a plausible candidate among the variety of alternate explanations so far put forward. The usual fallback position is that ecological requirements of reef species are controlling their range. Unfortunately, there is little information on this subject. Few reef ecologists have much enthusiasm for promoting ecological requirements as the elusive answer, since reef fishes in general appear to be highly catholic in their tastes. Leis (1986b) rejuvenated this explanation by extending it to planktonic larval ecological requirements. He found that taxa with larvae that are primarily distributed onshelf in the Great Barrier Reef (versus offshelf and oceanic) tended not to be found on islands on the Pacific Plate. He suggested that oceanic islands would not be able to provide the planktonic "habitat" required by these larvae. While this may yet prove to be valid, the analysis is compromised by the interaction of larval distribution with larval duration. It appears that most of the taxa with offshelf larval distributions have large and presumably long-lived larvae, while many onshelf taxa have small and probably short-lived larvae (see Leis and Rennis, 1983). At this point, these factors cannot be separated.

What is needed to evaluate these hypotheses are carefully controlled experiments. A set of useful experiments has been performed unwittingly by fisheries biologists over the years as they attempted to introduce valuable food fishes to islands where they did not previously exist. Many species, including the Marquesan sardine and some snappers and groupers, introduced to Hawaii over the years have established successfully (Randall and Kanayama, 1972; Williams and Clarke, 1983), proving that ecological requirements were not the reason for their absence. Even a mullet, an incidental species released with the sardine, has become a problem for Hawaiian fisherman. A few other species did not get established, although the fact that these were one-time releases mitigates the failure.

Historical factors may also play a role but are notoriously difficult to demonstrate. Springer (1982) has reviewed a number of dispersal hypotheses but personally favors a vicariance hypothesis to explain reef fish distributions. His final conclusion, which seems a little forced, was that some great barrier to dispersal existed along the entire western margin of the Pacific Plate. Unfortunately, that margin passes between island groups that are relatively close to each other and it is difficult to envision what the barrier really was. In any event, he suggests that the barrier may no longer exist. The hypothesis is driven by the presence of a relatively small number of widespread Pacific Plate endemics: species that occur widely over the scattered islands of the Pacific Plate and its margins. Whether this small set of species (about 1% of Indo-Pacific species) is truly that significant in explaining the complex distributions of Indo-Pacific reef fishes is certainly arguable.

There are other possible explanations for certain facets of the observed distributional patterns. There could be a "hybridization barrier" to dispersal of species with close relatives in adjoining yet distant areas. If an occasional recruit of a distant species arrived and mated with its resident relative, there would be a very effective barrier to colonization. As long as colonizers were few and residents were many there would be little chance of spread by the colonizing species as well as minuscule gene flow between species. This explanation may account for curious patterns of presence and absence of some species with very long-lived larvae, such as the wrasses of the genus *Thalassoma* in the Indo-Pacific. Indeed, hybrids of a Hawaiian endemic and a widespread species not from Hawaii (*T. duperrey* × *T. lutescens*) are sufficiently common in the southernmost island of Hawaii to show up in routine collections (J. E. Randall, personal communication; Victor, 1986a).

It is likely that all of these factors, and perhaps a number of others yet to be discovered, play some role in determining the geographic range of species. At present, most of these ideas remain speculative, for the detailed studies to evaluate the relative importance of potential determinants of biogeography have yet even to be proposed. Rather than some unifying theory, it is likely that a complex weave of explanations will come together to explain the intricate tapestry of species boundaries that trace the map of the Indo-Pacific.

IV. Daily Otolith Increments and Early Life History Studies

A. Introduction

The discovery of daily increments on the otoliths of fishes by Panella (1971) introduced a powerful new method for studying the age and growth of fishes. This technique, which requires little more than a good microscope, can provide exceptionally precise and detailed age and growth estimates—a serendipitous finding that, no doubt, must excite the envy of ecologists working on other less cooperative organisms. Daily increments are often particularly clear on the otoliths of larval and juvenile fishes (Fig. 9), making this technique a valuable tool for early life history studies. While the general methodology has become well established (Campana and Neilson, 1985; G. P. Jones, 1986), many promising applications have yet to be explored. This is especially true for tropical fishes, where the potential applications of this technique have only begun to be appreciated.

B. Review of the Methodology

Otoliths are accretions of calcium carbonate within the semicircular canals of the bony fishes. They function in both balance and hearing (Popper and Coombs, 1980). The production of visible daily increments depends on a daily cycle of differing rates of accretion of a protein matrix (darker under transmitted light microscopy) and crystalline inclusions (lighter) (Watabe *et al.*, 1982). There are three pairs of otoliths, each with different characteristics influencing their usefulness for aging and varying in quality between species. The largest pair, the sagittae, and the next largest, the lapilli, are used for aging. The sagitta usually has wider increments, making it more useful in slower-growing fishes, where increments on the lapillus rapidly become too narrow to resolve accurately. In faster-growing fishes, the sagitta often has prominent subdaily arrays of increments, making interpretation more difficult. In addition, the sagittae more often require grinding and polishing, which makes the technique much more labor-intensive.

Otoliths are typically removed with fine forceps from the sides and base of the braincase. They can be extracted from all size classes of fish, including embryos within the egg. Finding the otoliths in tiny fishes can be facilitated by the use of two polarizing filters, since otoliths are strongly birefringent (rotate polarized light) and glow brightly against a dark background when the two filters are placed in opposition. After removal, the otoliths are cleaned, allowed to dry, and placed into a drop of immersion oil on a microscope slide. For many small fishes this is sufficient preparation and the otolith increments can

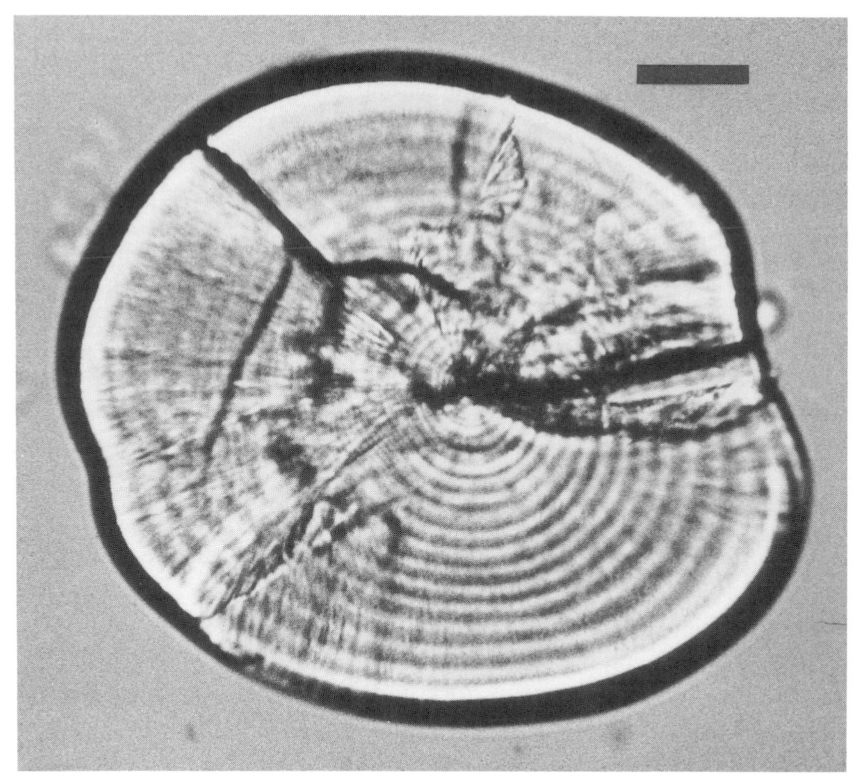

Figure 9 The sagitta from a larva of *Halichoeres bivittatus*. Note the subdaily array of increments to the left of the center. Scale bar, 40 microns.

be counted directly. Transmitted light that is polarized improves the contrast of the daily increments. If the otoliths are relatively large and opaque, some grinding is often necessary. Alternatively, for a more detailed view of the otolith increments, one can embed, grind, polish, and etch the surface with either acid or a calcium chelator (EDTA) and examine them under a scanning electron microscope.

C. Applications of the Daily Otolith Increment Technique

The most basic information available from the otolith is the total number of daily increments for a fish of a given size. With these data, one can derive an age–growth curve (Ralston, 1976a; Brothers and McFarland, 1981; Victor, 1983b, 1986c, 1987a, 1991; Fowler, 1989) or document a growth rate difference between populations (Victor, 1986b). Given the ease of this analysis and the importance of basic growth data, it is surprising that this application has not become more common in studies of the early life history of reef fishes.

A more commonly used application of this technique is the estimation of the duration of the planktonic larval phase. This is done either by estimating the total age of new recruits (Brothers and McFarland, 1981; Brothers *et al.*, 1983; Victor, 1986c, 1991; Robertson, 1988b; Robertson *et al.*, 1988; Wellington and Victor, 1989) or by estimating the age up to a mark corresponding to settlement (Brothers *et al.*, 1983; Brothers and Thresher, 1985; Thresher and Brothers, 1985; Victor, 1986a,c; Fowler, 1989; Thresher *et al.*, 1989; Wellington and Victor, 1989). Estimates of the planktonic larval duration are invaluable for biogeographic studies and can also be used to calculate rough estimates of growth rates for larvae when the size at settlement is known (Victor, 1986c, 1987a, 1991).

Transitions in the characteristics of the increments on otoliths can be very useful markers for major events in the early life history of individual fishes (Panella, 1971, 1980; Brothers and McFarland, 1981; Victor, 1982). On reef fishes, the most prominent transition is typically associated with settlement (Fig. 10). The presence of a settlement mark permits one to easily calculate the date of settlement, simply by subtracting the number of daily increments between the mark and the edge of the otolith from the date of capture. This technique has been used to reconstruct the daily pattern of settlement (Victor, 1982, 1984; Pitcher, 1988a), even up to a year after the settlement episode (Victor, 1983a). Comparisons of reconstructed settlement patterns have been used to measure the persistence of age cohorts over a season (Pitcher, 1988a), to assess the effect of recruitment on subsequent population sizes (Victor, 1983a), and to infer the patch size of planktonic fish larvae by estimating the spatial scale of settlement events (Victor, 1984).

Similarly, the daily pattern of spawning can be back-calculated by subtracting total age from the collection date (McFarland *et al.*, 1985; Robertson *et al.*, 1988). Of course, this would only reflect the ultimately successful spawning patterns, which may well have little correspondence to true spawning patterns. Robertson *et al.* (1988), however, found a close match between such back-calculated spawning patterns and the actual spawning patterns observed on the reef.

There have been few attempts to extend the analysis of daily otolith increments beyond the simple procedures described here. Perhaps the most promising application is deriving a detailed history of daily growth rates from the pattern of widths of the increments. Since there is some predictable relationship between some function of otolith size and fish size (e.g., Brothers and McFarland, 1981; Victor, 1987a; Fowler, 1989), the width of each increment should reflect the daily growth of the individual. Although this relationship has yet to be validated for reef fishes, there is some evidence for a direct

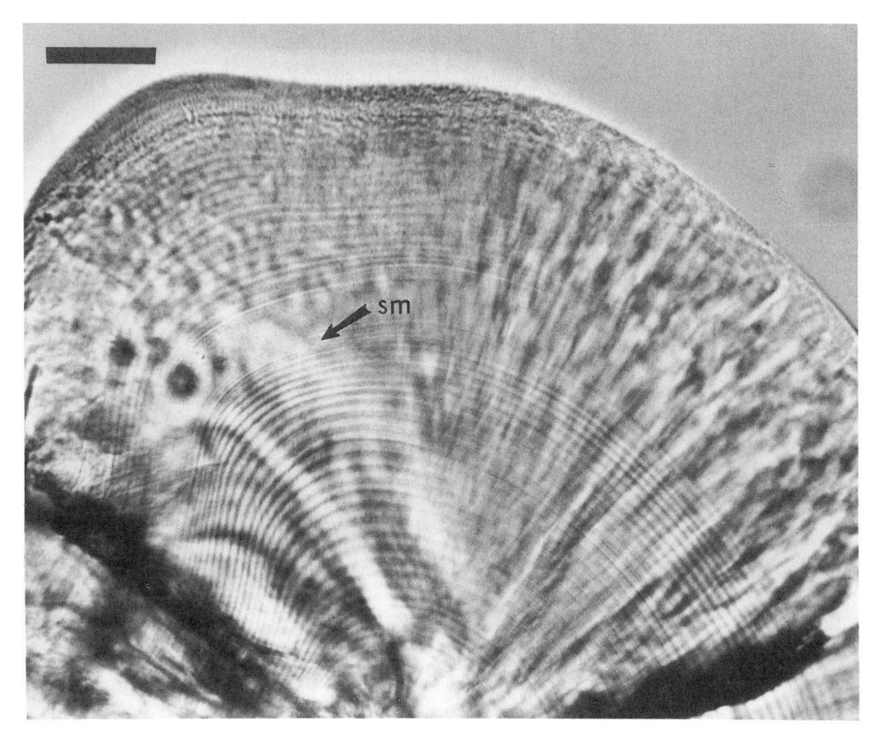

Figure 10 The settlement mark (sm) on the sagitta of a juvenile bluehead wrasse, *Thalassoma bifasciatum*. Scale bar, 40 microns.

relationship. Daily otolith increments on wrasse larvae dropped to half their former width during a period when the growth rate fell by about two-thirds (Victor, 1986c). Whether there is a one-to-one correspondence between daily increment width and growth rate or a running average is not known (Campana and Neilson, 1985), although five days of extra feeding of a wild population of reef wrasses resulted in five wide otolith increments, with an indication of wider increments on days with more feeding time (Victor, 1982).

Changes in the width and appearance of daily increments are common on the otoliths of young fishes. These transitions are sometimes correlated with settlement, as in the damselfishes (Pitcher, 1988a), although they often occur earlier in the life of the larva for other reasons (Brothers and McFarland, 1981; Victor, 1986c). In the bluehead wrasse, and probably other wrasse species, a transition to narrow increments near the end of the larval period signals the onset of a period of delayed metamorphosis that can last up to several weeks, during which growth is reduced (Victor, 1986c). Many other changes in increment width occur on the otoliths of young fishes of a variety of taxa, presumably reflecting changes in behavior, habitat, or life-style (Fig. 11).

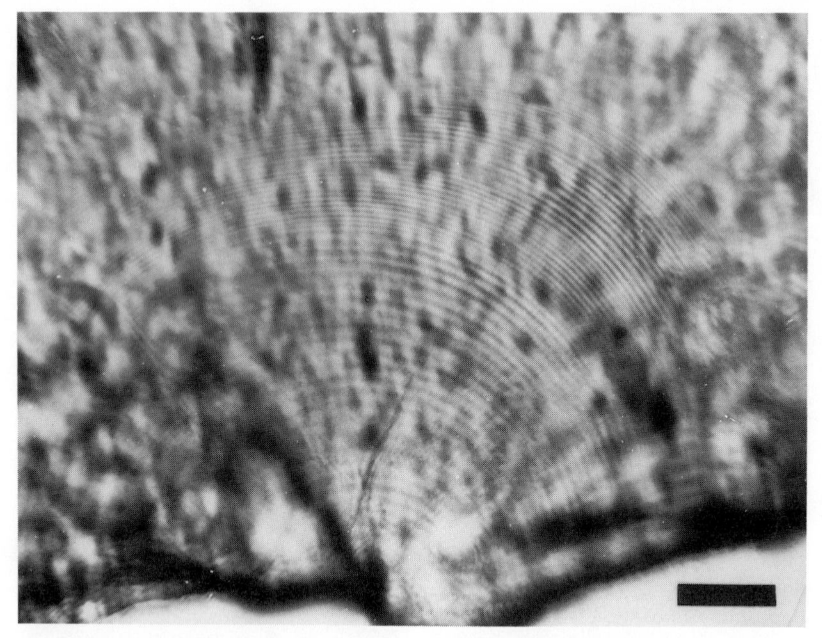

Figure 11 Unexplained transitions in the width of otolith increments during the larval period on the otolith of a juvenile Indo-Pacific wrasse, *Pseudocheilinus evanidus*. Scale bar, 40 microns.

Since daily increments are preserved unmodified within the otolith, it is possible to derive information about past environmental conditions. To date, it has proven feasible to infer past temperature exposures by analyses of stable isotope compositions, such as oxygen and carbon isotopes (Radtke *et al.*, 1987), or by strontium-to-calcium ratios (Radtke *et al.*, 1988). These techniques may make it possible to discover where in the plankton the larvae are traveling at various ages or from what oceanic region larvae are derived. Analyses of the elemental composition of sections of the otolith may be particularly valuable if they could localize the source population of larval recruits (Campana and Neilson, 1985).

Otolith increments vary in shape and pattern from species to species and perhaps even between populations of the same species (Brothers, 1984; Victor, 1986a, 1987a). Since each incremental ring preserves the outline of the otolith at the time the ring was formed, the morphology of otoliths of larvae can be easily derived from settled fishes of all ages. This method has been used to identify unknown wrasse larvae by comparing their otolith shape to that back-calculated from known juveniles (Victor, 1987a). Clearly identifiable characteristics of increments in the core region of the otolith should permit the identification of very early stage larvae of different families and even some genera (Brothers, 1984; Victor, 1986a).

Without a doubt, many other applications of the otolith increment technique await development in the future. The array of increments on the otolith preserves the record of the past in such unparalleled detail that the potential of the technique may be limited only by the imagination of reef fish ecologists.

D. Potential Pitfalls and the Need for Validation

Our knowledge of the basic science behind the otolith increment technique is still in its infancy and there is no question that exceptions do occur and assumptions are often unwarranted. Anyone who has examined a variety of otoliths knows that bizarre and unexpected increment patterns are commonplace and that few rules are transferrable from one taxon to another. It is therefore imperative that researchers not yield to temptation and run amok in their interpretations of otolith increments and transitions without thorough validation. At present, unfortunately, the only reef fish taxa for which clear experimental validation in the field is available are wrasses (Victor, 1982, 1983b), damselfishes (Robertson *et al.,* 1988; Pitcher, 1988a; Wellington and Victor, 1989), and chaetodontids (Fowler, 1989).

The first and most basic problem is whether the observed increments on the otolith are really formed daily. Despite widespread use of the adjective "daily," there are painfully few direct demonstrations of daily production of increments for reef fishes. Since Geffen (1982) first documented that increments were not always produced daily on the otoliths of larval herring, there have been a number of reports of increments being formed at a rate less than one per day (Campana *et al.,* 1987). This phenomenon has so far been reported only for temperate fish larvae, probably because of their slow growth rate. Campana *et al.* (1987) point out that the resolution limits of light microscopy are about one micron, and that this could account for the apparent nondaily formation of increments in slow-growing larvae. Clearly, the presence of narrow increments requires a thorough validation of their periodicity.

The otoliths of many reef fish larvae present a very different problem. There often appears to be additional arrays of "subdaily" increments within the larval period (see center left of Fig. 9). The presence of different superimposed sequences of increments on the same otolith is not new or unusual: both Panella (1980) and Brothers and McFarland (1981) mention extra "subdaily" increments. Despite this appellation, there is no experimental evidence demonstrating that these arrays are subdaily. I have proposed two criteria for the subjective presumption of subdaily increments (Victor, 1986a). The first criterion was based on the assumption that the largest repeating cycle of increment formation was daily. If, as the focus under the microscope is changed, each increment splits into two or three, resulting in an array of exactly twice or three times as many increments, one should presume that the fewer and larger increments are daily. The basis for this presumption rests on the fact that otolith increments are thought to be the product of physiological cycles entrained by light, temperature, feeding, or activity cycles influencing the metabolism of the fish (Panella, 1980). While there is some evidence for subdaily cycles of feeding and activity in fishes (e.g., vertical migrations of planktonic larvae), there are no reports of supradaily cycles occurring on two- or three-day cycles.

The higher-order array (with fewer increments) is usually visible over more quadrants of the otolith and at wider planes of focus than the finer array of "subincrements," which tend to occur over a smaller section of the otolith and at a single plane of focus (Victor, 1986a). These "subincrements" are typically found on otoliths with relatively wide higher-order increments. The relative prominence of the wider array of increments supports the assumption that they are formed on a daily cycle. It may be, as Panella (1980) suggested, that the arrays of "subincrements" occur on all fish otoliths, but are only detectable on those that are fast-growing. Despite their prevalence on larval otoliths, it remains an open question whether these arrays are truly structural or some form of optical artifact.

All of these complications only serve to emphasize the importance of verifying the assumption that increments are daily. The classic method for validating daily increments in the field is to mark the otolith either by isolation of the fish in the dark or by immersion in tetracycline, both of which interfere with the normal process of otolith accretion. The subject is later recaptured and the number of increments between the mark and the edge is compared to the number of days since release (Victor, 1982; Pitcher, 1988a; Robertson *et al.*, 1988; Fowler, 1989; Wellington and Victor, 1989). These methods have consistently affirmed the daily periodicity of increments on the otoliths of juvenile wrasses and damselfishes. Several studies have verified the formation of daily increments on reef fish larvae and juveniles under laboratory conditions (Schmitt, 1984b; Radtke *et al.*, 1988; Thresher *et al.*, 1989). A major drawback of validations in the laboratory is that artificial daily cycles imposed by captivity could induce daily increments that may not be analogous to those found on wild fishes.

A more pressing question is whether the increments observed on otoliths of planktonic larvae are also daily. In many ways, these increments closely resemble those on the otoliths of juvenile fishes, and therefore it is reasonable to assume that they are also daily. Nevertheless, it is imperative that this assumption be verified, especially considering the widespread occurrence of presumed subdaily increment arrays on larval otoliths. This is a difficult experiment to perform in the field, especially given the necessity that the larvae be permitted normal planktonic behaviors and migrations. Marking individuals within a very large enclosure drifting with the plankton is probably the only practical method possible at present.

Many studies that employ otolith increment counts have resorted to assumptions about when the first increment is formed. While it is probably true that errors in these assumptions would only change estimates by a few days, it is important for accuracy that the age and size at which the first increment is formed be documented so that the correct adjustment can be made to estimates of planktonic age and growth. Otoliths first appear at hatching in bluehead wrasse larvae, about one day after fertilization and release into the plankton (Victor, 1986c). Wellington and Victor (1989) found that larvae of a Caribbean damselfish have developed the primordium of the otolith by hatching (the start of their planktonic phase and five days after fertilization). In damselfishes, therefore, it seems that the count of increments on the otolith is a direct estimate of the planktonic duration, while for wrasses it is necessary to add about two days to the count. Whether these findings hold true for other families of reef fishes is unknown and needs to be examined.

When estimating planktonic larval duration from new recruits it is important to be sure that the recruits are recent arrivals from the plankton and not newly appeared from hiding places in the reef. The best way to avoid that complication is to capture new recruits from artificial reef substrates that are suitably isolated from nearby reefs and surveyed daily. To further verify that recruits are indeed new, it helps to compare competent larvae of the same species collected just prior to settlement (Fig. 4).

The presence of a variety of transitions on the otoliths of many species can easily lead to mistaken estimates of the duration of the planktonic phase or other stages. Settlement marks therefore should be verified in the field, a process that is, fortunately, relatively simple. The clearest demonstration of a settlement mark is to document the absence of the mark on planktonic larvae just before settlement and the presence of the mark on newly settled recruits (Victor, 1982; Pitcher, 1988a). Less rigorous methods include relying on the presence of the mark at the edge of the otolith of newly recruited fish (Wellington and Victor, 1989) or the smallest individuals captured (Radtke *et al.*, 1988; B. C. Victor, unpublished observations on the reef drum), back-calculating the transition from the otoliths of fishes of a known age since settlement (Fowler, 1989), or by statistical comparison of increment counts of newly recruited fishes with pretransition counts from older fishes (Thresher and Brothers, 1985; Thresher *et al.*, 1989).

Estimates of larval durations (and other otolith counts) may become biased by the exclusion of data. Fowler (1989) found that a fraction of the fish he studied did not have a clear transition at settlement and could not be included in his estimates of larval duration. Furthermore, some studies are forced to exclude otoliths from which accurate replicate counts could not be obtained. Since there can be some correlation between otolith quality and growth rate (in my experience, slow-growing individuals tend to have otoliths that are harder to read), it is imperative to use another method to corroborate one's findings.

The validity of the otolith increment technique is still untried for most species of reef fishes. It would not be surprising if some of our assumptions proved to be incorrect in some fishes that have yet to be examined. Clearly, a vigilant and skeptical attitude needs to be cultivated among workers in the field, and thorough verification of techniques (e.g., Pitcher, 1988a) should be strongly encouraged.

CHAPTER 10

Spatial and Temporal Patterns in Recruitment[1]

Peter J. Doherty
Australian Institute of Marine Science
Townsville, Queensland, Australia

I. Introduction to Nonequilibrial Systems

The chapters in this section show that coral reef fishes have complex life cycles, that is, two or more developmental stages that live in spatially distinct habitats (Roughgarden *et al.*, 1988). With rare exceptions (Thresher, 1984), reef fishes do not protect offspring through their full development so that most species spend some part of their early life history in a planktonic refuge when they may be dispersed to new locations (Johannes, 1978a). The effect of these bipartite life histories (exemplified by a relatively sedentary reef-associated phase recruited from a more mobile pelagic phase) is that individuals have to negotiate the hazards of two totally different environments within their lifetime, which increases the number of factors with potential to limit their abundance. Research done since 1975 has demonstrated that the recruitment of juveniles from larval sources into benthic populations is often variable and unpredictable (Doherty and Williams, 1988a); this has been the basis for an active debate about whether the size of an adult population is determined primarily by events occurring before, during, or after the settlement of juveniles into the reef environment (Jones, Chapter 11).

Variable replenishment of populations is not unique to coral reef fishes; it is a hallmark of marine organisms that disperse pelagic propagules, especially those with lengthy larval development. Known examples include plants (Hoffmann and Ugarte, 1985), a wide range of invertebrates (Coe, 1953; Loosanoff, 1964; Dayton, 1984; Caffey, 1985; Jamieson, 1986), and many temperate (Sissenwine, 1984; Steele, 1984; Cowen, 1985) and tropical (Doherty and Williams, 1988a) fishes.

[1] AIMS Contribution No. 509.

The commercial catches of temperate fisheries provide some rare long-term data sets that show how variations in the recruitment of juvenile cohorts can profoundly affect the subsequent abundance and demography of adult stocks (Fig. 1). Furthermore, these time-series include examples of high- (inter-annual) and low- (decades or longer) frequency signals in year-class size that are often coherent over large spatial domains and covariable among species (Koslow, 1984; Sherman *et al.*, 1984; Hollowed *et al.*, 1987). Both characteristics suggest that fish recruitment can be driven by climatic effects on the survival and/or dispersal of larvae. Since there is little evidence of compensatory mortality in postlarval populations (Gulland, 1982), it seems that there is a substantial component of environmental forcing in the dynamics of at least some benthic fish populations (Cushing, 1982; Longhurst, 1984; Sissenwine, 1984).

Despite historical precedents stretching back to the start of this century (Hjort, 1914), researchers initially gave little attention to settlement as an important source of variation in coral reef populations (Ehrlich, 1975). This lack of interest in recruitment seems to have derived from a widespread expectation among ecologists that coral reef communities should be paramount examples of biologically accommodated assemblages, exhibiting equilibrial properties of stability and self-regulation (Sale, 1980a). This perspective has been challenged during the last decade by empirical studies showing that traditional competition theory with its resource-based concepts of equi-

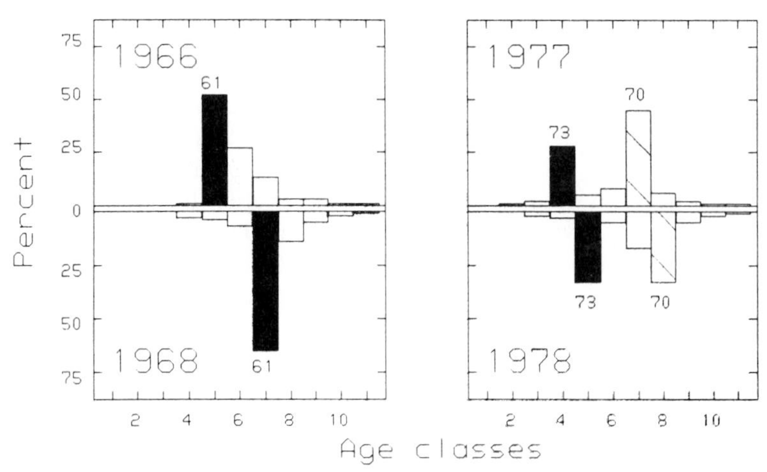

Figure 1 Age composition in four samples of *Merluccius productus* from the Pacific Northwest showing the disproportionate contributions and persistence of dominant year-classes. (Redrawn from Bailey, 1981.)

librium cannot explain all the complex spatial and temporal variability of abundance found in reef fish populations (Doherty and Williams, 1988a).

The most compelling evidence that fish do not track the carrying capacity of the benthic environment has been the apparent inertia of some populations after widespread natural disturbances like El Niño and predation by crown-of-thorns starfish (Wellington and Victor, 1985, 1988; Williams, 1986a; Guzman and Robertson, 1989). The most telling result has been the failure of herbivore populations to increase despite an obvious expansion of their food supply, apparently due to limiting larval supply (Wellington and Victor, 1988). While part of the inertia of populations after major environmental perturbations could be attributed to the longevity of individuals, some assemblages have been followed for almost a generation after the event without evidence of directional change in community structure (D. McB. Williams, personal communication). Although others have reached different conclusions about the importance of resource limitation, their results were all derived from tiny controlled manipulations (e.g., Shulman, 1984) that may be unrealistic simulacra of processes operating at larger scales.

Manipulative experiments generally have found little evidence of density dependence in the colonization and persistence of recruits from larval sources at spatial scales above 1 m^2 (Doherty and Williams, 1988a). The exceptions may be those fishes that reside on discrete objects like anemones, coral heads, or tiny artificial reefs where small size and lack of refugia make it possible for dominant individuals to exercise control over other individuals in the same patch of habitat (Doherty, 1982). Most natural reef habitats, however, occupy much larger volumes and present species with vastly greater opportunities for coexistence because of the fractal nature of reef surfaces (Bradbury and Reichelt, 1983). Thus, I suggest that much of the experimentation done to date (cf. Jones, Chapter 11) will be of little use in resolving questions about population regulation at the scales of greatest human interest. This is because the behavioral interactions among individuals that influence resource allocation in tiny experiments may be swamped by other influences, such as recruitment, at these larger scales (Lawton, 1987). Equally, it must also be acknowledged that failure to observe population regulation in small-scale experiments, as many have done, does not preclude the possibility of compensatory relationships at larger scales through the global stock-recruitment effects that are so much a part of classic fisheries science (Cushing, 1977). I conclude that one must be very cautious about extrapolating the results of experiments beyond their particular format, which is why I believe that "natural" experiments provide immensely valuable opportunities to understand large-scale system dynamics even though they rarely satisfy the accepted protocols of experimental design.

A third line of evidence that argues against equilibrium dynamics operating in populations of coral reef fishes, at least at local scales, has been provided by longitudinal studies on populations of damselfishes on the Great Barrier Reef (D. McB. Williams, 1979; Doherty, 1980) and wrasses in the Caribbean (Victor, 1983a, 1986b). In both oceans, local changes in abundance and/or demography of these common fishes have been shown to be driven by historical patterns of recruitment interacting with density-independent mortalities (Sale and Ferrell, 1988), which implies that these assemblages are not always saturated with individuals. While some researchers have emphasized low input (i.e., limiting recruitment) as the cause of undersaturated populations, others have emphasized high turnover (i.e., intense predation). Either way, the traditional competition paradigm is repudiated in favor of a simpler alternative in which local densities reflect the balance between settlement and survivorship. This model has two important characteristics; first, the balance is achieved by nonequilibrial mechanisms and hence can be stabilized over a range of densities, and second, the system is "open" in the sense that colonization rates are independent of local spawning effort.

Sale (1978b) demonstrated that the benthic stages of many coral reef fishes are unusually sedentary compared with those of other vertebrates of similar size. In contrast, there is much circumstantial evidence (e.g., biogeographic distributions, genetic patterns) that larvae of these same species are dispersed over spatial scales much greater than the patchiness in adult populations. Direct proof of larval dispersal between distant localities is provided by anomalous events such as recruitments in areas remote from possible spawning sources (Pillai *et al.,* 1983) and the successful colonization of the Hawaiian archipelago by introduced species (Randall and Kanayama, 1973). While long-distance dispersal may be the exception, it seems that the genetic stocks of most coral reef fishes are fragmented into numerous subpopulations linked to various degrees by larval dispersal ["metapopulations" in the current vernacular (e.g., Roughgarden and Iwasa, 1986; Roughgarden *et al.,* 1988)]. Some of the most challenging and hence most interesting questions about reef fish dynamics concern the patterns of dispersal and the connectivity of individual elements within these metapopulations.

The potential for transport, mixing, and death in the plankton is such that it seems certain that most of the presettlement fish carried to a local patch of habitat will have been spawned elsewhere. The lack of direct feedback between fecundity and recruitment over spatial scales from meters to kilometers (and possibly larger) means that homeostasis within local assemblages can only be accomplished through density-dependent control of settlement (e.g., Sale, 1977) and/or compensatory mortality after settlement (e.g., Shulman, 1985a); but there is little empirical proof that either mechanism is important in regulating natural populations of coral reef fishes (Doherty and Williams, 1988a). Consequently, it has become popular to regard these populations as "open non-equilibrial" systems (Talbot *et al.*, 1978) in which gains from external sources (recruitment across the boundaries of the local population) and losses to internal sinks (mortality) exert more control over abundance and community structure than the carrying capacity of the local environment (D. McB. Williams, 1980).

A. Models

The recent literature on population dynamics of coral reef fishes has been dominated by discussion about the merits of a particular form of nonequilibrial control: limitation by settlement. It could be argued that this situation occurs whenever population densities exist below carrying capacity since greater settlement ought to be able to drive densities to higher levels regardless of turnover. Victor (1986b), however, distinguished two qualitatively different scenarios of recruitment–limitation in order to emphasize the dual contributions of settlement and postsettlement mortality to adult abundance.

Primary recruitment-limitation (*sensu* Victor) exists whenever initial settlement is less than the turnover of adults between recruitment events. By definition, juvenile mortality is of secondary importance since it can only lower the population further below its potential carrying capacity. If postsettlement mortality is constant, variations in the intensity of settlement should directly impact both population size and age structure (Victor, 1983a). Because of the "storage effect" (=inertia) in populations composed of many age classes (Warner and Chesson, 1985), these correlations will be most obvious in short-lived species with high turnover of individuals. Warner and Hughes (1988) showed, however, that the absence of such correlations is not sufficient evidence to reject nonequilibrial dynamics since they will be masked if recruitment is low relative to total population size and they will be further weakened by random variations in the rate of mortality.

Secondary recruitment-limitation (*sensu* Victor) exists whenever initial settlement is greater than the turnover of adults between recruitment events. In this situation, mortality is of primary importance in preventing populations from running out of resources. The effect of variable recruitment on these populations will depend on both the constancy and magnitude of mortality. Even modest variations in the rate of mortality will destroy the correlations between settlement and population size, and between settlement and demographic structure (Warner and Hughes, 1988).

Shulman and Ogden (1987) suggested that populations of grunts in the Caribbean experience such high rates of juvenile mortality that their local abundance is determined primarily by patterns of postsettlement predation. Although this can be regarded as a case of secondary recruitment-limitation (*sensu* Victor), they argued that it is inappropriate to classify these systems as recruitment-limited since small changes in mortality will have greater effect on population size than changes of similar proportions in settlement.

In functional terms, these three scenarios describe a continuum of possible nonequilibrial outcomes ranging from primary control by initial settlement to primary control by postsettlement mortality, as defined by the sensitivity criterion of Shulman and Ogden (1987). This paradigm requires no modification to admit density dependence, which can arise through reduced predation on a population showing secondary recruitment-limitation and is thus most likely in populations with high and variable rates of mortality. One of the benefits of this simple model of local abundance is that it provides a parsimonious explanation for the full range of dynamics seen in natural populations (Warner and Hughes, 1988) and removes the false dichotomy between density-dependent and density-independent states. Questions about whether reef fish populations are one thing or the other are less appropriate than questions about the frequency of different states under varying regimes of recruitment and mortality.

Finally, the potential dispersal of larval fish over distances much greater than those of adult fishes may mean that population dynamics are qualitatively different when monitored over local and global scales. Metapopulations may exhibit density-dependent dynamics even when this condition is rare in local populations because the former are self-recruiting entities whereas the latter are not (see the earlier discussion). Whether reproduction actually acts in this manner to regulate global population size will depend on the variance in stock-recruitment relationships (Sale, 1982a). As yet, there are no published data on these relationships for any species of coral reef fish; this is identified in Section VI as a priority for future research and one of the justifications for studying patterns of recruitment.

II. Why Study Recruitment?

In recruitment-limited systems, recruitment statistics can be used to forecast the dynamics and demographic structures of benthic populations provided (1) that postsettlement mortality is relatively constant (Shulman and Ogden, 1987) and (2) that individuals do not migrate out of the area where recruitment is monitored (Robertson, 1988a). Both conditions restrict this application to a subset of all tropical reef fishes, notably those with sedentary lifestyles (or mobile populations contained by effective barriers to dispersal) and relatively low rates of mortality, which explains why a disproportionate amount of attention has been given to small site-attached species like damselfishes (Sale, 1980a). For this suite of species, at least, recruitment data can be used as the basis for empirical management of local fish stocks without any need to understand the processes that determine variability in settlement (Doherty, 1983b). While it sounds faintly ridiculous to speak of managing damselfishes, many small sedentary species are collected intensively for food and/or the aquarium trade; they may be even more vulnerable to overexploitation than larger species because of their sedentary nature and the efficient methods used to target them.

In populations that meet the two conditions of sedentary juveniles and constant postsettlement mortality, recruitment statistics can also be used to hindcast the relative abundance of fish immediately prior to settlement. Although this inferential approach is less satisfactory than direct sampling of presettlement fishes, the latter has only recently been made possible by the development of new sampling technologies (Kingsford and Choat, 1985; Doherty, 1987b; Milicich, 1988). In addition, there are situations where recruitment surveys may be more cost-efficient (see the following).

Even in populations with moderately variable mortality, recruitment surveys may still be used, at least for the purpose of hindcasting, if the data are collected from very young juveniles and if one is interested only in fairly gross spatial and temporal differences. One of the advantages of working with fish in the coral reef environment is that the juveniles of many species are conspicuous and can be counted immediately after settlement (Sale, 1980a). Another advantage is that the restricted movements of most newly settled juveniles and the hierarchical patchiness of reef environments allows pattern to be resolved over a variety of spatial scales (Sale *et al.,* 1984b).

Inferences about the abundance of presettlement fish have considerable strategic value (Doherty, 1983b). First, they can be used to define the scales of pelagic processes causing major variations in the supply of juveniles to benthic populations and hence to focus research into the proximate mechanisms underlying that variation. Second, they can be used to determine the stock-recruitment relationships essential for realistic modeling of global population dynamics (Sale, 1982a). Finally, they can be used to monitor directional changes in pelagic ecosystems. Since larval fishes occupy high positions in pelagic food webs, they integrate information about the lower levels of the ecosystem. Thus, systematic large-scale variations in year-class strength could provide valuable early warning of climate change (e.g., the greenhouse effect) and lesser forms of pollution.

III. Methods

Quantitative recruitment data are collected most often by nondestructive visual censuses, either from whole counts of discrete patches of habitat or from partial counts of larger areas using randomly located quadrats. There is a large literature on this group of techniques (Sale and Douglas, 1981; Andrew and Mapstone, 1987; Greene and Alevizon, 1989) that shows how the accuracy and precision of such counts diminishes with the dimensions of the sampling unit. This is especially true of newly settled fishes that are easily overlooked in wider transects. Although this bias can be reduced by careful attention to sampling design (Sale and Sharp, 1983), recruitment surveys will always underestimate settlement because of mortality. Although mortality schedules have only been determined for a handful of species, the consistent pattern so far is for greatest mortalities to occur immediately after settlement when losses may be greater than 10% per day (Doherty and Sale, 1986; Victor, 1986b; Eckert, 1987). Naturally, such high losses cannot be sustained for long and survivorship in perennial species often improves dramatically after just a few days in the benthic habitat (Sale and Ferrell, 1988). Some species, however, sustain high mortality for longer periods (Shulman and Ogden, 1987) and these may not be good candidates for the following techniques.

Even for the most suitable species, recruitment surveys based on visual censuses suffer from a major logistic constraint that makes it difficult to resolve a broad range of spatial and temporal variations simultaneously. Since each census requires the presence of a trained observer and is often time-consuming, visual surveys are best suited either to frequent monitoring of adjacent sites (Sale and Douglas, 1984) or to infrequent monitoring of distant sites (Eckert, 1984; Sale *et al.,* 1984b). It is usually impractical to resolve the fine-scale temporal pattern of recruitment on distant sites using any form of visual count.

This methodological limitation has been overcome to a large degree by back-calculating the age of an individual fish and the date of its settlement from growth increments deposited daily in various sites of calcification, most notably the otoliths (Victor, Chapter 9). Daily aging has revolutionized the study of recruitment phenomena because it removes the requirement for continuous *in situ* observation of populations. Instead, it is possible to reconstruct an approximation of the historical sequence of daily settlement at any site from the ages of fish collected in a single visit.

The back-calculation method offers several advantages. First, over short periods, it may resolve the temporal pattern of recruitment even better than most visual surveys for a lower investment of valuable field time (Fig. 2). Second, it can be used to resolve settlement in species with cryptic juveniles that cannot be counted reliably. Third, and most importantly, major variations in settlement intensity can still be detected in samples collected several months after the event (Fig. 2). Because of this, it becomes possible to examine the temporal coherence of recruitment variations over multiple spatial scales (Fig. 3) and hence to resolve the spatial coherence of transient phenomena.

The back-calculations do have limitations. First, like any indirect aging method, there is room for measurement error either because increments are

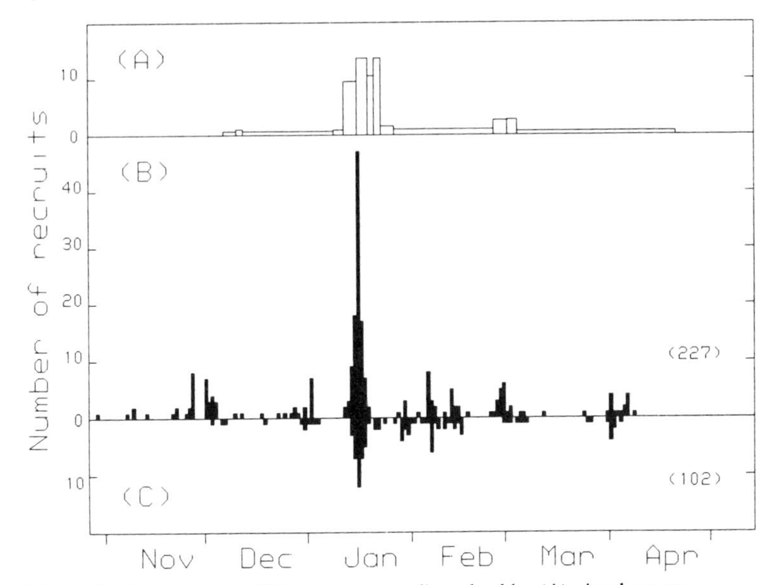

Figure 2 Recruitment of *Pomacentrus wardi* resolved by (A) visual census, (B) otolith back-calculations based on frequent sampling, and (C) otolith back-calculations based on a single sample. (Redrawn from Pitcher, 1988a.)

misclassified or because fish may not always lay down just one increment per day. This problem can be overcome with proper validation of each new species (Victor, 1982; Pitcher, 1988a; Fowler, 1989) and the most authoritative data sets will always be those backed by such evidence.

Second, all reconstructions are based on the ages of survivors and hence will reflect past settlement only to the extent that mortality follows fixed schedules. Pitcher (1988a) found reasonable correlations between continuous visual censuses and reconstructions based on both short and long periods of time (Fig. 2) so that the assumption of uniform mortality seems reasonable in this case. Interspecific comparisons of back-calculated recruitment data will often be confounded with differential turnover (Eckert, 1987; Shulman and Ogden, 1987; Sale and Ferrell, 1988) and should be treated with particular caution. However, it will become apparent that useful information about the timing of recruitment events can be obtained from such comparisons even though it might be inappropriate to compare the magnitudes of these events.

Third, even with constant mortality, it is obvious that the relative abundance of cohorts settling at different times must be systematically reduced by the cumulative losses over different time periods. In other words, uncorrected back-calculations will underestimate the strength at settlement of older cohorts relative to younger ones. To some extent, this is offset by the greater

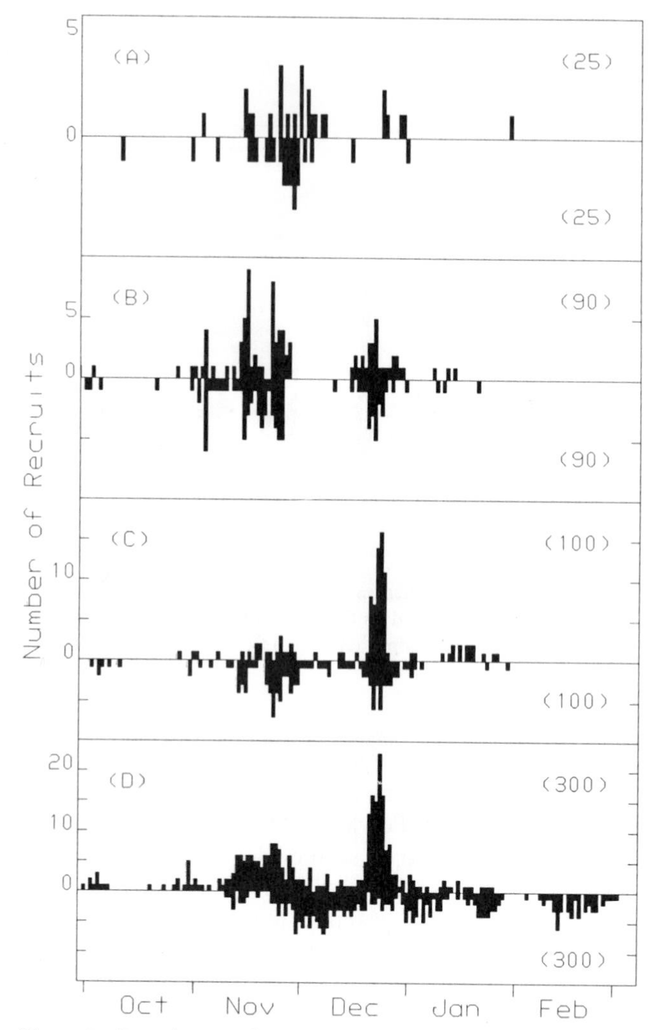

Figure 3 Recruitment of one year-class of *Pomacentrus* sp. (Allen, 1975) resolved by otolith back-calculations of collections from (A) two patch reefs at Lizard Island, northern GBR, (B) two sites, each of five patch reefs, at Lizard Island, (C) all sites at Lizard Island (lower graph) and similar sites on Waining Reef, and (D) four northern reefs (upper graph) and four southern reefs. (Redrawn from Pitcher, 1988b.)

rates of mortality suffered by the youngest individuals but it is one further complication. While it may be possible with accurate mortality schedules to correct such data back to time zero, this approach risks compounding sources of error unless very large sample sizes are available.

Finally, then, a logistic constraint on the back-calculation method is the time and effort required to age a large number of individuals. Pitcher (1988b) claimed that a sample of 100 individuals was adequate to resolve temporal pattern over 3–4 months at one location, but I believe that this will be marginal even assuming moderate levels of variability in settlement and survivorship.

For these four reasons, visual censuses probably remain the most efficient method for comparing a large number of closely located sites. Similarly, visual methods may be efficient when comparing a large number of distant sites where the researcher is only interested in year-class strength which would be the case in many monitoring studies (e.g., Sale *et al.*, 1984b). However, when the object is to gather as much detail as possible about both the temporal and spatial dimensions of recruitment, otolith reconstructions may be the only practical procedure despite the numerous qualifications that must be placed on the data.

IV. Spatial and Temporal Patterns in Recruitment

It is not intended that this short essay on recruitment should substitute for a thorough review of the phenomenon; interested readers can find further case studies in the work by Doherty and Williams (1988a). For obvious reasons, I have given priority here to data sets that allow simultaneous resolution of spatial and temporal variations in recruitment. Since otolith back-calculations generally allow better temporal and richer spatial resolution of these patterns, such data sets tend to predominate despite being a relatively recent development in the literature.

To build up a structured picture of recruitment variation in coral reef fishes, I have selected case histories that fit within a hierarchical classification of five spatial and two orthogonal temporal scales. The latter (variations "within" and "among" year-classes) are discussed first without explicit reference to spatial dimensions. When discussing temporal data, I will often use the term "cohort" as shorthand to refer to fishes that have settled together during a brief recruitment episode regardless of whether they originated from the same spawnings and the term "year-class" to refer to those fishes recruited within seasonal limits. Spatial patterns are considered within these two temporal frameworks at the following levels: (1) variations among replicates of habitat within one site (maximum separation of 0.1–1 km), (2) variations among different sites within the same reef system (separated by >1 km), (3) and (4) variations among whole coral reefs arranged both perpendicular and parallel to adjacent coastline (separated by >10 km), and (5) variations among different groups of reefs within the same biogeographic region (separated by >100 km).

I do not explicitly consider differences among biogeographic regions but I have attempted to illustrate each of the smaller-scale comparisons with data from more than one such region. Furthermore, I recognize that the classification of coral reefs into cross-shelf and longshore categories really applies only to reef systems on continental margins such as the Great Barrier Reef (hereafter GBR). However, I am not aware of any data sets that allow extended comparisons of recruitment among oceanic reefs. This is a significant gap in our present knowledge since isolated reefs may have qualitatively different dynamics from shelf reefs given the likelihood of different connectivities in such systems. Finally, there is one other category of recruitment phenomenon that will also be given little explicit attention. This is the selection of particular habitats by individual larvae. There is plenty of evidence that such behaviors occur and that they account for much of the structural differences in communities among different habitats and microhabitats (Sale *et al.*, 1980; Doherty and Williams, 1988a), but I suggest that the significance of these patterns lies in evolutionary and not ecological time.

A. Within Patches

1. Within Year-Classes

Observations on fixed sites have shown that recruitment is far from uniform in time. Seasonality has been observed in spawning and recruitment at most geographic locations (reviewed by Doherty and Williams, 1988a) and is particularly obvious on high-latitude reefs like those in Japan and the southern Great Barrier Reef (Fig. 4). The majority of fishes in the latter region only recruit during summer months (Russell *et al.*, 1977) and year-classes are typically formed from fewer than five consecutive months of settlement (Fig. 4E and F). In contrast, some Caribbean reef fishes have been reported as spawning in all months of the year (McFarland *et al.*, 1985) and there may be no period without some settlement (Fig. 4A and B). Nonetheless, even in equatorial regions, there have been many reports of seasonal variations in reproductive effort (Johannes, 1978a). No universal environmental correlate of seasonal spawning has been identified, although the most interesting idea is that spawning activity has been selectively modified to match temporal windows of favorable oceanographic conditions for larval survival and dispersal (Johannes, 1978a; Lobel and Robinson, 1988).

In all the studies for which comprehensive records of daily settlement are available, year-classes have been formed from multiple pulses of settlement, each lasting a few days to a week, superimposed upon a background of very low and more continuous settlement (Fig. 4A–F). There is evidence, however, of regional and species-specific differences in the timing and intensity of these recruitment episodes. For example, the Caribbean grunt has been

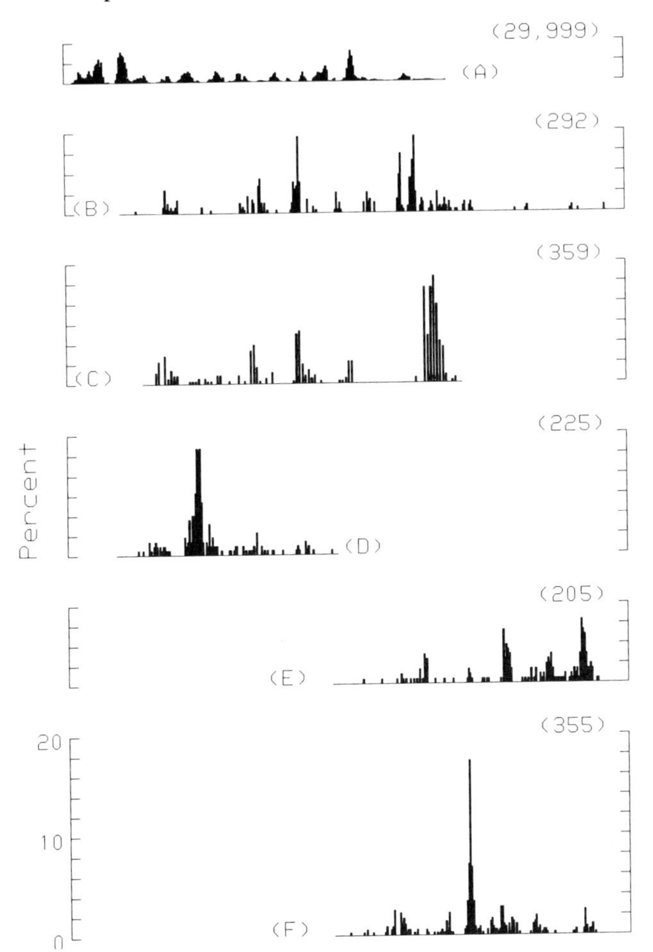

Figure 4 Recruitment scenarios for (A) *Haemulon flavolineatum* from St. Croix, U.S. Virgin Islands, (B) *Thalassoma bifasciatum* and (C) *Stegastes partitus* from the San Blas Islands, Panama, (D) *Amphiprion clarki* from southern Japan, (E) *Pomacentrus* sp. and (F) *Pomacentrus wardi* from the southern GBR. All data have been redrawn to the same scale (including lunar alignment) from the following sources: (A) McFarland *et al.*, 1985, (B) Victor, 1986b, (C) Robertson *et al.*, 1988, (D) Ochi, 1985, (E) and (F) Pitcher, 1987. All, except (A) and (C), are claimed to represent entire year-classes.

shown to settle in most months at regular intervals without huge variations in intensity among cohorts (Fig. 4A). In contrast, both of the GBR damselfishes settled in relatively few months with greater variability in the timing and intensity of individual cohorts (Fig. 4E and F).

Despite the difficulty of detecting cyclic trends in year-classes composed of relatively few cohorts, the data sets for five of the six species in Fig. 4 suggest lunar cycles in settlement (Fig. 5). These five include examples of uni- and bimodal patterns but, with the exception of *Pomacentrus* sp. (Fig. 5e), which shows no clear pattern, all species show low settlement during the days around full moon. It is assumed that settlement during the darker parts of the month is an adaptation to reduce predation on settling reef fishes and some species are known to employ olfactory cues to choose among sites during nocturnal settlement (Sweatman, 1988).

The linkage between spawning activity and settlement can be examined for four of the same species. McFarland *et al.* (1985) back-calculated the fertilization dates of newly settled Caribbean grunts from otolith records and found a dominant semilunar cycle with smaller weekly peaks (Fig. 5g). Larval duration was calculated at a relatively fixed 15 days. Because spawning was not observed directly and the fertilization dates were calculated only from successful recruits, the authors could not be sure that the apparent periodicity in

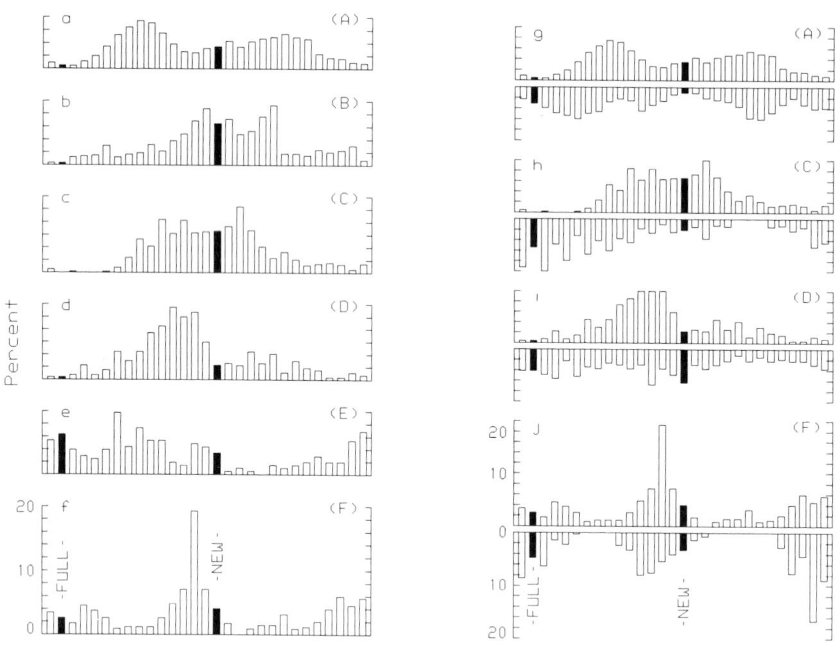

Figure 5 Average monthly settlement cycles from Fig.4 (left series) and average settlement (upper graphs) versus average production cycles (lower graphs) for selected species (right series). Letters in parentheses identify species from the previous figure. Letters on the opposite corner are used as identification in the text.

spawning was a true reflection of breeding effort. They hypothesized that spawning may actually be constant over time and that the bimodality found in settlement may reflect regular periods when favorable currents allow nocturnal transport of presettlement fishes into juvenile nursery areas.

Robertson *et al.* (1988) examined the linkage between spawning activity and settlement based on contemporaneous measurements of clutch size and recruitment in a Panamanian damselfish. These authors conceded that it was unlikely that they had measured recruitment that was sourced from local reproduction but assumed that their observations of both phenomena were typical of other places in the same year. They found that the form of the average monthly settlement cycle matched that of the average larval production cycle (Fig. 5h) with the following deviations. Monthly recruitment episodes were shorter and more variably timed than production cycles. Most of the recruitment in each month was produced by larvae spawned on a few consecutive days that recruited together during a settlement episode of similar duration after a fixed 5-week larval period. Settlement episodes showed four times more variability among months than the production cycles that generated them and the magnitude of settlement in any month could not be predicted from the spawning activity in the previous month. This suggests that both production and planktonic processes control the timing and magnitude of settlement.

Ochi (1985) examined the linkage between spawning and settlement in Japanese populations of an anemonefish (Fig. 5i). The correlation was not impressive but this owes more to the methods used than to any difference in biology. Ochi characterized larval production only by the number of hatching clutches and took no account of clutch size. Others have shown that systematic variations in mean clutch size during the lunar month (Doherty, 1983c) contribute to the cycles of egg and larval production.

Finally, an extended albeit not contemporaneous comparison of larval production and settlement is available for an Australian damselfish (Fig. 5j). Doherty (1983c) showed that this species spawns in semilunar cycles during a relatively short season of three months. Pitcher (1987) reconstructed the settlement of one year-class of this species on a nearby reef in a subsequent season (Fig. 4F). Although these data sets are matched neither spatially nor temporally, there is evidence that both are representative of other reefs in the region in most years (Doherty, 1987a). Figure 5j suggests similar cycling in both larval output and settlement, although the phase relationships are inconsistent with the estimated larval duration of 20 days (Thresher *et al.*, 1989). In any case, close comparison may be inappropriate given that the average settlement cycle (Fig. 5j) was constructed from relatively few cycles and was influenced disproportionately by a single dominant cohort (Fig. 4F). Although not conclusive, the matching periodicity of both spawning and settlement cycles in this species suggests that production processes may control the timing of settlement even in systems where planktonic processes introduce substantial temporal variation into monthly settlement dynamics.

While production processes may explain some of the timing of settlement, they do not explain the large temporal variations in cohort size observed within year-classes (see Addendum for one possible exception). Each of the recruitment series in Fig. 4 contains differences of at least an order of magnitude among the sizes of individual cohorts. A significant portion of this variation undoubtedly represents the random temporal influence of planktonic processes on larval supply. In species where year-classes are formed from relatively few cohorts, these variations may have a large influence on total year-class strength. For example, in Fig. 4F, the year-class was formed from six episodes of settlement of which one contributed greater than 40% of the individuals. In contrast, in Fig. 4A, the year-class contained at least a dozen cohorts of which the largest contributed only 15% to the total. When considering these differences, it must also be remembered that Fig. 4A was generated from a sample size that was two orders of magnitude greater than the others and this alone will reduce variation (Shulman and Ogden, 1987). However, it is important to note that the extreme temporal variation among cohorts revealed in the data series for GBR damselfishes is not an artifact of inadequate sampling size. As discussed in the following, these variations are often coherent among units of habitat and do not disappear when more units are aggregated to form a larger sample.

A final feature of the episodic recruitment of cohorts within year-classes is that pulses of settlement may be multispecific (Fig. 6), with different species, genera, and even families settling during the same brief period (Milicich, 1988). This trend may reflect no more than the lunar entrainment of settlement cycles and tidal flows may also be important in some places. Figure 4 includes two contemporaneous data sets for pairs of species: a wrasse and a damselfish from Panama (Fig. 4B and C) and two damselfishes from the Great Barrier Reef (Fig. 4E and F). Both pairs show evidence of synchronized cycles but no correlations between the magnitudes of contemporaneous episodes of settlement. Mindful of the caveats that must be applied to such interspecific comparisons (see Section III), this suggests that the presettlement stages of different species may have independent dynamics in the plankton even if they share the same water immediately before settlement. This hypothesis has been given additional support by Radtke (1988), who showed significant differences between the isotopic proxies of temperature recorded in the otoliths of two damselfishes that recruited on the same day. His interpretation of these differences was that the two individuals had inhabited different water masses until a week before settlement. If this is the

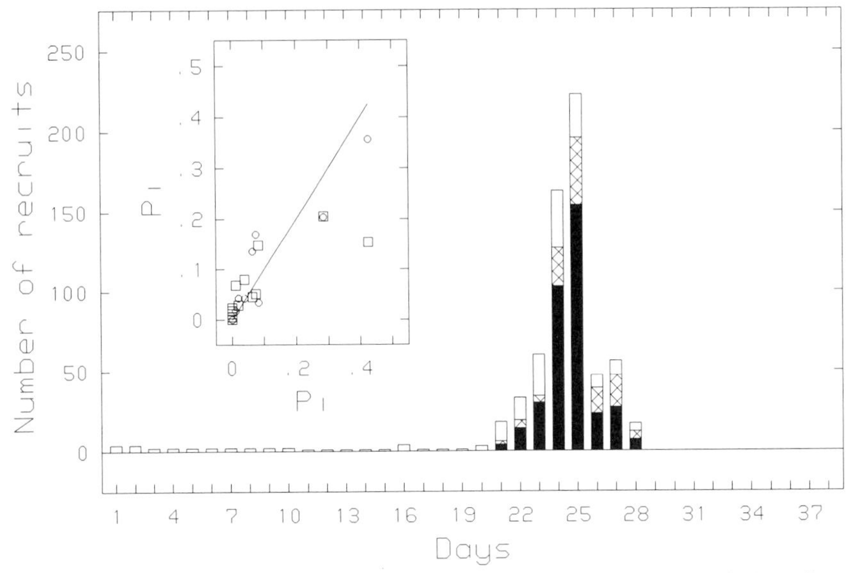

Figure 6 Settlement of *Pomacentrus amboinensis* (shaded), *Pomacentrus* sp. (hatched), and *P. wardi* (unshaded) to the same reef during one short period. Inset compares the proportional daily settlements of the rarer species against that of the most common recruit; the straight line indicates perfect correlation. (Redrawn from Williams, 1983a.)

correct interpretation, it suggests that planktonic mixing processes may have a substantial impact on the composition of settlement.

2. *Among Year-Classes*

While seasonal limits and lunar synchronies detected within year-classes generally remain stable over time, the fine details of settlement typically show little correlation among years (Fig. 7). This confirms that cohort formation is controlled by transient stochastic phenomena and reveals the potential for such phenomena to affect year-class size.

Many, if not most, long-term studies of recruitment have recorded interannual variations in year-class strength on fixed sites of at least one order of magnitude (Doherty and Williams, 1988a), which is similar to the level of temporal variation observed among cohorts within most year-classes. In a 5-year study of Hawaiian fishes, Walsh (1987) showed that all the most heavily recruiting species experienced significant interannual variations in juvenile abundance as a result of fluctuating recruitment and that most of the species experienced at least one exceptionally good or bad year-class during this period. His records of colonization emphasized the independence of recruitment dynamics among species but also suggested some community-

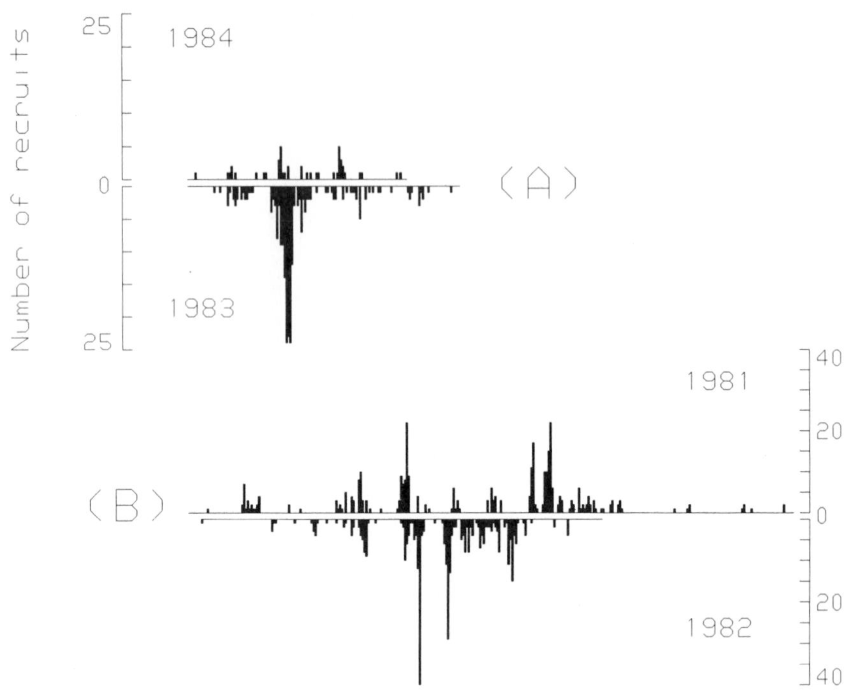

Figure 7 Interannual comparisons of recruitment for consecutive year-classes of (A) *Amphiprion clarki* and (B) *Thalassoma bifasciatum*. Each pair is aligned by moon phase. (Redrawn from Ochi, 1985, and Victor, 1986b.)

wide patterns (Fig. 8). In the first year, colonization was dominated by exceptional recruitment of three surgeonfishes, which all had relatively poor years for the remainder of the study. Three years later, a damselfish and a wrasse had exceptionally good recruitment in the same season while at least five other species experienced almost total failure. This is just one of a number of studies to show that major changes in the relative abundance of juveniles can result from unpredictable interannual fluctuations in the composition of settlement (D. McB. Williams, 1980; Schroeder, 1985).

Interannual variability in recruitment is not just associated with comparisons at small spatial scales. Kami and Ikehara (1976) reported interannual variations of two orders of magnitude in the total harvest of newly settled rabbitfishes from the reef flats of Guam. This suggests that the processes controlling year-class formation on small spatial scales are embedded within, or are a manifestation of, processes operating on larger scales. The lack of autocorrelation in the 13-year record of the siganid harvest (Runs test, $P > 0.9$) suggests that these processes vary independently among years.

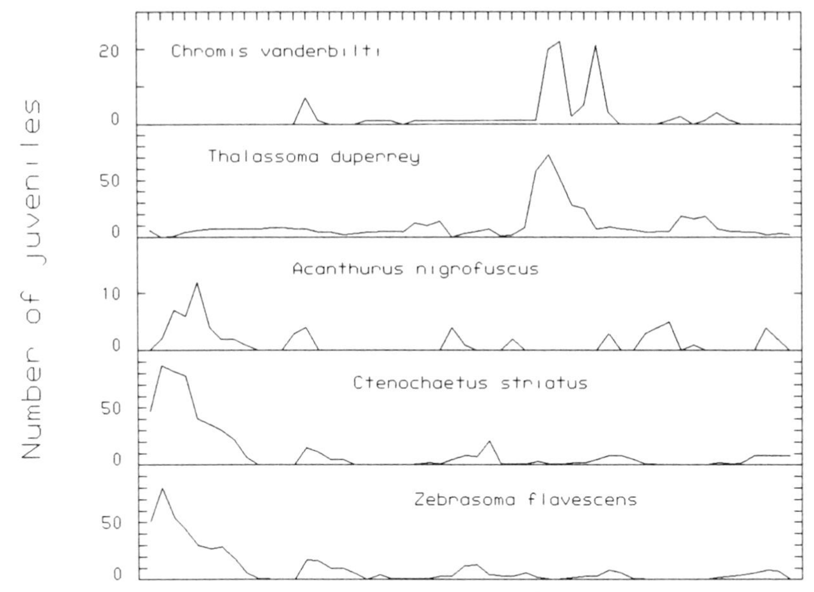

Figure 8 Colonization records spanning 55 months for five of the most common recruits to artificial reefs at Kona, Hawaii. (Redrawn from Walsh, 1987.)

B. Among Replicates within Sites (<1 km)

1. Within Year-Classes

Most species of coral reef fish exercise some habitat discrimination during settlement and some have very precise habitat requirements (e.g., anemonefishes and obligate associates with live coral). Eckert (1987) has shown that choice of settlement substratum may have immediate consequences for postsettlement mortality and there is little doubt that habitat selection by presettlement fishes is an adaptive behavior with a powerful role in shaping community structure among different types of habitat. This active selection should not be confused with variations caused by uneven larval supply but it is often difficult to separate the two processes when making comparisons among natural units of habitat. Because I wish to focus only on larval supply, I have drawn heavily on studies using artificial reefs and other uniform recruitment collectors.

The major result from studies at this scale is that temporal variations (i.e., pulses, episodes) of settlement are often significantly correlated among replicates (Fig. 3A). This indicates that the processes controlling cohort formation are coherent over distances greater than the average separation of those replicates. However, while the timing of such events may be generally synchronized, their magnitude may vary considerably among the replicates. In some species, the spatial variations in the intensity of settlement in one 24-hour period can be of similar magnitude to the daily variations observed on individual reefs during an entire settlement season (Doherty and Williams,1988a).

In some cases, the dispersion of daily settlement among replicates is such that it is most unlikely to have been produced by the independent random settlement of individuals. Instead, it seems likely that some species are aggregated in the presettlement phase and settle to reefs in groups (Shapiro, 1987b). Doherty (1981) showed that negative binomial models provided better simulations of the spatial variance in such situations than simple Poisson alternatives. In biological terms, the former models are equivalent to randomly located clumps of settlement and logarithmic distributions of group size.

In species with contagious settlement, the relative abundance of recruitment among replicates, indicated by the ranking of reefs, has been observed to change unpredictably from one settlement episode to the next within the same year-class (Doherty, 1983a; Victor, 1986b); this confirms that these spatial variations are produced by random encounters between presettlement fishes and reefs rather than by deterministic processes such as habitat selection. One consequence of such dispersion patterns is that spatial variation among replicates will change with replicate size since larger units should be colonized more frequently.

At the end of a recruitment season, spatial variability in year-class strength among replicate reefs is at least partly determined by the number of substantial recruitment episodes forming that year-class. This is because the spatial variations among cohorts should be independent and cancel. Consequently, spatial variations at this scale should be inversely proportional to the number of cohorts within a year-class. Shulman and Ogden (1987) have pointed out that researchers have historically tended to monitor recruitment for short periods and to use small sampling units, both of which are likely to capture small-scale variations in larval supply. They suggested that this has given a false impression of the real importance of recruitment variability. The answer surely is species specific because some fishes have short breeding seasons and never reside on large patches while others have long seasons and/or very mobile juveniles that respond to a different environmental grain. In the Caribbean, reef habitats are often surrounded by lush beds of seagrass that provide the juvenile nursery for many reef fishes. Since these species show ontogenetic migration onto rocky habitats, it is not surprising that they colonize replicate reefs with less spatial variability than species that settle directly into the adult habitat. In other reef systems, seagrass is rarer and ontogenetic migrations are less common. This may be a factor underlying the historical differences of opinion between ecologists working in different systems about the relative importance of recruitment processes to local population dynamics.

Finally, I have treated small-scale recruitment variations among replicates as though the only two causes are habitat selection and variations in larval supply. This overlooks the contributions of predators and other forms of disturbance between the time of settlement and the first recruitment census. Unfortunately, there are few data on this subject but enough to suggest that such factors can be important. Lassig (1983) showed that storm surge produced a substantial redistribution and mortality of newly settled fishes. Although this would normally be classified as a density-independent catastrophe, Andrewartha and Birch (1954) have argued that weather may produce density dependent outcomes when a resource like shelter influences survival.

Predators are likely to be more important than abiotic influences because their effects will be more continuous. Two recent studies have shown significant negative correlations between recruitment patterns of some small reef fishes among artificial reefs and the abundance of resident predators (Hixon and Beets, 1989; G. R. V. Anderson, personal communication). It is important to note that these correlations involve only some of all possible pairwise combinations of species and size classes of prey on the reefs as would be expected when predation is selective. Also, because the settlement of the predators should be controlled by factors similar to those affecting their prey, these effects are unlikely to be predictable among replicates at the start of each recruitment season or stable over time. Thus, they will be difficult to distinguish from variations caused by larval supply without direct measurement of actual settlement.

2. *Among Year-Classes*

Comparisons among fixed replicates over consecutive years confirm that settlement rates are controlled by both deterministic and stochastic processes. In the case of complex natural habitats, fixed differences among years often predominate indicating consistent habitat selections by presettlement fishes (e.g., Sale, Chapter 19). In the case of uniform recruitment collectors, fixed differences are the exception at this scale, confirming the stochastic nature of larval delivery and/or early postsettlement mortality both within and among year-classes.

As a result of the spatial coherence of strong cohorts within year-classes, interannual variations in year-class size tend to be correlated among replicate units of habitat. Thus, in a poor year, most local populations will experience below average recruitment. In a good year, most but not all will gain additional recruits. One of the obvious differences at this scale is that the degree of spatial variation in recruitment among patches rises as a function of average recruitment intensity especially in species that settle in groups (see Fig. 2 in Doherty, 1987a).

C. Among Sites within Coral Reefs (>1 km)

1. Within Year-Classes

At spatial scales greater than replicates within sites, there are few published data sets for which recruitment has been well resolved in time. The two most comprehensive studies both show that transient recruitment pulses may be coherent over distances of several kilometers within the same reef system (Victor, 1986b; Pitcher, 1987). The patterns at this scale are similar to those observed among replicates within sites (Fig. 9) except that the variability reported is often less than that from comparisons of individual units (cf. Fig. 3B versus Fig. 3A). This is purely a matter of technique in that sites are usually characterized by larger samples derived from multiple sampling units so that there is considerable smoothing of the small-scale random variations identified in the previous section. The most important lesson from these comparisons is that the processes controlling cohort formation on individual patch reefs are coherent across distances of several kilometers. This means not only that the timing of such events is consistent over these distances but also the relative magnitude of different cohorts within year-classes is often similar (Fig. 9). In situations where recruitment intensity has been found to vary over kilometer scales, there has been evidence of differential flow among the sites, indicating further the importance of larval delivery systems (Gladfelter *et al.,* 1980; Williams and Sale, 1981). Most studies agree that the abundances of different species are independent at this scale (Fig. 4E and F).

2. Among Year-Classes

When Williams and Sale (1981) monitored recruitment to sites separated by 1–3 km in one coral reef lagoon, they found significant interactions between site and year-class for some species. This suggests that recruitment of a single species may vary among sites within years as a result of transient conditions such as weather and tidal flow experienced at the time of settlement. It also suggests that the independence of different species observed at these scales may be brought about because they are settling at different times under the influence of different conditions.

D. Among Coral Reefs within Regions (Longshore)

1. Within Year-Classes

Victor (1984) presented data for the daily settlement of a wrasse on 18 reefs off the Atlantic coast of Panama. Two major recruitment pulses in midsummer

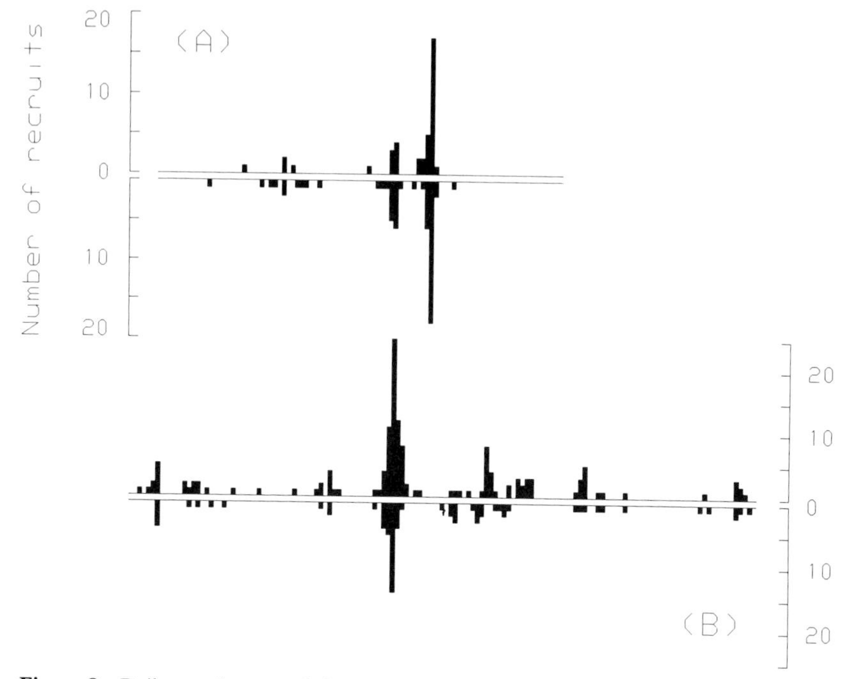

Figure 9 Daily recruitments of (A) *Thalassoma bifasciatum* to sites 1 km apart in the San Blas Islands, Panama, and (B) *Pomacentrus wardi* to sites 4 km apart on the southern GBR. (Redrawn from Victor, 1986b, and Pitcher, 1987.)

were observed to be coherent on many of these reefs representing synchronized settlement on reefs separated by up to 46 km (Fig. 10, reefs A,N,P, and R). On 11 of 14 reefs in a chain offshore and parallel to the coast, settlement peaked within 1–2 days of the same date on both occasions, well within the error of the method used to back-calculate settlement date. Another four reefs lying much closer to the coast did not participate in these events (see the following). Since settlement peaks were symmetrical around the new moon, it is possible that the inferred coherence was simply the result of lunar entrainment and not a mass transport phenomenon, although Victor suggested coordinated delivery of presettlement fish through the region by oceanographic fronts. If the latter hypothesis is correct, the variations in settlement intensity among neighboring reefs suggest that the presettlement wrasses were patchily distributed at kilometer scales along the front (Fig.10, reefs C,D,F, and G versus A,N,P, and R).

In the only comparable study, Pitcher (1987) resolved the daily settlement patterns of a damselfish on four coral reefs within northern and southern regions of the GBR. Within each of these regions, he was able to show that strong recruitment episodes were generally coherent on neighboring reefs,

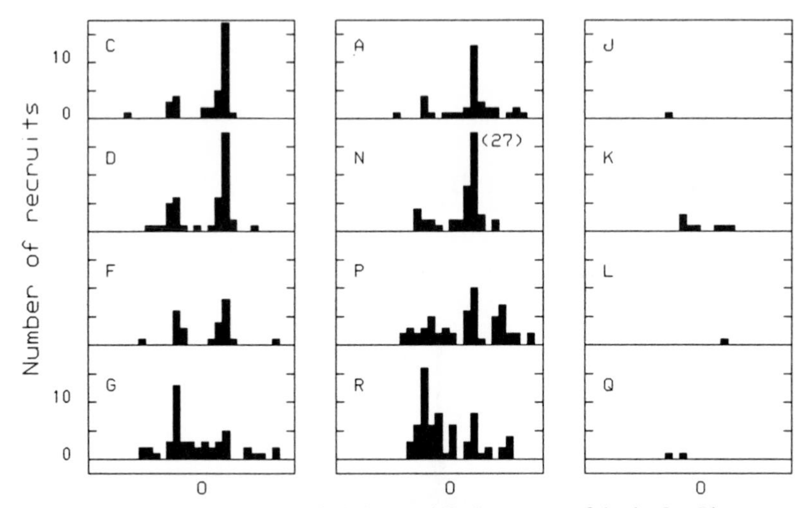

Figure 10 Daily recruitments of *Thalassoma bifasciatum* on reefs in the San Blas archipelago: offshore reefs separated by >1 km (left), offshore reefs separated by <10 km (center), and inshore reefs (right). Letters identify reefs in the original figure and the symbol on the x axis locates the new moon. (Redrawn from Victor, 1984.)

some of which were separated by up to 70 km. This was not always the case, however, and some events either lagged by several days among reefs or were not manifested on all reefs (see Fig. 18 in Doherty and Williams, 1988a). Although Pitcher's study was not designed to allow quantitative comparisons of recruitment strength among neighboring reefs, Fig. 3C shows substantial independence in the relative magnitude of consecutive cohorts recruited to two coral reefs separated by less than 30 km.

Two earlier studies compared year-class size among seven coral reefs from the southern GBR, including the four later studied by Pitcher. Eckert (1984) showed that the recruitment of labroid fishes varied significantly among similar habitats on these coral reefs; that is, there was greater variability over scales of 10 km than 1 km. A similar result was found by Sale *et al.* (1984b), who studied the recruitment patterns of a wide range of fishes to both lagoonal and reef-slope habitats on these same seven reefs. One of the most interesting results to emerge from their study was the different pattern of variability recorded in the two habitats. Only one species of cleaner wrasse was found to show similar spatial variations among reefs in both habitats. Some species were found to be more spatially variable in one habitat than the other, and some were relatively invariable in both. These results show that we need to know more about the habitat preferences and presettlement ecologies of different species to make sense of such different responses.

2. *Among Year-Classes*

The lagoonal censuses reported by Sale *et al.* (1984b) were started in 1981 and continue to the present. To my knowledge, they represent the longest time-series of fish recruitment from any coral reef region. A similar monitoring effort was started in the central GBR in 1983, this one consisting of annual censuses of 0+ fishes on reef-slope habitats from three midshelf reefs along a 30 km transect (D. McB. Williams and S. A. English, unpublished observations). Doherty (1988) summarized data from both of these programs for one common damselfish and showed that interannual variations in year-class size were more coherent among the northern reefs than among the southern ones (Fig. 11). The southern study also revealed the first possible evidence of a long-term change in recruitment of a species at regional scales, although similar change was not observed in the central section (Doherty, 1988). Both differences should caution us not to expect uniform dynamics from reefs in distant regions. Neither study suggested that spatial or temporal variations in year-class size at these scales were coherent among different species.

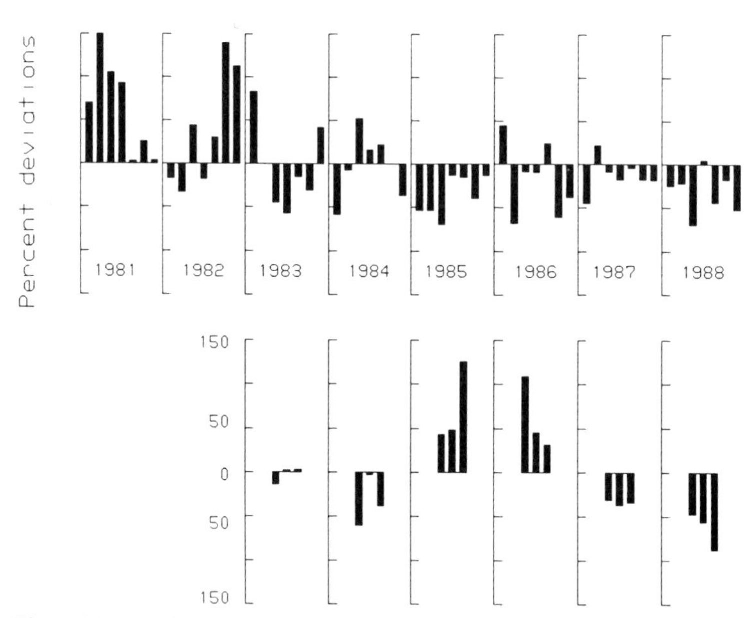

Figure 11 Residuals $(x_i - \bar{x})$ of recruitment time-series for eight year-classes of *Pomacentrus wardi* on seven reefs in the southern GBR (upper series) and six year-classes on three reefs in the central GBR. (Redrawn from Doherty, 1988.)

E. Among Coral Reefs within Regions (Cross-Shelf)

1. *Within Year-Classes*

In addition to the three midshelf reefs, the monitoring program of Williams and English in the central GBR also includes one inshore and one offshore reef so that recruitment can be examined along a 100-km cross-shelf transect. In a preliminary report, Williams (1986b) showed that no species colonized all cross-shelf zones and that the majority of species recruited among zones in similar proportions to the abundance of their adults (see Fig. 14 of Williams, Chapter 16).

The only other relevant study is that by Victor (1984) that was discussed earlier in the context of two synchronized settlements of a wrasse along a 46-km chain of reefs lying roughly parallel to the coast. Four reefs located much closer to the coast received little or no replenishment from these same events (Fig. 10, reefs J,K,L, and Q) and Victor (1986b) subsequently showed that the recruitment levels of this wrasse declined monotonically toward the coast. As in the GBR cross-shelf study, the density of adults and juveniles was directly correlated and Victor found that the best predictor of the abundance of both stages was the degree of exposure to the onshore current. Because growth rates of juvenile wrasse were actually higher in inshore habitats, Victor (1986b) argued against explanations based on active habitat selection and suggested that the outer reefs simply got more exposure to water containing the appropriate presettlement stages.

2. *Among Year-Classes*

Persistent correlations between adult abundance and settlement across the central GBR transect suggest that larval distributions are also the proximate mechanism determining the cross-shelf patterns in that fauna. An alternative hypothesis that the correlation is driven by the distributions of spawners is not supported by hydrodynamic modeling of the region, which suggests that planktonic propagules should have significant cross-shelf dispersal at each generation (Williams *et al.*, 1984). From the stability of the recruitment patterns, Williams (1986b) concluded that the quality of the larval habitat (defined by food/predators/water parameters) is likely to be the ultimate factor limiting further shoreward penetration of this fauna.

In his preliminary report, Williams (1986b) found no evidence of cross-shelf linkages in year-class size for five of six species. The exception was one wrasse found on both mid- and outer-shelf reefs that displayed coherent interannual variations in three consecutive seasons. Doherty (1988) analyzed three additional years of data for one damselfish and found a high correlation between year-class size of this species on inner- and midshelf reefs for five out of six seasons. Since the hydrodynamics of the GBR lagoon make it most unlikely that these reefs were exchanging larvae directly, such correlations are the best evidence yet of broad-scale climatic influence on year-class formation. The exceptional year, when recruitment failed on the inshore reef against the trend on the midshelf, simply illustrates the possibility for local effects to dominate regional trends. All analyses done to date have yet to show any evidence of coherent fluctuations involving more than one species.

F. Among Regions within Biogeographic Provinces

1. *Within Year-Classes*

Information about daily patterns of recruitment at very large spatial scales is limited so far to just one study (Pitcher, 1987). Some of the findings in this study were surprising and counterintuitive. As expected, populations in the warmer north started recruiting earlier in the year than their southern counterparts (Fig. 3D). But, contrary to the conventional wisdom that says that breeding cycles should be extended in more tropical latitudes, the duration of recruitment to southern populations was obviously longer. Furthermore, while settlement within each of these locations was broadly organized into cycles with similar period, these cycles were out of phase, implying settlement on different lunar–tidal cues in the two regions. Since genetic isolation over this distance seems unlikely (Shaklee, 1984), the result must mean that settlement, or more likely egg production, is entrained by local cues that differ between the regions. It is not obvious what these might be.

The timing of recruitment is not the only difference between populations in these two regions. Sweatman (1985a) suggested that average rates of recruitment of common damselfishes may be 10–20 times higher in the north. This may be more than just a latitudinal gradient of abundance. Initial surveys of 25 coral reefs representing five regions of the GBR and a linear distance of greater than 1000 km showed that very few species recruited uniformly among the five regions and that there were examples of many different types of distribution pattern (Doherty, 1987a, 1988; Doherty and Williams, 1988a). Doherty and Williams (1988a) suggested that these regional differences included a component due to different cross-shelf effects at the chosen latitudes and pointed to the impossibility of isolating distance effects from habitat differences at such large scales.

2. *Among Year-Classes*

Interannual comparisons of recruitment on the 25 coral reefs have revealed both deterministic and stochastic influences (P. J. Doherty, D. McB. Williams, and P. F. Sale, unpublished observations). Figure 12 illustrates this point with data from two congeneric damselfishes. In the case of *Pomacentrus*

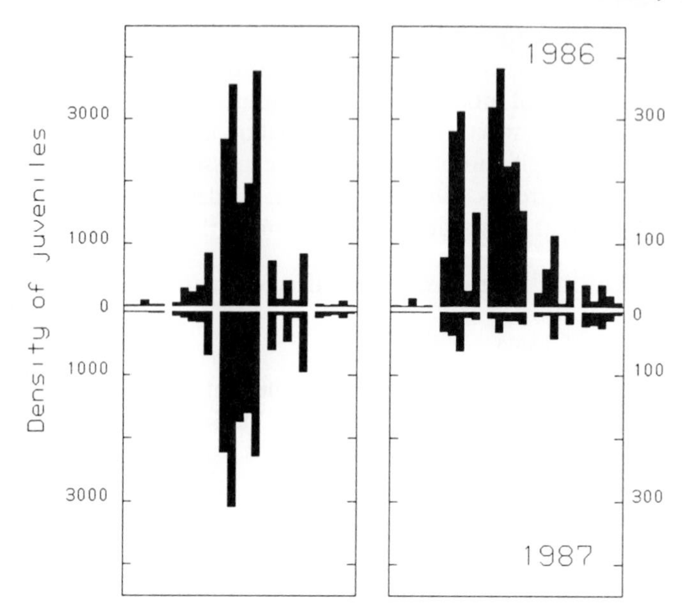

Figure 12 Two consecutive year-classes of recruitment of *Pomacentrus moluccensis* (left series) and *P. wardi* (right series) on five coral reefs in each of five geographic localities spanning >1000 km of the GBR. (Redrawn from Doherty and Williams, 1988a.)

moluccensis, which is a hard coral specialist, recruitment among the 25 reefs was remarkably consistent between years with persistent differences among individual coral reefs within regions as well as persistent differences among regional averages. The total numbers of this species counted in 600 transects each year varied less than 20% among three consecutive year-classes. In contrast, *Pomacentrus wardi,* which is a generalist herbivore, showed much greater instability in its spatial distributions at both scales and a sevenfold change in its total abundance among years. Once again, differences among species are emphasized.

Figure 12 shows that some of the interannual changes affecting *P. wardi* were manifested on a grand scale. In 1987, recruitment of this species collapsed on all the reefs monitored in at least two adjacent regions. This implies recruitment failure over 300–500 km and gives further support to the hypothesis that climate influences year-class size.

V. Synthesis

This chapter was prefaced with questions about the nature of numerical regulation in fish populations, specifically the choice between equilibrial and nonequilibrial alternatives. This is not an esoteric debating point but one in which the outcome affects practical issues like predicting the effects of fishing and the value of artificial reef placements (Bohnsack, 1989). The nonequilibrial paradigm also provides the strongest justification for studying the spatial and temporal patterns of reef fish recruitment.

The sedentary and conspicuous nature of many newly settled reef fishes allows a degree of surveillance over the replenishment process in these animals that is rare among marine fishes. In particular, the spatial information that can be gained by recruitment surveys potentially allows a synoptic view of the formation of a year-class. This pattern can be mapped onto oceanographic variability as measured through remote-sensing in the hope of finding environmental correlates of abundance (Svejkovsky,1988). By including the use of daily aging techniques, this search can be narrowed to patterns correlated with cohort formation, hence my earlier claim that coral reef fishes provide a model system for detecting and understanding the causes of variability in fish populations, an objective that is still a central occupation of fisheries science despite 90 years of effort (Rothschild, 1986).

The potential for future understanding of the recruitment process is revealed by how far we have progressed in a very short period, although it could be argued that the steepest part of the learning curve has been surpassed with the present available methods. Systematic surveys of the recruitment of coral reef fishes are often identified with Sale and Dybdahl (1975), who studied colonization by denuding replicate coral heads every two months from an area of just 12 m^2. Ten years later, results have been gathered on daily events from sites more than 1000 km apart.

Critics have cited the small size of Sale and Dybdahl's (1975) study as a reason for rejecting their conclusions about the general significance of recruitment variability. However, while subsequent surveys of larger structures have revealed different patterns of variation, they certainly have not eliminated recruitment as an agent of change in reef fish communities. Meanwhile, the debate about scale has contributed to a general awareness among ecologists of the hierarchical nature of many ecological processes (Dayton and Tegner, 1984b; Lawton, 1987). The outcome has been a realization that complex ecological phenomena like recruitment are multidimensional problems in which no one scale has logical primacy for all questions.

While this argument explains the hierarchical approach to scale taken in this chapter, it should be stressed that my choice of scales has been fairly arbitrary (essentially following those obvious in the GBR system) and that the underlying distributions of presettlement fish may be nearly continuous over a wide range of space and time. Haury *et al.* (1978) have graphically illustrated this for the processes causing variability of biomass in zooplankton assemblages. One of their major conclusions was that the important processes have covariable spatial and temporal dimensions, that is, that the spatial scales of the major sources of variability increase with the period of observation. A number of authors have discussed the relationship between the scaling of biological variability in marine populations and circulation processes driven by the dissipation of heat in the ocean (Longhurst, 1984; Steele, 1984, 1985).

What evidence is there that physical processes affect the recruitment of tropical reef fishes? It is clear that reproduction in most species is not random. Many of the seasonal and lunar synchronies have been explained as strategies by which animals that do not nurture their offspring try to optimize their survival by spawning gametes into the most favorable oceanographic conditions (Johannes, 1978a). The variability observed in recruitment surveys suggests that there is a substantial component of chance, despite the nonrandom spawnings, in matching reproductive output with favorable conditions for larval survival and/or transport. This unpredictability is also reflected in the small investment that most parents make in individual progeny and the preponderance of iteroparous spawning strategies among species.

By their nature, recruitment surveys reveal spatial information about successful matches between spawning and environment, although recruitment failures can potentially be detected if spawning effort is known. Year-classes of recruitment for many species have been shown to be formed from discrete pulses of settlement that stand out against a background of low continuous colonization. These pulses are unpredictable in exact timing and magnitude. While individual pulses are often multispecific, relative abundance changes from one pulse to the next. Within species, pulses tend to be temporally coherent over spatial scales from meters to tens of kilometers albeit with substantial variability at every scale. At the low end, this variability may represent the spacing among presettlement fish and, in some cases, schooling is indicated. At the high end, this variability may indicate a large heterogeneous patch of larval fishes (Victor,1984) but often a low density of sampling points means that other explanations cannot be excluded, that is, smaller independent patches entrained by similar temporal cues.

While a range of species may recruit during the same narrow period, no consistent associations of species have been detected, especially in the magnitude of settlement pulses. Closely related species are no more likely to be correlated than distant taxa. The independence of species dynamics may reflect the occupation of different water masses in space and time or different performance by species in multispecies assemblages. Almost nothing is known about the comparative ecologies of different ichthyoplankton, although the isotopic tools used by Radtke (1988) have exciting potential to access new information.

At larger spatial scales, it becomes more difficult to sustain the kinds of spatial comparisons that are possible at smaller scales. Quite apart from the greater logistic difficulties, there is greater likelihood of confounding differences in environment with straight isolation by distance effects, an uncertainty that is usually compounded by a lower density of sampling points. Nonetheless, some large-scale trends have been detected, notably the steep gradients associated with distance from land. These species replacements should remind us that dispersal is as much a problem of larval biology as it is of physics.

Within regions, coherent interannual variations in year-class size of single species have been detected over tens of kilometers. Since settlement may be nearly synchronous on widely separated reefs, this suggests that food distributions and/or physical transport mechanisms are simultaneously favorable over large areas. Either hypothesis is likely to reflect the influence of the underlying physical environment. The independence of recruitment variations in different species can be explained if they occupy the same water column at different times or respond in different ways to the same climatic influences. As more comparisons become available at large scales, it is likely that regional differences in hydrodynamics and biological oceanography will also be accentuated.

Finally, at very large scales (>100–1000 km), isolation by distance may become intractably confounded by local and regional habitat differences to the extent that close temporal comparisons of recruitment are unwarranted. The major benefits of monitoring the replenishment over such large scales is likely to be the detection of major environmental change that will only be revealed by long-term monitoring. In such cases, annual surveys of year-class size may be the most cost-effective way of obtaining such data (Sale *et al.*, 1984b).

VI. Future Directions

While recruitment surveys still have much to offer to the understanding of why fish numbers vary, there may also be limits to how well this question can be resolved following the approach described here (hence my comment about how far we have progressed along the learning curve). This limitation arises from the distortion of recruitment data by postsettlement mortality. Shulman and Ogden (1987) reported that more than 90% of newly settled grunts in a Caribbean location were lost within 12 months. While this is one of the highest rates of turnover on record, mortalities of 25–50% in the first month of settled life are commonplace (Doherty and Sale, 1986). Since predation rates are unlikely to be uniform (Shulman, 1985a; Hixon and Beets, 1989), recruitment surveys will at best be imprecise records of settlement. When recruitment data are extrapolated even further to estimate the abundance of presettlement stages, spatial comparisons may also be distorted by habitat selection.

Historically, there has been little choice about using recruitment surveys to estimate the abundance of presettlement fishes despite the lack of precision. Conventional ichthyoplankton surveys have been of little help because plankton nets are very size-selective and they grossly undersample the larger pelagic stages (Choat *et al.*, in press). Instead, net samples are dominated by tiny preflexion larvae that have not been exposed to more than a fraction of the planktonic processes thought to decouple egg production and settlement. Even without this bias, the nature of net sampling means that few studies can afford to resolve both spatial and temporal dimensions of larval assemblages (Doherty, 1987b).

Recently, there have been a number of experiments with alternative sampling methods. One such development has been the plankton purse seine net, which is potentially useful for targeting the surface features (flotsam, slicks) exploited by some species during the final stages of pelagic life (Kingsford and Choat, 1985). Another approach has been the attraction of large net-shy stages with lights during nocturnal sampling and the development of this concept into various automated trap designs (Doherty, 1987b).

Early results from the deployment of light-traps have been very encouraging. Milicich (1988) has used these devices in nearshore waters around one high continental island to monitor the composition and abundance of presettlement fishes within and among years. Her records of presettlement abundance have the same temporal characteristics as recruitment data collected at this scale. Another study has closed this link by demonstrating linear relationships between light-trap catches and daily settlement into nearby habitats (M. G. Meekan, unpublished observations).Consequently, these devices have considerable potential to replace recruitment surveys as the method of choice for monitoring presettlement fish and estimating replenishment. Because they are automated, it is possible to sample many locations simultaneously, including positions away from the reef where recruitment surveys are not possible. While these new tools will make it easier to tackle traditionally daunting questions about larval fishes (e.g., dispersal and connectivity, environment and larval production), recruitment surveys have done much to set the current agenda.

Addendum

In keeping with the rapid evolution in this field over the last 5–10 years, there have been several significant developments between the first and last draft of this chapter.

Contemporaneous sampling of egg production, presettlement abundance, and recruitment for a common damselfish at Lizard Island, northern GBR, has demonstrated tight temporal coupling among all three variables (M. G. Meekan, unpublished observations). Of greater interest, Meekan has found that weekly estimates of egg production spanning two seasons explained 65% of the subsequent variations in settlement intensity. Taken at face value, this result is the first evidence that local stock-recruitment relationships may be possible in coral reef fishes with pelagic larvae.

In a different development, P. J. Doherty and A. J. Fowler have sought to make an empirical test of the recruitment-limitation hypotheses by examining the age structures of populations after monitoring their replenishment for 10 years (see Section IV,D) and developing protocols for the annual aging of tropical fishes (Fowler, 1990b). Preliminary results from one damselfish have shown that the final densities of fish removed from seven coral reefs were directly proportional to the cumulative recruitment inputs witnessed over the 10-year period. Detailed comparisons between the initial and final densities of individual year-classes on the reef with highest densities have shown density-independent mortality and preservation of the recruitment signal in the age structure for up to 10 years (i.e., the "dominant cohort" effect of Fig. 1). These observations add substantially to the view that the population dynamics of coral reef fishes are not qualitatively different from those forming the basis of major fisheries in the temperate zone.

In order not to paint too rosy a picture of progress, I prefer to end on a cautionary note. B. C. Victor (personal communication) has drawn my attention to the enigma posed by two Caribbean wrasses, *Halichoeres poeyi* and *H. bivittatus*. Both have similar early life histories: they are morphologically almost identical, spend the same time in the plankton growing at the same rate, settle at the same size in the same habitat, and thereafter have similar juvenile characteristics. Despite these similarities, one species recruits only during the wet season and the other recruits only during the dry season. The enigma is that the dry season species also spawns during the wet season and planktonic processes appear to act differentially on these two very similar larvae to produce radically different settlement patterns. I include this example to remind readers that the problems discussed in this chapter owe as much to the nuances of larval biology as they do to hydrography and biological oceanography.

CHAPTER 11

Postrecruitment Processes in the Ecology of Coral Reef Fish Populations: A Multifactorial Perspective

G. P. Jones
Department of Zoology
University of Auckland
Auckland, New Zealand

I. INTRODUCTION

No ecologist studying coral reef fishes would doubt that there is life after recruitment. Afterall, it was the brightly colored adults and diverse reef-based assemblages that first captured their imagination (Ehrlich, 1975). However, is life after recruitment important? A cursory reading of the recent literature on the dynamics of coral reef fish populations could lead the beginner to have serious doubts (see reviews by Sale, 1980a, 1984; Munro and Williams, 1985; Doherty and Williams, 1988a,b). "Recruitment" usually refers to the input of juveniles to the observed reef-based population. The possibility that patterns of recruitment explain the observed distribution and abundance of adults has recently been rediscovered and reemphasized in all branches of marine ecology (Butman, 1987; Doherty and Williams 1988a; Roughgarden *et al.,* 1988; Keough, 1988; Underwood and Fairweather, 1989; Sale, 1990).

Although the recruitment message inspired a flood of recruitment studies, turning attention away from postrecruitment events, one can sense that the "new" testament is under revision. The dichotomy between pre- and postrecruitment events has been rejected, and both phases are being examined in a different light (Jones, 1987a; Richards and Lindeman, 1987; Keough, 1988; Mapstone and Fowler, 1988; Warner and Hughes, 1988). To what degree are patterns of recruitment modified by postrecruitment events? If this modification is extensive, what processes are important? This review primarily examines these two questions, in the spirit of compromise (C. L. Smith, 1978), in spite of Occam, and in the light of empirical research.

The current preoccupation with recruitment follows a tradition of fashionable explanations and controversies. The short history of reef fish study has been punctuated by a series of relatively simple models—a recurring pattern in ecological studies (Krebs, 1978; McIntosh, 1986). One can separate the four main theoretical arguments on the basis of only two criteria (Fig. 1). First, there are those that postulate that populations are at the carrying capacity (invoking competition as the major structuring process) and those that postulate that populations are below capacity (Fig. 1, columns). The second criterion distinguishes those that postulate that the organizing processes occur during the postrecruitment phase from those arguing that such processes are unimportant. These models predict that changes are driven by prerecruitment events (Fig. 1, rows).

The first model (Fig. 1, box 1) claimed that competition during the postrecruitment phase structures these diverse communities (and by implication its constituent populations) (Smith and Tyler, 1972; C. L. Smith, 1978). The assumption was that living space (including shelter) on the reef is in short supply, and that there is always an excess of potential recruits attempting to

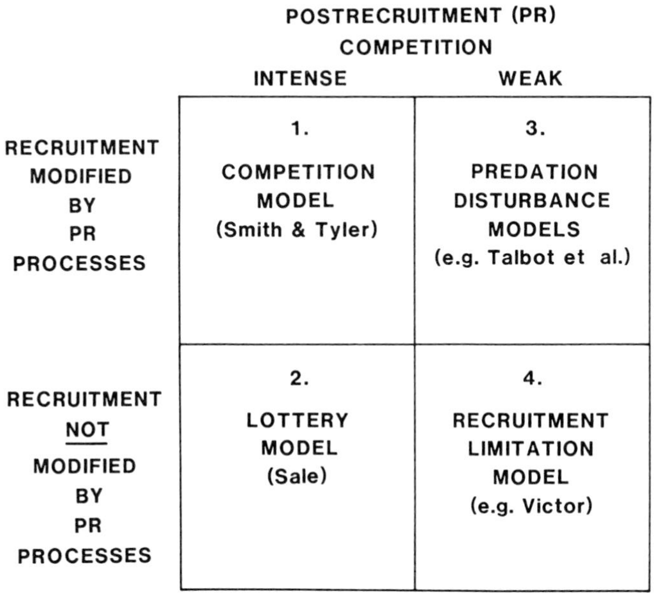

	POSTRECRUITMENT (PR) COMPETITION	
	INTENSE	WEAK
RECRUITMENT MODIFIED BY PR PROCESSES	1. COMPETITION MODEL (Smith & Tyler)	3. PREDATION DISTURBANCE MODELS (e.g. Talbot et al.)
RECRUITMENT NOT MODIFIED BY PR PROCESSES	2. LOTTERY MODEL (Sale)	4. RECRUITMENT LIMITATION MODEL (e.g. Victor)

Figure 1 Classification of simple models that make predictions about factors limiting adult numbers in coral reef fishes. Models can be distinguished on the basis of the importance of competition (intense or weak) and the importance of postrecruitment processes (recruitment modified or unmodified). Competition model: Smith and Tyler (1972). Lottery model: Sale (1977). Predation/Disturbance Model: Talbot *et al.* (1978). Recruitment-limitation model: Victor (1983).

colonize any free space. Therefore, the magnitude of recruitment is determined by competition for space on the reef. As long as resource levels remain constant, the community will remain in equilibrium and population numbers would remain constant.

The first departure from the traditional ecological principle being advocated here was the "lottery hypothesis," proposed in a series of papers by Sale (1974, 1977, 1978a). Sale was the first to realize that patterns in community structure are potentially driven by patterns of recruitment (Fig. 1, box 2). He retained the assumption that coral reef fishes compete for space, but assumed that competition would only influence overall numbers, not the relative abundances of individual species. The central tenet of the lottery hypothesis is that patterns of change in community structure are driven by stochastic recruitment events.

A third group (Fig. 1, box 3) of ideas can be distinguished that have in common the assumption that factors influencing the mortality of adults during the postrecruitment phase keep populations below the level at which resources become limiting. This includes the contention that predation pressure is higher in the tropics than in temperate regions (Goldman and Talbot, 1976; Johannes, 1978a) and is the major process limiting populations and structuring reef fish assemblages, and the major selective agent affecting reproductive strategies. Talbot *et al.* (1978) modified this view by arguing that populations vary not only as a response to predation pressure, but also to the uncertain nature of recruitment. Bohnsack and Talbot (1980) incuded physical disturbance in this nonequilibrial view of reef fish communities, suggesting that global patterns of successional change in community structure reflect differences in disturbance regimes.

The most recent proposal has been termed the "recruitment-limitation" model (Victor, 1983a, 1986b; Doherty, 1982, 1983a,b). The argument is that larval supply is normally insufficient for total population size to ever reach a carrying capacity determined by resource levels (Fig. 1, box 4). Population changes and age structure should then reflect variation in input and be relatively independent of postrecruitment processes. If this central tenet is true, then studies on larval rather than postrecruitment events will shed more light on the processes limiting the abundance of reef fish populations.

The fact that recruitment is variable in space and time is arguably the only undisputable fact to come out of 20 years of research on the numerical structure of coral reef fish assemblages (e.g., Williams and Sale, 1981; Williams, 1983a, 1986b; Sale and Douglas, 1984; Sale *et al.*, 1984b; Shulman, 1985b; Doherty, 1987a). Variable recruitment is an inevitable consequence of species producing many small dispersed propagules, which may suffer on the order of 99% mortality prior to reaching the reef (Fig. 2). Because numbers settling to the reef determine the initial abundance of each juvenile

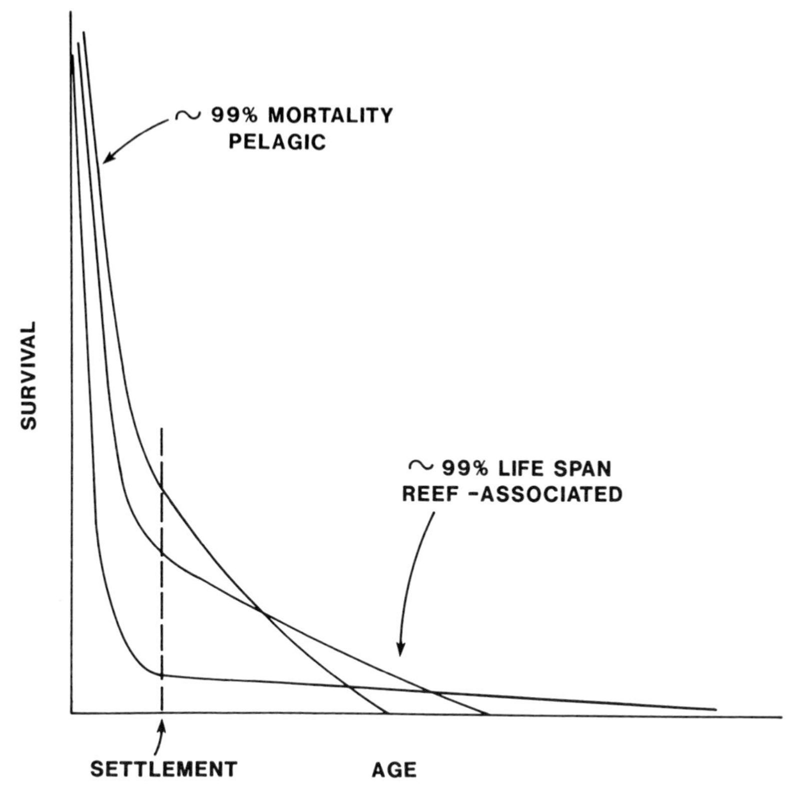

Figure 2 Hypothetical survivorship curves for three cohorts of a coral reef fish species. Despite large variation in settlement, the relative sizes of the cohorts at recruitment can be modified by differences in postrecruitment mortality over the life span.

cohort observed on the reef, recruitment must be an important value in marine population models. But is it all-important? It is frequently forgotten that on the order of 99% of the potential life span occurs after settlement to the reef, although few fish live to enjoy it (Fig. 2). Integrated over tens of years, small changes in postrecruitment parameters can dramatically alter the patterns set up at the start of reef life in long-lived species.

This chapter is a message to the uninitiated. Life after recruitment IS important. Postrecruitment processes are reexamined in a multifactorial framework. I take a critical view of single-process explanations (e.g., competition) or those based on single demographic events (e.g., recruitment). Comprehensive and useful models of coral reef fish populations must take into account the demographic changes and ecological processes occurring through the entire life history. Such models must be based on a clear quantitative description of the patterns in question, the demographic parameters establishing these patterns, and the variety of ecological processes that may be the ultimate cause of patterns observed.

II. Patterns in the Postrecruitment Phase

It must always be borne in mind that population ecologists are trying to explain the patterns in the distribution and abundance of the reef-associated phase. The relative importance of different demographic parameters and contributing processes cannot be assessed without reference to these patterns (Jones, 1987b). This raises the question as to what variables are important in population ecology. What patterns do we choose to explain and why?

The present emphasis on total population numbers (all individuals of the species in a defined area) stems from the community-diversity oriented history of reef fish studies. In early publications one can find only occasional reference to the abundance of an individual species. Information on important secondary characteristics such as breeding population size or age structure is completely absent. The first review of the population biology of coral reef fishes (Ehrlich, 1975) is most notable for the virtual absence of any information on population biology. It has been widely recognized that many of the patterns observed for complex assemblages reflect changes in the abundance of the constituent species (Strong *et al.,* 1984b). Many of the questions relating to species diversity cannot be answered until the necessary information on the population ecology of individual species accumulates.

One has to seriously question whether population studies should primarily focus on total numbers (Doherty and Williams 1988a). Total numbers, adult numbers, relative proportions of the mature sexes, and age and size distributions are all equally valid variables to consider in numerical population studies and should logically be considered in a hierarchical manner. By dividing the population into smaller, more homogeneous subunits, the primary factors affecting patterns may be obtained.

Doherty and Williams (1988a) have argued that factors limiting adult numbers are of little importance, because recruitment to reef populations is independent of local variations in breeding stock size. Even if this assertion turns out to be true, their argument misses the point completely. Total population numbers (adults + juveniles) appear not to determine recruitment to the reef either, yet this variable has been the major focus of these workers. Total population size is relevant, but why is it any more relevant than focusing on adult numbers? Since the main emphasis of terrestrial population studies has been the breeding population size (Begon, 1984; Crawford, 1984), there is enormous comparative value in examining this variable. The historic focus on total numbers in reef fish studies has led to a situation in which important processes affecting substantial proportions of the reef population have been completely glossed over (Jones, 1987a).

If the recruitment-limitation model is interpreted literally, it can become a trivial statement when applied to the distribution and abundance of all members of a species. Populations tend not to exist in places that the larvae never reach. In species in which juveniles make up a large proportion of the population (i.e., short-lived species), total densities will almost invariably fluctuate in response to recruitment variation. However, in these situations it does not follow that adult numbers, or the age structure, will be primarily determined by variation in recruitment.

Current advances in reef fish ecology are being hampered by the incomplete description of population patterns. Few studies have provided information on the distribution and abundance of adults (but see Thresher, 1985a; Jones, 1987a). Use of the term "adults" is frequently arbitrary and unrelated to breeding status (e.g., D. McB. Williams, 1979). Some studies have focused on completely nonreproductive populations (e.g., Ralston, 1976b), which are likely to reveal totally different processes than those limiting breeding populations.

Information on secondary characteristics such as age structure is scarce (Williams, 1978). Constancy in overall densities can mask an underlying population instability, which can only be revealed by a description of the dynamics in age structure. It is recognized that short-lived species may be limited by different processes than long-lived ones (Jones, 1987a; Warner and Hughes, 1988). Knowledge of the age structure provides the appropriate time frame for observations and experiments examining these processes. Without such data it is difficult to examine historical responses of populations to recruitment fluctuations or to make predictions about how populations will change in the future. As technological difficulties in aging tropical fish are overcome, using both annual and daily annuli on otoliths (e.g., Samuel *et al.*, 1987; Thresher, 1988), and information relating size and age accumulates, we will see a greater application of such data to this important ecological problem (Sale *et al.*, 1985).

Data on size distributions have been easier to collect and have revealed a number of important processes not evident when body size is ignored (Doherty, 1983a; Thresher, 1985a; Jones, 1987a; Gladstone and Westoby, 1988). Among these processes has been the importance of variation in growth parameters in determining adult population size (Jones, 1984b, 1987a). Consideration of patterns in the biomass of reef fishes is at an early stage (Doherty, 1982; Williams and Hatcher, 1983; Roberts, 1987), though the comparison of density and biomass responses to fluctuations in recruitment should be a productive line of research.

Another major problem of interest in population studies concerns the factor(s) affecting reproductive output. Again, a different set of processes may be involved than those limiting total numbers of individuals, etc. Thresher's (1985a) study on *Acanthochromis polyacanthus* contains the only detailed description of spatial and temporal variation in reproductive characteristics of a reef fish. Although future change in adult numbers appears not to relate closely to the production of offspring, the latter must be the starting point in any multifactorial description of the fate of each cohort.

III. Evidence for Recruitment Modification

Two important criteria must apply if adult populations are controlled by recruitment. First, recruitment must be independent of resident adult population size. One of the perceived inadequacies of the models based on competition for space was the seeming absence of any effects of resident populations on the recruitment of juveniles to the reef habitat (D. McB. Williams, 1980; Doherty, 1983a). However, resident removal experiments have produced the complete range of results, including positive effects of residents on recruitment (Sweatman 1983, 1985a; Jones, 1987b), negative effects (Sale, 1976; Shulman *et al.,* 1983; Sweatman, 1985a; Jones, 1987b), and no measurable effects (D. McB. Williams, 1980; Doherty, 1983a; Sweatman, 1985a; Jones, 1987b). Where strong interactions exist, the magnitude of recruitment is clearly being modified, and it cannot be said that populations at that scale are recruitment-limited.

The second criterion is that it must follow that adult density is dependent on recruitment levels, that is, the converse of the preceding statement must be true. However, the only study that has examined the relationship between recruitment levels and numbers subsequently reaching reproductive maturity indicated that the two variables were not closely related (Jones, 1987a). Focus on the resident–recruit interactions is probably misplaced, since most species undergo profound changes in their use of resources during their early years on the reef. Interactions between juveniles and adults are more likely to develop as juveniles grow to sizes at which they begin to use the same shelter and food resources as adults.

The most compelling arguments for the recruitment-limitation model are based on correlations between recruitment levels and long-term changes in adult abundance. For example, Doherty and Williams (1988a) describe the temporal relationship between "adult" (fish > 1 year) and juvenile densities in *Pomacentrus amboinensis* (Fig. 3). Increases in adults "appear" to follow successive increases in peaks of recruitment. However, the magnitude of the increase in adult numbers is not obviously proportional to the size of the recruitment pulse.

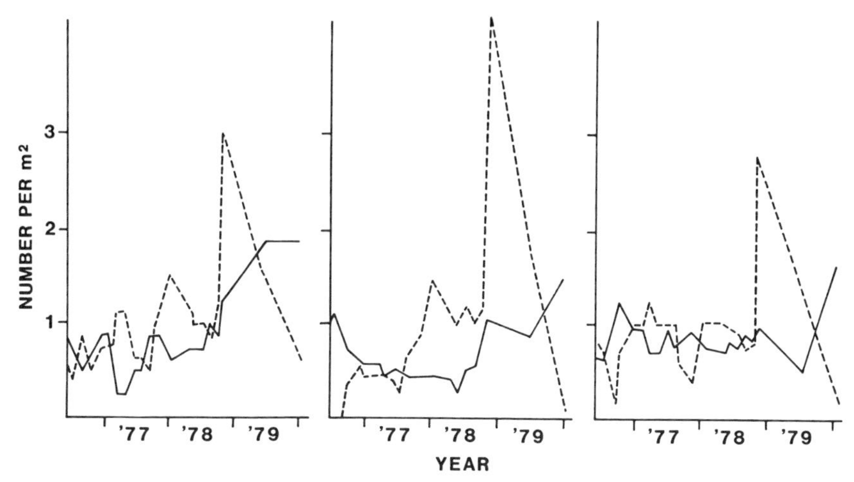

Figure 3 Temporal relationship between "adult" (solid line) and "juvenile" (dashed line) densities in *Pomacentrus amboinensis* on patch reefs within One Tree Lagoon. (Modified from D. McB. Williams, 1979.)

There are considerably more examples accumulating for which changes in adult numbers do not correspond to the immediate history of recruitment events. Jones (1990) looked at natural changes in "actual" adult numbers (fish > 50 mm) in *P. amboinensis* and found that breeding population size plateaued above a certain minimum recruitment (Fig. 4). This was supported by an experiment in which recruitment levels were manipulated over a 3-year period (Jones, 1990). The contrasting results of Williams and Jones do not necessarily represent a conflict, but do emphasize how a slight change in the pattern of focus can lead to totally different conclusions.

Robertson (1988a–c) examined the relationship between adult numbers and recruitment in three completely different taxa. Densities of adults in three surgeonfishes were not correlated with settlement patterns over a 6-year period (Robertson, 1988a). A large pulse in the recruitment of the triggerfish *Balistes vetula* (50–100 times normal) resulted in only a slight increase in adult numbers (1.5 times normal) (Robertson, 1988b). Adult populations of the damselfish *Abudefduf saxatilis* were unrelated to settlement over a 6-year period (Robertson, 1988c). More recently, Forrester (1990) found that initial patterns of recruit abundance in *Dascyllus aruanus* were highly modified by a variety of processes.

Clearly there are sufficient data to implicate the importance of postrecruitment processes in many species. Obviously it is not a simple problem to determine which processes are important, and at what stage in the pre- and postrecruitment phases they operate. If one takes adult population size as a

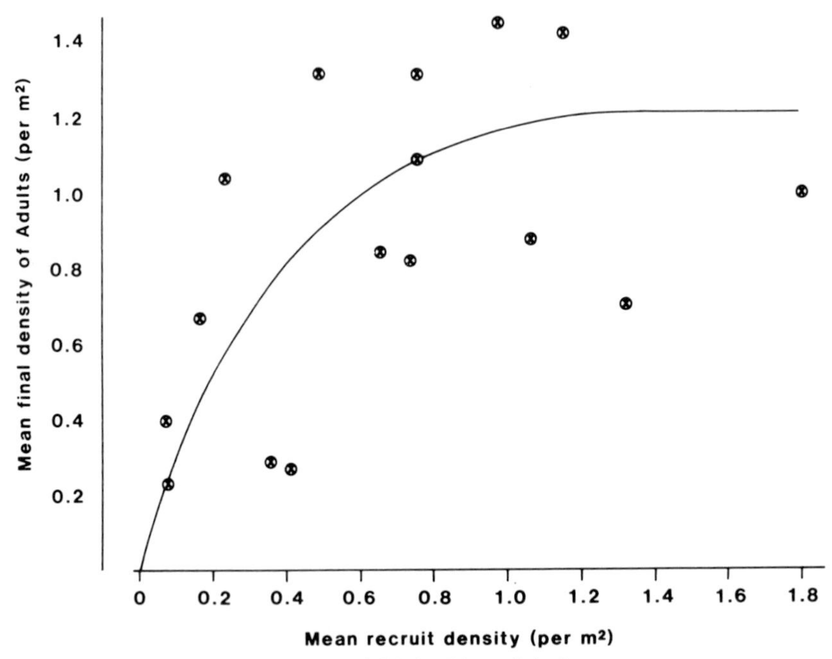

Figure 4 Relationship between true "adult" densities and the between-year average recruitment success for *Pomacentrus amboinensis* on 16 natural patch reefs monitored over three years (Jones, 1990).

temporary focus, then there is a complex series of demographic parameters that may all ultimately contribute to changes in adult population size. Figure 5 traces the fate of a cohort through the pelagic (top half) and benthic (bottom half) stages of the life cycle. Movement, mortality, and development time, during larval and juvenile phases, can contribute to the numbers of individuals reaching maturity from any given cohort. The size of the adult population may also be modified by adult mortality and movement patterns. A further complication is that there may be large differences among generations or cohorts. Despite the considerable attention population ecologists have given to a small number of species, there is none for which all this information is available, even for one location. And just how typical any one location is for any one species has never been ascertained.

The slow task in accumulating and comparing the magnitudes of different processes at different stages is well under way. For the postrecruitment phase, I first review patterns in the demographic parameters that directly determine changes in the primary (abundance) and secondary (structure) characteristics of populations. I then consider the ecological processes (competition, pre-

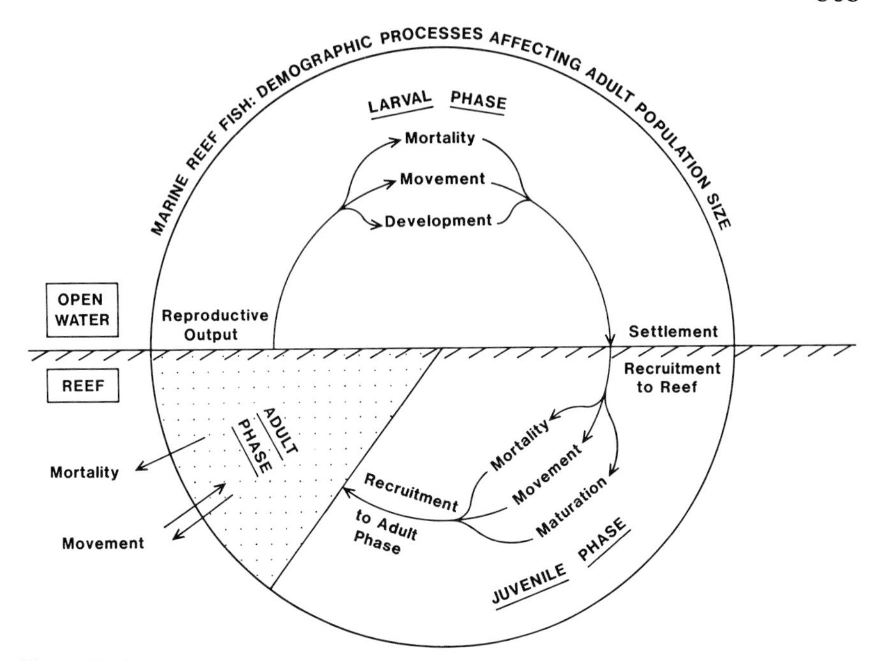

Figure 5 Complex sequence of demographic processes influencing a single cohort that will ultimately determine adult population numbers in a coral reef fish.

dation, etc.) that, through their effects on the demographic parameters, produce the patterns of abundance and structure.

IV. Demographic Parameters

Where the information exists, spatial and temporal variation in mortality, movement, growth, and maturation will be examined at five different levels: (1) among species, (2) among places and times within a species, (3) among sexual or reproductive categories, (4) among age classes, and (5) among individuals.

A. Mortality

Data on mortality are enormously difficult to obtain and are variable in quality because of problems with detecting the loss of known individuals, partitioning losses from mortality versus emigration, differences in methodologies, lack of statistical confidence limits, and a host of other problems. However, since mortality is the most direct way in which recruitment patterns can be numerically modified, it is critical that we overcome these difficulties.

Munro and Williams (1985) were the first to assemble mortality estimates for a variety of different reef fish species and taxa. Mortality rates of adults appear to differ considerably among species (Munro and Williams, 1985; Eckert, 1987; Sale and Ferrell, 1988). For example, Eckert (1987) found that adult survival varied from 4.9 to 69.5% per annum in a guild of wrasses, which extrapolates to projected life spans from 1.6 to 11.5 years. Such data may not represent real interspecific differences in mortality rates, however, because they do not take into account temporal (within and between generations) and spatial (within and between habitats) variation in mortality rates within each.

Since many recent studies indicate that age-specific differences in mortality rates are common, single mortality estimates for a species are clearly inadequate (Doherty and Sale, 1985; Victor, 1986b; Eckert, 1987; Shulman and Ogden, 1987; Sale and Ferrell, 1988). Most species exhibit extremely high mortality rates in the first days or weeks after settlement, after which survival gradually improves. This has two important consequences: (1) slight variation in survivorship during this time may produce substantial spatial and temporal variation in the size and structure of populations; and (2) settlement and early mortality patterns are inextricably confounded in most recruitment estimates.

Different cohorts have been shown to suffer very different mortality rates. For example, adults of *Halichoeres melanurus* exhibited 33% mortality in one year and 81% in the next (Eckert, 1987). Thresher (1983b) observed considerable differences in juvenile survival among broods of *Acanthochromis*, although this did not alter the correlation between numbers of offspring produced and numbers "fledged."

Mortality rates may differ widely among locations. Aldenhoven (1986b) monitored losses of *Centropyge bicolor* at four sites near Lizard Island, where life expectancy estimates varied from 1 to 5–13 years. Survivorship estimates for juvenile *Pomacentrus wardi* and *P. amboinensis* at One Tree Reef were much greater on deep patch reefs (>3 m depth) than on shallow reefs (Doherty, 1980; G. P. Jones, 1986), and for *P. amboinensis*, they were also greater near the perimeter of the lagoon than in the center (Jones, unpublished). On a larger scale, Eckert (1985b) found major differences in survival between lagoon and reef-slope populations of *Thalassoma lunare*.

Although data are still very limited, variation in mortality rates, whether correlated with or independent of density, is likely to be of major importance to any study on the structure and dynamics of fish populations. However, since recruitment fluctuations can be large, it may take some time before the input patterns are completely modified (Fig. 2). The importance of mortality may not be evident from short-term studies of one or two months in duration (e.g., Meekan, 1988).

B. Movement

The focus of reef fish studies on relatively sedentary taxa (e.g., Pomacentridae) has diverted attention from the issue of movement and its effect on spatial and temporal patterns in population size and structure. However, even in site-attached species, estimates of mortality have an undisclosed component due to the balance of immigration and emigration. It is expected of more mobile taxa (e.g., Lethrinidae) that the patterns of distribution across entire reef systems will be to some degree determined by postrecruitment habitat selection.

Difficulties in quantifying movement make this the most challenging demographic parameter to assess. It can only be measured directly by observing individually recognized or tagged fish shifting from one place to another and then calculating the net effects on the population. Since relocation may be unpredictable and occur almost instantaneously, it may be difficult to detect without continuous observation. Given these difficulties, it is not surprising that the suggested importance of movement comes from indirect sources.

Ontogenetic changes in depth and habitat preferences are evident from information on the size–age structures of populations in different places (Helfman, 1978; Werner and Gilliam, 1984; Jones, 1988a). For example, juvenile *Haemulon flavolineatum* settle onto sand and seagrass beds and migrate to nearby reefs within a few weeks (Shulman and Ogden, 1987). Different habitats are also occupied by juveniles and adults in five species of Indian Ocean surgeonfishes (Robertson *et al.,* 1979). Adults and juveniles of *Ctenochaetus hawaiiensis* and *Thalassoma duperrey* occupy lava reefs of different ages in Hawaii (Godwin and Kosaki, 1989).

Strong indications of the importance of movement among isolated patch reefs have been obtained for surgeonfishes (Robertson, 1988a) and the pomacentrid *Abedufdef saxatilis* (Robertson, 1988c). The lack of any relationship between adult numbers and settlement over a 6-year period was partially attributed to relocation of adults among the six patch reefs of the study. If movement is important among isolated patch reefs, then it is likely to be even more important on large habitat mosaics, which constitute the greater part of most coral reef systems.

To get a balanced picture of reef fish dynamics, we need to establish spatial scales over which the different demographic parameters are of greater or lesser importance for taxa of different relative mobilities, and for species occupying different types of habitat. The importance of recruitment variation in determining adult numbers must be examined over larger scales for more mobile species, but this cannot be done without accounting for the patterns of movement and redistribution within these large areas.

C. Growth

The rate at which individuals increase in weight and length is probably the demographic parameter that is most sensitive to environmental conditions and also the easiest to measure. Variable patterns in growth can have a major and direct influence on the size structure of reef fish populations. Maturation, the size of the adult population, and its reproductive output can vary as a direct consequence of growth (Jones, 1984b, 1987a), since these variables are usually more dependent on size than age in fishes. Indirect effects of growth on absolute population numbers may also occur. If predation on juveniles is heavy, the slower-growing individuals may be subject to higher age-specific mortality rates. Although this possibility has been recognized (Doherty and Williams, 1988a; Mapstone and Fowler, 1988), it has not yet been tested.

Munro and Williams (1985) summarized the growth information for 125 tropical species spread across 22 families (not all reef fish). Growth patterns varied greatly among species within families, with no apparent familywide patterns or similarities among ecologically similar species. However, with no estimate of the variation in the growth patterns within individual species, such comparisons are suspect.

Variation in the growth of a species among areas may be extreme (Thresher, 1983b, 1985a; G. P. Jones, 1986; Gladstone and Westoby, 1988). Thresher (1985a) compared the growth of juvenile *Acanthochromis polyacanthus* at 10 locations in One Tree Lagoon over two summers (Fig. 6). He found that variation in growth rates among places was considerably higher than variation between years. Two sites exhibited substantially higher growth rates in both years. Vertical spatial patterns also exist. G. P. Jones (1986) showed that growth in *Pomacentrus amboinensis* was faster in the deeper water of One Tree lagoon. Growth may be faster during warmer months (Jones, 1987a; Gladstone and Westoby, 1988), and males often grow faster than females [e.g., *Canthigaster valentini* (Gladstone and Westoby, 1988)]. Individual variation in growth rates is also very important, as evidenced by the increase in the variance in the mean size of juveniles as they grow (Doherty, 1982; Thresher, 1983b).

D. Maturation and Fecundity

The degree to which size and age to reproductive maturity vary for coral reef fishes is not well known. If maturation is more closely related to size than age [as in *Pomacentrus amboinensis* (Jones, 1987a)], then age to maturity will be

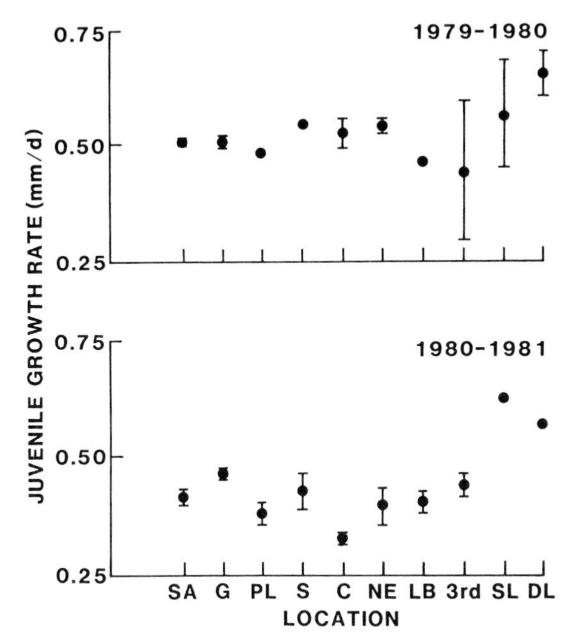

Figure 6 Spatial variation in the growth rate of two cohorts of juvenile *Acanthochromis polyacanthus* in One Tree Lagoon. Error bars = 2 × SE. (Redrawn from R. E. Thresher, 1985. Distribution, abundance, and reproductive success in the coral reef fish *Acanthochromis polyacanthus*. *Ecology* **66,** 1139–1150.)

very dependent on patterns in growth. Gladstone and Westoby (1988) provide the only account of site-specific variation in female maturity. The effect of differences in maturation time on input to the adult population has been pointed out by Jones (1984b, 1987a). The longer maturation time is, the fewer individuals survive to join the breeding population. Some species appear not to mature at all in some habitats [e.g., lagoon populations of *Chaetodon rainfordi* at One Tree Reef (A. J. Fowler, personal, communication)].

Since female fecundity increases with body size in coral reef fishes (e.g., Gladstone and Westoby, 1988), fecundity must also be influenced by any variation in growth rates. For *Acanthochromis polyacanthus*, Thresher (1985a) described considerable variation among locations in One Tree Lagoon, in both female fecundity and reproductive success or numbers fledged (Fig. 7). While the differences in female fecundity among sites were consistent between years, reproductive success was considerably more variable. The differences must be due to processes affecting juveniles during the brood stage.

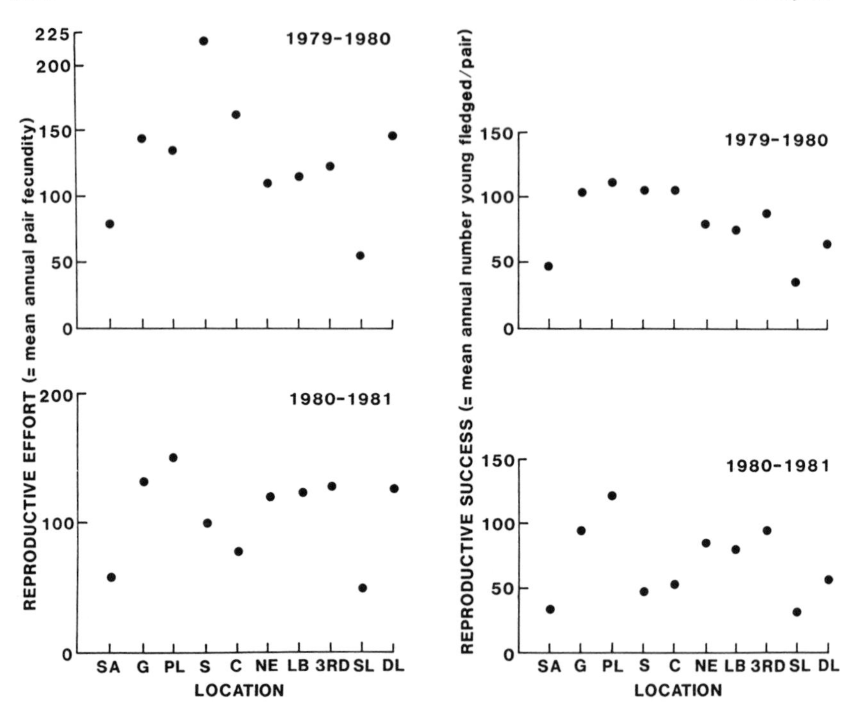

Figure 7 Spatial variation in the fecundity and reproductive success of pairs of *Acanthochromis polyacanthus* in One Tree Lagoon over two spawning seasons. (Redrawn from R. E. Thresher, 1985. Distribution, abundance, and reproductive success in the coral reef fish *Acanthochromis polyacanthus*. *Ecology* **66,** 1139–1150.)

V. Ecological Processes

A. Habitat Structure

Fish populations may respond in predictable ways to spatial and temporal changes in the structure of the reef with which they are associated. The term "habitat" includes such things as the physical and biological characteristics of the substratum (e.g., coral cover, substratum complexity, degree of siltation, etc.). A population response to changes in habitat structure is consistent with, but does not imply, resource limitation (Doherty and Williams, 1988a,b). However, fish–habitat interactions may be a key factor in predicting changes in overall abundance, breeding population size, and other aspects of population structure (Jones, 1988a).

Adult densities have been correlated with reef height (Thresher, 1983a), reef size (Warner and Hoffman, 1980a), depth (Thresher, 1983a), coral cover

(Nagelkerken, 1977; Bell and Galzin, 1984; Bell *et al.,* 1985), topographic complexity (Luckhurst and Luckhurst, 1978c; Thresher, 1983a; Roberts and Ormond, 1987), and distance from emergent reef (Thresher, 1983b). Roberts and Ormond (1987) have drawn attention to the importance of spatial scale in fish–habitat associations. Surface area index was a good predictor of the abundance of pomacentrids along 200-m transects in the Red Sea, but not at a smaller scale (10 × 2-m transects).

Many of these fish–habitat associations appear to be established at the time of settlement (Sale *et al.,* 1984a; Eckert, 1985a). The extent to which habitat influences postrecruitment processes has only recently been considered, and in most cases we do not know whether it reinforces or alters patterns established at settlement. *Pomacentrus amboinensis* prefers to settle in deeper parts of lagoons (Eckert, 1985a), where subsequently it grows faster and survives better than in shallow water (G. P. Jones, 1986). While juvenile growth and mortality in *Pomacentrus wardi* and *P. amboinensis* are affected by the depth of reefs and their positions in the lagoon (Doherty, 1980; G. P. Jones, 1986), the habitat characteristics responsible for this remain unknown. G. P. Jones (1986 and unpublished observations) established that such patterns exhibited by *Pomacentrus amboinensis* and *Dascyllus aruanus* were evident for juveniles transplanted to identical coral substrata at different depths and locations. This suggests that these places had a different potential for growth and survival unrelated to changes in the distribution of coral substrata.

Other work has clearly implicated the importance of coral substratum. Jones (1988b) examined the growth and survival of two planktivorous damselfishes, *Pomacentrus amboinensis* and *Dascyllus aruanus*, on reefs made from two different coral species, *Pocillopora damicornis* and *Porites* sp. Figure 8 shows the survival of *D. aruanus* on the two coral types, at different densities, and in the presence and absence of a potential competitor (*P. amboinensis*). The effects of substratum far exceeded those due to competitive interactions. The complete analysis showed that both fish species exhibited lower mortality rates and grew faster on *Pocillopora* reefs, perhaps because this coral provides better shelter where the fish can devote more time to feeding.

Establishing the relative importance of recruitment and postrecruitment processes in determining fish–habitat associations at different scales should be a priority area for future research. Postsettlement habitat selection and the influence of habitat on adult mortality are particularly in need of study.

B. Disturbance

The importance of disturbance is an integral part of the nonequilibrial view of reef fish communities, although early proponents seldom provided any concrete evidence for the assertion (e.g., Talbot *et al.,* 1978; Bohnsack and

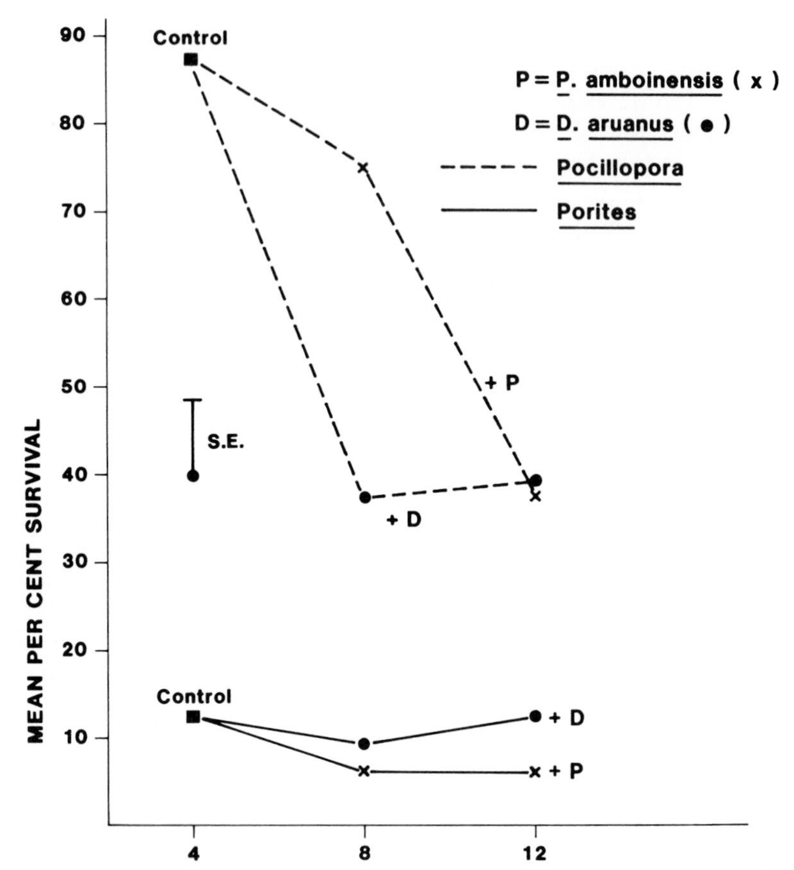

Figure 8 Influence of of coral substratum (*Porites, Pocillopora*), *Pomacentrus amboinensis* density (+P), and conspecific density (+D) on the percentage survival of juvenile *Dascyllus aruanus* over one year. Pooled SE indicated. (Redrawn from G. P. Jones, 1988. Experimental evaluation of the effects of habitat structure and competitive interactions on the juveniles of two coral reef fishes. *J. Exp. Mar. Biol. Ecol.* **123,** 115–126.)

Talbot, 1980). The impact of physical and biological disturbance on reef fish populations depends on its frequency and magnitude, relative to the longevity of the species. At present, the data base is restricted to isolated incidences of quite dramatic disturbances, with no appropriate time scale provided to judge their overall importance. If disturbances occur at frequent intervals relative to life span, the impact on the age structure could be large. Disturbance may directly influence populations after recruitment by increasing mortality, retarding growth, or inducing movement to or from affected areas. Indirect or coincident effects due to habitat destruction are also likely. Physical mechanisms include cyclones and temperature extremes, while biological agents include crown-of-thorns starfish outbreaks and sea urchin plagues.

Considerable attention has been given to the immediate response of fish populations to single-event, large-scale habitat destruction resulting from the foregoing disturbances. These are the so-called "natural experiments" of Doherty and Williams (1988a,b). The temporal patterns in fish populations following habitat destruction are many and varied.

There are reports of substantial effects in different parts of the world stemming from a variety of disturbances. Abnormal cold conditions caused fish kills on patch reefs at Big Pine Key, Florida (Bohnsack, 1983a). Declines in the total number of individuals and species were recorded, and long-term effects on population structure were suggested, but data on individual species are lacking. Kaufman (1983) monitored hurricane-induced destruction of coral near Discovery Bay in Jamaica, which transformed branching, live coral habitat into low rubble reefs. Densities of the corallivore *Stegastes planifrons* declined in disturbed areas, but increased in surviving coral refuges, suggesting a degree of redistribution. Species naturally common in rubble areas (e.g., *Stegastes diencaeus* and *Centropyge argi*) increased. There were short-term effects on the abundance of benthic carnivores and predators. Hurricane-induced damage at Oahu, Hawaii, destroyed corals and severely affected the catches of *Zebrasoma flavescens* and other aquarium fishes (Pfeffer and Tribble, 1985). Sano *et al.* (1987) described a major impact of coral destruction by *Acanthaster* on reef fish communities in Japan, where coral-polyp feeders died out completely. Numbers of other resident species declined over a 4-year period, perhaps as a result of the change in the structural complexity of the reef.

A number of studies have emphasized the similarity of fish communities before and after disturbances and have interpreted the data as evidence against resource-limitation models (Walsh, 1983; Williams, 1985; Wellington and Victor, 1985). Walsh (1983) quantified substantial short-term effects of a catastrophic storm on the abundance of *Thalassoma duperrey* juveniles, whose numbers increased immediately after the storm (Fig. 9). However, the effect was short-lived, with numbers returning to predisturbance levels four or five months later, apparently due to predation. There is some suggestion of a lesser increase in adult densities following the storm. However, most other species were not affected, and if they were, it was mainly in terms of redistribution rather than mortality. Using a resource-limitation model, Wellington and Victor (1985) predicted an increase in the abundance of the herbivore *Stegastes acapulcoensis* following the massive mortality of corals during the 1982–1983 El Niño, but this did not occur. It is significant that no species directly feeding on corals was considered. A massive disturbance by *Acanthaster* on

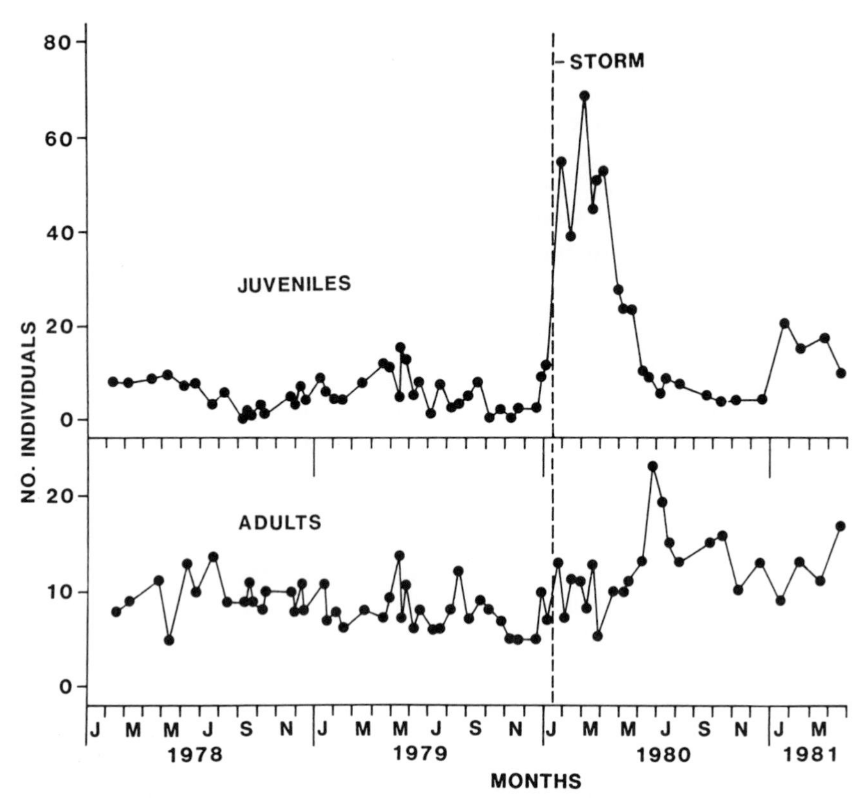

Figure 9 The numbers of *Thalassoma duperrey* juveniles and adults in fixed quadrats before and after a storm on the Kona Coast of Hawaii. (Redrawn from W. J. Walsh, 1983. Stability of a coral reef fish community following a catastrophic Storm. *Coral Reefs* **2**, 49–63.)

midshelf reefs on the Great Barrier Reef resulted in readily observable changes in the abundance of chaetodontid species that feed on scleractinian corals, but not for the majority of noncorallivorous species (Williams, 1985).

Clearly, some fish are susceptible to disturbance and others are not. Whether this is due to differences in robustness, dependence on the reef, or the degree of specialization is unkown. The issue of resource limitation should not be raised in the context of disturbance (Doherty and Williams, 1988a), unless direct effects of disturbance on populations have been discounted. What we really need are comprehensive measurements of just how and how often disturbance modifies population structure, whether directly or as a result of habitat destruction.

C. Resource Availability

1. Space

Populations may be limited by the availability of suitable living space, particularly if space is defended or mutual avoidance occurs. Space requirements usually relate to the individual's territory or home range, which obviously includes more specific resources such as food supply or shelter. The early belief that reef fish populations saturated their available space (Smith and Tyler, 1972; Sale, 1974) derived from observations of high density and diversity of coral reef fish assemblages, and overt aggression in many species (see D. McB. Williams, 1980, for a discussion). Evidence against this belief came in the form of observations on population fluctuations and the absence of effects of residents on recruitment, implying that new settlers always find space.

In the debate over the importance of space, certain basic facts appear to have been overlooked. First, territoriality is not universal and even in species that do exhibit this behavior, it is frequently restricted to certain age or sex classes. Therefore, space limitation may only apply to subcomponents of a population. Second, different age classes tend to use different types of space (Helfman, 1978). Experiments examining resident–recruit interactions may be an inappropriate test for space limitation, since they focus on life history stages that interact very little. The interactions between adults and juveniles may occur much later during the postrecruitment phase, when space requirements begin to overlap. A third point that should be noted is that competition for space can occur even if there is plenty of space if it varies in quality.

Few have attempted to test the importance of space limitation by manipulating the availability of the critical resource—space. Robertson *et al.* (1981) reduced patch reefs to half their original size to examine the effect of a reduction in space on *Eupomacentrus planifrons*. The manipulation had no effect on abundance, body weight, fat reserves, or the development of the ovaries. They argued that space per se did not influence the population biology of this species. It should be acknowledged, however, that a significant effect in this experiment would not have established space limitation as such, because of the technical difficulty of altering the availability of space without also affecting the availability of shelter and food.

In summary, the role of territorial behavior in limiting adult numbers or structuring populations is uncertain, despite the fact that this is one of the longest-standing assumptions in reef fish studies. Well-crafted experiments, in which space and other resources are manipulated independently, are urgently required to resolve this issue.

2. *Shelter*

The availability of holes or crevices in which to shelter during storms, or from predators, could limit numbers at any stage during the reef-associated part of the life history. Rough correlations between reef topography and fish abundance (Luckhurst and Luckhurst, 1978c) and observations on the defense of shelter sites (Robertson and Sheldon, 1979; Shulman, 1985a) are only suggestive of the importance of shelter as a limiting resource. Distinguishing the role of shelter from that of other resources requires an assessment of what shelter is and how it varies in availability, and experimentally demonstrating how it effects populations. Attempts to do this are few and vary in quality. Correlative studies include that by de Boer (1978), who found that the abundance of *Chromis cyanea* correlated with the number of hiding places. Also, Roberts and Ormond (1987) found that the numbers of holes of three different sizes together accounted for 77% of the variation in overall fish abundance in their Red Sea study sites.

Three comprehensive experimental studies, which set out to test hypotheses about the importance of shelter, illustrate the two possible extremes. Adult *Thalassoma bifasciatum* do not appear to be limited by the availability of sleeping holes (Robertson and Sheldon, 1979). These workers destroyed known shelter holes and could not detect an effect of the survival of blue head wrasse. Fish added to reefs already occupied also did not suffer mortality from failure to find sleeping sites. From these comprehensive experiments, Robertson and Sheldon (1979) concluded that sleeping holes were not in short supply. Shulman (1984) examined recruitment and/or survival of different species on experimental substrata with differing numbers of shelter holes. A series of six separate experiments indicated that early survivorship was increased by providing a greater number of refuges. Hixon and Beets (1989) manipulated the sizes of shelter holes on artificial reefs. Increasing the abundance of "large" shelters caused an increase in the abundance of large piscivorous fishes, which in turn resulted in a decrease in the local abundance of small prey fishes (see also Hixon, Chapter 17).

The value of these experiments would be greater if they were placed in the context of information on the ontogeny of shelter dependence and selection, and its effects on a range of demographic parameters and population characteristics. An obvious direction is toward larger-scale and longer-term descriptions and manipulations, so that the influence of shelter on adult population size and structure can be assessed.

3. *Food*

For many years food was considered the resource least likely to be in short supply on coral reefs (Sale, 1977, 1978a, 1980a; C. L. Smith, 1978). It was argued that territorial species would defend sufficient area to buffer themselves from fluctuations in food supply. However, observations suggesting that food may have considerable influence on postrecruitment events are increasing, mainly for herbivorous and planktivorous species. These observations are as follows: (1) Territorial herbivorous fish primarily defend areas from species that are most similar in terms of algae consumed (Low, 1971; Ebersole, 1977). (2) Herbivorous fish that usually feed on algae of low calorific value consume and digest feces of fish that consume algae of high calorific value (Robertson, 1982; Bailey and Robertson, 1982). (3) Tsuda and Bryan (1973) reported mass death of juvenile siganids due to starvation. (4) The growth and abundance of planktivorous species are often correlated with current speed, hence the supply of food to reefs (Hobson and Chess, 1978; Thresher, 1983a,b).

Despite the early consensus that food was not important and the evidence accumulating to the contrary, there have only been a few attempts to examine this problem experimentally (G. P. Jones, 1986). While there are difficulties with manipulating space or shelter without affecting food, it is quite feasible to supplement or reduce food supply without affecting other resources. Such food manipulations have produced contrasting results, depending on the variables examined. Robertson *et al.* (1981) reduced the amount of food available to *Eupomacentrus planifrons* by removal of substratum that contained algal mats. This had no measurable effects on body weights or fat reserves. Settlement or early survival of a group of coral reef fishes was not affected by manipulations of algal and invertebrate abundances on small patch reefs at St. Croix (Shulman, 1984).

Experimental studies focusing on juveniles have indicated that food has a substantial effect on growth and related parameters. Growth of juvenile *Pomacentrus amboinensis* was enhanced substantially in both shallow and deep habitats by supplementing their diet with plankton collections over a one-month period (G. P. Jones, 1986) (Fig. 10). Forrester (1990) obtained similar effects for the humbug *Dascyllus aruanus* using a unique feeding device. In contrast, survival was not affected by availability of food. When additional food was supplied to larger juveniles of *P. amboinensis,* their maturation was advanced and fecundity increased relative to controls (Jones, unpublished). The sexual maturation of *Thalassoma lunare* was advanced a full year by supplying juveniles with additional food (Eckert, 1985b).

D. Intraspecific Competition

There is strong evidence that intraspecific competition affects some of the characteristics of reef fish populations. Competition is normally evaluated in experiments designed to detect negative effects of density on either survival,

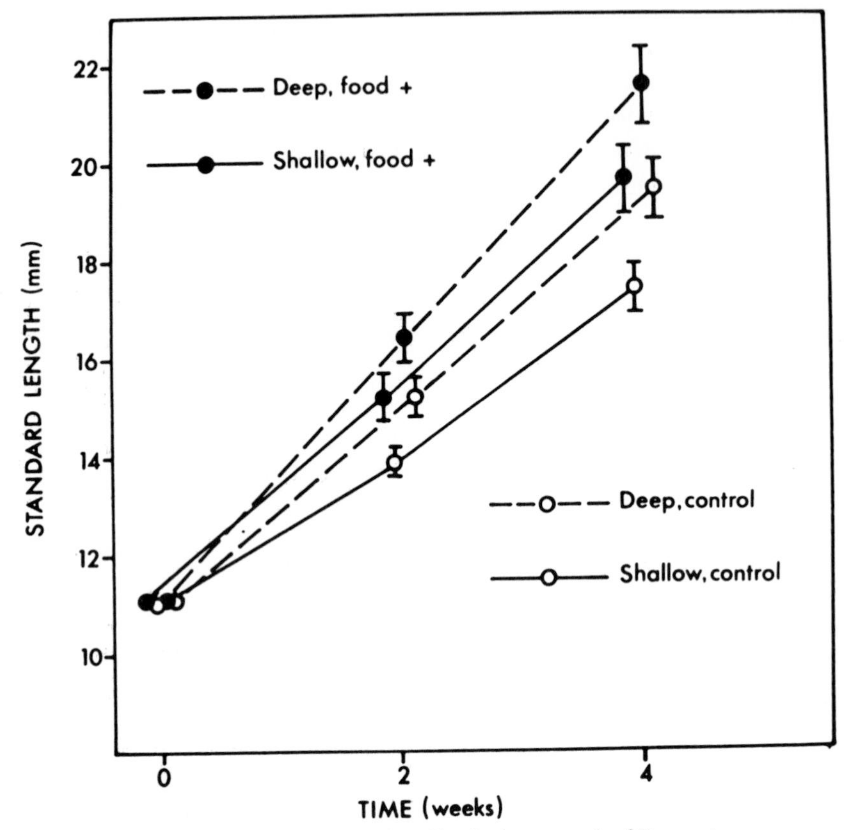

Figure 10 Effect of food supplementation (Food +) on growth of *Pomacentrus amboinensis* in shallow (<1 m) and deep (>3 m) habitats in One Tree Lagoon. Error bars = SE. (Redrawn from G. P. Jones, 1986. Food availability affects growth in a coral reef fish. *Oecologia* **70,** 136–139.)

growth, maturation, or fecundity. Such effects establish that competition is occurring but do not indicate whether it is due to limited resources (exploitation) or behavioral interactions (interference).

Intense intraspecific competition is primarily evident from effects on growth rates, with fish growing slower at higher densities (Doherty, 1982, 1983a; Victor, 1986b; Jones, 1987a,b, 1988b; Forrester, 1990). Experimental evidence indicates that the effects are large enough to have a major impact on the structure of reef fish populations. Competition has been detected both between and within cohorts, although the intensity of competitive influences on a cohort clearly changes as fish grow. Jones (1987a and unpublished observations), for example, examined the effects of varying adult (>50 mm) and juvenile densities on the growth of juvenile *Pomacentrus amboinensis* over a 2-year period (Fig. 11). Weak effects of juvenile density on growth were observed, causing the mean size of juveniles in three different density treatments to diverge over the 2-year period. The presence of pairs of adults (male + female) did not affect the growth of juveniles during their first six months.

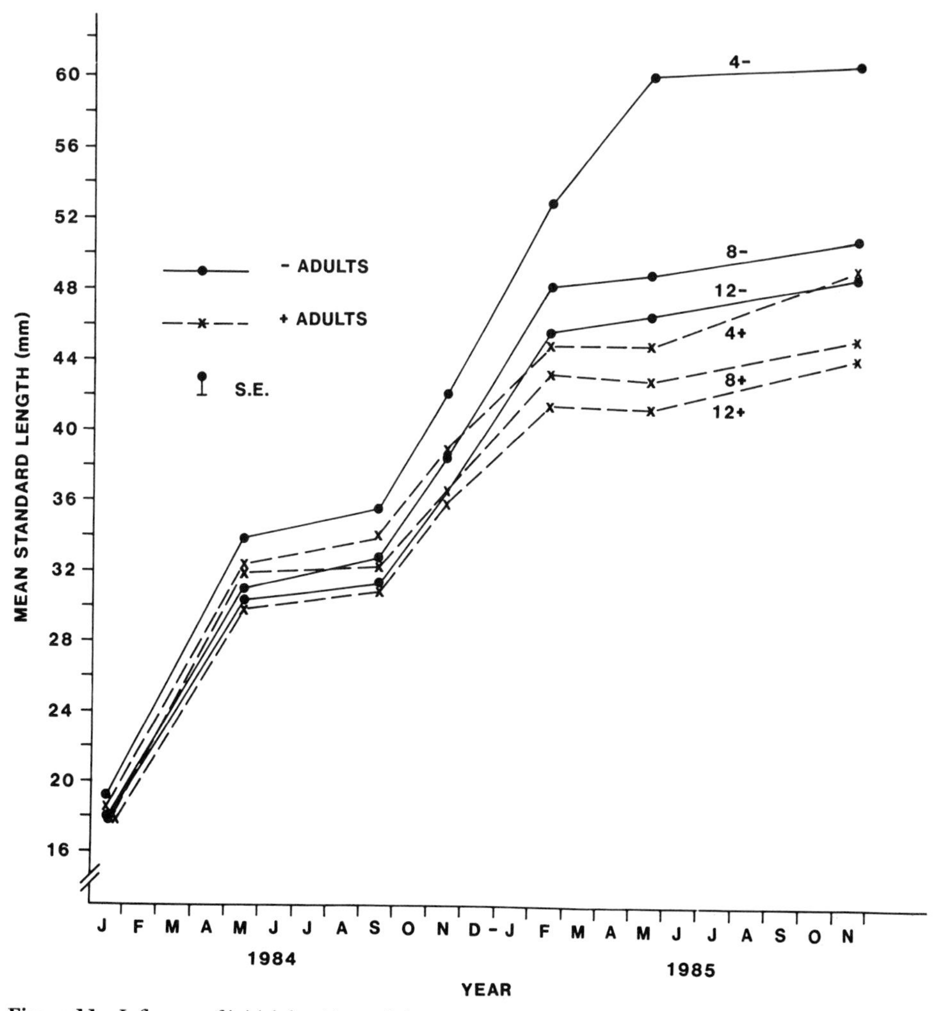

Figure 11 Influence of initial densities and the presence of adults on the growth of juvenile *Pomacentrus amboinensis* over two years following recruitment. Maturity is reached at 50 mm standard length. Pooled SE indicated. (Modified from G. P. Jones, 1987. Competitive interactions among adults and juveniles in a coral reef fish. *Ecology* **68,** 1534–1547.)

However, once juveniles reached a size of 30 mm, growth was suppressed in the presence of adults. This suggests that juveniles only compete with adults once they reach a certain size.

The degree to which competitive interactions are sex-related is unknown. I carried out an adult removal experiment at One Tree Reef to determine whether it was the adult male or the adult female that primarily inhibits the growth of large juveniles in *P. amboinensis* (previously unpublished information). Growth was monitored on natural patch reefs in which (a) both an adult male and female were present, (b) only a female was present (male removed), and (c) no adults were present (both removed). Growth was suppressed on reefs that had a male present (Fig. 12). Juveniles on reefs with an adult female present reached a similar size to juveniles in isolation from adults over the 7 months of the study. Hence, the interactions are sex-related. Since this is a sex-changing species (all juveniles are female), the inhibition of juvenile growth by a male may relate to suppression of sex change. Social control of growth and sex change appears to be common in species occupying isolated social units (Fricke and Fricke, 1977; Moyer and Nakazono, 1978).

Competition has been implicated in explaining differences in the growth rates of early and late settlers. Ochi (1986a) suggested that the growth of late-settling *Amphiprion* was limited by short growing time and by competitive interactions with early settlers. Such effects have been established experimentally for *Pomacentrus amboinensis* (Jones, 1987b). Late settlers grow considerably faster on reefs that do not also harbor individuals that settled earlier.

Changes in the intensity of competition among individuals of different absolute and relative sizes have a large effect on the variance in growth rates and the size structure of the populations (Doherty, 1982, 1983a; Jones, 1987a). Intracohort variation in growth increases with increasing density (Doherty, 1982, 1983a). However, the presence of adults can reduce the variation in juvenile growth, because interactions mainly affect faster-growing fish (Jones, 1987a). One important consequence of this asymmetric competition (*sensu* Begon, 1984) is a density-dependent reduction in the numbers of juveniles reaching maturity, not as a consequence of mortality but as a consequence of the reduced growth of large juveniles. In *P. amboinensis*, for example, juveniles do not reach maturity in the presence of adults (Jones, 1987a). Numbers reaching maturity are independent of the starting densities of a cohort. Similar processes appear to limit breeding population size in anemonefishes (Ochi, 1986a).

Fishes appear to be very resilient to competition in terms of mortality. Those species exhibiting density-dependent growth, for example, do not appear to exhibit density-dependent losses (Doherty, 1982, 1983a; Jones, 1987a; Victor, 1986b). However, there are indications of increased mortality

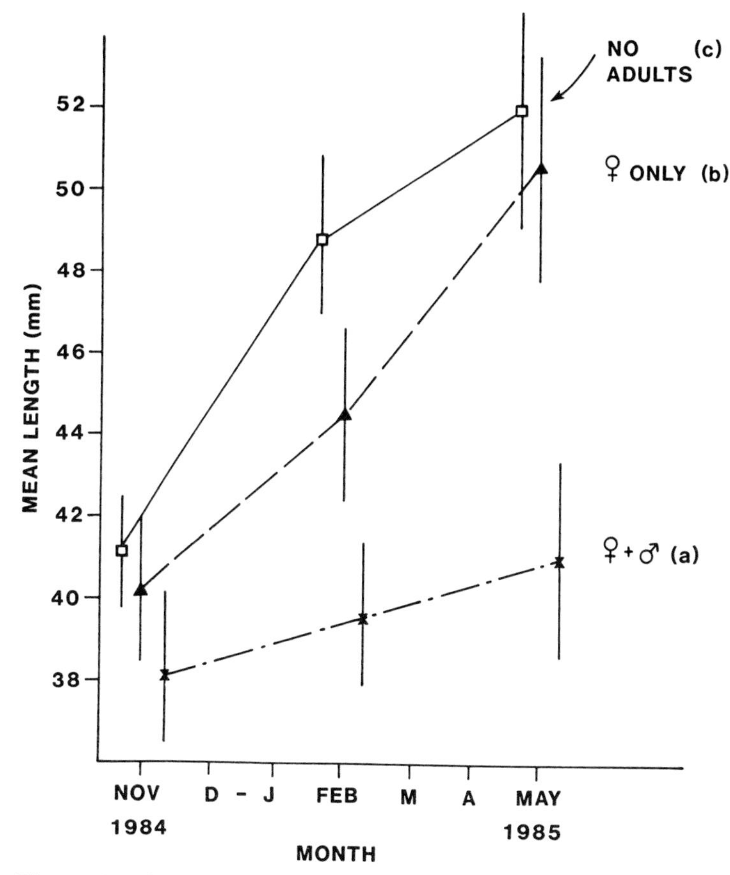

Figure 12 Effect of the presence of a pair of adults (a), mature females only (b), and no adults (c) on the growth of juvenile *P. amboinensis* on natural patch reefs in One Tree Lagoon (unpublished observations). Error bars = SE.

at high densities in some species. Thresher (1983b) found that mortality rates in *Acanthochromis polyacanthus* were positively related to brood size. In *Dascyllus aruanus,* Jones (1987b, 1988b) and Forrester (1990) detected negative effects of juvenile density on their survival, and in *Ophioblennius atlanticus*, Hunte and Cote (1989) found that mortality appeared to be density dependent following a major pulse of juveniles. Survival of surgeonfishes in the Caribbean is unrelated to the density of adults, but is negatively related to the density of settlers per unit of adult habitat (Robertson, 1988a).

E. Competition with Other Fishes

The dispute over the role of interspecific competition in determining the distribution and abundance of reef fishes can be traced back to the earliest studies, in which its importance was either assumed or denied (Smith and Tyler, 1972; Sale, 1974). Although it is far from over today, the present debate is at least "informed," as the number of manipulative experiments designed to test for effects of one species on another is increasing.

In the majority of these experimental studies, pairs of species are selected that are most likely to interact, given basic data on their diets and use of space. The majority of studies focusing on effects on demographic parameters have not detected important effects of interspecific competition. Interactions between *Pomacentrus flavicauda* and *P. wardi* (Doherty, 1982, 1983a), between *Pomacentrus amboinensis* and *Dascyllus aruanus* (Jones, 1987b, 1988b), and between *Pomacentrus flavicauda* and *Salarias fasciata* (Roberts, 1987) do not influence basic parameters such as abundance, growth, or mortality. The implication is that if competition is not important among these ecologically similar species, it is not likely to be important at all.

However, some demographic effects have been detected. Thresher (1983b) found that juvenile mortality in *Acanthochromis polyacanthus* was correlated with the collective density of all planktivores. Adult size was smaller, adult brood size was smaller, and spawning date was progressively delayed with increased density of pooled planktivores. Experimental clearances either of all fish or of all planktivores only resulted in (1) an advance in spawning date, (2) an increase in brood size, and (3) better juvenile survival, (4) but no effect on juvenile growth (Fig. 13). The relative importance of intra- and interspecific competition cannot be determined from this experiment. In the Caribbean, the herbivore *Microspathodon chrysurus* has a negative effect on the body mass and fat deposits of *Stegastes planifrons*, a smaller herbivorous species cohabiting the same reefs (Robertson, 1984).

There is greater evidence for an ecological role of interspecific competition from studies focusing on foraging activities and habitat use. Ebersole (1985) examined differences in microhabitat utilization by *Eupomacentrus leucostictus* and *E. planifrons*. He attributed the observed differences partially to a reversal of dominance, based on aggressive interactions between paired fish on two different substrata. No evidence of interspecific dominance, or a reversal in dominance on different substrata, was found in my study on *Pomacentrus amboinensis* and *Dascyllus aruanus* on their preferred coral habitats (Jones, 1988b). However, Robertson and Gaines (1986) presented comparative data that suggest that interference competition for food among surgeonfishes affects habitat use by subordinate species. Also, in two species of *Acanthemblemaria*, the dominant species competitively excludes the other from its pre-

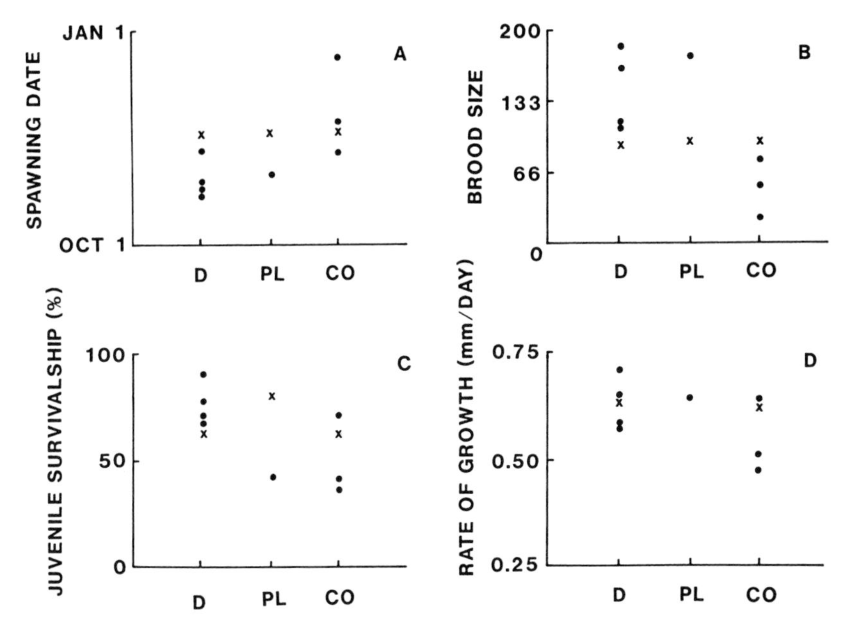

Figure 13 Results of experiment in testing for the effects of all other fish species and planktivores only, on four demographic parameters in *Acanthochromis polyacanthus*: spawning date, brood size, juvenile survival, and juvenile growth. D, totally denuded reefs; Pl, planktivore-denuded reefs; Co, Controls. Circles represent 1980–1981 experimental results; crosses represent values prior to manipulations, 1979–1980. (Redrawn from R. E. Thresher, 1983. Habitat effects on reproductive success in the coral reef fish, *Acauthochronsis polyacanthus* (Pomacentridae). *Ecology* **64,** 1184–1199.)

ferred depth range (Clarke, 1989). Collectively, these observations suggest that interspecific interactions among fishes may alter distributional patterns among habitats, but not overall abundances of species populations.

F. Competition with Invertebrates

Potential competitive effects of invertebrates on fish populations have received very little attention. Most of the published work concerns interactions between herbivorous fish and sea urchins. In an experimental removal of *Diadema*, Hay and Taylor (1985) observed an increase in grazing rates of parrotfish and surgeonfish. A massive increase in algal biomass and fish feeding rates followed mass mortality of *Diadema* at St Croix in 1983–1984 (Carpenter, 1985, 1988). Conversely, removal of damselfish potentially enhances the abundance of *Diadema* (A. H. Williams, 1980, 1981; Sammarco and Williams, 1982). More work is required to determine if such indirect interactions are widespread and have a major effect on the dynamics of reef fish populations.

G. Facilitation

Sweatman (1983, 1985a) was the first to detect a positive influence of conspecifics on the recruitment of some coral reef fishes, particularly *Dascyllus aruanus*. The degree to which the presence of conspecifics may facilitate the survival and growth of these settlers requires further investigation. Jones (1987b) had indications from experimental manipulations that the survival of juvenile *D. aruanus* was enhanced when adults were present, although this may be offset by a negative affect of juvenile density on mortality. This was later demonstrated by Forrester (1990). In another experiment, the mortality of juvenile *Pomacentrus amboinensis* decreased as their density increased (Jones, 1988b). Growth of *Acanthochromis polyacanthus* appears to be positively related to brood size (Thresher, 1983b). The processes that bring about these "positive" effects of density are unknown, but they will be important if this phenomenon turns out to be widespread.

H. Predation

Our understanding of the influence of predation on the dynamics of reef fish populations has been hampered by the lack of information on what the predators are. Although piscivores often make up a large proportion of reef fish communities (e.g., Goldman and Talbot, 1976; Williams and Hatcher, 1983), we know little of their rates of consumption of individual prey species at the different stages in their life histories (but see Sweatman, 1984; Hixon, Chapter 17). Associations between fish and shelter cannot be extrapolated to conclude that predator pressure controls overall abundance (e.g., Shulman, 1985a), since shelter may be important for other reasons (e.g., protection during storms and protection of demersal eggs).

It is common practice to infer patterns in predator pressure from patterns in mortality (Doherty and Williams, 1988a,b). This approach is unjustifiable in the absence of any data establishing that predation is the main cause of mortality.

One can find references to correlations between the abundance or mortality of particular reef fish and known predators. For example, adult density and survival of *Acanthochromis polyacanthus* is inversely related to the density of the piscivore *Plectropomis leopardus* (Thresher, 1983a,b). However, the link is rather tenuous without information establishing that this predator consumes the species of interest.

Direct observations of predator behavior can provide useful information on consumption rates of particular species, and some assessment can be made of how much this contributes to mortality rates. Using this method Sweatman (1984) calculated that predation by the lizardfish, *Synodus englemani*, could effect a 65% mortality rate on planktivorous fishes.

Alternative approaches to experimentally testing the importance of predation are predator removal experiments and exclusion experiments employing cages. Bohnsack (1982) surveyed reefs in which large predators had been reduced in abundance by spearfishing. The abundance of several small species (e.g., *Thalassoma bifasciatum*) was higher on such reefs, when compared with controls, suggesting that piscivory may be reducing their numbers.

Doherty and Sale (1985) detected an impact of predator exclusion cages on the rate of accumulation of juveniles to the One Tree Reef slope (Fig. 14). The real effect was restricted to soiitary, sedentary species, with other species

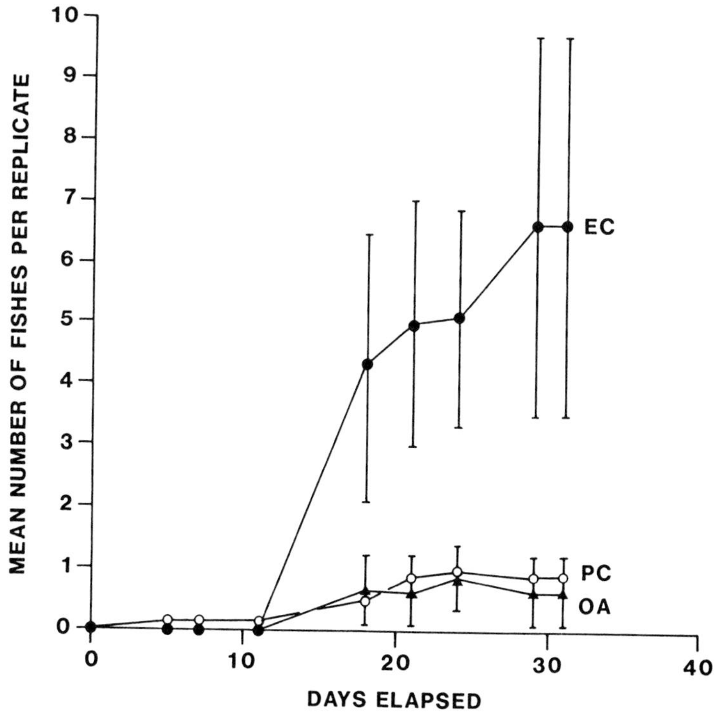

Figure 14 Effect of the experimental exclusion of "unknown" predators on the accumulation of recruit fishes (EC, exclusion cages; PC, partial cages; OA, open areas) on the reef slope at One Tree Reef. Error bars = 95% confidence limits. (Redrawn from P. J. Doherty and P. F. Sale, 1986. Predation on juvenile coral reef fishes: An exclusion experiment. *Coral Reefs* **4,** 225–234.)

showing effects due to mobility and cage artifacts. Problems with cages have been widely discussed (Lassig, 1982; Doherty and Sale, 1986), however, it is not so much the technique that is flawed, but a lack of ancillary data to assist in interpreting the results. In no case have data been provided on the predators involved, prey being consumed, and the behavior of predators and prey in relation to cages. A combination of approaches, including cage exclusions, may still be the best plan to increase our somewhat limited understanding of this potentially important problem (see also Hixon, Chapter 17).

VI. Multiple Causality and Ecological Importance

This survey of postrecruitment processes has highlighted two important facts. First, there is ample and accumulating evidence that recruitment patterns can be modified by postrecruitment processes—sometimes substantially. The list of processes limiting population numbers must include those acting on the planktonic (prerecruitment) and the benthic (postrecruitment) phases of the life cycle (Sale *et al.,* 1985; Jones 1987a, 1990; Richards and Lindeman, 1987; Mapstone and Fowler, 1988; Keough, 1988; Warner and Hughes, 1988). Questions must concern the circumstances that shift the relative importance from one end of the scale to another. Ideally, we should consider changes in the magnitude of different processes as they influence different stages in a cohort's existence (Fig. 5).

In a recent review covering similar material, Doherty and Williams (1988a) concluded that reef fish populations seldom reach carrying capacities. Further, they argued that whether this was due to inadequate recruitment or patterns of turnover on the reef, the dynamics of populations will be strongly influenced by recruitment variation. While the data support the observation that reef fish populations are clearly variable in space and time, that this variability is primarily due to recruitment is not evident from the work published to date. Our future understanding of the importance of postrecruitment processes could be compromised by any false impression that such processes, however complex, produce only small effects relative to recruitment.

The second point is that during the postrecruitment phase itself, a wide range of processes affect different species, and different demographic parameters act on each species, in many different ways. This range of factors is especially noticeable as one considers a hierarchy of population patterns, ranging from total numbers and adult numbers to the sex, age, and size structure of populations. Our understanding of these factors has been hindered in the past by a preoccupation with only total numbers and with a single factor—competition. Models based on the presence or absence of competition (Fig. 1) are far too simple. While recent work has indicated that competition can occur, it does not act in the way that was previously assumed. That is, it does not preclude larvae from settling on the reef, but does affect their subsequent growth and reproduction.

The foregoing limitations of the historical perspective apply to single-cause models and hypothesis testing in general (Hilborn and Stearns, 1982). Response surface descriptions of pattern and multifactorial experiments represent more realistic ways to cope with complex systems with multiple causes. The relative importance of different processes and their interactions are readily revealed, given two constraints on the use of the term "importance." First, it can only be defined in relation to the patterns in question. For example, competition may be intense (as measured by effects on growth rates), but unimportant because the pattern of interest is species composition (Welden and Slauson, 1986). Second, it can only be defined in the context of the other factors being tested (N. L. Andrew, personal communication). Since it is not normally possible to test all factors in one experiment, we are unable to assess the importance of all factors relative to one another.

These points are illustrated by a comparison of the results of a series of multifactorial experiments on *Dascyllus aruanus* and *Pomacentrus amboinensis*, which focused on the factors affecting their growth and survival (Fig. 15). The main effects considered were competition among juveniles, competition with adults, the effects of the other fish species, food supply, coral substratum, depth, and location (the latter independently of coral substratum, but not necessarily food). At the risk of being oversimplistic, interaction terms are omitted. Major differences in the factors found to be significant statistically

	P. amboinensis		D. aruanus	
	Growth	Loss	Growth	Loss
Density	yes (-)	no	no	yes (-)
Adults	yes (-)	no(-?)	no(+?)	no
Species	no	no	no	no
Food supply	yes (+)	no	x	x
Location	yes	yes	yes	no
Depth	yes	yes	yes	no
Coral substratum	no	yes	yes	yes

Figure 15 Summary of the results of a series of multifactorial experiments examining the effects of different factors and processes on the growth and loss of juveniles of *Pomacentrus amboinensis* and *Dascyllus aruanus* from patch reefs. Yes = significant effect (0.05 level); no = no significant effect; (+) = positive effect; (−) = negative effect; x = untested.

were found between the two species and between growth and loss rates within species. Hence, importance was very much dependent on the species and the variable examined. A maximum of only three factors were tested at any one time, and the particular combinations were chosen rather arbitrarily. A factor that was highly significant when tested against one factor in one experiment may have been only marginally significant when tested against another factor in a later experiment.

Ultimately the yes/no result of the significance test is less valuable than information on the variance attributable to each factor in relation to the total variation observed. Unquestionably, the initial experimental phase designed to detect contributing factors must be followed by a second phase in which only statistically significant factors are compared. At this stage, the emphasis will be on variance components in any analysis of variance rather than on statistical hypothesis testing. For example, further work is required on mortality in *P. amboinensis* to examine the relative importance of location, depth, and coral substratum, but competition could probably be ignored. Other untested factors (e.g., predation) must also be included.

The "failure" for any generalizations to emerge for comparisons of similar species may be due to the fact that too few species have been examined or that no generalizations are possible. The latter would not overly concern the researcher primarily inspired by the discovery of new detail. However, for those interested in similarities rather than differences (repeated patterns in the sense of Sale, 1988a), I believe the former situation is most likely to apply. It is encouraging, for example, that in an analysis of density dependence of survival, growth, and recruitment in young and adult fish stocks in many commercial temperate fishes, some general patterns emerged (Valiela, 1984). The survival and growth of juveniles were invariably negatively related to juvenile density. However, survival, growth, and recruitment into the adult population were not related to the density of adult fish.

In other systems in which the detailed local dynamics of small groups of species has been shown to be complex, broad patterns have emerged. For example, in bracken herbivores, local population abundance and variability are correlated with the proportion of sites occupied and body size (Gaston and Lawton, 1988). Such a comparative approach has not yet been applied to coral reef fishes. The preoccupation that coral reef fish ecologists have with outwardly similar species (e.g., pomacentrids) may be restricting their ability to detect the patterns that will form the basis of useful generalizations (Doherty and Williams, 1988a). Whatever their form, these generalizations will be based on concepts of relative importance of different processes rather than the historical emphasis on single causes, such as competition or recruitment. Many workers have stressed the need for observations on species with very different life spans, and those exhibiting a range of variation in recruitment and postrecruitment juvenile mortality (Shulman and Ogden, 1987;

Doherty and Williams, 1988a; Warner and Hughes, 1988). To these pleas I would add variation in maturation time, movement, and adult mortality, which are still largely unknown.

In pursuing different population patterns and causative processes, attention must be given to the consequences of changing the spatial and temporal scales of focus (Mapstone and Fowler, 1988). While different workers have often focused on problems at different scales (with much confusion and debate; see Ogden and Ebersole, 1981; Mapstone and Fowler, 1988), it is unusual for single studies to build in this very important factor. There are many sampling designs that enable the spatial scale at which individuals are responding to their environment to be detected (Grieg-Smith, 1983; Andrew and Mapstone, 1987; McArdle and Blackwell, 1989; Yoshioka and Yoshioka, 1989).

While the description of spatial and temporal pattern is necessary, it should not constrain studies focusing on the importance of different processes to particular scales, since the effects of opposing processes may cancel one another out. Different processes will have different sets of appropriate scales and, intuitively, more than one should be considered. For example, the intensity of competition will be evident on extremely local scales, emphasizing the importance of interactions among individuals, within and between patches (De Jong, 1979; Jones, 1987a). Disturbance may occur at a wide range of scales, although at present large-scale disturbances due to hurricanes and outbreaks of crown-of-thorns starfish have dominated discussions. Our understanding of the roles of predation and habitat structure are also restricted to a limited range of scales in space and time, and we do not know to what degree fishes respond to or covary with changes in the distribution of habitat-forming organisms or predators.

VII. Looking to the Future

In an attempt to highlight areas for future research effort, I suggest the following as "emerging" but unconfirmed generalizations about postrecruitment processes:

1. Spatial and temporal changes in habitat structure, and associated changes in shelter availability and predator pressure, have a major impact on population size and structure, primarily through influences on movement and mortality.
2. Intraspecific competition and food availability have major effects on adult numbers and population structure through influences on growth and reproduction.
3. Interspecific competition is of minimal importance in determining

abundance, but associated behavioral interactions may govern the local-scale use of habitats.

4. Disturbances affecting fish populations directly, or indirectly via changes in habitat structure, will also be important at a wide range of spatial scales.
5. Variations in recruitment are decreasingly important in high-recruiting populations, and longer-lived species, whose population sizes are more closely related to adult mortality and movement patterns.

Perhaps the most instructive conclusion from this discussion is that recruitment-based models (supply-side ecology—*sensu* Roughgarden *et al.*, 1987) are too simple because they single out one life history event from a series of potentially important demographic parameters. However, these models were the catalyst for us to place the study of postrecruitment events in a complete demographic framework, and in doing so they stimulated many of the recent advances in this field. Competition-based models (surplus-side ecology?) fail because they make assumptions about the operation of competition that are incorrect and ignore a wide range of other potentially important ecological processes. However, it was the discovery that these assumptions were incorrect that stimulated research on the potential importance of variation in recruitment. It is to be hoped that the realistic models of the future will be just as stimulating to empirical research and discussion. It is also hoped that they recognize that the observed patterns of distribution and abundance of reef fish populations are shaped by a complex array of interacting processes, acting differentially on different growth stages and on different scales of space and time. If so, the importance of "life after recruitment" will finally be recognized.

Acknowledgments

Many thanks to those with whom I have discussed these ideas and to those who have commented on the manuscript, including N. L. Andrew, J. H. Choat, R. Cole, A. W. Ebeling, U. L. Kaly, B. D. Mapstone, and N. Moltschaniwskyj. Special thanks to P. F. Sale, whose encouragement started the wheels in motion, and whose patience allowed me to get there in the end.

Part IV

Reproductive and Life History Patterns

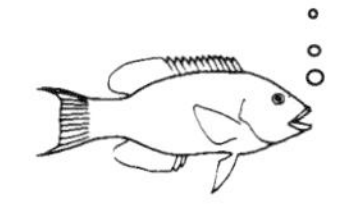

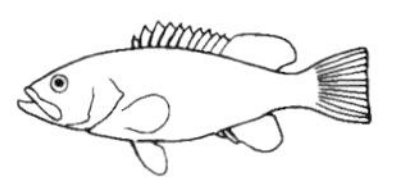

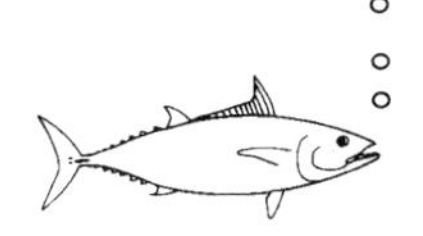

CHAPTER 12

Intraspecific Variability in Social Systems of Coral Reef Fishes

Douglas Y. Shapiro
Department of Marine Sciences
University of Puerto Rico
Mayagüez, Puerto Rico

I. INTRODUCTION

The social structure of terrestrial animals often has been related to the temporal and spatial distribution of food, water, shelter, mates, and other needed resources (Emlen and Oring, 1977; Rubenstein and Wrangham, 1986). Theoretical explanations of variability in social systems have generally relied on interspecific comparisons within genera or families (Rubenstein and Wrangham, 1986; Slobodchikoff, 1988). When intraspecific variability has been examined, explanations have been very similar to those for interspecific differences.

In comparison with terrestrial vertebrates, the social systems of coral reef fishes have been studied for a relatively small number of years by comparatively few workers. Yet intraspecific variation in social organization is already striking (e.g., Barlow, 1975; Colin, 1978; Ogden and Buckman, 1973; Popper and Fishelson, 1973; Robertson, 1981; Shapiro, 1979, 1987c; Thresher, 1979). How do we explain such variability?

Part of the explanation will undoubtedly lie with the complexity and variability of resource distribution within and among coral reefs. At the simplest physical level, reefs themselves vary enormously in size, shape, exposure to currents and waves, geological history, and vertical depth contours. Even within similar ecological zones, the actual spatial distribution of physical structures, sessile and mobile animals, and plants, at a scale relevant to fishes, is likely to vary widely within and among reefs. Fishes can be expected to adapt their social behavior and discrete features of their social system to these local variations (Robertson and Hoffman, 1977).

Another part of the explanation will lie with the opportunistic and generalist feeding strategies of many species, which enable them to occupy a range of microscale habitats (Randall, 1967). If the distribution of resources varies from one habitable area to another, aspects of the social system of a species are likely to vary in kind. To take one example, spawning sites are probably selected, at least in part, on the basis of how pelagic eggs and larvae are influenced by local water movement at those sites (Barlow, 1981; Doherty *et al.*, 1985; Johannes, 1978a; Shapiro *et al.*, 1988). Consequently, the location of suitable spawning sites will vary widely from one reef to another as the reefs vary in size, shape, and exposure to wind, current, and waves, that is, in all of those factors influencing local water movement. Variations in the degree of separation between foraging and spawning sites will force the fish to vary the extent to which they do or do not migrate out of foraging zones to spawn.

A third part of the explanation for social system variation derives from variability in the time, place, and rate of recruitment of pelagic juveniles onto reefs. Variable recruitment will place juveniles on reefs of different size, shape, and water-movement regimes and possibly in varying microhabitats. Subsequently, particular features of social systems may alter to fit the local pattern of resource distribution, as suggested earlier. This argument implies that a cohort occupying one part of a reef may form a social system different from a subsequent cohort settling later on a different portion of the same reef. Variable recruitment and mortality rates will also create large variations in local population density, which may, in turn, influence spatial and behavioral aspects of social systems with no alterations by the fish themselves in the way they respond behaviorally to the nonsocial environment (Shapiro, 1987c).

Juveniles settling initially on one type of microhabitat may move into new microhabitats as they grow and mature. Consequently, there may be ontogenetic changes in the form taken by the social system at different stages in life history (Helfman *et al.*, 1982; McFarland and Hillis, 1982).

In some cases, social systems will be seen to vary or not depending on how we classify them. Groups containing one male and several females may be interpreted as qualitatively different from groups containing many males and females (Fricke, 1975), or both could be seen as representing minor differences in group size and composition within a single type of social structure, namely, stable bisexual groups. Populations in which individuals alter the form of social units they occupy as they mature may be viewed as containing several different types of social system (Ogden and Buckman, 1973) or as having a single system composed of several types of units, with individuals moving among the different units as they develop (Warner and Downs, 1977).

Finally, social systems may be responsive to local demographic history. Where features within the social system are maintained by tradition, for example, particular migratory pathways (Helfman and Schultz, 1984) or spawning sites (Colin *et al.,* 1987; Warner, 1987), the social history of a population will strongly influence its social system. Even in the absence of traditionality, recruitment of juveniles into a local population may produce variable change in the social system depending on the nature of the social system prior to recruitment (Shapiro, 1987c).

A traditional approach to differences in social structure would treat the social system as a response to local variations in predation, interspecific competition, and particularly to the distribution of limited resources (Emlen and Oring, 1977; Rubenstein and Wrangham, 1986). This approach has been used successfully to explain interspecific differences in the social systems of butterfly fishes (Hourigan, 1989). When applied to intraspecific variation, this approach might argue as follows. If individual behavior, which produces a social system, has evolved to match a particular distribution pattern of restricted resources, variations in the pattern of distribution should select for sufficient intraspecific plasticity in behavior to maintain the match by varying the social system (e.g., Gill and Wolf, 1975; Rubenstein, 1986).

In this chapter, I review the literature on intraspecific variation in social systems of coral reef fishes and consider how those variations have been explained. Many of the explanations employ the traditional approach mentioned earlier. However, this literature review also suggests that some variations in social systems could result from variations in recruitment and from the spatial distribution patterns resulting from the settlement behavior of juveniles, that is, from factors not operating strongly in many terrestrial systems. Consequently, I then discuss recent models illustrating how settlement by juveniles onto reefs can influence spatial distribution and I discuss lessons from these models that help us understand variable social systems.

II. Types of Variation within Social Systems

A. Alternative Reproductive Behaviors

A relatively simple form of variation lies with the behavior of individuals within the system. Many protogynous species (in which adult females change sex and become males), particularly wrasses, have two types of males (Warner and Robertson, 1978). Initial-phase (IP) males are generally small, bear the same coloration as females, and spawn in groups—several males release sperm simultaneously with the egg release of an individual female. Some IP males also spawn by "streaking" or "sneaking" (Warner *et al.,* 1975). In contrast, terminal-phase (TP) males are larger, bear a brighter coloration pattern, often defend temporary or permanent spawning territories, and spawn one-on-one with individual females.

The coloration and spawning behavior of IP and TP males are closely related to two different developmental sequences for these males. Most IP males in most populations developed initially into adult males. These are called primary males. TP males are either sex-changed females and are called secondary males, or they are primary males that have changed color and behavior. Explanations for these alternative male reproductive tactics and developmental sequences have generally relied on theories of alternative life history strategies (Leigh *et al.*, 1976; Shapiro, 1989b; Warner, 1985; Warner *et al.*, 1975).

These behavioral alternatives relate to broader variations in social systems when the proportions of primary males in the population vary among reefs (Warner and Hoffman, 1980a). When reefs contain few or no IP males, then the reproductive system reduces simply to territorial pair-spawning by TP males. When reefs contain large numbers of IP males, then the reproductive system contains both territorial pair-spawning and nonterritorial group-spawning fish. These variants can be interpreted to represent different social systems.

The explanations for these variants have, in fact, viewed the social system as resulting directly from the proportion of IP and TP males on the reef (Warner and Hoffman, 1980a). Alternative explanations, for example, based on differences in substrate availability or water movement among reefs, have not yet been provided for variant social systems in these circumstances.

B. Ontogenetic Changes in Social Organization

In some species, individuals change the type of social unit they form or occupy as they grow (Helfman, 1978). A clearly documented case concerned Caribbean populations of the French grunt, *Haemulon flavolineatum* (Ogden and Zieman, 1977). After settling onto the reef, the smallest postlarvae formed mixed schools with mysid shrimps over sea urchins or coral promontories (McFarland and Kotchian, 1982). Slightly larger juveniles assorted by size into daytime resting schools over specific coral heads (Helfman *et al.*, 1982; McFarland and Hillis, 1982). The largest juveniles defended small territories over crevices or other locations on the coral head. At night, all individuals migrated off the reef along traditional paths to feed in nearby sea grass beds (Helfman and Schultz, 1984).

Segregation by size into separate schools was accomplished by size-related variations in agonistic behavior (McFarland and Hillis, 1982). The resulting uniformity in the size of school members was thought to increase their locomotor efficiency and reduce the risk of predation. Territories held by the largest juveniles were also thought to enhance protection against predators (McFarland and Hillis, 1982).

In the protogynous Caribbean redband parrotfish, *Sparisoma aurofrenatum*, juveniles matured as females in territories, where most reproduction occurred (Clavijo, 1982). Most females remained within their territories at all times, but a small portion of territorial females moved to deeper water at night, presumably for increased shelter. Later in life, females left a territory and wandered, changed sex, and then quickly became wandering males. Eventually, wandering males obtained a territory where they spawned with females in the territory.

Thus, in the redband parrotfish, individuals changed, as they developed, from occupying territories to wandering and then again to occupying territories. Individuals shifted from territorial to wandering partly because leaving a territory is probably causally related to the onset of sex change and partly because newly sex-changed males can substantially improve their reproductive success by locating and occupying a vacant territory already containing two to five females (Clavijo, 1982). Consequently, it is presumably advantageous for developing individuals, either just before, during, or after they have changed sex, to leave their original territory rather than remain and compete with the original and much larger male. This conclusion is not a full explanation for this ontogenetic sequence because it fails to distinguish this case from other sex-changing species in which males share space with a larger territory holder (Thresher, 1979).

A similar, but more complex ontogenetic sequence has been proposed for a population of protogynous *Scarus iserti* (formerly *S. croicensis*) in Panama (Warner and Downs, 1977). Three forms of social units occurred at this site: stationary groups, which showed no evidence of territorial behavior; territorial groups; and roving groups, which moved widely over the reef feeding on algae (Ogden and Buckman, 1973). The size of individuals in these groups suggested that females and primary males occupied stationary and roving groups when small and territorial groups when larger (Warner and Downs, 1977). Once a female changed sex, the newly transformed male repeated the basic sequence (Clifton, 1989; Warner and Downs, 1977): small secondary males occupied stationary groups, slightly larger secondary males were in roving groups, and the largest secondary males occupied territorial groups.

The shift from roving to territorial groups was understood as the result of attaining a size allowing the individual to enter a territory successfully, either as a female or as a male. Since a newly sex-changed male is presumably less competitive than older, larger males, sexual transition might be accompanied by an inability to compete with a territorial male. The new male would thus be relegated to roving or stationary groups again. The ontogenetic explanation (Warner and Downs, 1977), while complex, appears satisfactory, although it does not explain the coexistence of stationary and roving groups.

C. Intrasite, Intersite, and Temporal Variations

1. *Qualitative Variation*

While ontogenetic variations in social systems have been explained primarily in terms of life history strategies, explanations of intra- and intersite variations have relied more on the availability and distribution of resources, including food and mating sites.

The striped parrotfish, *S. iserti*, illustrates not only ontogenetic variation in social systems but intrasite and intersite variation as well. On shallow reefs in the San Blas Islands of Panama, this fish occupied stationary units, wandering groups of foragers, and territorial harems, with fish spawning both in pairs and in groups (Buckman and Ogden, 1973; Clifton, 1989; Ogden and Buckman, 1973; Robertson *et al.*, 1976). On a deeper reef in Puerto Rico, individuals migrated to the edge of the reef to spawn and the social system was described as being "fundamentally different" from that in Panama (Barlow, 1975). In Discovery Bay, Jamaica, all individuals foraged in roving groups in the morning and migrated in the afternoon to a specific site where they spawned exclusively in aggregations of IP individuals (Colin, 1978). TP males did not spawn and no stationary groups or territorial harems were seen. In Barbados, although territorial groups were found at offshore sites, neither stationary, territorial, nor roving groups occurred at an inshore site, where individuals occupied loosely packed home ranges (Dubin, 1981).

Explanations for the varying social system of this fish have been almost as diverse as the social systems. Wandering groups of foragers were sufficiently successful at overcoming the defense of algal food by territorial groups and by damselfishes that feeding rates were equivalent for individuals feeding in roving groups and in territories (Clifton, 1989; Robertson *et al.*, 1976). Consequently, roving groups may represent an alternative feeding strategy to maintaining all-purpose territories, although roving groups can probably only form when there are sufficient individuals to swamp territorial group defenses (Clifton, 1989). The need for a minimum number of fish may relate to the observation that roving groups in two studies only formed on relatively large reefs (Colin, 1978; Ogden and Buckman, 1973). Thus, intersite variation in the presence or absence of roving groups might simply reflect differences in local population density or reef size. Barlow (1975) felt that the difference between the Puerto Rico and Panama populations related to differences in the patchiness of the food supply between shallow and deep reef systems, while Dubin (1981) felt that the differences in social system among her reefs were not attributable to differences in food supply or population density, but were related to availability of suitable spawning sites. Thus, authors have not agreed on which environmental variables influence the types of social units at a site nor on the direction that the social system is likely to be pushed by variations in those variables.

A similar quandary remains over variations in the social system of *Halichoeres maculipinna* between sites in Florida and Panama. In Florida, two or three herds of IP females foraged and spawned in a territory occupied by one dominant and several subordinant TP males (Thresher, 1979). Thresher felt that the substrate-dwelling invertebrate food of this fish was dispersed indefensibly, so females foraged over large areas rather than defending feeding territories, but that female mates could be defended by TP males. He did not explain why females foraged in herds rather than as individuals.

In Panama, IP fish foraged within individual territories and migrated elsewhere to spawn, while TP males fed and spawned within individual, all-purpose territories (Robertson, 1981). There was no consistent spatial relationship between the location of IP and TP territories. Robertson (1981) attributed the differences between the social systems in Florida and Panama to differences in reef topography of the occupied sites. He suggested that fish in Florida fed in areas with abundant spawning sites and consequently spawned and fed in the same area; fish in Panama fed on shallow reef tops lacking good spawning sites and thus migrated to the reef edge to spawn. It was not explained why IP fish defended feeding territories in Panama but foraged in large, wandering herds in Florida.

Robertson's (1981) explanation based on proximity between suitable feeding and spawning sites, has recently been invoked to explain social system variations in the bluehead wrasse, *Thalassoma bifasciatum* (Fitch and Shapiro, 1990). In the San Blas reefs of Panama, fish fed on zooplankton on the upcurrent ends of reefs during the morning and migrated during the afternoon to downcurrent regions where large males defended temporary spawning territories (Robertson and Hoffman, 1977; Warner, 1987; Warner and Hoffman, 1980a,b; Warner *et al.,* 1975). However, in the back reef areas of large patch reefs on the southwest coast of Puerto Rico, males and females occupied all-purpose home ranges throughout the day, fed on benthic invertebrates, and spawned in the afternoon within or very close to their morning home ranges (Fitch and Shapiro, 1990). There was no migration to distant, downcurrent spawning areas and there was no convincing evidence of male territoriality in these back reef areas.

Fitch and Shapiro (1990) argued that *T. bifasciatum* is sufficiently flexible in its feeding ecology that it can settle successfully both in fore reef areas where it feeds on zooplankton and in back reef areas where it feeds on benthic invertebrates. In fore reef areas, there is a wide separation between sites rich in zooplankton and sites where planktonically spawned eggs would not be swept over the reef flat. Consequently, individuals occupying the fore reef migrate away from the feeding area to spawn. Where particular spawning sites are concentrated and relatively few in number, they become defensible and form the center of temporary spawning territories defended by large males. In back reef areas, eggs spawned planktonically would already be on the downcurrent side of the reef flat. Consequently, there is no advantage in migrating to spawn and much of the morning feeding area constitutes an adequate spawning site. The result is that individuals spawn and feed within the same location. The numerous spawning sites are seemingly indefensible, so male–male aggression is less intense and not noticeably site-related.

Intrasite variation in the bluehead wrasse is of particular importance because natural selection arguments, based on the size advantage model for the evolution of female-to-male sex change, have relied heavily for support on the highly competitive nature of the migratory spawning system of this fish (Charnov, 1982; Warner *et al.,* 1975). Although the size advantage model is controversial for many reasons (Policansky, 1987; Shapiro, 1987a, 1988a, 1989a,b; Warner, 1988a,b), this particular evidence in support of the model is weakened to the extent that a competitive, migratory spawning system is joined by nonmigratory, less competitive social variants for the same species.

Several species of butterflyfish form social units that vary geographically or within subpopulations or are related to foraging behavior (Hourigan, 1989; Neudecker and Lobel, 1982). For example, *Hemitaurichthys zoster* and *Chaetodon sancthelenae* occupied large, stable groups when they fed on current-borne zooplankton in the water column and formed pairs when feeding benthically (Fricke, 1986; Hourigan, 1989). In these cases, as in interspecific comparisons, differences in social structure appeared to be primarily related to differences in feeding ecology (Hourigan, 1989). However, *C. capistratus* also formed pairs for foraging and spawning in some locations, but formed groups of 15 individuals or less elsewhere (Colin, 1989; Neudecker and Lobel, 1982). Change from pairs to groups was attributed not to differences in feeding ecology but to increased density of adults, which broke down the underlying tendency for pair formation (Colin, 1989).

2. *Quantitative Variation*

Some coral reef fishes form social groups that vary in size and sexual composition (e.g., Donaldson, 1989, 1990; Moyer, 1990; Reese, 1975; Sikkel, 1990). Such variations have been considered either as the usual, expected biological variants within a single social system (Robertson, 1972, 1974), as quantitative variants of a magnitude requiring special explanation (Shapiro, 1988b), or as separate social systems or separate subdivisions within a single overall social system (Fricke, 1980; Kuwamura, 1984).

The cleaner wrasse, *Labroides dimidiatus,* formed adult, male–female pairs, with juveniles occupying separate home ranges from adults, in some locations (Potts, 1973), and social groups containing one male and up to 16 juvenile and adult females at other places or times (Robertson, 1972, 1973a, 1974; Robertson and Hoffman, 1977). Robertson basically considered these variations to comprise a single form of social structure.

In Shirahama, southern Japan, two types of *L. dimidiatus* groups were distinguished (Kuwamura, 1984). "Linear" types contained females that shared overlapping home ranges and feeding and mating sites. "Branching" types contained females divided into two portions or branches, with all females within each branch sharing overlapping home ranges but excluding females of other branches from their territory. The behavioral relationships of individuals in linear groups differed from the relationships of individuals in branching groups.

Kuwamura (1984) considered linear groups to be the primary form of social group in this species but saw both types as adaptations facilitating the orderly behavioral control of sex change. Although he had no experimental evidence, he suggested that the formation of linear or branching groups depended on whether there are several females of equal size within a group. If so, then similar-sized individuals would segregate into different spatial and behavioral branches within the group. If not, then the group would remain linear.

The anemonefish *Amphiprion clarkii* forms colonies on clusters of sea anemones (Dunn, 1981). On southern Japanese islands, all colonies contained one mating pair and up to 10 juveniles; individuals remained close to their home anemone and pairs were stable (Moyer, 1976). On more northerly islands, colonies often contained several mating pairs, each pair occupying a clearly defined territory, as well as up to 16 nonmating, subdominant adults. Individuals in northern populations often moved into and out of colonies, resulting in instability of mated pairs and occasional polygyny (Moyer, 1980).

Moyer attributed these differences in colony size, structure, and behavior to the spatial distribution of the host anemone, which occurred singly in the south and in large clusters in the north, and to lower predation rates in the north, which permitted greater movement of fish outside of home colonies (Moyer, 1980). Since adult anemonefish, including *A. clarkii,* are generally highly territorial and recognize and prefer to remain with their mating partner (Fricke, 1973a, 1979; Moyer and Bell, 1976; Ross, 1978a), dense clustering of host anemones could only produce greater movement among colonies, lower stability of mated pairs, and more exchange of colony members if territorial tolerance and group cohesion by these fish were plastic.

Indo-Pacific *Dascyllus* species live in stationary groups, usually in discrete colonies of highly branching corals. Group size ranges between 1 and 80 adults and 0 and 23 juveniles, both within and between species (Coates, 1982; Fricke, 1973b; Fricke and Holzberg, 1974; Katzir, 1981; Sale, 1972b). Fricke produced experimental evidence that group size in *D. marginatus* is constrained by the size of the coral head occupied by the group (Fricke, 1980). The limiting factor might be the number of shelter holes in the coral (Sale, 1972b). As group size of *D. aruanus* increased, intragroup aggression among individuals crowding together for shelter rose, presumably resulting in expulsion of some individuals from the group (Sale, 1972a). However, group size in *D. aruanus* did not correlate significantly with coral area when data from low- and high-density populations were included (Sale, 1972b). Thus, it seems likely that actual group size is influenced by additional factors, such as local population density, recruitment and mortality rates, and number of available coral heads.

The structure of *Dascyllus* groups also varies widely. In *D. aruanus,* individuals mated in one or several pairs, in a one-male, polygynous cluster, or promiscuously, depending on the group (Fricke and Holzberg, 1974). Small, all-female groups have been seen in some populations (Coates, 1982). In *D. marginatus,* the adult male-to-female sex ratio varied from 1 : 1 in small groups to 1:6 in intermediate-sized groups and back to 1:1 in large groups. Fricke considered these variations to represent entirely different mating systems: pairs represented monogamy in small groups; single-male, multifemale groups represented harems; and large groups represented promiscuity (Fricke, 1980). To explain this variability, one must postulate not only a relation between group size and coral area but also a degree of behavioral plasticity on the part of group members to permit or deny varying access to the group to members of each sex. Finally, in some populations, *D. aruanus* and *D. marginatus* formed mixed-species social groups, with only one male reproducing in the group (Shpigel and Fishelson, 1986). This male could be of either species. Females of the other species were said not to mate.

3. Grading between Qualitative and Quantitative Variation

One of the most variable social systems in coral reef fishes is that of the small, Indo-Pacific serranid, *Anthias squamipinnis* (called *Pseudanthias* by Katayama and Amaoka, 1986). Social variants of this fish illustrate the extent to which recruitment and mortality produce dynamic changes in the social system through time and the importance of behavioral plasticity and local history in mediating those changes.

At many locations, individuals formed discrete groups on coral aggregates surrounded by open sand or on unbroken stretches of reef face (Avise and Shapiro, 1986; Shapiro, 1987b; Shapiro and Boulon, 1982, 1987; Yogo, 1986). Most groups contained 1–400 individuals, with adult female-to-male sex ratios averaging 5.9–8.4 (Shapiro, 1980, 1988b). Males occupied discrete home ranges within the group, while females occupied home ranges that overlapped heavily with those of other females and with one or several males (Shapiro, 1986a). A few groups contained 1000 or more individuals, had a low female-to-male sex ratio, and two types of males. "Active" males occupied space in the upper portion of the group and spawned with females, while "inactive" males remained in lower regions of the group, failed to maintain spermatogenesis, and did not spawn (Popper and Fishelson, 1973).

While the volume of the occupied coral aggregate accounted for 33% of the variation in group size, the internal structure and composition of groups also varied widely and required additional explanation (Shapiro, 1988b; Shapiro and Lubbock, 1980). Groups were either spatially unitary or divided into two or three subgroups, with frequent movement by individuals between neighboring subgroups. The tendency for groups to be subdivided increased with group size (Shapiro, 1988b). Unitary groups closely resembled large subgroups and differed from small and medium subgroups in sex ratio, relation to the substrate, and juvenile membership. These patterns suggested that groups tended to subdivide as they increased in size, with large subgroups representing the later stage of a unitary group after group splitting (Shapiro, 1977, 1988b).

The possibility that groups might vary in time, as well as in space, by splitting as they grow was supported by two types of observation. First, if division into subgroups were the first step in the production of a new, independent group, then different groups could be expected to be at different stages of this developmental process. In conformity with this expectation, the five subgroups of two intensely studied groups varied widely in degree of independence (Shapiro, 1986b). Second, group splitting and seasonal variation in group size, composition, and structure were observed directly in a detailed study of 24 social groups over a 31-month period (D. Y. Shapiro, D. Goulet, D. Forestier, and M. Tacher, unpublished observations).

The one evolutionary explanation of group splitting suggested to date is that in consequence of the behavioral control of sex change (this species is a protogynous hermaphrodite), males are expected to have greater net reproductive access to females under some conditions in small- and medium-sized groups than in large groups (Shapiro and Lubbock, 1980). As groups grow, it would become increasingly advantageous for males to leave the group and enter or form a new, smaller group. Since large females can expect to change sex in the near future, they would reap the same advantage as males from departing from large groups. The result of these departures, if made simultaneously, would be group splitting. Splitting would be possible only if suitable, vacant substrate adjoined existing groups and only if group cohesion altered with group size.

This species' social system also varies in ways other than simply in group size through time. On reefs with relatively continuous coral cover, huge, unbroken "clouds" of *Anthias,* containing many thousands of fish and stretching dozens of meters along vertical reef faces, sometimes coexist with the more typical, discrete groups just described. In areas where scattered coral aggregates and uneven coral substrate are interspersed with large sand patches, a type of social grouping intermediate between discrete groups and continuous clouds is seen (Shapiro, 1987c). Intermediate groupings generally contained 100–200 individuals, which moved irregularly in space and time over relatively large areas. Consequently, these groupings were not readily definable spatially and were less dense internally than discrete groups.

Patterns of space use by individuals in intermediate groupings proved to be similar to those within discrete groups (Shapiro, 1986a, 1987c). Females occupied larger home ranges than males; most female home ranges overlapped those of several males; and female movements over the substrate and among portions of the discrete group or intermediate grouping depended on the continued presence of a male.

On the basis of this behavioral similarity, Shapiro (1987c) suggested that variations in the social system resulted from a combination of (1) fixed patterns of spatial behavior; (2) plasticity of other behaviors, such as aggressiveness toward nongroup members; (3) local substrate distribution; and (4) recent demographic history. The argument, in effect, described how one variant of the social system could grade into another variant. For example, sparse recruitment of juveniles onto a scattered, patchy substrate would produce a number of small, scattered, discrete groups. Juveniles would mature with the limited set of conspecifics within their group and would consequently distinguish group members from others, that is, would become xenophobic. If recruitment continued to exceed mortality, groups would increase in number and in size but would remain discrete and would follow the developmental sequence of splitting at characteristic sizes.

Alternatively, moderate recruitment onto the same habitat might place enough juveniles in close proximity to each other that aggressive intolerance would develop weakly or not at all. Lessened xenophobia would permit females to roam more widely, particularly after the death of a male and before the emergence of a newly sex-changed male, when females could move from one nearby set of individuals to another in search of mates (Shapiro, 1987c). The result would be an intermediate grouping in the same habitat that under different recruitment conditions produced discrete groups. Finally, another possible explanation would be that predation pressure is lower at some sites, thereby reducing the cost of moving between substrate patches and permitting the formation of intermediate groupings.

D. Conclusions from Empirical Studies

Several forms of variation in social structure are apparent from the empirical studies. First are variant social structures resulting clearly from two or more behavioral types of individuals coexisting in the same population. When the relative proportion of behavioral phenotypes varies, the social system may vary with it, as in the diandric wrasses with variant spawning systems on reefs with different proportions of IP and TP males (Warner and Hoffman, 1980a). When the behavioral types of individuals correspond to life history stages, variant social structures are formed by individuals at different sizes or ages, as in the grunt and parrotfish examples (Clavijo, 1982; Clifton, 1989; Helfman, 1978; Helfman and Schultz, 1984; Helfman *et al.,* 1982; McFarland and Hillis, 1982; McFarland and Kotchian, 1982; Ogden and Zieman, 1977; Warner and Downs, 1977). In general, different behavioral types of individuals may represent alternative ways of responding to an identical environment or separate responses by individuals sampling different environments.

The second type of variation in social systems are those occurring within or between sites and that are possibly, but not necessarily, the result of different behavioral phenotypes. In the literature most of these variations have been attributed to similar individuals sampling different microhabitats, that is, sites where the distribution and availability of food, shelter, mates, spawning sites, and competitors differ. Presumably, selective pressures for various forms of spatial and social behavior vary among microhabitats, and individuals respond by forming variant social structures. It should be noted that, in many of these studies, the actual distribution of environmental factors was not carefully measured.

Another explanation for this type of variation would attribute separate social structures to different behavioral genotypes, each occupying separate but similar sites and forming alternative social structures at those sites, or each selecting or surviving differentially in different microenvironments in accordance with their behavioral differences. This type of explanation might be invoked to explain cases like *S. iserti,* where some individuals defend feeding territories and others form roving groups of foragers, and *T. bifasciatum,* where individuals on fore reefs feed pelagically and form a different social system from individuals that feed benthically on back reefs (Buckman and Ogden, 1973; Clifton, 1989; Fitch and Shapiro, 1990; Ogden and Buckman, 1973; Robertson *et al.,* 1976). The possibility of having behaviorally different genotypes at one or several sites has not been examined, probably in part because of the widespread assumption of complete mixing of the gene pool during the prolonged pelagic phase of these fishes (Avise and Shapiro, 1986; Shapiro, 1983).

Finally, some intersite differences in social structure probably result from different rates and temporal patterns of recruitment onto a particular habitat type. This possibility was raised for *A. squamipinnis* when it was suggested that sparse recruitment would produce discrete groups at the same site at which moderate recruitment would produce intermediate groupings (Shapiro, 1987c).

The third type of variation in social systems is change occurring over time.

Group splitting in *A. squamipinnis* illustrates the dynamic nature of social units, which respond not only to static features of conspecific distribution and availability of environmental resources, but also to local demographic changes wrought by successive waves of recruitment and mortality (Shapiro, 1977, 1979, 1986b, 1988b). As additional long-term studies of coral reef fish social systems are conducted, more of this type of temporal variability within sites will undoubtedly emerge.

Several points stand out from the studies reviewed here. First, no single type of explanation is likely to apply to all cases of variation in social systems. Several types of hypotheses and many factors will need to be considered, including the influence of recruitment and mortality over time, and local demographic history. Second, for intra- or intersite variation in the social system to result, either several behavioral genotypes must be present within the population or behavior must be sufficiently plastic to produce those variations under different environmental and demographic conditions. Finally, the way in which individuals use space and the rigidity or plasticity of that behavior are of prime importance in determining the form taken by the social system.

In the next section, I consider several approaches to the theory of space use under conditions of recruitment similar to those typifying coral reef fishes. I then discuss how these approaches help us to understand intrapopulational variations in social systems.

III. Settlement Models for Individual and Group Spacing and Their Application to Social System Variation

One basic feature of social systems is the pattern of spacing by individuals within the population (Waser and Wiley, 1979). Variations in social systems may result from different patterns of spacing, for example, individuals foraging solitarily or in clusters, or from the superposition of different patterns of behavioral interaction on a relatively fixed system of spacing. For example, individuals spaced aggregately within colonies might mate as monogamous pairs or as polygynous groups within the colony.

In many coral reef fishes, particularly those with sedentary juveniles and adults, the basic pattern of spacing is strongly influenced by the spatial behavior of juveniles during settlement onto the reef. Thus, a good starting point for understanding variations in social systems of reef fishes is to examine spatial aspects of recruitment. In this section, I examine several theoretical models that predict that recruitment behavior will produce colonies of individuals in territorial species, and evenly dispersed social groups in group-living, protogynous species. I then discuss the application of these models to coral reef fishes and the instruction they provide for understanding variations in social systems.

A. Individual Spacing in Territorial Species

1. Theory

In territorial coral reef fishes, individuals defend temporary or permanent territories for feeding, spawning, shelter, or a combination of factors. In several theoretical models, the density and pattern of dispersion of territories depend on the sequence of events associated with the establishment and turnover of territories within a local habitat area (Getty, 1981; Maynard Smith, 1974; Stamps and Krishnan, 1990). If one assumes that mean territory size is fixed within a species and that new settlers arriving simultaneously at a site space themselves evenly apart, then the maximum number of individuals that can fit within a given habitat varies widely depending on whether all territories are established synchronously or successively (Maynard Smith, 1974). In this model, density can be three times as high with synchronous as with successive settlement.

If territories are not fixed in size, but may be compressed by the aggression of individuals settling close to preexisting territories, then the choice of settlement site should be governed by behavior called "preemption" (Getty, 1981). Preempting individuals settle closer to an existing territory than is required by initial density conditions. This behavior leaves such a small space between the settler and the preexisting territory holder that no subsequent settler can later intervene and force a compression of territory size. Individuals that preempt are expected eventually to have territories larger than those that do not preempt (Getty, 1981).

In Getty's model, a consequence of preemption is that territories are clustered around early "seed territories" (Getty, 1981). Within clusters, territories are uniformly distributed in space and nearest-neighbor distances remain constant through time as the cluster grows by accretion of new territories around the periphery. Saturation density may be much less than the maximum possible density since saturation can occur without having contiguous territories. As a result, aggression between neighbors may be less than in non-preempting populations.

Conditions necessary for preemption are strikingly similar to those suggested by some recent studies on behavior and recruitment in territorial coral reef fishes. The main conditions are that the system be nonequilibrial, with sequential settlement by new individuals onto territories, and that there be a potential for future new recruits (Getty, 1981; Sale, 1980a; Sale and Douglas, 1984).

Getty's (1981) model applies where territories are compressible and later arrivals insert themselves in the interstices between neighboring territories and subsequently expand the borders of their territories. Juveniles of coral reef fishes, however, are likely to settle onto reefs under a much wider range of conditions. For example, settlers of some species are likely to be able to settle within adult territories as well as in the interstices between neighboring territories (Sale, 1974). Recently, Stamps and Krishnan (1990) investigated the effect of a variety of behavioral settlement rules on the final size of territories of settling individuals, using computer simulations. The results of these simulations were unambiguous. In every case, average territory sizes were larger when individuals settled near a previously settled territory holder than when they settled randomly in space. In all simulations, a lower density of territories resulted from settlement near neighbors than from random settlement.

Thus, the intuitive model of Getty (1981) and the simulations of Stamps and Krishnan (1990) predict that settling near neighbors is advantageous. We expect, then, that any species settling by this rule would generate territories in clusters or colonies.

2. *Application to Coral Reef Fishes*

These models should apply to territorial coral reef fishes in which newly settled juveniles remain at the initial site of recruitment throughout much or all of their lives. They should also apply to species in which juveniles or adults spend part of their life wandering or switching territories, provided the rules for territory establishment remain the same as those assumed in the models for initially settling juveniles.

Degree of synchrony of arrival corresponds roughly with the extent to which a habitat patch is settled simultaneously by a large wave of recruitment or gradually over many small recruitment pulses. The correspondence is not exact because the models assume that early and later settlers are of comparable size and competitive abilities and have comparable resource requirements. In coral reef fish, early recruits will have grown by the time later recruits arrive; thus, successive settlers will not be exactly of comparable size. Both simultaneous and successive recruitment have been observed: massive recruitment during short periods in the triggerfish *Balistes vetula* (Robertson, 1988b) and the bluehead wrasse *T. bifasciatum* (Victor, 1986b), and gradual, successive recruitment in these and in many other species (Doherty, 1987a; Williams, 1983a; Williams and Sale, 1981).

A clear prediction of the models by Getty (1981) and Stamps and Krishnan (1990) is that new territories should form closer to neighbors than expected at random, and consequently that territories will be clustered. To evaluate the role of conspecific attraction in the formation of clusters in a real population, one must find examples where clusters occupy less than an entire patch of suitable habitat. Otherwise, territorial clusters may simply be the result of filling a habitat patch (Stamps, 1988).

Territorial adults of a number of coral reef fishes are said to cluster in colonies (Baird, 1988; Clifton, 1989; Fricke and Kacher, 1982; MacDonald, 1981; Myrberg, 1972; Roberts, 1985; Robertson and Lassig, 1980; Robertson *et al.*, 1979; Sadovy, 1986; Schmale, 1981; Victor, 1987b; Vine, 1974; A. H. Williams, 1979). In only a few cases, however, have the territories of individuals been mapped in a sufficiently large area to evaluate clumped dispersion visually or statistically. For example, the frequency distribution of individual *Xyrichtys martinicensis* in a large square grid of sand bottom revealed strong clustering (Victor, 1987b). When the locations of individual *Acanthurus lineatus* were plotted over a 400-m section of fringing reef, adults and juveniles clearly formed clusters, with apparently similar but unoccupied substrate surrounding the colonies (Robertson *et al.*, 1979).

To my knowledge, clumping has been evaluated statistically only for the territorial Caribbean damselfish *Stegastes partitus* (Sadovy, 1986). All individuals within a 14 × 30-m area of homogeneous habitat were each followed for 20 min and the centers of their activity were mapped (Fig. 1). Nearest-neighbor (NN) analysis was performed on all points remaining after those within 1 m of the borders of the study area were discarded (to avoid edge effects). The observed mean NN distance (0.73 m) was significantly lower ($p < 0.001$) than the mean NN distance expected (1.18 m) if territories were randomly dispersed. Sadovy (1986) concluded that individuals were aggregated. Aggregations were called "colonies" and were considered discrete if members of one aggregation did not interact behaviorally with members of another aggregation (Sadovy, 1986). In most other cases in the literature, it is not clear whether the titular colonies represented a greater degree of aggregation than would be expected by saturating a local habitat patch, that is, whether they were true colonies or not.

In no case have colonies been observed during their formation. Consequently, it is not certain whether individuals behave preemptively or in the various ways assumed in the settlement models. Furthermore, clustering of territories could result from other processes not included in these models, for example, postsettlement predation or exclusion of individuals by members of competing species. However, several observations suggest that colonies might have formed by accretion of juveniles around the periphery of older individuals, in accordance with the view of sequential settlement in the models. In *S. planifrons* and *A. lineatus*, large adults occupied the center of colonies, while smaller adults and juveniles occupied the periphery (Itzkowitz, 1978; Robertson *et al.*, 1979). While this pattern might result from large individuals dominating smaller ones to occupy the most beneficial, central sites, as these authors suggested, they might also be the natural consequence of colony

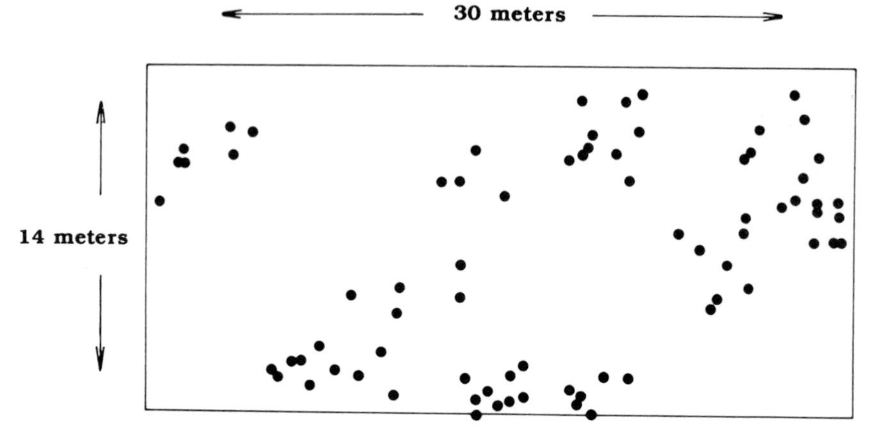

Figure 1 The location of territory centers (·) of 75 *Stegastes partitus* in a 14 × 30-m study area on Mario Reef, Puerto Rico. Territories were statistically clumped. (From Sadovy, 1986.)

growth. If later-arriving juveniles settled near the territorial borders of earlier-arriving individuals, then the oldest individuals would always occupy the center of the colony.

On the Great Barrier Reef, the damselfish *Plectroglyphidodon lachrymatus* lived in colonies in staghorn coral (Robertson and Lassig, 1980). At the center of the colony, most of the coral was dead and displayed signs of having died well before dead or moribund coral at the edges of the colony. Robertson and Lassig hypothesized that this pattern resulted from the formation and growth of the fish colony: "the coral occupied first by the damselfishes had been dead longest and the most recently dead material, mixed with living coral, was that into which the fishes had most recently moved." These remarks suggest that colonies grew by accretion of territories around the periphery.

These observations, by seemingly conforming to the models' predictions, tend to support the assumption that settlement near previous territory holders is selectively advantageous. Why might this be so? In the models, preemptive behavior produces two main consequences: colonies of territory holders and an increase in average territory size (Getty, 1981; Stamps and Krishnan, 1990). Preemptive behavior would be selected for if one or both of these consequences were selectively advantageous.

Colony-living may offer various advantages to territorial coral reef fish. Adults and young may obtain increased protection against predation (Baird, 1988; Foster, 1989; Victor, 1987b), and increased defense against interspecific food competitors because "some avenues of entry (into a territory by a competitor) would be blocked by neighboring territories" (Robertson *et al.*,

1979). Colony formation might also function to reduce the distance traveled and time to locate a mate, which would reduce time off the territory (Itzkowitz, 1978). Consequently, food loss to competitors would decline, allowing individuals in colonies to hold smaller territories than if they were isolated. Robertson *et al.* (1979) suggested that this effect might become stronger as colonies grew, resulting in progressive declines in individual territory size and gradual condensation of the colony. Colonies of territories might also be beneficial if neighbors can provide valuable information about the habitat to new settlers (Stamps, 1988) and if neighbors form alliances to help defend each other's territory against incursions from potential territorial usurpers (Getty, 1987).

On the other hand, the models of Getty (1981) and Stamps and Krishnan (1990) imply that the main advantage of preemption is to increase average territory size. The value of any particular territory size depends largely on the way the costs and benefits of territories change with territory size and on whether the species involved is a "time minimizer" or an "energy or area maximizer" (Ebersole, 1980; Harvey and Mace, 1983; Hixon, 1980b; Schoener, 1983a). Most conditions that have been examined theoretically favor an intermediate, optimal territory size (Hixon, 1980b). However, individuals should maximize the size of their territory where males, but not females, are territorial and females choose mates on the basis of the area of the male's territory, and when larger areas provide excess resources as a buffer against periods of scarcity (Hixon, 1980b, 1987). Preemptive behavior would also be favored whenever nonpreemptive settlement produced suboptimal territory size in consequence of conspecific competitor density. Under these conditions, preemptive behavior would provide individuals with a closer approximation of optimal territory size. Finally, if territories are established while the individual is small, but resource requirements increase with body size, then it may be advantageous for small settlers to behave preemptively because their final territory size will be larger, by the time they grow to need it, than it would have been had they not behaved preemptively.

If the primary advantage of settling close to neighbors is colonial life, then preemption can be viewed simply as the mechanism for the formation of this advantageous form of social organization. If the primary advantage is increased territory size, then the formation of colonies becomes an epiphenomenon resulting from selective pressure for increased territory size. Thus, the models illustrate that recruits into a population might occupy a particular type of social structure either because selection favors the social structure itself or because selection favors some other consequence, for example, increased territory size, of the behavior that produces the social structure as a by-product.

These models are important for two other reasons. First, they provide insight into how and why settlement behavior by individuals leads to a particular type of spacing pattern, that is, colonies, which in turn might be shared by several types of social systems. For example, once individuals hold territories, males could expand their movements to incorporate the territories of one or several females. If females tolerated this male behavior, the underlying spatial arrangment would become a one-male, one-female pair or a one-male, multifemale group, respectively (Hixon, 1987). If the initial territory site were suitable for both feeding and spawning, the social system would consist of small, all-purpose groups. Alternatively, males and females might retain individual territories and seek mates during short excursions out of their territory; if sites of initial territories were suitable for feeding but not spawning, these excursions might result in migrations to better spawning sites.

Second, these models illustrate that a fixed settlement strategy, preemption, can produce the same pattern of space use under varying recruitment conditions. Although population density and average territory size may vary, depending on whether settlement on territories is synchronous or successive, preemption will lead to spatial aggregation of individuals in both cases.

Thus, if intraspecific variation in social systems involves fixed and plastic behaviors, then these models help us to understand at least those fixed components that concern colonial patterns of space use. Once that underlying pattern is established, superposition of other, variable behaviors produces variant social systems. In the next section, I consider how models of spacing behavior apply to individuals living in groups rather than in individual territories and how those models aid in explaining variant social systems in group-living species.

B. Local Spacing in Group-Living Species

The advantages of settling close to preexisting neighbors would accrue both to individuals each defending its own territory and to individuals in territorial social groups whose defense of space is shared by some or all of its members, provided new groups form by juveniles settling together at a new site. When groups defend territories we can expect newly settling groups of juveniles to follow the neighbor rule. The result will be clusters of groups, with even dispersion of groups and nearly contiguous group territories within each cluster. Little is known about the spatial dispersion of territorial, group-living coral reef fishes (Sale, 1972b). Anemonefishes occupying separate groups within large fields of contiguous anemones are likely candidates to show this dispersion pattern of groups (Fricke, 1979).

Since many group-living fishes do not defend group space, however, territorial models would not apply to them. For these species, other considerations may be paramount in determining spatial dispersion. One such consideration is the role of sex change.

Many group-living coral reef fishes are protogynous hermaphrodites in which sex change is controlled behaviorally (Shapiro, 1987a). Adult females change sex in response to alterations in their behavioral interactions with members of both sexes within the group. These alterations can be produced by sudden separation of females from males, for example, by the death of a male, or by the maturation or entry of new females into the group (Aldenhoven, 1986a; Shapiro and Lubbock, 1980). The larger the number of females living together at the time of separation of the sexes, the greater is the probability that one or more of them will change sex (Cole and Robertson, 1988; Ross *et al.*, 1983; Shapiro and Boulon, 1982). These observations suggest the existence of a sex ratio threshold, such that whenever the threshold is exceeded, an individual will change sex (Shapiro and Lubbock, 1980). Sex change mediated by exceeding a sex ratio threshold forms the mechanism for a selective pressure favoring even dispersion of social groups, in the following way.

The sex ratio in protogynous fishes generally exceeds 1.5 females per male (Sadovy and Shapiro, 1987) and may be as high as 5–8 females per male (Shapiro, 1979; Yogo, 1986). The effective sex ratio at spawning may be even higher, particularly for successful males (Hoffman *et al.*, 1985; Warner and Hoffman, 1980a). As long as the sex ratio remains high within a mating unit, it should be advantageous for a female to change sex as soon as possible, provided, once having changed sex, its reproductive value as a male exceeds the expected reproductive value had it remained a female (Charnov, 1982; Ghiselin, 1969; Shapiro, 1989b; Warner, 1975). The faster new juveniles enter a group and mature, the sooner the sex ratio threshold will be exceeded and the sooner a large female will change sex. Females seem to change sex roughly in the order of their size rank within the unit (Fishelson, 1970; Robertson, 1972; Shapiro, 1981). Consequently, as the rate at which large females change sex increases, smaller females will move up in the size rank faster and they also will change sex sooner. Since successive sex changes occur when the sex ratio threshold is exceeded, the average sex ratio will remain stable (Shapiro and Lubbock, 1980) and the advantage of changing sex will continue through time over successive sex changes.

Thus, an individual can increase its lifetime reproductive value by occupying a group that is more attractive to juveniles than other groups. The result of this process is selective pressure for individuals to form or enter groups that are as attractive to juveniles as possible.

The attractiveness of a group to a settling juvenile is likely to be a function of many factors: group size, availability of food and shelter, and local predator density, although which of these factors settling juveniles can assess is not known. Certainly one factor determining group attractiveness, measured empirically as the rate at which juveniles enter it, is the distance from that group to its nearest neighbor. With a few assumptions we can model the influence of nearest-neighbor distance on recruitment of juveniles (Shapiro and Boulon, 1987).

Let us assume (1) that juveniles are randomly or homogeneously distributed above the reef just prior to settlement, (2) that recruitment rate onto a reef is independent of the density of adults occupying it, as was found in some recent studies (Doherty, 1983a; Victor, 1986b), (3) that all juveniles enter established groups at the time of settlement, and (4) that juveniles settle into whichever preexisting group is closest to them when they leave the water column.

Where these assumptions hold, an established group that is far from its neighbors will attract all juveniles that are in its vicinity at the time of settlement. A group that is close to a neighbor, which resembles it in size and resource availability, will attract roughly half of the juveniles settling in its vicinity. The other half will enter the neighboring group. The result is that a preexisting group attracts settling juveniles in proportion to the distance separating the group from its neighbors. The farther a group is from a neighbor, the larger the number of juveniles that will enter it during settlement.

Let us now expand the model and assume that juveniles either enter preexisting groups or settle aggregately and form new groups at unoccupied sites at the time of settlement. Let us also assume that new groups can form at any unoccupied site. Since an individual can expect to change sex earlier in its life and reap the reproductive advantage of being a male sooner by forming a group that is highly attractive to other juveniles, locations for new groups that are far from neighbors will be favored over locations that are close to neighbors. Consequently, new groups should form at sites that are as far from preexisting groups as possible. If new groups consistently formed in this way, the result would be statistically even dispersion of groups over the reef. This model could easily be expanded to include other factors that are likely to influence the attractiveness of a group to settling juveniles, that is, group size or local resource availability (Shulman, 1985a; Sweatman, 1985a).

To my knowledge, no published study of a group-living species has examined spatial aspects of settlement strategies relevant to this model and only one study has evaluated spatial dispersion among nonterritorial social groups. In two populations of *A. squamipinnis,* one occupying individual coral aggregates scattered over open sand and the other occupying relatively uniform, vertical reef face, social groups were mapped and nearest-neighbor analysis was performed. In both habitats, groups were statistically evenly dispersed (Shapiro and Boulon, 1987). In the habitat where groups occupied discrete coral aggregates, there was much unoccupied space between neighboring groups and more than four times as many coral aggregates of inhabitable size as existing groups. Along the vertical reef face, there was no discernible difference between unoccupied substrate separating neighboring groups and the portions of reef face that were occupied. Thus, the even dispersion pattern was probably a reflection of the behavioral strategy governing formation and spacing of groups, rather than even dispersion of suitable habitat (Shapiro and Boulon, 1987). It remains to be discovered, of course, whether this dispersion pattern was generated by the fish's settlement strategy or by later events, such as mortality or group migration.

The relevance of this model of group spacing for social system variation can be revealed by a brief examination of the model's assumptions. In the model, recruitment behavior follows a simple rule: juveniles either enter preexisting groups or aggregate at new sites as far from preexisting groups as possible. When preexisting groups are relatively sparse and are separated by much unoccupied habitat, and when recruitment rates are low or moderate (conditions implicitly assumed in the model), this rule produces new social groups evenly spaced among preexisting groups. However, when preexisting groups are dense (but separated by available habitat) and recruitment is moderate or heavy, or when recruitment is massive, the settlement rule results in a large, contiguous mass of individuals, filling all available space. Discrete groups will form or maintain their identity in this situation depending on the extent to which individuals display aggression and affiliative behaviors differentially among near and far neighbors. Thus, a fixed behavioral rule could produce discrete groups under one set of recruitment conditions and a continuous cloud of individuals not broken into discrete groups under other conditions.

In other words, the model helps us understand variations in social structure by distinguishing fixed from plastic behaviors and by implicitly specifying the environmental conditions under which those behaviors might be expected to produce social variants. This model and those concerning settlement behavior in individually territorial species are heuristically useful for thinking through the problems of how and why a mixture of spatial and social behavioral elements combine to form variant social systems.

IV. Conclusions

Social systems of coral reef fishes can be highly plastic and may differ intraspecifically from one general location to another or from one part of a reef to another. The social structure on one reef may change through time for several reasons. Changing mortality and recruitment patterns may alter which portions of the reef are occupied or change the demographic structure of the local population, thereby favoring new patterns of individual space use and mutual interaction. The basic social structure itself may incorporate processes of social development and change, driven by selective pressures related to other features of the species' basic biology, such as protogynous sex change. Finally, individuals may change their ecological priorities as they grow, resulting in selective pressure for different social structures, ontogenetic movement between those structures, and intrapopulational differences within the social system itself.

We have seen that there are good theoretical reasons to expect individuals to follow concrete rules governing spatial dispersion at the time of settlement and later. These rules are likely to produce particular spatial patterns on the reef, for example, clustering in colonies for individually territorial species or evenly dispersed groups for group-living fishes. The pattern of space use represents a fundamental aspect of the social system (Waser and Wiley, 1979), but it constitutes neither a complete description nor a sufficient force to generate a final form of social organization. Any particular pattern of spatial dispersion may support a variety of social systems.

It is important to remember that social systems are produced by the behavioral tendencies of its members. Natural selection can act separately, in producing variations in the social system, at two places: on the behavioral tendencies themselves or on the final form of the social structure generated by those tendencies. It might be advantageous under some conditions, for example, for individuals to relax their tendency to chase conspecifics regardless of whether or not such an alteration changes the social structure from individual territoriality to small groups. If a variant social form resulted and were selectively neutral or not strongly disadvantageous, then the alteration in social system would be an epiphenomenon contingent on the advantages of a change in the underlying behavioral tendency. On the other hand, if strong selective pressures made it advantageous to alter the form of social structure, then a change in underlying behavioral tendency would become the proximate mechanism generating the newly favored structure. Only a careful analysis of the way in which fixed and plastic behavioral tendencies generate particular social structures will enable us to begin to make such explanatory distinctions.

The studies on *L. dimidiatus* and *A. squamipinnis* demonstrate that not all behavioral tendencies are equally plastic. In *L. dimidiatus,* individuals at all sites formed single-male, multifemale groups of restricted size. Thus, all females moved only within a small area and all had a very limited tendency to interact with more than one male. These relatively fixed behavioral tendencies constrain the general form of the social system. In branching groups, however, females have a greater tendency to discriminate between two classes of group members, allowing one set to move within their common territory while chasing others a short distance away, but not out of the group (Kuwamura, 1984). The tendency to chase other females is thus relatively plastic and intrapopulational variations in that tendency generate the observed differences in group structure.

In *A. squamipinnis,* the large variation in size and internal structure of discrete groups and the range of social systems from discrete groups through intermediate groupings to enormous clouds mask an underlying similarity in all group types in the way individual males and females use space. Females move over larger home ranges than males, and female rates of movement exceed male rates and depend on the continued presence or absence of males in the immediate vicinity (Shapiro, 1986a,b, 1987c). Thus, these aspects of movement in space remain constant while other behaviors, including tendency to discriminate between group members and strangers and tendency to roam over small or large areas, are plastic.

I suspect that comparatively small changes in the tendency to perform critical behaviors can have apparently large effects on the form of the social system. For example, imagine a set of individuals each holding a separate territory contiguous to that of neighbors. Individual territoriality remains as long as the tendency to chase intruders is high for all individuals. If the chasing tendency altered even slightly, say, becoming lower for individuals that were familiar while remaining high for strangers, then immediate neighbors could begin to share space. If there were small-scale patchiness in the distribution of original territories, sets of immediate neighbors within small patches would tend to share space and coalesce. The social system would change from individual territoriality to small groups. One can easily imagine other examples where small changes by individuals in the tendency to perform one or a small number of simple behaviors generate qualitative alterations in the form of the social system.

Thus, to understand variations in social systems we need to know (1) how particular behavioral tendencies combine to produce each variation; (2) to what extent different tendencies are fixed or plastic; and (3) how those tendencies are influenced by individual growth, juvenile recruitment, adult mortality, the distribution and availability of mates and needed resources, local water movement (for marine fishes), and the preexisting social system, that is, local social history. The literature is just beginning to fill in these points for some vertebrates at empirical and theoretical levels (Chase, 1982, 1985; Hinde, 1983; Hogeweg and Hesper, 1983; Rubenstein and Wrangham, 1986; Slobodchikoff, 1988; Wrangham, 1986).

Acknowledgments

During the preparation of this chapter, I was supported by NSF Grant RII-8610677 and NIH Grant SO6RR-08103, whose backing I gratefully acknowledge. Thanks go to Y. Sadovy for permission to use a figure from her Ph.D. thesis and to M. A. Hixon, M. E. Leighton, and J. A. Stamps for extensive comments on earlier versions of the manuscript.

CHAPTER 13

The Role of Adult Biology in the Timing of Spawning of Tropical Reef Fishes

D. Ross Robertson
Smithsonian Tropical Research Institute
Balboa, Republic of Panama

I. INTRODUCTION

Most marine fishes, including tropical reef species (Sale, 1980a), have two-phase life cycles that include a planktonic larval stage. The observation of considerable fluctuations in temperate region stocks of marine invertebrates and fishes that have this type of life cycle led to the proposition that their adult population sizes are controlled primarily by variation in the larval supply (e.g., Thorson, 1950). A logical development of this idea, which continues to play a major role in the population biology of such organisms (for reviews see, e.g., Cushing, 1982; Doherty and Williams, 1988a; Roughgarden *et al.,* 1988), is that their temporal patterns of reproduction are tailored to larval requirements and have been selected to maximize larval recruitment and cope with both predictable and unpredictable variation in larval mortality. The hypothesis that larval biology ultimately controls the scheduling of reproduction has dominated analyses not only of seasonal patterns of spawning of temperate marine fishes (e.g., Qasim, 1956; Parrish *et al.,* 1981; Bakun *et al.,* 1982; Lambert and Ware, 1984; Sherman *et al.,* 1984; Checkley *et al.,* 1988), but also of the timing of reproduction of tropical reef fishes on a range of temporal scales (for reviews see Johannes, 1978a; Thresher, 1984; Walsh, 1987; Gladstone and Westoby, 1988).

There is a great variety of patterns of spawning by reef fishes on the daily, lunar and seasonal time scales. Some species have fixed, short diel periods when they spawn, while others spawn throughout much of the day. The timing of discrete diel spawning periods may also vary and track changes in the daily timing of the tides. On the lunar scale, spawning of some local populations may be synchronized, either sporadically or cyclically at lunar, semilunar, and other frequencies. In other cases, reproduction of a population is acyclic and unsynchronized, although its individual members may have their own cycles of activity. The level of activity varies from individuals spawning only once or twice a month to spawning every day or two. Seasonal spawning patterns among the members of a single reef fish community may range from situations in which an entire local population spawns more or less synchronously once a year, through uni- and bimodal seasonal cycles that vary in their strength and timing, to apparently nonseasonal, year-round activity.

This temporal variation is compounded by variation in where and how species spawn, what types of eggs they produce, and the extent of parental care. Some fishes spawn inside their small permanent feeding areas while others migrate up to tens of kilometers to traditional spawning grounds. When spawning, many species abruptly release a cloud of pelagic eggs high in the water column and pay no further attention to them. In some species, benthic eggs are scattered more or less indiscriminately on the substratum, while in others they are carefully laid in discrete, dense, monolayered clutches in prepared, permanent nests where they are guarded intensively by their parent(s) for several days until they hatch and the larvae disperse.

Although much attention has been focused on the ways in which larval biology may influence the evolution of reproductive patterns of tropical species, we have tended to ignore ways in which adult biology also may do so. While recognizing the central role that larval biology plays in the life histories and reproduction of these organisms, the purpose of this chapter is to consider how adult biology factors may control reproduction on the diel, lunar and seasonal time scales. In it I will review existing larval biology and adult biology hypotheses, evaluate how different hypotheses cope with the available data, and, in some cases, indicate what types of data are lacking. I will point out how adult biology may affect spawning patterns in ways not previously considered, and how adults may sometimes be constrained by their own biological limitations from scheduling reproduction in a manner that is most appropriate for maximizing larval recruitment. In doing so I hope to provide a more balanced perspective of factors that control reproduction in reef fishes and show how adult biology explanations may represent viable alternatives to many larval biology explanations.

II. Diel Spawning Patterns

Almost all the available data are for day-active fishes that spawn during daylight or the crepuscular periods. Analyses of diel spawning patterns have concentrated on predator–prey interactions involving both spawners and their eggs and larvae, although some attention also has been paid to the energetic requirements of newly hatched larvae, and to the effects of parental care of eggs on parents.

A. Larval Biology Hypotheses

Existing larval biology hypotheses deal mainly with risks of predation on planktonic propagules as they are released from a reef. Pelagic spawners produce planktonic eggs that disperse immediately after spawning, while eggs of benthic spawners develop on the substratum for a day or more before the planktonic larvae disperse from them. Even though spawning and propagule dispersal are well-separated events in benthic spawners, it has been proposed that diel spawning periods of both pelagic and benthic spawners represent adaptations to larval requirements.

1. Tidal Influences on Propagule Dispersal

Johannes (1978a) developed the hypothesis that the tidal regime is the primary factor controlling the diel spawning periodicity of reef fishes. He proposed that dispersal of planktonic propagules away from adult habitat is advantageous and that adults facilitate dispersal and minimize predation on propagules as they leave the reef by spawning at the stage of the tide most appropriate for dispersal, typically the beginning of the ebb tide.

If the tidal regime is a simple one, high tides of a particular relative size consistently occur at the same time of the day. In such a system, restriction of spawning to a particular time of the day could represent a response solely to the tidal regime. Alternatively, spawning during a particular time of the day could be advantageous per se, and fish may spawn at a particular tidal stage only when it occurs during the particular diel period because they are responding to both diel and tidal factors (Robertson, 1983). Time of day per se does appear to be important to some species. Many that live either at localities with almost no tidal influence or in habitats that are not strongly influenced by tidal effects restrict their spawning to a particular time of day but not to a particular tidal stage (see examples in Robertson, 1981; Gladstone and Westoby, 1988; Colin and Clavijo, 1988; Myrberg *et al.,* 1989). Colin and Bell's (1991) observations of intraspecific variation in diel spawning activity of pelagic spawners at a single locality show that although both time of day and the tidal cycle affect spawning patterns, the relative importance of each varies in different habitats. In one habitat with minimal tidal flow a species has a short, fixed, daily spawning period, regardless of the stage of the tide, while in another habitat with strong tidal flows the same species spawns at a fixed tidal stage at a variety of different times during the day.

2. *Risks from Mobile Egg-Predators*

There are at least three mechanisms by which the restriction of pelagic spawning to a certain period of the day might reduce losses to mobile egg-predators. Spawning could be restricted (1) to periods when egg-predators are inactive, or (2) to periods when egg-predators, although active, are likely to be satiated by prior feeding. Alternatively, (3) synchronization of spawning could simply oversaturate active egg-predators regardless of their feeding status (Robertson and Hoffman, 1977; Johannes, 1978a; Lobel, 1978; Robertson, 1983; Colin and Clavijo, 1988).

Thresher (1984) criticized this egg-predation hypothesis and claimed that rates of egg predation are too low to be of real significance. Whether any general pattern exists is uncertain. While rates of egg predation are very low at some localities (Colin and Clavijo, 1988), they are known to be high at others (Robertson, 1983; Moyer, 1987). Rates of egg predation also can vary considerably among sites on the same reef (Colin and Bell, 1991) and vary in their impact on different species of spawners at the same site (Colin and Clavijo, 1988).

Many pelagic spawners spawn in the afternoon or around dusk (see Thresher, 1984; Gladstone and Westoby, 1988; Colin and Clavijo, 1988). This diel pattern would be expected if they were restricting spawning to a time when diurnal egg-predators would be expected to be relatively satiated. However, so little is known about diel and tidal patterns of feeding activity of mobile egg-predators (I cannot find any directly relevant published studies) that the various forms of the egg-predation hypothesis cannot be seriously evaluated at present.

3. *Initiation of Larval Feeding*

Prompt initiation of feeding is thought to be critical to the survival of newly hatched fish larvae (e.g., Rothschild, 1986). A restricted, fixed daily spawning period could be advantageous if spawning time affects the survivorship of newly hatched larvae by determining the size of their energy reserves and when they begin feeding.

Diel dispersal windows of benthic larvae appear to be quite narrow—in all known cases larvae are released shortly after sunset (Thresher, 1984; Gladstone and Westoby, 1988). Doherty (1983c) proposed that, because benthic eggs that are not ready to hatch during one evening period must wait an additional 24 hours, a failure to hatch during the evening closest to the time when they complete development could result in unhatched but fully developed larvae depleting energy reserves they will need at the start of their planktonic life. Consequently, spawning early in the day may be advantageous for such species because it maximizes the proportion of larvae that are ready to disperse during the nearest suitable hatching window. This hypothesis could be tested by examining relationships between intraspecific seasonal variation in (1) the duration and timing of diel spawning periods, (2) egg size (which affects development rates), and (3) temperature-mediated egg development times.

Colin (1989) suggested that spawning may be timed so that eggs hatch in the morning and larvae have a full daylight period in which to begin feeding. Such a relationship seems unlikely to be of general importance, for two reasons. First, since benthic eggs typically hatch shortly after sunset, their larvae are released during the worst period from the point of view of diurnal feeding. Second, diel hatching periods of pelagic eggs of various species are likely to be broad, since they spawn throughout most of the day (e.g., Robertson, 1983; Colin and Clavijo, 1988; Colin and Bell, 1991) and there can be significant intraspecific variation in development rates of eggs at a constant temperature (e.g., Colin, 1989).

B. Adult Biology Hypotheses

1. Predation on Spawners

Hobson (1968, 1974) observed that some predatory fishes are most active in attacking particular types of prey during crepuscular periods, and he developed the idea that aspects of the diel patterns of activity of prey fishes represent risk-minimizing responses to the diel feeding cycles of predatory fishes. This led to the suggestion that, by spawning during the day, pelagic spawners reduce the risk of attack on themselves by crepuscular predators (Robertson and Hoffman, 1977; Johannes, 1978a). However, Thresher (1984) assembled data from a wide variety of sources, which show that crepuscular spawning is very common among pelagic spawners. He suggested that most predatory fishes actually are least active during crepuscular periods, noted that crepuscular spawners indulge in what appear to be more risky types of spawning behavior than do day spawners, and concluded that by spawning at dusk, pelagic spawners could be avoiding high-risk periods. This analysis did not take into account potentially confounding effects of several variables, since it combined data on species in a wide variety of taxa, and from different localities and habitats. However, it is a useful starting point that indicates aspects of the reproductive behavior of prey fishes that should be examined in comprehensive comparative analyses of the activity of entire communities of predatory and prey fishes.

Rates of attempted predation on pelagic spawning fishes are very low in some areas (Colin and Clavijo, 1988; Colin and Bell, 1991) but can be high in others (Moyer, 1987). Effects of attempted predation range from serious to minimal disruptions of spawning (Colin and Bell, 1991; Robertson, 1983). Large fishes may be relatively immune to predation, and size-mediated variation in predation risks may influence not only the rapidity of spawning and the propensity for spawners to move away from shelter (Thresher, 1984; Colin and Bell, 1991) but also the diel timing of spawning.

The biggest deficiency in any arguments about diel variation in predation risks is that there are *very* few studies of the diel patterns of behavior of predators. Most of the data derive from stomach content analyses (see e.g., Nagelkerken, 1979; Shpigel and Fishelson, 1989, and references therein). Those data and the few observations that have been made of predation indicate that some predators are crepuscular while others feed mainly during the day. It is far from clear whether there are any general diel patterns of predator activity (Sweatman, 1984). Piscivorous fishes sometimes have quite restricted diets (e.g., Nagelkerken, 1979; Shpigel and Fishelson, 1989), and predator–prey interactions that are important to spawning fishes may be quite specific. Future analyses will need to be quite precisely focused to take all of these sources of variability into account.

2. The Duration of the Spawning Act

In many pelagic spawners a female releases her entire clutch in one act that lasts only a few seconds. In contrast, a female benthic spawner may release eggs individually and take half an hour or so to deposit her entire clutch (e.g., Pressley, 1980; Kohda, 1988; Hunte and Cote, 1989). A consequence of this difference may be that spawning of benthic spawners is more vulnerable to disruption and premature termination than that of pelagic spawners. Risks of disruption of spawning by diurnally active egg-predators could have selected for crepuscular spawning by benthic spawning fishes (Pressley, 1980), many of which spawn around dawn, but not in the evening. Pelagic spawners, on the other hand, spawn at dusk much more often than they do at dawn (Gladstone and Westoby, 1988; Kohda, 1988). This difference could reflect differences in the diel feeding cycles of different types of predators on eggs and on spawners, including asymmetries in the crepuscular activity of those predators (see, e.g., Sweatman, 1984; Nagelkerken, 1979). However, the difference in spawning times also could simply reflect constraints arising from the structure of the spawning act of benthic spawners. While a female pelagic spawner could complete a spawning at dusk even if light levels decline prematurely due to bad weather, a female benthic spawner could have her spawning curtailed by the same event. A similar situation would be very unlikely to occur at dawn.

3. *Effects of Cost of Parental Care*

Doherty (1983c) proposed that the timing of spawning of brood-guarding species might represent a response to the cost of parental care. In brooders whose eggs complete development within several days, each additional day a clutch spends in the nest represents a substantial percentage increase in the duration and cost of brood care of that clutch. For day-spawning species whose benthic eggs hatch only during the early evening, it could be advantageous to spawn early in the day so that larvae have the greatest chance of being ready to leave during the first available hatching period. Such a response to the cost of care could account for the preponderance of morning spawning among benthic brooders. Effects of variation in that cost would potentially be most important for tropical species whose eggs develop quickly and at fairly predictable rates at stable high temperatures. Comparison of the diel spawning patterns of temperate and tropical species, or, preferably, of different populations of the same species at different latitudes or different seasons (i.e. under different temperature regimes) could be used to test this hypothesis.

4. *Feeding Biology of Spawners*

One area that has been largely neglected is the relationship of diel patterns of spawning to the feeding biology of spawners. Examination of this relationship could be instructive because there may be both benefits and costs to feeding biology from particular patterns of spawning activity. Since reproductive output depends on food intake, diel patterns of spawning could represent adaptive responses that minimize feeding losses. Conover and Kynard (1984) pointed out that, by hydrating its clutch overnight and spawning in the early morning, a diurnally active fish should minimize feeding losses due to the time spent in reproduction, including the period when the body cavity is preempted by the greatly expanded gonads.

The potential influence of feeding costs as a factor selecting for particular diel spawning periods of reef fishes has not been examined. That cost should depend on a variety of factors, including the relative volumes of the clutch and food mass, the amount of time and energy spent on each spawning, the spawning frequency of individuals, the percentage of the day spent feeding, and the distribution of feeding over the day. For example, feeding losses might be expected to have relatively strong effects on herbivorous reef fishes that spend much time migrating substantial distances between feeding and spawning sites (e.g., Myrberg *et al.*, 1989), spawn frequently (Robertson *et al.*, 1990), process relatively large volumes of food during daily feeding periods (e.g., Polunin and Klump, 1989), and spend much time feeding each day and have afternoon peaks of feeding activity (e.g., Robertson, 1984;

Polunin and Klump, 1989). If feeding costs are of general importance, we would expect that herbivorous reef fishes would tend to spawn in the early morning. Herbivorous damselfishes generally do spawn in the morning (Kohda, 1988), whereas herbivorous surgeonfishes generally do not (Robertson, 1983; Myrberg *et al.*, 1989; Colin and Bell, 1991).

The tidal regime also controls diel feeding patterns of many species independently of time of day. Tidal currents provide food for many planktivores. In some benthic feeding species, feeding activity may be restricted regularly by tidally mediated limitations on their access to intertidal feeding habitats. Consequently, in species in which diel patterns of feeding are tidally controlled, particular diel spawning patterns might minimize loss of feeding opportunities, or the development of a clutch in the female (and hence the timing of the spawning period) might be controlled directly by the diel cycle of food intake.

The risk of loss of food to competitors also could influence diel patterns of spawning, and some fishes may restrict spawning to periods when their food competitors are least active. Robertson (1983) was able to relate some of the interspecific variation in spawning times of herbivorous surgeonfishes at a single site to interspecific variation in potential pressure from food competitors. Interspecific variation in spawning times among damselfishes also follows a pattern predicted by this hypothesis, since territorial herbivorous species generally spawn at dawn, when their herbivorous competitors are inactive, while mobile nonterritorial planktivorous species spawn at various times of the day (Kohda, 1988). However, some of the difference Kohda (1988) ascribed to food competition pressures could simply reflect phylogenetic relationships and constraints, since the species he compared belong to a mix of genera and subfamilies.

C. Conclusions

Although most of the attention directed at factors that select for diel patterns of spawning of reef fishes has focused on the dispersal of planktonic propagules and the feeding activity of predatory fishes on eggs and spawners, so little is known about the feeding patterns of these predators that we cannot realistically assess the absolute or relative importance of either form of predation at this time.

The focus of attention needs to be broadened to include diel patterns of the full range of activities of spawners, particularly those relating to feeding biology, since the timing of spawning may affect the opportunity for adults to engage in other activities. Interspecific variation in diel spawning patterns may also be due to constraints that an extended spawning act might place on benthic spawners, and to mechanisms that minimize the cost of parental care of benthic eggs.

III. Lunar Spawning Patterns

Lunar and semilunar spawning cycles occur commonly in reef fishes. However, such traditionally emphasized cycles represent only a part of the range of patterns of spawning that occur on this time scale, patterns that include not only cycles at higher frequencies, but also continuous, sporadic, and intermittent acyclic activity. Lunar cycles in the return of juveniles to adult habitat at the end of the planktonic phase also occur in many species (for reviews see Johannes, 1978a; Thresher, 1984; Gladstone and Westoby, 1988; Robertson *et al.,* 1990).

Larval biology hypotheses concerning lunar reproductive patterns deal with both the beginning and the end of the planktonic phase. Propagule dispersal hypotheses consider factors that may affect rates of predation on planktonic eggs and larvae as they leave the spawning site. The settlement-linkage hypothesis considers relationships between spawning and the return of pelagic juveniles into adult habitat. Adult biology hypotheses consider not only environmental factors that may directly affect the success of reproductive activities, but also a variety of intrinsic advantages to population synchronization of spawning for both benthic brooding and pelagic spawning species. They also propose that environmental and intrinsic constraints influence the ability of adults to develop population cycles of spawning.

A. Larval Biology Hypotheses

1. The Beginning of the Planktonic Phase: Dispersal of Eggs and Larvae

The propagule dispersal hypotheses propose that success of dispersal of planktonic eggs or larvae is affected by tidal heights and flows or moonlight levels, or that synchronization of spawning is intrinsically advantageous, because of predator oversaturation (for reviews see Johannes, 1978a; Thresher, 1984; Gladstone and Westoby, 1988).

Thresher (1984) reviewed the literature and found that lunar spawning cycles occur more commonly among species that produce benthic eggs than among pelagic spawners. He suggested that this difference is related to propagule dispersal mechanisms: Diel windows during which pelagic eggs can be successfully released are relatively broad and, by adjusting the time of day of release to coincide with appropriate tidal stage, many pelagic spawners can reproduce throughout the month. Benthic eggs, on the other hand, have much narrower diel dispersal (hatching) windows, and those windows coincide with tides that are appropriate for dispersal only during two short periods each lunar month. A recent, more comprehensive analysis by Gladstone and Westoby (1988) supported Thresher's conclusions. The ability of broad comparative analyses such as these to test hypotheses is limited by potentially confounding effects of many other variables. These include effects of phylogenetic relationships, of differences in sampling effort among taxa, of geographic variation in the tidal regime, and of differences in tidal influences in different habitats at the same site. Such problems can be minimized in analyses of intraspecific variability in spawning patterns in relation to, for example, seasonal and geographic variation in the tidal regime (see, e.g., Conover and Kynard, 1984). These effects also can be reduced if the comparative analysis considers spawning patterns of sets of closely related species living in different habitats at several sites with different tidal regimes.

Robertson *et al.* (1990) performed such an analysis of the reproductive patterns of 17 benthic brooding fishes (15 damselfishes and 2 blennies) at three neotropical sites that have very different, and seasonally variable, tidal regimes. Those fishes exhibit a great variety of lunar spawning patterns, including not only uni- and bimodal lunar cycles of varying strength and lunar timing, but also high-frequency spawning cycles and sporadically synchronized, intermittently variable and continuous spawning. Variation in spawning patterns was examined in relation to sets of predictions from the various dispersal hypotheses. As they are currently framed, those hypotheses assume that larvae of different species have essentially the same requirements for dispersal, and that dispersal mechanisms affect them in the same manner. Lunar hatching patterns of those neotropical fishes followed various predictions of the different propagule dispersal hypotheses, including the tidal control hypothesis, in a small minority of cases. Further, none of the eight species for which data were available consistently conformed to multiple predictions from the tidal control hypothesis. Robertson *et al.* (1990) concluded that, although many damselfishes and blennies from a broad range of habitats have lunar and semilunar spawning cycles, only species that live in habitats from which little or no dispersal may be possible during certain lunar phases consistently exhibit spawning patterns predicted by the tidal control hypothesis. Similar analyses of lunar patterns of spawning of assemblages of pelagic spawning fishes have yet to be made.

2. *The End of the Larval Life: Linkage of Spawning and Settlement*

Christy (1978) and Kingsford (1980) proposed that lunar reproductive patterns of some shallow-water invertebrates and fishes may have been selected to maximize the numbers of larvae that are competent to settle during lunar periods that are most favorable for settlement.

Some data support this settlement-linkage hypothesis. Two damselfishes that have lunar spawning and settlement cycles also have relatively uniform-age settlers (Kingsford, 1980; Robertson *et al.,* 1988). Further, at the same Caribbean site as one of those damselfishes, (1) a wrasse whose lunar settlement cycle coincides with that of the damselfish (a peak at new moon) lacks a lunar spawning cycle but has variable-age settlers (Victor, 1986b), and (2) several other damselfishes that have (differently timed) lunar spawning cycles also have settlement peaks around the new moon. This suggests that, at that site at least, there may be a single lunar period that is most favorable for settlement of various species, and that the lunar spawning cycles of some of them could represent a response to the constraint of uniform-age settlers (Robertson *et al.,* 1990). This begs the question why variation in age at settlement might constrain some species and not others.

More equivocal data are available from other localities. In a Japanese damselfish that has a fixed lunar settlement cycle, variation in its spawning (semilunar and acyclic in different years) is contrary to what the settlement-linkage hypothesis would predict. Also, data are beginning to emerge that show that there may be intraspecific geographic variation not only in lunar spawning and settlement patterns, but also in the age at settlement (Doherty and Williams, 1988a; Wellington and Victor, 1989; Robertson *et al.,* 1990). Although such data are not necessarily inconsistent with settlement-linkage, they do reduce its likelihood. There is sufficient variation in lunar spawning patterns among damselfishes, for example (see Robertson *et al.,* 1990), that it should be possible to make a reasonable test of this hypothesis by examining relationships between such variation and variation in settlement patterns and the duration of the planktonic phase in a suite of closely related species at multiple sites.

B. Adult Biology Hypotheses

1. Moonlight and Nocturnal Activities

Unimodal lunar spawning cycles could be selected for if moonlight affects the efficiency of some aspect of reproduction. Colin *et al.* (1987) suggested that moonlight could facilitate the long-distance nocturnal spawning migrations of a grouper, while Allen (1972) proposed that moonlight might enhance parental ability in a benthic brooding damselfish. Different groupers spawn at different times of the lunar month (Johannes, 1978a) and a comparison of their diel migration patterns could be used in a simple, preliminary test of the first suggestion. Interspecific variation in lunar spawning patterns of damselfishes at a single site is such that, although some spawn at the time predicted by Allen's hypothesis, many others spawn either when moonlight levels are lowest or at both extremes of moonlight conditions (Robertson *et al.*, 1990). There are no data available that show whether brooding success is dependent on moonlight conditions in any species.

2. *Effects of the Tidal Regime on Adult Activities*

Besides having effects on the success of the dispersal of propagules from certain habitats, the tidal regime also may have direct effects on the spawning ability of adults. Conover and Kynard (1984) proposed that a semilunar cycle in the availability of intertidal spawning habitat could be directly driving the semilunar spawning cycle of a temperate, benthic spawning atherinid. Lunar cycles of spawning of some reef fishes may represent responses to tidal influences on various aspects of adult biology.

The tidal regime must affect the feeding of many reef fishes and produce a variety of lunar cyclic patterns of food intake. Intertidal feeding habitats will be available for differing percentages and times of the day over the course of the lunar month. Tidal currents will bring planktonic food in lunar cyclic patterns. Risks of predation on adults that engage in particular activities in intertidal habitat may also depend on tidally mediated water depth. These effects of the tidal regime are likely to be strongest in species in which propagule dispersal also is most likely to be affected by the tidal regime, that is, nonmigratory benthic spawners that live in laterally extensive intertidal habitats.

3. *Intrinsic Advantages to Spawning Synchronization*

Population synchronization of spawning may provide intrinsic advantages to individuals, and synchronization may often be lunar cyclic simply because the lunar cycle provides the strongest, most universally available set of environmental cues.

a. Synchronization of Aggregating Pelagic Spawners Many pelagic spawning fishes that live in subtidal habitats and have lunar spawning cycles also migrate to and aggregate at traditional spawning grounds [e.g., some serranids (Shapiro, 1987d; Carter, 1989)]. The association of lunar spawning cycles with migration and aggregation may be related to adult biology constraints.

Colin and Clavijo (1988) observed spawning aggregations of two Caribbean surgeonfishes, one that spawned daily and one that spawned on a lunar cycle. They suggested that this difference is related to differences in the costs of spawning migrations, which appeared to be higher in the lunar cyclic species. Extending this line of reasoning, it can be seen that whether or not lunar synchronization of spawning is likely to facilitate the formation of spawning aggregations may depend not only on the cost of migration but also on the frequency with which individuals spawn and their population density (see, e.g., Carter, 1989). In an unsynchronized species, the chance that sufficient individuals would arrive at the spawning ground to form an aggregation of some minimum size on any randomly chosen day will be low if individuals spawn infrequently and occur at a low density. With lunar synchronization of activity, an individual of such a species would have a much greater chance of encountering an aggregation whenever it is ready to spawn.

The most extreme examples of lunar synchronized spawning aggregations occur among the groupers. In some species, spawning occurs only during a fixed 1- to 2-month period each year, and individuals of a dispersed population migrate tens of kilometers and aggregate and spawn at a traditional spawning ground during a fixed lunar phase (e.g., Colin *et al.,* 1987; Carter, 1989).

b. Spawning Synchronization of Benthic Brooders Two adult biology hypotheses propose that there are intrinsic advantages to population synchronization of spawning in species that brood benthic eggs. Those advantages include enhanced effectiveness of brood defense and reduction in mortality of eggs that results from the cost of parental care.

(i) Synchronized Colonial Spawning In a number of freshwater fishes and at least one damselfish (Foster, 1989), the effectiveness of defense of benthic eggs by males is increased when nesting is colonial, because of the collective activities of the aggregated males. Lunar cyclic synchronization of spawning should facilitate colony formation in such species.

As expected from the enhanced-defense hypothesis, the degree of development of this synchronized colonial spawning in two cogeneric damselfishes correlates with differences in pressures from egg-predators, which are high in the colonial species and low in the noncolonial one (Foster, 1989). However, the occurrence of this pattern of spawning is not related only to variation in pressure from egg-predators. Highly synchronized colonial spawning, although common among damselfishes, is restricted to planktivorous species that are mobile and normally live in schools, and that temporarily defend territories only when breeding. This spawning pattern is lacking among benthic feeding herbivorous damselfishes whose adults are sedentary and singly defend permanent feeding territories that are dispersed on the substratum. Such species lack the social and ecological potential to form highly aggregated colonies (Robertson *et al.,* 1990).

(ii) The Cost of Parental Care Intensive guarding of benthic eggs by parental male fishes, including tropical damselfishes, can be energetically costly, and that cost may induce nest abandonment and filial egg cannibalism by parent males (Robertson *et al.,* 1990). Dominey and Blumer (1984)

proposed that population synchronization of spawning should result in increased parental investment in each clutch, since clutches will be larger and males will be less likely to receive additional eggs if they abandon or eat first-laid clutches. Lunar cycles in brooding activity might reduce egg mortality not only by having effects on parental investment, but also by providing regular, frequent, nonbrooding periods during which males can recuperate from the cost of care (Robertson *et al.*, 1990).

Neotropical damselfishes and blennies show not only different patterns of population synchronization of brooding (sporadic activity, and lunar and higher-frequency cycles), but also unsynchronized cycles of brooding activity in individual nests in some species (Robertson *et al.*, 1990). As expected from the cost-of-care hypothesis, continuous brooding appears to be associated with a negligible cost of care in at least one blenny. However, there are no data that help explain the existence of the variety of patterns of brood cycling in terms of this hypothesis. Different patterns might represent independently evolved, alternative solutions to the same problem (Clutton-Brock and Harvey, 1984), or graded responses to varying intensities of the problem. Population cycles might arise when individual females cannot predict or control brooding activity in unsynchronized nests. Some of the variation also may be due to limitations on the options that different species have. For example, population synchronization of spawning appears to be both less strongly linked to the lunar cycle and more intraspecifically variable in its timing among the schooling planktivorous damselfishes than among the territorial herbivorous species in this family. Two factors may contribute to this difference. First, schooling species may have greater potential to synchronize their activity independently of lunar cues (i.e., by using social cues). Second, variability in spawning output may be produced by unpredictable fluctuations in the planktonic food supply, while a greater constancy in the availability of benthic algae may allow more regular reproductive activity (Robertson *et al.*, 1990). These possibilities remain to be examined in the field.

The cost-of-care hypothesis predicts that lunar cycles of spawning should be more common among benthic brooding species than among nonbrooders that produce pelagic eggs, and such does occur among Caribbean reef fishes (Robertson *et al.*, 1990). Since those brooders and nonbrooders belong to different families, this comparison is limited by potentially confounding phylogenetic effects. However, since both the brooding and pelagic spawning families include both lunar cyclic and acyclic species, it is unlikely that the differences between brooders and nonbrooders are due to evolutionary conservatism in one family or another. There are insufficient data available on benthic spawners that lack parental care to include them in any analysis at this point.

C. Conclusions

The hypothesis that lunar patterns of spawning are related primarily to the success of dispersal of planktonic propagules away from adult habitat is most strongly supported by data from fishes (particularly benthic spawning species) that live in laterally extensive intertidal habitats, habitats from which dispersal may be severely limited during particular lunar phases. However, the lunar cycle also may control reproductive patterns of species living in this type of habitat by way of effects of the tidal regime on a range of adult activities.

The hypothesis that lunar patterns of spawning represent adaptations that facilitate the entry of juveniles into adult habitat at the end of their pelagic existence is supported by some, but not all, of the few sets of data available.

Adult biology may be a major determinant of lunar spawning patterns of both intertidal and subtidal species. Moonlight might facilitate nocturnal spawning migrations and nocturnal eggcare, although there is no direct evidence for either effect. In many cases the lunar cycle may be most important as a source of environmental cues for spawning synchronization. Lunar spawning cycles may facilitate the formation of spawning aggregations in pelagic spawners that occur at low densities and spawn infrequently. Various patterns of spawning synchronization among benthic brooders, including lunar cyclic activity, may not only facilitate the formation of nesting colonies that increase the effectiveness of nest defense, but also may reduce egg losses due to the costs of parental care. Some of the variation in lunar patterns of spawning among benthic spawners may be related to differences in social system structure and to temporal patterns of food availability.

Hypotheses that invoke adult biology advantages or adult biology constraints appear to be as or more successful than propagule dispersal hypotheses in their ability to account for variation in lunar spawning patterns of reef fishes. Some of the success of adult biology hypotheses may be due to their making less restrictive predictions than the propagule dispersal hypotheses about the structure and lunar timing of cycles of activity. In addition the adult biology hypotheses accommodate interspecific variation that we know exists in adult capabilities and ecology. In the absence of any data that indicate otherwise, the propagule dispersal hypotheses assume that there is great interspecific similarity in requirements and capabilities of those propagules. Thus they may be oversimplified.

IV. Seasonal Spawning Patterns

Most of the discussion of factors that control seasonal spawning patterns of reef fishes has centered on the hypothesis that spawning output tracks the seasonal cycle of change in larval survivorship. Data are only now becoming available that provide estimates of seasonal change in larval survivorship and enable us to begin to assess its relationship to the spawning pattern. The main existing alternative to the larval survivorship hypothesis proposes that the timing of reproduction ultimately is controlled by the requirements of juvenile fishes in the beginning of their benthic existence. Some thought also has been given to how the reproductive capacity of adults might be controlled by seasonal change in environmental factors that affect them directly.

A. Larval Biology Hypotheses

1. Seasonality in the Larval Environment

Previous analyses of the role of larval biology as a factor controlling seasonal patterns of spawning of reef fishes have been limited to attempts to correlate community-level seasonality of spawning with gross seasonality of environmental variables that are thought to affect the food supply, growth, and dispersal of larvae. These variables include water currents, which might retain larvae near or transport them away from suitable habitat; wind patterns, which may affect not only currents but also the density and stability of food patches; the seasonal cycle of primary production, which might determine the types and average abundance of larval foods; and temperature and day length, which may affect development rates of larvae (Munro *et al.*, 1973; Russell *et al.*, 1977; Johannes, 1978a; Kock, 1982; Doherty, 1983c; Bakun, 1986; Walsh, 1987; Lobel, 1989).

This type of correlative analysis can be extended by examining several predictions from this "gross environmental seasonality" hypothesis. (1) Since the proposed mechanisms of environmental control of larval survivorship are fairly simple, they should affect communities at different sites in the same way and one should see consistency in relationships between environmental seasonality and community-level spawning patterns. (2) Further, closely related species with similar reproductive biology probably have similar larval biology, and, if they live and spawn in the same small area of a reef, they should have quite similar seasonal spawning patterns. (3) The spawning output of individual species should track seasonal gradient(s) of environmental change. (4) Intraspecific latitudinal variation in spawning seasonality should follow consistent patterns of change in relation to environmental change. There are data available that relate to each of these predictions.

a. Geographic Variation in Community-Level Spawning Seasonality There are three tropical sites from which sufficient data are available to make a comparative analysis: the Australian Great Barrier Reef, the Caribbean, and Hawaii (Table 1).

Table 1 Community-Level Spawning Seasonality and Environmental Seasonality at Three Tropical Sites

	Site		
	Great Barrier Reef	Hawaii	Jamaica
Latitude	23°S	20°N	18°N
Temperature range	20–28.5°C	22–28°C	25–30°C
Spawning seasonality	Strongly unimodal	Moderately unimodal	Weakly bimodal
	Environmental state during spawning peak(s)		
Temperature	Near maximum	Intermediate	Min & max
Day length	Near maximum	Maximum	Min & max
Prevailing winds	Minimum	Submaximal	Max & min
Hurricanes/cyclones	Maximum	Intermediate	Min & max
Rainfall and runoff	Maximum	Intermediate	Min & max
Dispersive currents	Maximum	Intermediate?	No data
Primary production	Weak peak?[a]	No pattern	Intermediate?[b]

[a] There is a summer peak (due to river runoff) inshore, but no seasonal cycle on offshore reefs.
[b] There may be little interannual consistency in the cycle.
Sources: Russell *et al.*, 1977; Doherty, 1983c; Munro, 1983; Sammarco and Crenshaw, 1984; Walsh, 1987; Furnas and Mitchell, 1987; Doherty and Williams, 1988a; and the U.S. Department of Commerce, 1982a,b, 1989.

Both seasonal patterns of spawning and the extent and nature of seasonal change in various environmental parameters differ among those sites. There is a trend for decreasing strength of spawning seasonality with decreasing latitude (see also Munro, 1983). However, although there are latitudinal changes in some environmental variables (e.g., day length and minimum water temperatures), spawning peaks are not consistently associated with a particular relative or absolute state of any of them (Table 1). Thus, community-level spawning seasonality does not appear to be related in a consistent manner to gross seasonality in any of the environmental factors that are thought to directly or indirectly influence larval survivorship.

This lack of consistency could be due to differences in which factors have most influence on larval survivorship at the different sites. The Hawaiian archipelago is a well-isolated cluster of small islands. Since larvae that move away from the islands may be very unlikely to survive, a premium could be placed on retention of larvae near adult habitats (Johannes, 1978a; Lobel, 1978). Although it has been proposed that spawning peaks in Hawaii when current systems are most likely to retain larvae near the islands, it is unclear whether such a coincidence does occur (for opposing views see Walsh, 1987; Lobel, 1989). In addition, there are many exceptions to the "ideal" pattern, including among endemic species (data in Walsh, 1987), which might be expected to strongly conform to the predicted pattern.

A special-case argument can be used to explain differences between Hawaii and the other two sites, since both of the latter are large reef systems in which larval retention is less likely to be of overriding importance. However, it is not obvious how such a line of reasoning could be applied to differences between the Great Barrier Reef and the Caribbean (see Table 1).

b. Spawning Seasons of Closely Related Species at the Same Site The Russell *et al.* (1977) data from the Great Barrier Reef indicate that species in each of at least three families (Apogonidae, Pomacentridae, and Labridae) spawn at opposite seasonal environmental extremes. Similar differences in spawning seasons occur among both labrids and pomacentrids at Hawaii (data in Walsh, 1987). For example, among the herbivorous damselfishes, *Stegastes fasciolatus* has a short spawning season that peaks in late winter (MacDonald, 1981), while *Abudefduf sordidus* spawns at a uniform level from spring through autumn (Stanton, 1985). This diversity of patterns among closely related species at two sites is not predicted by simple environmental seasonality hypotheses.

c. Local Spawning Cycles and Environmental Gradients Contrary to expectations, seasonal patterns of spawning of many species at different sites do not track environmental gradients. Jamaican fishes include species that appear to be nonseasonal spawners, and others in which the peaks of a bimodal spawning cycle coincide with both extremes of environmental conditions (see Munro, 1983). Further, some Hawaiian fishes, including species in the same family, show either bimodal spawning activity with peaks at both extremes of environmental conditions (Walsh, 1987) or uniform reproduction during all but one period of extreme conditions (Stanton, 1985).

d. Intraspecific Geographic Variation in Spawning Seasonality Data on lutjanids and serranids at sites scattered throughout the tropical West Atlantic (Fig. 1) show that simple, consistent patterns of geographic variation in spawning cycles are lacking. First, although three serranids show a pattern of winter spawning at low latitude and spring-summer spawning at high latitude, one other serranid and five lutjanids clearly do not (Fig. 1). Colin and Clavijo (1988) have suggested that a change from winter to summer spawning with increasing latitude arises because these fishes are adapted to spawning at the summer temperatures of the last glacial period, and those temperatures now occur in different seasons at different latitudes. However, there is no obvious reason why lutjanids and serranids at the same series of sites should not show similar patterns of latitudinal variation in spawning seasons (Fig. 1) or why

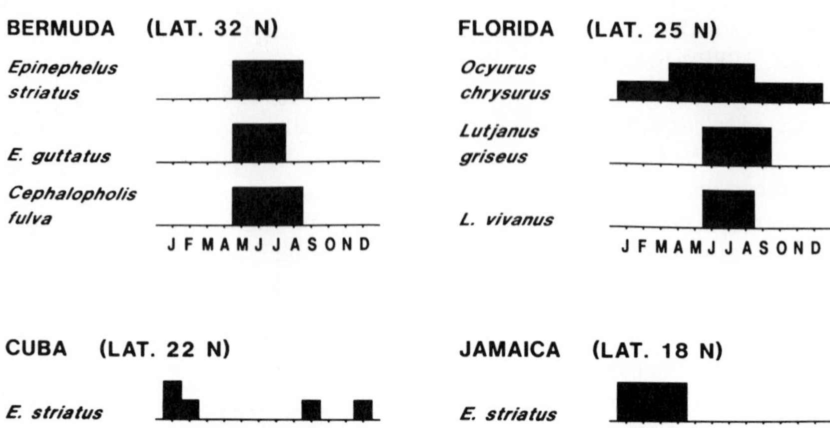
BERMUDA (LAT. 32 N)
Epinephelus striatus
E. guttatus
Cephalopholis fulva
J F M A M J J A S O N D
FLORIDA (LAT. 25 N)
Ocyurus chrysurus
Lutjanus griseus
L. vivanus
J F M A M J J A S O N D

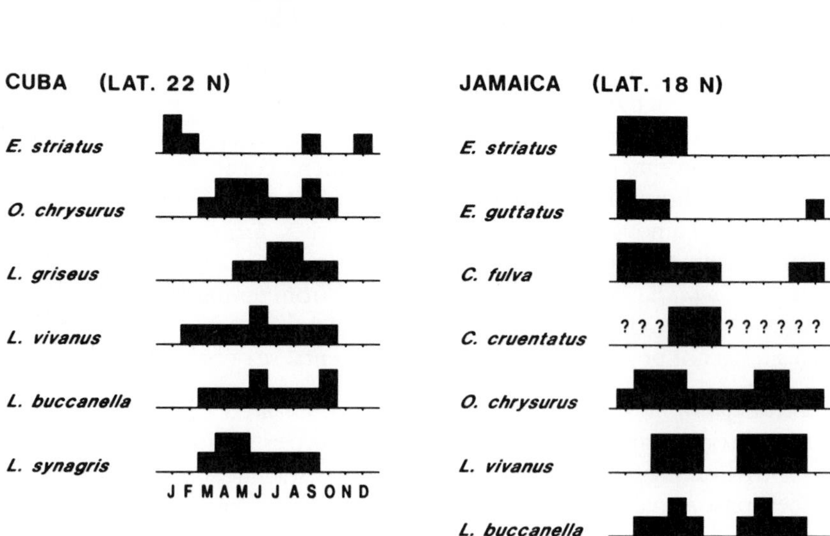
CUBA (LAT. 22 N)
E. striatus
O. chrysurus
L. griseus
L. vivanus
L. buccanella
L. synagris
J F M A M J J A S O N D
JAMAICA (LAT. 18 N)
E. striatus
E. guttatus
C. fulva
C. cruentatus
? ? ? ? ? ? ? ? ?
O. chrysurus
L. vivanus
L. buccanella
J F M A M J J A S O N D

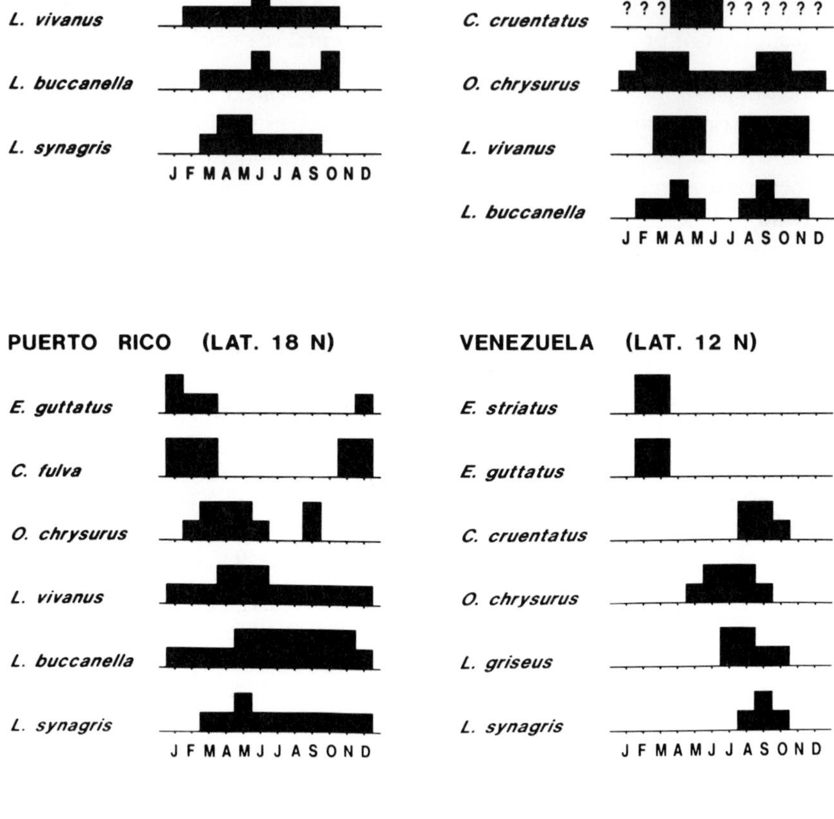
PUERTO RICO (LAT. 18 N)
E. guttatus
C. fulva
O. chrysurus
L. vivanus
L. buccanella
L. synagris
J F M A M J J A S O N D
VENEZUELA (LAT. 12 N)
E. striatus
E. guttatus
C. cruentatus
O. chrysurus
L. griseus
L. synagris
J F M A M J J A S O N D

different serranids should show different patterns. For example, the serranid *Cephalopholis cruentatus* spawns in late spring at high and intermediate latitudes (i.e., Bermuda and Jamaica) and in summer at low latitude (Venezuela) [Fig. 1, and *anecdotal* information from Bermuda (B. E. Luckhurst, personal communication, 1989)]. These latitudinal changes in the timing of spawning do not correlate with patterns of change in other variables [e.g., the relative and absolute states of day length and of the annual cycle of primary production (see Burnett-Herkes, 1975; Nagelkerken, 1979)].

Second, the spawning cycles of different serranids and lutjanids vary independently at the different low-latitude sites. While spawning seasons of some species may not differ between two sites, those of others do, and often do in different ways (Fig. 1 and Table 2).

Third, there are indications of mesoscale geographic variation in seasonal patterns of spawning of several Caribbean reef fishes. *Scarus iserti* (Scaridae) appears to have summer versus winter/spring spawning peaks on the north and south coasts of Jamaica (Munro, 1983; Colin and Clavijo, 1988). Two surgeonfishes, *Acanthurus bahianus* and *A. coeruleus,* have different seasonal spawning patterns at two sites off the south coast of Jamaica—a restricted, strongly unimodal pattern inshore, and a more extended, weakly bimodal pattern 100 km offshore (data in Munro, 1983).

These patterns of intraspecific geographic variation in spawning seasons are very difficult to reconcile with larval biology hypotheses that invoke simple, gross seasonal changes in one or another variable in the larval environment. If spawning patterns are responses to larval requirements then changes in the larval environment must be subtle and complex and follow independently varying, species-specific patterns at different sites within the Caribbean.

2. *Matching of Seasonal Patterns of Spawning and Recruitment at the Same Site*

Correlative analyses such as the preceding offer limited insight into factors that control spawning seasonality because they do not incorporate information on seasonal patterns of larval survivorship. Direct measurements of larval survivorship have not been made for any tropical species, and at this stage we are limited to measuring spawning output and larval recruitment at the same site and using changes in the ratio of recruitment to spawning effort to

Figure 1 Spawning seasons of lutjanids and serranids at six sites in the tropical West Atlantic. Sources: references in Grimes (1987) and Shapiro (1987d); also Munro (1983), Starck (1971), Colin *et al.* (1987), Bardach *et al.* (1958), Alcala (1987), Mendez (1989), Perez-Villanoel (1982), Y. Sadovy (personal communication, 1989) for Puerto Rico, and R. Claro (personal communication, 1989) for southwest Cuba. Note: "Venezuela" also includes the Netherlands Antilles.

Table 2 Similarity of Spawning Seasons of Lutjanids and Serranids at Different Sites in the Caribbean (see also Fig. 1)

	Cuba/Jamaica	Cuba/Puerto Rico	Cuba/Venezuela	Puerto Rico/Jamaica	Jamaica/Venezuela
Lutjanidae					
Ocyurus chrysurus	Similar, J longer	≈ Same	Bi-/unimodal and timing difference	≈ Same, J longer	Bi-/unimodal and timing difference
Lutjanus griseus	ND[a]	ND	≈ Same	ND	ND
L. buccanella	Timing difference	Bi-/unimodal and longer	ND	Uni-/bimodal and timing difference	ND
L. vivanus	Uni-/bimodal and timing difference	≈ Same, PR longer	ND	Uni-/bimodal, PR longer	ND
L. synagris	ND	≈ Same, PR longer	Timing difference, C longer	ND	ND
Serranidae					
Epinephelus striatus	Similar	ND	≈ Same	ND	≈ Same
E. guttatus	ND	ND	ND	≈ Same	Similar
Cephalopholis fulva	ND	ND	ND	Similar	ND
C. cruentatus	ND	ND	ND	ND	Timing difference

[a] ND = No data.

estimate how relative larval survivorship changes over the year. This method assumes that the recruitment pattern observed at the study site is representative of that experienced by the larvae that were produced there. Seasonal patterns of larval survivorship can be estimated in this way for 12 species from four sites in the Caribbean and the East and West Pacific.

a. Neotropical Fishes Robertson (1990) examined spawning and larval recruitment of nine neotropical fishes (eight damselfishes and one blenny) from the Caribbean and Pacific coasts of Panama. The average seasonal patterns of spawning and settlement did not match in any of those species. The seasonality of settlement was stronger than that of spawning in most cases, and/or peaks of spawning and settlement were out of synchrony. The seasonal pattern of spawning differed from the seasonal pattern of larval survivorship in all but one of the nine species (*A. saxatilis;* Fig. 2).

b. Western Pacific Fishes Gladstone and Westoby (1988) measured seasonal changes in both spawning effort and recruitment of a small pufferfish, *Canthigaster valentini,* on the northern part of the Australian Great Barrier Reef. Spawning and settlement occur year-round, but both are reduced during the cool half of the year. Their data indicate that since the reduction in spawning is disproportionately greater than the decrease in settlement during the cool season, larval survivorship may be highest during that season (Table 3). That is, spawning effort may be lowest during the period of highest larval survivorship. Gladstone and Westoby (1988) suggested that cool-season spawning may be a bet-hedging strategy to cope with occasional catastrophic losses in recruitment due to warm-season cyclones.

Doherty (1980, 1983c) studied two damselfishes, *Pomacentrus wardi* and *P. chrysurus* (=*P. flavicauda*), at One Tree Island at the southern limit of the Great Barrier Reef. Both of those species showed consistent seasonal patterns of spawning over a 5-year period. Since the study population of *P.wardi* spawns for 2.5 months (October to December), and the larval duration of this species is ≈20 days (Brothers *et al.,* 1983; Thresher *et al.,* 1989), most larvae produced by that population settle in November and December. However, the great bulk of each year's settlement of *P.wardi* in the vicinity of One Tree Island occurs in one short episode each year, and those episodes have occurred in January or February of each of the five years for which data are available (Russell *et al.,* 1977; D. M. Williams, 1979; Doherty, 1980; Pitcher, 1988a). Thus the consistent spawning season of the study population of *P.wardi* is distinctly asynchronous with the strong, consistent seasonal peak of settlement. Most *P.wardi* that settle at that site must be produced by other populations of adults that have either differently timed or more extended spawning seasons. *Pomacentrus chrysurus,* on the other hand (which has a similar larval duration to *P.wardi;* see Thresher *et al.,* 1989, where *P.chrysurus* = *P.rhodono-*

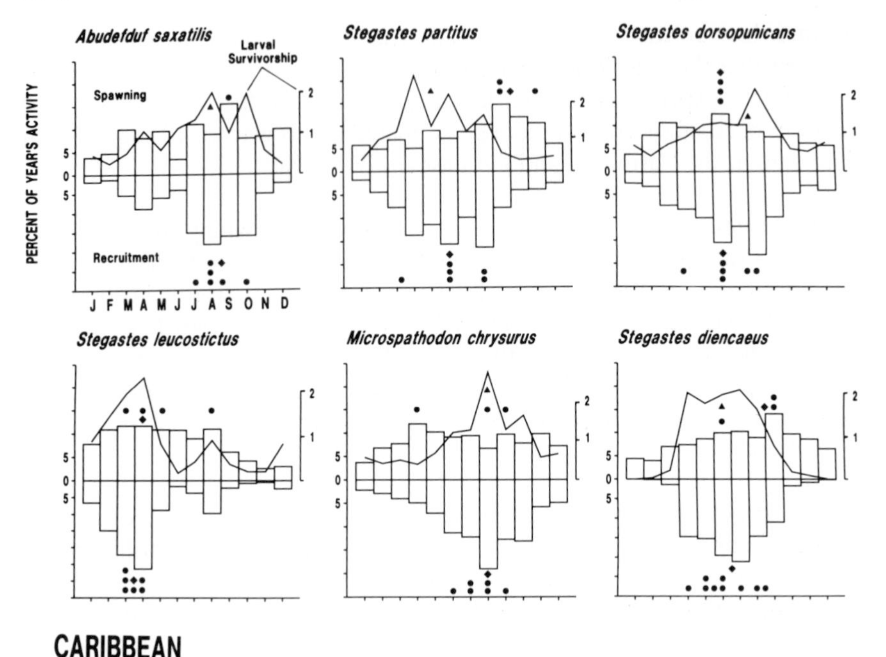

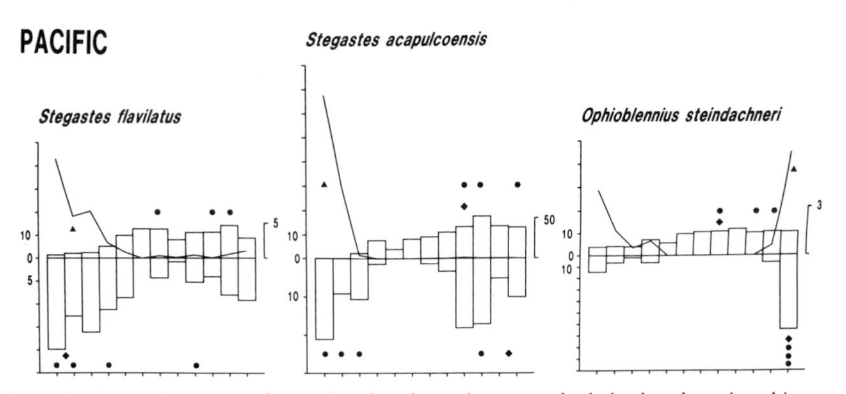

Figure 2 Seasonal patterns of spawning, larval recruitment, and relative larval survivorship of nine neotropical fishes. Histograms show the mean percentage of the year's spawning and recruitment that occur in each month. The recruitment pattern is appropriately lagged (e.g., January's recruits are larvae that were spawned in January). Relative larval survivorship is the recruitment %/spawning % for each month. Diamonds indicate average timing of annual peaks (where present) of spawning and recruitment. Triangles indicate peak of larval survivorship cycle (not calculated for *S.leucostictus* because the distribution is bimodal). Circles indicate the peak spawning or recruitment month of each year sampled. (Reprinted with permission from D. R. Robertson, 1990. Differences in the seasonalities of spawning and recruitment of some small neotropical reef fishes. *J. Exp. Mar. Biol. Ecol.* **144,** 49–62.)

Table 3 Life History Parameters of *Canthigaster valentini*[a]

Average adult female
Weight: 8.6 g
Survivorship: < 6 months
Estimated annual spawning output: 72,000 eggs (≈ 1.5 times body weight)
Larval life: Mean duration of 92 days (range 64–113 days)
Seasonal changes in activity
Seasons: Cool—sea temperatures < 24–26°C; Warm—27–28.5°C
Female mortality rate: cool > warm
Female growth rate: cool = warm × 0.25
Spawning output: cool = warm × 0.40[b]
Larval recruitment: cool = warm × 0.78 to 0.88[c]
Relative larval survivorship[d]: cool = warm × 1.9 to 2.2[c]

[a] Data or calculations are based on data from Gladstone (1985, 1991), Gladstone and Westoby (1988), and Stroud *et al.* (1989).
[b] Since no consistent patterns of seasonal change in adult density are evident (W. Gladstone, personal communication), I used seasonal change in per capita output.
[c] Range is for values based on the season during which larvae arrive and the season during which larvae are produced (assuming a larval life of 3 months).
[d] Recruitment/spawning output.

tatus), does not show any apparent discrepancy between its seasonal patterns of spawning and settlement (data in Doherty, 1980).

Thus the data from 12 species at four sites indicate that spawning may not be well matched to the seasonal pattern of larval survivorship in most cases. The degree of mismatching ranges from slight between-season differences to a strong within-season discrepancy in a species (*P.wardi*) in which both the spawning and settlement seasons are short.

Since the larvae of reef fishes typically spend weeks in the plankton (e.g., Victor, 1986a; Wellington and Victor, 1989) it is possible that most individuals settle well away from their source, and that local mismatching of spawning and recruitment could be due to geographic variation in either spawning patterns or larval survivorship patterns. Sufficient microgeographic variation in spawning patterns does appear to exist in some species (see Section IV, A, 1, d) to warrant serious consideration of spatial variation in spawning as a potential causal factor. MacDonald (1985) found local seasonal mismatching of reproduction and recruitment of a Hawaiian lobster at various sites in the Hawaiian archipelago and showed how spatial variation in seasonal recruitment patterns was correlated with variation in currents that could transport pelagic juveniles. Thus, local mismatching in this case could reflect spatial variation in the seasonal cycle of pelagic-phase survivorship. However, since this lobster's pelagic phase is long and variable (6 to 11 months), dissimilarity in spawning and recruitment cycles could be due to variation in the duration of that phase.

There are no data available on intraspecific geographic variation in seasonal recruitment patterns in reef fishes. In addition, there are too few data to show whether variation in larval durations of many species could be sufficient to account for local mismatching of spawning and settlement. The larval life of damselfishes, among others, appears sufficiently short [a few weeks (Thresher *et al.*, 1989; Wellington and Victor, 1989)] that seasonal variation in larval duration is unlikely to be involved in the mismatches described here (Robertson, 1990). The larval duration of *C.valentini* is, however, longer and more variable than in damselfishes (Table 3), and variation in that duration could have been involved in its spawning/recruitment mismatch.

Plausible explanations can be constructed that take into account what we know about larval durations and the dispersive potential of water currents and that attribute local differences between spawning and recruitment to geographic variation in spawning and larval survivorship cycles. However, based on what we know about complexity and the retentive potential of currents and the ability of fish larvae to migrate between different currents, equally plausible arguments can be made that many larvae of many species are retained near their source (e.g., Bakun, 1986) and that mismatches between spawning and larval survivorship cycles do occur. Since there are pronounced interspecific differences in larval durations, morphology, and size at recruitment, and in larval distributions across both vertical and horizontal profiles around reefs (see, e.g., Leis, Chapter 8), we should not expect the larvae of all species to be equally susceptible to a set of dispersal mechanisms. Only more intensive studies of variation in spawning and recruitment cycles, and of larval distributions and ecology, will help resolve the question of how much connection there is between local larval sources and sinks and show whether spawning is tracking larval survivorship cycles.

B. Juvenile Biology Hypotheses

Russell *et al.* (1977), Stanton (1985), and Walsh (1987) suggested that spawning output could be tracking seasonal change in the suitability of the benthic environment (due to changes in food availability, temperature, and the physical stress) for the growth and survival of juvenile fishes after they arrive in a reef.

Analyses of this possibility have been limited to a search for correlations between community-level spawning seasonality and gross environmental seasonality. It has been shown that differences in the within-season timing of recruitment can strongly affect juvenile growth and maturation rates in temperate reef fishes (Jones and Thompson, 1980; Ochi, 1986a). Similar studies have not been made of seasonal variation in juvenile growth, survival and maturation rates in tropical species. Species that differ in terms of the extent to which their recruitment cycles match the environmental cycle could make useful test organisms in a comparative study of this question.

C. Adult Biology Hypotheses

1. *Hawaiian and Red Sea Fishes*

MacDonald (1981) examined seasonal patterns of spawning, food acquisition, and fat storage in two Hawaiian damselfishes and proposed that both use fat reserves built up during the period of peak food availability to support later spawning (but see Walsh, 1987, regarding seasonal change in food availability). He also suggested that interspecific differences in the duration of their spawning seasons could reflect differences in the risk of delaying reproduction, since the two species have quite different average longevities. Fishelson *et al.* (1987) also proposed that a Red Sea surgeonfish stores resources during a strong seasonal peak in availability of algal food and uses those reserves to support spawning immediately afterward. However, in neither study was any estimate made of the proportion of a female's reproductive output that could have been derived from the stored reserves. The Fishelson *et al.* (1987) data indicate that reserves may not contribute much to the surgeonfish's output, since they are depleted in the first month of a five-month breeding season, well prior to the peak of spawning.

Stanton (1985) found that reproduction of the Hawaiian damselfish *Abudefduf sordidus* is uniformly high throughout most of the year but ceases in winter. He suggested that this hiatus could be due to this species' nesting activities being particularly vulnerable to disruption by increased wave action in winter. Since Stanton studied this species at a site protected from such wave stress, direct inhibition seems unlikely to be involved in producing the pattern he observed.

2. *Neotropical Fishes*

Robertson (1990) found that environmental stresses acting on the adults of several Caribbean and eastern Pacific fishes influence the strength of seasonality of their spawning cycles. He proposed that a variety of environmental and intrinsic adult biology constraints control the seasonal pattern of spawning and indirectly determine the extent to which adults match the flux of their spawning output to the seasonal cycle of larval survivorship.

a. Extrinsic Constraints In Panama there are two main seasons, one wet and calm and the other dry and windy. Levels of spawning of both Caribbean and

Pacific species during the dry season are negatively correlated with the degree of exposure to physical conditions (rough water, sediment movements, and currents in the Caribbean; low temperatures in the Pacific) in different habitats and different years. Consequently those conditions evidently are stressful for adults and depress their reproductive output. This inhibition of reproduction occurs during different parts of the seasonal larval survivorship cycles of different species. In two of the Pacific species, spawning is minimal but larval survivorship is maximal during the early dry season (Robertson, 1990) (see Fig. 2).

b. Intrinsic Constraints

(i) Risks of delaying reproduction Although fecundity increases with age in fishes, the value of early spawning may exceed that of delayed spawning in short-lived species because of the risk involved in the delay (Williams, 1966). The magnitude of any risk of a seasonal delay in spawning will depend on both adult longevity and the length of the delay between the peak of spawning potential and the peak of larval survivorship. The potential importance of this constraint varies among the neotropical species and may be quite high in some [e.g., a delay of ≈9 months in a species whose adults have a half-life of ≈1 year (Robertson, 1990)].

(ii) Body size and storage capacity In short-lived species, reproductive effort is expended early and is large in relation to body size (Williams, 1966). A female of a small species can achieve a large absolute reproductive output only by spawning numerous times and releasing in excess of her body weight in eggs each year (DeMartini and Fountain, 1981; Burt *et al.,* 1988). Published information on sizes of the ovaries and clutches of fishes whose females spawn only once or twice a year (e.g., Williams, 1966; Thompson and Munro, 1983a; DeMartini and Fountain, 1981) indicate that storage can allow a female to produce a clutch weighing up to about one-third of her body weight. In two neotropical damselfishes the average female's annual spawning output is equivalent to 3–4.7 times her body weight (unpublished observations). Species such as these would be able to make only minor increases in spawning output during the season of peak larval survivorship by relying on previously stored resources. Thus body size may strongly limit the ability of small fishes to delay reproduction and favor continued spawning during periods of submaximal larval survivorship.

(iii) Physiology of reproduction Since fecundity increases with body size, a fish potentially could enhance reproductive output during the season of highest larval survivorship if it reallocated resources from current reproduction into growth during the preceding period when larval survivorship is low. Whether such a reallocation strategy would produce an increase in spawning

during the period of peak larval survivorship sufficient to overcompensate for spawning foregone at other times of the year might depend on whether current levels of egg production are controlled by intrinsic physiological limitations and on the relative costs of growth and reproduction and allometries in those costs.

Per capita rates of egg production can reach about twice a female's body weight during a 2.5-month season in damselfishes (my calculations from data in Doherty, 1980), and six to eight times a female's body weight during a 3.5-month season in other small fishes (Hubbs, 1976). Since both of these rates are considerably higher than those achieved by two of the neotropical damselfishes that Robertson (1990) studied, it seems unlikely that their current flux of output is determined by intrinsic physiological constraints.

Because of differences in the efficiency of conversion of food to growth versus reproduction (Wootton, 1979), growth may be relatively more costly than reproduction. Further, as a result of allometries in metabolic rates, in the rate of growth of reproductive tissue, and in the efficiency of conversion of food (Burt *et al.*, 1988; Pauly, 1986), there may be a decline in the relative rate of reproductive output and an increase in the unit cost of reproduction as body size increases. Even in the absence of any reduction in efficiency with increasing size, an adult that had grown at the expense of prior reproduction would need many more resources for the increase in spawning required of it during the period of peak larval survivorship. Resource limitation on adults during the season of peak larval survivorship could prevent a reallocation strategy from producing increased annual recruitment.

It is possible to see how various combinations of adult biology constraints might have produced the varying degrees of mismatching between the seasonal patterns of spawning and larval survivorship of the neotropical fishes. They are short-lived, small species that have a limited ability to delay reproduction. Their reproductive output is partly controlled by physical environmental stresses, the nature of which differs in the two oceans, and the intensity of which varies in different habitats. Whether their output also could be limited by food availability remains to be determined.

3. *Australian Fishes*

Both spawning and recruitment of the pufferfish *Canthigaster valentini* peak during the warm half of the year, although larval survivorship appears to be higher during the cool half. The fact that adult mortality is highest and adult growth and reproduction lowest during the cool season indicates that environmental conditions are unfavorable for adults during that period (Table 3). Delaying reproduction from the warm to the cool period would be very risky, because of low longevity, and the potential to increase output during the cool period could be limited by both small body size and reduced metabolic potential (Table 3). Thus a mismatch between *C. valentini's* spawning and the larval survivorship cycle could derive from a combination of environmental, longevity, and size constraints.

Pomacentrus wardi and *P.chrysurus* (=*P.flavicauda*) are benthic feeding herbivores that Doherty (1980, 1983c) studied in a lagoon. *Pomacentrus chrysurus* is a habitat specialist, and the study habitat is typical for it. *Pomacentrus wardi* is a habitat generalist that occupies a range of other habitats in addition to lagoons (Doherty, 1980, 1983c; Robertson and Lassig, 1980). Differences between the degree of matching of the spawning and settlement seasons of those two species could be a reflection of the extent to which the study habitat is representative for each. Synchrony may have been observed in *P.chrysurus* because the study population was in the habitat in which most individuals of this species occur. Asynchrony may have been observed in *P.wardi* because populations in different types of habitats have different spawning patterns, and only one habitat type was sampled. The One Tree Island lagoon is semienclosed, has limited water exchange with the sea, and experiences physical and biological conditions different than in the remainder of the reef (Russell *et al.*, 1977; Hatcher and Hatcher, 1981). Breeding of both *Pomacentrus* species occurs during the warmest, calmest part of the year (Doherty, 1980). During that period the lagoon could experience a unique pattern of change in environmental variables (e.g., temperature?) that cue or drive the onset and termination of reproduction. For example, there is seasonal bloom of benthic algae (in response to changes in nutrient levels) that is restricted to the lagoon (Hatcher and Hatcher, 1981) and occurs during the onset of *P.wardi's* spawning season.

Although the asynchrony that exists between spawning and recruitment shows that *P.wardi* in the lagoon must have a different spawning season from that of other populations of conspecifics in the region, the significance of this difference is unclear. The few settlers that arrive before the annual settlement peak could result from many populations spawning at the same time as the study population, but few of their larvae surviving to settle. They also could result from high larval survivorship but few populations spawning. Regardless of the precise mechanism, it seems clear that the lagoon population of *P.wardi* is not matching its spawning output to the overall seasonal pattern of larval survivorship since most of that species' successful settlers are spawned at a different time of the year. A simple, testable explanation for this situation is that environmental controls on spawning vary among *P.wardi's* habitats.

D. Conclusions

Attempts to relate spawning seasonality to larval survivorship cycles by way of correlations between spawning patterns and gross seasonality in the larval environment can tell us very little about factors that determine spawning seasonality. The complexity of patterns of inter- and intraspecific geographic variation in spawning seasons and within-site interspecific differences in relationships between spawning and environmental gradients show that, if spawning is tracking change in larval survivorship, then environmental factors that determine larval survivorship must be much more complex, subtle, species-specific in their action and spatially variable than has generally been thought. The simple explanations simply cannot account for the variability that exists.

Data that have recently become available indicate that there are significant differences between local spawning and larval recruitment cycles in many species. These differences could be due to adults employing a bet-hedging strategy of spawning over an extended period to cope with unpredictable spatial variation in the larval survivorship cycle. These differences also could arise because larvae that settle at one site were produced elsewhere and their settlement cycle is a direct reflection of the (different) spawning cycle at their source. If so, the differences in spawning cycles at different sites presumably would be due to adult biology mechanisms since a lack of feedback (i.e., larvae not settling at their source) would prevent selection from tailoring each local population's spawning cycle to a specific local larval survivorship cycle.

The possibility that spawning of tropical reef fishes tracks seasonal regimes of change in environmental conditions for benthic juveniles remains untested. There is evidence for such tracking in temperate reef fishes.

The combined action of a variety of environmental and intrinsic constraints on adults of some species may prevent them from matching their spawning output to the seasonal pattern of larval survivorship. There is no reason to assume that the season that is best for spawning also is best for larval survivorship, and there is evidence to the contrary for some species. The response of many fishes to adult biology constraints may be to spawn continuously, at whatever level environmental conditions permit at the moment, so long as there is some minimal level of return for effort. Relatively benign tropical conditions may permit year-round spawning with little seasonal change or may allow adults of different species to respond to a range of factors that show a variety of seasonal patterns of change. On higher-latitude coral reefs, seasonal environmental constraints on both adults and larvae may increase in strength and produce not only more sharply defined spawning seasons, but also a stronger tendency for those seasons to coincide in a broad range of species. The latitudinal differences in spawning seasons of West Atlantic reef fishes could reflect a latitudinal gradient in the intensity of such constraints.

To assess the relative importance of larval biology, juvenile biology, and adult biology factors as determinants of seasonal patterns of spawning, we need much additional information about the most basic aspects of these fishes' life histories. How much spatial variation is there in seasonal patterns of spawning and recruitment? What factors affect juvenile growth and maturation? How do environmental factors such as food availability and physical stresses affect adults' ability to reproduce? How do effects of those factors vary seasonally and how do adults partition resources between reproduction, growth, and storage in different seasons? What influence do intrinsic limitations such as low longevity and low storage capacity have on the scheduling of spawning?

If constraints are preventing many small, short-lived reef fishes from releasing their spawning effort during a season of peak larval survivorship, one might expect this to have consequences for their adult populations. There is evidence that adult populations of some reef fishes are determined by larval supply and that these populations are below levels that would be set by resources (reviewed by Doherty and Williams, 1988a). A pattern of consistent partial reproductive "failure" during a season of high larval survivorship could contribute to maintaining the larval supply below levels that would enable adult populations to reach the point where they are controlled by resources.

CHAPTER 14

The Use of Phenotypic Plasticity in Coral Reef Fishes as Tests of Theory in Evolutionary Ecology

Robert R. Warner
Department of Biological Sciences and Marine Science Institute
University of California at Santa Barbara
Santa Barbara, California

I. INTRODUCTION

Coral reef fishes show an immense variety of social and mating systems, including solitary territorial species, monogamous pairs, small groups or harems with a single dominant male, and on up to schooling species that mate en masse in group spawns. Life histories and sexual patterns are equally diverse. Mature individuals may occur over a great range of sizes, and average longevities range from a few months to many years. Some species have permanently separate sexes, while others demonstrate sex change or the presence of both sexes simultaneously in a single individual.

This tremendous interspecific variability is mirrored by a large amount of variation within certain species. I suggest that we can utilize this intraspecific variability in tests of hypotheses in evolutionary ecology. Past discussions have stressed how phenotypic plasticity allows local adaptation to diverse physical and social environments, particularly in organisms that are widely dispersed but have little control over their present location (Gause, 1942; Bradshaw, 1965; Fagen, 1987). Here I extend this idea of local adaptation in order to explore hypotheses concerning what environmental factors have shaped the diversity of particular behavioral and life history characteristics. Interspecific diversity in a trait represents the genetic response of separate species over evolutionary time to particular challenges posed by the environment. Similarly, phenotypic plasticity of a trait within a species represents a set of functional adaptive responses to short-term variation in the environment. A study of phenotypic plasticity can identify just which aspects of the environment appear important in eliciting a functional response. Competing hypotheses about the evolution and maintenance of a trait usually differ in terms of the particular aspects of the environment that they consider to be important, and the information from a study of phenotypic plasticity can be used as a test of these hypotheses (Warner, 1980; Stearns and Koella, 1986; West-Eberhard, 1988). Given a particular trait, the environmental factors to which a particular species responds in the short term should be the same as those that shape differences between species over evolutionary time.

Phenotypic plasticity should be expected in situations where some aspect of the environment, critical to fitness, varies in a pronounced but unpredictable way. The variation should be over relatively large temporal or spatial scales, so that a particular individual experiences only one or a few expressions of the environmental parameter (Bradshaw, 1965).

There are compelling reasons to suppose that coral reef fishes should show phenotypic plasticity and respond flexibly and adaptively to their environment. First, a highly dispersive larval stage makes local genetic adaptation of populations much less likely, and some recent studies have supported the idea that gene flow is extensive and local populations result from highly mixed populations of larvae (Victor, 1984; Avise and Shapiro, 1986).

Second, the physical environments into which individuals settle are highly variable in extent of area, in substrate, in food availability and distribution, and in depth, and these factors can directly affect life history characteristics (Sale, 1984; Victor, 1986b). The low and sporadic recruitment rates seen for many coral reef fishes (see Doherty and Williams, 1988a) suggest that larvae have little opportunity to assess several environments before settling: they may be fortunate to find a reef at all. In addition, many coral reef fishes are restricted to the local population into which they settle (Sale, 1984); migration at a later stage to a more appropriate environment is limited to those habitats available in the immediate area. Thus reef fishes may not often be distributed among resources in an equitable manner.

Third, the social environment for reef fishes can be extremely unpredictable in space and time, so that behaviors and life history allocations appropriate in one population may be quite maladaptive in another. The filtering effect of successive areas for settlement along a current can result in consistently higher population densities in upstream locales versus those downcurrent (Victor, 1986b; Gaines and Roughgarden, 1987). In addition to density, the actual size of a local population often varies with local reef size. Unpredictable temporal variation in recruitment can have another major effect on the social environment: age and size structure can vary dramatically as cohorts of different initial abundances grow older in a local population. For example, a single successfully recruited age class can predominate in a population over considerable periods (e.g., Victor, 1983a).

Thus the unpredictability of settlement and varied reef environments create the proper conditions for the evolution of phenotypic plasticity. Given enough time, even relatively rare environments may evoke appropriate responses (Via and Lande, 1985). This plasticity in turn may be used to explore theory in evolutionary ecology. Because the comparisons can be made intraspecifically, the problem of differing evolutionary histories does not cloud an analysis of response to the environment (Clutton-Brock and Harvey, 1984).

In the literature on coral reef fishes, descriptions of a particular species' mating system or life history in one location are often countered by a contrary report by another worker in another location (see Thresher, 1984, for several examples). I suggest that the reported discrepancies are due less to differences in technique or expertise, and more to functional responses of the same species to different environments.

I have no intention to review variability in social systems and life histories of coral reef fishes. Within- and between-species variation in mating systems were comprehensively covered recently by Thresher (1984); for most of these examples, we have little or no information on accompanying differences in the environment. Instead, I want to focus on three examples of how between-population differences can be used to identify the critical environmental parameters shaping the evolution and maintenance of particular traits. The hypothesis is simple: those environmental parameters that lead to a response in some trait within a species are likely to be the same parameters that have led to species differences in fixed traits (Warner, 1980).

II. Examples of Intraspecific Variability

A. Population Density, Economic Defendability, and Mating Systems

Variability in recruitment can lead to major differences in density both between local populations and within a single population over time. In addition, larger reefs may have relatively higher densities at certain sites where fishes gather to mate, feed, or seek shelter. Differences in densities, in turn, can affect a myriad of environmental factors, including food distribution and abundance, defendability of resources, and future prospects for growth and reproduction (e.g., Jones, 1987a).

The concept of economic defendability (Brown, 1964) suggests that territoriality is favored under conditions where the benefits conveyed by the resource exceed the costs of defense. It is often used to explain differences between species in the degree of territoriality. Most often, the proposed critical environmental factor is the dispersion of the resource being defended: at equivalent mean densities, resources that are uniformly distributed are not worth defending, while resources that are concentrated in only a few areas are too costly to defend. Only when resources are distributed in a moderately concentrated fashion do the benefits outweigh the costs of defense (Brown and Orians, 1970). An additional factor that has received less attention is how population density affects defendability (Hixon, 1980b): obviously, an increase in potential intruders raises the costs of defense.

Past work on the bluehead wrasse (*Thalassoma bifasciatum*) illustrates how an intraspecific comparison, followed by experimentation, can sort through the relative importance of resource distribution versus population density in the expression of territorial behavior. Like many reef fishes, the bluehead wrasse exists in local populations with little or no interchange of adults between reefs, but with extensive mixing at the larval stage. Mating sites are the resources that are defended by large males. These sites tend to be upward projections on the downcurrent periphery of a reef and serve as egg-launching sites for the pelagic eggs; the same sites remain in use for generations (Warner, 1988c, 1990). Females that are ready to spawn migrate to a mating site from their upcurrent feeding areas, release their eggs, and return to feeding. Large males attempt to mate singly with these arriving females by defending the area surrounding a mating site from other males.

Some mating sites are not defended. Instead, they are occupied by large aggregations of smaller males who mate in groups with the arriving females (Warner *et al.*, 1975). On larger reefs, the majority of matings occur on these undefended sites, while on the smaller reefs all sites are defended by single large males (Warner and Hoffman, 1980a). What leads to the differences in the expression of territorial behavior among local populations? Because larval dispersal is extensive in this species and the populations being studied are within a few hundred meters of one another, the differences in territorial behavior are not likely due to differences in evolutionary history or local genetic adaptation.

Let us deal with the two hypotheses on economic defendability in turn, beginning with spatial distribution. It could be that mating sites are either more uniformly distributed or more highly aggregated on larger reefs relative to small reefs. However, measurements of mating site characteristics do not indicate any differences in the dispersion or physical aspects of sites among reefs. On all sizes of reefs, the distances between mating sites are approximately the same, and they always occur in the same general downcurrent area.

Overall population densities are constant with reef size, but an important difference appears when mating densities are noted (Warner and Hoffman, 1980a). Because area increases more rapidly than periphery, and because the mating population tends to converge on the same downcurrent extremities regardless of reef size, the density of mating individuals (both females and potentially intruding males) is much higher on larger reefs. Increased numbers of females mean that a particular site is a richer resource, but the increased numbers of intruders that this resource draws may make a site uneconomical to defend (see Hixon, 1987, for a model). This leads to the nonintuitive prediction that the most successful sites should be undefended, and it is just this pattern that is seen on large reefs (Warner and Hoffman, 1980a). Only sites less frequented by females, slightly upcurrent from the reef end, are included in large male territories.

These results suggest a major role for local population size and mating density in the expression of territoriality. It also suggests experiments, which is one of the great advantages of the plasticity seen in reef fishes: one can further test a hypothesis arising from a comparative study by manipulating specific variables. Warner and Hoffman (1980b) used transplants from one reef to another to alter population density of mating individuals. They showed that 40–65% increases in intruding male numbers did indeed lower the mating success associated with particular territories to less than half of their original value. Corresponding reductions in density on other reefs to a third of the original levels raised mating success threefold.

Other studies of the bluehead wrasse in different geographical areas suggest that the pattern seen in Panama is a general one. Fitch and Shapiro (1990) studied a low-density population in a back reef area of Puerto Rico and noted only pair-spawning. In a very comprehensive project, von Herbing (1988) observed a dense population of blueheads on a large fringing reef in Barbados where both types of matings occurred; the predominant mode was group-spawning. The differences between these reports of mating systems are much more likely due to the size of the population studied than to large-scale geographical effects. In St. Croix, U.S. Virgin Islands, for example, the population densities and mating patterns are similar to that seen in Panama: pair-mating predominates on small patch reefs but is rare on the fore reef, where most matings occur in groups (unpublished observations).

In this species, a change in the effectiveness of territoriality has a dramatic effect on the mating system and distribution of male reproductive success: on small reefs we see resource defense polygyny and mate monopolization by large males, while large reefs are characterized by extreme promiscuity and a much more equitable distribution of mating success among males. This means that we can relate mating system expression to local environmental factors such as geographical location, current direction, and reef area, all of which are beyond the control of natural selection. Even more important, comparative observations and experiments have identified out of a set of alternatives the most probable *mechanism* by which changes in mating system occur in this species.

B. Population Size, Opportunity for Accession, and Life Histories

In many species there exist groups of "bachelor" males or "floaters" that are often slightly smaller or younger than more reproductively successful males. The existence of these males is most often ascribed as a default rather than a strategy: larger males monopolize reproduction, so smaller males must wait their turn (see Wiley, 1981). As in the previous discussion, mate monopolization may arise from a shortage of economically defendable mating resources (see Emlen and Oring, 1977).

The possibility that bachelor status may actually represent a life history tactic arose when such bachelors were noted in species that have reasonably successful alternative reproductive modes (Hoffman *et al.,* 1985). For example, in sex-changing species, females could refrain from sex change and continue egg production instead of becoming bachelors; in species such as the bluehead wrasse, smaller males could remain in group-spawning aggregations. But in these species individuals abandon previously successful modes of reproduction and suffer a decrement in their current reproductive success in order to become a bachelor male. Why?

This problem was first attacked through interspecific comparisons. Hoffman *et al.* (1985) suggested that bachelors were individuals that reallocated resources away from reproduction in order to grow rapidly. If large size conveys a higher probability of attaining successful male status, it can be adaptive at a certain point in the life history to cease current reproduction in order to grow more quickly. If this view is correct, those sex-changing species with very high levels of reproductive success among large males should also have more bachelor males, because the possibility of gaining greater rewards permits the accrual of greater costs. Conversely, species with lower large male success should have no bachelor males at all, since the potential loss of reproduction as a female is relatively large. This was the pattern seen by Hoffman *et al.* (1985) in three species of sex-changing wrasses: in one species in which maximum male mating success was about 12 matings a day, the largest females remained as such until the disappearance of a local dominant male. In two other species in which the largest males mate 25 to 50 times a day, females become bachelor males and do not reproduce for long periods.

While these observations are consistent with a hypothesis of tactical reallocation of energy, the study suffers from the limitations of interspecific comparisons: the differences seen may be due to other factors now lost in separate evolutionary histories, or the bachelors seen in the two species may be the result of imprecise sex-changing mechanisms. Intraspecific comparisons would provide a more definitive test of the reallocation hypothesis.

The best example of such an intraspecific comparison is the work by Aldenhoven (1986a) on the angelfish *Centropyge bicolor,* in which females mate in harems defended by large males. She pointed out that several factors could increase the value of an early sex change in addition to high mating success of dominant males. These include factors that shorten the wait to achieve dominant status, such as higher death rates of dominant males and larger numbers of harems among which to sample as a bachelor. Aldenhoven studied four populations within a few kilometers of each other. While the mating success of large males did not differ among the populations, one population was about twice as densely distributed and had a higher death rate than the others. As predicted, many individuals underwent earlier sex change and became bachelors in this dense population but not in the others.

There is no particular reason to expect that mortality rates or large male mating success will be consistently higher in larger populations of any particular haremic species. However, the opportunity for accession to a harem by a bachelor will vary consistently with population size. For example, on smaller reefs, there may be room for only four harems, and the waiting time for any one of these to come available will be correspondingly long. In this case, it may profit a large female to simply remain in her present sex and attempt to take over her own harem when the present owner dies. On a larger reef with forty harems, the average waiting time for an opportunity to present itself is ten times less, and it may become worthwhile for some individuals to lose reproduction as a female in order to become a roving bachelor that can sample widely.

Here we see a major shift in life histories predicted from the changes in opportunity that arise from variation in local population size and/or mortality rate. The existence of intraspecific phenotypic plasticity in this trait, and the fact that the trait varies in a fashion consistent with a reallocation hypothesis, is strong evidence against a hypothesis based solely on males making the best of a bad situation. There is now need for manipulative experiments designed to elicit bachelor behavior in response to alterations in local population size.

Studies of other species also suggest that the bachelor strategy is more common on larger reefs where opportunities are greater. The parrotfish *Sparisoma aurofrenatum* has no bachelors among the harems on small patch reefs in the San Blas Islands of Panama (S. G. Hoffman, personal communication), but bachelors were common in the larger reef populations studied by Clavijo (1982) in Puerto Rico.

C. Sex Allocation, Local Group Size, and Mating System

Another important aspect of life history theory is sex allocation, which can take the form of offspring sex ratio, the timing of sex change, or the relative contribution to male and female function in individuals that can perform both roles (Charnov, 1982). Allocation in simultaneous hermaphrodites is particularly labile in fishes, and it forms fertile ground for an investigation of the factors that might lead to biases toward one sexual expression or the other.

Interspecific comparisons among the simultaneously hermaphroditic serranine sea basses have revealed a close relationship between mating system and sex allocation (Fischer, 1984; Fischer and Petersen, 1987). Those species that are monogamous tend to have strongly female-biased gonads; these pair-mating individuals reciprocate in male and female roles, each contributing eggs. Species in which there is a substantial amount of sperm competition from other individuals during mating are still female-biased and reciprocate, but have a higher male allocation than in purely monogamous species. Finally, species that exist in harem-like groups have a mixed pattern: smaller individuals are strongly female-biased, but the largest individuals are pure males and do not reciprocate.

This overall pattern suggests that while egg production forms the ultimate limit to fitness in certain mating systems, the returns for increases in male function can exceed those for increases in female function. In monogamous species, there is little to be gained from increased sperm production and both partners benefit from a heavy bias toward eggs. Proximity to other matings opens the possibility of participating as an extra male; this and the resulting sperm competition elicit a somewhat higher male allocation than the previous situation. Finally, in species where a large number of individuals occur in a relatively small area, a large individual can profit by spending his energy on monopolizing the egg producers and not functioning as a female at all. The subordinates in the group, with reduced opportunity as males, should in turn bias their allocation toward female function.

The interspecific pattern is consistent with mating system theory (e.g., Emlen and Oring, 1977). The species that show polygyny are at moderate densities and defend all-purpose territories, which allows a male to dominate a group of hermaphrodites (Fischer and Petersen, 1987). However, consistency does not necessarily indicate causation: the serranine sea basses are extremely diverse and numerous, and many other factors (including different recent phylogenies) could lead to differences in sex allocation among species.

To resolve this issue, Petersen (1990a) has studied intraspecific variation in the polygynous species *Serranus fasciatus*. The species was investigated at three different population densities, and the social system differed dramatically among the three locales. In low-density areas the fishes formed monogamous pairs that reciprocally spawned with each other, and thus each individual gained about equally from male and female function. Larger groups (up to 11 individuals) were haremic: the smaller members of the group were hermaphrodites who achieved nearly all of their reproductive success as females, while the largest individual functioned only as a male. All the hermaphrodites spawned once nightly with the dominant male, and only occasionally (6.6% of the total spawns observed) did they interfere as males in nearby matings.

At the highest densities, Petersen (1990a) observed a further transition in mating system. In three groups ranging from 11 to 23 individuals, some of the larger hermaphrodites not only spawned in a female role with the dominant male each night, but also pair-spawned regularly in a male role with small hermaphrodites. The rate of interference in other spawns by small hermaphrodites acting as males was also much higher (22.8%) in these high-density populations. The dominant male directed his aggression and spawning activity toward the larger hermaphrodites. Because of these mating patterns, all hermaphrodites had more reproductive success as males than did similarly sized individuals in smaller harems.

The pattern of actual allocation of gonadal tissue to male and female function paralleled these behavioral observations (Petersen, 1990b). In a year that hermaphrodites had relatively high reproductive success as males they also had larger allocations to testes, and male reproductive success and testis allocation were positively correlated across individuals.

The changes in sexual allocation with changes in mating system are a particularly compelling confirmation of the idea that population density and group size control the ability of a male to act polygynously (Emlen and Oring, 1977), because each individual has the opportunity to regulate allocation to male function. In monogamous situations, it profits both members of the pair to act as both males and females, with most of the investment devoted to egg production. In small groups, a dominant individual can gain more by monopolizing the female functions of the other group members, even if this means ceasing reproduction as a female entirely. In the largest groups the ability of the largest male to monopolize matings is reduced, and the opportunity to successfully function as a male extends to the subordinates as well (Petersen, 1990a).

The decrease in mate monopolization and increase in opportunity for smaller males that occur with increasing density in this species are similar to the pattern seen in the previous two examples. But hermaphroditic species are capable of much finer adjustments in both mating system and physical sex allocation (Lloyd and Bawa, 1984). Not only does social group size change within this species, but sex allocation can vary within an individual in a precise and measurable way. Field experiments with such variable hermaphrodites provide an important opportunity for further research.

III. Discussion

I have described three instances where a study of intraspecific variation has identified the critical environmental parameters affecting the expression of behavioral or life history traits. The inherent difficulty associated with the comparative method is separating ancestral effects from adaptive genetic responses (Clutton-Brock and Harvey, 1984; Bell, 1989), and this problem is removed here. An additional advantage is that the prospective parameter can then be experimentally manipulated in further studies to confirm or refute its importance.

Very few studies exist that document both intraspecific variation and the associated environmental parameters. To encourage further studies of this nature, I here suggest what are, in my opinion, the important environmental parameters to investigate. The environment influences the life histories and social systems of animals in three basic ways. The first aspect has been much discussed, the second forms the theme of this paper, and the third is a focus for future research.

The *spatial distribution of resources required by females for food and reproduction* probably applies to reef fishes in ways similar to that proposed for other animals. A male's ability to monopolize the matings of many females will often depend on their home range size and whether they are willing to mate within that home range. There are several excellent reviews of the relationship among species between general feeding mode and social systems (e.g., Robertson and Hoffman, 1977; Thresher, 1984; Hourigan, 1989), but there is surprisingly little information on intraspecific variation in mating system that can be specifically ascribed to differences in resource distribution.

In the wrasse *Halichoeres maculipinna*, resource defense polygyny has been reported in studies made in Panama (Warner and Robertson, 1978; Robertson, 1981), but Thresher (1979) claimed that the same species formed harems in Florida. These differences may have been caused by different resource distributions, because females in the Panama populations migrate from their home ranges to mating sites, but in Florida they both reside and spawn within a male's territory. In Panama, females feed on shallow reef tops that may be unsuitable for spawning, while feeding in Florida occurs in deeper areas where mating can also take place (Robertson, 1981). I suggest that the different mating systems may be found quite near each other under the right conditions of local environmental variability.

Many food items on reefs should exhibit variability in space and time, similar to that seen in fish populations themselves. Thus studies of the effect of spatial differences in resources on intraspecific differences in mating systems are likely to be fruitful. Fishes that feed on benthic invertebrates and other patchy prey are especially good candidates for such a study.

The effect of the *size and density of the population* has been a recurrent theme in this chapter. Density can of course affect female home range size directly and thus have an impact on a male's ability to monopolize mates, as mentioned earlier. But here I have stressed two other aspects of density: male numbers and their effect on defendability, and population size and the resultant opportunity for accession to dominant male status.

Changes in economic defendability can lead to abrupt shifts in the mating system, as seen in the bluehead wrasse. Changes in opportunity can determine the presence or absence of a "bachelor" phase in the life history, as seen in *Centropyge bicolor*. Since size and density of coral reef fish populations are prone to very wide variation, it would not be surprising to see similar intraspecific variability in mating systems when other species are studied in detail.

An aspect of coral reef fish biology that could have profound effects on social systems and life histories is *variation in age and size structure*. Many aspects of life history depend on the relative values of current versus future reproduction (e.g., Pianka and Parker, 1975). In life history theory, these values are fixed by extrinsic forces of mortality and intrinsic limitations on growth, with the population usually assumed to be in a stable age distribution. However, for many species, and coral reef fish in particular, populations are highly variable in age and size structure because of recruitment fluctuations.

Different size structures in populations offer different optimal life history solutions (Schultz and Warner, 1989). For example, consider a species with a mating system in which large males monopolize reproduction and small males have alternative reproductive strategies. A male of intermediate size would increase his fitness by maximizing growth and reaching a large size quickly if there were few individuals larger than he and many smaller females with whom he could potentially mate as a large male. In contrast, a male of the same size with few individuals smaller than he has diminished future prospects. Resources should be allocated more toward alternative reproductive strategies presently available and less toward growth and future reproduction. Similar predictions can be made for females in protogynous species, because they also can eventually become large males; in this case, the allocation decision is between growth and egg production (Schultz and Warner, 1989).

The idea that individuals might alter their life history trajectories in response to local conditions is not new (see Stearns, 1982). The important point is that coral reef fishes offer particularly likely candidates to show such responses because of extreme temporal variability in recruitment. As such, they offer an opportunity to test life history theory in a noninvasive and experimental fashion (Schultz and Warner, 1989). Past experiments have directly altered the physiology of target organisms (e.g., by changing temperature; see Hirschfield, 1980) and then sought evidence for adaptive reallocation. However, changes in other life history parameters such as growth could be directly due to the manipulation. Size and age structures can be altered in reef fish populations without directly manipulating the target individuals, and the predictions for shifts in life history trajectories are straightforward. I think that this approach is likely to be a valuable avenue for future research.

Studies of intraspecific variation in coral reef fishes are just beginning. We already know that reef fishes show great interspecific variety in social and mating systems as well as in life histories. As studies proliferate, we are becoming aware that similar variety can often be found within single species. It is not surprising that organisms faced with wide dispersal and great variability in social and physical environments should show phenotypic plasticity. We now need to use this plasticity in further tests of evolutionary ecology.

Acknowledgments

Thanks to Chris Petersen for supplying me with his unpublished manuscripts and a continual flow of ideas. J. Endler, C. W. Petersen, D. R. Robertson, P. F. Sale, C. St.Mary, E. T. Schultz, and E. van den Berghe all made helpful comments on an earlier draft. Preparation of this paper was supported by the National Science Foundation (BSR 87-04351).

PART V

Community Organization

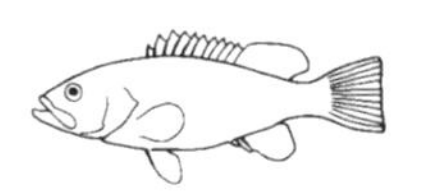

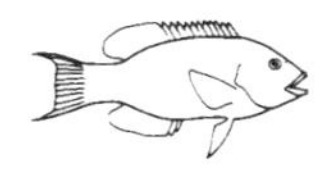

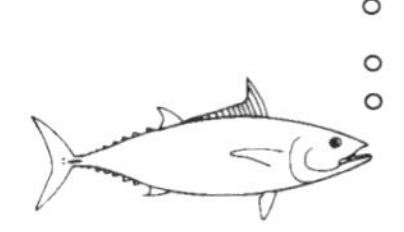

CHAPTER 15

Geographic Variability in the Ecology of Coral Reef Fishes: Evidence, Evolution, and Possible Implications

R. E. Thresher
CSIRO Division of Fisheries
Hobart, Tasmania, Australia

I. Introduction

Fish families found on tropical coral reefs have a virtually worldwide distribution on both coral and noncoral, rocky substrata. The consistency of this fauna at the family level, often noted (Ekman, 1953; Briggs, 1974; Longhurst and Pauly, 1987; Sale, Chapter 19), underlies the integration of studies in different geographic areas into comprehensive theories about the community as a whole (see reviews by Sale, 1980a; Doherty and Williams, 1988a). Yet there are not many data to justify this integration. The few quantitative ecological comparisons between areas are usually based on small samples at a few sites, suggesting considerable caution in generalizing their implications. The easy integration of results from studies in different areas is also worrying in the light of well-documented, often fundamental differences between areas in their oceanographic and geographic settings (e.g., Philander, 1989) and consistent differences between workers in the western Atlantic and western Pacific on the importance of resource limitation and recruitment-related processes on the structuring of reef fish assemblages (see reviews by Helfman, 1978; Doherty and Williams, 1988a). For these reasons, a detailed analysis of possible large-scale geographic differences in reef fish ecology is called for.

This chapter addresses the following question: Are there consistent differences between reef fishes and their assemblages in different geographic areas that bear on the extent to which studies from these areas can be integrated into comprehensive theories? The question is addressed at two spatial scales: interoceanic or New World–Old World comparisons, and "regional" comparisons among island groups and similar-sized units in the same ocean. With few exceptions, data for comparisons at smaller spatial scales are too sparse to warrant analysis.

The chapter proceeds in four stages: (1) a review of comparisons that have already been made at interoceanic and regional spatial scales, (2) an analysis of geographic variations in the taxonomic composition and reproductive characteristics of the reef fish fauna, (3) an examination of hypotheses that have been proposed to account for these variations, and (4) a discussion of the possible implications of apparent large-scale geographic differences in ecology, with some suggestions as to how to test their generality and significance.

II. Previous Geographic Comparisons of Reef Fish Assemblages

Analyses of geographic variations in reef fish communities have a long history but, until recently, have been restricted to relatively gross features. It has long been recognized, for example, that there are more species of fishes (and other marine organisms) on coral reefs in the Indo-Pacific than in the western Atlantic and more in the Atlantic than in the "depauperate" eastern Pacific (see Ekman, 1953; Sale, 1980a). In the Pacific, there has also been widespread recognition of a west-to-east gradient in species richness, with "centers of distribution" of various marine organisms, including fishes, in the general region of the Philippines and the Malayan Archipelago. Kay (1980) summarized previous descriptive analyses of the west and central Pacific as yielding three generalizations: (1) the central Pacific biota has a strong western affinity, (2) species richness declines from west to east across the Pacific, and (3) major taxonomic groups gradually disappear from west to east (however, see Springer, 1982, for evidence of an abrupt change in species composition at the western margin of the Pacific Plate). Good examples of these patterns among reef-associated fishes are provided by Allen (1975), Goldman and Talbot (1976), and Woodland (1983).

Recent debate about the "order" and "lottery" hypotheses (see Sale, 1980a; Mapstone and Fowler, 1988; Doherty and Williams, 1988a) has stimulated more detailed analyses of geographic variations in community characteristics. To a very large extent, the evidence supporting each hypothesis was drawn from work in different faunal provinces (western Atlantic = "order"; western Pacific = "lottery"). This geographic component to the debate prompted speculation that the factors structuring the fish communities differ in some fundamental way in the two areas (Helfman, 1978; Thresher, 1982) and inspired several interoceanic comparisons of fish assemblages in similar habitats. C. L. Smith (1978) compared the numbers of individuals, species, genera and families and overall diversity of fishes on sets of similarly sized patch reefs in the Bahamas and the Society Islands. Although the number of species was slightly higher in the Pacific, fish assemblages at the two sites were generally similar, based on data for each site pooled across reefs. Essentially the same conclusion—that there are no striking differences between areas—has generally been drawn in subsequent studies. Rates of recolonization following defaunation and final assemblage composition are much the same on similar artificial reefs off Florida and in the lagoon at One Tree Reef (Great Barrier Reef) (Bohnsack and Talbot, 1980). The predictability of assemblage structure based on environmental factors and the degree to which assemblages differ among neighboring patch reefs appear to be roughly equivalent at St. Croix (U.S. Virgin Islands) and Enewetak (Marshall Islands), which Gladfelter *et al.* (1980) concluded was likely to reflect similar structuring mechanisms. Species diversity within habitats (alpha-diversity) does not differ significantly for lagoonal and back reef sites between the western Atlantic and Indo-west Pacific (Sale, 1980a). Nor is there any indication that the numbers of species and individuals in another habitat, the reef front, vary substantially among widely separated sites in the Indian and Pacific Oceans (Talbot and Gilbert, 1981).

Geographic comparisons involving data other than counting the number of fishes in an assemblage are not common. Several studies have noted that congeneric species in different areas often behave similarly (e.g., Robertson and Hoffman, 1977; Thresher, 1984) or interact with the reef community in much the same way, for example, "algal farming" by damselfishes in both the Atlantic and Pacific (see papers in Emery and Thresher, 1980). In one of the first geographic comparisons of trophic ecology, Bakus (1967) suggested that grazing herbivores were relatively more common on tidal flats in Oceania than in other parts of the Pacific or Caribbean. Subsequent studies present a mixed bag of observations, some reporting geographic differences and others suggesting that differences are slight. Losey (1974) found higher densities of cleaning species at Puerto Rico than at several sites in the Indo-Pacific. Goldman and Talbot (1976) noted that scarids that eat mainly corals were present in the Pacific but not in the Atlantic, while Bouchon-Navaro and Harmelin-Vivien (1981) report that among herbivores, scarids appeared to be relatively more abundant than acanthurids in the Caribbean than in the Indo-Pacific. In contrast, both Gladfelter *et al.* (1980) and Thresher and Colin (1986) reported similar trophic structures for fish assemblages on reefs in the Caribbean and the Marshall Islands. Within the Pacific, Anderson *et al.* (1981) noted an increase from west to east in the proportion of planktivorous chaetodontid species, which they contrasted with a decline in the proportion of planktivorous pomacentrids across the same sites. Similarly, reported herbivore densities tend to be higher in the central Pacific than at other sites in either the Indo-Pacific or the Atlantic (Bouchon-Navaro and Harmelin-Vivien, 1981).

Much of the information available on trophic features of reef fish assemblages has recently been reviewed by Parrish (1989). Parrish cautions against generalizing about broad-scale geographic differences, particularly for trophic features, on the basis of sparse data from a few areas. He notes that it is often difficult to assign a species to a particular trophic level, either because of lack of data or because it has a broad or flexible diet. Nonetheless, Parrish (1989) made a few tentative observations based on a comparison of twelve sites scattered across the Indo-Pacific and Atlanto-east Pacific: (1) trophic structures often vary widely between sites within oceans; (2) species that feed on sea grasses directly are probably more common in the Caribbean than in the Indo-Pacific (probably because sea grass beds are generally closer to reefs in the Caribbean); (3) overall levels of herbivory are likely to be similar in the different oceans, though scarids appear to be more abundant than acanthurids on Caribbean island reefs and vice versa on Pacific island reefs; (4) zooplanktivores appear to be more abundant and constitute a larger portion of the community on Pacific reefs than on Atlantic reefs; and (5) there appear to be some geographic trends in abundance of carnivorous groups, such as more feeders on sessile invertebrates in Oceania than in the Atlantic. Parrish suggests that patterns of carnivory are particularly difficult to discern, however, because of often large variations within regions, the effects of human impact on larger carnivores, and difficulties in assessing the abundance of such species.

To summarize, a number of studies have attempted interoceanic and regional comparisons of the structure and ecology of reef fish assemblages. In most cases, however, the data were too sparse to draw any robust conclusions, aside from obvious trends in species richness and diversity. The interest in quantitative comparisons stimulated by the debate over the order and lottery hypotheses has largely waned, at least in part because no compelling evidence was found that the numbers of individuals, numbers of species, rates of recolonization, or densities of fishes differed markedly between assemblages in similar habitats in the Indo-Pacific and in the western Atlantic. However, it is premature to conclude that interoceanic differences in assemblage structure are slight. The variance within regions has not been properly assessed, few habitats and sites have been examined quantitatively, and the few data lack the power to resolve anything other than very conspicuous differences between areas. The few comparative studies of the dynamics of the assemblages are generally of too short a duration to be useful. As Parrish pointed out, comparisons involving other ecological features are usually casual adjuncts to ongoing, very specific studies. There are intriguing hints of regional differences—for example, persistent comments that the trophic structure of Oceania differs from that in the rest of the Pacific—but these have not, as yet, been followed up by detailed comparative studies.

III. An Analysis of Interoceanic and Regional Variation in Assemblage Structure

A. Taxonomic Composition

Although there are insufficient data to test for consistent geographic variations in population or community ecology of reef-associated fishes, it is possible to test for geographic variations in the taxonomic composition of reef fish assemblages, which may be relevant to their ecology. To this end, I took data from 21 checklists of species from sites in the Indo-Pacific and western and central Atlantic (for locations and literature sources see Fig. 1) and tested for geographic trends in the species composition of their assemblages. To minimize problems of uneven data quality, I included a checklist only if it was the most recent checklist available for a site; it was the most comprehensive for that site, did not conspicuously underrepresent the fauna, and was intended as reasonably comprehensive; and it was produced by or in cooperation with

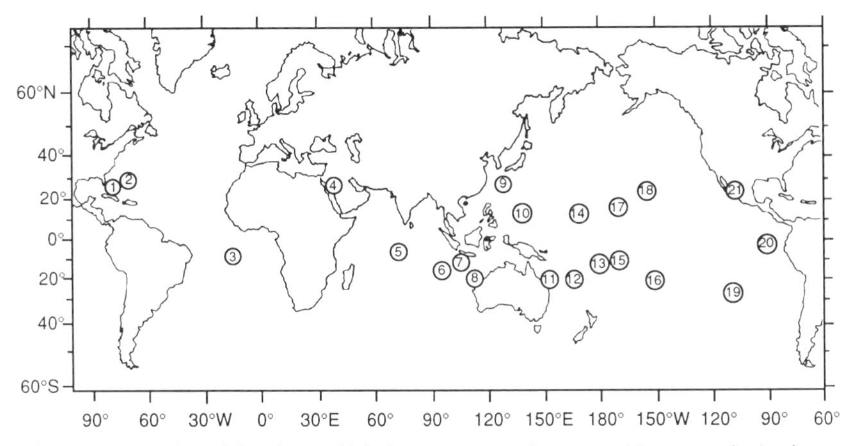

Figure 1 Location of sites from which data on community composition were obtained. Key and references: 1. Florida Keys (Starck, 1968); 2. Bahamas (Bohlke and Chaplin, 1968); 3. Ascension Island (Lubbock, 1980); 4. Red Sea (Randall, 1983); 5. Chagos Archipelago (Winterbottom *et al.,* 1989); 6. Cocos (Keeling) Islands (Smith-Vaniz and Allen, 1991); 7. Christmas Island (Allen and Steene, 1979); 8. northwest Australia (Allen and Swainston, 1988); 9. Ryukyu Islands (Yoshino and Nishijima, 1981); 10. Guam (Shepard and Myers, 1981; Amesbury and Myers, 1982); 11. Great Barrier Reef (Russell, 1983b); 12. New Caledonia (Kulbicki, ms.); 13. Fiji (Carlson, 1975); 14. Marshall Islands (Randall and Randall, 1987); 15. Samoa (Wass, 1984); 16. Tahiti (Randall, 1985a); 17. Johnston Island (Randall *et al.,* 1985b); 18. Hawaiian Islands (Randall, 1985b); 19. Easter Island (Randall and Egana, 1984); 20. Galapagos Islands (Snodgrass and Heller, 1905); 21. Sea of Cortez (Thomson *et al.,* 1979).

observers with strong taxonomic backgrounds (as evidenced by their treatment and associated literature).

From the checklists, I extracted the number of species in each of 46 teleost families of fishes I consider to be primarily reef-associated, based on the literature and personal observations (Table 1). Similarly, I excluded species in these families that are characteristically associated with nonreef habitats.

The number of species and families declines from west to east with distance from the western continental margin of the Pacific Ocean, with a minor peak at continental margin sites in the New World (Fig. 2). For Pacific sites, both species and family richness correlate significantly with distance along latitudes from the western margin of the ocean, after excluding the New World reefs (rs = -0.84, $p < 0.01$, and rs = -0.91, $p < 0.01$, respectively). There is an indication of a decline east to west across the Indian Ocean, but it is weak, driven mainly by the difference in species richness between northwest Australia and more westerly Indian Ocean sites. New World reefs (sites 21, 1, and 2) are inhabited by fewer species (approximately 60% less) and families (12% less) than reefs near the western continental margin of the Pacific (sites 8–12), but are similar in both species and family richness to reefs in the central Pacific (sites 16–18). Ascension Island (site 3), in the central Atlantic, has approximately the same number of species and families as isolated areas of the central and eastern Pacific, such as Easter Island (site 19) and the Galapagos (site 20).

Table 1 List of Families Included in the Geographical Analysis of Species Checklists[a]

Acanthuridae (P)	Gobiesocidae (D)	Pinquepididae (P)
Antennariidae (P)	Gobiidae (D)	Pomacanthidae (P)
Apogonidae (D)	Grammistidae (P) (includes *Liopropoma* and *Pseudogramma*)	Pomacentridae (D)
Aulostomidae (P)	Haemulidae (P)	Priacanthidae (P)
Balistidae (B)	Holocentridae (P)	Pseudocromid-like fishes (D) (includes Acanthoclinidae, Grammidae, Plesiopidae, and Pseudochromidae)
Blenniidae (D)	Kuhliidae (P)	Scaridae (P)
Bothidae (P)	Kyphosidae (P)	Scorpaenidae (P)
Caesionidae (P)	Labridae (P)	Scorpidae (P)
Chaetodontidae (P)	Lethrinidae (P)	Serranidae (P)
Cheilodactylidae (P)	Lutjanidae (P)	Siganidae (B)
Cirrhitidae (P)	Monacanthidae (B)	Syngnathidae (D)
Clinidae (D) (includes Labrisomidae and Chaenopsidae)	Mullidae (P)	Synodontidae (P)
Congridae (P)	Muraenidae (P)	Tetraodontidae (B)
Diodontidae (P)	Opistognathidae (D)	Zanclidae (P)
Ephippidae (P)	Ostraciidae (P)	
Fistulariidae (P)	Pempheridae (P)	

[a] Key to spawning mode (see text): D = normal demersal; P = pelagic; B = balistid-type demersal.

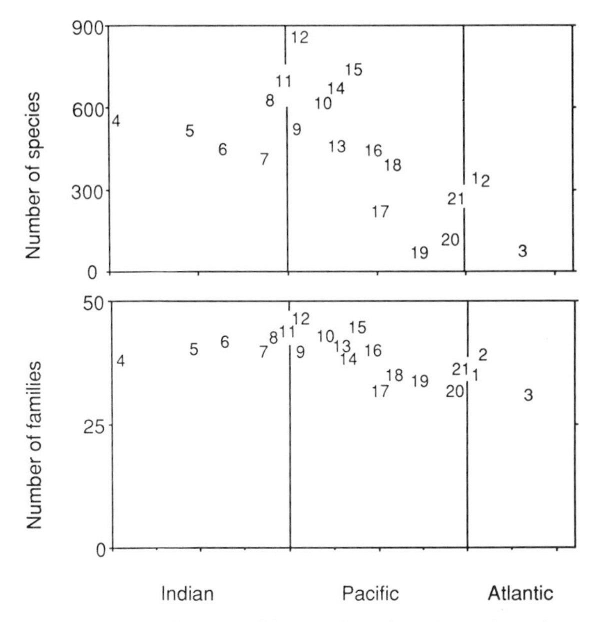

Figure 2 Distribution of the number of species and number of families of reef-associated fishes reported at sites in the Indian, Pacific, and Atlantic oceans. Numbers refer to sites identified in Fig. 2. The horizontal axis is scaled in terms of proportional distance east to west across each ocean basin (see text).

The ranking of families at each site in order of species richness provides more detail about the geographic affiliations among sites. A matrix of nonparametric correlations among the 21 sites, based on these rankings, is everywhere positive and significant. Fundamentally, some families, such as damselfishes and wrasses, are relatively species-rich everywhere, whereas others, such as ephippids and priacanthids, are everywhere species-poor. These global differences among families unite all tropical reef assemblages. Nonetheless, families do not rank identically by species richness across sites. A nonparametric cluster analysis divides the 21 sites into three groups (Fig. 3): (1) a "New World" group, consisting of the Florida Keys, the Bahamas, the Sea of Cortez, and the Galapagos Islands; (2) a "Central Oceanic" group, consisting of Johnston Island, Easter Island, Hawaii, and, in the Atlantic, Ascension Island, and (3) an "Indo-nonoceanic Pacific" group, composed of the remaining Pacific and all Indian ocean sites. Varying the clustering routine has no effect on these major groupings of sites. Nonparametric clustering further divides the Indo-nonoceanic Pacific into two subgroups, apparently according to

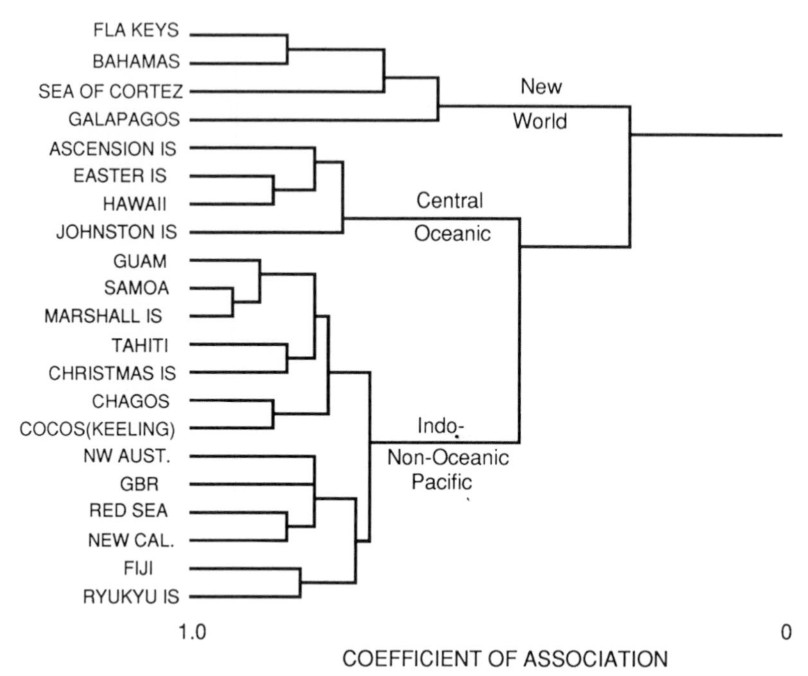

Figure 3 Results of a cluster analysis of the 21 reef sites, based on similarities and differences in the relative species richness of 46 families of reef-associated fishes. The clustering routine used is based on farthest-neighbor distances and the gamma coefficient of association (see Goodman and Kruskal, 1954), a nonparametric measure of predictability.

their proximity to a continental margin. One subgroup consists entirely of island sites in the Pacific and Indian oceans [Guam, Samoa, the Marshall Islands,Tahiti, Chagos Archipelago, Cocos (Keeling), and Christmas Island], whereas the other consists of sites near continental margins (the Rea Sea, Great Barrier Reef, and northwest Australia) or island groups near these margins (New Caledonia, Fiji, and the Ryukyu Islands). Sites jump back and forth a bit between the two subgroups depending on the clustering routine used, which suggests that the distinction between continental margin and "offshore" sites is fuzzy.

I could not find comprehensive checklists of fishes for sites close to the apparent center of diversity of reef fishes (the Philippines-Indo-Malayan region), but analysis of species lists not intended to be comprehensive for the Philippines (Murdy *et al.*, 1981) and Thailand (Monkolprasit, 1981) align them closely with the continental margin subgroup. The order of family richness for the Philippines site is similar to that of the Great Barrier Reef

(rs = 0.93) and that for Thailand is most similar to those of sites in the Ryukyus and Fiji (rs = 0.84 for both). At the same time, two sites in the primary data set show nearly equal levels of similarity to more than one group. The Galapagos is most similar to the New World sites (rs = 0.65 to 0.72) with which it was grouped, but it is only slightly less similar to the Pacific reefs in the Central Oceanic group (rs = 0.61 to 0.65), while Tahiti shows affinities with both the offshore subgroup of the Indo-nonoceanic Pacific (rs = 0.90 to 0.93) and Pacific sites in the Central Oceanic group (rs = 0.87 to 0.91).

The division between Central Oceanic and the Indo-nonoceanic Pacific sites is also clearly evident in the distributions of species within families (Fig. 4). Four families that are sufficiently conspicuous and taxonomically well worked out as to be unlikely to be misidentified in checklists—pomacentrids, acanthurids, chaetodontids, and pomacanthids—were analyzed. In each, a cluster analysis based on presence–absence data for each species at each site identifies Johnston Island, Hawaii, and Easter Island as one group and Tahiti, Samoa, Guam, and the Marshall Islands as a second faunistically similar group. These correspond exactly with the Pacific elements of the Central Oceanic group and the "offshore" subgroup of the Indo-nonoceanic Pacific group, as identified in the family-level analysis. The relationship between the continental margin and offshore subgroups of the Indo-nonoceanic Pacific group differs slightly

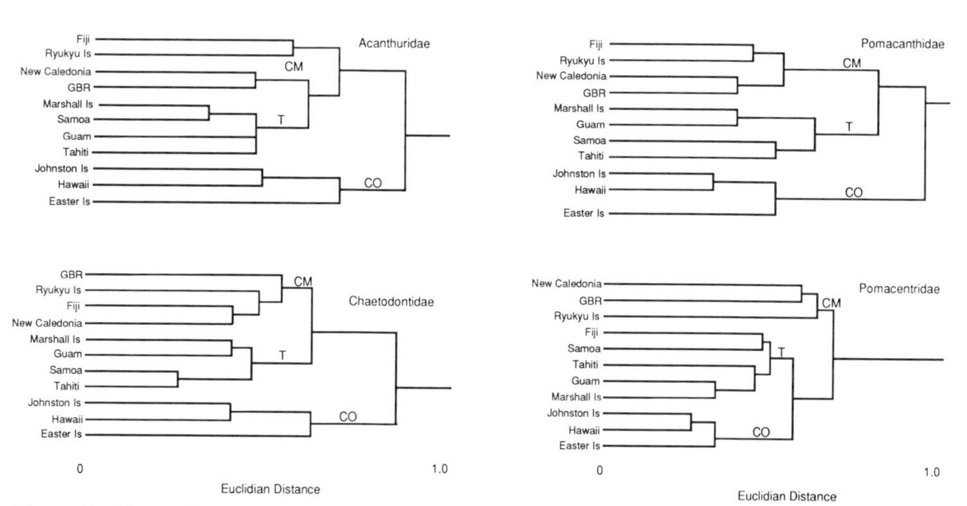

Figure 4 Cluster diagrams showing affiliations between sites in the Pacific Ocean, based on the presence/absence of shared species in the Acanthuridae, Chaetodontidae, Pomacanthidae, and Pomacentridae. The clustering routine used is based on farthest-neighbor distances and Euclidean distances as a measure of dissimilarity. CM = continental margin sites; T = transitional sites; CO = central oceanic sites, all as previously identified based on analysis of geographic patterns of species richness among coral reef fish families in general (Fig. 4).

among families. The four continental margin sites form a distinct cluster in the pomacanthids and chaetodontids, but Fiji is excluded and linked more closely to the offshore sites in the pomacentrids, and both New Caledonia and the Great Barrier Reef cluster, albeit distantly, more with the offshore sites than they do with the two other continental margin sites in the acanthurids. This inconsistency among these families parallels the shifting of sites among the two subgroups in the family-level analysis, when subjected to different clustering paradigms, and presumably reflects the same "fuzziness" of the boundary between the two faunal regions.

The nature of the differences between geographic groups is shown by the distributions of the number of species in each family. Compared with the Great Barrier Reef—an arbitrary standard for the Indo-nonoceanic Pacific group—the distribution of species richness across families is quite different for a New World fish assemblage and only broadly similar for a site in the Central Oceanic group (Fig. 5). Endemic families have little effect on these differences; excluding endemics from the analyses has almost no effect on the comparison between the Great Barrier Reef and a representative New World site (the Florida Keys) (rs = 0.56 and 0.58, with and without endemic families, respectively) and improves only slightly the comparison with the representative Central Oceanic site (Hawaii) (rs = 0.72 and 0.84).

Rather, differences in the species richness of the families common to all sites separate the geographic groups. Only two families—pomacentrids and labrids—rank in the ten richest families at all sites (Table 2). Otherwise, the three major geographic groups have little in common. The three New World sites (not including the Galapagos, see below) are consistently rich in species of serranids, gobies, clinids, haemulids, and lutjanids. There are no equivalent, consistently speciose families in the Indo-Pacific (other than pomacentrids and labrids). Continental margin reefs and those in the Central Oceanic group are dominated by wholly different sets of families. At continental margin sites, serranids, gobies, blennies, chaetodontids, and apogonids are consistently rich in species. None of these consistently ranks in the top ten at the Pacific Central Oceanic sites, which are dominated consistently by muraenids, acanthurids, and holocentrids.

The offshore subgroup of the Indo-nonoceanic Pacific group is taxonomically and geographically intermediate between continental margin and Central Oceanic assemblages, throughout which there is an overlapping of the domi-

Figure 5 The distribution of species richness per family for sites in New World, Central Oceanic, Indo-Pacific transitional, and the Indo-Pacific continental margin subgroup assemblages. The ranking of families at all sites follows that of the Great Barrier Reef. Note that the vertical axes have different scales.

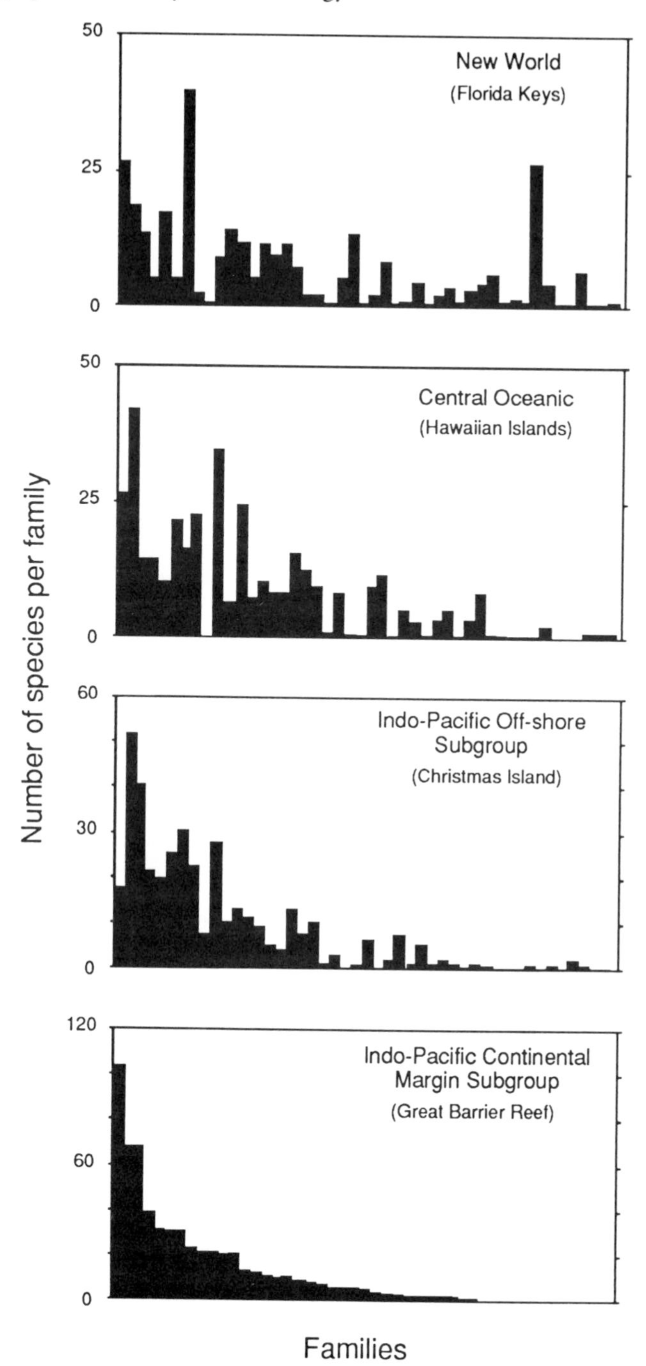
50
25
0
New World
(Florida Keys)
50
25
0
Central Oceanic
(Hawaiian Islands)
60
30
0
Indo-Pacific Off-shore
Subgroup
(Christmas Island)
120
60
0
Indo-Pacific Continental
Margin Subgroup
(Great Barrier Reef)
Number of species per family
Families

Table 2 List of the Ten Most Species-Rich Families at Each Site Used in the Geographical Analysis of the Structure of Reef Fish Assemblages[a]

Site	Rank in species richness									
	1	2	3	4	5	6	7	8	9	10
Ascension Island	Mu	La	Bo	Po	Se	Ba	Bl	Sp	D	(Ac = Ho = Pa)
Bahamas	Cl	Go	Se	Ap	La	Po	Lu	Sc	Ha	Mu
Florida Keys	Se	Go	Cl	La	Ap	Ha	Sc	Po	Sy	Lu
Sea of Cortez	Cl	Go	Se	Bo	Lu	Mu	La	Ha	Po	Gb
Galapagos Islands	Se	La	Po	Ha	Mu	Go	Ac	Lu	Ho	Bl
Easter Island	La	Mu	Ho	Go	Ml	Bl	D	Sp	Po	Ac
Johnston Island	Mu	La	Ac	Ch	Ho	Sp	Ml	Po	Lu	Se
Hawaiian Islands	La	Mu	Go	Sp	Ac	Ch	Se	Ho	Po	Bl
Tahiti	La	Go	Mu	Se	Po	Ac	Ch	Bl	Ap	Sc
Samoa	Go	La	Mu	Bl	Po	Se	Ap	Ac	Ho	Ch
Marshall Islands	La	Go	Po	Se	Mu	Ac	Bl	Ch	Ap	Sc
Guam	Go	La	Po	Mu	Bl	Se	Ap	Ac	Ch	Lu
Fiji	Po	La	Ch	Se	Bl	Ap	Go	Mu	La	Ac
New Caledonia	La	Go	Se	Po	Ap	Lu	Bl	Sp	Ch	Mu
Great Barrier Reef	Go	Po	La	Bl	Ap	Ch	Se	Ac	Ps	Mu
Ryukyu Islands	Po	La	Go	Bl	Ch	Se	Ap	Ac	Sc	Pa
Western Australia	Go	La	Ap	Po	Se	Bl	Ch	Lu	T	Sy
Christmas Island	La	Po	Se	Mu	Ch	Ac	Bl	Ap	Go	Ho
Cocos (Keeling)	La	Go	Po	Ap	Se	Mu	Ch	Ac	Bl	Ho
Chagos Arch.	Go	La	Po	Mu	Ap	Se	Ac	Ho	Bl	Sp
Red Sea	Go	La	Bl	Sp	Ap	Po	Se	Sy	Lu	Ch

[a] Data sources are given in Fig. 2. Key to families: Ac = Acanthuridae; Ap = Apogonidae; Ba = Balistidae; Bl = Blenniidae; Bo = Bothidae; Ch = Chaetodontidae; Cl = Clinidae; D = Diodontidae; Gb = Gobiesocidae; Go = Gobiidae; Ha = Haemulidae; Ho = Holocentridae; La = Labridae; Lu = Lutjanidae; Ml = Mullidae; Mu = Muraenidae; Pa = Pomacanthidae; Po = Pomacentridae; Ps = Pseudochromid-like fishes; Sc = Scaridae; Se = Serranidae; Sp = Scorpaenidae; Sy = Syngnathidae; T = Tetraodontidae.

nant families and a gradual shift from one set of dominants to the other (Fig. 6). The ten most species-rich families in Tahiti, Samoa, Guam, and the Marshall Islands (sites 10,14,15, and 16) include both the five families that are abundant along continental margin sites and two of the three from the Central Oceanic group (acanthurids and muraenids; holocentrids just miss, ranking tenth at Samoa and eleventh at Tahiti, the Marshall Islands, and Guam). A similar melding of geographically adjacent faunas is evident at the Galapagos (site 20). Its dominant families include both the three Central Oceanic dominants and four of the five families that are species-rich on New World reefs (the missing family is the Clinidae).

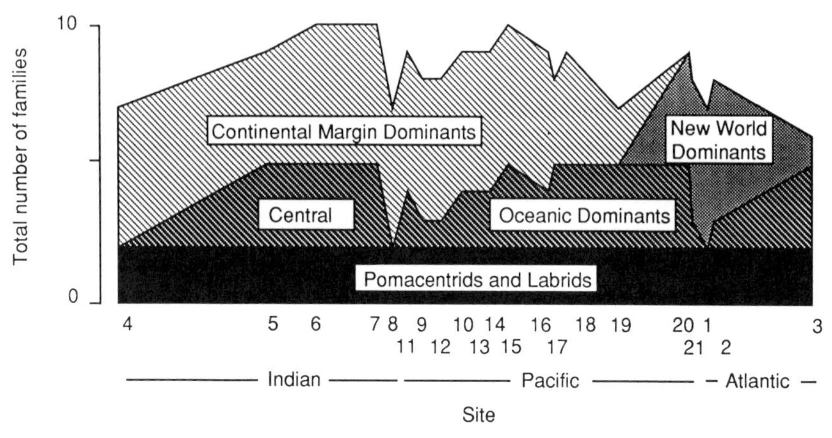

Figure 6 The proportion of the ten most species-rich families constituted by the New World, Indo-Pacific continental margin, and Central Oceanic "dominant" families at sites distributed circumtropically. Sites are identified and arrayed as in Fig. 3. Pomacentrid and labrids are species-rich everywhere. However, Central Oceanic dominant families consistently decline in importance along continental margins (in the western Indian Ocean, eastern Indian/western Pacific rim, and both the eastern Pacific and western Atlantic), whereas continental margin dominant families, in both the Indo-Pacific and New World, decline in relative importance toward the center of all three oceans.

Finally, the geographic patterning of species-rich families in the Indian Ocean is similar to that in the western and central Pacific. Sites along the continental margin (northwest Australia and the Red Sea) have the same five consistantly dominant families as similar sites in the western Pacific, as well as one other family (Lutjanide) (Table 2), whereas the three offshore sites (Christmas Island, Chagos Archipelago, and Cocos-Keeling) (sites 5–7) show an overlapping of continental margin and oceanic families similar to that of offshore sites in the Pacific (Fig. 6).

B. Reproductive Strategies

Two spawning modes predominate in reef-associated fishes: pelagic-spawning, in which buoyant, usually spherical eggs are shed directly into the water column, where they disperse; and demersal-spawning, in which adhesive eggs, heavier than seawater, are usually tended by one or both parents until hatching into planktonic larvae. Demersal-spawning species can be further divided into two rather different groups: "balistid-type" species (mainly balistids, tetraodontids, and siganids), which tend eggs for only a short period, if at all, before they hatch quickly into yolk-sac larvae at roughly the same stage of development as larvae hatched from pelagic eggs; and "normal demersal-spawning" species (all other demersal-spawning families), which incubate and tend the eggs for a relatively long time and whose larvae are large and well developed at hatching.

The geographic affiliations among reef fish assemblages evident within and among families is reflected in the distribution of the principal spawning modes. The three families that are consistently species-rich at Central Oceanic sites are all pelagic-spawners, whereas three of the five along continental margins are normal demersal-spawners. Reflecting this, the relative abundance of demersal-spawning fish species declines significantly ($rs = 0.83, p < 0.01$) west to east across the Pacific until reefs in the New World group are reached (Fig. 7). Geographic variation in the proportion of demersal-spawning in the Indian Ocean varies inconsistently around a mean close to that for Pacific sites in the offshore subgroup.

However, the shift in the proportions of pelagic- and demersal-spawning species across the Pacific involves only the normal demersal-spawners. The proportion of balistid-type spawners varies throughout the Indo-Pacific about a mean of around 7% of the total number of species present. By comparison, the proportion of normal demersal-spawning species in the assemblage varies from about 35 to 40% along the western continental margin of the Pacific to 10–15% in Oceania.

The relative proportions of the three spawning modes differ again at New World sites (Fig. 7). The proportion of normal demersal-spawners is similar to that of continental margin sites in the Indo-Pacific, and hence is higher than characteristic of offshore and Central Oceanic sites (including Ascension Island). However, balistid-type spawners are very poorly represented in the New World; the proportion averages only about half that of Pacific sites in general. This difference reflects more than just the absence of siganids in the New World. If the comparison is restricted only to the balistids and tetraodontids, differences between New World and Indo-Pacific reefs are still significant ($p < 0.01$, Mann–Whitney U test) and of about the same order. Again, the Galapagos appears to be intermediate between Central Oceanic and New World assemblages.

C. Synthesis and Comparison with Other Proposed Schemes for the Geographic Classification of Reef Sites

To summarize these sections, the assemblage of fishes associated with coral reefs shows a high level of global similarity at the family level. Nonetheless, there are at least four statistically definable regional assemblages: (1) a New World assemblage (western Atlantic-eastern Pacific) that is intermediate in terms of species richness between continental margin and oceanic reefs in the Pacific, has a high proportion of normal demersal-spawning species and a low

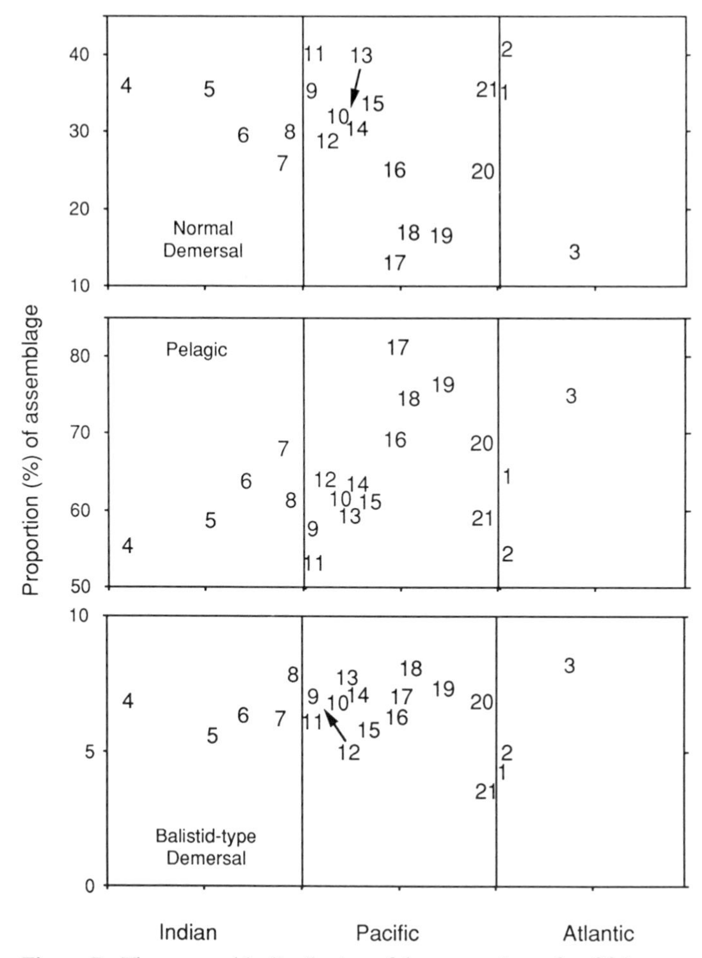

Figure 7 The geographic distribution of the proportions of reef fish assemblages in each of the three major spawning modes (see text). Site identification and horizontal array of sites as in Fig. 3. Note that the vertical axes are different.

proportion of balistid-type spawners, and is dominated by families many of which are not species-rich on Indo-Pacific reefs; (2) a Central Oceanic assemblage, with sites in both the Atlantic and Pacific, that is poor in species and dominated by pelagic-spawning families; (3) an Indo-Pacific continental margin assemblage, rich in species, with a high proportion of both balistid-type and normal demersal-spawning species; and (4) a transitional, offshore zone that, in the Pacific, reaches at least from Guam and Samoa to Tahiti and, in the Indian Ocean, from Christmas Island to the Chagos Archipelago. The dividing line between the last two is less well defined than the major faunal divisions. The transitional zone includes some families that are species-rich on continental margin sites and and others that are rich at Central Oceanic sites; it is also intermediate between these sites in terms of the richness of the fauna and proportional representation of the spawning modes. The Galapagos Islands appear to be a similar transition point between the Central Oceanic assemblage and New World reefs.

Schemes for dividing the tropical reef biota into biogeographic units have been proposed since at least the mid-1800s (see Kay, 1980). With the noteworthy exception of Vermeij (1987), these divisions have not been based on ecological features, but on patterns of endemism, that is, the presence of species broadly distributed throughout a particular faunal province, but restricted to that province. Nonetheless, there are clear similarities between the scheme outlined here and those proposed by taxonomists. Every marine biogeographer, for example, distinguishes between a New World (Atlanto-eastern Pacific) province and an Indo-Pacific province. Most also distinguish between a Pacific central oceanic biota (usually focused around Hawaii) and an Indo-Malayan biota broadly distributed in the western Pacific (see Kay, 1980). These are the first and major splits between sites produced by my analysis. In most instances, however, biogeographic affiliations among sites based on my analysis differ in detail from those defined on the basis of endemism. For example, taxonomists justifiably consider Ascension Island to be part of the Atlantic province, on the basis of species shared with western and eastern Atlantic sites; at the levels of family structure and reproductive characteristics, however, it shows much greater affinities with isolated sites in the tropical Pacific.

The differences between groupings of sites based on ecology (to the extent that it is reflected in my analysis) and endemism are particularly clear in the Indo-Pacific. Numerous schemes have been proposed to divide the Indo-Pacific reef biota geographically. Early schemes are reviewed and compared by Kay (1980). The only major scheme proposed since Kay's review is Springer's (1982), who focused attention on the distinctive nature of the Pacific Plate biota. None of these schemes matches the "ecological" one I propose, though that outlined by Ekman (1953) comes close. Ekman provisionally identified three major regions in the Indo-Pacific: an Indo-Australian region stretching from tropical Japan to the Great Barrier Reef with strong affinities to continental reefs in the Indian Ocean; a Hawaiian region (including Johnston Island); and an "islands of the central Pacific" group. Ekman noted that the boundary of the Indo-Australian (Indo-Malayan) fauna was unclear, but might occur east of Fiji and Melanesia. This scheme is close to what I propose, with the codicil that the "Hawaiian group" also includes Easter Island.

My division of the Indo-Pacific also supports Springer's (1982) hypothesis that the western margin of the Pacific Plate (approximately the Andesite Line) is a conspicuous boundary in the distributions of many fish groups and that there are affiliations between the Hawaiian Islands and southeast Oceania (if you include Easter Island). Springer's grouping of continental sites with island groups west of the Pacific Plate boundary (Fiji, New Caledonia) differs from the results of my analysis only in our treament of Guam. Springer considers it a marginal site off the Pacific Plate, whereas I find that its strongest affinities ecologically are with island sites on the Pacific Plate. We also differ on the relative strength of the link between Hawaii and other Pacific Plate sites. Although clearly distinguishing them and noting Hawaii's affinities with southeast Oceania, Springer nonetheless felt that Hawaii was an integral part of the Pacific Plate biota. In contrast, my analysis suggests that the sharpest faunal break is that between the Central Oceanic group, including Hawaii, and the transitional sites scattered elsewhere about the Pacific Plate.

IV. Life History Features and the Evolution of Geographic Variation in Indo-Pacific Reef Fish Communities

The division of the global reef biota into continental margin assemblages (both Indo-west Pacific and New World), Central Oceanic assemblages, and transitional sites places the schemes proposed by Ekman and Springer for the Pacific into a wider context. It also implies that factors more general than a particular geological fault line running down the western Pacific are involved in the evolution and maintenance of this biogeographic structure. One factor that has not previously been considered in this context is geographic variation in the representation of the spawning modes.

Among reef fishes, the demersal- and pelagic-spawning modes differ in several respects (Barlow, 1981; Thresher, 1984), as summarized in Table 3. Two of these differences could logically bear directly on the distributions of the spawning modes. First, the larvae of normal demersal-spawning fish families typically occur closer to shore than those hatched from pelagic eggs, off Hawaii (Leis and Miller, 1976; Leis, 1982a), the Great Barrier Reef (Leis and Goldman, 1984) and the Australian Northwest Shelf (Young *et al.*, 1986; see Leis, Chapter 8), and also in temperate waters (Richardson and Pearcy, 1977; Furlani *et al.*, 1991). Where larvae of balistid-type spawners occur along an inshore–offshore axis is not yet clear (Leis, 1986b), as they are generally caught only in small numbers. However, the large prejuveniles of at least some balistids commonly occur well offshore. Second, the larvae of normal demersal-spawning species tend to remain in the plankton for less time

Table 3 Reported Characteristic Differences Between Reef-Associated Species in Each of the Three Major Spawning Modes[a]

	Normal demersal	Balistid-type demersal	Pelagic	Reference
Egg characteristics	Adhesive, heavier than seawater	Small, adhesive, heavier than seawater	Buoyant	Barlow (1981), Thresher (1984)
Incubation period	>48 hours, often 3–7 days	<12 hours	<48 hours	Thresher (1984)
Parental care	Male or biparental until hatching	None (Tetradontids, Siganids) or female only (Balistids)	None	Barlow (1981), Thresher (1984)
Stage of larval development at hatching	Small yolk, open gut, pigmented eyes, active swimmer	Large yolk, pro-larva, apparently passive	Large yolk, pro-larva, apparently passive	Leis and Rennis (1983)
Adult size	Typically <15 cm TL	Varies widely	Varies widely, but predominantly >15 cm	Barlow (1981), Thresher (1984)
Spawning time (diel)	Daytime, early morning peak, hatch at dusk	Dawn, eggs hatch at dusk	Tidally entrained or dusk	Thresher (1984)
Larval distribution	Predominantly inshore	Predominantly offshore, often neustonic	Predominantly offshore	See text
Larval planktonic duration	Typically <25 days	Typically 20–50 days	Typically 20–40 days	See text

[a] Leis and Goldman (1984) also suggest that demersal-spawners typically spawn in spring and summer, and pelagic-spawners typically in summer and autumn. This difference is not obviously supported by the literature (e.g., Young *et al.*, 1986; Munro *et al.*, 1973; Erdman, 1977), nor by subsequent work by J. M. Leis (personal communication).

than the larvae of pelagic or balistid-type spawners (Fig. 8). Inter- and intraspecific differences in planktonic duration within spawning modes are considerable, but overall the normal demersal-spawning species tend to average between 15 and 25 days, whereas both pelagic-spawning and balistid-type species average about 10 days longer, most falling in the range of 20 to 40 days. Planktonic durations in excess of 50 days are commonly reported for pelagic and balistid-type demersal-spawners (Brothers and Thresher, 1985; Stroud *et al.,* 1989), but have not been reported for any normal demersal-spawning species.

These data imply that the likelihood of long-range larval dispersal varies between spawning modes. The short planktonic durations and inshore development of normal demersal-spawners imply relatively short average dispersal distances, whereas the long planktonic duration and offshore development of many pelagic and balistid-type species suggest high probabilities of extensive dispersal. I emphasize that these are logical and seemingly reasonable, but not inevitable, conclusions; there are no direct data on the mean or maximum dispersal distances of any species of reef fish (other than the few that lack a planktonic larval stage). They also are, by intent, generalizations about the spawning modes. Both larval distributions and planktonic durations vary widely within each mode, and comparable variability in the prospects of long-distance dispersal is inevitable.

Three general hypotheses, summarized by Springer (1982) and Vermeij (1987), have been proposed to explain the biogeography of the Indo-Pacific. (1) "Habitat-based" hypothesis: ecological differences between sites dictate the nature of the species assemblage present at each (see, e.g., Leis, 1986b; Vermeij, 1987). (2) "Extinction" hypothesis: the decline in the number of species and families from west to east across the Pacific is the result of partial extinction in Oceania of the Tethyan reef fauna, which is now largely restricted to the Philippine/Indo-Malayan region (Ekman, 1953). (3) "Dispersal" hypothesis: regional variation derives from differences in the abilities of species to disperse from the western continental margin (or thereabouts), which has acted as a center of speciation and colonization.

"Habitat-based" hypotheses are supported by, on the one hand, documented associations between particular habitat elements and characteristics of fish communities (e.g., Carpenter *et al.,* 1981; Bell and Galzin, 1984; Leis, 1986b) and, on the other, by taxonomic and trophic features of some reef-associated invertebrate taxa that vary across the Pacific in a pattern similar to the distributional patterns of the fishes (Kay, 1980; Vermeij, 1987). Habitat-based hypotheses to account for geographic patterns of distribution are, however, unsatisfying in two respects. First, it is usually impossible to determine whether the differences in the invertebrate taxa underlie variations in the fish community or are parallel responses to biogeographical or historical factors. Experimental studies on habitat manipulation could distinguish be-

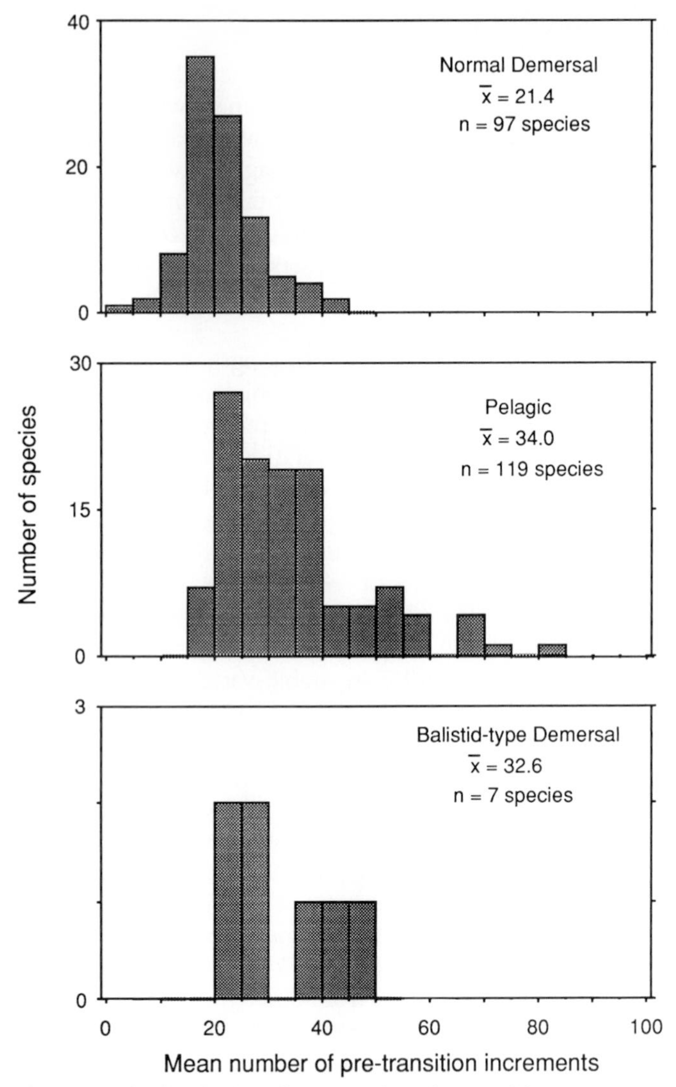

Figure 8 The distribution of mean number of pretransition increments in the otoliths of species of Pacific reef fishes in each of the three major spawning modes. The data are based primarily on personal observations and the work of E. Brothers. The family Gobiidae is poorly represented in the normal demersal-spawners and would be likely to broaden the peak slightly to the right. Planktonic larval durations (in days) are assumed to be equivalent to pretransition counts for demersal-spawning species, whereas they are assumed to be 2–3 days longer than pretransition counts in the pelagic and balistid-type demersal-spawning species, because of the interval between hatching and the development of the otoliths in pro-larvae in each group (see Thresher *et al.*, 1989, and Thresher and Brothers, 1985, for

tween these alternatives, but they are logistically difficult on the spatial and temporal scales required. Second, falsifying the general habitat-based hypotheses is impossible; there is always the possibility of some, as yet unidentified, habitat feature that dictates shifts in the composition of fish assemblages. For these reasons, habitat-based hypotheses cannot be ruled out, but also may not have much explanatory power. Parenthetically, the success of at least some reef-associated fishes (Randall, 1987) and invertebrates (Carlton, 1987) introduced to Hawaii does not appear consistent with habitat-based hypotheses. Similarly, the presence of a healthy expatriate population in Fiji of a species of *Omobranchus*, otherwise found only in continental margin regions (Springer and Gomon, 1975; Thresher, 1991), demonstrates that Fiji meets the species' ecological requirements, and that some other factors account for its previous absence at the site. Finally, there is no apparent association between the groupings of sites in the Pacific described earlier and any conspicuous habitat feature. Transitional sites include both high and low islands, and Central Oceanic sites include both those with healthy coral populations and those lacking corals altogether.

The "extinction" hypothesis is discussed by Springer (1982) with regard to the shorefishes and by Vermeij (1986) with regard to marine invertebrates. The hypothesis has been tested by examining the molluscan fossil record, which is fairly extensive for the Pacific. Springer found little evidence of the large-scale extinctions of the central Pacific biota required by the hypothesis and suggested that the extinctions that had occurred were local and could be ascribed to events specific to the sites [such as differential effects of sealevel changes on high and low islands (see Grigg and Epp, 1989)]. Vermeij, however, concluded that the fossil records of several molluscan taxa were consistent with the extinction hypothesis and that within the Pacific, high islands and the western continental margin, in particular, have served as refugia for taxa now extinct elsewhere. He suggests that high levels of coastal productivity near each buffered the effects of environmental stress on larval and adult survival.

The "dispersal" hypothesis, and particularly the role of irregular dispersal events in determining the biogeography of the Indo-Pacific, has typically been invoked with particular reference to sites remote from the centers of faunal diversity, such as the Hawaiian and Easter islands (Gosline, 1968; Scheltema, 1986) and the eastern Pacific (Rosenblatt, 1967b; Zinmeister and Emerson, 1979; Brothers and Thresher, 1985). Several studies report that species whose

---

data on demersal- and pelagic-spawning families, respectively). Differences among the spawning modes are significant at $p < 0.001$ (Kruskal–Wallis test). Note that the vertical axes are different.

distributions include these sites have unusually long planktonic durations (e.g., Thorson, 1961; Brothers and Thresher, 1985; Richmond, 1987), which seems to lend credence to the dispersal hypothesis. At the same time, several mechanisms have been proposed to account for high rates of speciation in the western Pacific [see Potts, 1983; McManus, 1985; Rosen (cited in Potts, 1985); see also Kohn, 1985, for evidence of high rates of speciation in peripheral island groups]. All the proposed mechanisms combine some element of geographic isolation due to vicariant events (usually resulting from changes in sea level) with subsequent dispersal of newly evolved taxa. On a broader scale, Springer (1982) has argued that vicariant events have been important in structuring the western and central Pacific fish fauna and underlie the distinctive nature of the Pacific Plate fauna.

The differences between the spawning modes in distribution and likelihood of long-distance dispersal are superficially consistent with the "dispersal" hypothesis. Specifically, the hypothesis is that demersal-spawning species are relatively uncommon at Central Oceanic sites because they are less likely to disperse from the western continental region than either the pelagic or balistid-type demersal-spawning fishes abundant at those sites. The exceptions appear to prove the rule. Not all the species-rich families along continental margins are demersal-spawners. Most of those that produce pelagic eggs, however, develop inshore (Leis, 1986b) and either have relatively short planktonic durations (lutjanids) or include large numbers of species with short durations (chaetodontids, serranids) (Brothers and Thresher, 1985). A relationship between apparent dispersal abilities and distribution also holds within families: species broadly distributed across the Pacific have longer mean planktonic durations (and hence presumably higher likelihood of long-distance dispersal) than those found only in the western Pacific (see the review by Thresher and Brothers, 1989). No broadly distributed species that occurs in the Hawaiian Islands ("Hawaiian-inclusive" species), for example, has a mean planktonic duration less than 35–40 days (Brothers and Thresher, 1985). This is longer than the mean duration of the planktonic larval stages of nearly 60% of the non-Hawaiian-inclusive species studied by Brothers and Thresher. On that basis, we suggested that the larvae of most extant western Pacific fish species may not normally be capable of reaching the Hawaiian Islands. This conclusion, and those of subsequent studies, I have in the past interpreted as providing strong indirect evidence for the "dispersal" hypothesis (Thresher *et al.*, 1989).

This conclusion is premature. The problem is essentially philosophical. All three hypotheses can be fitted to any given set of distributions by the addition of appropriate ancillary hypotheses. For example, as currently articulated, extinction hypotheses do not specifically predict a community bias at Central Oceanic sites toward pelagic-spawners or species with longer than average planktonic durations. These observations can be incorporated into an extinction framework, however, by assuming that species with broadly dispersed larvae are less likely to become extinct than those with limited dispersal ranges. Such an assumption is even supported conceptually by studies on marine invertebrates (Hansen, 1978; Jablonski and Lutz, 1983; Jablonski, 1986; see, however, Bouchet, 1981). Alternatively, one can turn the argument around and argue that the shift in the relative dominance of the major spawning modes is due to a higher probability of extinction of these pelagic-spawning families along continental margins when their offshore developing larvae were "trapped" in isolated regions by sea-level changes during glacial periods. Plausible extinction hypotheses can be framed in either context; neither is easily falsified; and both can lead logically to a predominance of pelagic-spawners in Oceania and demersal-spawners at continental sites. Much the same can be done to the habitat-based hypotheses, such as Leis' (1986b) quite logical extension of this set of hypotheses into the pelagic larval realm; such modifications render the core hypothesis even more difficult to tackle, even if it could be falsified. It is difficult to specify direct tests of alternative hypotheses to account for the distribution of the spawning modes or the dominance of some families along continental margins and others at central oceanic sites.

Still, some limits can be set on the range of workable hypotheses. The biogeographic structure of Pacific reef fish assemblages is the consequence of two trends: a west-to-east decline in the species richness of continental margin dominant families (as well as the ubiquitous pomacentrids and labrids) and a contrasting increase in the absolute species richness of Central Oceanic dominants (Fig. 9). This appears to rule out simple versions of at least some hypotheses, that is, both groups speciating along continental margins, but dispersing east at different rates. One must either invoke separate mechanisms to account for the distribution of each set of families or propose a single mechanism that has directly opposite effects on the two groups. The latter is certainly possible. Sea-level changes, for example, could have simultaneously enhanced speciation of inshore developing fishes along continental margins while locally obliterating the predominantly pelagic-spawning groups that require offshore development. Dispersal abilities seems to be a key to understanding the different geographic distributions of the spawning modes and the biogeographic structure of the Indo-Pacific, but they must be combined with some extinction or speciation mechanism that affects continental margin and central oceanic families differently.

As alluded to in the foregoing, one mechanism commonly invoked is sea-level changes associated with tectonic activity and glaciation [e.g., Potts, 1983; McManus, 1985; Rosen (cited in Potts, 1985); Springer and Williams, 1990)]. Rosenblatt (1963) provides an alternative hypothesis. He suggests that the richness of demersal-spawning taxa along continental margins can be

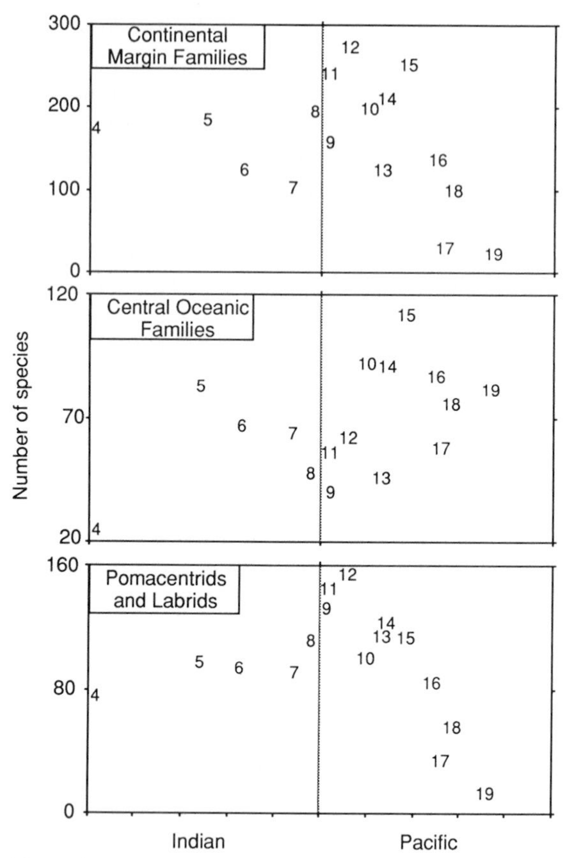

Figure 9 Geographic distribution across the Indo-Pacific in the absolute number of species in continental margin dominant and Central Oceanic dominant families (see text), as well as for the pomacentrids and labrids. Sites are identified and arrayed as in Fig. 3.

attributed to their small size, consequent low individual mobility and low fecundity, and, hence, low rates of genetic exchange between what are often small and physically isolated coastal habitat patches. Rosenblatt's hypothesis is quite consistent with the inshore development and short planktonic durations of the larvae of most demersal-spawning taxa, but it is difficult to see how this mechanism could result in a dearth of pelagic-spawning species in the same environments. By comparison, changes in coastal topography, sea-surface temperatures, circulation patterns (Overpeck *et al.*, 1989), and oceanic productivity (Berger *et al.*, 1989) associated with sea-level changes in general, and changes associated with glaciation in particular, would have far more dramatic effects on larvae in continental regions than in Oceania (e.g., Grigg, 1988) and could well account for differential speciation and extinction by the two spawning modes.

However, hypotheses based on the effects of sea-level changes have two weaknesses. First, as pointed out by V. G. Springer (personal communication), it is difficult to reconcile a hypothesis that critically depends on speciation near, and differential dispersal from, continental margins with the presence in the central Pacific of small, demersal-spawning, usually endemic species with inshore developing larvae. These are the species the dispersal hypothesis says should not be there. Some of these small blennies, damselfishes, apogonids, and the like are in genera with planktonic durations longer than is typical for these families (Thresher *et al.*, 1989), but others are not and cannot be fitted easily into the dispersal and extinction hypotheses. A suggestion that they are remnants of some preexisting fauna that has somehow avoided extinction begs the question of how they got into the central Pacific in the first place. Second, hypotheses based on sea-level changes and their effect on continental margins ignore other, perhaps episodic, variations in oceanic conditions that could also determine speciation, extinction, and dispersal patterns. Oceanographic conditions vary on a range of time scales, from diurnal and seasonal through short-term interannual (e.g., El Niño–Southern Oscillation Events) to those spanning thousands and millions of years (e.g., Milankovich cycles), the consequences of which are not well worked out.

One such cycle that I think deserves wider attention is described by Fischer and Arthur (1977) and, independently, by Grigg (1988). Both note that oceanographic conditions in the Pacific have changed markedly over the last 20–30 million years, from a relatively sluggish circulation in the early Tertiary, dominated by east-to-west transport, to a more rapid circulation with strongly developed gyral currents. Grigg suggests that the recent colonization of the Hawaiian Islands by corals was the result of these strengthening currents and consequent increased likelihood of long-range dispersal by coral larvae. Fischer and Arthur propose a more general scheme. They note that the pelagic biota, from phytoplankton to "mega-carnivores," varies in diversity over about a 30-million-year cycle, and that this cycle corresponds strongly with variations in the strength of oceanic current regimes. During periods of apparently strong gyral currents, biotic diversity is low ("oligotaxic" periods); when currents are relatively sluggish, diversity is high ("polytaxic" periods). They suggest that the link between current regimes and biotic diversity is mediated by larval ecology. During strong current periods, long-range transport of larvae is common, previously allopatric species mix widely, and diversity declines as a result of competition. As currents become more sluggish, however, rates of larval transport are reduced, local isolation occurs, and rates of speciation increase.

Aside from Grigg's (1988) work, this scenario has not been applied to the tropical reef biota. If it is correct, then one can envisage the biotic history of the Pacific as constituting cycles of vigorous speciation throughout the region followed by waves of predominantly west-to-east dispersal and mixing as oceanic gyral currents accelerate, with communities thereafter slowly sorting themselves out and approaching stability prior to the onset of relatively sluggish currents and a polytaxic period. This hypothesis has three points to recommend it, aside from its fitting nicely with the history of corals at Hawaii. First, it relates directly to dispersal abilities, and hence would have markedly different consequences for demersal- and pelagic-spawning taxa. Second, it can account for the apparently anomalous "poor dispersal" species in Oceania; they are the surviving and now-isolated remnants of previous "dispersal waves" (a hypothesis, by the way, that should be very amenable to direct testing by biochemical genetic techniques). And third, it might also account for the marked overlapping of niches among reef fishes (see Sale, 1977). Fischer and Arthur (1977) suggest that we appear to be entering an oligotaxic period, that is, currents have recently (geologically speaking) accelerated, previously isolated faunas have mixed (hence the correlation between planktonic duration and breadth of distribution west-to-east across the Pacific?), and the sorting-out process is just starting. Recent mixing of allopatric faunas could account for the diversity of ecologically similar fish species on western Pacific reefs, a diversity that appears to be inconsistent with theories of competitive exclusion.

V. Ecological Consequences of Regional Variation in Assemblage Structure

A. The Indo-Pacific

The shift in the dominance of the spawning modes across the Pacific and the division of the Indo-Pacific into clusters of sites with similar assemblage structures could have important consequences for ecological studies in the region, as the shift in dominance is likely to be accompanied by two changes in community characteristics. The first is an increased likelihood of long-distance dispersal, as discussed earlier. The second is that the mean size of fishes present at a site is likely to increase west to east across the western and central Pacific. The reason for this is that pelagic-spawning fishes are characteristically larger than those that produce demersal eggs (Barlow, 1981; Thresher, 1984), a trend that may be common among marine organisms (e.g., Menge, 1975; Jablonski and Lutz, 1983). Hence, as the proportion of demersal-spawners declines across the Pacific, the proportion of small-bodied species in the assemblage is also likely to decline. Testing this hypothesis statistically is difficult, since none of the checklists provides data on body sizes. However, it is consistent with my own observations at, for example, the Great Barrier Reef and Hawaii.

Body size is an important feature in life histories, not only for fishes but for most organisms (Blueweiss *et al.,* 1978; Wootton, 1984; Saether, 1988). Larger fishes produce more eggs per spawning than smaller fishes, irrespective of spawning mode (see Thresher, 1984, for data on reef fishes), and, depending on spawning frequencies and length of reproductive life, may be individually more fecund. Large animals also live longer and mature more slowly than small ones (Saether, 1988) (Fig. 10). Correlations of correlations provide weak tests of hypotheses, but constitute a reasonable basis for developing them. Hence, I hypothesize that as the mean size of fishes in an assemblage

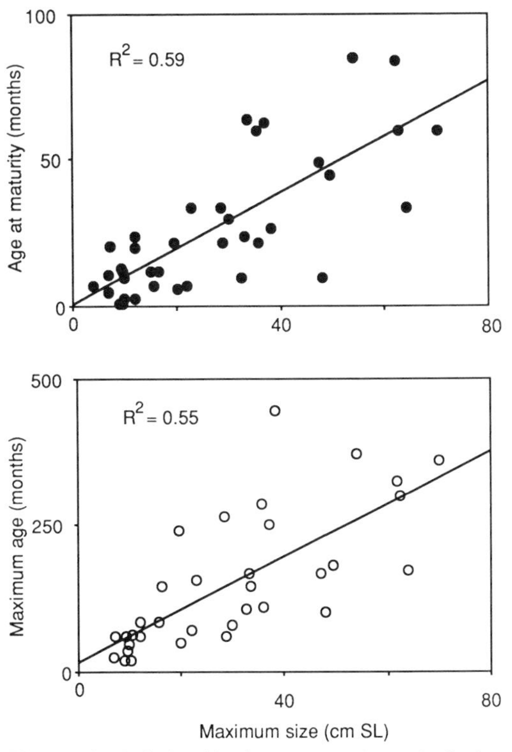

Figure 10 Relationships between maximum body size (cm Standard Length) and minimum age at sexual maturity (of females, if the sexes differ) and maximum ages for 38 species in 15 families of reef-associated fishes. The data are gleaned from a scattered literature; most are cited by Thresher (1984). Both correlations are significant at $p < 0.001$.

increases west to east across the Pacific there are parallel increases in mean fecundity, mean and maximum ages, and age at first reproduction.

This hypothesis could be tested directly, but data to do so have not yet been assembled. If it is correct, however, then it implies that the population and community ecology of reef fishes varies across the Pacific in two important respects. First, because of regional variations in mean fecundity and dispersal capabilities, the rates of exchange of offspring between habitat patches ("inter-reefal connectivity") should increase west to east across the Pacific, all else being equal. Second, because the fishes live longer and mature more slowly, populations and communities in the Central Pacific assemblage should react more slowly (or not at all) to variance in settlement (e.g., variations in year-class strength) and should be less affected by extremes in settlement levels (the "storage effect" of Warner and Chesson, 1985; see also Warner and Hughes, 1988) than populations and communities in the western Pacific. It would follow, then, that populations and communities in the central Pacific are more resilient, more stable, and less prone to local and global extinction (see Vermeij, 1986; Jablonski, 1986; Pimm *et al.,* 1988) than those in the western Pacific, but also less able to respond quickly to locally favorable conditions. By analogy, the central Pacific is inhabited mainly by the piscine equivalents of elephants and lions, whereas the western Pacific also has its complement of lemmings and shrews. Both assemblages will have long-lived, slow-maturing species, found in relatively stable, genetically well-mixed populations, but western Pacific assemblages have, as well, large numbers of species with generation times that may often be as short as a few months (e.g., Eyberg, 1984; Thresher *et al.,* 1986), populations of which are likely to be genetically relatively discrete, temporally unstable, and, perhaps, prone to local extinction.

As yet, there are no data to test these predictions directly. Sale (1980a) noted that three types of field data were available to assess the temporal stability and persistence of reef fish assemblages: studies of colonization of artificial and natural reefs, monitoring of select species in natural assemblages, and monitoring of entire assemblages. He further noted that the available data were too few to deduce patterns of stability. Little has changed since 1980.

Many studies have examined the relative stability of fish assemblages on small habitat patches, either artificial or natural. However, these provide a poor test of geographic variability in stability and persistence, for two reasons. First, they are rarely conducted for more than three or four years, which is not sufficient to assess variations in the numbers of large-bodied species, many of which live for twenty years or more. And second, the small spatial scales and the small habitat patches almost invariably examined in such studies bias them toward results based on the short-lived fishes that, being relatively small, dominate on such structures. For example, normal demersal-spawning species constitute 27% of the species studied by Walsh (1985) on artificial reefs in Hawaii, 41% of those monitored by Nolan (1975) at Enewetak Atoll, and a massive 80% of the 25 most abundant species on the patch reefs studied by Sale and Douglas (1984) on the Great Barrier Reef; such species constitute only 16, 31, and 40% of the fish assemblages in these areas, respectively. Consequently, studies of temporal stability on these small habitat units provide what may be a distorted indication of regional variations in community characteristics. The literature contains hints of the predicted regional differences in stability and persistence: "Temporal stability of the communities on the Kona reefs [Hawaii] was significantly greater than on the Great Barrier Reef" (Walsh, 1985) and "the similarity of the two collections [of a Hawaiian patch reef sampled at 11-year intervals by Brock *et al.*, 1979] . . . is considerably greater than that among our much smaller reefs [on the Great Barrier Reef]" (Sale and Douglas, 1984). Unfortunately, the problems arising from the effects of small-scale spatial heterogeneity, different sampling intervals, scales, and methods, and unrepresentative proportions of fishes in the two main spawning modes at the sites studied render regional comparisons based on studies to date of doubtful value. We need better-quality data on community dynamics before we can attempt regional comparisons within the Pacific.

Finally, the declining proportion of small-bodied, demersal-spawners across the Pacific could also underlie some interesting and previously noted shifts in trophic ecology and taxonomic diversity within groups. Bakus (1967), for example, suggested there were relatively more grazing herbivores on tidal flats in Oceania than in the western Pacific. This could be a predictable consequence of the regional differences in community composition discussed above if mobile, grazing herbivory correlates with large body size in reef fishes (which is likely), given the relative dearth of small, site-restricted herbivores, such as blennies and damselfishes, in Oceania to crop stocks of intertidal algae. Similarly, Anderson *et al.* (1981) noted an increase in the relative diversity of planktivorous chaetodontids west to east across the Pacific, paralleled by a decline in the numbers of planktivorous pomacentrids. In the context of the current discussion, I interpret this as ecological diversification of the relatively large chaetodontids in the relative absence of the small, demersal-spawning pomacentrids.

B. Interoceanic Differences: The Western Pacific versus the New World

Reef fish assemblages in the New World resemble those of continental margins in the Indo-Pacific, in that both are characterized by high proportions of small-bodied, demersal-spawning species. Hence it would be reasonable to expect that the dynamics of the two assemblages are similar and directly comparable. However, the assemblages in the two regions differ in two respects that may bear on their community and population ecology.

First, different families are species-rich in the two regions. In New World assemblages, there are numerous species in the demersal-spawning family Clinidae [all in what is treated by V. G. Springer (personal communication) as the families Labrisomidae and Chaenopsidae] and the pelagic-spawning families Haemulidae and Lutjanidae. None of these is characteristically species-rich on Pacific reefs. In contrast, blenniids, apogonids, and chaetodontids are abundant in the western Pacific, but are poorly represented in the New World. Pomacentrids, labrids, and an entire spawning mode—balistid-type spawners—are also relatively species-poor on New World reefs as compared with the western Pacific. To the extent that species in these families have characteristic ecologies, for example, clinids eat mainly small benthic invertebrates (Randall, 1967) whereas blennies are mainly herbivores (Sano *et al.*, 1984b), these differences in species richness could affect community characteristics.

Second, there are consistent "interoceanic" differences in life history features in at least some families of reef fishes. These differences, discussed by Thresher (1982, 1985b) and Victor (1986a), are reviewed and summarized by Thresher and Brothers (1989). While noting that sample sizes are small and that geographic variation is not well documented in any species, Thresher and Brothers draw several tentative conclusions. (1) Most demersal-spawning fishes in the western Atlantic have smaller eggs, smaller larvae at hatching, longer planktonic larval durations, and slower rates of larval growth than their relatives in the western Pacific. The pomacentrid genus *Stegastes* is a noteworthy exception. (2) Interspecific differences in early life history features are less between confamilial species in the western Atlantic than in the Indo-Pacific. (3) These differences appear to be limited entirely to normal demersal-spawning families; there is no indication of a comparable interoceanic difference in any pelagic-spawning fish family yet examined. The authors note, however, that there are also many fewer data on pelagic-spawners, and the power of statistical tests to detect small differences is slight. There are not enough data as yet to draw any conclusions about balistid-type demersal-spawners. (4) Early life history features covary strongly both between sites and between species within families (Fig. 11), which suggests they are not evolutionarily independent (for an example of similar covariance in a temperate fish species, see Sinclair and Tremblay, 1984). Hence, regional differences are likely to be the integrated effect of selective pressures operating on several elements of the life histories of the animals. (5) Eastern Pacific species tend to be reproductively similar to those in the Western Atlantic and hence have smaller eggs, etc., than their relatives in the western and central Pacific.

These interoceanic differences in early life history features may, Thresher

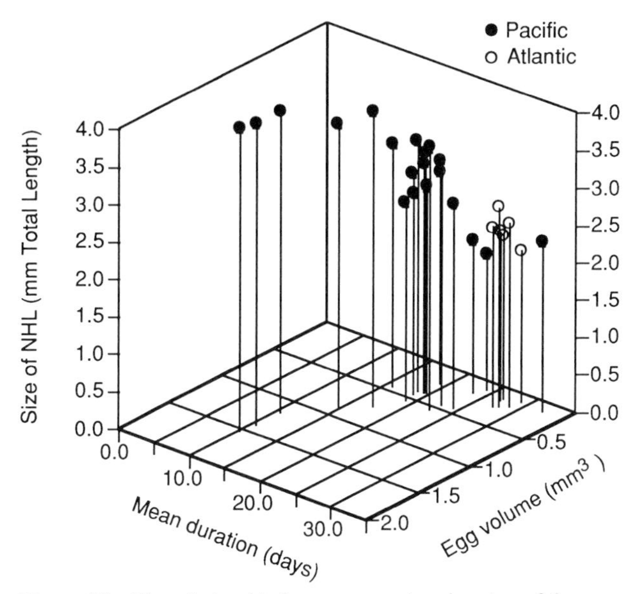

Figure 11 The relationship between egg size, duration of the planktonic larval stage, and the size of newly hatched larvae (NHL) among Pacific and western Atlantic species in the reef-associated family Pomacentridae. Points are mean values for each variable for each species. Pairwise correlations among the three reproductive parameters are all significant at $p < 0.01$.

and Brothers (1989) suggest, be the result of either local selection for extreme values of one parameter, such as long planktonic durations among Atlantic species, or regional differences in the optimal balance between fecundity and the allocation of resources to individual offspring. As yet, there are few data to test either hypothesis. The second, however, is consistent with evidence of higher mean levels of water column productivity in the western Atlantic (Birkeland, 1987; Wilkinson, 1987; see, however, Sinclair and Tremblay, 1984). Whatever underlies them, interoceanic differences in early life history features combine with perhaps slight taxonomic differences between assemblages to constitute a combination of community and reproductive features unique to New World reefs. Because the New World and continental Indo-Pacific assemblages both have a high proporton of small-bodied, demersal-spawning species, their dynamics should be similar. But these species differ reproductively between the two sites: New World species have relatively long planktonic durations and are more similar in that regard to the relatively few demersal-spawning species in Central Oceanic assemblages than they are to related species along Indo-Pacific continental margins. Consequently, reef fish assemblages in the New World should have the high turnover rates of Indo-Pacific continental margin assemblages, but also the high levels of exchange of offspring between habitat patches, the low genetic divergence between sites, and high resilience and low levels of local and global extinction of asssemblages in Central Oceanic assemblages.

Finally, other elements of the life history of demersal-spawning fishes may also differ consistently between New World and Indo-Pacific sites. Smaller egg size suggests that New World species (and probably those at Central Oceanic sites) produce clutches that, relative to body size, are larger than those of their Indo-Pacific continental margin counterparts. Among pomacentrids at least, the predominant mode of reproductive periodicity also appears to differ between the two regions (Table 4). Western Atlantic species, with one

Table 4 Reproductive Periodicity of Indo-Pacific and Western Atlantic Damselfishes (Pomacentridae)

Species	Location	Periodicity	Reference
Indo-Pacific			
Abudefduf sordidus	Hawaii	Irregular synchronized peaks	Stanton (1985)
Abudefduf abdominalis	Hawaii	Irregular synchronized peaks	May (1967)
Amphiprion bicinctus	Red Sea	Lunar	Fricke (1974)
Amphiprion melanopus	Guam	Semilunar	Ross (1978b)
Amphiprion species	Marshall Islands	Continuous, but peak near full moon	Allen (1972)
Chromis notata	Japan	Irregular synchronized peaks	Ochi (1986b)
Chrysiptera biocellatus	Great Barrier Reef	Semilunar	Thresher and Moyer (1983)
Chrysiptera cyanea	Great Barrier Reef	Arhythmic	Thresher and Moyer (1983)
Chrysiptera rollandi	Great Barrier Reef	6-day rhythm	Thresher and Moyer (1983)
Dascyllus marginatus	Red Sea	Semilunar	Holzberg (1973)
Dascyllus trimaculatus	Red Sea	Irregular synchronized peaks	Fricke (1973b)
Plectroglyphidodon johnstonianus	Hawaii	Arhythmic	MacDonald (1976)
Pomacentrus nagasakiensis	Japan	Lunar	Moyer (1975)
Pomacentrus rhodonotus	Great Barrier Reef	Semilunar	Doherty (1983c)
Pomacentrus wardi	Great Barrier Reef	Semilunar	Doherty (1983c)
Western Atlantic			
Abudefduf saxatilis	Panama	Continuous	Foster (1987b)
	Bahamas	Lunar (?)	Cummings (1968)
Chromis cyanea	Curaçao	Lunar	de Boer (1978)
Microspathodon chrysurus	Panama	Lunar	Pressley (1980)
Stegastes leucostictus	Jamaica	Lunar	Itzkowitz (1985)
Stegastes partitus	Florida	Lunar	Schmale (1981)
Stegastes planifrons	Jamaica	Lunar	Williams (1978)

possible exception, have unimodal peaks of spawning activity close to or immediately after the full moon, whereas Indo-Pacific species have more diverse spawning cycles, which are, if anything, modally semilunar. A similar regional difference is evident in the Blenniidae. The western Atlantic blenny, *Ophioblennius atlanticus*, like the pomacentrids, spawns only during the period near the full moon (Marraro and Nursall, 1983), whereas Indo-Pacific blenniids are variously reported to be asynchronous [*Exallias brevis* at Hawaii, by B. A. Carlson (personal communication)], semilunar [several species at One Tree Island, Great Barrier Reef (unpublished observations; see also Wickler, 1965)], and lunar spawners [*Istiblennius zebra*, at Hawaii, by Strasburg (1953)].

I bring up these two possible differences for two reasons. First, consistent regional differences within families might be useful in discovering the adaptive significance of lunar spawning cycles, about which there is much speculation and not much data (see the review by Thresher, 1984). Second and more important, clutch size and spawning frequency in part determine per capita rate of larval production, which in turn could bear on settlement rates. As noted in the Introduction, the vigorous discussion over the relative importance of competitive interactions and recruitment limitation in structuring reef fish assemblages has a large geographic component: advocates of the importance of competitive interactions based most of their evidence on data from New World reefs, whereas the lottery and recruitment-limitation advocates draw most of their evidence from what I would now identify as Pacific continental margin reefs. Logically, differences between assemblages in per capita settlement rates could well bear on this debate. Although there are insufficient data as yet to determine whether settlement rates differ consistently between areas (Shulman, 1985b; Shulman and Ogden, 1987), that two components of reproductive effort apparently differ between members of the same families at New World and Pacific continental margin sites suggests that the debate may reflect as much real differences between assemblages as it does the paradigm preferences of individual scientists (see Dayton, 1979).

VI. Summary, Conclusions, and Even More Unsubstantiated Speculations

"The great tragedy of science—the slaying of a beautiful hypothesis by an ugly fact."

—Thomas Henry Huxley (*The Method of Zadig*, 1878)

This review is already too long. Even so, it does not begin to encompass the range of spatial scales and elements of ecology and behavior that could well vary geographically. I suspect, for example, that pomacentrin damselfishes in the western Atlantic are more habitat generalists and recruit to a broader range of habitat types than in the Indo-Pacific. The social organizations of other groups, such as angelfishes, also seem to vary more within and between species in the Atlantic than in the Pacific. The data base for such comparisons, however, is still so sparse that even I am not prepared to expatiate on them in print. The comparisons that I have braved also suffer from the difficulties of working with data from widely separate areas and different habitat types, collected by different sampling techniques and small sample sizes. The geographic structuring of sites that my analyses detect is real, but its causes and the implications for community and population ecology must be couched in terms of hypotheses and predictions. If subsequent work proves them wrong, then so be it.

I suggest three lines of evidence that could prove particularly useful in destroying my hypotheses.

First, hypotheses based on a comparison of a particular pair of areas (e.g., western Atlantic versus western Pacific) should be tested against data from other areas. If interoceanic differences in early life history features result from geographic differences in oceanic productivity, then they should also be different at other sites of high and low production (the Red Sea and the eastern Pacific versus Ascension Island?). At the moment, the sparse data available for the eastern Pacific do not unambiguously support the productivity hypothesis, and it may well be that a complete data set will negate that hypothesis altogether.

A second line that may be worth pursuing is seeking evidence of parallel variations between sites in unrelated taxa. That a range of demersal-spawning fish families shows much the same pattern of differences in egg sizes between the Indo-Pacific and western Atlantic lessens the bias from having small samples for any one family and implies that a mechanism is operating with pervasive effects across families. Similarly, evidence that Hawaiian endemic species in such different animals as reef fishes and cone shells have smaller eggs than their relatives elsewhere in the Pacific (see Thresher and Brothers, 1989) strengthens the argument that Hawaiian endemic species are doing something consistently different. More important, however, are the exceptions to what may otherwise be broad parallel variation. Why doesn't the damselfish genus *Stegastes* have smaller eggs in the western Atlantic than in the western Pacific, if such a difference is a general rule for other demersal-spawning taxa? Are the population dynamics of Atlantic and Pacific species of *Stegastes* more alike than those of comparably distributed species of *Abudefduf*, for example? "Exceptions" like *Stegastes* could provide powerful tests of hypotheses raised to account for regional differences.

And third, introduced species could provide a similar, but potentially even more powerful test of these hypotheses. Over the last few centuries, numerous species of marine organisms have been accidentally introduced at sites well outside their normal ranges, primarily by courtesy of ballast water (Carlton, 1985). In tropical regions, Hawaii in particular has been a "receiver" of introduced species (Carlton, 1987). Most are plants and sedentary marine invertebrates, which tolerate ship-board transport readily. Few species of tropical marine fishes have successfully invaded areas outside their native ranges, though at least a couple have navigated the Panama Canal (McCosker and Dawson, 1975), others have been successfully introduced to the Hawaiian Islands (Randall, 1987), and at least one species of blenny, *Omobranchus punctatus,* which is native to the Indo-Pacific continental margin assemblage, has established expatriate populations in Fiji, southern Africa, and two sites in the tropical western Atlantic (Springer and Gomon, 1975). Depending on the date of the introduction, generation times, and the intensity of local selection, the reproductive and population ecologies of these introduced species should converge on those of native species if regional differences in life history traits are adaptive. Introduced invertebrates are likely to be a more fertile field of study than fishes, because of their generally shorter generation times and the greater number and diversity of species introduced. To my knowledge, no such study has been done, though a comparison of native and expatriate populations of *O. punctatus* is in progress (Thresher, ms.).

To conclude, I suggest that the data available strongly indicate that there is substantial geographic patterning of elements of the population and community ecology and behavior of reef-associated fishes. The evidence of geographic differences in, for example, early life history features and the relative dominance of the different spawning modes is sufficiently compelling that ascribing them solely to the "value-laden, theory-driven, experience-dominated definitions of hypotheses and experimental designs" of scientists working in different areas is probably not warranted (see Underwood and Fairweather, 1985). Although the cautious approach suggested by Underwood and Fairweather (1985) to geographic comparisons remains justified, the problems at this point are to determine the limits of these geographic differences and to understand their implications for population and community dynamics. The essential need is for well-designed comparative studies in different areas, carried out over a long enough period that some understanding of population dynamics at time scales relevant to the species studied is possible. Two- to three-year-long field studies are not sufficient for any species. In the meantime, generalizations about the factors that are or are not important in determining the composition and dynamics of geographically diverse reef fish assemblages should be assayed with caution.

Acknowledgments

S. Blaber, A. W. Ebeling, R. E. Johannes, J. M. Leis, J. E. Randall, K. J. Sainsbury, V. G. Springer, and P. C. Young reviewed early drafts of this paper. I very much appreciate their vigorously critical comments and their willingness to plow through my, at times, difficult prose. Editorial comments by V. Mawson and F. R. Harden Jones were particularly helpful. Discussions and/or correspondence with A. Gronell, K. J. Sainsbury, V. G. Springer, and J. Williams were also extremely valuable in forming my ideas. Neither they nor the reviewers, however, share any blame for the chapter's contents. This chapter developed and benefited greatly from discussions held at a UNESCO-sponsored workshop on interregional differences in coral reef ecology, organized by C. Birkeland and held in Fiji in 1986, and a subsequent symposium on the same topic, organized by C. Birkeland and C. R. Wilkinson for the Sixth International Coral Reef Symposium (1988). This paper is in part based on field work funded by a grant (#3486-87) from the National Geographic Society.

CHAPTER 16

Patterns and Processes in the Distribution of Coral Reef Fishes

David McB. Williams
Australian Institute of Marine Science
Townsville, Queensland, Australia

I. INTRODUCTION

The warm waters of coral reefs, the general clarity of the water, and the relative lack of reaction of fishes to a diver's presence make coral reef fishes ideal subjects for *in situ* studies by SCUBA divers. The consequence has been a considerable number of descriptive studies of the distributions of coral reef fishes and the composition of assemblages of reef fishes.

The spatial heterogeneity of coral reefs over a wide range of scales, from meters to thousands of kilometers, is one of their most striking features (Fig. 1). Together with the fact that reef fishes tend to be more site-attached than similar-sized terrestrial vertebrates (Sale, 1978b), this heterogeneity and how fish respond to it are of major concern in the study of reef fish ecology. Assemblages of reef fishes and their physical and biological environment vary greatly among habitat patches at all spatial scales. Generalizations concerning the population dynamics of these fishes and management strategies for these populations must take into account this spatial variation in abundances and in processes determining abundances.

Much of the research on coral reef fishes in the last decade and a half has focused on mechanisms of coexistence and a relatively fruitless argument as to whether or not reef fishes partition their resources in ways consistent with niche theory (e.g., Anderson *et al.*, 1981; Sale and Williams, 1982). Many of these studies have provided good descriptive data of patterns of distributions but have done little to increase our knowledge of factors determining the distribution and abundance of reef fishes. This is because the spatial scales of most studies have been greatly restricted and process-oriented studies have been rare.

The factors determining the abundances of reef fishes within a habitat patch

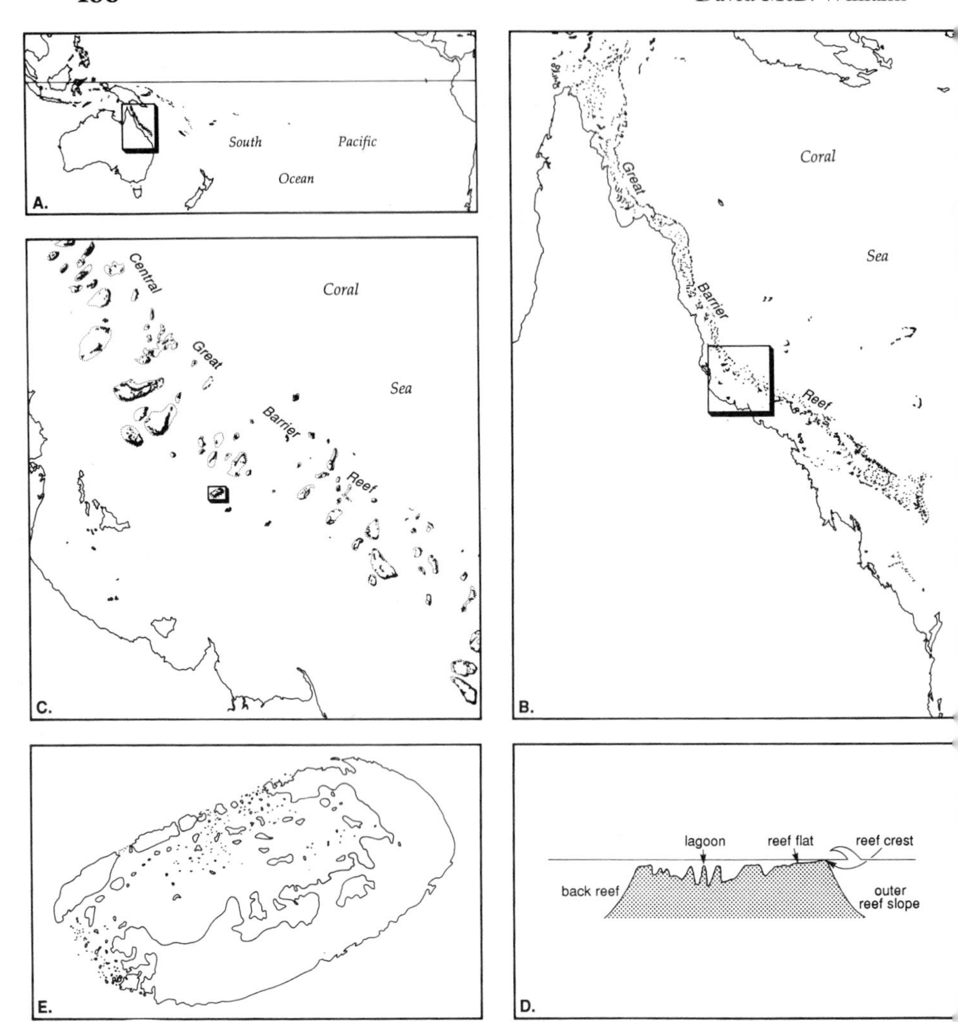

Figure 1 The spatial heterogeneity of coral reefs over a wide range of scales, from meters to thousands of kilometers, is one of their most striking features. (A) The South Pacific Ocean contains many regions of coral reefs; (B) The Great Barrier Reef contains more than 2900 individual reefs; (C) a detail of the Central GBR; (D) a lateral view of a single reef with many reef zones; (E) a vertical view of a reef lagoon with many patch reefs.

have recently been reviewed in depth by Doherty and Williams (1988a). The aim of this chapter is to synthesize our knowledge of patterns of distributions of reef fishes among habitat patches over a wide range of spatial scales and to review our understanding of the processes determining the observed patterns. Emphasis is placed shamelessly on studies on the Great Barrier Reef (hereafter

GBR), because of the existence of a number of studies using comparable data sets over a wide range of spatial scales in this region. An introduction to the ecology of the major taxa discussed in this chapter is given in Table 1.

II. Patterns

A. Within Reefs

Physiographic zonation is a striking feature of all coral reefs (Done, 1983). Most reefs can be divided at least into reef flat, crest, and outer slope zones with characteristic depths and exposure to wave environments. Surge zones, channels, variously developed lagoons, and back reef areas can also be readily recognized on many reefs. Because of logistic constraints and the obvious nature of this environmental zonation, most studies of the distributions of fishes of coral reefs have either examined distributions within one of these zones or compared abundances of species among the zones.

1. Within Zones

Reef zones may vary in width from tens of meters in the case of reef crests and surge zones to lagoons kilometers across. A species' perception of the zone (as a coarse- or fine-grained habitat) is largely dependent on the home range of individuals. Some species spend their entire postsettlement lives in a single coral head tens of centimeters in diameter (e.g., Sale, 1971). Others roam across zones in a day's activities (e.g., Choat, 1969; Vivien, 1973) (Fig. 2) or during spawning migrations (references in Johannes, 1978a; Robertson, 1983). The size of the home range tends to be larger in species with larger individuals and to increase with the size of individuals within species (Sale, 1978b), but considerable variation may occur among closely related species. For example, home ranges of surgeonfishes at Aldabra range from less than 40 m^2 to hectares (Robertson and Gaines, 1986).

Most detailed studies of within-zone distributions of reef fishes have tended to concentrate on the smaller ($<$200 mm length), more site-attached species. The preferred habitat of these smaller species can usually be readily defined in terms of substratum type or physical relief (e.g. coral life-forms or species, patch reefs or rocks of varying size and height, rubble or sand, vertical walls, ledges or caves) and water depth (e.g., Table 2; Sale, 1969, 1972b; Fishelson *et al.,* 1974; Allen, 1975; Itzkowitz, 1977a; Chave, 1978; Gladfelter and Gladfelter, 1978; Luckhurst and Luckhurst, 1978c; Doherty, 1980, 1983a; Robertson and Lassig, 1980; Waldner and Robertson, 1980; D. McB. Williams, 1980; Robertson and Polunin, 1981; Thresher, 1983a; Harmelin-Vivien, 1989). These microhabitats or habitat patches are rarely distributed

Table 1 Ecological Summary of Major Taxa Discussed in This Chapter

Family:	Acanthuridae	Chaetodontidae	Labridae	Pomacentridae	Scaridae
Common name:	Surgeonfishes	Butterflyfishes	Wrasses	Damselfishes	Parrotfishes
Typical maximum length:	15–40 cm	10–20 cm	8–30 cm	5–15 cm	15–50 cm
Approximate number of coral reef species:	60	110	320	400	80
Major diet:	Herbivorous, few detritivorous or planktivorous	Corals, sessile invertebrates	Motile invertebrates	Herbivorous or planktivorous	Herbivorous
Morphologically specialized larvae:	Yes (*Acronurus*)	Yes (*Tholichthys*)	No	No	No
Relative length of larval life:	Long	Short–long	Short–long	Short	Medium

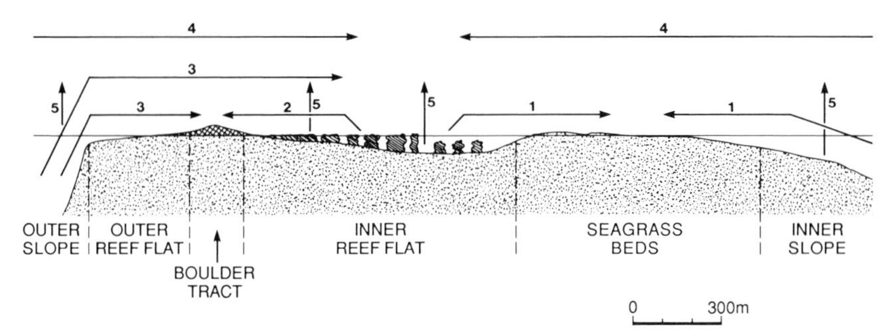

Figure 2 Movements of different groups of fishes within the fringing reef of Tulear (Madagascar) during a rising tide. 1,2: from the inner reef flat to sea grass beds and rubble bank, respectively, to feed; 3: from the outer slopes to feed in shallower waters; 4: movement of pelagic fishes from deeper waters to feed on inner reef flat; 5: general movement up into water column. [Redrawn from M. L. Vivien (1973). Contribution a la connaissance de l'ethologie alimentaire de l'ichthyofaune de platier interne des recifs coralliens de Tolear (Madagascar). *Tethys, Suppl.* **5**, 221–308.]

uniformly within a zone, nor are fish uniformly distributed among them where they do occur (e.g., Sale, 1972a). The distribution of fishes among these patches has variously been related to topographic complexity (Luckhust and Luckhurst, 1978c), food availability (D. McB. Williams, 1979; Thresher, 1983b), current flow and water quality (Hobson and Chess, 1978; D. McB. Williams, 1979; Thresher, 1983a), exposure to wave action (Talbot, 1965; Williams, 1982), availability of hiding places (de Boer, 1978; Roberts and Ormond, 1987), and live coral cover (Bell and Galzin, 1984; Bouchon-Navaro *et al.,* 1985; Hourigan *et al.,* 1988).

Except for an increased knowledge of some spawning migrations (see the foregoing) and studies of diel movements, quantitative knowledge of the home ranges of the larger (>200 mm length) and more mobile fishes has advanced little since some of the pioneering studies in reef fish ecology (Bardach, 1958; Randall, 1962; Springer and McErlean, 1962). Overall, these studies concluded that "reef fishes, in general, are non-migratory. Such movements that are made are usually more-or-less localized. Over long periods, however, some fishes may make more extensive movements" (Randall, 1962). Clearly more quantitative studies are required in this area.

Many of these larger species undergo striking diel migrations from resting places to feeding areas. These have been described in particular detail for striped parrotfish (Ogden and Buckman, 1973), for French and white grunts (e.g., McFarland *et al.,* 1979; Helfman *et al.,* 1982), and for a wide range of species by Hobson (e.g., Hobson, 1965, 1968, 1972, 1973). Distances traveled vary from at least several kilometers to just a few meters and, in many species, timing and routes taken are predictable (Hobson, 1973). During these feeding migrations, fishes characteristically move in shoals. Those that

Table 2 Summary of Habitats That Eight Species of Damselfishes Are Most Characteristically Associated with in Various Sites in the West Indies[a]

Species	Substrate or habitat type[b]	Depth range
Stegastes planifrons	Staghorn (1,2,4,5,6,7), elkhorn (1), *Agaricia* (2), massive corals (1,4,5,6,8), *Millepora* (1)	Shallow–deep (1,2,9)
S. dorsopunicans	Elkhorn (1,2,6), beach rock (1), rocky back reefs (2,6)	Very shallow–moderate (1,2,8,9)
S. variabilis	Back reefs with sand and massive corals (1,2,8)	Moderate–deep (1,2,8,9), shallow (9)
S. leucostictus	Bare back reefs (2,6), low-profile rubble (1)	Very shallow (1,2,8,9)
S. partitus	Sandy slopes (1,2), low-profile rubble (1,2,6,8), beach rock (1)	Moderate–deep (1,2,8,9), shallow (9)
S. diencaeus	Rocky back reefs (2)	Very shallow–shallow (2)
S. mellis	Beach rock (1)	Very shallow (1), shallow–deep (8,9)
Microspathodon chrysurus	Elkhorn and *Millepora* (1,3,6,8,9)	Very shallow–moderate (1,3,8,9)

[a] Modified from Waldner and Robertson (1980).
[b] Categorization of habitats by Emery (1973), Clarke (1977), and Itzkowitz (1977a) differed in some respects from ours. This makes direct comparisons difficult in some cases. Authority: 1 = present study, Puerto Rico; 2 = present study, Panama; 3 = D. R. Robertson, unpublished observations; 4 = Williams, 1978 (Jamaica); 5 = Kaufman, 1977 (Jamaica); 6 = Itzkowitz, 1977a (Jamaica); 7 = Itzkowitz, 1978 (Jamaica); 8 = Clarke, 1977 (Bahamas); 9 = Emery, 1973 (Florida).

swim away from the reef at night (often to feed on sandflats or seagrass beds) tend to return to the same location each morning, with the daytime schooling sites close to prominent topographical features (Hobson, 1973). At twilight there is a characteristic "changeover period" between diurnal species entering their resting sites and the emergence of nocturnal species (Hobson, 1965, 1972).

Factors determining distributions of larger, more mobile species within zones are less well studied than those of smaller species. Presumably the behavior of these species will respond closely to the availability of food and, at least for the nonschooling species, will be influenced by the availability of suitable shelter (Randall, 1963; Talbot, 1965; Goldman and Talbot, 1976; Hixon and Beets, 1989).

2. *Among Zones*

a. Within Species Juvenile reef fishes tend to have similar among-zone distributions to their adult conspecifics or to occur in shallower areas than the adults.

On the Great Barrier Reef, juvenile damselfish generally settle directly into the adult habitat (e.g., D. McB. Williams, 1979), as do common species of wrasse (Eckert, 1985b). A different situation occurs for two of the common damselfishes in the Bahamas, however, where juveniles predominate in shallow water and adults in deeper water (Clarke, 1977). Juvenile butterflyfishes in Moorea typically settle within the adult habitat, as they do on the GBR (Fowler, 1990a), but some occur in shallower water (Bouchon-Navaro, 1981). Distributions of juvenile acanthurids and scarids in Aqaba are more restricted than those of the adults, with both juvenile scarids and acanthurids occurring primarily on the shallow reef front down to 10 m depth. This is the preferred habitat of the adult scarids but not that of all acanthurids, many of which occur further inshore (Bouchon-Navaro and Harmelin-Vivien, 1981). In Hawaii, juvenile manini (a surgeonfish) select shallow-water habitats but move into deeper water as they mature (Sale, 1969). Robertson *et al.* (1979) reported habitat segregation of juveniles and adults of the same species in five species of surgeonfish at Aldabra. Of these, juvenile *Acanthurus lineatus* occurred shallower than the adults (distributions of other juveniles are not made explicit). In Tulear (Madagascar), juveniles of all species of grouper (Serranidae), several wrasses including *Cheilinus diagramma, C.trilobatus,* and *C.undulatus,* and several parrotfishes are found on the inner reef flat but adults are found primarily the outer reef slopes (Vivien, 1973).

The role of seagrass beds as a nursery habitat for reef species has been emphasized in Caribbean studies (e.g., Ogden and Zieman, 1977; Ogden, 1980; Ogden and Gladfelter, 1983; Shulman, 1985a) but these habitats do not, in general, appear to have the same role in the Indo-Pacific region (Quinn and Kojis, 1985; Parrish, 1987a; Birkeland, 1987, 1988). Clearly high-diversity communities of reef fishes exist on the GBR, for example, in the general absence of seagrass beds in close association with reefs. A detailed understanding of the extent to which the proximity of seagrasses or mangroves to coral reefs affects the presence or abundance of species of reef fish remains to be determined in both the Caribbean and the Indo-Pacific (Quinn and Kojis, 1985; Parrish, 1987a).

For many deeper-water species of the outer reef slope as well as for some of the shallower species, greater numbers of smaller fish are found in the upper part of a species depth range and fewer and larger fish in the lower part of the range (Mead, 1979; Munro, 1983; Brouard and Grandperrin, 1985; Wright *et al.*, 1986; Kulbicki and Grandperrin, 1988). This phenomenon is not restricted to coral reefs (Johannes, 1981; Longhurst and Pauly, 1987; McCormick, 1989). Mead (1979) further suggests that the best yield to fishing (=highest biomass of the population?) is usually at intermediate levels of the species depth range.

Limited data suggest that some of the larger species of the outer reef slopes make significant diel migrations between different depths. In Vanuatu this migration was a nocturnal upward movement of 40 to 80 m, particularly marked in the species living at intermediate depths of 80 to 240 m (Brouard and Grandperrin, 1985).

b. Among Species The uneven distributions of species among different reef zones is obvious even to the most casual observer. Species richness of microhabitats within reef zones may differ little between the Caribbean and equivalent habitats in the Indo-Pacific (e.g., Bohnsack and Talbot, 1980; but see Gladfelter *et al.*, 1980), but total richness over all zones is obviously less in the Caribbean (Sale, 1980a). The strong implication is that among-zone differences in species composition are smaller in the Caribbean than in the (regionally more diverse) Indo-Pacific. This difference remains to be quantified (Sale, 1980a). Harmelin-Vivien (1989) has quantified a similar difference between a central Pacific reef (Moorea) and an Indian Ocean reef (Tulear). She did find that among-zone differences in species composition are significantly smaller on the reef with regionally lower species richness (Moorea) than on the reef with higher regional diversity (Tulear).

For all families examined in relative detail, including damselfishes (Allen, 1975; Clarke, 1977), butterflyfishes (Bouchon-Navaro, 1980, 1981, 1986; Fowler, 1990a), scorpaenids and serranids (Harmelin-Vivien and Bouchon, 1976), and surgeonfishes, parrotfishes, and siganids (Russ, 1984a,b), some species are relatively restricted in their distribution among zones and others widely distributed (Fig. 3).

The best data on among-zone distributions are for the butterflyfishes. At One Tree Reef (GBR), the 8 abundant species of butterflyfish are widely

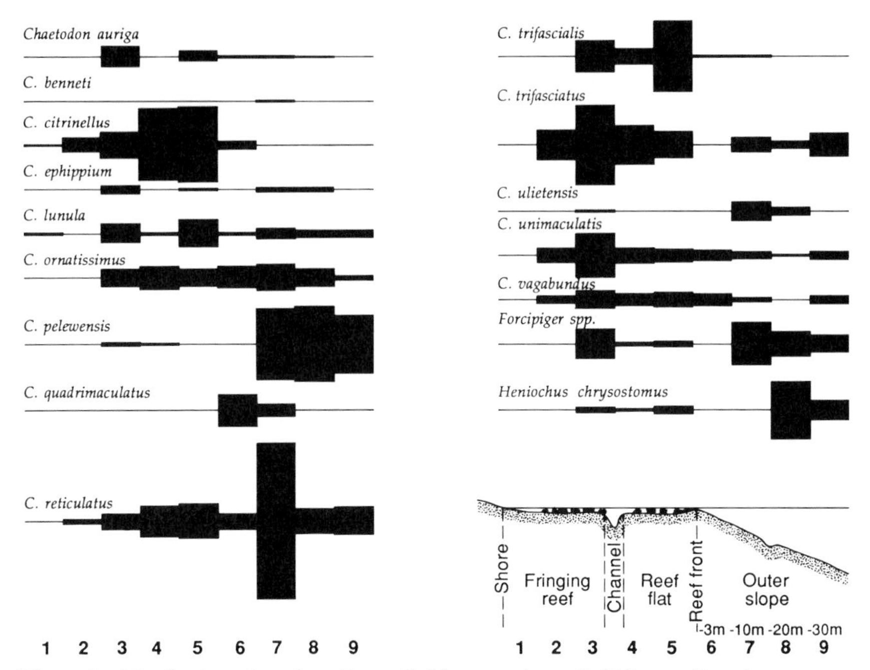

Figure 3 Distribution of species of butterflyfish across the reef of Moorea (French Polynesia). Height of columns is proportional to density of individuals in each zone. [Redrawn from Y. Bouchon-Navaro (1981). Quantitative distribution of the Chaetodontidae on a reef of Moorea Island (French Polynesia). *J. Exp. Mar. Biol. Ecol.* **55,** 145–157.]

distributed among zones, while 11 out of the 14 rarer species are highly restricted in distribution (Table 3). At Aqaba the two species that comprise 92% of all individuals are widely distributed and rarer chaetodonts are more restricted. In Moorea, French Polynesia, there is no clear relationship between distribution and abundance.

Species that are restricted or widely distributed at one geographic location are not necessarily similarly distributed on other reefs. Six butterflyfishes restricted at One Tree Reef (GBR) are relatively widely distributed in Moorea (French Polynesia). One of the species restricted in Moorea is relatively widely distributed at One Tree. Two of the species relatively restricted in Aqaba are widely distributed at One Tree (Table 3). In all nine cases, species that are relatively widely distributed at one geographic location have considerably higher densities at that site (Table 3). In four cases, species' distributions differ little between Moorea and One Tree Reef. In two of these there are approximately twofold differences in densities, and in the other two there is

Table 3 Dispersion of Butterflyfishes Among Reef Zones at Three Geographic Locations[a]

	One Tree			Moorea			Aqaba		
	W	Mean Density	Max. Density	W	Mean Density	Max. Density	W	Mean Density	Max. Density
Chaetodon rainfordi	0.60	62.1	167.6						
C. plebius	0.88	16.7	23.8						
Chelmon rostratus	0.47	15.7	29.2						
Chaetodon flavirostris	0.83	11.9	20.6						
C. trifasciatus	0.40	10.4	32.2	0.42	6.1	19.5			
C. melannotus	0.53	10.3	28.9				0.26	1.0	4.5
C. auriga	0.33	5.7	11.4	0.23	1.2	5.5	0.07	0.4	2.5
C. lineolatus	0.48	4.1	7.5						
C. trifascialis	0.10	1.3	7.2	0.10	3.4	18.0			
C. aureofasciatus	0.34	0.9	2.8						
C. ulietensis	0.06	0.9	6.0	0.10	0.8	4.5			
C. speculum	0.54	0.7	1.7						
C. ephippium	0.08	0.6	2.8	0.27	0.4	1.5			
C. citrinellus	0.10	0.6	4.2	0.14	5.6	19.0			
C. baronessa	0.07	0.5	4.7						
C. vagabundus	0.02	0.4	3.1	0.53	2.3	4.5			
C. pelewensis	0.05	0.3	2.8	0.17	6.1	19.5			
C. lunula	0.02	0.3	1.4	0.49	2.2	6.5			
C. bennetti	0.31	0.2	0.6	0.03	0.0	0.5			
C. ornatissimus	0.05	0.2	2.1	0.71	3.7	6.5			
Forcipiger spp.	0.05	0.2	2.1	0.33	3.7	11.5			
C. unimaculatus	0.05	0.2	1.5	0.52	3.3	11.0			
C. reticulatus				0.33	8.7	40.0			
C. quadrimaculatus				0.06	1.1	7.5			
C. pauscifasciatus							0.83		73.0
C. austriacus							0.76		35.5
C. fasciatus							0.22		4.0

[a] *W* = a weighted measure of dispersion across zones [based on the Shannon–Weiner formula for niche breadth (*W'* of Clarke, 1977)]. Mean density = mean density across all zones. Max. density = density in zone in maximum abundance. [Recalculated from data in Fowler, 1988 (One Tree); Bouchon-Navaro, 1981 (Moorea); and Bouchon-Navaro, 1980 (Aqaba).]

little apparent difference. A similar positive correlation between local densities and breadth of distribution has been found for a number of terrestrial groups (Brown, 1984).

Relationships between the distributions of butterflyfishes and their preferred prey are difficult to determine because of the lack of detail on feeding preferences, particularly species of coral, and on the distribution of the prey. Most of the studies of distribution have divided species into relatively broad categories such as obligate coral feeders, facultative coral feeders, noncoral benthic invertebrate feeders, and planktivores (e.g., Anderson *et al.*, 1981; Bell *et al.*, 1985; Findley and Findley, 1985). Hourigan *et al.* (1988; along with Reese, 1977, and other work by Reese), working in Hawaii, have divided the obligate coral feeders further into specialists and generalists. They found that at the within-reef scale, the distribution of generalist butterflyfishes tended to follow the total coral cover but that the distribution of the specialists more closely followed the distribution of their preferred prey species. The extent to which the distributions of butterflyfishes in more diverse areas can be related to their preferred food awaits more detailed studies of diets versus prey distributions.

Some groups of species within families exhibit characteristic among-zone distributions. For example, plankton-feeding species of the damselfish genus *Chromis* are largely restricted to the outer reef slopes (Allen, 1975). Algal-grazing damselfishes (such as *Stegastes* spp.) are largely restricted to shallow reef areas (Allen, 1975).

All of these studies have been carried out by direct observation in shallow reef waters (generally <20 m depth). Fisheries investigations of the deeper waters of the outer reef slopes down to 400 m have also found distinct distributions of species that can be characterized according to depth and bottom type. Three distinct bottom types have been recognized throughout the Pacific: level bottom, mixed sand and coral; gradual slope, mixed sand, coral, and rock; sharp drop-off, rock and coral (Mead, 1979). Within any reef, the depth range of maximum concentration of a species can be characterized. For example, in Vanuatu, species can be characterized as shallow (<120 m), intermediate (120–240 m), or deep (>240 m) (Brouard and Grandperrin, 1985), but absolute ranges vary among locations (Mead, 1979; Ralston and Polovina, 1982; Brouard and Grandperrin, 1985). Variations in depths of sharp thermoclines may be a major factor influencing these vertical distributions (Fourmanoir, 1980; Brouard and Grandperrin, 1985).

c. Among Families Few data are available to compare the distributions of families within reefs. Clarke (1977) clearly demonstrated that butterflyfishes in the Bahamas are more broadly distributed across reef zones than damselfishes on the same reef and cited similar data from Fanning Island in the Pacific (Chave and Eckert, 1974).

Bouchon-Navaro and Harmelin-Vivien (1981) demonstrated that among herbivorous fish, surgeonfishes dominate the reef flat in Aqaba and parrotfishes dominate the outer slope. They further suggested that this phenomenon seemed to be generally valid for other reefs. In contrast, Robertson *et al.* (1979) found parrotfishes at Aldabra (Indian Ocean) to be most abundant in the intertidal area of the reef crest and surgeonfish more abundant on the deeper reef slope. Russ (1984a,b), working on a number of reefs in the

Central GBR, found that surgeonfishes and parrotfishes both have higher numbers of species and individuals on reef crests and in lagoons than on reef flats and outer slopes; rabbitfishes (Siganidae) (a third family of herbivores) have higher numbers of species and individuals in lagoons and back reefs than in the other three zones studied (Fig. 4).

d. Among Trophic Groups Russ (1984b) divided the herbivorous fishes of the Central GBR into a series of guilds (based on Hatcher, 1983) and found that these were distributed differently among zones. "Suckers" feeding on fine

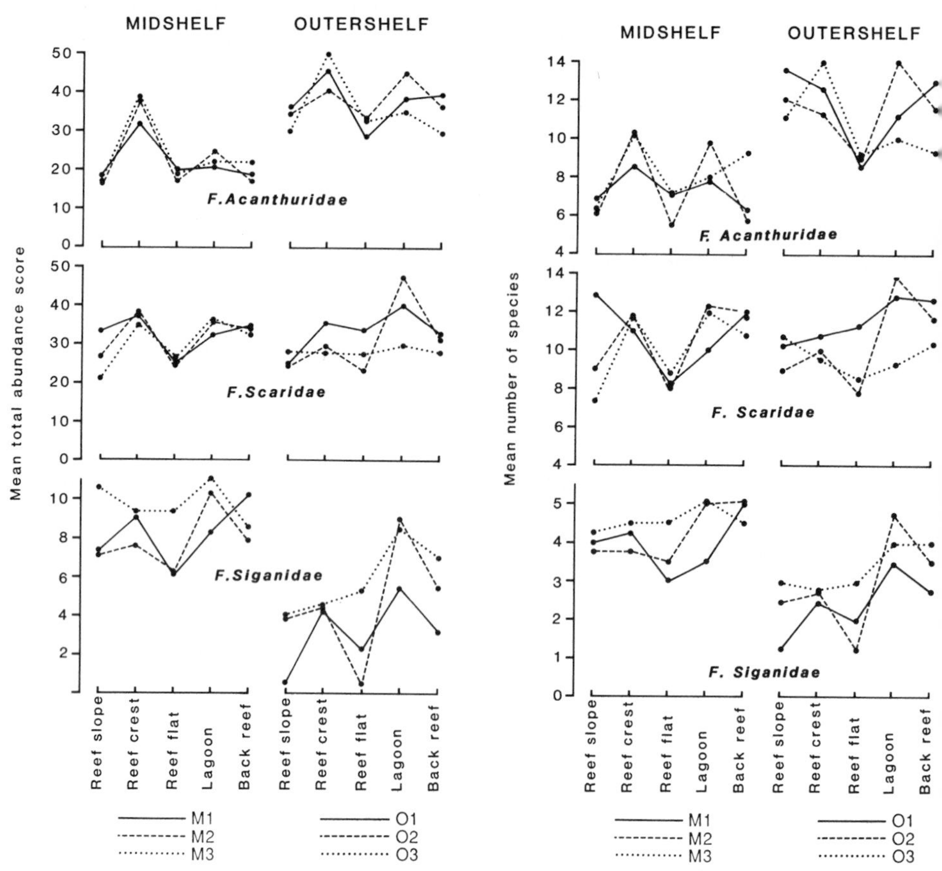

Figure 4 Distributions of surgeonfishes (Acanthuridae), parrotfishes (Scaridae), and rabbitfishes (Siganidae) among reef zones of three midshelf (M1–M3) and three outer shelf (O1–O3) reefs in the Central Great Barrier Reef. Locations of reefs are given in Fig. 6. [From G. R. Russ (1984b). Distribution and abundance of herbivorous grazing fishes in the central Great Barrier Reef. II. Patterns of zonation of mid-shelf and outershelf reefs. *Mar. Ecol. Prog. Ser.* **20,** 35–44.]

sediments were most abundant near windward and leeward edges of reefs. Suckers feeding over sand were most abundant in the back reefs and lagoons. Large and small "croppers" and "scrapers" were more abundant in shallow zones (reef crest, reef flat, lagoon) than in deep zones (reef slope, back reef) (Fig. 5).

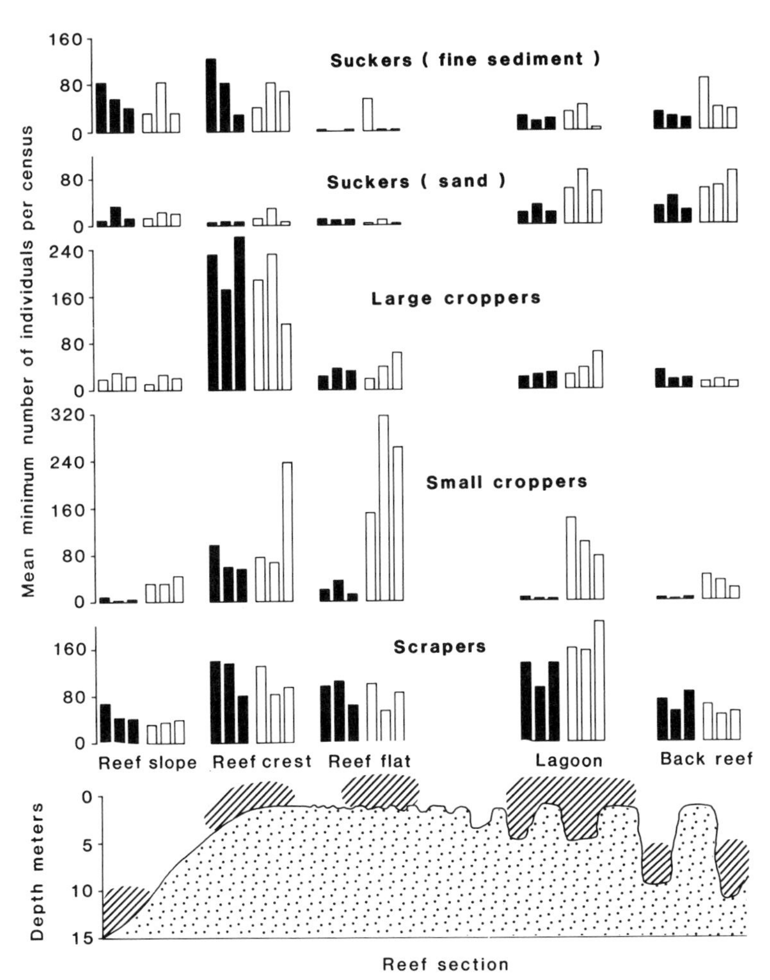

Figure 5 Distributions of trophic groups of herbivorous fishes among reef zones on midshelf and outer shelf reefs. Shaded and unshaded columns represent midshelf and outer shelf reefs, respectively, of Fig. 4. [From G. R. Russ (1984). Distribution and abundance of herbivorous grazing fishes in the Central Great Barrier Reef. II. Patterns of zonation of mid-shelf and outershelf reefs. *Mar. Ecol. Prog. Ser.* **20,** 35–44.]

The proportional abundance (of total biomass) of four major feeding types of fishes in three main habitat types has been examined at One Tree Reef (Goldman and Talbot, 1976). Benthic invertebrate feeders predominated (44%) in the lagoon where few planktivores were present (6%). In the two reef slope habitats, there was about half the proportion by weight of benthic invertebrate feeders and two to four times the proportion of plankton feeders. The authors suggested that the greater proportion of planktivores on the outer slopes might be related to a greater availability of food due to stronger currents. They also suggested that the greater abundance of planktivores on the leeward slope relative to the windward was due to the considerably greater shelter present on the leeward slopes. Sixty-four percent of the biomass of fishes on the windward slopes (sampled during daylight) comprised piscivores. They attributed this to the presence, in the deeper regions, of many large carnivorous species that were considered to be in a daytime resting phase but probably fed elsewhere at night.

Harmelin-Vivien (1989) compared the distribution of trophic groups between reef flats and outer slopes on both Tulear and Moorea. She found that on both reefs, omnivores were more abundant on the reef flat and zooplanktivores more abundant on the outer slope. Piscivores were more abundant on the reef flat at Moorea and on the outer slopes in Tulear.

B. Among Reefs

The traditional view of coral reefs as energetically self-contained ecosystems occurring only in clear oceanic waters (Odum and Odum, 1955) suggests little environmental variability and relatively little variation in the structure of fish communities among reefs. This view is grossly in error. As far as fishes are concerned, coral reefs are not energetically self-contained. A major proportion of the fish biomass feeds on zooplankton derived from an external source—the waters surrounding the reef (Russ, 1984c; Munro and Williams, 1985; Hamner *et al.*, 1988; Williams *et al.*, 1988). Also in contrast to the traditional view, coral reefs occur in a wide range of tropical environments (Done, 1982) and significant changes in the structure of resident fish communities are associated with these different environments (Munro and Williams, 1985).

Major problems in understanding among-reef variation in reef fish communities have been uncertainty in the similarity of reef environments and lack of comparable sampling strategies among studies. These difficulties have been overcome in a series of studies on the GBR, which comprises over 2900 individual reefs spread over 2000 km from north to south (lat. 10°S to 23.5°S) and ranging from continental fringing reefs to platform reefs perched on the edge of the continental shelf more than 200 km from shore (Williams, 1982, 1983b, 1986a; Williams and Hatcher, 1983; Russ, 1984a,b). In the Central

GBR, coral reefs are found in a wide range of environments; from turbid, relatively productive nearshore waters exposed to limited wave energy to reefs on the edge of the continental shelf surrounded by extremely clear and very low productivity Coral Sea waters and subject to severe wave action (Done, 1982).

The census technique used in the latter studies of Russ and myself (Williams, 1982, 1983b, 1986a; Russ, 1984a,b) involves visual, underwater counts by divers using SCUBA gear. My census dives involved a 45-min swim along the outer reef slope, swimming in a zigzag pattern from the surface to a depth of 13 m and recording the presence of species and their abundance (on a log abundance scale) along oblique transects stretching approximately 5 m to either side of the diver. All data were recorded on prepared census forms of waterproof paper. Five censuses of nonoverlapping areas of reef slope were made on each reef. Each 45-min swim covered approximately 150 m of reef front in the horizontal direction. A species list of approximately 150 species was used, including all species of butterflyfish, the majority of common damselfishes and surgeonfishes, virtually all parrotfishes, and selected wrasses, fusiliers (Caesionidae), and rabbitfishes (Siganidae). The species were selected to represent a wide range of ecological types, including: coral feeders, algal grazers, plankton feeders, carnivores, species with short larval lives, species with long larval lives, species typically associated with nearshore habitats, and those associated with isolated oceanic reefs. Russ used a similar technique in a wide range of habitats, although he counted only herbivorous species.

1. Similar Locations

These among-reef studies were initiated in the Central GBR (Fig. 6) along a strong cross-shelf environmental gradient (Done, 1982), but were later extended to the entire length of the GBR. Significant differences were found in the abundance of species on reefs kilometers and tens of kilometers apart at similar locations (e.g., differences between "nearshore" or between "midshelf" reefs in Fig. 6) along the environmental gradient (Fig. 7). For some species, these differences have been maintained over 8 years (Williams, 1986a and unpublished observations).

Following on my work, Russ (1984a,b) compared distributions of herbivorous fishes (parrot-, surgeon-, and rabbitfishes) across the same Central GBR transect (Fig. 6) but extended the study to compare abundances among five different zones (reef slope, crest, flat, lagoon, and back reef) on each reef. Analyzing abundance and number of species at the family level by ANOVA, he found significant differences among reefs at the same cross-shelf location and significant Reef (Locations) × Zone interactions, indicating that zonation patterns differed among reefs at similar locations. Except for the parrotfishes, however, these sources of variation in distributions were small compared to those associated with among-zones or across the shelf (Table 4).

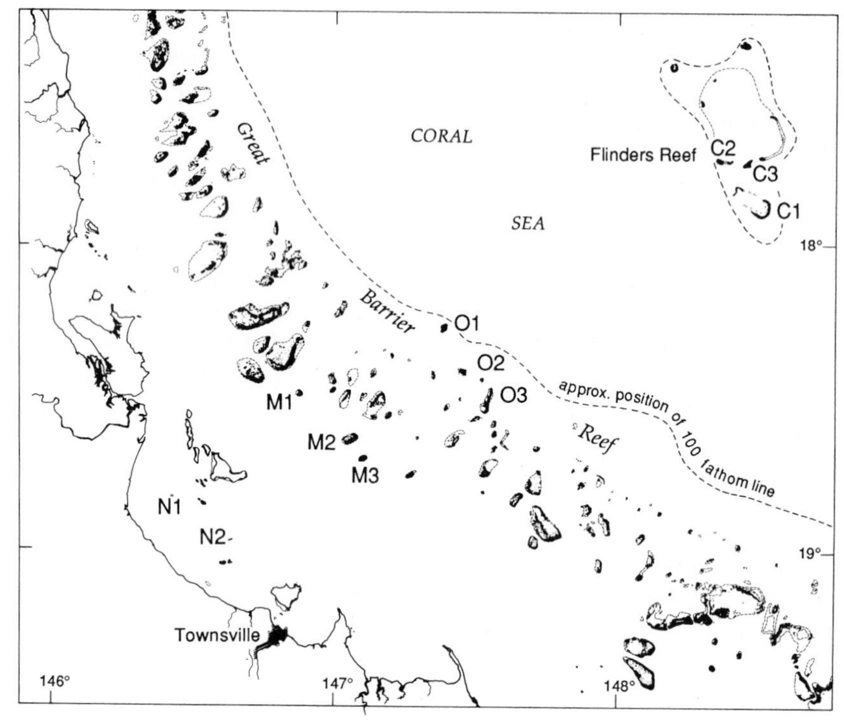

Figure 6 Location of two nearshore (N1, N2), three midshelf (M1–M3), three outer shelf (O1–O3), and three Coral Sea reefs (C1–C3) along cross-shelf transect in the Central Great Barrier Reef.

2. *Across Continental Shelf*

Williams (1982) compared the structure of reef fish communities on the outer slopes of the Central GBR reefs indicated in Fig. 6 and found major changes in the abundance of fishes across the cross-shelf transect (Fig. 7). There were significant differences in species abundances among replicate censuses within reefs in this study and, as indicated in the previous section, there were significant differences among reefs at similar cross-shelf locations. These differences

---

Figure 7 Distribution of damselfishes across Central Great Barrier Reef transect. Letter codes are given in Fig. 6. For each species, each of 11 reefs is represented by a circle. Each circle is divided into five sectors, representing five nonoverlapping, replicate censuses of shallow outer reef slope on each reef. The radius of each sector is proportional to a log abundance category. (Modified from Williams, 1982.)

POMACENTRIDAE

Inshore — N2 N1
Mid-shelf — M1 M2 M3
Outer-shelf — O1 O2 O3
Coral Sea — C1 C2 C3

Neopomacentrus sp.
N. cyanomus
Pomacentrus popei
P. wardi
Acanthochromis polyacanthus
Chromis nitida
Stegastes apicalis
Pomacentrus sp.
Glyphidodontops rollandi
Pomacentrus amboinensis
Glyphidodontops talboti
Pomacentrus coelestis
Chromis atripectoralis
Pomacentrus lepidogenys
Neopomacentrus azysron
Pomacentrus philippinus
Abudefduf whitleyi
A. sexfasciatus
A. saxatilis
Dascyllus reticulatus
Chromis ternatensis
Plectroglyphidodon lachrymatus
P. dickii
Stegastes fasciolatus
Chromis margaritifer
C. weberi
C. atripes
C. xanthura
Pomacentrus 'emarginalis'
Chromis chrysurus
C. lomelas
C. amboinensis

Table 4 Summary of Three-Factor Analyses of Variance Involving Number of Species and Abundance of Individuals of Three Families of Herbivorous Fishes (Acanthuridae, Scaridae, Siganidae) in the Central Great Barrier Reef[a]

Variation	Degrees of freedom of F ratio	Number of species			Abundance			All three families
		Acanthuridae	Scaridae	Siganidae	Acanthuridae	Scaridae	Siganidae	
Location on shelf	1,4	** (53%)	ns (0%)	* (38%)	*** (56%)	ns (0%)	* (46%)	** (21%)
Reefs (Locations)	4,90	ns (1%)	*** (20%)	** (5%)	ns (0%)	** (6%)	*** (8%)	ns (0%)
Zones	4,90	*** (12%)	*** (20%)	*** (16%)	*** (25%)	*** (34%)	*** (13%)	*** (43%)
Location × Zones	4,16	ns (2%)	ns (2%)	ns (2%)	ns (2%)	ns (5%)	ns (1%)	ns (5%)
Reefs (Locations) × Zones	16,90	** (9%)	*** (22%)	ns (5%)	*** (7%)	*** (20%)	ns (4%)	*** (10%)
Residual		(22%)	(36%)	(34%)	(10%)	(35%)	(28%)	(21%)

Significance levels: *p.05 > P > 0.01; **0.01 > P > 0.001; ***P < 0.001; ns P > 0.05

[a] Significance of the F-value and the variance component (parentheses) expressed as percentage of the sum of variances for each analysis are given for each factor. Values from these analyses are plotted in Fig. 5. (From Russ, 1984b.)

were small, however, compared to cross-shelf variation in abundances (Fig. 7). The nature of the distribution of species across the transect varied markedly among families. In particular, damselfishes and butterflyfishes were significantly more restricted in distribution than surgeonfishes, parrotfishes, and wrasses. Cross-shelf patterns in distributions have remained constant for more than 8 years (Williams, 1986a and unpublished observations).

Following this study, Williams and Hatcher (1983) carried out an intensive comparison of the fish communities on one nearshore, one midshelf, and one outer shelf reef using quantitative explosive stations. This greatly increased the number of taxa shown to vary in abundance cross-shelf (virtually all species) and also demonstrated major cross-shelf changes in diversity and trophic structure. Species richness was greatest on the midshelf, intermediate on the outer shelf, and lowest nearshore (Table 5). Species' evenness (Pielou's *J'*) increased markedly with distance of the reef from shore. There was a significantly lower biomass of algal grazers on nearshore reefs than elsewhere and a considerably greater biomass of planktivores on the midshelf than elsewhere (Table 5). Russ (in Williams *et al.,* 1986) correlated the distribution of algal grazers with algal productivity (Fig. 8). The greater biomass of planktivores on midshelf reefs than outer shelf reefs may be related to differences in the availability of food, but the lower biomass of planktivores on the nearshore reef cannot be so related (Williams *et al.,* 1986, 1988) (Fig. 8).

Russ's (1984a,b) studies confirmed that marked cross-shelf variation in the abundance of taxa occurred over all reef zones, not just the outer reef slope. In a nested analysis, he found the two major sources of variation in distribution of surgeonfishes, parrotfishes, and rabbitfishes to be among zones within reefs, and among locations cross-shelf (Table 4 and Fig. 9). When zones were examined individually and compared among reefs across the shelf, the majority of species displayed significant cross-shelf variation in abundance, irrespective of the zone examined, with cross-shelf change accounting for as much as

Table 5 Mean Number of Species, Mean Value of Pielou's Index of Evenness (J'), and Mean Biomass (kg) of Herbivores and Planktivores per 150-m^2 Site on Nearshore (N), Midshelf (M), and Outer shelf (O) reefs in the Central Great Barrier Reef[a]

Parameter	Nearshore	Midshelf	Outer shelf	SNK
Number of species	38	70	56	M > O > N
J'	0.431	0.562	0.748	O > M > N
Herbivore biomass	1.3	3.4	6.8	O = M > N
Planktivore biomass	6.3	25.3	10.1	M > O = N

[a] N, M, and O represent N1, M1, and O1 of Fig. 7, respectively. SNK: Student–Newman–Keuls test for ranking of means ($P < 0.05$). (Modified from Williams and Hatcher, 1983.)

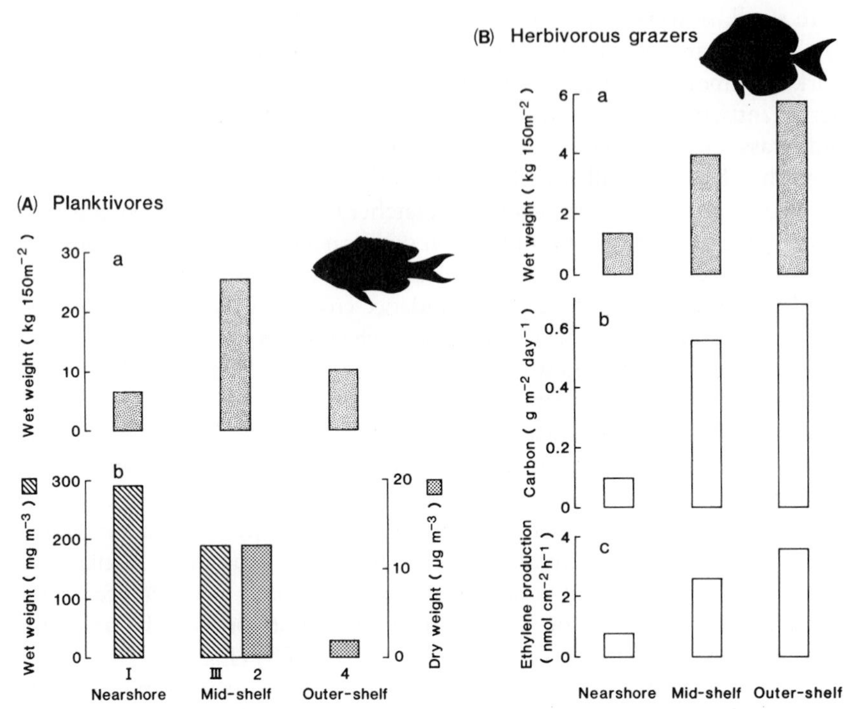

Figure 8 Cross-shelf distribution, in the Central GBR, of: (A) (a) standing crop of planktivorous fishes and (b) mean summer biomass of plankton; (B) (a) standing crop of algal-grazing fishes, (b) rates of trophic exchange from algae to grazers, and (c) rates of nitrogen fixation by algae. Nearshore, midshelf, and outer shelf reefs are N1, M1, and O1 of Fig. 6, respectively. [Modified from D. McB. Williams, G. Russ, and P. J. Doherty (1986). Reef fishes: Large-scale distributions, trophic interactions and life cycles. *Oceanus* **29,** 76–82.]

80–90% of the variability in abundance. For most species this was due to an absence or low abundance on nearshore reefs (Russ, 1984a). The three families differed significantly in their distributions across the shelf. Surgeonfishes increased in abundance and diversity with distance of the reef from shore. There was no significant difference in diversity or abundance of parrotfishes between outer and midshelf reefs but both variates were much less on the nearshore reefs. Rabbitfishes were more abundant and diverse on midshelf than on outer shelf reefs (Russ, 1984a,b) (Fig. 4).

3. *North–South*

An obvious extension of the cross-shelf studies was to determine whether similar changes occur along the length of the GBR and to relate the extent of north–south to east–west variation. Anderson *et al.* (1981) found a cross-

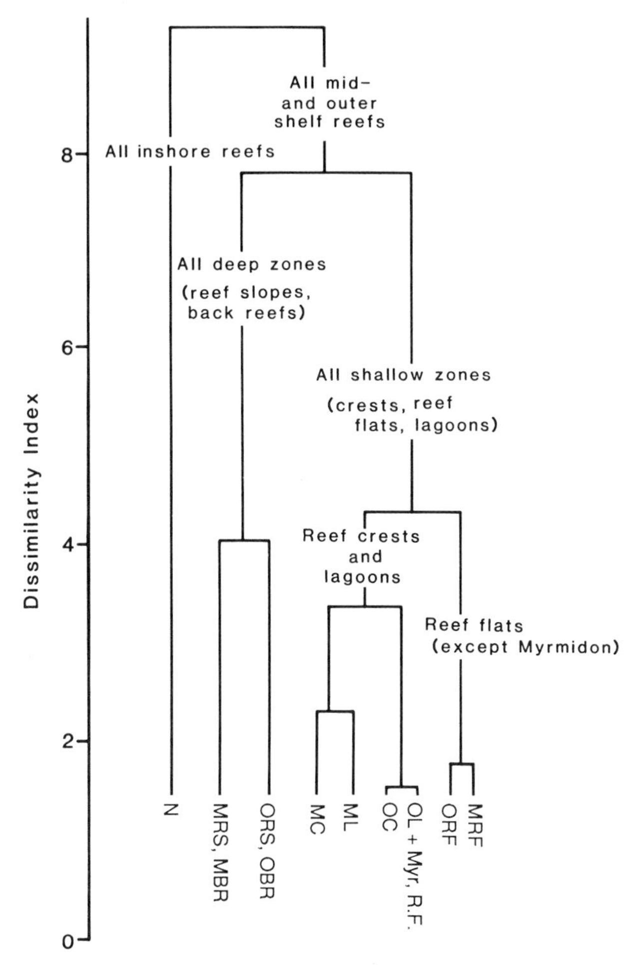

Figure 9 Dendrogram from classification analysis (incremental sum-of-squares sorting strategy, Bray–Curtis similarity coefficients) of abundance of surgeonfishes, parrotfishes, and rabbitfishes across all reefs and zones in Fig. 4 plus reef slope sites on three nearshore reefs (N1, N2, and one other not shown in Fig. 6). [Modified from Russ (1984a). Distribution and abundance of herbivorus grazing fishes in the Central Great Barrier Reef. I. Levels of variability across the entire continental shelf: *Mar. Ecol. Prog. Ser.* **20,** 23–34.

shelf pattern in the distribution of butterflyfishes at 14.5°S that was very similar to that found in the Central GBR (18.5°S) (Williams, 1982; Sale and Williams, 1982).

Williams and English (in prep.) have repeated cross-shelf transects similar to those in the Central GBR at eight different latitudes from 11°S to 23.5°S

(Fig. 10A). Preliminary results (Abel *et al.,* 1985; Williams, 1983b) emphasize a consistent cross-shelf variation in community composition, that is, communities on outer shelf reefs are more similar among latitudes than to midshelf communities at the same latitude and viceversa (groupings 3 and 2, respectively, in Fig. 10B). A notable exception to this pattern is that communities on the far northern midshelf reefs are more similar to nearshore communities elsewhere than to other midshelf communities (grouping 1 in Fig. 10B). The preliminary analysis also indicates significant latitudinal variation in the composition of communities. In particular, the midshelf (and outer shelf) communities of the Pompeys and Swains appear distinct from the midshelf (and outer shelf) communities of the more northern latitudes (Fig. 10B).

The study (Williams and English, in prep.) on geographic variation throughout the GBR indicates that north–south variation is much less than cross-shelf variation and that the cross-shelf continuum of reef fish communities in the Central GBR from nearshore to western Coral Sea reefs encompasses most of the variation observed throughout the GBR. It also shows that the seaward extent of the distribution of nearshore and midshelf communities is related to the extent of exchange of oceanic water across the shelf at any latitude and that the nature of outer shelf communities is dependent on the distance of the reef from the edge of the continental shelf.

Despite these generalizations, the complexity of the geographic patterns within the GBR is indicated by the distributions of two butterflyfishes, *Chaetodon rainfordi* and *C. plebius* (Fig. 11). These are the two most abundant species of butterflyfish at One Tree Reef (Table 3), where they are ecologically indistinguishable on the basis of work to date (Reese, 1977; Fowler, 1988). They differ greatly, however, in their geographic distributions.

A study of among-reef variation in the structure of reef fish communities along the 800-km length of the Vanuatu island chain indicates significant among-reef variation but no systematic latitudinal trends. The among-reef variation, including that between fringing reefs around high islands and platform reefs, is small compared to cross-shelf variation over far smaller distances (tens of kilometers) in the GBR (Williams, 1989).

Much greater latitudinal variation in the structure of reef fish communities appears to occur along the Hawaiian chain of reefs and islands (1500 km long) than through Vanuatu or the GBR (Hobson, 1984). This reflects, at least in part, a greater temperature variation along the chain (cooling from southeast to northwest) and a greater latitudinal variation in gross habitats, from large volcanic islands in the southeast to isolated atolls in the northwest. The species in the southeastern end of the Hawaiian archipelago show close relationships with the fishes of more tropical regions of the Pacific. The northwestward spread of some of these species is probably limited by decreasing water temperature, but others may be limited by reduced protection from prevailing winds and seas northwestward and the decreased occurrence of exposed basalt

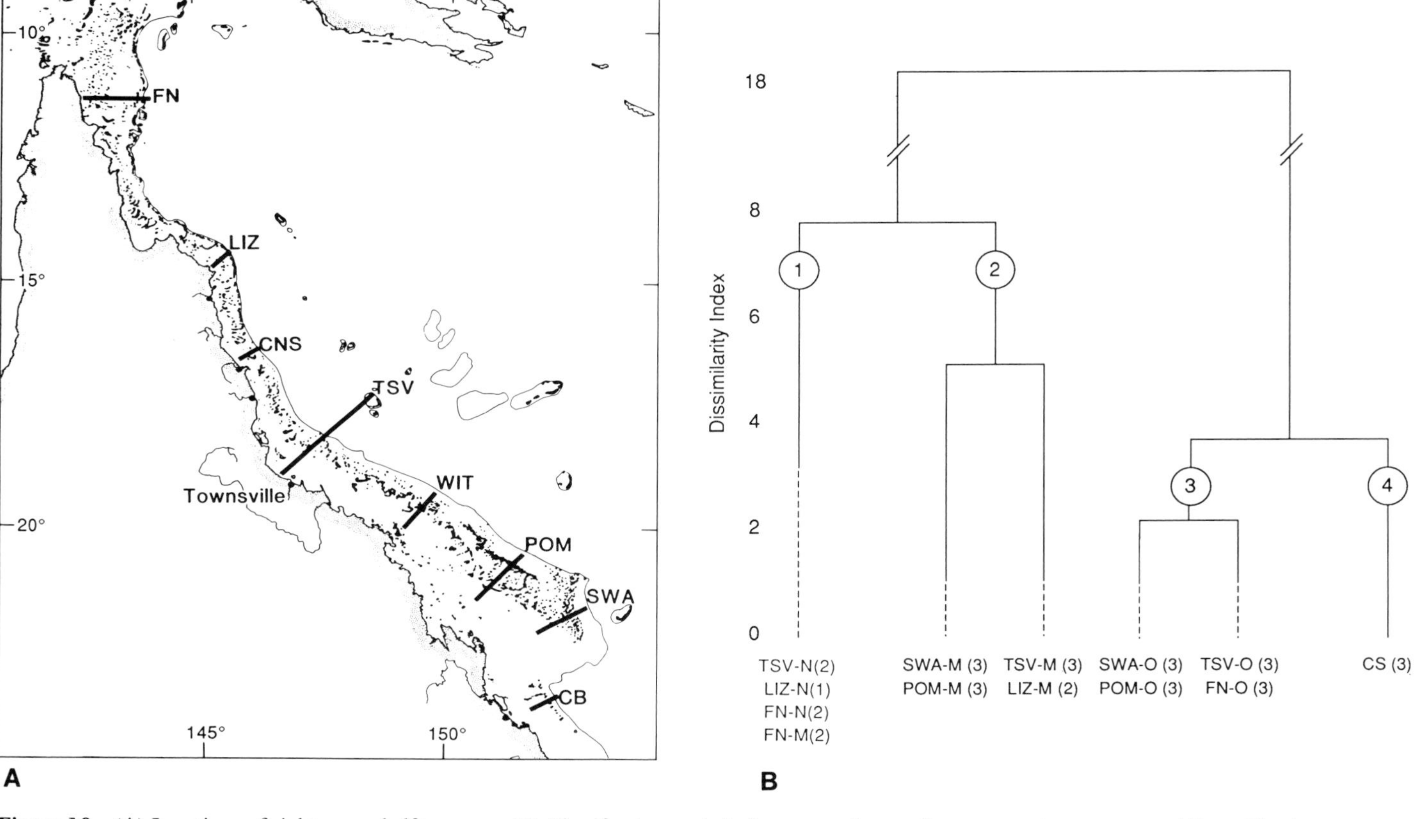

Figure 10 (A) Locations of eight cross-shelf transects. (B) Classification analysis (incremental sum-of squares sorting strategy and Bray–Curtis similarity coefficients) of 125 species of fish on 33 reefs distributed among five of the cross-shelf transects in (A). Transect abbreviations: -N, nearshore reef; -M, midshelf reef; -O, outer shelf reef; -C, Coral Sea reef. Numbers of reefs at each location are given in brackets. (Modified from Williams, 1983b.)

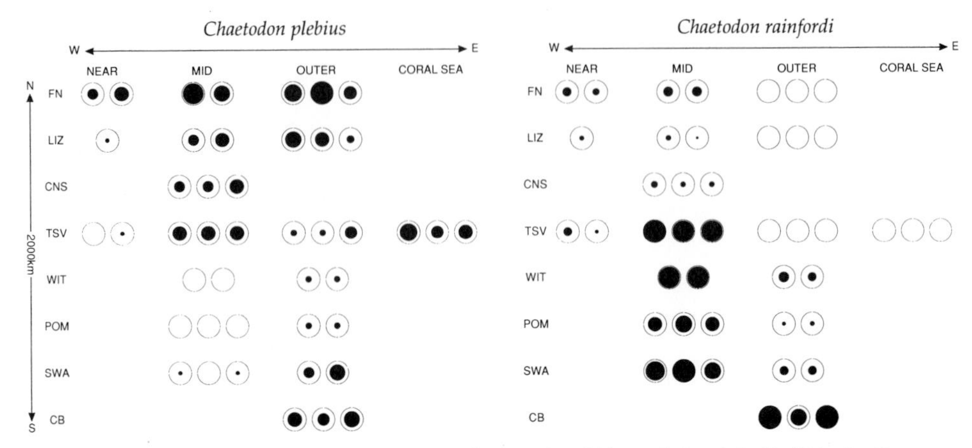

Figure 11 Schematic geographic distributions of *Chaetodon plebius* and *C. rainfordi* within the Great Barrier Reef. Each open circle represents a surveyed reef. The four columns represent four cross-shelf locations, and the eight rows represent eight different latitudes (as in Fig. 10A) arranged from north to south. Areas of solid circles are proportional to the mean relative abundance of each species on the reef slopes of each reef.

in this direction (Hobson, 1984). Certain species of reef fish that are widespread in the central Pacific are more abundant at French Frigate Shoals in the center of the Hawaiian Archipelago than elswhere in the chain. Hobson (1984) cites the study by Grigg *et al.* (1981) on the abundance of the coral genus *Acropora* at French Frigate Shoals and suggests that the same mechanism indicated by Grigg *et al.* for *Acropora* (colonization by larval transport in the subtropical countercurrent) may explain the fish distributions.

4. Geographic Locations

Exactly the same survey technique used throughout the GBR (Williams, 1982) has also been used elsewhere in the South Pacific in the Western Coral Sea, Vanuatu, and French Polynesia and on Elizabeth and Middleton reefs (Williams, 1989). The former three sites are at similar latitudes to the GBR reefs but lie progressively farther eastward in the Pacific (Fig. 12). The latter two reefs are south of the GBR and form the southernmost platform coral reefs in the world. A classification analysis of these data highlights the distinctiveness of Elizabeth and Middleton reefs; the similarity of the Vanuatu communities to those of the outer shelf of the GBR; the distinctiveness of the French Polynesian reefs from those of the western Pacific; and the difference between the two Polynesian reefs sampled (Fig. 13).

Relative species richness of four major families (surgeonfishes, butterflyfishes, damselfishes, and parrotfishes) on the Vanuatu reefs is similar to, or

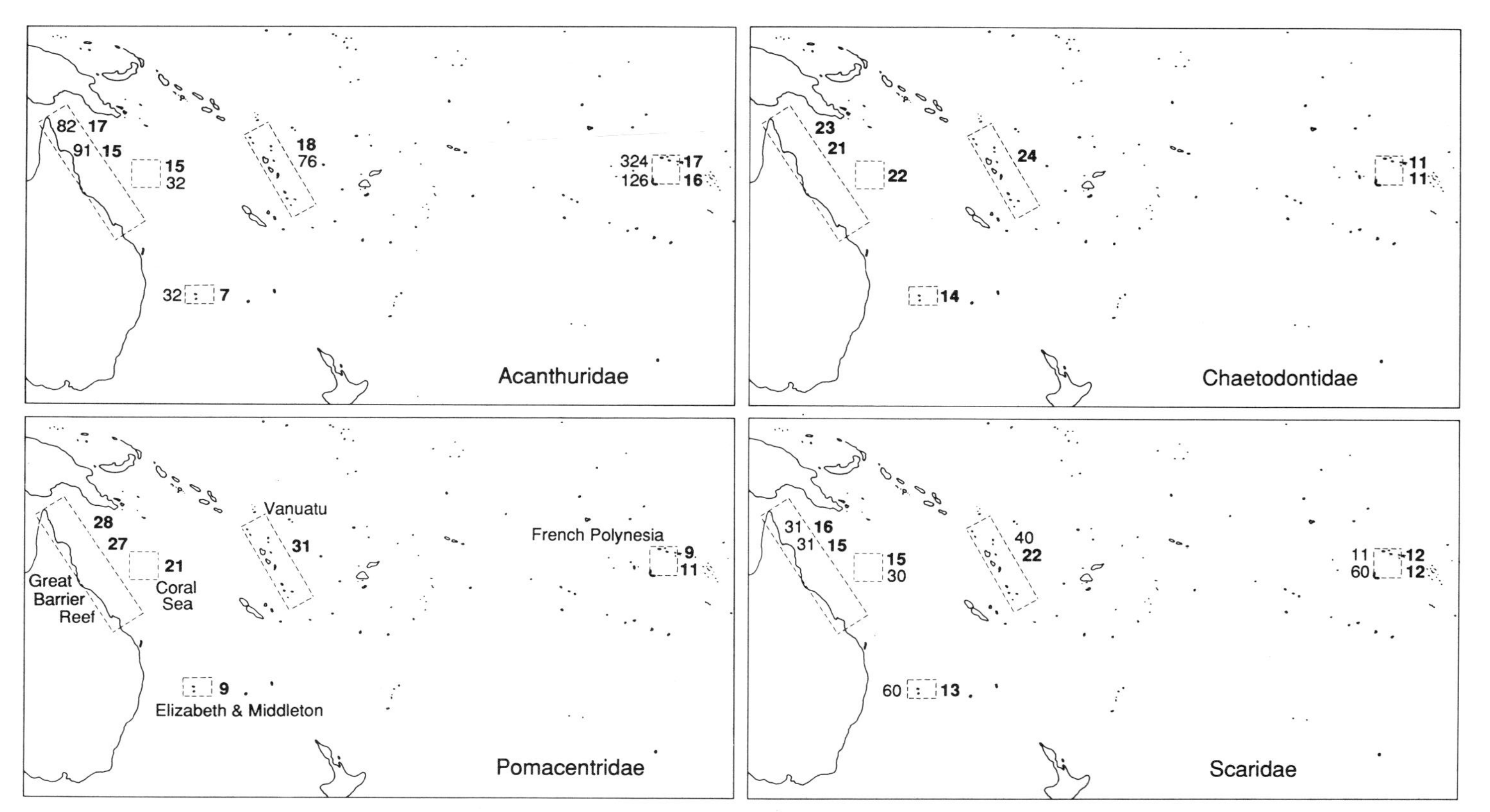

Figure 12 Distribution of four families of reef fish among four South Pacific sites. Bold figures represent total number of species observed in surveys. Other figures represent mean relative abundance of the family in surveys.

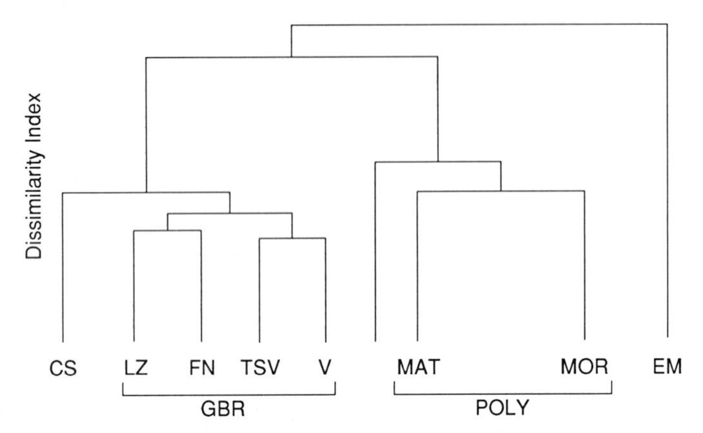

Figure 13 Dendrogram of classification (UPGMA sorting, Bray–Curtis similarity coefficients) of abundances of reef fishes in Coral Sea (CS), Great Barrier Reef (LZ, FN, TSV), Vanuatu (V), French Polynesia (MAT = Mataiva, MOR = Moorea) and Elizabeth and Middleton reefs (EM). The unlabeled group is an outlier from an aberrant Vanuatu site of extremely low diversity.

greater than, that on the outer reefs of the GBR. Species richness on the Coral Sea reefs is similar to that on the outer GBR except for the damselfishes, which are reduced. Relative species richness is generally greatly reduced in French Polynesia except for the surgeonfishes (Fig. 12). Although species richness at Elizabeth and Middleton reefs is high compared to nonreef environments, almost all the dominant species in terms of abundance are species endemic to the general region of the northern Tasman Sea and not found on the GBR (unpublished observations).

The overall changes in species richness observed here are correlated with the distances of the sites from the "center of Indo-Pacific diversity" in the vicinity of the Philippines and Indonesia (Sale, 1980a). The dominance of "endemic" species at Elizabeth/Middleton can be related to the relative isolation of these reefs and the fact that they are at the very limits of reef growth, experiencing significantly lower minimum temperatures than is normal on coral reefs.

Patterns in abundance (densities of fish) of the herbivorous surgeonfishes and parrotfishes are of particular interest. Surgeonfishes are relatively low in abundance in the western Coral Sea and at Elizabeth/Middleton but extremely abundant on the Polynesian reefs. Parrotfishes are particularly abundant at Elizabeth/Middleton and at Moorea in Polynesia but low in abundance on Mataiva in Polynesia. Moorea is a high volcanic island, whereas Mataiva, 300 km away, is an oceanic atoll. The major differences in fishes between the two reefs is the much greater abundance of surgeonfishes on the atoll and a much greater abundance of parrotfishes around the high island (Fig. 12).

The reason for the pronounced switch in relative abundances of herbivorous surgeonfishes and parrotfishes on the two Polynesian reefs is unclear. Mataiva is considerably more exposed to oceanic swells and wave action than Moorea and surgeonfishes are similarly more abundant on the exposed outer shelf platform reefs of the GBR than on the more sheltered midshelf reefs (Fig. 4). Parallel changes in abundance of parrotfishes do not occur, however, across the GBR (Williams and Hatcher, 1983; Russ, 1984a).

I originally proposed (unpublished research proposal) that because of the similarity of the oceanic waters surrounding the two Polynesian reefs, the differences in fish fauna between these two reefs should be less than that between outer and midshelf reefs of the GBR that occur closer together but in quite different pelagic environments. However, the differences are of a similar magnitude and it is not clear to what extent the differences between the two Polynesian reefs are due to one being an atoll and the other a high volcanic island or to other reasons.

Bell *et al.* (1985) compared the abundance of butterflyfishes on the outer slopes of four French Polynesian reefs, including Moorea and Mataiva, spread over 600 km. Densities of fish and species diversity differed significantly among the reefs. The extent of live coral cover explained some, but only a moderate amount, of this variance. Sites with different coral cover often held similar numbers of individuals, and others with comparable cover had different numbers of individuals. Diverse and abundant assemblages, however, were only found at sites rich in coral.

Galzin (1985, 1987a) compared the structure of communities on the outer slopes of the four reefs examined by Bell *et al.* (1985) plus that on Mehetia, approximately 200 km east of Moorea. His analysis showed that the communities on the outer slopes of the atolls with a pass through their outer rim (including Mataiva) were significantly more similar to each other than they were to the community of the atoll without a pass. Further, the community on the high island without a lagoon was more similar to that of atolls than to those on the high island with a lagoon (Moorea). Galzin hypothesized that the presence or absence of passes in atolls and of lagoons around high islands are major determinants of spatial variability in the structure of outer slope communities on these reefs.

III. Processes

Considerable pattern exists in the distribution of reef fishes over a wide range of spatial scales. These patterns are particularly strong along major environmental gradients such as among physiographic reef zones and across continental shelves from nearshore to oceanic waters. A common theme over most scales and many taxonomic or functional groupings is that distributions can often be correlated to the availability of "suitable" food and shelter. However, there is a problem of tautology here in that suitable food and shelter are to a large extent defined by the distribution of the taxon. Although the availability of these two resources does not necessarily limit the local densities of fish in a suitable habitat patch (Doherty and Williams, 1988a), they can clearly influence the distribution of fishes among habitat patches.

Identification of correlations between resources and fish distributions does not in itself identify the processes that cause the observed patterns. This section examines the role of dynamic processes, such as dispersal, habitat selection, settlement, competition, mortality, and reproduction, in determining patterns of distribution. Emphasis is on the proximate causes of distributions rather than the ultimate, evolutionary causes. The former are both more readily testable than the latter and more relevant to fisheries-related problems. I have chosen to discuss these processes under spatial headings to emphasize the scale dependence of their relative importance. While a number of experiments and studies have explored the processes determining local densities of reef fishes (reviewed by Doherty and Williams, 1988a), few studies of the distributions of reef fishes have gone beyond the descriptive stage. One reason for this, of course, is the greater difficulty of studying species over larger spatial scales.

A. Within Reefs

Given the typical length of the larval life of reef fishes of approximately 3 weeks to 3 months (e.g., Brothers *et al.,* 1983), the apparent dispersal of larvae of reef fishes, the sizes of individual reefs (generally hundreds of meters to kilometers, sometimes tens of kilometers), and generally poor stock-recuitment relationships in fecund marine fishes (references in Doherty and Williams, 1988a), it is most unlikely that local variation in reproductive output is a major process determining the distribution of fishes within a reef. Recruit availability, habitat selection, interference competition, and differential mortality are, a priori, more likely to be significant processes at this scale.

1. Within Zones

Habitat selection by presettling fishes, independent of interactions with resident fish assemblages, appears to be a major process determining the substratum type and depth distribution of many (but not all; see Doherty, 1980) small, site-attached damselfishes within reefs (D. McB. Williams, 1980), as well as their distribution among these habitat patches within zones (D. McB. Williams, 1979, 1980; Williams and Sale, 1981). Patterns of settlement can also explain most of the variation in distribution of wrasses among sites of different depth and coral cover within the One Tree Lagoon (Eckert,

1985a,b). Fowler (1988) suggested that the distribution of recruits of butterflyfish in One Tree Lagoon may relate to the distribution of their larvae, with sites farther upstream with respect to inflowing water being more likely to receive recruits. Many other species, however, such as the cardinalfishes (Apogonidae), show much more variable patterns of settlement at widely separated sites within the One Tree Lagoon and no clear relationship between the distributions of adults and new recruits (Williams and Sale, 1981). The latter species characteristically settle in groups or aggregate very quickly after settlement, are often less site-attached than most damselfishes, and suffer relatively high postsettlement rates of mortality (Williams and Sale, 1981).

Habitat selection of large shelter sites may be a major factor determining the distribution of larger fishes within reefs. Correlations between such sites and the distribution of large fish within reefs is well known to divers and fishermen. Experimental support comes from manipulations of artificial reefs in which the number of large fishes inhabiting the reefs was clearly correlated with the number of large holes present (Hixon and Beets, 1989).

The earliest observations of coral reef fishes indicated that many of the small species were overtly aggressive toward one another and maintained defended territories. Although only a relatively small proportion of all species are so territorial, interference competition clearly does affect the distributions of some species at the scale of individual territories: meters or less in the case of territorial damselfishes (e.g., Sale, 1974; Robertson, 1984) and meters to tens of meters in the case of some butterfly- and surgeonfishes (e.g., Tricas, 1985; Fricke, 1986; Robertson and Gaines, 1986). For example, the experimental removal of a territorial algal-grazing damselfish or surgeonfish is often followed by the entry of other grazers usually excluded (e.g., Sale, 1976; Robertson *et al.*, 1976, 1979; Doherty, 1983a). Where multiple territories form large continuous mosaics such as in colonies of *Acanthurus lineatus*, positive (Robertson and Polunin, 1981) and negative (Robertson *et al.*, 1979) effects on the distributions of other species may be more extensive. Few data are available on the proportion of reef habitat influenced by territorial species. Klumpp *et al.* (1987) measured the proportion of substratum occupied by damselfish territories on one reef in the central GBR. They found that territories occupied a very small proportion of the substratum on the reef slopes but could occupy more than 50% of all substratum in some parts of the reef flat.

Interference competition between resident fishes and juvenile fishes settling from the the plankton was once thought to be a major determinant of the distribution and abundance of reef fishes (Sale, 1980a). Most recent experimental studies have, however, concluded that there has been either no effect of residents on settling fishes or else a positive interaction (reviewed by Doherty and Williams, 1988a).

The possible role of differential mortality rates (due to predation) in determining within-zone distributions has not been well explored although several recent studies have documented spatial variability in rates of mortality at this scale. Aldenhoven (1986b) found that mortality rates of an angelfish varied significantly among four sites separated by 0.3 to 6.0 km. Differences between two of the areas varied 10-fold and all site-specific differences were maintained over the 5.5-year study period. Eckert (1987) found high variability in rates of mortality of wrasses within a reef zone but found few consistent spatial patterns.

In experimental studies of predation using artificial reefs, Shulman *et al.* (1983) found that successful recruitment of two taxa, including grunts, was significantly affected by the prior settlement of a juvenile piscivore to that reef. Similarly, Hixon and Beets (1989) found a negative relationship between the abundance of large fishes and the maximum number of small fishes (again mostly small grunts) present on artificial reefs. Shulman (1985a) also found that recruitment of juveniles of almost all species in a rubble/sand assemblage occurred more heavily on artificial reefs built 20–40 m away from the main reef (but in a similar depth of water) than on those built at the edge of the reef. She then convincingly demonstrated in a series of experiments that this pattern was probably due to two factors: differences in available shelter (algae and sea grass) and differences in encounter rates with predatory fish.

2. *Among Zones*

Despite the many descriptions of zonation of fishes across the physiographic zones of reefs, and correlations with apparently suitable food and shelter, little is known of the proximate causes of the observed distributions.

Habitat selection by settling fish again appears important at this scale for at least some of the damselfishes (D. McB. Williams, 1979). For example, an initial survey of the richest area of One Tree Lagoon for damselfishes found adults of 18 species, all of which are characteristic of a lagoon environment. Intensive monitoring of recruitment to this area over the next 2.5 years resulted in 1690 individuals comprising only 21 species of the more than 90 recorded from One Tree Reef (Allen, 1975). Only 9 individual recruits (4 species) were of species not in the original adult count, and of the extra 4 species, only one was not a characteristic inhabitant of adjacent lagoon habitats. Eckert (1985b) demonstrated that large differences in abundances of wrasses between lagoon and reef slope sites could also be attributed to differential patterns of settlement. Meekan (1988) found that patterns of settlement of a pomacentrid and a serranid differed among zones at Lizard Island and that rates of early postsettlement mortality did not differ significantly among the zones. He concluded that settlement patterns, rather than postsettlement mortality, would have the greatest influence on patterns of distribution of these species, at least of juvenile stages.

The relative importance of habitat selection, competition, and predation in determining among-zone distributions of species whose adults are less site-attached than damsels and/or whose juveniles recruit to different reef zones to the adults has been little studied (e.g., Robertson, 1988a). Robertson and Gaines (1986), however, convincingly argue the significance of interference competition in determining zonation patterns of many species of surgeonfishes on contiguous habitat at Aldabra. In a careful assessment they conclude that although agonistic interactions (probably due to competition for food) are a major force influencing distributions, other factors, in particular habitat selection, may also be important. Earlier I observed that species that are restricted or widely distributed (among reef zones) at one geographic location are not necessarily similarly distributed on other reefs. Robertson (in Robertson and Gaines, 1986) makes the important observation that not only do habitat-use relationships of agonistic species-pairs at Aldabra vary at different geographic locations but also their agonistic interactions vary. For example, a particular species-pair (*Acanthurus lineatus* and *Ctenochaetus striatus*) cohabits and shows no agonistic interactions at Aldabra and two other Pacific locations. At Moorea and Bora Bora (South Pacific), similar cohabitation was exceptional (and apparently habitat specific) and most *A. lineatus* defended their territories against and did not cohabit with *C. striatus*.

Given apparent correlations between distributions of species and their food, habitat selection for areas of greatest availability of preferred food is a likely proximate factor determining among-zone distributions. As a caveat, the potential problem of tautology mentioned earlier must be kept in mind. I know of no studies that have adequately tested the role of habitat selection in determining the among-zone distributions of adult fishes on coral reefs. Diel feeding migrations among zones provide circumstantial evidence.

The pronounced diel feeding migrations of many reef fish, both within and between reef zones, and the associated twilight "changeover periods" between diurnal and nocturnal species, have often been related to predation pressure (references in Hobson, 1973; Helfman *et al.,* 1982). Hobson (1972) clearly sees predation as an ultimate, evolutionary process determining these patterns: "the well-ordered pattern of events that characterizes twilight in Kona [Hawaii] also occurs on other reefs widespread in tropical seas, whether or not there exists in each of these areas today a severe threat from large piscivorous predators. The twilight pattern of actions is the result of a long evolution that in any one area transcends the existing situation and species" (Hobson, 1972).

These behavioral responses (habitat selection) are presumably reactions to both food availability and the need for shelter. Hobson's (1972) assessment begs the question of the flexibility of migration patterns in response to spatially varying food supply and predator pressure. Fishelson *et al.* (1987) provide an example for the surgeonfish *Acanthurus nigrofuscus* that indicates considerable flexibility in behavior. Two sites 1.8 km apart in the Gulf of Aqaba were studied. At the first site there was plenty of food but no hiding sites. Here there were pronounced feeding migrations over 400–500 m between shelter sites and feeding areas. At the second site, however, both food and adequate shelter sites were found side by side. Here fish did not migrate and the home range of a group did not extend more than 10 to 20 m alongshore.

I know of no studies that have demonstrated that among-zone patterns in distributions are the result of differential mortality, of either juveniles or adults, in different zones. However, given Aldenhoven's (1986b) observations of consistent, large differences in mortality rates among sites within a zone, the possibility cannot be dismissed out of hand.

The processes determining "bigger fish in deeper water" are not clear. The pattern could be the result of recruitment to shallow waters followed by migration to greater depths with age and growth, differential rates of mortality, or simply different growth rates at different depths. Different processes may determine the pattern in different species. Pauly (in Longhurst and Pauly, 1987) has suggested a simple physiological explanation of faster rates of growth in deeper water due to a decrease in routine metabolism because of decreased water temperature.

B. Among Reefs

The distribution and availability of larvae may be critical in determining the distribution of species among reefs—more so than within reefs. After settlement, few species of shallow-water reef fish (i.e., most species) are likely to move between reefs separated by kilometers or by depths greater than, say, 100 m. During the larval phase, however, such movement is both possible and in many situations likely (Williams *et al.,* 1984). Interference competition among postrecruits, on the other hand, is far less likely to be of significance as a proximate factor at this scale.

Numerous recent studies of spatial variability in the recruitment of juvenile reef fishes (Doherty and Williams, 1988a; Doherty, Chapter 10), have permitted us to determine at what stage in the life cycle various patterns are determined, and whether observed patterns of distribution are determined prerecruitment (during larval stages and at the time of settlement) or postrecruitment (involving reef-established juveniles and adults).

1. Similar Locations

Interannual variability in recruitment patterns among reefs at similar locations shows both deterministic and stochastic patterns (Doherty, 1987a; Eckert, 1984; Sale *et al.*, 1984b; Williams, 1986b; reviewed by Doherty and Williams, 1988a; Doherty, Chapter 10). While such variability may explain, at least in part, differences in abundance among reefs at similar locations (Eckert, 1984; Russ, 1984b; Williams, 1982), no studies have yet been published that demonstrably relate differences in adult densities among reefs at similar locations to differences in recruitment patterns among these reefs. The near extinction of two species (*Chromis atripectoralis* and *Pomacentrus popei*) on several reefs following extensive coral mortality due to crown-of-thorns outbreaks has, however, been related to recruitment failure as a result of habitat selection by larvae of these species against dead coral habitat (Williams, 1986a and unpublished observations).

Studies of early postsettlement survivorship of different species among similarly located reefs demonstrate significant interannual variation in rates of mortality among reefs but remarkably consistent patterns of survivorship among reefs within any one year (D. McB. Williams and S. English, unpublished observations). On this basis, differential rates of mortality of newly settled fishes do not appear to be significant in explaining among-reef differences at this scale (but see Doherty, Chapter 10).

Eckert (1984) found that the observed distributions of adult wrasses among seven reefs were relatively even despite the recruitment of larvae of six species showing consistent differences among the reefs. As she points out, however, her study did not cover the generation length of many of the study species and she could not be certain that patterns of recruitment prior to her study were similar to those observed during the study. In species where there were very large differences in recruitment among reefs, the distribution of adults tended to follow the same trend.

2. Across Continental Shelf

Cross-shelf patterns of recruitment of the majority of species examined are remarkably consistent from one year to the next and are strongly correlated with distributions of adults at this scale (Williams and English, in prep.) (Fig. 14). It thus appears that distributions of adults at this scale are determined at the time of settlement or earlier. Two sets of data suggest that the patterns are determined prior to settlement, that is, by the distribution of larvae. First, cross-shelf patterns of recruitment to identical coral heads from which all fishes were periodically removed are as distinct as patterns to areas of natural reef slope (P. J. Doherty, unpublished observations). Second, sampling of

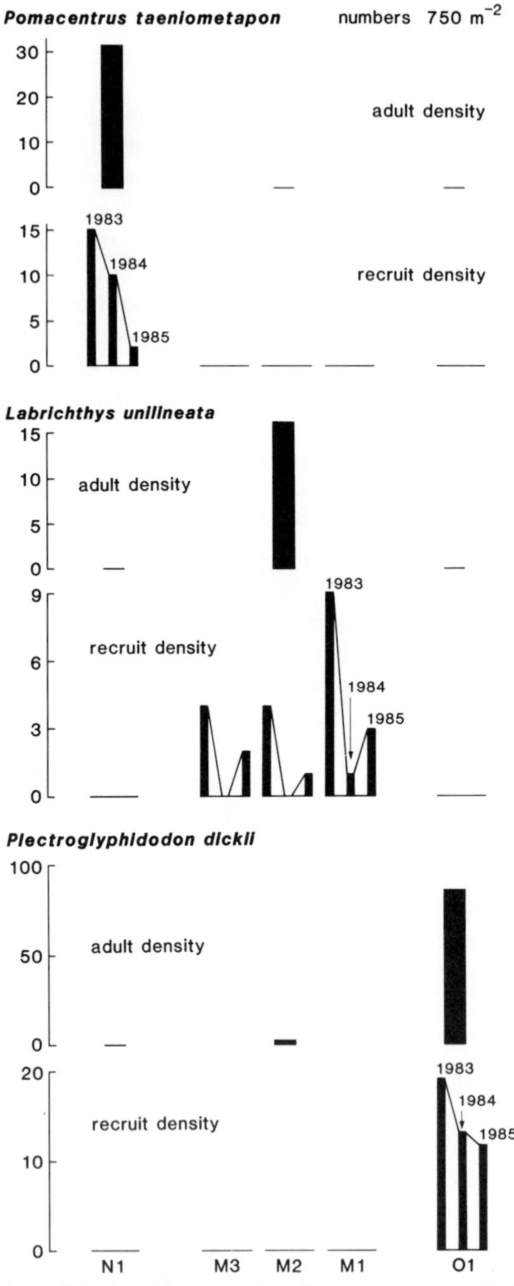

Figure 14 Densities of recruits of three species of reef fish for three successive years on one nearshore (N1), three midshelf (M1–M3), and one outer shelf (O1) reef of the Central Great Barrier Reef Transect (Fig. 7), together with

settlement stage larvae attracted to lights at night yielded primarily larvae of nearshore species in nearshore waters and midshelf species in midshelf waters (Williams *et al.,* 1986). Further validation of the hypothesis that cross-shelf patterns of recruitment, and hence adult distributions, are determined by the distribution of larvae requires more extensive sampling of the cross-shelf distribution of larvae (Williams *et al.,* 1988).

If adult distributions at this scale are determined by the distributions of larvae, what determines the distribution of larvae? Cross-shelf distributions of larvae could be determined by:

1. adult spawning sites and restricted dispersal
2. behavior of larvae, or
3. differential mortality of larvae.

Physical mixing processes, such as channelization of longshore currents by reefs, and wind, tide, and turbulent mixing by reefs (Parslow and Grubic, 1989; Williams *et al.,* 1984), suggest that the observed patterns of recruitment could not be maintained by passive drift from spawning sites alone. This leaves the probability that cross-shelf distributions of larvae are determined by the availability of suitable pelagic habitats, which vary greatly from nearshore to oceanic waters (Williams *et al.,* 1988; cf. Leis, 1986b, for a similar conclusion at the zoogeographic scale). The relative extent to which the proximate cause of larval distributions along this gradient is larval behavior or differential mortality is presently unknown.

While this scenario is speculative, it is supported by limited circumstantial evidence. We know relatively little about the life histories of the larvae of coral reef fishes (see Leis, Chapter 8), but one particular observation has been made a number of times: species with relatively small, morphologically "unspecialized" larvae tend to have nearshore distributions and those with larger, highly modified larvae tend to have offshore distributions (e.g., Leis and Miller, 1976). The large and bizarrely modified larvae of many surgeonfishes, butterflyfishes, and squirrelfishes (Holocentridae) are frequently found, for example, in the stomach contents of tuna feeding in offshore waters but are rarely found nearshore. The small and relatively unspecialized larvae of many (but not all) damselfishes and parrotfishes are often caught nearshore but not offshore (e.g., Leis, 1986b). A striking pattern occurs in cross-shelf distribu-

---

densities of adults (1981) of these species on one reef at each of these locations (N1, M1, O1). [From D. McB. Williams, G. Russ, and P. J. Doherty (1986). Reef fishes: Large-scale distribution, trophic interactions and life cycles. *Oceanus* **29,** 76–82.]

tions on the central GBR. If one looks at patterns of species richness across the shelf of the nine most diverse families of reef fish (Table 2 in Williams and Hatcher, 1983), only three families increase markedly with distance offshore: surgeonfishes, butterflyfishes and squirrelfishes. Each of these families, and only these families of all those listed, tends to have large, highly modified larvae with presumed offshore distributions. Further circumstantial evidence comes from cross-shelf distributions of species and genera within families known to vary in their larval life histories (Williams and English, in prep.) and from latitudinal variation in cross-shelf distributions (see the following).

3. North–South

Two studies have demonstrated major north–south variability in the availability of recruits in the GBR. Sweatman (1985a) indicated that rates of recruitment of two species of damselfish at Lizard Island were, on average, 10–20 times greater than in similar habitats at One Tree Reef, 1200 km to the south. Doherty, Williams, and Sale (in Doherty and Williams, 1988a) present data on regional variation in recruitment of five species of damselfishes and wrasses to five midshelf reefs in each of five regions spread over a north–south transect of 1000 km. These data indicate very strong, species-specific, regional variation in the availability of recruits. However, the authors point out that the data partially reflect differences among regions in the relative positions of reefs on the cross-shelf axis (see also Doherty, Chapter 10).

No published studies to date have attempted to compare latitudinal variation in recruitment to latitudinal variation in distributions of adult fishes. Williams and English (in prep.), however, compare cross-shelf patterns in distributions of adults of 125 species (including each of the families in Table 1 plus rabbitfishes and fusiliers (Caesionidae)) at eight different latitudes of the GBR spread over 1800 km. The oceanward or landward extent of the distribution of taxa varied with latitude and appeared closely related to water quality, in particular the exchange of oceanic water across the continental shelf. There was thus a correlation between variation in cross-shelf distributions and the width of the continental shelf (which varies from <50 to >250 km) and the extent of barrier reefs on the outer edge of the shelf.

The distributions observed by Williams and English (in prep.) provide further circumstantial evidence of the potential importance of larval habitats in determining distributions at this scale. Despite the general patterns observed, all taxa did not vary in a similar manner with latitude. At the family level, distributions of similar trophic groups did not covary as might be expected. For example, the algal-grazing surgeonfishes and parrotfishes had markedly different, rather than similar, distributions. Instead there were striking similarities in the distributions of families with similar inferred larval life histories and larval distributions (see the earlier discussion) but very different postsettlement life histories. For example, surgeonfishes and butterflyfishes had very similar distributions, as did parrotfishes and damselfishes, but the the distributions of these two pairs of families were quite different to one another and were apparently related to water quality.

I have already indicated that the considerable north–south variation in fish communities within the Hawaiian chain may relate to major variation in reef habitat. The proximate factors determining these distributions are not clear, although possibly lethal temperatures have been inferred (Hobson, 1984). The possible role of larval dispersal in the oceanic subtropical countercurrent in determining some of this pattern (Hobson, 1984) has also been discussed earlier.

4. *Geographic Locations*

While the spatial scales of some of the patterns in this section approach zoogeographic scales, the studies were aimed at preliminary comparisons of the structure of fish communities on widely separated reefs rather than testing zoogeographic hypotheses. For discussions of zoogeographic processes in reef fishes, the reader is referred elsewhere [Briggs, 1974; Springer, 1982; Rosenblatt and Waples, 1986; Hourigan and Reese, 1987; see Leis (Chapter 8), Thresher (Chapter 15), and Victor (Chapter 9) and references therein).

The observed patterns of diversity on the reefs discussed here are consistent with previous studies but the lack of a significant reduction in richness of surgeonfishes in Polynesia is curious. The data highlight differences in the fish fauna among oceanic reefs, particularly in the relative abundances of the herbivorous surgeonfishes and parrotfishes. A significant analysis of the potential factors determining these patterns awaits further research and will require a larger sample size of oceanic reefs and more intensive studies of factors determining abundances within each reef.

IV. Conclusions

It is trite to suggest that adequate explanations of distributions, as of abundances (e.g., C. L. Smith, 1978; D. McB. Williams, 1980; Thresher, 1983b; Aldenhoven, 1986b; Eckert, 1987; Jones, 1987a; Shulman and Ogden, 1987), of coral reef fishes are likely to be pluralistic and taxon- and site-specific. The challenge in a review such as this is to search for nontrivial generalizations and to generate hypotheses that might direct and stimulate further productive research.

Studies to date do suggest one generalization with respect to within-reef distributions. Patterns of recruitment are likely to be of major importance in determining the distributions of site-attached species with relatively low postsettlement mortality (such as many damselfishes). For less site-attached species and those with relatively high postsettlement rates of mortality (such as some grunts), however, postsettlement processes such as predation will be of relatively greater importance in determining distributions (cf. Williams and Sale, 1981). Parallel arguments have been made for the relative importance of presettlement and postsettlement events in determining local abundances (Shulman and Ogden, 1987; Robertson, 1988a,c; Doherty and Williams, 1988a,b). The validity of this generalization can only be determined by rigorous studies of factors determining the within-reef distributions of an ecologically diverse range of species over a diverse range of habitats.

Patterns of recruitment are likely to have a more general effect on distributions at the among-reef scale than within reefs ("reefs" being defined as separated from other reefs by kilometers or surrounded by water too deep to provide suitable habitat for postsettlement reef fishes) because at the former scale, relocation (Robertson, 1988a) of fishes is not generally possible. Shortfalls in recruitment cannot be compensated by migration from adjacent patches. The role of interannual variability in recruitment in determining distributions among reefs at similar locations on a continental shelf remains to be determined. For the species examined to date, however, patterns of recruitment do appear to be the major determinant of at least cross-shelf distributions of reef fishes. Circumstantial evidence suggests that these patterns in recruitment are likely to be determined primarily by the distribution of larvae at this scale, and I have hypothesized that the distribution of larvae is determined to a large extent by the distribution of suitable pelagic habitats for the larvae. Further testing of these hypotheses will require improved means of sampling the distribution of ichthyoplankton (e.g., Doherty, 1987b), major advances in the taxonomy of larvae at the species level, and a better understanding of the comparative life histories of larvae of different species.

Most of the data on cross-shelf patterns of recruitment have, again, concerned species with relatively low postsettlement rates of mortality. To the extent that the larval habitat hypothesis is correct, however, I would expect species with relatively higher rates of postsettlement mortality to be similarly affected. The large-scale distribution of adults of species that move kilometers or tens of kilometers between reefs during normal postsettlement ontogeny will not be similarly determined. Relatively few species have such life histories but some of these are species of major commercial importance. An example is the large "interreefal" snappers (Lutjanidae) that may settle in nearshore habitats and move offshore with age and growth, the distance moved presumably determined by the depth and width of the continental shelf.

CHAPTER 17

Predation as a Process Structuring Coral Reef Fish Communities

Mark A. Hixon
Department of Zoology and
College of Oceanography
Oregon State University
Corvallis, Oregon

I. INTRODUCTION

A. Background

A long-standing controversy on the major processes structuring communities of coral reef fishes has centered mainly around two alternative hypotheses. First, the **Competition Hypothesis** states that competition is the predominant interaction determining the abundance (and ultimately the distribution and local diversity) of fishes. The underlying assumption is that the population densities of adult fish are sufficiently high that resources are limiting. Consistent with this hypothesis are observations of resource partitioning among species (e.g., Smith and Tyler, 1972, 1973b, 1975; C. L. Smith, 1978; Anderson *et al.,* 1981; Gladfelter and Johnson, 1983; Ebersole, 1985; reviewed by Ross, 1986; Ebeling and Hixon, Chapter 18). Second, the **Recruitment Limitation Hypothesis,** as originally formalized by Doherty (1981), maintains that presettlement mortality of eggs (zygotes) and larvae determines adult patterns of abundance. This hypothesis asserts that mortality in the meroplankton results in such low recruitment that adult populations never reach levels at which resources become severely limiting, thus precluding significant competitive interactions (e.g., D. McB. Williams, 1980; Doherty, 1981, 1982, 1983a; Victor, 1983a, 1986b; Sale *et al.,* 1984a; Wellington and Victor, 1985; reviewed by Doherty and Williams, 1988a; Doherty, Chapter 10). (Here I equate "recruitment" with postlarval/prejuvenile settlement for fish that settle in adult habitat.)

Relatively little attention has been paid to a third alternative, the **Predation**

Hypothesis, which states that postsettlement mortality due to piscivory determines adult patterns of abundance. This hypothesis asserts that predation on new recruits, juveniles, and adults results in such low population sizes that severe resource limitation and competition are precluded. The relatively low level of interest in the predation hypothesis among reef fish ecologists is surprising (but see C. L. Smith, 1978; Doherty and Williams, 1988b; Glynn, 1988; and data papers cited below), especially considering the increasing attention given to predation in behavioral and community ecology in general (e.g., recent symposia: Feder and Lauder, 1986; Simenstad and Cailliet, 1986; Kerfoot and Sih, 1987), and in marine fisheries biology (e.g., Sissenwine, 1984; Rothschild, 1986).

The goals of this chapter are threefold: (1) to review and evaluate both direct and indirect evidence for the predation hypothesis; then (2) to show that the competition, recruitment limitation, and predation hypotheses are not so much alternatives as overlapping regions along a continuum of structuring processes; and finally (3) to suggest general means of testing the role of predation relative to other processes, adopting methods that have been used widely in other systems. I intentionally focus on evidence supporting (rather than falsifying) the predation hypothesis to provide some balance in a literature that is dominated by advocacy of other hypotheses. I personally advocate no particular hypothesis.

B. The Predation Hypothesis

Field studies have demonstrated the importance of predation in structuring various ecological systems. There is particularly strong evidence for predation effects in freshwater and rocky intertidal communities (reviewed by Connell, 1975; Clepper, 1979; Zaret, 1980; Paine, 1984; Sih *et al.,* 1985; Kerfoot and Sih, 1987). Previous studies have generated two general classes of predictions from the predation hypothesis, one evolutionary and the other ecological. First, if the predation hypothesis has operated through evolutionary time, then the morphology and behavior of prey fishes should exhibit patterns consistent with minimizing the risk of predation. Second, if predation is presently an important process in structuring reef fish communities, then the abundance (and consequently the distribution and local diversity) of prey fishes should shift in predictable ways as the density of piscivores or prey refuges changes through time or space.

From the ecological perspective of this review, the evolutionary class of predictions is relatively weak because its verification provides only circumstantial evidence that predation presently structures communities. However, documenting antipredatory patterns in the morphology and behavior of prey fishes is essential for formulating more directly ecological predictions, which can then be evaluated by either correlative or experimental approaches. As will become obvious, circumstantial evidence that the predation hypothesis may be true for reef fish assemblages is abundant, correlative evidence is uncommon, and experimental evidence is rare.

II. Circumstantial Evidence

Indirect evidence that piscivory has exerted a strong selective force on reef fishes is plentiful, and previous reviews have compiled numerous examples (Ehrlich, 1975; Hobson, 1975, 1979; Thompson, 1976; Huntsman, 1979; Helfman, 1986a). Such circumstantial evidence indicates the widespread occurrence of both piscivorous fishes and antipredatory mechanisms among prey fishes. The evidence falls into three categories: the ubiquity of piscivores; morphological/chemical prey defenses; and behavioral prey defenses.

A. The Ubiquity of Piscivores

Piscivores are an ever-present component of all coral reef systems. At first glance, one might argue that this assertion is false for exploited systems. However, fishing disproportionally removes large piscivores, such as groupers, snappers, and jacks (Bohnsack, 1982; Munro, 1983; Koslow *et al.*, 1988; Russ and Alcala, 1989). In fact, many smaller generalized predators, which are often not the focus of fisheries, consume new recruits, juveniles, and other small fish. Table 1 summarizes major regional surveys of reef fish trophic categories and shows that 8 to 53% of the species in an area consume other fishes. This wide range of values can be partially attributed to the relative detail of the study; the higher values are from studies that included extensive analyses of gut contents (e.g., Hiatt and Strasburg, 1960; Randall, 1967; Hobson, 1974; Parrish *et al.*, 1986). Fishes from the studies in Table 1 that ate small fish at least occasionally included members of families not normally associated with piscivory, such as squirrelfishes, cardinalfishes, goatfishes, damselfishes, and wrasses.

Piscivorous reef fishes are diverse in behavior as well as taxonomy. Hobson (1975, 1979) reviewed the predatory modes of piscivores in detail and distinguished five major categories: (1) open-water species that pursue their prey, such as jacks (Potts, 1981); (2) cryptic species that ambush their prey, such as lizardfishes (Sweatman, 1984); (3) species that apparently habituate prey to an illusion that they are nonpredatory, such as groupers and snappers (Harmelin-Vivien and Bouchon, 1976); (4) species that slowly stalk their prey, such as trumpetfishes (Kaufman, 1976); and (5) species that attack prey within crevices, such as moray eels (Bardach *et al.*, 1959). Of course, a given

Table 1 Regional Surveys on the Abundance of Piscivorous versus Nonpiscivorous Coral Reef Fishes[a]

Location	Percentage piscivore abundance by			Reference
	Species number	Fish number	Fish biomass	
Caribbean Sea				
Virgin Islands	52.8	—	—	Randall (1967)
Indian Ocean				
East Africa	13.2	—	11.0	Talbot (1965)
Madagascar	13.4	2.8	—	Harmelin-Vivien (1981)
Pacific Ocean				
Great Barrier Reef	—	—	54.0	Goldman and Talbot (1976)
Great Barrier Reef	8.0	1.0	5.7	Williams and Hatcher (1983)
Hawaii (Hawaii)	28.0	—	—	Hobson (1974)
Hawaii (Oahu)	18.4	4.0	8.3	Brock *et al.* (1979)
Hawaii (NW)	41.3	6.0	19.4	Parrish *et al.* (1986)
Marshall Islands	47.6	—	—	Hiatt and Strasburg (1960)

[a] Expanded from Parrish *et al.* (1986).

piscivore may use more than one of these modes, and the final attack by all piscivores involves a rapid strike.

Some predators employ various levels of aggressive mimicry, ranging from merely hiding among nonpredatory fishes [e.g., trumpetfishes (Aronson, 1983)] to "fishing" with modified lures resembling various prey items [e.g., anglerfishes (Pietsch and Grobecker, 1987)]. In the latter case, piscivores may actually attract and consume other piscivores by utilizing fishlike lures. For example, the anterior dorsal fin of the scorpionfish *Iracundus signifer* resembles a small fish (complete with an eyespot), which undulates as if hovering (Shallenberger and Madden, 1973). Even more remarkable is the anglerfish *Antennarius maculatus*, whose lure is extremely fishlike (Pietsch and Grobecker, 1978) (Fig. 1).

The abundance as well as the diversity of piscivorous fishes is high (Table 1). At the extreme, Goldman and Talbot (1976) reported that piscivores accounted for 54% of the total fish biomass at One Tree Island on the Great Barrier Reef. Parrish *et al.* (1986) and Norris and Parrish (1988) reported over 30% piscivores by weight and over 8% by number at one of their stations in the northwestern Hawaiian Islands.

In addition to predatory fishes, a considerable host of invertebrate and tetrapod piscivores also occur on coral reefs. Invertebrate piscivores include some anemones (Gudger, 1941), cone snails (Kohn, 1956), mantis shrimps

Figure 1 The anglerfish *Antennarius maculatus* (58.5 mm SL). The fishlike lure is a modified first dorsal spine, which moves in a circular pattern resembling swimming motions. The highly cryptic body provides this fish with an effective sit-and-wait ambush mode of piscivory. [From Pietsch and Grobecker (1978); copyright © 1978 by the American Association for the Advancement of Science.]

(Steger and Benis-Steger, 1988), asteroid seastars (Robilliard, 1971), and even ophiuroid brittlestars (Morin, 1988) (Fig. 2). Seasnakes (Voris and Voris, 1983) and seabirds (Hulsman, 1988) can also be major predators of reef fishes. Overall, there are probably many more piscivores on any given reef than most researchers suspect.

B. Morphological/Chemical Prey Defenses

The general mechanisms whereby prey minimize the risk of predation have been the subject of numerous books (Cott, 1940; Wickler, 1968; Edmunds, 1974; Curio, 1976) and review articles (Bertram, 1978; Harvey and Greenwood, 1978; Endler, 1986; Ydenberg and Dill, 1986). Because such mechanisms do not bear directly on the question of community structure, my goal

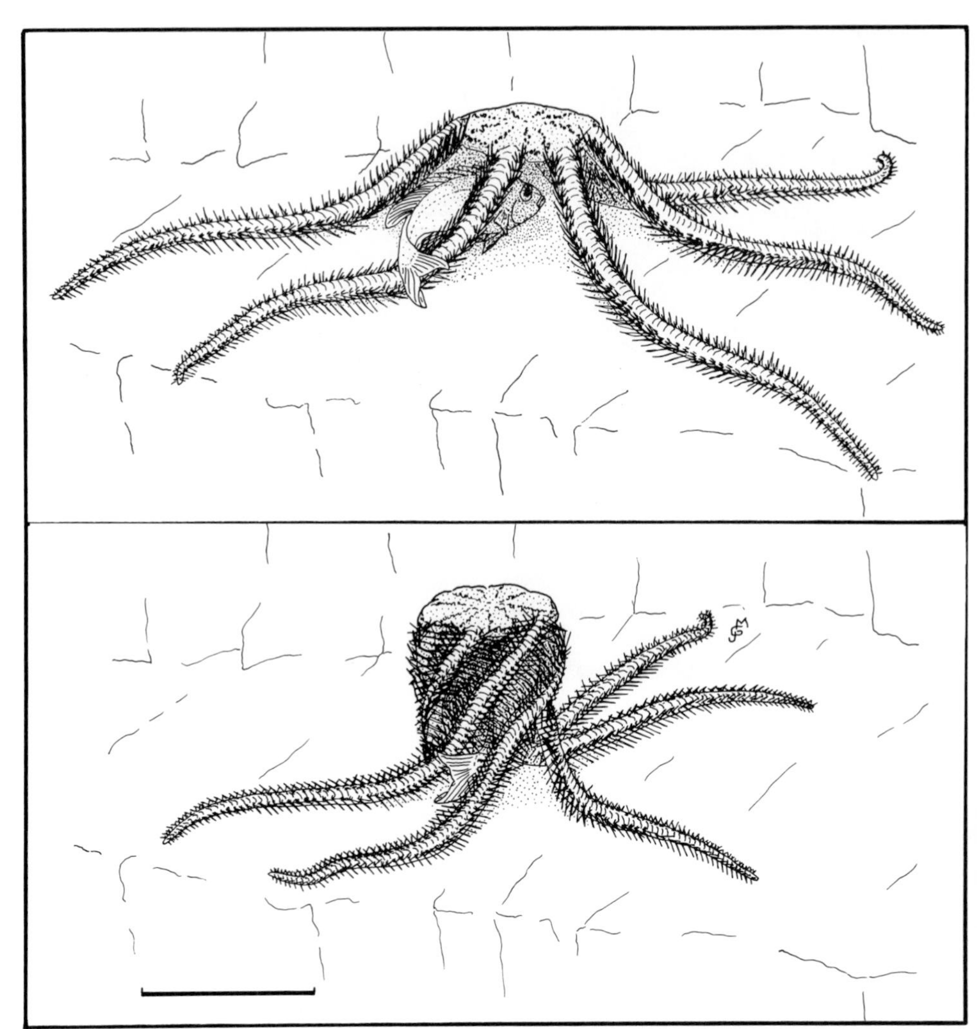

Figure 2 Piscivory by the brittlestar *Ophiarachna incrassata* (bar indicates 5 cm). *Above:* ambush posture, with a fish entering the "shelter" formed between the disc and the reef. *Below:* spiral posture, with the fish imprisoned within a helical cylinder of spines. [Reprinted from J. G. Morin (1988), in "Echinoderm Biology—Proceedings of the Sixth International Echinoderm Conference, Victoria, 23–28 August 1987" (R. D. Burke, Ph. V. Wadenor, Ph. Lambert, and R. L. Parsley, eds.), 832 pp., Hfl.155/US$80.00. A.A. Balkema, P.O. Box 1675, 3000 BR Rotterdam, Netherlands/A.A. Balkema, Old Post Road, Brookfield, Vermont 05036, U.S.A. Copyright © 1988 by A.A. Balkema Publishers.]

here is not to provide an exhaustive review, but simply to emphasize that prey defenses are both abundant and diverse among reef fishes.

1. Structures

Reef fishes exhibit a wide variety of body shapes and structures that are clearly useful in discouraging attack. Tough skin (e.g., boxfishes), fin spines (e.g., scorpionfishes), exceptionally deep bodies (e.g., angelfishes) that can be expanded in some cases (e.g., triggerfishes), and the ability to inflate (e.g., puffers) interfere with a piscivore's ability to grasp and consume its prey. The mucous envelope produced by some parrotfishes during nocturnal inactivity may inhibit olfactory detection by moray eels (Winn and Bardach, 1959). Beyond obvious structures, the threat of predation has also undoubtedly selected for the swimming morphology of prey fishes that allows for rapid escape responses (Hobson and Chess, 1978).

2. Colors

Cryptic coloration, often associated with cirri and other structural modifications, is widespread among benthic reef fishes (e.g., clinid blennies). Crypsis may involve masquerade mimicry of inedible objects (Randall and Randall, 1960), often associated with special behaviors (e.g., pipefishes resembling seagrass). Prey fish may even resemble piscivores [Batesian mimicry (reviewed by Russell *et al.,* 1976)]. For example, a harmless cardinalfish of the genus *Fowleria* strongly resembles the venomous and piscivorous scorpionfish *Scorpaenodes guamensis* (Seigel and Adamson, 1983). The plesiopid *Calloplesiops altivelis* dives into a hole when frightened, but leaves its tail exposed. The tail strongly resembles the head of the moray eel *Gymnothorax meleagris*, complete with an appropriately placed eyespot (McCosker, 1977) (Fig. 3). An eyespot on the tail, especially when combined with obliterative coloration around the eye (e.g., some butterflyfishes), may also cause piscivores to misdirect their attacks (Neudecker, 1989). Finally, conspicuous "warning" (aposematic) coloration may advertise fish that are toxic (e.g., some puffers) or otherwise unpalatable (e.g., some butterflyfishes) (Neudecker, 1989).

3. Toxins

The slowest swimming and presumably most vulnerable reef fishes, the Tetraodontiformes, include many toxic species (Halstead, 1978). Puffers contain tetrodotoxin, among the most potent of neurotoxins, and the demersal eggs and larvae of some species are unpalatable to predators (Gladstone, 1987). Trunkfishes excrete ostracitoxin, which can kill other fish in confined areas. Skin toxins (crinotoxins) have also been identified in the moray eels *Gymnothorax nudivomer* and *Muraena helena*, soapfishes, some gobies, and soles of the

genus *Pardachirus* (Halstead, 1978; Randall *et al.*, 1981; Tachibana *et al.*, 1984). The blenniid *Meiacanthus atrodorsalis* not only has toxic buccal glands providing a defensive venomous bite, but also two nonvenomous mimics (the blenniids *Ecsenius bicolor* and *Runula laudanus*) that are apparent examples of Batesian mimicry (Losey, 1972b).

C. Behavioral Prey Defenses

The most obvious ways that reef fishes avoid predation are, first, by remaining close to shelter [including urchins and anemones (reviewed by Ehrlich, 1975)] and, second, by dodging a piscivore's final attack. Helfman (1986a, 1989) reviewed the behavioral interactions between prey fish and approaching piscivores. Three more subtle, but clearly important means of avoiding predation involve schooling behavior, spawning patterns, and daily activity patterns. The ubiquity of these behaviors bolsters the conclusion that the risk of predation is severe and widespread on coral reefs.

1. Schooling

That prey derive antipredatory benefits from living in groups is well documented among animals in general (reviewed by Bertram, 1978; Harvey and Greenwood, 1978), as well as fishes in particular (reviewed by Radakov, 1973; Hobson, 1978; Shaw, 1978; Partridge, 1982; Pitcher, 1986). Various mechanisms have been proposed to explain how schooling (polarized or nonpolarized) lowers the risk of predation. Besides social and foraging advantages, there is general agreement that avoiding predation is a major reason why so many reef fishes school or otherwise occur in groups. Heterospecific schools are common, although the foraging advantages of such groups are difficult to separate from the safety advantages (e.g., Ehrlich and Ehrlich, 1973; Ormond, 1980; Wolf, 1987). In an unusual case, postlarval haemulid grunts school on reefs with similarly sized mysid "shrimp," which they apparently use as a source of both safety and food (McFarland and Kotchian, 1982).

There is considerable evidence that groups of reef fishes may actually mob and otherwise harass piscivores (Johannes, 1981; Dubin, 1982; Motta, 1983; Donaldson, 1984; Sweatman, 1984; Ishihara, 1987). Reciprocally, by attacking in groups, some piscivores effectively isolate individual prey from schools (Major, 1978; Schmitt and Strand, 1982). In any case, large spawning

Figure 3 Batesian mimicry by the plesiopid *Calloplesiops altivelis*. *Above:* the model moray eel *Gymnothorax meleagris,* in typical posture with its head exposed from a reef crevice (ca. 15 cm head length). *Below:* intimidation posture by the mimic, which enters crevices when frightened but leaves its tail exposed (ca. 15 cm TL). (Photos courtesy of Tom McHugh, Steinhart Aquarium.)

aggregations of reef fishes appear to suffer little predation, although their eggs are often not so lucky (Colin, 1976, 1978, 1982, 1989; Colin and Clavijo, 1978, 1988; Bell and Colin, 1986).

2. *Spawning Patterns*

Johannes (1978a, 1981) compiled a large body of evidence suggesting that the timing, location, and behavior of spawning in reef fishes are strongly affected by the risk of predation on adults and their eggs (see also Thresher, 1984). Table 2 summarizes Johannes' interpretation of various patterns. Overall, reef fishes appear to spawn when and where both the risk of predation on adults is minimized and the probability of eggs and larvae drifting safely away from the reef is maximized. Johannes (1978a, 1981) argued convincingly that the planktonic dispersal of reef fish larvae evolved primarily as a refuge from the severe risk of predation by planktivores on reefs (documented by Hobson and Chess, 1978; Leis, 1981; Hamner *et al.,* 1988). Corroborating these ideas, Gladstone and Westoby (1988) showed that relatively invulnerable toxic reef fishes do not display the patterns listed in Table 2. Instead, toxic fishes unhurriedly court and spawn throughout the day, do not defend their toxic demersal eggs, and spawning and hatching are unrelated to tidal cycles. Note, however, that one can provide adaptive explanations for virtually any observed behavior. Johannes' (1978a) hypotheses remain largely untested (Shapiro *et al.,* 1988), and other hypotheses for larval dispersal have been proposed (Barlow, 1981; Doherty *et al.,* 1985).

3. *Daily Activity Patterns*

Reef fishes worldwide exhibit pronounced behavioral shifts associated with daily cycles of sunlight. Detailed studies of this phenomenon have been made

Table 2 Hypothesized Ways That Spawning Coral Reef Fishes Minimize the Risk of Predation from Reef-Based Piscivores and Planktivores[a]

Reproductive behavior	Hypothesized advantage
Broadcast Spawners (most species)	
Offshore spawning migration	Reduces predation on eggs and larvae
Spawning near shelter	Reduces predation on spawners
Vertical spawning rush	a. Reduces time spawners are exposed b. Reduces predation on eggs (off bottom)
Spawning during ebbing spring tides	Reduces predation on eggs (offshore)
Spawning at night	Reduces predation on spawners and eggs
Demersal Spawners	
Brood defense	Reduces predation on eggs
Live-bearers	Eliminates predation on eggs and larvae

[a] Extracted from Johannes (1978a).

in the Gulf of California (Hobson, 1965, 1968), Hawaii (Hobson, 1972), the Caribbean (Collette and Talbot, 1972), and the Great Barrier Reef (Domm and Domm, 1973). Hobson (1975, 1979) and Helfman (1978, 1986b) provide general reviews of these patterns. Crepuscular periods (dawn and dusk) are times when the specialized visual systems of both diurnal and nocturnal prey fishes, suited for detecting their own small prey, are ineffective for detecting piscivores (Munz and McFarland, 1973; McFarland and Munz, 1975c; McFarland, Chapter 2). Piscivores, on the other hand, have visual systems with maximum sensitivity in twilight, presumably because they do not require the specialized vision needed by strictly diurnal or nocturnal species. These relative constraints leave prey fishes particularly vulnerable to predation at twilight, a period when virtually all such species seek shelter. This behavior results in a brief "quiet period" when neither diurnal nor nocturnal species occupy the water column. Consistent with the risk of predation causing these patterns, smaller and more vulnerable fish seek shelter earlier and emerge later than larger individuals.

Associated with day–night shifts in activity are daily migrations between safe resting areas and relatively exposed feeding areas. Grunts spend the day schooling inactively on reefs, and after dusk migrate to nearby seagrass beds and feed (Ogden and Ehrlich, 1977). The fish migrate along predictable routes, where they are frequently intercepted by waiting lizardfish (McFarland *et al.*, 1979; Helfman *et al.*, 1982). Helfman's (1986a) manipulations of lizardfish models suggested that the timing of migration is influenced by the activity of these piscivores.

III. Correlative Evidence

The evidence reviewed in the previous section suggests that the risk of predation on coral reefs is great and that reef fishes have evolved a variety of mechanisms that minimize this risk. Such evidence is consistent with but not sufficient for demonstrating the importance of predation in structuring reef fish communities. Predation structures a prey assemblage by altering the absolute and relative abundances of species, thus affecting the distributions and perhaps the local diversity of the prey. There are both indirect and direct sources of evidence for such population- and community-level effects. In this section, I first review relevant predictions generated by the predation hypothesis, then evaluate the correlative evidence for each.

A. Predictions from Theory

Although a wide variety of specific predictions can be generated from the predation hypothesis, only a handful can presently be evaluated by data on reef fishes. Two indirect predictions can be derived from general concepts, one concerning patterns of survivorship of prey cohorts and the other concerning prey refuge space. Two more direct predictions involve relationships between the abundance of predators and prey, and between predator abundance and local prey diversity.

1. Prey Survivorship

A general pattern from the literature is that small, young, or otherwise naive animals are more susceptible to predation than larger, older, experienced adults (Murdoch and Oaten, 1975; Taylor, 1984; Werner and Gilliam, 1984). Therefore, if predation affects prey abundance, then mortality is likely to be more severe for early-aged individuals than for adults, as has been argued for reef fish eggs, larvae, and juveniles (Johannes, 1978a). The resulting prediction is that cohorts of reef fishes under intense predation should suffer disproportionally high mortality early in life. Such a pattern is called "type III" survivorship (Pearl and Miner, 1935; Deevey, 1947) (Fig. 4A).

The inferential power of this prediction is weakened (but not destroyed) by two facts. First, mortality is not necessarily equivalent to death by predation, and acts of predation among reef fishes are notoriously difficult to observe. However, other sources of mortality (physical disturbances and pathogens) can be detected by careful monitoring. In the case of physical disturbances, storms may (Lassig, 1983) or may not (Walsh, 1983) be a source of mortality for reef fishes, while hypothermal events can be locally catastrophic (Bohnsack, 1983a). Considering pathogens, I know of no evidence of fatal epidemics among reef fishes, except perhaps the occasional mass mortalities of fantail filefish (*Pervagor spilosoma*) in Hawaii (unpublished observations), the causes of which have not been determined.

Second, as discussed in detail in Section V,B,6, type III survivorship provides evidence for the predation hypothesis only if initial population densities exceed levels at which resources become limiting. However, of the three types of survivorship curves depicted in Fig. 4A, type III is the pattern that would most quickly lower population size below some competitive threshold, and is the pattern most consistent with population size being limited by predation (Deevey, 1947).

2. Prey Refuge Availability

If the predation hypothesis is true, then the possibility exists that predators cause prey to compete for refuge space (Holt, 1984, 1987; Jeffries and Lawton, 1984). In this scenario, the competition and predation hypotheses intersect. Predation limits prey population sizes, with refuge availability setting the extent of this limit, as mediated by competition for refuges. Applied to reef fishes, such predation-induced competition has been advocated as the

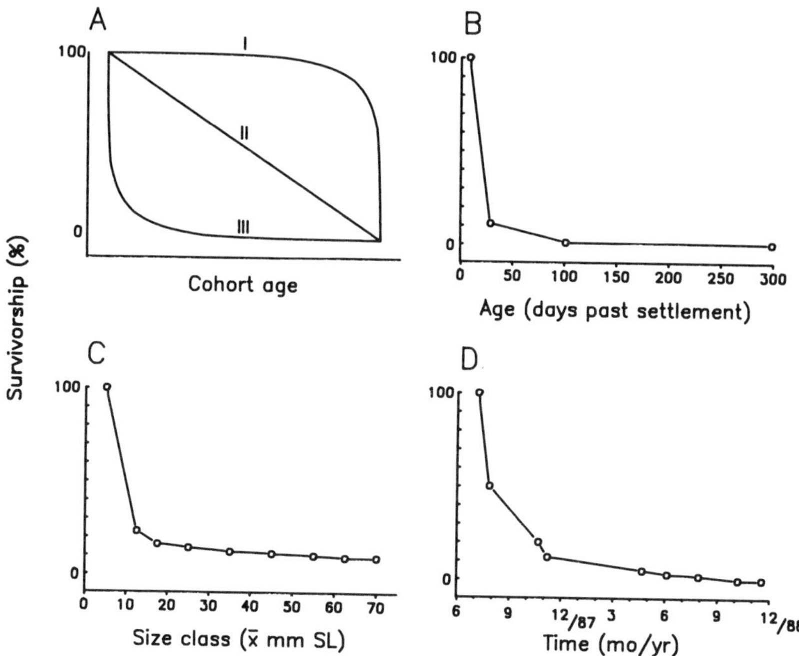

Figure 4 (*A*) The three basic types of survivorship curves, type III being most consistent with the predation hypothesis. Actual survivorship curves for: (*B*) French grunt (*Haemulon flavolineatum*) at St. Croix, U.S. Virgin Islands (extracted from Shulman and Ogden, 1987); (*C*) bluehead wrasse (*Thalassoma bifasciatum*) at the San Blas Islands, Panama (extracted from Warner and Hughes, 1988); and (*D*) blackbar soldierfish (*Myripristis jacobus*) at St. Thomas, U.S. Virgin Islands (M. A. Hixon and J. P. Beets, previously unpublished information).

hypothesis that structural shelters (holes) of the appropriate size are a primary limiting resource for reef fishes (e.g., Randall, 1963; Smith and Tyler, 1972, 1973b, 1975; Luckhurst and Luckhurst, 1978b; C. L. Smith, 1978). The correlative prediction is that comparing different reef systems should result in a positive relationship between refuge availability and prey fish densities.

Two problems in testing this prediction are, first, the potential difficulty of accurately characterizing and measuring refuge availability and, second, the fact that refuge availability may be correlated with other factors affecting fish densities independent of piscivory. In any case, the relationship between refuge availability and prey-fish densities is likely to change through time as the prey grow. In particular, prey fish may outgrow juvenile refuges and/or face different suites of predators, resulting in "ontogenetic niche shifts" between different subhabitats (reviewed by Werner and Gilliam, 1984).

3. *Predation Effects on Prey Density*

A more forthright means of detecting predation effects on community structure involves predictions concerning the relative abundances of piscivores and their prey. Applying predictions from general models of predator–prey population dynamics (reviewed by Taylor, 1984) is not directly possible because such models assume closed populations, where predator and prey birth and death rates are interactively linked. Local reef fish populations are generally open systems, where adult reproductive output is probably not linked to subsequent larval recruitment at the same site.

Nonetheless, the most basic prediction from the predation hypothesis is easy to derive without mathematics (see Warner and Hughes, 1988, for a more formal approach). First, in the absence of piscivores, the density of prey fish occupying a particular reef would reach a limit imposed by other processes, such as recruitment limitation or competition. Second, if the density of piscivores increased to a level at which they reduced the density of prey, then predation would become the predominant process limiting prey-fish abundance. The resulting correlative prediction is that, if the predation hypothesis is true, then a comparison of otherwise similar reef systems with broadly different piscivore densities should result in a negative relationship between piscivore and prey-fish densities.

4. *Predation Effects on Prey Diversity*

If the predation hypothesis is true, then local prey species richness and/or evenness should shift in predictable ways as predation intensity increases over a broad range. The two major patterns observed in other systems and predicted by various models are that prey diversity should either decline monotonically or initially increase then subsequently decrease as predation intensity increases from zero to high levels (reviewed by Hixon, 1986; Ebeling and Hixon, Chapter 18).

B. Patterns of Prey-Fish Survivorship

Having derived specific correlative predictions from the predation hypothesis, what is the evidence? As explained in Section III,A,1, cohorts of reef fishes regulated by predation are likely to exhibit type III survivorship curves (Fig. 4A). Unfortunately, field survivorship data for reef fishes have been virtually nonexistent until recently. Most studies have investigated survival only during the first few weeks after settlement, and in most cases, mortality is quite high during this period (Doherty and Sale, 1986; Victor, 1986b; Meekan, 1988; Sale and Ferrell, 1988). However, for the type III survivorship model to be tested adequately, survival must be monitored into adulthood. I am aware of seven studies that have followed survivorship for at least the first year of postsettlement life.

Four of these studies, all conducted on the Great Barrier Reef, did not report sufficient data to construct lifelong survivorship curves. Doherty (1982) found that approximately 20% of transplanted recruits of *Pomacentrus flavicauda* survived a year at One Tree Island. At the same site, Doherty (1983a) estimated 75% survival of transplanted *P. wardi* recruits after one year. This is by far the highest survivorship yet documented, yet this value may be somewhat inflated because Doherty transplanted as many fish as necessary to establish a set stable number over the first few days of each experiment (P. F. Sale, personal communication). Thus, while the pattern of *P. flavicauda* is difficult to categorize, *P. wardi* apparently exhibited either type I or II survivorship (Fig. 4A).

Aldenhoven (1986b) monitored harems of the angelfish *Centropyge bicolor* over three years at four sites at Lizard Island. She found 10-fold differences in mortality between two sites, but did not investigate causation. Although she found that mortality did not vary significantly with size, she noted that "a significant decreasing trend in mortality with increasing size may have been found in each area had more data been available" (p. 239). Such a pattern would have indicated type III survivorship. Finally, Eckert (1987) monitored survival of various wrasses on ten patch reefs at One Tree Island. She followed multiple cohorts of new recruits of 11 species for one year. Of 27 cohorts that included at least five fish initially, 24 showed monthly mortality better described statistically as declining exponentially rather than linearly, indicating type III survivorship. Moreover, in comparing 9 species for which Eckert gathered data on both new recruits and adults, I calculated that the weighted average annual mortality for recruits was 78.0%, while that of adults was 20.6%, again indicating type III survivorship.

Three more complete data sets also support the prediction of type III survivorship. Shulman and Ogden (1987) estimated the survival of French grunt (*Haemulon flavolineatum*) from settlement to about one year of age at St. Croix, U.S. Virgin Islands (Fig. 4B). Warner and Hughes (1988) summarized postsettlement, size-based survivorship data from Warner (1984) and Victor (1986b) on bluehead wrasse (*Thalassoma bifasciatum*) at the San Blas Islands off the Caribbean coast of Panama (Fig. 4C). In this case, size was approximately linearly related to age for fish up to 75 mm TL (at least one year past settlement). Finally, J. P. Beets and I are following the survivorship of cohorts of various species on isolated artificial reefs at St. Thomas, U.S. Virgin Islands. Figure 4D is a typical pattern from our study, showing the fate of a single cohort of about 100 blackbar soldierfish (*Myripristis jacobus*) on one such reef. Logarithmic plots of the three data sets in Fig. 4 are still hyperbolic, indicating that cohorts of all three species exhibited type III survivorship and suffered disproportionally high mortality early in life (Deevey, 1947). While mortality can have several sources, the investigators in all of these studies could detect no source other than predation. At the same time, virtually all studies to date have equated disappearance with mortality. The role of juvenile emigration remains largely unknown (Robertson, 1988a), indicating a need for monitoring tagged or otherwise recognizable individuals.

C. Prey-Fish Density versus Refuge Availability

Consistent with the prediction developed in Section III,A,2, there is considerable circumstantial evidence that reef fishes compete for shelter holes as refuges from predation. First, observations and experiments indicate that fish select shelter holes closely matching their body sizes, which would minimize the risk of predation within those holes (Robertson and Sheldon, 1979; Shulman, 1984; Hixon and Beets, 1989). Second, fish often defend shelter sites, suggesting that suitable holes are in short supply (Low, 1971; Hobson, 1972; Ebersole, 1977; McFarland and Hillis, 1982; Shulman, 1985a). Third, settlement (by postlarvae) and/or colonization (by juveniles and adults) is often more rapid to empty sites than to similar sites already occupied by fish, consistent with there being competition for shelter [Sale, 1976; Talbot *et al.*, 1978; Sweatman, 1985a (one of two species)]. Fourth, juveniles and adults (in addition to postlarvae) rapidly colonize artificial reefs or denuded natural reefs, suggesting that nearby natural reefs are crowded (Randall, 1963; Sale and Dybdahl, 1975; Molles, 1978; Talbot *et al.*, 1978; Bohnsack and Talbot, 1980; Shulman *et al.*, 1983; Walsh, 1985; Hixon and Beets, 1989; Bohnsack, 1990).

There have been a considerable number of community-level studies relevant to the prediction of a positive correlation between refuge availability and prey-fish densities. Unfortunately, all but one of these studies examined fishes in general, including predators and prey, as well as habitat complexity in general, which is not necessarily equivalent to refuge availability. Perhaps as a consequence, the evidence has been mixed. On one hand, both Luckhurst and Luckhurst (1978c) and Carpenter *et al.* (1981) found significant correlations between reef-substrate complexity and fish abundance in the Caribbean and the Philippines, respectively. Temporally, the abundance of resident fishes decreased dramatically following the collapse of a Japanese reef caused by the coral-eating seastar *Acanthaster planci* (Sano *et al.*, 1987). The authors associated this decline to the associated decrease in shelter availability, although food abundance was also undoubtedly affected. On the other hand, both Risk (1972) and Sale and Douglas (1984) detected no correlation between habitat complexity and fish abundance when comparing reefs in the Caribbean and the Great Barrier Reef, respectively. In a study examining a single guild, Thresher (1983a,b) similarly found no correlation between an index of topographical relief and the abundance of planktivorous fishes at the Great Barrier Reef. However, in the most thorough such study to date, Roberts and Ormond (1987) examined both structural complexity in general and hole density per se in Red Sea reefs. A stepwise multiple regression of these and other habitat variables showed that the density of holes in the reefs accounted for 77% of the variance in fish abundance.

Concerning ontogenetic niche shifts involving prey refuges, Shulman (1985b) documented that juvenile grunts in the Caribbean refuge effectively in seagrass beds, where piscivores are relatively rare. The grunts shift to sheltering in nearby reefs only when they become too large to hide among seagrass blades. This spatiotemporal correlation is more consistent with the predation hypothesis than with other explanations, especially given that adult grunts continue to forage in seagrass beds nocturnally (see Section II,C,3).

D. Prey-Fish Density versus Piscivore Density

Several studies provide correlative evidence testing the predicted negative relationship between the densities of piscivores and their prey (Section III,A,3). At One Tree Island on the Great Barrier Reef, Thresher (1983a) compared 26 patch reefs and documented a significant inverse relationship between the abundance of the piscivorous serranid *Plectropomus leopardus* and the abundances of both the diurnal planktivorous pomacentrid *Acanthochromis polyacanthus* and a group of four nocturnally planktivorous cardinalfishes. Interestingly, there was no correlation between the abundance of these potential prey and the total abundance of all piscivores on the reefs. There was also no correlation involving a group of 12 diurnally planktivorous species. In the same study, Thresher (1983b) noted that, over a year-long period, adult *Acanthochromis* disappeared on 3 of 4 reefs where *Plectropomus* occurred, but on only 1 of 20 reefs where the piscivore was absent. He also noted that the percentage mortality of 27 broods of juvenile *Acanthochromis* over 30 days was positively correlated with the mean total of all fish present, but was not correlated with the number of piscivores per se. Overall, such contradictory patterns are difficult to interpret.

At St. Croix in the Caribbean, Shulman *et al.* (1983) examined colonization and recruitment patterns on 30 concrete-block reefs (11 blocks per reef) over two months. On reefs with early immigration by small snappers (*Lutjanus* spp.), subsequent recruitment and/or survival of grunts (*Haemulon* and *Equetus* spp.) was significantly lower than on reefs where grunts settled without such piscivores. Off Miami, Florida, Bohnsack (1990) observed a similar pattern with respect to immigration by the apparently piscivorous serranid *Diplectrum formosum* and recruitment and/or survival of various species on larger artificial reefs.

In another study at St. Croix, Shulman (1985b) followed recruitment to meter-square plots of conch shells placed various distances from a reef. She found increasing rates of recruitment with increasing distance from the reef. Typical of the Caribbean, the reef was surrounded by a grazed bare "halo zone" that gave way to a seagrass bed farther from the reef. In a series of experiments, Shulman manipulated natural and artificial seagrass, which provided shelter for new recruits, and followed the fates of juvenile grunts tethered at various distances from the reef. Her results suggested that the risk of predation facing new recruits decreased with increasing distance from the reef, following the predicted pattern.

Finally, Hixon and Beets (1989) examined fish assemblages that developed on isolated, cubic-meter, concrete-block reefs (48–72 blocks per reef) at St. Thomas in the Caribbean. Over the 30-month duration of their study, they detected a highly significant negative regression between the number of resident piscivores (moray eels, squirrelfishes, and groupers) and the maximum number of potential prey fish (defined as fishes small enough to be consumed by the piscivores, mostly juvenile grunts) occupying a reef (Fig. 5A).

At the level of individual species, Hixon and Beets (unpublished data) followed the fates of recruit cohorts of the damselfish *Chromis cyaneus,* which settled simultaneously and in nearly equal abundance among three adjacent reefs. Survivorship of these cohorts over the first 3 months following settlement was inversely correlated with the abundance of piscivores on these reefs. Taken together, these correlative studies support the idea that piscivores do affect the density of their prey.

E. Prey-Fish Diversity versus Piscivore Density

If predation structures reef fish assemblages, then local prey-fish diversity should respond to broad ranges in predation intensity (Section III,A,4). An ongoing study by J. P. Beets and myself on fish assemblages occupying artificial and natural reefs at St. Thomas provides the only relevant data of which I am aware. Figure 5B illustrates the same reef censuses as Fig. 5A (see above), showing the number of potential prey species as a function of the number of resident piscivores occupying a reef. Assuming that predation intensity increased with piscivore density, this statistically significant pattern suggests that predation negatively affected local prey species richness. While there are a number of possible mechanisms that could produce this pattern (reviewed by Hixon, 1986; Ebeling and Hixon, Chapter 18), it appears that piscivores in this system simply extirpated locally rare species.

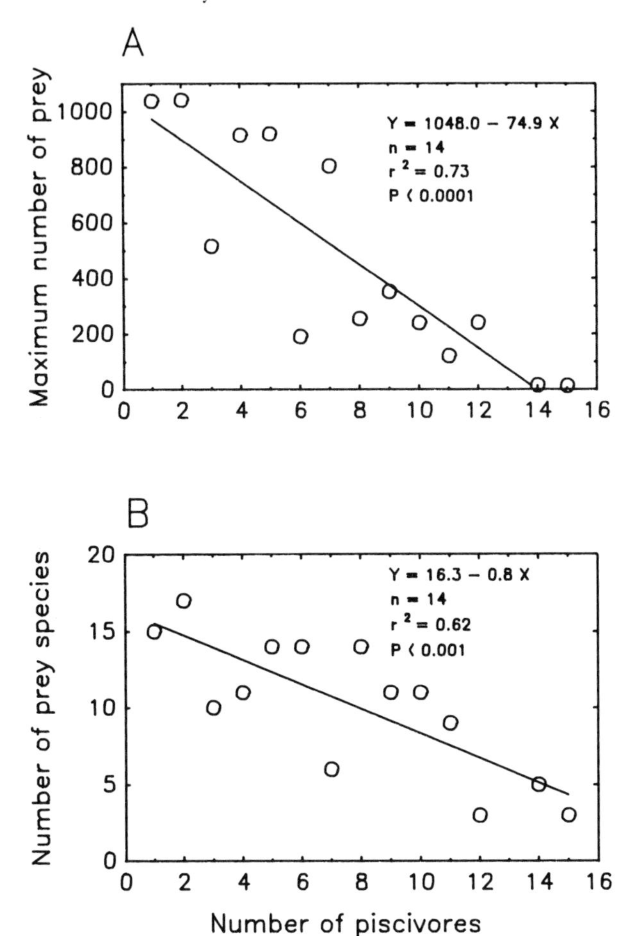

Figure 5 (*A*) Maximum observed number of potential prey fish as a function of the number of piscivorous fish occupying artificial reefs at St. Thomas, U.S. Virgin Islands. (Modified from Hixon and Beets, 1989.) (*B*) Same censuses as the previous graph, examining the number of potential prey species as a function of piscivore abundance (M. A. Hixon and J. P. Beets, previously unpublished information).

IV. Experimental Evidence

Two predictions developed in Section III,A have been evaluated experimentally. With suitable controls, the relatively indirect prediction of a positive relationship between refuge availability and prey-fish densities can be tested by manipulating shelter holes or fish densities. Similarly, the more direct prediction of a negative relationship between piscivore and prey-fish densities can be tested by population manipulations of either predators or prey. Unfortunately, few such experiments have been attempted, and those manipulating piscivore densities have encountered problems in design or implementation.

A. Prey-Fish Density versus Refuge Availability

Five separate studies have manipulated refuge availability with suitable controls, three involving natural shelters and two involving artificial reefs.

1. Natural Refuges

At the San Blas Islands in the Caribbean, Robertson and Sheldon (1979) investigated possible nocturnal shelter limitation for bluehead wrasse. When natural shelter holes were removed, displaced fish found unoccupied shelters nearby. When fish were removed, few of the vacated shelters were used by other fish. When fish were added to a reef, the immigrants readily found unoccupied shelters and survived for the 2-month duration of the study. At the same site, Robertson *et al.* (1981) manipulated territory sites, including shelter holes, of the damselfish *Eupomacentrus* (now *Stegastes*) *planifrons* by, first, removing all substrate within territories (leaving bare sand) and, second, removing half of three entire patch reefs (originally measuring 5–11 m in diameter). These manipulations effectively increased local population densities by 50%, yet the displaced fish readily reestablished territories near their old sites and persisted through the 3-month duration of the former experiment and the year-long duration of the latter. Clearly, refuge availability did not limit local population size for either the wrasse or the damselfish.

At the Red Sea, Fricke (1980) manipulated shelter for social groups of the damselfish *Dascyllus marginatus*, which permanently occupy distinct coral heads. Translocating equal-sized heads that normally supported about three to four fish, he established isolated "blocks" of one, three, and six heads and seeded each with various combinations of six fish. Following the prediction, subsequent survival was significantly greater on the larger blocks. Fricke noted that more fish were aggressively expelled from the smaller, crowded blocks by dominant group members, and that many of those expelled were eaten by lizardfish and groupers.

2. Artificial Refuges

At St. Croix, Shulman (1984) monitored recruitment to small arrays of conch shells modified to provide zero, one, or two holes per shell. Over periods ranging between 3 and 8 weeks, she found greater recruitment to shells providing more holes. Similarly, comparing arrays of two concrete blocks to meter-square arrays of conch shells or branching coral, thus providing a gradient of increasing shelter, Shulman found the predicted positive relationship between the abundance of recruits and shelter availability.

Only one study of larger artificial reefs has included adequate controls for testing shelter limitation. Previous studies (e.g., Talbot *et al.,* 1978) sometimes included holeless control reefs, but invariably confounded hole size and hole number (i.e., reefs with more holes had larger holes). At St. Thomas, Hixon and Beets (1989) compared fish assemblages occupying isolated concrete-block reefs of identical size (about 1 m^3) varying only in the number (0, 12, or 24) of identically sized holes (12×14 cm). We found that, as predicted, reefs with more holes supported more fish of that size over a 30-month period. The fact that the 24-hole reefs provided more (yet less variable) refuges than comparably sized natural reefs near this site indicated that experimental refuge limitation at this scale was realistic.

B. Prey-Fish Density versus Piscivore Density

The most direct test for a causal relationship between predator and prey abundances is to manipulate experimentally either the predator or the prey populations, including both controls and replication (Connell, 1974, 1975). To my knowledge, only predator manipulations have been attempted for reef fishes, and only four such studies have been published. The first of these was more a correlative study than an experiment (Bohnsack, 1982); the second lacked replication and failed to document a significant manipulation (Stimson *et al.,* 1982); the third lacked sufficient replication (Thresher, 1983b); and the fourth included a variety of artifacts (Doherty and Sale, 1986). Because two of these studies appeared in publications that were not widely circulated (Bohnsack, 1982; Stimson *et al.,* 1982), I provide detailed summaries.

1. Removal by Spearfishing

Bohnsack (1982) compared piscivore and potential prey-fish abundances on heavily fished versus relatively unfished reefs in the Florida Keys from 1979 to 1981. Looe Key Reef ("removal") had been spearfished by sports divers for years, while some 100 km to the northeast, Molasses and French reefs ("controls") had been protected from spearfishing since 1960. Bohnsack used 20-min random-point censuses to estimate abundances of all fishes, mostly potential prey ($n=130$ censuses at the removal site, 63 and 40 at the controls), and 15-min searches to estimate only piscivore abundances ($n=33$ searches at the removal site, 17 and 12 at the controls).

The removal reef supported significantly fewer (and smaller) piscivorous fishes ($\bar{X}$ per census$=124$ piscivores at the removal site, 757 and 204 at the controls), especially snappers, which accounted for over 75% of the total piscivores. However, there was no significant difference among reefs in the total number of fish censused ($\overline{X}$ per census=273 at the removal site, 309 and 212 at the controls). Examining the 25 most abundant species, Bohnsack (1982) noted six species that were both significantly different in abundance between the removal and control reefs and not significantly different between the control reefs. In four of these cases, including the most abundant species overall (bluehead wrasse), the removal reef supported significantly more fish than the control reefs. (The other three species were the grunt *Haemulon aurolineatum*, the damselfish *Eupomacentrus planifrons*, and the wrasse *Halichoeres garnoti*.) Bohnsack concluded forthrightly: "stating that piscivorous predation is an important factor controlling community structure of reef fishes based on present evidence would be premature" (1982, p. 266). On the basis of a similar study involving various fishing methods in the Philippines, Russ (1985) reached the same conclusion (see also Russ and Alcala, 1989).

More recently, R. E. Schroeder (personal communication) attempted an experimental spearfishing removal at Midway Lagoon, Hawaii. Following one year of baseline observations, more than 2500 piscivores were removed from four of eight patch reefs over a three-year period. Despite this marathon effort, virtually no change in the prey-fish fauna could be detected; continuous immigration resulted in no net change in piscivore abundance or biomass. M. J. Shulman (personal communication) had encountered the same problem in attempting to remove moray eels from Randall's (1963) artificial reef in the Virgin Islands.

2. *Removal by Trapping*

Stimson *et al.* (1982) attempted an eel removal experiment in Kaneohe Bay, Oahu, Hawaii. Muraenids and congrids were trapped throughout the one-year study on two isolated 30-m-diameter patch reefs separated by 0.5 km. Eels captured at the control reef were tagged, measured, and released, while those captured at the removal reef were translocated to nonexperimental reefs. The gut contents of the eels were not examined. Unfortunately, the "catch per trap night" of eels did not decline appreciably on the removal reef throughout the study, bringing into question whether or not this study actually manipulated piscivore densities significantly. The abundances of potential prey fishes were estimated by counting the number of fish swimming over fixed 6-m lines per unit time. This method conceivably resulted in fish whose home ranges overlapped the lines being counted repeatedly, yet such biases were at least consistent between the reefs. Eels were not censused visually.

The results of this experiment were presented somewhat obscurely. Data on changes in the abundance of only one species (the butterflyfish *Chaetodon miliaris*) were reported, this being the third most abundant species at the beginning of the experiment (326 "line crossings"). [Note that this species is mislabeled as *Dascyllus albisella* in the second figure of the publication (J. Stimson, personal communication).] For the two most abundant taxa (initially 735 and 834 line crossings, respectively), Stimson *et al.* (1982) simply stated that "no obvious differences in relative densities or size distributions were found for *Thalassoma duperrey* [a wrasse] and *Scarus* sp. [parrotfish] between reefs over time" (p. 3). There was no appreciable change in the density of *C. miliaris* throughout the experiment at the control reef. At the removal reef, there was no change until the very last census of the experiment, when a sudden immigration of adult *C. miliaris* (>10 cm TL) occurred. Stimson *et al.* examined their census data with a discriminant function analysis, which indicated that the similarity of the two fish assemblages had diverged through time. The authors concluded that "eels alone can evidently alter community structure of reef fishes" (p. 5). Given the results as presented in this paper, this conclusion seems unwarranted.

3. *Removal by Poisoning*

Thresher (1983b) examined the effects of other fishes on the survivorship of juvenile *Acanthochromis* at One Tree Island, Great Barrier Reef. Unlike other reef fishes, the larvae and juveniles of this species remain with their parents. In June 1980 (between spawning seasons), Thresher subjected small patch reefs to one of three treatments, each of which left the resident pair of *Acanthochromis* in place: (1) unmanipulated controls (three reefs); (2) removal of all fishes, that is, all potential predators and competitors (four reefs); and (3) removal of all planktivores, presumably only competitors for food (three reefs, but only one was subsequently studied because the *Acanthochromis* disappeared). The reefs he selected had exhibited comparable juvenile survivorships the previous spawning season. The manipulations were accomplished by capturing the resident *Acanthochromis* with an anesthetic and holding them upstream while either all the remaining fish were poisoned or only planktivores were collected with anesthetic. From October 1980 to January 1981, Thresher recensused these reefs an unreported number of times. The small sample sizes did not allow meaningful statistical comparisons.

Thresher predicted that the survivorship of juvenile *Acanthochromis* over 30 days would be highest on the reefs where all other fishes were removed, intermediate on the planktivore-cleared reefs, and lowest on the control reefs. While average survivorship was indeed greatest on the completely denuded reefs, there was little difference between the planktivore-cleared and control reefs (Fig. 6). However, the patterns on the denuded reefs were similar to those on the same reefs the year before the experiment, when no manipulations were made. Thus, these results are suggestive of a predation effect, yet somewhat equivocal.

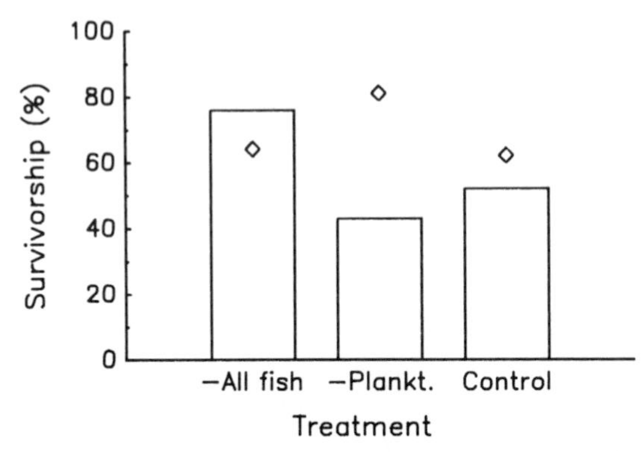

Figure 6 Mean survivorship over 30 days of juveniles of the damselfish *Acanthochromis polyacanthus* on patch reefs under three treatments: all fish except parents removed (*−All fish*; n=4 reefs), only planktivores removed (*−Plankt.*; n=1), and unmanipulated control (*Control*; n=3). The diamonds represent the mean values for the same reefs during the previous year, when no manipulations were made. (Modified from Thresher, 1983b.)

4. *Removal by Caging*

In the most carefully designed and executed study to date, Doherty and Sale (1986) monitored recruitment of fishes to caged and uncaged 3.24-m^2 plots at One Tree Island. They designed their experiment to avoid a major artifact encountered during previous unpublished and apparently unsuccessful studies: leaving piscivore-exclusion cages in place too long resulted in abundant algal growth inside (due to exclusion of herbivores), which left the results impossible to interpret accurately. In a series of four independent trials, each lasting from about 15 to 30 days and involving 7 to 11 complete visual censuses, Doherty and Sale studied 8 to 10 replicate plots subjected to one of three treatments: (1) fully caged with 10-mm wire mesh (2 m^3 per 0.6-m-high cage), which effectively excluded piscivores, but not settling postlarvae; (2) partially caged with a full roof and half of each wall; and (3) uncaged. The partial cage was designed to control for secondary cage effects, presumably allowing access by piscivores. All cages were cleaned regularly to prevent fouling.

The average outcome of all trials was that there were always more juvenile fish in the full cages than the partial cages or open plots, superficially supporting the prediction of the predation hypothesis. However, this trend was never statistically significant because of high variances; an *a posteriori* power analysis determined that reducing the chance of a type II error to 5% would have required about 90 replicates per treatment. Moreover, some fishes showed unanticipated evidence of: partial cages providing partial protection from predation; fish settling differentially on caged plots, apparently selecting the high relief offered by the cages; postsettlement movement of fish among plots; and an apparent preference by settlers for the disturbed substrate bordering the plots.

Despite these problems, which Doherty and Sale (1986) forthrightly acknowledged, one category of fishes during one trial appeared to provide a convincing pattern. Recruits of these fishes (solitary and sedentary butterflyfishes, damselfishes, and wrasses) could be recognized and followed as individuals, and they apparently did not produce the problems listed above. As a group, these fishes settled evenly among the three treatments and remained in the plots where they settled. During the first 15 days following settlement, apparent survivorship in fully caged plots was considerably greater than that in both the partially caged and open plots (Fig. 7). Surprisingly, Doherty and Sale (1986) interpreted this pattern as not so much supporting the predation hypothesis as indicating the relative importance of the recruitment limitation hypothesis: "it seems likely that density limitation in these populations occurs before settlement ... predation on recruited individuals simply widens the gap between potential and realized densities in this system" (p. 233).

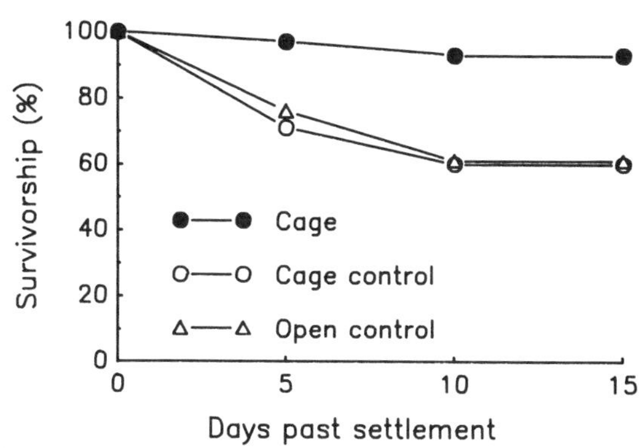

Figure 7 Survivorship of juveniles of solitary, sedentary reef fishes under three treatments: within piscivore-exclusion cages (*Cage*), within partial cages (*Cage control*), and within open plots (*Open control*). (Modified from Doherty and Sale, 1986.)

V. Synthesis

A. Conclusions from the Evidence

Overall, there is insufficient evidence to conclude unequivocally that piscivores strongly affect the absolute and relative abundances and, by extension, the community structure of reef fishes in many systems. The circumstantial evidence that piscivory has been a powerful selective agent molding the morphology and behavior of reef fishes is diverse and convincing (Section II). This evidence strongly suggests that researchers should pay more attention to the possibility that piscivory is a major process structuring reef fish assemblages.

Correlative evidence evaluating the predation hypothesis has been less abundant, yet largely supportive. All species studied through adulthood suffer disproportionally high mortality shortly after settlement (type III survivorship), apparently due to severe predation on new recruits (Section III,B). Evidence for a positive correlation between shelter availability and prey-fish density, based on the hypothesis that the risk of predation forces prey fishes to compete for refuge space, has been mixed (Section III,C). This may be because all but one of these nonexperimental studies examined fishes in general (not prey fish per se) and habitat complexity in general (not shelter availability per se). Finally, several studies have documented a significant negative relationship between piscivore and prey-fish densities, the strongest correlative inference supporting the predation hypothesis to date (Section III,D). Considering piscivore effects on local prey-fish diversity, only a single study has provided relevant data: a significant inverse relationship between piscivore abundance and prey species richness (Section III,E).

There have been very few field experiments relevant to the predation hypothesis, and these have provided mixed and ambiguous results. Tests for predation-induced competition for refuge space have demonstrated excess shelter for two local populations, yet limited shelter for three other systems (Section IV,A). Direct piscivore manipulations have faced various problems in design and implementation, and to date have produced equivocal results (Section IV,B).

B. Predation in Context: A Continuum of Processes

In this section, I offer my ideas on how we can best approach the question of what processes, including predation, structure assemblages of reef fishes. In so doing, I synthesize and extend the work of many researchers. Basically, I suggest that we stop treating different processes as being mutually exclusive, and instead examine their relative contributions to structuring communities (see also Jones, Chapter 11).

In the introduction to this Chapter (Section I,A), I simplistically state the competition, recruitment limitation, and predation hypotheses as mutually exclusive alternatives. Unfortunately, this approach has been embraced by most reef fish ecologists, resulting in an artificial controversy that has persisted for over a decade (Sale, 1984, 1988a). Despite similar controversies among community ecologists in general (see Salt, 1984; Strong *et al.,* 1984a; Diamond and Case, 1986), various researchers have reached the conclusion that most communities are not structured by a single predominant process (e.g., Strong *et al.,* 1984a; Diamond and Case, 1986; Menge and Sutherland, 1987; Hixon and Menge, 1991). Only recently have some reef fish ecologists followed suit (e.g., Shulman and Ogden, 1987; Warner and Hughes, 1988). Even the most avid proponents of recruitment limitation have lately adopted this new attitude, and now acknowledge the potential importance of predation-induced mortality after recruitment (Victor, 1986b; Doherty and Williams, 1988a,b).

Considering only the three processes treated here, it is easy to envision a combination of low larval abundance (recruitment limitation) and high postsettlement predation precluding competition for food (Victor, 1986b), yet forcing prey fishes to compete for refuge space (Holt, 1984, 1987; Jeffries and Lawton, 1984). Clearly, such complex reality would render controversy over the predominant process meaningless. A more realistic controversy would address the relative contribution of each process. Once this determination was made for all local populations belonging to a particular guild, one could then draw cogent conclusions concerning the determinants of community structure *at a particular place and time*. I stress the last part of the previous sentence because repeated overgeneralizing from one study of one species at one site over one period has unnecessarily aggravated the ongoing controversy. Indeed, given that most of the evidence for recruitment limitation has come from damselfishes on the Great Barrier Reef (reviewed by Doherty and Williams, 1988a), while most evidence favoring the predation hypothesis has come from grunts in the Virgin Islands (Shulman, 1984, 1985a; Shulman and Ogden, 1987; Hixon and Beets, 1989), we are hardly in a position to generalize either hypothesis.

As a prelude to examining experimental designs, consider the circumstances under which presettlement mortality, postsettlement predation, and competition would each be the major process structuring a local reef fish population. Extending Victor's (1986b) graphical model, Fig. 8 illustrates a variety of hypothetical survivorship curves for an average cohort, from the time of spawning (fertilization), through settlement (**S**), to the death of the longest-lived individual. The initial abundance at spawning represents the number of postlarvae that would settle on a particular reef (following dilution due to dispersal) if there was zero mortality in the plankton. I define $\mathbf{N_c}$ as the mean postsettlement population density (over the life span of the cohort) above which competition for nonrefuge resources becomes detectable through experimentation. Victor (1986b) used adult carrying capacity here, but there are problems with applying this concept (Peters, 1976; Sale, 1979a). Note that $\mathbf{N_c}$ in reality is not so much a threshold density as a range of densities, resulting in a gradient of increasing resource limitation and varying at different life stages. (Note also that the relative position of $\mathbf{N_c}$ along the ordinates and **S** along the abscissas is purely arbitrary to allow comparison between different scenarios.) In each plot on Fig. 8, the upper curve represents extreme type I postsettlement survivorship (low early-life mortality), while the lower curve represents extreme type III survivorship (high mortality, presumably due to predation; see Fig. 4A).

1. *Competition*

Figure 8A depicts the circumstances where either competition or predation is the predominant process determining the abundance and distribution of the cohort. As a result of relatively high recruitment, the initial density of recruits at settlement is above $\mathbf{N_c}$, setting the stage for competitive interactions. Subsequently, if the intensity of predation is low (upper curve region, **C**), then resources will remain limiting, and intraspecific (and possibly interspecific) competition will be the primary process determining resource use (niche breadth, etc.). Here I assume that competition will not be so severe as to grossly reduce survivorship to the point where the population density drops below $\mathbf{N_c}$. [In fact, studies of simple laboratory systems have demonstrated that severe competition can shift survivorship from type I to type II [see Fig. 4A], but virtually never to type III (Pearl, 1928; Deevey, 1947).]

2. *Predation*

If, on the other hand, postsettlement mortality due to predation is severe (lower curve region, **P,** in Fig. 8A), then piscivory has reduced the density of prey fish below $\mathbf{N_c}$, precluding competition for nonrefuge resources. This is the purest manifestation of the predation hypothesis. The possibility still exists that refuges from predation may be limiting, so that competition for shelter occurs (Holt, 1984, 1987; Jeffries and Lawton, 1984). Note that if survivorship was linear (Fig. 8A, type II), the predominant structuring process would shift from competition for nonrefuge resources early in life to predation later (perhaps including competition for refuges).

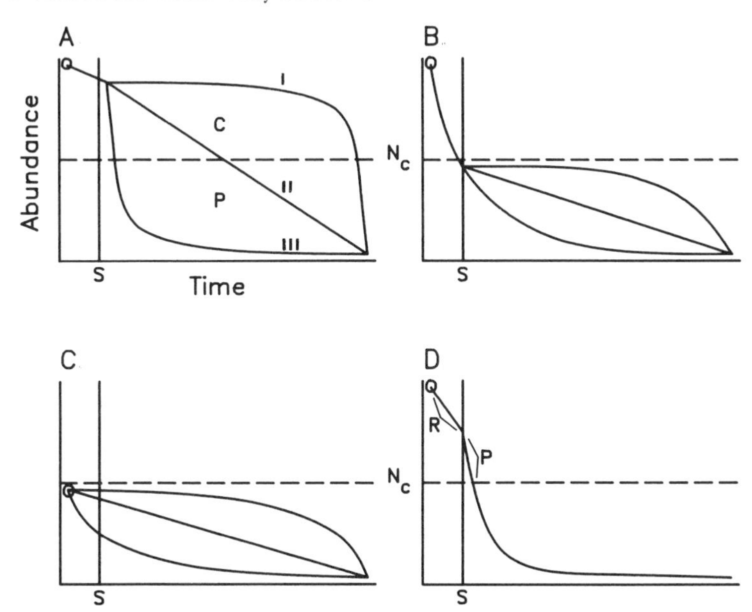

Figure 8 Hypothetical survivorship curves for a cohort of coral reef fish, with postsettlement survivorship ranging from extreme type I (upper curves) to extreme type III (lower curves). Each curve extends from the time of spawning (open circle), through planktonic larval life, postlarval settlement (**S**), juvenile, and adult life. The initial abundance (open circle) reflects the number of postlarvae that would settle on the reef if there was no mortality in the plankton. $\mathbf{N_c}$ is the mean postsettlement population size above which nonrefuge resources become limiting. The different sets of curves represent cases where the major process affecting the population is: **A,** competition (**C,** above $\mathbf{N_c}$) or predation (**P,** below $\mathbf{N_c}$); **B,** recruitment limitation; **C,** zygote limitation; or **D,** a nearly equal combination of presettlement mortality (the mechanism of recruitment limitation, **R**) and postsettlement predation (**P**). See text for discussion.

3. Recruitment Limitation

Figure 8B illustrates the purest manifestation of recruitment limitation [Victor's (1986b) "primary recruitment limitation"]. At settlement and subsequently independent of survivorship, the number of recruits is less than $\mathbf{N_c}$, precluding competition for nonrefuge resources. As before, if predation is severe, then the prey may still be forced to compete for refuge space. Moreover, to the extent that mortality in the plankton is caused by predation (reviewed by Frank and Leggett, 1985; Richards and Lindeman, 1987; Bailey and Houde, 1989), recruitment limitation can actually be considered a subset of a broader predation hypothesis.

4. *Zygote Limitation*

Figure 8C represents a possibility that, to my knowledge, has not been proposed for reef fishes. This is the situation where the initial number of zygotes at spawning is lower than $\mathbf{N_c}$. While some may consider this pattern simply a subset of recruitment limitation, Victor (1986b) asserted in his formulation of recruitment limitation that "the shortage of planktonic larvae certainly does not reflect the production of zygotes by spawning adults" (p. 145). Therefore, I believe that the distinction is important for separating causative mechanisms. If the initial number of zygotes is less than $\mathbf{N_c}$, which may be caused by any number of processes affecting adult abundance, spawning success, and/or larval dispersal, then subsequent mortality of eggs, larvae, juveniles, and adults is clearly irrelevant as far as competition for nonrefuge resources is concerned—such competition does not occur in any case. Of course, without data on the production and fate of zygotes, this "zygote limitation" hypothesis is indistinguishable from the recruitment limitation hypothesis, and therefore untestable.

5. *Combined Processes*

Thus far, I have considered mostly cases in which competition, predation, recruitment limitation, or zygote limitation is the predominant process structuring the population. Figure 8D represents what I hypothesize may approximate a typical pattern for reef fishes. As drawn, this survivorship curve shows presettlement mortality (the mechanism of recruitment limitation, **R**) and postsettlement mortality (predation, **P**) both contributing substantially to bringing the population density below $\mathbf{N_c}$. Victor (1986b) called such patterns "secondary recruitment limitation" when they occur before a cohort reaches sexual maturity. In fact, this pattern resembles the predation hypothesis (Fig. 8A, type III curve) more than the recruitment limitation hypothesis (Fig. 8B). Indeed, without the effect of predation or other postsettlement mortality, the population density exceeds $\mathbf{N_c}$ in this case, manifesting the competition hypothesis. Actually, arguing that one process is more important than the other is meaningless in this scenario. The point is that predation, especially on new recruits, can be every bit as important as presettlement recruitment limitation in precluding competition for nonrefuge resources (see also Talbot *et al.*, 1978).

6. *Conclusions*

The foregoing exercise provides three lessons for determining what structures reef fish assemblages. First, two or more processes may operate simultaneously in structuring a population or assemblage. Asserting that data consistent with one hypothesis falsify all alternatives is mere advocacy. For example, Doherty and Sale's (1986) statement that predation serves only to limit population densities below levels that are already recruitment limited (see Section IV,B,4) asserts that the type III pattern in Fig. 8B is the truth. In fact, the type III patterns in Figs. 8A, C, and D are equally as viable given their data set.

Second, predation can play an important role regardless of whether it is the predominant process structuring a population or assemblage. In all cases illustrated in Fig. 8, predation may cause important secondary effects, having ramifications for community-level interactions. In particular, whether or not population densities are sufficiently high to cause competition for food or other nonrefuge resources, predation can conceivably cause prey to compete for refuge space (Holt, 1984, 1987; Jeffries and Lawton, 1984). Further, predation can reduce prey densities beyond reductions due to other factors. In particular, if two prey species are competing for food, differential predation on the dominant may tilt the effective competitive asymmetry to the otherwise subordinate species (reviewed by Sih *et al.,* 1985).

Third, *postsettlement* survivorship patterns can suggest whether or not predation is severe, but cannot indicate the predominant process structuring a population or assemblage. That is, given that $\mathbf{N_c}$ is unknown without experimentation, the various curves in Fig. 8 are qualitatively identical to the right of the "settlement line" (**S**). What occurs before settlement remains largely an unknown "black box" (Richards and Lindeman, 1987; Doherty and Williams, 1988a). To test the models in Fig. 8 completely, one would need to document survivorship from spawning onward, which is currently an impossibility.

C. Field Experiments and Community Structure

Given the present impossibility of testing the models in Fig. 8 directly, is it possible to determine what structures assemblages of reef fishes? I believe so. If reef fish ecologists can adopt a pluralistic attitude toward community structure, then they can take advantage of multifactorial experimental designs that have proven very useful in other systems. Although all field experiments are necessarily limited in spatial and temporal scale, carefully executed experiments would provide a much more rigorous determination of structuring processes than the indirect approaches emphasized thus far. A detailed review of such experimental designs is beyond the scope of this chapter; the basic designs have been reviewed by Sih *et al.* (1985).

Given that manipulating postlarval settlement of reef fishes may be impossible in most systems, we are left with manipulating potential competitors (including new recruits) and predators by locally increasing or decreasing their densities. The simplest factorial design would involve four treatments replicated at similar yet isolated reefs: predator manipulations only; competitor manipulations only; predator and competitor manipulations; and controls. The relative response of the study species among treatments would indicate the relative importance of predation versus competition in structuring the study populations (Sih *et al.*, 1985). The less the difference in responses between control and manipulated sites, especially if the responses to all four treatments were identical, the stronger the inference that recruitment limitation (or zygote limitation) was operating. Particularly valuable for determining the effects of predation would be to monitor and compare the survivorship of local recruit cohorts among different treatments, as in Doherty and Sale's (1986) pilot study (Section IV,B,4). More complex designs would provide greater resolution and allow tests of more detailed hypotheses (Sih *et al.*, 1985), including interactions between predation, competition, and refuges in determining local prey diversity (Holt, 1987; Hixon and Menge, 1991). Of course, such designs would be limited by logistic trade-offs between the number of treatments and the number of replicates.

Surprisingly few experimental manipulations of entire local populations of coral reef fishes have been attempted, none of which have included factorial manipulations of both predators and competitors. Thresher's (1983b) small pilot experiment was perhaps the closest to date (Section IV,B,3). A major task from the outset is to identify potential predators and competitors. Predators can be inferred from food-habit studies. Conspecifics are obviously potential competitors; selecting potential heterospecific competitors requires documenting relative patterns of resource use, thus identifying guilds. Of course, complications such as individuals of one species eating the juveniles of a competitor ("intraguild predation") would strain the dichotomy between predation and competition (Werner and Gilliam, 1984; Polis *et al.*, 1989).

I reviewed four approaches to removing piscivores in Section IV,B. Caging must necessarily include adequate controls against a variety of possible artifacts (Doherty and Sale, 1986). Spearfishing, trapping, or selective poisoning of piscivores is certainly feasible (Bohnsack, 1982; Stimson *et al.*, 1982; Thresher, 1983b; R. E. Schroeder, personal communication; M. J. Shulman, personal communication), although more labor-intensive. Critical to any such manipulations would be verification that piscivore densities were actually reduced. This verification emphasizes the importance of adequate isolation of experimental reefs, which was insufficient in some previous studies (Section IV,B,1 and 2). Adding rather than removing large piscivores is probably not a viable manipulation given the extreme difficulty of capturing and translocating such predators unharmed, and the tendency of these fishes to home (M. A. Hixon and J. P. Beets, unpublished observations). Smaller nonhoming piscivores would be easier to translocate.

Regardless of the method used to manipulate piscivores, it is imperative that a substantial number of the fish capable of consuming the study species be manipulated. As reviewed in Section II,A, the impact of what can be called "diffuse predation," a prey species facing many species of predators, may be considerable for small reef fishes (including the new recruits and juveniles of any species). Analogous to the effect of "diffuse competition" (MacArthur, 1972), diffuse predation may result in the situation where the overall negative effect of all piscivore species on a prey population is substantial, even though the impact of each individual piscivore species is minor. Diffuse predation has recently been documented in the northwestern Hawaiian Islands (Parrish *et al.*, 1986; Norris and Parrish, 1988). For example, wrasses were consumed by 18 species of piscivores representing nine families. Additionally, where piscivorous species consume each other's juveniles, diffuse predation could possibly involve "predatory networks." Like "competitive networks" (Buss and Jackson, 1979), predatory networks could provide a mechanism for maintaining local diversity within a guild of piscivores. Testing such ideas will clearly require factorial experiments.

One probable reason that factorial designs have not been widely employed is that both logistic constraints imposed by working underwater and the very nature of reef fish systems inhibit their implementation. The studies reviewed here indicate two major problems. First, unless the experimental reefs are adequately isolated, immigration may negate the effects of removals, and apparent mortality may in fact be emigration. Tagging fish, including new recruits, is probably the best way to test the isolation of reefs (e.g., Hixon and Beets, 1989). Second, the reef framework inhibits a diver's ability to count, capture, or otherwise remove sheltering fish, especially piscivores like moray eels. If either or both of these problems render factorial experiments logistically impossible on natural reefs, artificial reefs could be employed to provide both sufficiently isolated replicates and shelters that are accessible to divers (M. A. Hixon and J. P. Beets, unpublished observations).

Whatever methods are used, it seems obvious that the time has come for more pluralistic experimental studies of the processes structuring assemblages of coral reef fishes. Predation is clearly one process that deserves more attention.

ACKNOWLEDGMENTS

Thanks to many colleagues for answering my pleas for information, especially J. A. Bohnsack, P. L. Colin, W. Gladstone, G. S. Helfman, J. A. Koslow, J. D. Parrish, E. S. Reese, G. R. Russ, J. St. John, and R. R. Warner. Special thanks to D. B. Grobecker, J. E. McCosker, J. G. Morin, and T. W. Pietsch for permission to reproduce their fascinating photos and drawings. Constructive comments by J. P. Beets, D. J. Booth, M. H. Carr, A. W. Ebeling, E. S. Hobson, B. A.

Menge, P. F. Sale, D. Y. Shapiro, M. J. Shulman, P. C. Sikkel, and B. N. Tissot improved the manuscript. My sincere apologies to authors whose relevant work I either overlooked or misinterpreted. Thanks to the National Science Foundation, the Center for Field Research, and the University of Hawaii for supporting my work in Hawaii, and to the Puerto Rico Sea Grant Program, the Oceanic Research Foundation, and the University of the Virgin Islands for supporting my work in the Caribbean. Ultimate thanks to M. Hixon, R. Hixon, R. Antoine, and L. Bolella, each of whom contributed in their own special ways.

CHAPTER 18

Tropical and Temperate Reef Fishes: Comparison of Community Structures

Alfred W. Ebeling
Department of Biological Sciences and
Marine Science Institute
University of California at Santa Barbara
Santa Barbara, California

Mark A. Hixon
Department of Zoology and
College of Oceanography
Oregon State University
Corvallis, Oregon

I. Introduction

The structure of a reef fish community—the local distribution, abundance, and functional relations of species—is the product of the regional pool of available species and the processes that influence the establishment and persistence of populations assembled locally from the pool (Sale, 1980a). A large body of literature describes community-level patterns of reef fishes in temperate and tropical habitats, and a more recent spate of papers provides experimental evidence of the processes that may create these patterns. Potentially important processes include competition, predation, recruitment, and abiotic disturbance, which, in some combination, interact in producing the observed structure of a local assemblage. If a reef is "saturated" with fish such that resources are limiting, then competition is a major structuring process, which may be expressed as niche diversification or a competitive lottery, or may be modified by compensatory mortality of superior competitors or shifting of species dominance in a changing environment (Connell, 1978, 1988). If the habitat is unsaturated, then the community may be structured by a host of possible mechanisms, such as the replenishment of individuals from "sources" to "sinks" (Pulliam, 1988), limitation of population sizes by low recruitment (Doherty and Williams, 1988a), or population limitation by physical disturbance or predation (Connell, 1978, 1988).

The rising tide of literature has prompted needed reviews of what is now known of these patterns and processes. Several of these reviews concentrate on the tropical communities associated with coral reefs (e.g., Ehrlich, 1975; Sale, 1980a; Doherty and Williams, 1988a); one treats the temperate communities in New Zealand (Jones, 1988a). As pointed out by Choat (1982), however, few works compare tropical and temperate reef fish communities, which are derived from noticeably different faunas and environments.

Our aim, therefore, is to compare fish communities of temperate rocky reefs with those of tropical coral reefs, in terms of their environments, structure, and possible regulatory mechanisms, to see if predictable differences occur between latitudes. Do tropical and temperate communities differ fundamentally in structure? Are these communities regulated in basically different ways? Are there enough data of the right kinds to answer these questions? A voyage in the sea of information is worthwhile if for no other reason than to set the course. We limit the scope of our review to systems occurring in analogous habitats. Tropical corals grow in shallow regions about the equator where minimum water temperatures seldom drop below 20°C during the winter; temperate reefs with characteristic forests of large brown algae occur poleward of these isotherms (Fig. 1). Coral reef fishes obviously live among living corals in relatively shallow tropical water. Since temperate reef fishes live on rocky reefs occurring at any depth, we restrict our coverage of temperate communities to those occurring in comparably shallow water where attached macroalgae can grow.

We begin by comparing the environments in which reef fishes assemble as communities. We examine the constraints and potential biases of sampling populations in temperate versus tropical environments as potential sources of systematic error in latitudinal comparisons. We then describe and compare the structure of these communities and evaluate the hypothesized models of regulation of community structure, given the known faunal and environmental differences. We conclude that, despite real differences in the structure of temperate and tropical reef fish communities, there is presently insufficient evidence for a universally appropriate model of community regulation in either system. We close by suggesting directions for future research.

II. Comparison of Environments

A. Habitat Structure

1. Hard Substrata

Temperate rocky reefs and tropical coral reefs are fundamentally different in the composition of their hard substrata. The coral reef is a complex organic matrix with many convolutions that increase spatial heterogeneity, microhabitat variety, and refuges from predation (Fig. 2). The coral itself is a source of food, although relatively few species are so specialized that they depend on

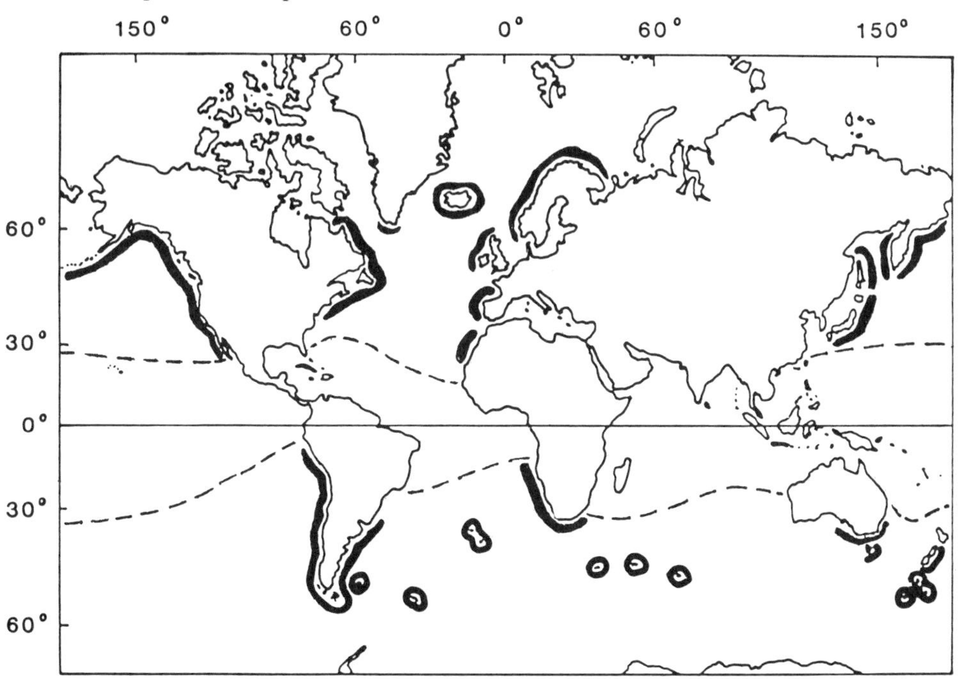

Figure 1 World distribution of kelps (Laminariales) relative to the 20°C isotherm for the coldest month of the year, which marks the boundary between tropical and temperate seas. [Adapted from M. S. Foster and D. R. Schiel (1985, Fig. 2, p. 5). The ecology of giant kelp forests in California: A community profile. U.S. Fish and Wildl. Serv. Biol. Rep. 85(7.2). Redrawn from Womersley (1954), with additional data from Chapman (1970), Briggs (1974), and Michanek (1975).]

the soft parts for their nutrition [some Chaetodontidae (Motta, 1988)]. More important, perhaps, are the interstices formed by coral skeletons that trap plankton and provide shelter for invertebrate food (Abel, 1972). Fish species diversity is often related to coral diversity or topographic complexity (reviewed by Roberts and Ormond, 1987). For example, both the diversity and abundance of Hawaiian reef fishes in Kaneohe Bay, Hawaii, were primarily correlated with coral diversity and bottom relief and only secondarily with such factors as water circulation, light, and substratum hardness (Smith *et al.*, 1973).

Temperate reefs consist of various kinds of rock, often including crevices, holes, and promontories, but lacking the enormous surface complexity of the corals (Fig. 3). Nonetheless, fish abundance and diversity may be linearly related to bottom relief (Ebeling *et al.*, 1980a) or reach an asymptote at intermediate levels of reef height and algal density (Patton *et al.*, 1985). As

Figure 2 View of a fringing coral reef along the coast of Jamaica, showing massive coral buttresses about 30 m long at 10 m depth and canyons between. [From T. F. Goreau and N. I. Goreau (1973, Fig. 2, p. 403). Sketch by P. D. Goreau. The ecology of Jamaican coral reefs. II. geomorphology, zonation, and sedimentary phases. *Bull. Mar. Sci.* **23,** No. 2, 399–464.]

parts of broad coastal headlands, often running virtually uninterrupted along continental coasts, temperate reefs are usually more continuous than coral reefs. With the important exceptions of the large barrier reefs associated with continental margins off Australia (the Great Barrier Reef) and Central America (the Belize Barrier Reef), coral reefs are relatively patchy along coasts and are usually associated with small tropical islands. Coral reef habitats may thereby comprise more patchy environments for fishes, with little or no migration of smaller individuals from one reef system to another (Barlow, 1981). This patchiness may increase the between-habitat component of fish diversity (Bradbury and Goeden, 1974; Goldman and Talbot, 1976; Sale, 1980a; Alevizon *et al.,* 1985).

2. Foliage

Besides the hard substrata, the most conspicuous difference between temperate and tropical habitat structures is the extension of macroalgae above temperate reefs (Fig. 4). Forests of large laminarian kelps and fucoid brown algae are the hallmark of temperate reefs throughout the world (Dayton, 1985; Schiel and Foster, 1986). In sharp contrast, most algae is cropped back in the tropics, leaving an unbroken view through clear water above the coral. Various vegetation layers or "canopies" in a temperate kelp forest play vital roles in structuring higher trophic levels (Dayton, 1984; Dayton *et al.,* 1984). The

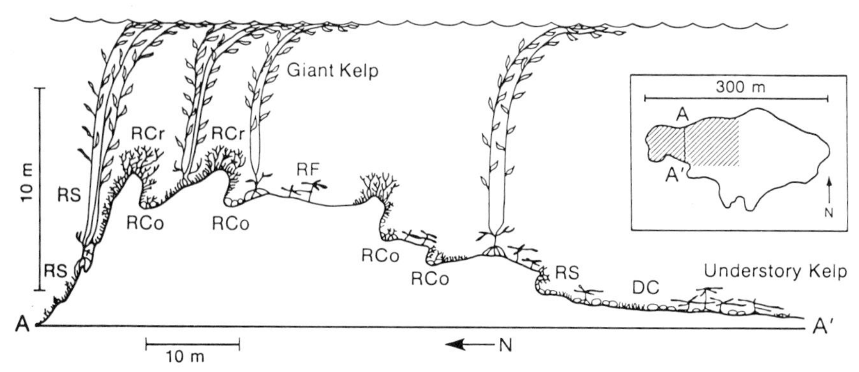

Figure 3 Section of a temperate reef (Naples Reef) off Santa Barbara, southern California, showing microhabitats of deep cobble (DC), depressions or rills with algal turf and kelp plants growing on a cobble base (RCo), the crest of the reef supporting dense bushy stands of red algae and scattered understory kelp plants (RCr), reef flat of relatively level rock base (RF), and reef slope with algal turf, bushy algae, and scattered kelp plants (RS). [From A. W. Ebeling and D. R. Laur (1986, Fig. 1, p. 126). Foraging in Surfperches: Resource partitioning or individualistic responses? *Environ. Biol. Fishes* **16,** 126. Reprinted by permission of Kluwer Academic Publishers.]

bottom or crustose layer may play a role in settlement of other organisms (Breitburg, 1984). The second or algal-turf layer, often overgrowing the prostrate crust, is composed of articulated coralline and filamentous or fleshy macroalgae. This canopy provides shelter and food for many small invertebrate grazers and detritivores, which in turn are the principal food of epibenthic fishes (Leum and Choat, 1980; Choat and Kingett, 1982; Laur and Ebeling, 1983; Holbrook and Schmitt, 1984). A third layer of understory kelp may form a dense subsurface canopy in which juvenile fishes shelter (Ebeling and Laur, 1985). In the North Pacific and some regions of the Southern Hemisphere, a fourth layer reaches upward to form a dense surface canopy. Layers of kelp are major contributors to the detritus-based food chain (Gerard, 1976; Mann, 1982; Ebeling *et al.,* 1985; Harrold and Reed, 1985; Duggins, 1988; Duggins *et al.,* 1989). Surface kelp deflects nearshore currents and thereby depresses through-transport of larvae and other plankton (Jackson and Winant, 1983; Bernstein and Jung, 1979; Jackson, 1986). Tropical reefs, on the other hand, have only the first two layers of algae at most and lack the complex filter composed of tall plants (Earle, 1972; Hay, 1981a).

Thus, temperate kelp beds provide a living vertical dimension that coral reefs lack; they extend the space available for substratum-oriented fishes seeking food and shelter among the blades and stipes (Quast, 1968b; Feder *et al.,* 1974; Coyer, 1979; Choat, 1982). Kelp beds generally support a larger biomass of fish than nearby areas without kelp (Quast, 1968c; Estes *et al.,*

Figure 4 Temperate–tropical contrast of reef structures. The upper photograph is of an area of temperate reef and understory kelp (*Eisenia arborea*), with a pile surfperch (*Damalichthys vacca*) in the foreground, off Santa Cruz Island, California; the lower photograph is of a similar area of coral reef, with a French angelfish (*Pomacanthus paru*) in the foreground, in the Caribbean Sea off Belize. (Photographs courtesy of William S. Alevizon.)

1978; Larson and DeMartini, 1984; Ebeling *et al.*, 1980b; Carr, 1989; but see Stephens *et al.*, 1984; Choat and Ayling, 1987). Kelp beds moderate wave action and form productive "edges" where reef residents and pelagic species mingle (Quast, 1968b; Miller and Geibel, 1973; Feder *et al.*, 1974; Ebeling *et al.*, 1980b; Leaman, 1980). The plant surfaces in midwater may provide essential substrata for egg attachment (Limbaugh, 1955; Quast, 1968b) or settlement of fish larvae and pelagic juveniles from out of the plankton (Quast, 1968d; Leaman, 1976; Jones, 1984a,c; Choat and Ayling, 1987; Carr, 1989). On the other hand, coral reef fishes must settle either directly onto the reef or nearby refuges such as seagrass beds (Shulman, 1984, 1985b).

Temperate nursery areas for juvenile fishes are often associated with areas of heavy foliage (Miller and Geibel, 1973; Larson and DeMartini, 1984; Carr, 1989; Holbrook *et al.*, 1989), analogous to peripheral seagrass beds in the tropics (Shulman, 1985b); in both systems, juveniles occur in shallow water, then move into deeper water as they grow older (Sale, 1969; Miller and Geibel, 1973; Jones and Chase, 1975; Choat and Kingett, 1982). Such nursery areas include tidal pools, back reef lagoons, and estuaries, but apparently not tropical mangrove systems to the extent as previously assumed (Quinn and Kojis, 1985). Temperate waters typically provide large offshore mats of drifting kelp and other algae, moving unidirectionally along the surface or bottom and used as rafts by a variety of juvenile fishes (Mitchell and Hunter, 1970; Boehlert, 1977; Jones, 1988a). A diverse assemblage of temperate fishes may shelter in accumulations of detached seaweeds nearshore as well (Lenanton, 1982). The dispersal stages in tropical systems usually drift free in the plankton (Johannes, 1978a; Thresher, 1984) and less frequently associate with detached algae (Robertson and Foster, 1982), unless swept into seaweed sinks such as the Sargasso Sea.

B. Environmental Conditions

1. Seasonality

As the well-traveled diver knows, temperate waters tend to be cold and murky and tropical waters are often warm and clear. This contrast varies seasonally and geographically. In temperate regions, water temperatures experienced by reef fishes may vary annually between 2°C and 20°C along coastal New England (Olla *et al.*, 1974), for example, but only between 10°C and 20°C off southern California (Quast, 1968a; Ebeling and Laur, 1988). Moreover, temperate waters along the western continental coasts are subject to intense upwelling, which may shift temperatures several degrees Celsius within a few hours. For example, the thermally unstable environment off southern California alters fish behavior by inducing fish to seek the bottom and shelter during exceptionally cold and turbid episodes (Quast, 1968a). Variation in thermal regimes—either seasonal or with depth—may enhance local diversity in transition zones between temperate and subtropical waters by promoting the co-occurrence of cool- and warm-water species (Stephens and Zerba, 1981). In general, temperate continental habitats are characterized by more turbid waters and fewer typical reef fishes than are habitats at offshore islands (Robins, 1971; Gilbert, 1972). Where temperate and tropical faunas overlap in the absence of well-developed coral systems, such as in the Gulf of California, the typical tropical reef species inhabit clear-water areas around precipitous rocky points (Gilligan, 1980).

Temperate fishes show varying degrees of bathymetric movements related to seasonal change in environmental conditions. In shallower water, densities may decrease as fish respond to loss of macroalgae and decreasing food supply during fall and winter (Leaman, 1980; Ebeling and Laur, 1988). Fish may migrate well offshore during the winter in more extreme climates as along New England (Olla *et al.,* 1975). In Puget Sound, where seasonal change is less severe, schooling midwater species such as rockfishes (*Sebastes,* Scorpaenidae) simply descend a few meters to deeper water, while some benthic species remain in the same shallow (<15 m deep) area of reef and kelp throughout the year (Moulton, 1977). However, in warmer waters off southern California, wherever level bottom separates inshore and offshore reefs, some species migrate inshore to more productive habitats during the winter (Stouder, 1987; Ebeling and Laur, 1988; Ebeling *et al.,* 1990).

Coral reefs grow where the minimum annual surface temperature exceeds about 20–21°C (e.g., Newell, 1971; Goreau and Goreau, 1973) (Fig. 1). Seasonal fluctuations in water temperature are slight compared to in temperate regions. While water clarity is often very high on oceanic reefs, coastal and lagoonal reefs can undergo reduced visibility due to terrestrial runoff, storms, and localized plankton blooms. Although less so than in temperate waters, seasonal change influences tropical systems. Over much of the tropics, two monsoon-related seasons replace the four solar seasons, but are characterized more by changes in patterns of wind, rain, and currents than temperature and day length (Johannes, 1978a). Compared with seasonal enrichment due to large-scale upwelling or overturn of temperate waters, however, open-ocean nutrients in the tropics are consistently low and relatively invariant in all but the eastern parts of the oceans.

In sum, seasonality increases from the tropics through subtropical and warm-temperate climates to cold-temperate zones. The stronger seasonal variation in the higher latitudes elicits greater, albeit predictable, responses in the reef fish assemblages.

2. Disturbance

Superimposed on seasonal events in both tropical and temperate regions are interannual fluctuations in climate. Of major importance on a global basis is the El Niño–Southern Oscillation (ENSO) phenomenon in the eastern Pacific Ocean. It is now thought that couplings between the tropical atmosphere and upper layers of the ocean bring about repeated appearances of warm or cool water in the equatorial eastern and central Pacific at 3- to 5-year intervals (Graham and White, 1988). Episodes may be accompanied by severe storms generated by marked changes in heat transfers between circulating air and water masses. Interannual variability in tropical environments has been ascribed to the quasi-periodic ENSO cycle (Longhurst and Pauly, 1987). During warm-water phases of a cycle, normally cool, upwelled water is displaced in the eastern Pacific. Displacements extending poleward and drastically affecting the temperate oceanographic regimes are called El Niño episodes. The intrusions of warm, nutrient-poor water during an extreme El Niño may cause deterioration of kelp forests and alterations in structure of subtidal reef communities off southern California (reviewed by Tegner and Dayton, 1987). Reduced levels of nitrates and phosphates combined with elevated seawater temperatures tend to disrupt the nitrogen and carbon metabolism of kelp and impair its growth processes (see, e.g., Germann, 1988).

Storms, however, are the major cause of physical disturbance on all shallow reefs. It seems that, at a specific location, storms are seasonally more predictable and frequent in temperate regions, but usually do not match the severity of sporadic tropical hurricanes and typhoons. In higher temperate latitudes, regular cycles of defoliation and regrowth of large algae, with associated shifts in migration and activity patterns of fishes, are correlated with seasonal storm cycles (e.g., Moulton, 1977). However, unusually strong turbulence may cause substantial mortality; for example, large numbers of dead fish from kelp beds were cast ashore with obvious injuries during a major winter storm off central California (Bodkin *et al.,* 1987). In lower temperate latitudes, major disturbances may occur less predictably and cause upheavals of community structure with indefinite recovery periods. In the Southern Hemisphere along coastal Chile, drastic environmental changes wrought by earthquakes, as well as El Niños and sporadic storms, can cause catastrophic mortalities in the intertidal and shallow subtidal reef communities (Castilla, 1988). In the warm-temperate region off southern California, unusually severe storms add a marked unpredictable element to a moderate seasonal regime (Dayton and Tegner, 1984a; Ebeling *et al.,* 1985; Estes and Harrold, 1988). Kelp forest communities were significantly altered by destruction of mature plants by storm surge during periods of unfavorable conditions for growth and altered patterns of recruitment of both plants and animals associated with a major El Niño episode in 1982–1984 (reviewed by Tegner and Dayton, 1987). Severe storms may trigger processes whereby either existing kelp forests are transformed into urchin-dominated barrens or vice versa, depending on the recent history of the community (Ebeling *et al.,* 1985; Ebeling and Laur, 1988). Thus, storm disturbances may have either deleterious or beneficial effects on kelp forests. The destruction of mature canopies opens space and admits sunlight for rejuvenation of other macroalgal assemblages by means of recolonization and succession (Dayton *et al.,* 1984; Reed and Foster, 1984; Kennelly, 1987a,b, 1989).

Where major storms occur in the tropics, cycles of community perturbation and recovery are very irregular. A host of major physical and biological disturbances, often related to storm activity, tend to perturb coral reef communities unpredictably (Endean, 1976). In the Caribbean region, for example, strong hurricanes may reset the course of ecological succession to alter the long-term patterns of coral distribution and sea urchin grazing (Woodley *et al.,* 1981; Hughes, 1989). Mobile fishes may constitute the least affected group and show most resiliency to perturbation (Springer and McErlean, 1962; Bortone, 1976; Tribble *et al.,* 1982; Walsh, 1983), although the turbulence of major tropical storms can kill fish, and population structure may be temporarily altered (Bortone, 1976; Lassig, 1983; Tribble *et al.,* 1982; Woodley *et al.,* 1981). A severe disturbance during the period of larval settlement may alter the age composition of species populations (Lassig, 1983) or chronically disrupt distributional patterns of sedentary species (Woodley *et al.,* 1981). Sporadic hypothermal events (Bohnsack, 1983a) or episodes of toxic red tides (Smith, 1979) may also be locally catastrophic for subtropical reef fish assemblages.

C. Forage Base

A reef community captures energy in two basic ways: attached plants fix light energy as primary production, and certain animals consume organic matter, including plankton, that is swept over the reef by currents (Odum and Odum, 1955). Some temperate beds of giant kelp (*Macrocystis pyrifera*) have among the highest productivities known for any natural ecosystem (Whittaker, 1970), a capacity equal to or greater than that of coral reefs and seagrass flats (McFarland and Prescott, 1959). This capacity is reflected in the large biomass of fish that temperate reefs support, about equal to or greater than that on tropical reefs (Brock, 1954; Odum and Odum, 1955; Bardach, 1959; Randall, 1963). On coral reefs, concentrations of plankton over coral constitute a major potential source of food (Herman and Beers, 1969; Goreau *et al.,* 1971). Reefs tend to act as highly efficient "filters" or "transducers" in removing and regenerating particulate organic matter by a combination of processes, including feeding by resident animals, adsorption on mucoid surfaces, trapping in eddies and backwaters, settlement of planktonic larvae, and becoming detritus upon dying (Tranter and George, 1969; Odum and Odum, 1955; Glynn, 1973; Hamner *et al.*, 1988; Hobson, Chapter 4). Recycling of nutrients within a large biomass prevents losses to the ocean and sustains a large benthic productivity (Froelich, 1983; Smith, 1983).

1. Plankton

The density and composition of tropical reef plankton vary with season, water flow over the reef, and time of day (Glynn, 1973). Nocturnal plankton includes not only the smaller daytime forms (copepods, larvaceans), but larger reef-generated forms (polychaetes, larger crustaceans) as well (Hobson, 1968, 1972; Hobson and Chess, 1986). Planktivorous fishes comprise a major means of diverting this wafted energy source into the reef community (Stevenson, 1972; Emery, 1968; Glynn, 1973; Bray, 1981; Hobson, Chapter 4), and as expected of the exploitation of a continuously renewable food supply, planktivory occurs both day and night on both temperate and tropical reefs (Schroeder and Starck, 1964; Ebeling and Bray, 1976; Hobson and Chess, 1976).

Planktivores appear to be efficient at removing plankton swept over reefs in both regions. Glynn (1973) observed that zooplankton volumes were reduced by some 60% in water streaming off a Caribbean coral reef, and Hamner *et al.* (1988) provided evidence that planktivorous fishes consumed most of the zooplankton passing over a reef on the Great Barrier Reef. Zooplankton was also depleted by resident planktivorous fishes over temperate rocky reefs off California (Bray, 1981) and New Zealand (Kingsford and MacDiarmid, 1988). Such depletion may reduce the supply of invertebrate larvae that would otherwise settle inshore (Gaines and Roughgarden, 1987). On both temperate and tropical reefs, feces generated by diurnal planktivores may be consumed directly by other fishes during the day (Robertson, 1982) or be deposited in reef crevices at night (Bray *et al.*, 1981, 1986), thus facilitating the transfer of allochthonous material and energy to the reef system.

2. Benthos

The most important source of reef-generated production providing food for fishes is the benthic community contained in a matrix of algae, sponges, tunicates, bryozoans, hydroids, and worm tubes covering much of the reef surface. Small animal inhabitants, especially amphipods, constitute the forage base for a host of demersal microcarnivores (Section IV,C). Thus, the well-being of a substantial part of the reef community depends on production associated with this carpet of "turf." On a southern Californian reef, for example, the distribution of abundant, microcarnivorous surfperches (Embiotocidae) is closely correlated with the distribution of turf, in both space and time (Ebeling *et al.,* 1985; Stouder, 1987). On coral reefs, the abundance of corallivorous butterflyfishes (Chaetodontidae) is correlated with that of the corals on which they feed (Bell and Galzin, 1984; Bouchon-Navaro and Bouchon, 1989). In the Caribbean, production from surrounding seagrass beds is imported to reefs by daily migrating grunts (Meyer *et al.,* 1983). Foraging over seagrass beds at night, the grunts defecate in their diurnal shelters on reefs, thereby fertilizing nearby corals (Meyer and Schultz, 1985a,b).

Beyond being a source of food, macroinvertebrates may interact directly and indirectly with fishes in ways that affect the benthic forage base. Sea urchins, rather than fishes, appear to be the principal macroherbivores on temperate reefs (e.g., Mann, 1982; but see Horn, 1989). As either small patches or extensive areas, barren grounds (barrens) created by overgrazing due to sea urchins have occasionally displaced kelp forests in temperate waters the world over (Lawrence, 1975; Schiel and Foster, 1986). Zones of barrens and stands of low kelp occur together naturally off New Zealand, where small microcarnivorous fishes forage in the kelp canopy and larger mesocarnivorous fishes eat macroinvertebrates from the barrens (Choat and Ayling, 1987). In the North Pacific, however, barrens often occupy areas where an important predator of urchins, such as the sea otter (*Enhydra lutris*), has been removed (reviewed by Estes and Harrold, 1988; Estes *et al.,* 1989), although the essential role of this predator in preserving the kelp forest under all circumstances is still questioned (Foster and Schiel, 1988). Indeed, urchin outbreaks decimate kelp forests in the North Atlantic, where otters have never occurred (e.g., Breen and Mann, 1976). Such outbreaks may be due to unusually strong recruitment cohorts of urchins combined with unusually favorable physical conditions (Hart and Scheibling, 1988). Although herbivorous fishes are thought to be the principal grazers in the tropics (e.g., Bakus, 1969; Hay, 1981a, 1984a), aggregations of urchins may also remove most macroalgae (Hay, 1984b, 1985; Carpenter, 1986; Foster, 1987a; Hughes *et al.,* 1987) or even living corals (Glynn *et al.,* 1979) in local patches or in the absence of their fish predators. In addition, periodic local outbreaks of crown-of-thorns starfish (Walbran *et al.,* 1989) alter fish community structure by consuming living coral used as food and shelter (Sano *et al.,* 1984a, 1987; Bouchon-Navaro *et al.,* 1985; Williams, 1986a).

On both temperate and tropical reefs, sea urchins may compete directly with fishes for algal food sources. Hay and Taylor (1985) experimentally removed urchins from a reef flat in the Virgin Islands and observed a subsequent increase in the abundance of parrotfishes and surgeonfishes. Damselfish are able to exclude some urchins from the algal mats within their defended territories off Jamaica (Williams, 1981), while territories of surfperch can be overrun and decimated by urchins off California (Hixon, 1981).

D. Summary and Conclusions

Temperate reefs lack the enormous structural complexity of coral substrates, but often comprise areas of relatively continuous rock or boulder fields along coastal headlands. Compared to temperate rocky reef fishes, therefore, coral reef fishes often inhabit a more heterogeneous and patchy reef substrate with a greater variety of refuges and microhabitats. On temperate reefs, however, seasonal upwelling of nutrient-rich water supports tall, dense stands of kelp and other macroalgae. Unlike coral reefs, where foliage is mostly restricted to seagrass beds in surrounding sand flats, temperate reefs provide an added dimension of plant canopy for food and refuge, especially for juvenile fishes. Tropical waters are often warm and clear with relatively little seasonal fluctuation in conditions. The greater magnitude of seasonal change experienced by temperate rocky reef fishes is reflected in sharper seasonal peaks in their offshore–onshore migratory behavior. The stronger seasonal variation in the higher latitudes causes progressively greater, but more predictable, responses by resident fishes.

Although El Niño–Southern Oscillation conditions periodically disturb kelp forests, storms are the major source of physical disturbance on both temperate and tropical reefs. At any specific location, storms may be more frequent at higher latitudes, but are usually less severe than the occasional hurricanes and typhoons of tropical regions. In lower latitudes, the worst disturbances cause upheavals of community structure with indefinite recovery periods. Even though assemblages of mobile fishes are less affected by powerful water motion than are the sessile plants and macroinvertebrates, the effects of environmental perturbation may eventually cascade through the entire community.

Allochthonous food energy is imported into both tropical and temperate systems in the form of oceanic plankton consumed by large numbers of planktivorous fishes. Autochthonous energy generated by the rich detrital food chain supports benthic communities associated with algal turf, which harbors the small invertebrate prey of microcarnivorous fishes. Detrital accumulations are much greater on temperate reefs, where, unlike on coral reefs, plant production greatly exceeds consumption. Macroinvertebrates rather than fishes are the principal grazers of benthic algae on temperate reefs. Sea urchins may create “barrens” with an associated decline in fish abundance if, for example, a major storm disturbance removes mature kelp and other detrital food sources for urchins. Urchins may also compete directly with fishes for access to benthic food sources on both temperate and tropical reefs. Crown-of-thorn starfish can decimate live corals, also affecting fish distributions.

III. Environment-Induced Sampling Constraints and Bias

The differences between temperate and tropical reef environments, summarized in the previous section, have ramifications for any attempted comparisons between the fish communities occupying these habitats. The logistic constraints confronting temperate reef fish ecologists are quite different from those facing their counterparts who work in the tropics. Hence, many of the apparent temperate–tropical differences in community structure may be as much or more a function of relative methodological constraints and bias as real differences. With a few notable exceptions, tropical reef systems are remote from the world's major population centers. Yet tropical reefs are eminently more workable because predictably benign conditions allow longer time in the field and greater experimental replication (Sale, 1980a).

A. Seasonal Accessibility and Working Conditions

Access to temperate sites may be severely limited during the winter and spring, when surge, turbidity, and cold curtail the monitoring of experiments and limit observations to periods of fortuitous breaks in the weather. Experimental constructs such as cages are often badly damaged or lost so that long-term data on ecological succession or grazing pressure are usually incomplete and reports must interpret the outcomes of procedural fits and starts (see Breitburg, 1984). Winter fish counts are suspect when made in frigid water of low visibility because the observer is likely disoriented, weak, and nauseous as he or she is tumbled back and forth in a heavy surge. Shallow sites become unworkable, and deeper ones are workable only during reasonably calm periods. This is unquestionably why many temperate sites are sampled mostly during the summer–fall period of maximum calm and water clarity (see Ebeling *et al.,* 1980a,b; Larson and DeMartini, 1984). Conversely, the benign tropics can be sampled in comparable comfort all year, except of course when tropical storms threaten (see Sale, 1980a; Doherty and Williams, 1988a).

Tropical conditions allow more productive time under-water than do temperate conditions. Over many coral reefs, observations can often be made in a shallow, calm, and crystal clear medium. This allows for long, comfortable scuba dives or almost unlimited time simply snorkling at the surface. It provides opportunities for extensive and careful censusing, experiment tending, or behavioral monitoring by an alert and enthusiastic observer. In contrast to the tropics, the intimidating temperate conditions reduce observational efficiency, accuracy, and precision. It is hard to see fish in a murky milieu. Visibility frequently drops below 3–4 m, the minimum for accurate counts, either by cinetransects or sight transects (Ebeling *et al.*, 1980b; Larson and DeMartini, 1984). And this is not to mention the discouraging prospects for correctly identifying drably colored temperate species in the shade of a kelp canopy, which may eliminate more than 90% of the incident sunlight at midday (Fei and Neushul, 1984).

Divers using the older and thinner neoprene wet suits suffer dyskinesia, distraction, mental disruption, attention loss, and lapse of memory when water temperatures fall below 10°C (Bowen, 1968). This tends to destroy any enthusiasm for the project, let alone a sense of rigorous scientific purpose (A. W. Ebeling, unpublished observations). Observational difficulties when conditions are particularly awful during winter and spring may exaggerate the seasonal effect. In such times and after 20–30 years of watching temperate fishes, an aging ichthyologist tends to "burn (freeze?) out" and become wrongly convinced that temperate systems are hopelessly disturbed and uninviting (A. W. Ebeling, unpublished observations).

B. Sampling Accuracy

The obvious fact that complex substrates contain more nooks and crannies to hide small cryptic species may introduce an unknown amount of bias in temperate–tropical comparisons of reef fish diversity and abundance. Being enormously variegated, convoluted, and ramified, reefs with high coral cover provide a vast potential for concealing fishes of all shapes and sizes (Talbot, 1965). Consequently, attempts have been made to design complete censusing techniques (e.g., Sale and Douglas, 1984). Since many species reveal themselves only at night, a complete census requires around-the-clock observations (Helfman, 1983). Alternatively, combinations of destructive sampling and daytime visual transecting provide total estimates of cryptic and overt coral reef species (Brock, 1954; Bardach, 1959). Yet neither is all revealed on temperate reefs, since they harbor cryptic species in a habitat elaborated by a complex of short and tall algae. Destructive sampling of kelp-bed fishes in California has yielded a host of small cryptic forms, such as clinids, gobies, and clingfishes, that are seldom recorded along standard belt transects (Quast, 1968c). Short of complete censuses, visual belt transects (Brock, 1954) are widely used to estimate fish densities on both temperate and tropical reefs, and much effort has gone into measuring and compensating for the bias of this method (Jones and Chase, 1975; Brock, 1982; Sale and Sharp, 1983; Fowler, 1987; McCormick and Choat, 1987; Lincoln-Smith, 1988).

Because temperate and tropical fish may behave differently, the same sampling techniques may bias samples in different ways. For example, the Rapid Visual Technique (RVT) of Jones and Thompson (1978), which ranks species according to their frequency of encounter, overemphasizes the importance of widespread rare species but underestimates that of patchy but abundant ones (DeMartini and Roberts, 1982). Thus, the technique may bias a tropical census toward evenly distributed species, which are perhaps more consistently visible within the areas censused, but bias a temperate census against abundant species, which tend to move into and out of a target area in large schools. DeMartini and Roberts also point out that human observers tend to consistently underestimate the number of objects in a three-dimensional target such as a dense school of fish. Davis and Anderson (1989) demonstrated this effect dramatically by comparing species abundances estimated by visual transect with "true" abundances determined by marking and resighting fish in systematic censuses of all individuals in a circumscribed area. Visual transect estimates of abundant and aggregated species measured less than half the estimates of "true" abundances. The two estimates of the brightly colored garibaldi (*Hypsypops rubicundus*), which tends to be solitary and spread more evenly, were more in accord.

An additional confounding problem that may render temperate–tropical comparisons difficult stems from the lower regional species richness and larger body sizes of temperate reef fishes (Sections IV,A and B). All else being equal, the same visual sampling method would probably result in fewer errors in species identification and fish abundance estimates in temperate systems.

Because of these biases, tropical–temperate comparisons are suspect even when the same transecting technique was used. For example, Table 1 compares diversity and abundance of tropical western Atlantic (Alevizon and Brooks, 1975) and temperate Californian (Ebeling *et al.,* 1980b) reef fish communities sampled in the exact same way from 2.5-min 8-mm movie strips ("cinetransects"). From this representation, the two assemblages appear similar in local abundance, diversity, and evenness. In total species recorded, the tropical reefs appear to be not much richer than the temperate reefs, although this comparison is probably invalid because of the much smaller number of transects (sample size) filmed from each tropical site. Sample means provide a better basis for comparison: the tropical samples average 67% more species per transect than those from California, even though regional pools of reef fishes contain an estimated 267% more species in the tropical western Atlantic off Venezuela or Florida (Emery, 1978) than off temperate southern California (Quast, 1968b). Although this may represent a real difference between tropical and temperate reef fish communities in their expression of local versus larger-scale patterns of diversity (see Sale, 1980a), the western Atlantic samples may simply be more biased in their underrepresentation of small,

Table 1 Comparison of Cinetransect Samples of Fish Abundance and Diversity between Two Tropical Western Atlantic Coral Reef Sites off Venezuela and Florida[a] and Two Temperate Eastern Pacific Rocky Reef and Kelp Habitats along Coastal California[b]

	Western Atlantic tropical reefs		Eastern Pacific temperate reefs	
	Aves Island, Venezuela	Key Largo, Florida	Santa Barbara, California	Santa Cruz Island, California
Number of transects	32	28	168	185
Mean number of fish	83.3	63.8	64.7	70.6
Total species recorded	44	53	36	34
Rare species (<1.0% of total fish), % of total species	75	64	50	56
Mean number of species	13.0	14.7	8.2	8.4
Mean H' diversity	2.50	2.90	2.42	2.58
Population J' evenness	0.66	0.73	0.74	0.78

[a] Adapted from Alevizon and Brooks (1975).
[b] Adapted from Ebeling *et al.* (1980b).

cryptic, and nocturnal, or rare and widespread species (see Brock, 1982; DeMartini and Roberts, 1982); the Rapid Visual Technique, wherein such species may be more thoroughly surveyed, yielded substantially higher species numbers in comparable sites off Florida (cf. Jones and Thompson, 1978). Using only the cinetransect samples, therefore, it is difficult to judge whether the tropical western Atlantic assemblage of reef fishes is locally much more diverse than the temperate Californian assemblage, even though the former is derived from a much larger species pool.

More generally, Sale (1980a, Chapter 19) pointed out how different sampling and analytic techniques may influence assessments of degree of similarity among reef fish assemblages in either space or time. We suggest that, in addition, even the identical visual or photographic sampling technique and analysis may tend to distort comparisons of structure between tropical and temperate reef fish communities because of the fundamental differences in their respective habitats.

C. Summary and Conclusions

The marked differences between temperate and tropical environments may contribute to differences in logistic constraints and sampling bias, confounding apparent similarities or differences in their reef fish communities. Counts of some species may be unrealistically low from many temperate sites where fishes are difficult to observe throughout the year in frequently frigid, turbid, and rough waters overlying bottoms shaded by dense stands of kelp. The usually clearer, warmer, and calmer waters bathing most coral reefs permit longer and more careful censusing and attention to field experimentation. On the other hand, the complex coral substrate (like the macroalgal cover on temperate reefs) may provide more opportunities for cryptic fish species to stay out of sight. All else being equal, moreover, visual species identification and estimates of fish abundances may be less error prone on temperate reefs, which support fewer species and larger fish. Thus, in both systems, a segment of the fish community may remain invisible to nondestructive sampling by transect or quadrat.

Perhaps more attention should be paid to possible behavioral differences between temperate and tropical species of reef fishes. Humans tend to underestimate the number of objects in dense, three-dimensional clusters. If common temperate reef fishes are more likely to move about in dense schools than are their tropical counterparts, for example, the numerical range between abundances of common and rare temperate species observed by nondestructive transect would be artificially small. Yet, the rank or logarithmic transformations of abundance arrays commonly used to obviate this problem tend to mask basic differences in dominance–diversity relations (see e.g., Sale, Chapter 19). Consequently, temperate–tropical comparisons may be suspect even when the same transecting technique was used.

IV. Description of Community Structure

A. Species Composition and Richness

Regional pools of coral reef fishes are much richer than temperate pools; the number of fish species decreases poleward from the Indo-West Pacific center of tropical diversity (supporting thousands of species), through temperate zones (hundreds of species), to boreal regions (tens of species) (Mead, 1970; Briggs, 1974; Ehrlich, 1975; Springer, 1982). At the low end, boreal rocky reefs of southern New Zealand are dominated by only eight fish species (Kingsford *et al.,* 1989). The latitudinal diversity gradient is correlated with increasing seasonality and frequency of winter storms, and with decreasing solar radiation, geographic area, continuity of productivity, and climatic stability over geologic time in higher latitudes (Emery, 1978). This broad geographic pattern is reflected in regional comparisons. For example, the well-studied tropical western Atlantic and temperate coastal California have regional pools of about 500–600 and 100–200 fish species, respectively

(Emery, 1978; Quast, 1968b). Regional fish diversity also decreases longitudinally on either side of the Indo-West Pacific, with fewer than one-third the species contributing to pools in the Atlantic and eastern Pacific tropics (Briggs, 1974, 1985; Sale, 1980a; Findley and Findley, 1989, for butterfly fishes).

Collectively, tropical reefs have more species per genus (mean = 2.3) and more genera per family (3.0) than temperate reefs (1.5 and 2.1, respectively) (Ross, 1986). Their communities are dominated by perciform and tetraodontiform fishes, which occupy a multitude of niches in their complex habitat. Primarily tropical families, such as the Serranidae, Pomacentridae, Labridae, Gobiidae, and Blenniidae, are among the most diverse for marine fishes (Nelson, 1984). In both the North and South Pacific, temperate reef faunas include warm-temperate or subtropical species derived from the tropics, mostly perciformes, as well as cool- or cold-water species of primarily temperate or boreal origin, such as scorpaeniforms (Mead, 1970). Distinctive tropically derived families occur in temperate regions across the Pacific Ocean. Besides distinct species of pomacentrids, kyphosids, and labrids in both the North and South Pacific, families such as the labridlike Odacidae and Chironemidae (kelpfishes) are typical of the Southern Hemisphere and families such as the Pholidae (eel blennies) and Embiotocidae (surfperches) are abundant in the North Pacific (Mead, 1970).

Relative to temperate reef fish assemblages, this greater diversity of coral reef fishes creates the theoretical potential for more intricate social structures and a greater number of symbiotic relationships (see, e.g., Robertson and Polunin, 1981). However, the role of these interactions in structuring reef fish communities remains unclear (Sale, 1980a), and from what is known of temperate reef fishes, especially of the more recent offshoots from tropical families, many behave much like their tropical counterparts (Sale, 1978a).

The latitudinal increase in richness of reef fish communities between temperate and tropical regions may reflect a decrease in tolerance of environmental variability between temperate and tropical species. Compared to tropical species, temperate species must tolerate much greater ranges of seasonal change and interannual fluctuation in climate. In regions of large seasonal variation, such as the North Atlantic, many temperate fishes pass the winter period of frigid cold, storm turbulence, and loss of cover or food in a state of hiding and torpor (Emery, 1978; Olla *et al.,* 1979). Where seasonal fluctuations are less, such as the eastern North Pacific, adaptations are less extreme. Some species acclimate to seasonal change in the same habitat; others undergo onshore–offshore migrations. Spawning seasons are adjusted, with the more northern species tending to spawn in the winter months and the more southern forms during spring and summer (Tarzwell, 1970; Moulton, 1977; Leaman, 1980). Reanalyzing others' data, Stevens (1989) suggested that tolerances for broader seasonal conditions account for "Rapoport's rule," which states that a poleward decrease in species richness is accompanied by an increase in species ranges, as among Californian coastal fishes (Horn and Allen, 1978). The greater yearly variation in, for example, temperature at higher latitudes favors the evolution of broader climatic tolerances of the resident species. This "preadapts" them to occupy wider latitudinal ranges as well. Stevens reasoned that if tropical species typically have much narrower environmental tolerances than their temperate counterparts, their equal or greater dispersal abilities would more frequently place them beyond their preferred ranges and into areas where they can compete less successfully. Thus, a steady input of inferior competitors may in itself augment local diversity in the tropics (see Section V,G).

Nonetheless, the richness of some temperate assemblages of reef fishes can be enhanced by the intermingling of species from different faunas historically adapted to different climatic regimes (Quast, 1968b; Choat, 1982; but see Moreno and Jara, 1984). Meridional ranges of warm- and cool-temperate species may fluctuate from year to year with shifting currents and climatic trends (Hubbs, 1974; Stephens and Zerba, 1981; Choat and Ayling, 1987; Ebeling and Laur, 1988; Jones, 1988a). Thus, temperate warm- and cool-water faunas tend to merge in transitional zones of abrupt changes in temperature and other environmental factors, which may vary in magnitude and extent from one year to the next. Warm interannual periods favor species more recently derived from tropical ancestors; cooler periods favor species historically adapted to seasonal fluctuations of temperature and other environmental conditions within a cooler range (Stephens *et al.,* 1970; Patton, 1985). Interannual climatic shifts have occurred, for example, off the western coast of North America over thousands of years, as recorded in long-term temperature records and through strata in aboriginal refuse heaps (Hubbs, 1948, 1960, 1967). In sum, a temperate reef assemblage in the middle latitudes may vary in its relative abundances of cool- versus warm-tolerant species and include summer or winter visitors as well as year-round residents (Moulton, 1977; Olla *et al.,* 1979; Ebeling *et al.,* 1980b; Stephens and Zerba, 1981; Patton, 1985).

B. Body Size and Reproduction

There is some evidence that average body size of adult marine teleosts is smaller in the tropics than in cooler temperate regions. Some two-thirds of all small-bodied (<10 cm long) marine teleosts live in the tropics, and the frequency of small-bodied species relative to the total teleostean fauna tends to decrease from about 18% in warm south-temperate waters to about 10% in the tropics, 8% in the warm north-temperate zone, and only 6% in cool north-temperate waters (Miller, 1979, using data from Lindsey, 1966). Within primarily tropical families such as Pomacentridae (damselfishes) and Labridae (wrasses), the relatively few derived subtropical or warm-temperate species are often larger than their tropical relatives (Choat, 1982). Furthermore, body size may increase with latitude within a single species, as in the hexagrammid *Oxylebius pictus* in the eastern North Pacific (DeMartini and Anderson, 1978).

While the causes of body-size differences are beyond the scope of this review (see Warburton, 1989), the ecological ramifications of this phenomenon may be considerable. Larger, nonterritorial individuals generally move longer distances with greater inclination to go from reef to reef than smaller fish, whether temperate or tropical (Bardach, 1958; Robertson, 1988a). This pattern reflects the fact that larger fish occupy larger home ranges than smaller fish (Sale, 1978a), perhaps due to decreasing food availability with increasing body size (Schoener, 1968). Within-family comparisons between latitudes bolster this conclusion. For example, small tropical damselfishes defend much smaller territories (only about 1 m^2) (Low, 1971; Myrberg and Thresher, 1974; Williams, 1978) than their large temperate counterparts (ca. 10 m^2) (Clarke, 1970; Moran and Sale, 1977; Norman and Jones, 1984).

Along with body size, a latitudinal increase in egg size but decrease in larval dispersal distinguish temperate from tropical demersal spawners (Thresher, 1988, Chapter 15). This difference is less well expressed in pelagic spawning species. Wide dispersal may be less important in temperate species, which need not be distributed among isolated patches of coral (Barlow, 1981) and would not benefit from offshore transport (Parrish *et al.,* 1981). Consequently, perhaps, coastal populations of temperate reef fishes off California seem to show greater genetic differentiation (Davis *et al.,* 1981) than do coral reef species in the western Pacific Ocean (Ehrlich, 1975; Doherty and Williams, 1988a). Temperate faunas also have a higher proportion of viviparous species, including the North Pacific Embiotocidae, whose females bear advanced young occupying the same general habitat type as their parents (Baltz, 1984).

Environmental seasonality (Section II,B) has its greatest effect on the reproductive cycle of reef fishes. Whereas many coral reef fishes spawn all year (Ehrlich, 1975; Johannes, 1978a; Sale, 1980a; McFarland and Ogden, 1985; Doherty and Williams, 1988a), most temperate species spawn during the spring and summer when planktonic productivity providing food for fish larvae is greatest (Lowe-McConnell, 1979; Miller and Geibel, 1973; Stephens and Zerba, 1981; Guillemot *et al.,* 1985; Wyllie Escheverria, 1987; Jones, 1988a). In cool-temperate Puget Sound, for example, spawning of most species peaks during late spring and summer, while only about 25% of the species spawn in fall and winter (Moulton, 1977). Spring and summer herald an influx of first-year juveniles onto temperate reefs (see also Kingett and

Choat, 1981; Stephens *et al.*, 1986; Ebeling and Laur, 1988). Although coral reef fishes have peak spawning periods that coincide with current patterns, productivity pulses, or moon phases and associated tidal cycles (Munro *et al.*, 1973; Johannes, 1978a; Sale, 1980a; Thresher, 1984; Lobel, 1989), they generally show extreme variation in spawning periodicities among species (McFarland and Ogden, 1985). If coral reef species, such as anemonefishes, inhabit subtropical or warm-temperate regions as well, their temperate populations may show more seasonality in their breeding cycles (Moyer, 1980).

C. Foraging Guilds

Temperate and tropical reef fish faunas may differ in trophic structure as well as taxonomic composition (Bakus, 1969; Choat, 1982; Jones, 1988a). Here, we consider five major foraging guilds (*sensu* Root, 1967), as summarized in Table 2. We do not consider several minor guilds, such as detritivores, which are relatively poorly studied.

Midwater microcarnivores pick small prey off various elevated surfaces. Certain small wrasses and other species often select prey from blades of macroalgae. As opportunistic "cleaner fishes," such species occasionally pick ectoparasites from larger fishes in temperate waters, but may be more specialized as cleaners in the tropics (Limbaugh, 1961; Hobson, 1971; Losey, 1972a; Bray and Ebeling, 1975). Experimental removals of the cleaner wrasse *Labroides dimidiatus* by Youngbluth (1968), Losey (1972a), and Gorlick *et al.* (1987) falsified previous suggestions that cleaning mutualism enhances the local abundance of host species (Limbaugh, 1961).

Demersal microcarnivores and *mesocarnivores* together constitute the principal carnivore biomass in both temperate and tropical faunas. The microcarnivores select tiny prey from the benthic community with their small mouths, while the mesocarnivores ambush larger motile prey, including small fish, or seek more sedentary macroinvertebrates, such as gastropods and echinoderms. At the family level, the microcarnivores were apparently derived phylogenetically from the more generalized, large-mouthed mesocarnivores (Kotrschal, 1988). Microcarnivores, whose main food is often amphipods in temperate and subtropical systems (Quast, 1968d; Choat and Kingett, 1982; Laur and Ebeling, 1983; Moreno and Jara, 1984; Hallacher and Roberts, 1985; Schmitt and Coyer, 1983; Kotrschal and Thomson, 1986), forage during the day and hide at night; mesocarnivores, such as serranids and scorpaenids, feed during dawn or dusk and at night (Hobson, 1973, 1974; Hobson and Chess, 1976; Ebeling and Bray, 1976; Hobson *et al.*, 1981; Helfman, 1986b). Eastern North Pacific kelp forests also support large numbers of mesocarnivores that may switch among prey types (small fishes, plankton, and substrate-oriented prey) either in midwater among algal fronds or on the bottom

Table 2 Comparison of Foraging-Guild Structure in Temperate Rocky and Tropical Coral Reef Fishes[a, b]

Guild[a]	Temperate reefs	Tropical reefs
Midwater microcarnivores	Small prey picked off large macroalgae and opportunistically off other fish (e.g., wrasses, various juveniles)	Ectoparasites picked off other fish exclusively (mostly wrasses)
Demersal microcarnivores	Most diverse guild in both systems; small invertebrate prey picked off primary and secondary substrates	
	(e.g., surfperches, various juveniles)	(e.g., butterflyfishes, gobies, various juveniles)
Demersal mesocarnivores	No major functional difference between systems; consume macroinvertebrates and small fish	
	(e.g., rockfishes, sculpins, greenlings)	(e.g., scorpionfishes, anglerfishes, hawkfishes)
Macrocarnivores	Mostly transient piscivores (e.g., sharks, electric rays)	Resident and transient piscivores (e.g., sharks, jacks)
Planktivores	Mostly diurnal; consume mostly transient zooplankton (e.g., damselfishes, various juveniles)	Diurnal and nocturnal; nocturnal species consume resident zooplankton (e.g., damselfishes, cardinalfishes)
Herbivores	Low diversity and abundance (e.g., kyphosids)	High diversity and abundance (e.g., damselfishes, parrotfishes, surgeonfishes)

[a] The examples of families in each guild refer to large adult fish, unless indicated otherwise.
[b] From G. S. Helfman (1978). Patterns of community structure in fishes: Summary and overview. *Environ. Biol. Fishes* **3,** 129–148. Reprinted by permission of Kluwer Academic Publishers.

(Love and Ebeling, 1978; Hallacher and Roberts, 1985). The kelp bass (*Paralabrax clathratus*) and various midwater rockfishes (*Sebastes*) make up a substantial proportion of the total mesocarnivore abundance and biomass on some reefs (Love and Ebeling, 1978; Hallacher and Roberts, 1985). There is probably greater predation by mesocarnivores on benthic invertebrates in the tropics (Vermeij, 1978; Choat, 1982). In contrast to temperate reefs, where fishes may have little effect on the overall abundance of benthic invertebrates (e.g., Choat and Kingett, 1982; Laur and Ebeling, 1983; reviewed by Choat, 1982), tropical reefs have prey restricted to small crevices, holes, and other refuges from fish predation (e.g., Menge *et al.,* 1985; Hixon and Brostoff, 1985; reviewed by Hixon, 1986; Choat, Chapter 6).

Macrocarnivores include large piscivorous species, such as sharks and jacks on coral reefs. Hixon (Chapter 17) devotes an entire chapter to this guild and its effects in coral reef fish communities. The evidence suggests that both piscivory and adaptations among prey fish to minimize the risk of predation are widespread, but that the community-level effects of such predation are largely unknown. On temperate reefs, the dominant piscivores include sharks, electric rays, and larger sea basses (Quast, 1968b; Bray and Hixon, 1978). Pinniped marine mammals join this guild in consuming reef fishes, although the impact of such predation is largely unknown (Section V,C).

Planktivores are either diurnal or nocturnal species that eat plankton imported by ocean currents or generated on the reef itself. Randall (1967) listed 26 families of Caribbean reef fishes containing species that eat zooplankton. Diurnal species that are planktivores as adults include various serranids, pomacentrids, and labrids (Davis and Birdsong, 1973). In addition, almost all the other species consume plankton during early life history stages on both temperate and tropical reefs (Davis and Birdsong, 1973; Hobson, 1982; Singer, 1985). Although numerous planktivorous species inhabit coral reefs, only a few warm-temperate or subtropical species (mostly pomacentrids) are present in higher latitudes (Springer, 1982). Even so, the planktivore biomass is about the same in temperate and tropical habitats (Talbot and Goldman, 1972; Goldman and Talbot, 1976; Russell, 1977; Ebeling *et al.,* 1980b; Hobson, 1982; Kingsford, 1989). On some reefs off California, for example, blacksmith (*Chromis punctipinnis*) (Bray, 1981) and blue rockfish (*Sebastes mystinus*) (Burge and Schultz, 1973; Miller and Geibel, 1973; Hallacher and Roberts, 1985) make up a substantial portion of the total reef fish biomass in southern and central regions, respectively. The relatively few species of obligate and part-time planktivores made up some 67% of the numbers and 50% of the biomass of fishes on a temperate reef off southern California (Ebeling *et al.,* 1980b). Resident nocturnal planktivores are more abundant in coral reef habitats (Helfman, 1986b), although juveniles and adults of some species off southern California enter the reef habitat at night and pick plankton (Hobson and Chess, 1976; Ebeling and Bray, 1976; Hobson *et al.,* 1981).

Herbivores comprise a major foraging guild on coral reefs in terms of both species diversity and biomass (reviewed by Ogden and Lobel, 1978; Horn, 1989). Despite their low standing crop, benthic algae on coral reefs support a diverse guild of herbivorous fishes absent from temperate reefs (see Section IV,C). The high productivity and turnover of closely cropped filamentous algae appear to compensate for low standing crops as a forage base for fishes (Montgomery, 1980). Indeed, the biomass of herbivores rivals that of planktivores in being the largest of any foraging guild (Brock *et al.,* 1979). Many damselfishes, blennies, parrotfishes, surgeonfishes, and rabbitfishes graze algae. Russ (1984a,b) provides a detailed analysis of the distribution and abundance of such fishes over the Great Barrier Reef in Australia, and the activities of these abundant grazers can have major effects on benthic community structure (reviewed by Hixon, 1983, 1986; Choat, Chapter 6).

On the other hand, herbivory is relatively uncommon among temperate reef fishes (Bakus, 1966, 1969; Gaines and Lubchenco, 1982; Choat, 1982; Jones, 1988a). The dearth of herbivorous fishes in cold waters may involve temperature-limited rates of digestion of plant material, which has a lower caloric density and is less easily degradable than animal food, relative to the energetic needs of such large and mobile species (see Mead, 1970; Montgomery and Gerking, 1980; Ralston and Horn, 1986; Horn, 1989). Among cool-temperate fishes off California, for example, the truly herbivorous category seems to be limited to a few, relatively inactive pricklebacks (Stichaeidae), which inhabit algal tufts in the intertidal zone and rarely move from these predictable food sources (Horn *et al.*, 1982; Ralston and Horn, 1986). The large and active plant eaters of temperate waters tend to be more generalized than their tropical counterparts and are more adaptable to a temporally variable and less predictable food supply (Horn, 1989). For example, plant-cropping subtidal species, such as the kyphosids *Girella nigricans* and *Medialuna californiensis,* consume substantial amounts of animal matter along with the algae (Quast, 1968d; Ebeling and Laur, 1988). It is not surprising that the reef fish communities of boreal regions contain no herbivorous species (Briggs, 1974; Moreno and Jara, 1984; Kingsford *et al.*, 1989).

D. Daily Activity Patterns

The structure of reef fish assemblages varies temporally as diurnal, crepuscular, and nocturnal species follow their daily cycles of activity. Hobson (1965, 1968, 1972, 1973, 1974, 1982, Hobson *et al.*, 1981), among others (e.g., Collette and Talbot, 1972), has described the diel changeover: In the evening, the diurnal fishes retire in orderly fashion to refuges in and about the reef, with smaller individuals preceding larger individuals both among and within species; in the morning, the reverse progression ensues. Following a twilight "quiet period" of little activity in the water column, nocturnal species emerge. In general, the diurnal species are more taxonomically derived with specializations of small, protrusible jaws for microcarnivory, or strong jaws and prolonged alimentation for herbivory; nocturnal species are more generalized for mesocarnivory. There are no nocturnal herbivores. Thus temporal partitioning is usually between representatives of different teleostean suborders or even orders. Larger piscivores tend to drive the system in that they are most active in and about the quiet period when the diurnal and nocturnal shifts are most vulnerable (Collette and Talbot, 1972; Hobson *et al.*, 1981; see also Hixon, Chapter 17).

Within this general scheme, the behavior of tropical reef fishes may be more tightly organized or programmed than that of their temperate counterparts (Choat, 1982). For example, Helfman's (1978) review of previous work indicates that a greater percentage of the tropical reef fish community has a structured cycle of diel activity (Table 3). The twilight changeover between diurnal and nocturnal assemblages may occur more quickly and orderly in the tropics than in temperate regions (see Hobson, 1972, 1973; Ebeling and Bray, 1976; Helfman, 1986b), although this apparent difference may simply reflect the activities of greater numbers of nonsedentary species in the tropical assemblages (Hobson *et al.*, 1981). Coral reef species may have longer and more consistent migratory routes between feeding and sleeping areas (Ogden

Table 3 Temporal Structure in Representative Communities of Temperate and Tropical Reef Fishes, Based on the Species' Feeding and/or General Activity Patterns[a]

Locality	Number of species with a known activity cycle	Percentage with a major period of activity	Percentage diurnal	Percentage nocturnal	Reference
TEMPERATE ROCKY REEFS					
Santa Barbara, California	23	70	57	13	Ebeling and Bray (1976)
Catalina Island, California	10	90	40	50	Hobson and Chess (1976) (planktivores only)
Catalina Island, California	27	93	63	30	Hobson and Chess (1976)
Average =		84.3[b]	53.3	31.0	
TROPICAL ROCKY OR CORAL REEFS					
Southern Gulf of California	53	96[c]	47	38	Hobson (1965, 1968)
Alligator Reef, Florida Keys	159	92[d]	43	44	Starck and Davis (1966)
Lameshur Bay, Virgin Islands	85	88[e]	46	33	Collette and Talbot (1972)
Lameshur Bay, Virgin Islands	61	97	67	30	Smith and Tyler (1972)
Kona, Hawaii	143	97[f]	72	24	Hobson (1972)
Average =		94.0[b]	55.0	33.8	

[a] Adapted and expanded from Helfman (1978, Table 3, p. 133).
[b] Significantly different (*t*-test of arcsine-transformed values, $P < 0.01$).
[c] Includes nine crepuscular and two nocturnal-crepuscular species.
[d] Includes four crepuscular species.
[e] Includes two crepuscular and four each of diurnal-crepuscular and nocturnal-crepuscular species.
[f] Includes one crepuscular species.

and Buckman, 1973; McFarland *et al.*, 1985; Helfman, 1986b). This trend of decreasing organization in diel activity pattern continues with increasing latitude. For example, the twilight changeover is least structured for kelp-bed fishes in cold-temperate Puget Sound, Washington, where there are no warm-temperate or subtropical species (Moulton, 1977) than off southern California where there are many of both kinds (Ebeling and Bray, 1976) (Fig. 5).

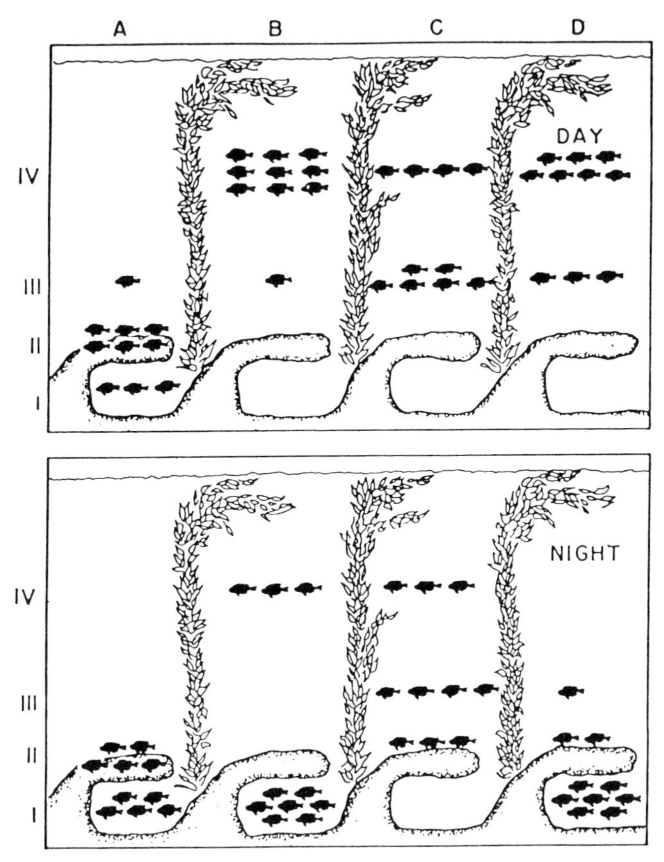

Figure 5 Day and night positions of reef fishes in a kelp forest off Santa Barbara, southern California, comparing (A) seven sedentary bottom species, (B) four mobile mesocarnivores, (C) six mobile demersal microcarnivores, most with primary temperate origins, and (D) five mobile demersal and/or midwater microcarnivores, all with close tropical relatives, observed in shelters (I), exposed on the bottom (II), within 1 m of the bottom (III), or in midwater (IV). Within each guild, one fish represents 10% of observations. [From Ebeling and Bray (1976, Fig. 2, p. 714).]

E. Summary and Conclusions

Tropical regions support more species of reef fishes than temperate areas by at least an order of magnitude. However, there is little evidence that this theoretically greater potential for interactions among coral reef species plays a major role in structuring their communities.

While the coral reef fish fauna is virtually circumglobal, the temperate array of faunas consists of independent elements separated longitudinally by the continental landmasses and latitudinally by the tropics. Along the continental coasts, north–south faunal boundaries fluctuate as warm-temperate and cold-temperate regions overlap in zones of transition. In these zones, species of different faunas mingle to various extents depending on yearly shifts in the oceanographic climate.

Temperate reef fishes tend to attain larger body sizes than their coral reef counterparts. The large numbers of small-bodied teleostean fishes in tropical marine habitats can exploit the diversity of small prey hiding in the myriad nooks and crannies of the coral substrate. Larger-bodied temperate species may tend to be somewhat more vagarious, having larger home ranges and territories. Temperate fish faunas include relatively more viviparous species and may, with some notable exceptions, have less dispersive larval and juvenile stages. Since ranges of temperate species are more one dimensional, there may be less advantage to disperse beyond the immediate coastal zone.

Temperate and tropical communities of reef fishes converge in many trophic characteristics but differ markedly in others. Carnivores are the most abundant and diverse foragers in both communities, but predation on benthic invertebrates may be greater in the tropics. Even though the total abundance and biomass of adult planktivores may be similar in temperate and tropical reef communities, the number of planktivore species is greater in the tropics. Most juveniles are planktivorous in both communities. The main difference in community trophic structure is in herbivory. Although a rich and abundant group of herbivorous species characterizes the coral reef fish communities, relatively few species are strictly herbivorous on cool-temperate rocky reefs. It has been suggested that the dearth of herbivorous reef fishes in cold marine waters may be due in part to restrictions on the rate of digestion, relative to other metabolic processes, at cooler temperatures.

On both temperate and tropical reefs, fishes show pronounced daily cycles in activity. A quiet period at twilight marks the transition between occupation of the water column by specialized species of microcarnivores and herbivores during the day and by generalized species of mesocarnivores at night. Large piscivores are most active in and about the quiet period when the smaller fishes are most vulnerable. This diel behavior pattern appears to be most structured in coral reef fish communities, and progressively less so in warm-temperate and cold-temperate communities, which are less diverse and contain fewer tropically derived species.

V. Regulation of Community Structure

Do temperate and tropical assemblages of reef fishes differ fundamentally in the mechanisms that regulate the patterns of distribution and abundance of their resident species? As members of more diverse assemblages in a presumably stable environment, do coral reef fishes usually coexist by partitioning space, food, and time resources as a result of present or past competitive interactions? As members of assemblages in more seasonally variable habitats, do fishes of temperate reefs co-occur below numbers where competition is important? These questions were often answered in the affirmative before the 1980s (reviewed by Sale, 1980a; Doherty and Williams, 1988a). Within the past decade, however, other models of community regulation have gained favor to the extent that, in a recent review, Doherty and Williams (1988a) averred: "The most important result [of the past decade's experiments] has been the repeated falsification of the hypothesis that populations of coral reef fishes are generally limited by the carrying capacities of reef environments, because this result removes the greatest difference separating studies of reef fishes from those of teleost populations in other neritic environments" (p. 544).

So foretold, we examine the evidence for the hypothesized mechanisms that may regulate the distribution and abundance of temperate and tropical reef fishes (see reviews by Doherty and Williams, 1988a; Mapstone and Fowler, 1988; Sale, Chapter 19). We do not attempt an exhaustive review of the literature. Rather we examine representative studies relevant to each model of community regulation. Thus, the number of studies tabulated in reviewing each model is not necessarily correlated with the validity of that hypothesis. This may be especially true of the circumstantial evidence for "niche diversification" (Section V,A), which was the prevailing research paradigm of the 1960s and into the 1970s (Sale, Chapter 19).

A considerable number of models of community regulation have been proposed. Our list mostly follows that of Connell's (1978) review, but also includes more recent ideas (recent in the sense of being applied to reef fishes), such as "recruitment limitation" (D. M. Williams, 1980; Doherty, 1980, 1982a, 1983a) and "sources and sinks" (Pulliam, 1988). Because there is presently no evidence from reef fish systems (either pro or con) concerning the "circular networks" model from Connell's (1978) review, we do not consider this hypothesis, although the possibility of "predatory networks" (Hixon, Chapter 17, this volume) is intriguing.

A. Competition and Niche Diversification

The "niche diversification hypothesis" maintains that past or present interspecific competition is the predominant process structuring an assemblage. There are basically two versions of this hypothesis. The first states that species have evolved specializations preventing competitive exclusion in an environment where available resources were limited through evolutionary time [the coevolutionary "ghost of competition past" (*sensu* Connell, 1980)]. The second version states that, despite ongoing present-day competition, the environment somehow provides refuges for subordinate competitors or situations where each competitor dominates a particular subhabitat or other subdivision of a limited resource base (reviewed by Colwell and Fuentes, 1975). Thus, in both versions, species are expected to exhibit "resource partitioning," occupying different microhabitats, eating different suites of prey, or being active in different time slots (e.g., Pianka, 1973; Schoener, 1974).

Because the niche diversification hypothesis was the most popular model of community regulation of the 1960s and 1970s, it has received far more attention than any other in studies of reef fish communities. We surveyed the primary literature to test the following predictions from this hypothesis for reef fishes (Table 4). If a community is structured by niche diversification, then: (1) The microhabitat distributions, diets, or perhaps times of activity (covered in Section IV,D) of the component species should show consistently low overlap ("resource partitioning"). Such partitioning should often be expressed as "niche complementarity," where co-occurring species that overlap greatly in one niche dimension (say, diet) separate along another dimension (say, microhabitat). (2) Distributions of potential competitors should be complementary at broader scales; they should displace one another geographically, as conditions favor one species over the other ("geographic complementarity"). Note that such patterns are also consistent with the sources and sinks model (see below). (3) Experimental removal of a species should result in expanded resource use by its competitors ("species complementarity"). (4) Vacated living space should be quickly regained, so patterns of distribution and abundance should be resilient when perturbed ("resiliency"). (5) Succession within newly created sites should culminate in the same equilibrial assemblage ("climax"). (6) The environment should be saturated with fish, to the extent that, for example, all holes and refuges are filled with appropriately sized individuals ("shelter saturation"). (7) An increase in the limiting resource, for example, shelter sites or food, should increase the abundance of fish ("resource limitation").

We present Table 4 as an incomplete and simplified overview of our current predicament regarding each of the preceding predictions. Table 4 does not include studies that focused on possible competition within species

Table 4 Outcomes of Representative Tests of the Predictions from the Niche Diversification Hypothesis of Regulation of Reef Fish Community Structure, Comparing Temperate Assemblages of the Eastern North Pacific with Tropical Assemblages of the Western North Atlantic/Caribbean and Indo-Pacific Regions[a]

Region	Prediction	Group	Prediction supported?	Reference
A. TEMPERATE ROCKY REEFS				
1. Circumstantial Evidence				
Eastern North Pacific	Resource partitioning	All	Yes	Quast (1968b)
		All	Maybe	Hobson and Chess (1976)
		All	Maybe	Yoshiyama (1980, 1981)
		All	Yes	Cross (1982)
		All	Yes	Gascon and Miller (1982)
		All	Yes	Larson and DeMartini (1984)
		All	Yes	Grossman (1986)
		Blennies	Yes	Stephens *et al.* (1970)
		Blennies	Yes	Kotrschal and Thomson (1986)
		Eel blennies	Yes	Barton (1982)
		"Picker-type" fishes	Yes	Bray and Ebeling (1975)
		Rockfishes	Yes	Love and Ebeling (1978)
		Rockfishes	Yes	Larson (1980)
		Rockfishes	Yes	Hallacher and Roberts (1985)
		Sculpins	No	Norton (1989)
		Surfperches	Yes	Alevizon (1975)
		Surfperches	Yes	Ellison *et al.* (1979)
		Surfperches	Yes	Hixon (1980a)
		Surfperches	No	Ebeling and Laur (1986)
		Surperches	Yes	Holbrook and Schmitt (1986)
	Geographic complementarity	Blennies	Yes	Stephens *et al.* (1970)
		Surfperches	Yes	Hixon (1980a)
		Surfperches	Yes	Schmitt and Coyer (1983)
	Resource limitation	All	No	Stephens and Zerba (1981)
		Surfperches	Yes	Schmitt and Holbrook (1986)
		Surfperches	Yes	Stouder (1987)
2. Experimental Evidence				
Eastern North Pacific	Species complementarity	Rockfishes	Yes	Larson (1980)
		Surfperches	Yes	Hixon (1980a)

(*continued*)

Table 4 *Continued*

Region	Prediction	Group	Prediction supported?	Reference
		Surperches	Yes	Schmitt and Holbrook (1986)
	Climax	All	Yes	Fager (1971)
		All	Yes	Gascon and Miller (1981)
		All	Yes	Carter *et al.* (1985)
	Resiliency	All	Yes	Thomson and Lehner (1976)
		All	Yes	Grossman (1986)
	Resource limitation	Gobies and midshipmen	Yes	Breitburg (1987a)
		Surfperches	Yes	Hixon (1980a)
		Rockfishes	Yes	Larson (1980)
		Wrasse and blenny	Yes	Thompson and Jones (1983)
B. TROPICAL CORAL REEFS				
1. Circumstantial Evidence				
Western Atlantic and Caribbean	Resource partitioning	All	Yes	Smith and Tyler (1973a)
		All	Yes	Parrish and Zimmerman (1977)
		All	Yes	Gladfelter and Gladfelter (1978)
		All	Yes	Luckhurst and Luckhurst (1978a,b)
		All	Maybe	Findley and Findley (1985)
		Angelfishes	Yes	Hourigan *et al.* (1989)
		Butterflyfishes	No	Clarke (1977)
		Butterflyfishes	Yes	Neudecker (1985)
		Damselfishes	Yes	Emery (1973)
		Damselfishes	Yes	Clarke (1977)
		Damselfishes	Yes	Ebersole (1985)
		Groupers	Yes	Roughgarden (1974)
		Squirrelfishes	Yes	Gladfelter and Johnson (1983)
		Surgeonfishes	No	Roughgarden (1974)
	Shelter saturation	All	Yes	Smith and Tyler (1972)
		All	Yes	C. L. Smith (1978)
Indo-Pacific	Resource partitioning	All	Yes	Hiatt and Strasburg (1960)
		All	Yes	Gosline (1965)
		All	Maybe	Talbot and Goldman (1972)

(*continued*)

Table 4 *Continued*

Region	Prediction	Group	Prediction supported?	Reference
		All	No	Bradbury and Goeden (1974)
		All	Maybe	Goldman and Talbot (1976)
		All	Yes	Harmelin-Vivien (1977)
		All	No	Russell *et al.* (1974)
		Butterflyfishes	Yes[b]	Anderson *et al.* (1981)
		Butterflyfishes	No[b]	Sale and Williams (1982)
		Butterflyfishes	Yes	Harmelin-Vivien and Bouchon-Navaro (1983)
		Butterflyfishes	Yes	Bouchon-Navaro (1986)
		Butterflyfishes	Maybe	Motta (1988)
		Butterflyfishes	Maybe	Findley and Findley (1989)
		Damselfishes	Yes	Fishelson *et al.* (1974)
		Damselfishes	No	Sale (1974)
		Damselfishes	Yes	Robertson and Lassig (1980)
		Damselfishes	No	Tribble and Nishikawa (1982)
		Damselfishes	Maybe	Sale *et al.* (1984a)
		Damselfishes	Yes	Jones (1988b)
		Squirrelfishes	Yes	Vivien and Peyrot-Clausade (1974)
		Surgeonfishes	Maybe	Jones (1968)
		Surgeonfishes	Maybe	Robertson *et al.* (1979)
		Surgeonfishes	Yes	Robertson and Gaines (1986)
		Surgeonfishes and damselfishes	Yes	Robertson and Polunin (1981)
	Resource limitation	Damselfishes	Yes	Thresher (1983a,b, 1985a)
		Planktivores	Yes	Hobson and Chess (1978)
		Planktivores	Yes	Hamner *et al.* (1988)
	Shelter saturation	Damselfishes	No	Sweatman (1983)
2. Experimental Evidence				
Western Atlantic and Caribbean	Climax	All	Maybe	Bohnsack and Talbot (1980)
	Resiliency	Damselfishes	Yes	Williams (1978)

(continued)

Table 4 *Continued*

Region	Prediction	Group	Prediction supported?	Reference
		Damselfishes	Yes	Waldner and Robertson (1980)
	Resource limitation	Herbivores	Yes	Robertson *et al.* (1976)
		All	Yes	Shulman (1984)
		All	Yes	Hixon and Beets (1989)
Indo-Pacific	Climax	All	Maybe	Nolan (1975)
		All	No	Sale and Dybdahl (1975)
		All	No	Talbot *et al.* (1978)
		All	No	Sale (1980b)
	Resiliency	All	Yes	Brock *et al.* (1979)
		All	Yes	Sale (1980b)
		Damselfishes	Yes	Belk (1975)
	Resource limitation	Herbivores	No	Roberts (1987)
		Surgeonfishes	Yes	Robertson and Gaines (1986)
		Damselfishes	No	Jones (1988b)

[a] Predictions of the hypothesis are tested by circumstantial or experimental evidence. Under circumstantial evidence it is indicated whether or not there is support that species within a guild: overlap little in their use of resources ("resource partitioning"), complement one another in geographical distribution ("geographic complementarity"), appear as a group to be limited by resources ("resource limitation"), or appear to saturate available refuges ("shelter saturation"). Under experimental evidence it is indicated whether or not there is support that species within a guild: complement one another when one is removed ("species complementarity"), colonize artificial reefs in similar relative abundances ("climax"), assume original relative abundances after perturbation ("resiliency"), or respond in distribution or abundance to manipulated resources ("resource limitation").
[b] Different interpretations of the same data set.

(e.g., Nursall, 1977; Robertson and Sheldon, 1979; Robertson *et al.*, 1981; Sweatman, 1985a; Jones, 1987a,b) because such interactions do not bear directly on niche diversification between species. Our attempt to categorize representative studies as supporting ("yes" in Table 4), partially supporting ("maybe"), or falsifying ("no") each of these predictions emphasizes only temperate systems along the Pacific coast of North America and coral reef systems of the Indo-Pacific and western Atlantic. So limited, this circumstantial and experimental evidence indicates that, contrary to the classic notion that the greater diversity of coral reef fish fauna reflects the narrower and more finely partitioned niches of its species (see, e.g., Mead, 1970; Briggs, 1974), the niche diversification hypothesis may actually apply equally well or even better to temperate reef fish communities. Besides the fact that our review is not exhaustive, however, we must add the following general caveats: descriptions of patterns of distribution, abundance, and resource use are only suggestive; the lingering results of past competition are not verifiable (Connell, 1980); and the number of positive outcomes in studies of competition is generally high because researchers tend to investigate this process in systems where they suspect that it is likely to occur (Connell, 1983; Schoener, 1983b).

The various kinds of circumstantial evidence for and against the existence of competition and niche diversification suggest that the hypothesis applies equally well to temperate and tropical systems (see "circumstantial evidence" in Table 4). The percentages of outcomes consistent with the hypothesis ("yes" and "maybe" categories) are quite high for "temperate rocky reefs" as well as "tropical coral reefs" (88% of 26 studies and 82% of 44 studies, respectively). The large majority of such evidence pertains to the relatively tenuous prediction of resource partitioning, however. A typical example is of six species of rockfishes (*Sebastes* spp.) co-occurring in kelp forests off central California [Larson (1980) and Hallacher and Roberts (1985) in section A, 1 of Table 4], wherein the species were shown to segregate vertically in the water column as well as by depth along the bottom (Fig. 6). Among all examples relating to resource partitioning in Table 4, percentages of outcomes consistent with the prediction are again very high for the temperate and tropical studies cited (90% of 20, and 82% of 38 studies, respectively). In addition, the partitioning of time as diel patterns of activity occurs in both systems, although the twilight changeover in species occupying the water column appears to be progressively less structured for the simpler reef fish communities at higher latitudes (Section IV,D). Overall, therefore, our results corroborate those of Ross's (1986) more extensive review and analysis of resource partitioning among fishes in general. He also found little difference in patterns between temperate and tropical reef fish assemblages; percentages of temperate and tropical species pairs showing at least one "substantial difference" in use of food, space, or time were 80 and 82%, respectively. Nonetheless, Ross cautioned that such differences may simply reflect chance divergences in the species' evolutionary histories, having nothing to do with competiton (see also Sale, 1979b; Connell, 1980). Peter Sale (personal communication) put the matter more bluntly: "Given that three species of fish on a reef will inevitably differ in some aspect of what they do, how useful is a triumphant documentation of resource partitioning anyway?" (see also Sale, Chapter 19).

The experimental evidence actually suggests that niche diversification may operate more frequently in temperate than in tropical reef fish communities. All 12 experimental outcomes from temperate areas provide evidence for ongoing competition, while only 69% of 16 tropical studies are corroborative (see "experimental evidence" in Table 4). This pattern is not consistent across

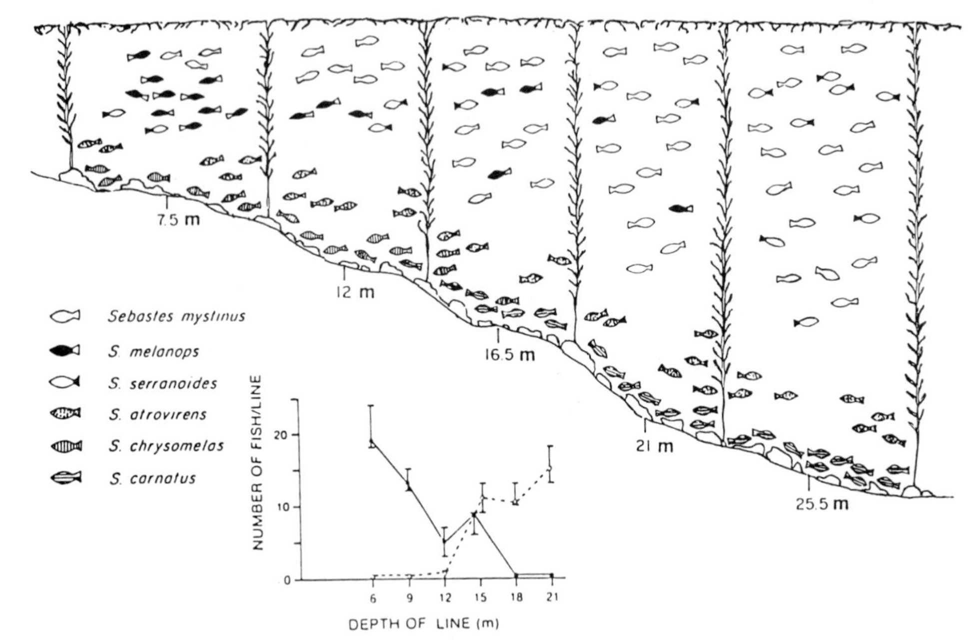

Figure 6 Depth distributions of rockfishes (*Sebastes*) in kelp forests off California. In the picture, species partly segregate by depth in the water column (first four species in the list) or along the bottom (last two species) off central California. [From L. E. Hallacher and D. A. Roberts (1985, Fig. 3, p. 98). Differential utilization of space and food by the inshore rockfishes (Scorpaenidae: *Sebastes*) of Carmel Bay, California. *Environ. Biol. Fishes* **12,** 91–110. Reprinted by permission of Kluwer Academic Publishers.] The inset graph shows the bathymetric distribution of the last two species (solid and dotted lines, respectively) off Santa Cruz Island, southern California. [Modified from R. J. Larson (1980, Fig. 2, p. 226). Competition, habitat selection, and the bathymetric segregation of two rockfish (*Sebastes*) species. Ecol. Monogr. **50,** No. 2, 221–239.]

all predictions of the hypothesis, however. For example, some assemblages of small species show considerable resilience between defaunations whether in temperate rocky intertidal zones or on coral heads in the tropics (Table 4, "resilience"). On the other hand, species assemblages colonizing small experimental reefs stabilized sooner and more predictably in temperate habitats, almost never doing so in tropical studies (Table 4, "climax"). This pattern may simply be a function of the fewer combinations of species available for settlement or to prioritize space in a less diverse fish fauna (see Sale, 1976; Helfman, 1978; Shulman *et al.,* 1983).

Although competition for food among coral reef fishes had often been discounted (Sale, 1980a, 1984), recent experimental and circumstantial evidence indicates otherwise for several tropical and temperate reef fishes (Table

4, "resource limitation"). For planktivorous damselfishes in the western tropical Pacific, data from Thresher (1983b) are suggestive that breeding pairs of *Acanthochromis polyacanthus* spawn earlier in the season and have larger broods of juveniles with greater survival rates on small patch reefs from which potential competitors were removed. G. P. Jones (1986) found that small juveniles of *Pomacentrus amboinensis* show increased growth rates when provided plankton in excess of natural supplies. However, Jones (1988b) could find no evidence of interspecific competition between juveniles of *P. amboinensis* and of another planktivorous damselfish, *Dascyllus aruanus,* manipulated to different densities on patch reefs, even though each species is associated with a different kind of coral and both species grew better on one kind than the other. Others have presented circumstantial evidence that distributions of such planktivores are correlated with plankton abundance (Hobson and Chess, 1978; Thresher, 1983a, 1985a). Using fish-removal experiments, Thompson and Jones (1983) demonstrated that at certain growth stages a wrasse and territorial blenny compete for food in temperate waters off New Zealand, although Roberts (1987) found that a territorial damselfish and blenny apparently did not compete for the benthic algae on tropical coral reefs off Australia. In other studies, foraging and/or shelter sites were shown to be limiting for temperate surfperches and gobies off California (Hixon, 1980a, 1981; Breitburg, 1987a; Behrents, 1987), but not for a tropical wrasse and a territorial damselfish off the Caribbean coast of Panama (Robertson and Sheldon, 1979; Robertson *et al.,* 1981).

Larger scale manipulations of potential competitors between habitat types have been done in temperate communities only. For example, controlled reciprocal removals of each member of closely related pairs of temperate surfperches (Hixon, 1980a; Holbrook and Schmitt, 1986, 1989) and rockfishes (Larson, 1980) off California have demonstrated the existence of present-day interspecific competition (Table 4, "species complementarity"). Predictably, the less aggressive congener of each pair expanded into the more productive habitat of the more aggressive congener when the latter was removed (e.g., Fig. 7). This was corroborated circumstantially for the surfperch studies by "natural experiments" on a geographic scale as well (Table 4, "geographic complementarity"). Wherever the dominant congener is rare or absent, the subordinate congener occupies the habitat range of both (Hixon, 1980a; Schmitt and Coyer, 1983). Comparable population-level experiments and geographical analyses are sorely needed for coral reef fishes.

In conclusion, there is no evidence, except perhaps from temporal partitioning of resources between day and night, that the more diverse communities of fishes inhabiting coral reefs are more competitively organized than the simpler communities of fishes living on temperate reefs. Indeed, the experimental evidence listed in Table 4 indicates that, if anything, temperate reef fishes may

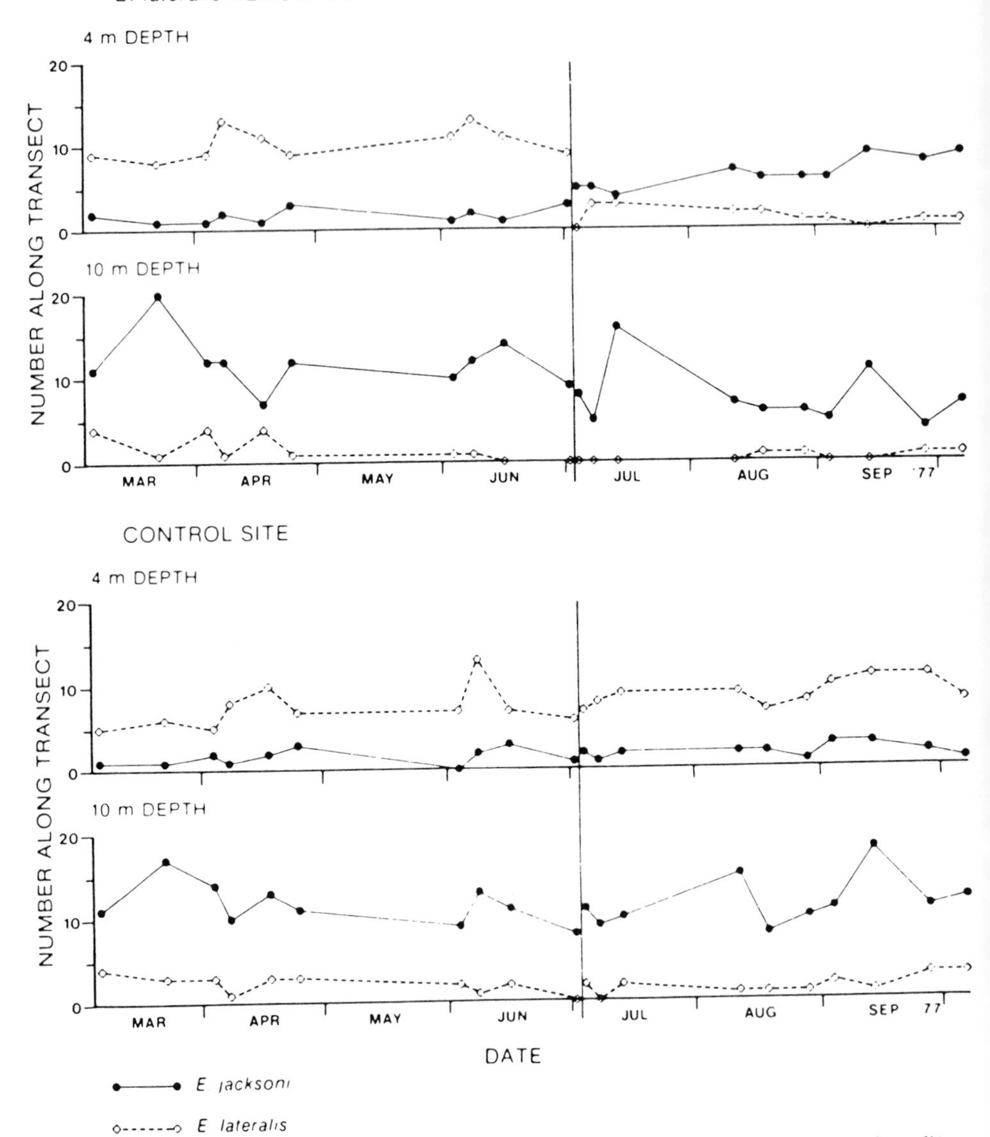

Figure 7 Counts in time of the black surfperch (*Embiotoca jacksoni*) and striped surfperch (*E. lateralis*) in shallow (4 m depth) and deep (10 m depth) habitats off Santa Cruz Island, southern California. Vertical lines mark the time when the striped surfperch, a specialist on the shallow habitat, was experimentally removed (upper pair of graphs) or left at natural densities (lower pair of graphs) in the presence of its congener. [Modified from M. A. Hixon (1980, Fig. 7, p. 926). Competitive interactions between California reef fishes of the genus *Embiotoca*. *Ecology* **61,** No. 4, 918–931.]

tend to be more organized by interspecific competition than are their their tropical counterparts. For both systems, the plethora of circumstantial evidence favors an equally large amount of niche diversification by means of resource partitioning due to present or past competitive interactions. Yet, most such evidence is inherently biased. All species must differ to greater or lesser degrees in their autecologies, for the trivial reason that all represent different evolutionary endpoints. It is noteworthy that of the 54 studies in Table 4 that examined only the observed overlap in resource use by different species, 46 (85%) concluded, though often equivocally, that the species partitioned one or more resources. However, four other studies—two each of tropical (Sale and Williams, 1982; Gladfelter and Johnson, 1983) and temperate (Ebeling and Laur, 1986; Norton, 1989) species—tested resource partitioning more objectively. They statistically compared mean observed overlap among species in a guild with a null value assuming species overlapped by amounts that varied randomly (as expected if species' differences in resource use evolved independently). Of these four analyses, only one (Gladfelter and Johnson, 1983) found significant cause to reject the null hypothesis of random overlap.

B. Competitive Lottery

Also referred to as "equal chance" (Connell, 1978) and "multispecies equilibrium" (Doherty, 1983a), the competitive lottery hypothesis of Sale (1974) obviates the requirement of niche diversification to sustain multispecies associations despite ongoing competition among individuals to secure space. Strictly speaking, each species need not have an "equal chance" at recruiting to each site, but only a chance equal to its probability of per capita loss (see Sale, 1977, 1978a, 1979a): "Each species . . . must simply win often enough to avoid going extinct" (P. F. Sale, personal communication). This hypothesis assumes that all species in a guild have similar resource requirements and larval settlement capabilities, and that the availability of settling postlarvae is locally unlimited. Species may differ in competitive ability, however (Sale, 1982a). Furthermore, local larval recruitment is independent of local adult stock, and a strong prior residency effect allows any established recruit to hold space against all comers. Since space is limited, chance vacancies are quickly filled by recruits (analogous to "lottery tickets"), and the unpredictability of the process prevents any one species from excluding another. Thus the species are distributed in a random or patchwork pattern, as vacated spaces are filled with new individuals recruited from the plankton in random order (Sale, 1974). Therefore, although local species populations may vary in size, the species composition and total abundance of fish within the guild should remain fairly constant (in "multispecies equilibrium"). This hypothesis has a quantitative theoretical basis if any species is able to increase its population size when it is rare and the others are common (Abrams, 1984). This condition is met if the co-occurring species have nonlinear (hump-shaped) stock-recruitment relations (Sale, 1982a), vary in their environmentally affected birth and death rates, and have overlapping generations for "storage" of strong cohorts over unfavorable years (Chesson and Warner, 1981).

Empirical support for a competitive lottery comes from a guild of territorial herbivorous damselfishes competing for sites on coral rubble patches off Australia (Sale, 1974, 1975, 1977, 1978a; Sale *et al.*, 1980) (Table 5, section B). Total available space remained occupied for more than 3 years, as space vacated by adults was reoccupied by other adults or new recruits apparently at random. Thus three species, *Eupomacentrus apicalis, Pomacentrus wardi,* and *Plectroglyphidodon lachrymatus,* could vary in relative abundance on the same patch while the total number of individuals remained fairly constant (Sale, 1975, 1979a). Observations of their distributions and foraging behaviors indicated that resource needs of these species were nearly equal and their competitive abilities appeared roughly similar, except that the most abundant species, *P. wardi,* tended to compensate for its somewhat inferior competitive ability by securing marginal spaces. Sale (1976) concluded from fish-removal experiments that recruitment of *P. wardi* is in fact limited by the availability of vacant space on which to settle.

Two studies suggest that the damselfish lottery may be a local phenomenon. Robertson and Lassig (1980) provide circumstantial evidence for spatial resource partitioning among the same species near Sale's site. At a different location on the Great Barrier Reef, Doherty (1982b) monitored recruitment and growth of *P. wardi* at different densities among patches. He summarily concluded that because neither recruitment rates nor post-recruitment survival were density dependent, population regulation was nonequilibrial and perhaps determined by the supply of recruits rather than the available space for recruits (Doherty, 1983a; see Section V,H). Therefore, the lottery may not have been operating during Doherty's study.

Our search for temperate examples of competitive lotteries was futile, perhaps because a multispecies equilibrium is difficult to recognize. Most available studies of resource allocation by temperate reef fishes imply that co-occurring species exploit different suites of resources to greater or lesser degrees (see Ross, 1986). In addition, temperate reef fish communities usually lack guilds of several territorial species whose discrete spaces can be monitored for long periods of time. Thus, it may be impossible to demonstrate that a constant number of individuals occupy all available spaces in a pattern that varies in the relative abundances of species through time.

C. Compensatory Mortality and Predation on Common Species

Species with similar requirements can coexist at equilibrium if the most common suffers the greatest mortality due to, for example, differential predation (Connell, 1978). Either the superior competitor is prevented from excluding inferiors by being more vulnerable to predators (or other sources of mortality), or, where competitive asymmetries among prey species are not pronounced, predators switch to the more abundant of alternative prey types in a "type III functional response" (Murdoch and Oaten, 1975). In both cases, the process of competitive exclusion is mitigated. Thus, if predation is the source of mortality, then the prediction of this hypothesis is that the local diversity of reef fishes should increase in the presence of piscivores.

Although predation pressure has often been assumed to be greater in the tropics where antipredator tactics appear to be more highly evolved (Ehrlich, 1975; Johannes, 1978a; Hixon, Chapter 17), there are very few data on how a compensatory response of predator to prey fishes may regulate either tropical or temperate communities (Table 5, Section C). In the only study we know of relating the number of potential prey species to the number of piscivorous fish, Hixon and Beets (1989) found that the local number of prey species declined with increasing piscivore abundance, opposite the prediction of compensatory mortality (Hixon, Chapter 17). However, density-dependent predation of reef fishes has not been investigated; choice experiments offering superior versus inferior competitors as prey have not been performed; and per capita mortality of prey species after altering predator or prey abundances has not been rigorously examined (reviewed by Hixon, Chapter 17).

D. Intermediate Disturbance

This hypothesis views local species diversity as a function of patchy mortality due to disturbance, be it physical (e.g., storms) or biological (e.g., nonselective predation). It states that low-disturbance areas have low diversity because late-successional "climax" species exclude early-successional "fugitive" species, and high-disturbance areas also have low diversity because most species are locally extirpated. However, areas disturbed at intermediate frequency or severity are augmented by both subordinate fugitives and dominant climax species (Connell, 1978). Time between disturbances is an important feature of this model. Succession within an empty habitat patch is seen as proceeding from a low-diversity assemblage of early colonists, to a high-diversity mix of early and late species, to a low-diversity climax assemblage. Each disturbance pushes the successional sequence partway or entirely back to the beginning.

Table 5 Some Key Studies Evaluating the Various Hypotheses on the Regulation of Reef Fish Community Structure[a]

Hypothesis	Species, location, and reference	
	Temperate reefs	Tropical reefs
A. Competition and niche diversification	---------------- see Table 4 ----------------	
B. Competitive lottery	+: none	+: damselfishes Heron Island, Australia (Sale, 1974, 1975)
	−: none	−: damselfishes Heron Island, Australia (Robertson and Lassig, 1980)
C. Compensatory mortality and predation on common species	+: none	+: none
	−: none	−: all species St. Thomas, U.S. Virgin Islands (Hixon, Chapter 17)
D. Intermediate disturbance	+: none	+: none
	−: none	−: none
E. Predation on rare species	+: none	+: all species St. Thomas, U.S. Virgin Islands (Hixon, Chapter 17)
	−: none	−: none
F. Gradual change	+: all species California, U.S.A. (Stephens *et al.*, 1984; Ebeling and Laur, 1988)	+: none
	−: none	−: none
G. Sources and sinks	+: wrasse California, U.S.A. (Cowen, 1985)	+: cardinalfishes Bahamas (Dale, 1978)
	−: none	−: none
H. Recruitment limitation	+: wrasse California, U.S.A. (Cowen, 1985)	+: damselfish One Tree Island, Australia (Doherty, 1983a); wrasse San Blas Islands, Panama (Victor, 1986b)
	−: none	−: grunt St. Croix, U.S. Virgin Islands (Shulman and Ogden, 1987)

[a] Major supportive studies follow a "+" sign; explicitly opposing studies follow a "−" sign; and "none" indicates no available information.

Thus, maximum local diversity occurs where intermediate levels of disturbance hold the assemblage occupying each patch at a midsuccessional stage.

Certainly both temperate and tropical reef fishes are displaced or killed by physical disturbances of various sorts, although the effect on community structure is difficult to assess (see Section II,B). Indeed, it seems unlikely that physical damage due to storms and other such events can affect typically mobile reef fishes severely enough to keep populations chronically low over a widespread area (Doherty and Williams, 1988a).

Predators probably have constituted a more important source of mortality, especially in the tropics, as evidenced by the widespread occurrence of specialized antipredator mechanisms among potential prey and various sources of correlative evidence (Hixon, Chapter 17). Nonetheless, temperate and boreal species may more likely fall prey to marine mammals and birds occurring in large concentrations (Moulton, 1977; Moreno and Jara, 1984) and locally abundant specialized predators such as electric rays (Bray and Hixon, 1978). Without proper shelter, juvenile fish become highly vulnerable to predators on both tropical (Shulman, 1984, 1985a; Hixon and Beets, 1989) and temperate reefs (Ebeling and Laur, 1985; Behrents, 1987). Talbot *et al.* (1978) attributed the high diversity of fish assemblages on small patch reefs to the maintenance of species abundances below equilibrium levels by means of predation, as well as seasonal variation and uncertain recruitment. They concluded that high within-habitat diversity may have been at least partly sustained by the intermediate-disturbance process.

However, none of these observations specifically supports the prediction that intermediate disturbances promote the diversity of prey fishes. The only study that we could find to supply pertinent data is that of Hixon and Beets (see Hixon, Chapter 17) of predatory fishes and their potential prey fish species inhabiting experimental reefs in the Virgin Islands. In this case, the relation between predator density and number of prey species was both inverse and monotonic, not the hump-shaped function predicted by the intermediate-disturbance hypothesis (Hixon, 1986; Chapter 17). In sum, we found little convincing evidence, pro or con, that disturbances maintain high levels of species diversity in either tropical or temperate reef fish communities (Table 5, section D).

E. Predation on Rare Species

Despite the ideas presented in the previous two sections, increasing levels of predation are not always predicted to increase local prey diversity. As explained previously, the intermediate-disturbance model predicts that prey diversity will decline at very high levels of predation. However, even predation at low levels can cause an immediate decline in prey diversity if predators

(1) selectively or disproportionately consume the competitively subordinate or otherwise rare species in an assemblage, or (2) consume all prey nonselectively, but rare prey are highly susceptible to predation and are locally extirpated before they can respond favorably to competitive release (Van Valen, 1974; Lubchenco, 1978; Hixon, 1986). These possibilities comprise an alternative hypothesis on the role of predation in regulating community structure. Note that this hypothesis is not mutually exclusive of all others; the initial rarity of a species can be caused by any number of processes, including competition and recruitment limitation (Section V,H). Yet, given that all communities include more rare than common species, predation on rare species can potentially be a major process in determining the diversity of a local assemblage.

The prediction of the predation-on-rare-species hypothesis is that local prey diversity will decline monotonically as the level of predation increases from zero. Here (as in the compensatory-mortality and intermediate-disturbance hypotheses), the level of predation is a measure of the extent to which predators decrease the local population sizes of prey species, and is presumably correlated with predator density.

As summarized previously, a study of fishes occupying artificial reefs in the Virgin Islands provides the only data we could find on the effects of piscivores on the local diversity of reef fishes (Hixon and Beets, 1989) (Table 5, section E). The strong negative relationship between the number of resident piscivores occupying a reef and the maximum number of resident prey species is consistent with the hypothesis of predators extirpating rare species (Hixon, Chapter 17, Fig. 5B). However, the gut contents of the piscivores in this study were not examined, so the precise mechanisms underlying this relationship remain unknown. Moreover, we could find no comparable data from any other tropical or temperate reef system, so this hypothesis remains largely untested.

F. Gradual Change

The gradual change hypothesis states that environmental conditions, which determine the dominance rankings of competitors, eventually change before any one species can exclude others and monopolize resources. The predicted result of this hypothesis is that a gradual change through time in species composition or relative abundances should be correlated with environmental variation (Hutchinson, 1961; Connell, 1978). As a seasonal or longer-term climatic trend alters their environment, species better adapted to the new conditions will increase in abundance but not to the point of excluding others before conditions change once again. Species track the environment but never come into stable equilibrium because of time lags between resource availability and the realization of the species' exploitative potentials (Boyce, 1979).

This model would seem to be potentially less important in tropical than in temperate regions, which undergo major latitudinal climatic shifts superimposed on more predictable seasonal fluctuations (Section II,B). On temperate reefs, seasonal variation in water temperature may promote a reversal of competitive abilities between species derived from different faunas. The gradual change model may be best expressed in ecotones supporting mixtures of cool- and warm-temperate faunas, where the cool-adapted species perform best during periods of low temperature and the warm-adapted ones are superior during warm episodes. For example, warm water temperatures appear to prevent the competitively dominant, cool-temperate surfperch *Embiotoca lateralis* from locally extirpating the subordinate, warm-temperate *E. jacksoni* at the southern edge of the geographical range of *E. lateralis;* north of this region, *E. lateralis* appears to have completely eliminated its congener from reef habitats (Hixon, 1980a). Although tropical assemblages show some seasonality in environmental conditions, the magnitude of seasonal variation does not approach that of temperate zones (Section II,B). Dominance shifts of tropical species may be less noticeable within the greater diversity of coral reef assemblages (Richards and Lindeman, 1987).

Necessary (but not sufficient) evidence supporting the gradual change hypothesis is variation in the distribution and abundance of reef fishes related to seasonal change and climatic trends. While there is ample evidence of such variation in temperate systems, we found none for assemblages of coral reef fishes. In Puget Sound in the U.S. Pacific Northwest, for example, densities of several species of rocky-reef fishes decrease during the fall and winter as productivity declines and kelp is torn away by strong wave action; a few more generalized carnivores remain abundant throughout the winter when they may eat a greater diversity of benthic prey than the declining specialists (Moulton, 1977). In the subtropical northern Gulf of California, Mexico, marked seasonal change had a greater effect on relative abundances of fish assemblages on small artificial reefs than did rate of colonization or succession (Molles, 1978). More generally, an influx of juveniles during the spring and summer contributes substantially to the predictable seasonal variation in temperate reef assemblages in the Northern Hemisphere (Olla *et al.,* 1975; Miller and Geibel, 1973; Leaman, 1980; Stephens *et al.,* 1986; Ebeling and Laur, 1988) and Southern Hemisphere off New Zealand (Choat and Kingett, 1982; Jones, 1984c). Because many species must adjust to losses of macroalgal cover and food in winter, they migrate to deeper water, become torpid, or seek more productive habitats. The degree of adjustment increases with latitude as seasonal cues and exigencies become more intense (Moulton, 1977). In warm-temperate New Zealand, however, the predictable spatial variation of reef fishes among habitat types usually exceeds the temporal variation (Kingett and Choat, 1981; Jones, 1988a).

At a given latitude, the composition of temperate reef fish assemblages changes with longer-term shifts in climate, ocean temperature, and current patterns. With episodes of poleward advection and warming, such as El Niño events in the eastern Pacific (Section II,B), recruitment and abundances of subtropical or warm-temperate species increase as those of cool-temperate fishes decrease (Section IV,A). Yet such climatic changes fluctuate with unpredictable frequency, and elements of the one fauna seldom completely displace elements of the other. Off northeastern New Zealand, abundances of subtropical species to the north and more temperate species to the south are correlated with latitudinal gradients in water temperature and current pattern (Choat and Ayling, 1987). At a particular latitude, temporal fluctuations in abundances of subtropical and temperate species reflect shifts in these gradients, and offshore islands support relatively more subtropical elements than the nearby mainland (Choat *et al.,* 1988). Analogous shifts occur off southern California as well. During a warming trend with periodic El Niño episodes of varying strength, abundances of southerly warm-water species increased, those of cool-water species decreased, and those of "central species," perhaps adapted to a more variable regime, remained unchanged (Stephens and Zerba, 1981; Stephens *et al.,* 1984; Patton, 1985; Ebeling and Laur, 1988). A recent shift to a cooler regime seems to have given northern species the advantage once again (A. W. Ebeling, unpublished observations). Since all three categories of species may live in one local area of reef and kelp, they gradually replace one another in numerical dominance.

Despite such evidence for seasonal variation in species composition and relative abundances, there is no evidence that this variation promotes coexistence. Thus, the gradual change hypothesis remains untested for reef fishes (Table 5, section F).

G. Sources and Sinks

Theoretically, inferior competitors can coexist with dominant species in a single locality if there are other areas where the subordinates are at an advantage and can supply surplus recruits to the locality (Abrams, 1984). For reef fishes, Dale (1978) offered an economic analogy called "money in the bank" (local "capital" augmented by foreign "interest"). The hypothesis was recently formalized as allowing coexistence in patches of an environmental mosaic where species vary in their ratios of birth/death rates (Pulliam, 1988). Thus any locality is at one time a "source" for some species populations and a "sink" for surplus individuals produced elsewhere; the local assemblage is as much a reflection of the conditions in other patches of the mosaic as the local availability of resources. If coexistence depends on this process, local populations should include species recruited from elsewhere along with any locally breeding populations.

Stevens (1989) suggested that this hypothesis is sufficient to account for the maintenance of greater local species richness in the tropics than in temperate areas. Tropical species, which have presumably adapted to more constant environments, may have more narrowly defined and closely approximated source areas where some species have a reproductive or competitive advantage, but have dispersal systems that are as "open" as those of temperate species. Among coral reef fishes, for example, there is some evidence that each species in an assemblage of western Atlantic cardinal fishes (Apogonidae) is most abundant in a different habitat (Dale, 1978). Similarly, a Caribbean damselfish has a refuge from multispecific habitat use by virtue of its broad geographic range (Waldner and Robertson, 1980). In addition, damselfishes recruiting to patches of different quality may form subpopulations varying in reproductive output (Doherty, 1980, 1982a; Wellington and Victor, 1988). Thus, if marginal populations can be maintained at low levels by a continued trickle of immigration, then tropical communities should be relatively sensitive to disturbances that create "habitat islands" with a larger proportion of very rare species (Stevens, 1989). The comparable censuses of temperate and tropical reefs in Table 1 indicate that a substantially larger proportion of tropical species are indeed "rare."

In temperate regions with broad gradients in temperature and other factors, sink populations may be created through expatriation from other sources over a much broader scale. Warm-temperate reefs and kelp beds off New Zealand and California support marginal or nonbreeding populations of species with more northern or southern centers of abundance along with central populations in optimal parts of their ranges (Choat and Ayling, 1987; Jones, 1988a; Patton, 1985; Ebeling and Laur, 1988; Choat *et al.*, 1988). Off southern California, for example, breeding of California sheepshead (*Semicossyphus pulcher*), a warm-temperate species, is usually more successful to the south (Cowen, 1985), while populations consisting mostly of adults occur commonly on northern reefs, where both successful spawning and recruitment appear to be sporadic (Cowen, 1985; Davis, 1986; Ebeling and Laur, 1988). At the northern sites, marked increases in sheepshead recruitment during 1983 coincided with a major El Niño climatic episode in 1982–1984, when northerly transport of larvae apparently increased (Fig. 8). Expatriate populations of large adults of some southern species inhabit reefs in Monterey Bay, far from their source along southern California (Miller and Geibel, 1973; Hubbs, 1974). On a smaller scale, offshore reefs may be sinks for certain relatively closed populations of viviparous surfperches, whose young are born

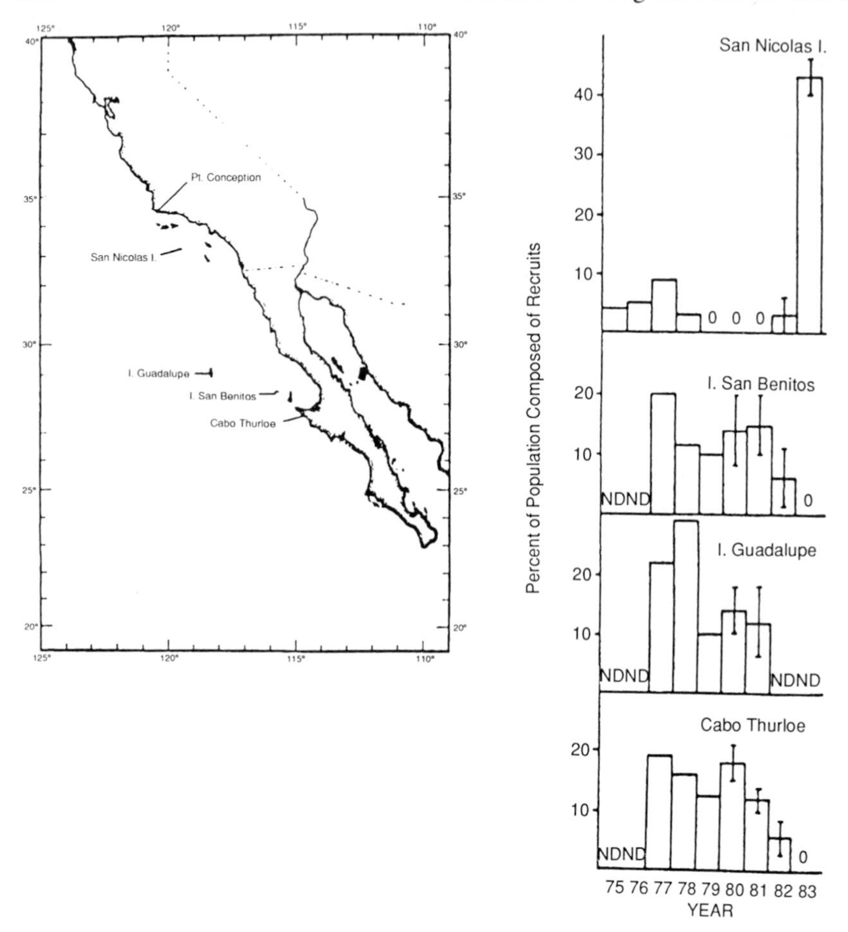

Figure 8 Annual recruitment success of the California sheepshead, *Semicossyphus pulcher,* a warm-temperate fish, at the four longitudinally arrayed study sites off southern California, U.S.A., and Baja California, Mexico, as indicated in the map (inset). Note that relative recruitment increased dramatically at the most northern site (San Nicolas Island) just south of Point Conception during the warm El Niño climatic event of 1982–1984. [Adapted and reassembled from R. K. Cowen (1985, Fig. 1, p. 722, and Fig. 2, p. 725). Large scale pattern of recruitment by the labrid, *Semicossyphus pulcher:* Causes and implications. *J. Mar. Res.* **43,** 719–742.]

mostly inshore where algal cover is dense (Ebeling and Laur, 1988; Ebeling *et al.,* 1990).

In sum, several lines of evidence from both temperate and tropical reefs are consistent with the sources and sinks model (Table 5, section G). As with the other models, however, rigorous tests have not been forthcoming.

H. Recruitment Limitation

The last model of community regulation that we consider seems to be the most popular at the moment. This hypothesis states that reef fish populations are undersaturated simply because the supply of postlarval recruits is limited (D. McB. Williams, 1980; Doherty, 1980, 1982a, 1983a; reviewed by Doherty and Williams, 1988a). Thus, population sizes virtually never reach levels where competition is important, and local species composition and relative abundances are determined simply by the species and number of postlarvae that happen to settle in a particular location. The idea is seductive. Since most species have planktonic larvae that spend varying periods adrift (Brothers *et al.*, 1983; Brothers and Thresher, 1985; Smith *et al.*, 1987; Victor, 1986a; Wellington and Victor, 1989) in irregular current patterns (Sette, 1960; Lobel and Robinson, 1983; Shapiro *et al.*, 1988), and since targets for settlement are small compared to the expanse of the open ocean (Richards and Lindeman, 1987), the odds that a particular locality is replete with competent larvae may be low. This does not preclude the possibility that some reefs occasionally become overpopulated by occasional dense settlement, perhaps from a particularly rich advected patch (see Doherty, 1983a; Victor, 1986b). One prediction, therefore, is that population density should vary widely among sites and times, without a concordant variation in resources. On reefs with the lowest density of new recruits, competitive interactions among juveniles and adults should be very rare, and changes in resource levels should not alter relative resource use or mortality rates among species. Removal of adults should not affect the recruitment rate, which may be very low; more strictly, residents should not inhibit settlers. [Some settlers may avoid conspecific adults (Talbot *et al.*, 1978; Sweatman, 1985a) or seek such adults (Williams and Sale, 1981; Sweatman, 1985a; Jones, 1987b) on smaller targets like patch reefs (Schroeder, 1987) or on certain substratum types (Sale *et al.*, 1984a).] Population density should be much more sensitive to the number of settlers than to postsettlement mortality. Therefore, adult density should be linearly related to the prior density of juveniles in short-lived species, and the age structure of adults should reflect interannual variation in recruitment in long-lived species (Jones, 1987a; Shulman and Ogden, 1987; Doherty and Williams, 1988a).

There is some debate over how strictly the criteria should be applied to demonstrate "adequately" that a population is recruitment limited (Jones, 1987a). Victor (1983a, 1986b) believes that the demonstration is adequate if the number of first-year adults equals the number of the previous year's recruits, minus a constant density-independent mortality of juveniles. This criterion ignores the role of postsettlement mortality (as opposed to presettlement larval availability) in keeping populations undersaturated. Others believe that, for recruitment limitation to occur, the density of newly settled postlarvae must be below that at which competition eventually occurs among juveniles or adults (Shulman and Ogden, 1987). Otherwise, competition, predation, or some other mortality factor will necessarily structure the assemblage (Hixon, Chapter 17).

Although momentum is gaining for generalizing the recruitment-limitation hypothesis as applied to coral reef fishes (Doherty and Williams, 1988a; Sale, 1988a), the evidence is mixed and few studies have met the stricter criteria (Table 5, section H). Hence, most results provide necessary but not sufficient evidence of recruitment limitation. Studies in recruitment patterns at different scales in time and space over the Great Barrier Reef show enormous variability in year-class composition of damselfishes, wrasses, and cardinalfishes. Episodic recruitment occurs over scales of a few meters up to 100 km (Williams, 1983a; Sale *et al.,* 1984b; Sale, 1985; Doherty, 1987a). Tremendous temporal variability in recruitment is also evident in the Caribbean (Shulman, 1985b) and Hawaii (Schroeder, 1985; Walsh, 1987). Manipulations of adult and juvenile densities on experimental reefs have tended to confirm the occurrence of population fluctuations and density-independent recruitment patterns (Sale, 1976; D. McB. Williams, 1980; Doherty, 1982, 1983a; but see Sale, 1979a). Jones (1987a) found that while juvenile survival in a planktivorous damselfish was independent of adult density, growth and time to maturity were not (see also Doherty, 1983a).

Some of the more compelling support for this hypothesis comes from studies on the Caribbean coast of Panama. The age composition of the adult population of a short-lived wrasse, *Thalassoma bifasciatum,* as determined by daily otolith increments, corresponded to sporadic recruitment pulses observed the previous year, and inputs of recruits were unrelated to either juvenile or adult mortalities on the reef (Victor, 1983a). Recruitment episodes occurred during periods of high species diversity of larvae in the plankton, and recruitment varied among reefs as correlated with reef exposure to prevailing currents (Victor, 1984, 1986b). Thus confirmation of a linear relationship between adults and prior recruits may be best sought in a short-lived species whose population structure is not confounded by several cohorts of adults (see Abrams, 1984; Warner and Chesson, 1985; Warner and Hughes, 1988). Additionally, Robertson and Sheldon (1979) provided experimental evidence that bluehead wrasse are not limited by shelter availability in this region.

Support for the recruitment-limitation hypothesis is not universal. In the Virgin Islands, French grunts (*Haemulon flavolineatum*) settled in such high densities that postrecruitment mortality, rather than recruitment, clearly determined population size (Shulman and Ogden, 1987). In other examples, either postsettlement mortality, competition, or relocation between reefs as juveniles and adults apparently overrode recruitment in determining population densities of surgeonfishes off Caribbean Panama (Robertson, 1988a) and blennies off Barbados (Hunte and Cote, 1989).

The fewer studies of possible recruitment limitation on temperate reefs indicate that this model may not generally apply over the same scales of space and time that the tropical studies have indicated. Although a subtropical wrasse (*Pseudolabrus celidotus*) from northeastern New Zealand varied spatially in recruitment density, this was more likely due to preference for algal habitats than to vagaries of settlement; density-dependent survival of juveniles, perhaps driven by intraspecific competition for food, limited adult densities (Jones, 1984b). In southern California and Mexico, local densities of a congeneric triplet of shallow-living blennies (*Hypsoblennius*), all with long-lived pelagic larvae, are probably determined more by interspecific competition for food and shelter than the supply of settlers (Stephens *et al.,* 1970). Stephens *et al.* (1986) found that recruitment of several species of reef fishes (Gobiidae, Labridae, Scorpaenidae, Serranidae) to a southern Californian breakwater was unrelated to densities of early-stage larvae sampled offshore, although later-stage competent larvae were inadequately sampled and limitation by presettlement mortality was not ruled out. Yet, postsettlement mortality due to predation apparently determined juvenile densities on a small, experimental reef nearby.

In addition, some temperate regions have relatively more live-bearing species without a pelagic dispersal stage. Twenty-six abundant species of viviparous surfperches (Embiotocidae) occur in the North Pacific (Baltz, 1984). Since precocious young are probably born in the same area as their parents, populations are unlikely to be recruitment limited on a local scale, and competition between surfperches has been demonstrated experimentally (Hixon, 1980a; Schmitt and Holbrook, 1986) (Section V,A).

During climatic shifts characteristic of temperate regions, rocky-reef fishes are more likely to show variation in recruitment over larger scales of space and time than their tropical counterparts. Off California, widespread differences in recruitment success may occur between cool- and warm-temperate species as the oceanographic regime changes unpredictably (Radovich, 1961; Hubbs, 1974; Stephens and Zerba, 1981; Stephens *et al.,* 1984, 1986; Ebeling and Laur, 1988). Recruitment of young-of-year blue rockfish (*Sebastes mystinus*), a cool-temperate planktivorous species, and blacksmith (*Chromis punctipinnis*), a warm-temperate planktivorous damselfish, provide examples on a southern Californian reef near Santa Barbara (Fig. 9). Blue rockfish young were abundant during cool years in the 1970s, but virtually disappeared during a warming trend in the 1980s. Densities of young blacksmith, on the other hand, increased dramatically during the climatic shift. Note that such patterns are also consistent with the gradual change model.

The most direct evidence of recruitment limitation on a regional scale is

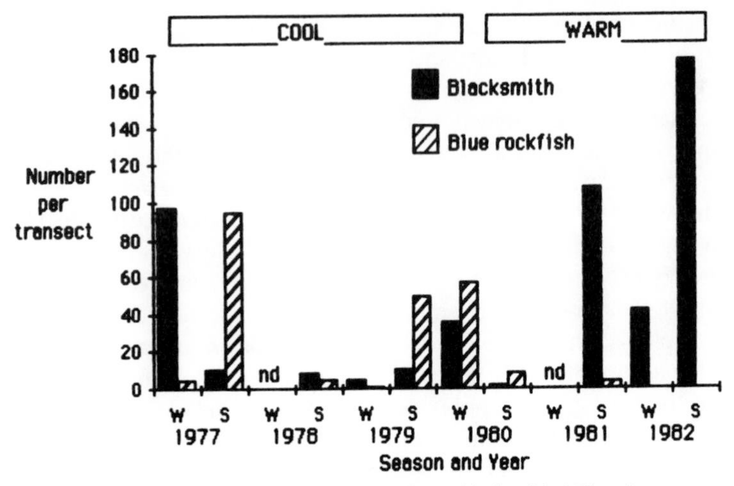

Figure 9 Semiannual densities of young-of-year blacksmith (*Chromis punctipinnis*) and blue rockfish (*Sebastes mystinus*) on Naples Reef near Santa Barbara, southern California, during periods of cool and warm water. Sample size is about 18–27 transects, each with a volume of about 240 × 3 × 3 m (2160 m^2) near the base at the ends of the long axis of the reef (see Fig. 3). S, July–December; W, January–June; nd, no data taken. [Transects have been described by Ebeling and Laur (1988).]

from an extensive analysis of unusually large northward recruitment of a warm-temperate wrasse during an intense El Niño episode of northerly transport (Cowen, 1985). Cowen suggested that, in general, recruitment limitation is observed locally whenever the major breeding population is located upcurrent. Thus, widespread recruitment of southern species to the north (and vice versa) is related to major climatic changes associated with reversals in prevailing currents. This causes "expatriation" of adult populations in geographic "sinks" (see Section V,G), where reproductive potential may be "stored" (*sensu* Warner and Chesson, 1985) in long-lived individuals until the next environmental shift (Cowen, 1985). Alternatively, of course, annual variability in recruitment may be due to temperature-related variation of local spawning success (Tarzwell, 1970).

Ultimately, demonstrating presettlement recruitment limitation requires one to show that postsettlement processes, including mortality and competition, do not significantly affect population sizes. Such demonstrations are largely wanting in both tropical and temperate reef fish systems.

I. Conclusions

A careful weighing of the literature on regulation of community structure provides few generalizations. Although there are real differences between temperate and tropical reef fish communities at a descriptive level, there appears to be no universally appropriate model of community regulation for either region. For models where evidence is available from both temperate and tropical systems (competition and niche diversification, sources and sinks, recruitment limitation), a sampling of key studies indicates that each model may apply more or less equally well to both regions. However, for most models (competitive lottery, compensatory mortality, intermediate disturbance, predation on rare species, gradual change), few data are available, the data are only from tropical reefs, and most results are negative. Thus, for the vast majority of reef fish assemblages, there are no data, either pro or con, directly relevant to each model of community regulation (Table 5). All models may operate to greater or lesser extents in temperate or tropical systems, depending on local environmental conditions during the study and during the system's recent history. Any search for "the factor" regulating reef fish community structure is thus doomed to failure and sweeping generalizations should be suspect (Sale, 1979b, Chapter 19; Hixon, Chapter 17; Jones, Chapter 11).

VI. Future Research

We believe that future research should be directed toward explicitly testing the different models of community regulation for reef fishes in both temperate and tropical regions (see gaps in Table 5). Along with more careful observation over longer periods, multifactorial experiments should be designed to manipulate different populations or resources to find out how processes such as competition and predation may interact to structure fish populations (reviewed by Sih *et al.*, 1985; Hixon, Chapter 17). There should be more effort at discovering when and where recruitment limitation is operative. Determining the relative distributions and abundances through time of postlarvae in the water column versus the substratum would provide valuable data on the mechanisms of recruitment (see McFarland and Ogden, 1985; Eckert, 1985a; Sweatman, 1985b; Kingsford and Choat, 1989; Kobayashi, 1989; Leis, Chapter 8). Testing recruitment limitation, compensatory mortality, intermediate disturbance, and predation on rare species requires measuring the relative mortality rates of different categories of common and rare species (in addition to monitoring recruits). Studies of relative breeding potential may reveal critical relationships between "source" and "sink" species populations in a particular locality. Little is known about patterns of distribution and abundance of transient species in either temperate or tropical regions. There should be continued effort to map home ranges of fish over their life spans to determine the scales over which processes requiring the interaction of individuals, such as competition and predation, operate.

A critical examination of latitude-associated biases in data gathering is long overdue. Perhaps a more objective comparison of environmental complexity and refuge availability should be made between coral and temperate reefs. More thorough searches for covert individuals would help in interpreting the apparent similarities and differences in species diversity (richness and evenness). More temperate–tropical pairs of studies using the same observational and experimental methods should be employed. Such studies would be particularly valuable if they involved the same personnel (and their associated personal biases) in both regions, such as Menge and Lubchenco's (1981) pioneering temperate–tropical comparison of intertidal systems. Studies comparing sampling methods, such as that by Davis and Anderson (1989), should be made to discover habitat-generated sources of bias, which may differ substantially between temperate and tropical reefs.

Finally, it would do us well to document losses in local diversity due to habitat destruction and overfishing by the human scourge. Natural processes that once structured fish communities may no longer prevail near heavily exploited urbanized areas. Loss of keystone species, such as the earlier depletion of sea otters in the North Pacific, may revolutionize community structure to the extent of indirectly changing the rank order of fish species abundances. Overfishing of predators in the tropics may upset relationships among less exploited prey species. Introduction of exotic species to geographically isolated sites such as Hawaii may destabilize community structure (Randall, 1987). Thus, outcomes of temperate–tropical comparisons should be read in light of the relentless human invasion. A particular example of an important problem for the tropical fish ecologist would be measuring the response of fish assemblages to the destruction of more than 70% of the Philippine coral reefs (McAllister and Rubee, 1986). This environmental disaster was inflicted by sedimentation due to excessive deforestation, and especially by dynamiting and cyanide poisoning of the fauna in response to the lucrative international market for coral habitat and tropical fish aquaria. This might lead one to suggest that the most crucial difference between coral reef and temperate reef fish assemblages is that the smaller and more brightly colored coral reef fishes constitute the more desirable object of such outrages.

Acknowledgments

We thank P. F. Sale for his direction, encouragement, and manuscript review; M. H. Carr, A. J. Fowler, S. J. Holbrook, D. C. Reed, and D. McB. Williams for further critiques; and R. Trench for help finding sources of illustrations. The Marine Science Institute at U. C. Santa Barbara provided administrative services. Some material is based on support by the National Science Foundation under Grants OCE-7925008, OCE-8208183, and OCE-8822931. Ultimately, we thank Jan Ebeling and Ursula Hixon for their love and patience.

CHAPTER 19

Reef Fish Communities: Open Nonequilibrial Systems

Peter F. Sale
Department of Zoology
University of New Hampshire
Durham, New Hampshire

I. Introduction

A. Reef Fishes as a Model System

This chapter concerns the ways in which the many species of reef fishes present at a location are organized into communities. This has been one of the primary foci of attention of reef fish ecologists from the start, for the simple reason that coral reefs typically support such a high diversity of fishes compared to temperate regions. Ecologists in the 1960s anticipated that there might be special ways in which communities would be organized in order to contain such richness. For the purposes of this chapter, I will consider community structure to consist of the patterns in species composition and species' abundances present in a defined habitat. The study of community structure includes description of such patterns and of changes to such patterns which may occur through time. It is also concerned with the causal factors responsible for determining these patterns and causing the changes.

Study of the organization of reef fish assemblages has, over the past 20 years, caused us to radically revise our understanding of the processes determining the structure of such communities. Originally, ecologists approached reef fishes in the expectation that their communities would be equilibrial assemblages organized through competitive processes among co-occurring species in the reef habitat. This expectation was called into question first when groups of successfully co-occurring species were not readily separated in terms of their resource requirements or abilities in competing for these resources. Second, it also became apparent that there was substantial change through time in the structure of local communities, and that this process of change was driven by a spatially and temporally variable recruitment of young fish to reefs at the close of their larval pelagic phase.

Recognition that reef fishes nearly universally had pelagic larval phases lasting from 10 to 100 or more days depending on the species, and that as juveniles and adults they were relatively sedentary in a spatially very heterogeneous habitat came during this same time (for details see Leis, Chapter 8; Victor, Chapter 9; Doherty, Chapter 10; Robertson, Chapter 13). The most widely accepted view at present (e.g., Mapstone and Fowler, 1988) is that reef fish species exist as excellent examples of spatially divided populations in a patchy environment (Fig. 1). The local subpopulations maintain their interconnection by exporting all gametes/larvae to the open ocean and receiving recruitment from that source alone—presumably drawing, in the process, on reproductive effort of other breeding groups. Under this view, local reef fish communities have a structure which is transient if the recruitment success of the component species is temporally variable. The expected rules governing the organization of communities of reef fish are much more subtle than was first anticipated.

To begin with, local recruitment of each species is quite independent of local reproductive success (in producing viable young) and is dependent, instead, on reproductive success in other sites. This decoupling means that even strong interspecific competition among local residents cannot lead to

Figure 1 Reef fish species exist as groups of subpopulations occupying disjunct patches of habitat. They communicate with one another by exporting all offspring to, and receiving all recruitment from, a surrounding larval bath of largely unknown structure. Reef fish assemblages are composed of the species which happen to share each patch.

permanent local extinction, nor to a coevolved partitioning of resources in the manner predicted by classical competition theory (Sale, 1977, 1985; Chesson and Warner, 1981; Chesson, 1986). Second, the very high mortality during the larval phase (presumably approaching 100%) ensures, in a real and variable world, that the rate of recruitment must vary strongly through time and among sites. (Since mortality plus survivorship must sum to 100%, the pattern of variance in larval mortality must mirror that of variance in recruitment to the reef. The latter will be proportionally much larger when compared to the low mean rate of recruitment.) Highly variable recruitment ensures that local history has to be important in determining the composition of an assemblage of fish.

If general, and if correct, current understanding means that reef fishes provide an easily manipulated empirical analogue of the interesting models of species coexistence in open, patchy and changeable environments developed by Caswell (1978), Levin (1976; Paine and Levin, 1981), Chesson (1985, 1986), and others. It also means that our understanding of the organization of diverse tropical communities is radically altered from our simple expectations of the 1960s.

B. Ecology and the Management of Reef Fisheries

In many regions of the world, reef fish provide a very important, usually artisanal fishery. Examples of the yields taken off reefs can be found in the work by Munro and Williams (1985) and Russ (1984c, Chapter 20). The principles on which we base our management of reef fish systems should depend on the ecology which characterizes them. Until recently, however, management of these fisheries has been largely reactive rather than proactive, and seldom based on current ecological ideas.

If reef fishes are organized as open nonequilibrial systems, there are important consequences for their management. Under these circumstances, local densities are not necessarily set by the limits of locally available resources, and local harvest may have little effect on future local recruitment. At the same time, overexploitation in one location may have severe effects on future recruitment to other, possibly quite distant sites. At present it is simply not possible to say whether the current ecological view of reef fish assemblages is correct, nor what spatial scale is correct on which to base management if this view is correct. There is some urgency to resolve these questions, because the pressure on reef fish stocks is high in many regions, and increasing. More effective management is needed now.

C. Structure of This Chapter

It is clear that views on the true nature of reef fish communities have been changing and that an informed consensus is needed both for fundamental understanding and for management.

In this chapter, I will explore the development of our current view of the organization of reef fish communities. In doing this I will look critically at the evidence, and the deficiencies of data which exist, and explore the possibility that different processes are important in different geographic regions. I will also consider the scale dependence of some of the conclusions which have been drawn. And I will suggest questions worthy of future attention by ecologists interested in community-level phenomena.

II. Difficulty in Evaluating Current Data

A. Geographic Differences in Reef Fish Ecology?

Our present understanding of the ecology of reef fish assemblages is based on a strongly biased set of data and may not be generally correct. While considerable information exists for the Caribbean, and some information is available for sites scattered through the Pacific, the greatest amount of information exists for the Great Barrier Reef. This situation has apparently arisen because scientists with interest in this question have had both good access and good support to pursue these questions on the Great Barrier Reef, while other questions have been of primary interest to ecologists working with reef fishes elsewhere. There is a serious paucity of ecological data on fish for reef systems in the Indian Ocean and the Red Sea.

It is not unreasonable to suppose that reef fish communities are organized differently in different biogeographic regions. The Great Barrier Reef, in particular, differs from many other reef systems in some potentially important respects. It is a continental system stretching 2500 km along the northeast coast of Australia and is bathed by a relatively consistent longshore current (Williams *et al.*, 1984). This feature may have the result that the advection of larvae away from the natal reef, with subsequent recruitment occurring at distant locations, may be a more important feature of reef fish ecology there than in more oceanic locations in both the Pacific and elsewhere.

The Great Barrier Reef also exists in a region of physically severe conditions because the tidal fluctuations (typically 2–3 m change) combine with the strong southeast trades to ensure strong interreefal currents and extensive movements of water masses among reefs. The Caribbean, by contrast, is in a region of much more modest tidal regime, as are many parts of the mid-Pacific. These differences may enhance the possibilities for widespread dispersal of larval fishes in the Great Barrier Reef system.

Certainly, some ecologists working elsewhere have concluded that the interpretations of reef fish ecology which have been derived from studies on the Great Barrier Reef may not apply in other places. Unfortunately, because different kinds of studies have been done by different scientists in different locations, it has not been easy to resolve whether the differences reported in the literature are real geographic differences in fish ecology or differences in interpretation made by different scientists with different experiences and data (see also Ebeling and Hixon, Chapter 18).

A primary need at present is for comparative studies using identical methods among geographic regions such as the Caribbean and the Great Barrier Reef. In this way it will be possible to discern whether there are fundamental geographic differences in the ecology of reef fish assemblages.

B. Different Research Designs

Reef fish ecology as practiced over the past 25 years has been characterized by a diversity of sampling methods applied by different scientists to fish populations in a wide range of locations. Work has often been purely descriptive (snapshot samples identifying patterns in distribution and abundance of fish species). Other studies have involved field experimentation using either natural or artificial units of habitat, or have been descriptive, but have tracked patterns over more or less long periods of time (monitoring of assemblages at defined sites over months or years). Locations used have ranged in size from single coral colonies to physiographically defined regions on reefs—windward slopes, back reefs, lagoonal walls. Many studies have made use of naturally occurring patch reefs, but these vary in size and character from geographic region to geographic region. Most studies have used small patch reefs 1–2 m in diameter, but studies using patch reefs ten times larger are also common. Under these circumstances, it is perhaps not surprising that a clear consensus on community structure has yet to develop. It is also possible that the different reef fish assemblages studied are actually different in the way they are organized.

The fact is that single-time "snapshot" studies cannot provide any information concerning the temporal persistence of community structure, nor the processes underlying that structure. While this should be an obvious limitation, the general faith in the constancy of the tropical environment led early workers to presume that their snapshots were samples of a persisting system (Sale, 1988b). Hiatt and Strasburg (1960, p. 66) wrote that the Enewetak coral reef was a

> complex ecosystem which apparently fluctuates in composition very little, if at all, from year to year, and has over a long period of time acquired a biota successfully adjusted competitively in the relatively constant environment of the tropical west-central Pacific Ocean.

This statement is taken from the introduction of a paper which in most ways is a landmark contribution to reef fish ecology. It was not a conclusion reached in the course of their study, but a statement of accepted faith—a faith in the constancy of nature which was very widely held at that time. Smith and Tyler (1972, 1975), working in Bimini, Bahamas, were perhaps the first reef fish ecologists to acknowledge the need to re-census assemblages if they were to draw conclusions about the temporal persistence of patterns in community organization.

Monitoring studies have their own limitations. They enable the detection of temporal change in assemblage structure, but provide no information about the cause or causes. (They also contain some analytical problems because many reef fishes are sufficiently long-lived that many of the individuals counted are the same ones census after census. Statistical purists are not the only ones who worry about nonindependence in repeat census data.) Experimental studies can test hypotheses about causes, but the hypotheses they can evaluate are usually severely limited by spatial extent of the manipulation and by duration of the postmanipulation study. No one of these approaches can provide a full answer to questions about community structure.

Differences among studies in the spatial scale at which they are executed can have profound effects on results for two different reasons. To begin with, studies done at a larger spatial scale frequently are done using broader-scale sampling procedures. Because of this, they yield a coarser-grained description of the assemblage studied. This problem is dealt with more fully in the following section when considering the problems caused by the diversity of sampling methods used.

Studies at different spatial scales can also yield different results because they detect phenomena which are themselves scale dependent (Powell, 1989; Wiens, 1989). For example, there are numerous documentations of the fact that different species of fish are assembled in different physiographic zones of a coral reef. These zonal differences show up again and again, on reefs in all geographic regions, and are persistent. We can generate lists of species found in particular zones, or a series of line drawings of typical scenes in each habitat as did Hiatt and Strasburg (1960). By contrast, if we sample different replicate sites within one of these zones, we find much more transient associations of species (discussed more fully in the following). Only some of the list of species appropriate to that zone are present in a given small site at any particular time, and species composition changes. These different results are scale dependent, and both are correct. Differences between zones are due to factors different than the ones responsible for the more transient differences among sites within zones.

C. Different Sampling Methods

Perhaps the most important difference among studies has been in the sampling methods used. A minority of studies is based on data obtained by conventional destructive sampling of fish, using chemical ichthyocides, various types of traps or nets, or explosive charges as the collecting method. All of these methods suffer biases due to their selective capture of particular types (ichthyocides or explosives) or sizes (traps) of fish. But more importantly, use of ichthyocides or explosives yields collections that are strongly influenced by sea conditions, topography, fish activity patterns, prevalence of larger scavenging fish in the area, and ability and industry of the collecting team. We need to remember that collections are not complete. They are but samples of the fauna that was present at the time and place the collection was made. They always underestimate what was present, because they always fail to include some individuals. That this comment is true does not negate the fact that some collecting methods can yield very comprehensive collections. That it needs to be stated emphasizes how sloppily we (and many other ecologists) confuse samples and the community from which they are taken.

If we can think sloppily about samples of fish taken by ichthyocide, net, or explosive, reef fish ecologists have usually been far too casual in interpreting the results of nondestructive visual sampling schemes. These also have to underestimate, to a greater or lesser extent, what was present at the time and place of sampling. In addition, while most studies of community structure have relied on nondestructive visual census procedures, rather than on destructive sampling, the details of the procedure used vary greatly among studies, with important consequences for the data collected (Sale, 1980a).

Most studies using patch reefs, small artificial reefs, or similarly well-defined small patches of habitat rely on a "total" census of all fish present. Only occasionally have authors tested the completeness of their total censuses (e.g., Sale and Douglas, 1981), although most state explicit criteria for deciding when a census is completed—usually observing for a set number of minutes without recording a new species. Workers using large patch reefs as sample units have frequently classified the species present into abundance categories rather than presume that an accurate total count is possible (e.g., Gladfelter *et al.,* 1980). This appropriate approach, however, increases the difficulty of comparing across studies. Abundance categories chosen differ among workers, the accuracy with which species are assigned to abundance categories must vary, and it is not easy to compare these studies with those on smaller habitat units where "total" counts have been obtained (Sale, 1980a).

Where studies have used large areas of contiguous (and homogeneous) habitat as sample units, the preferred approach to sampling fish abundances has been to use strip or belt transects. Transects have differed in length and width and in method of laying down and of surveying. In many cases width has been only estimated, and in some length is also estimated by duration of swim during the survey. Results obtained from such surveys are strongly influenced by factors such as transect width, rate of swimming, number of species being counted at one time, and behavior of the fishes being censused (Fowler, 1987; Lincoln-Smith, 1988). Variation in any of these compounds the difficulty in comparing among studies.

The imprecision in such methods has also led to problems in interpretation of results. Anderson *et al.* (1981) used a "snapshot" study sampling chaetodontid fishes on a broad scale and a wide geographic area by means of essentially unmeasured transects. They then used a variety of arguments in comparing the (spatial) patterns they found to the expectations of the classical competitive paradigm and concluded that these reef fishes fit this paradigm well. Their paper had a strong impact (because it confirmed generally widely held expectations) despite the methodological problems in their sampling, the limitations of the "snapshot" approach, and the lack of statistical significance of many of the patterns they had claimed were present (Sale and Williams, 1982).

As an alternative to transects, a number of studies on contiguous reef areas have used sampling methods which yield relative abundances of species in the vicinity. Some are based on the relative likelihood of seeing a species during a randomly directed swim in the area (Jones and Thompson, 1978). Others are based on a direct count of fish moving near a stationary diver during a set time period (Bohnsack and Bannerot, 1986). While such methods have definite advantages in certain circumstances, they have rarely been tested against other sampling methods, and thus the comparison among studies is further complicated. With different scientists favoring different sampling methods, applying them to different units of habitat, and working in different geographic locations, it was perhaps inevitable that clear general patterns and principles have been slow to emerge.

D. Different Analyses, Different Indices, and Different Criteria

A major concern of scientists examining the structure of reef fish communities has been the degree to which the observed structure is persistent through time. It was the demonstration of (or perhaps the claim to have demonstrated) temporally variable structure in local fish communities in Great Barrier Reef studies which first raised the possibility that these assemblages were not organized as internally regulated equilibrial solutions to the problem of coexistence (Sale and Dybdahl, 1975). This demonstration was counterintuitive as well as counter to prevailing conventional wisdom.

The concept of constancy of community structure is not absolute at least when applied to real (and therefore somewhat variable) data, and scientists are far more hostage to their preconceptions than they may believe when evaluating such attributes (Sale, 1988a). Suppose that on two successive occasions a patch reef is found to contain approximately but not exactly the same assemblage of fish. This result can be described by emphasizing the ways in which it has changed, or by emphasizing the ways in which it has remained the same. And one can choose indices to help do either of these.

For example, the data in Table 1 can be described by emphasizing that all but 1 of 10 species are present in both censuses. A numerical index of similarity which uses such presence/absence data only, such as the Jaccard Coefficient of Community, can be calculated to show high similarity. Smith (1973) did this to quantitatively express the similarity of assemblages in

Table 1 Two Hypothetical Censuses of Fish Occupying a Patch Reef at Different Times, and the Values of Selected Similarity Indices Used to Compare Them[a]

	Census	
Species	A	B
Apogon doederleini	52	28
Chromis caerulea	23	3
Pomacentrus wardi	2	0
Pomacentrus amboinensis	5	1
Thalassoma lunare	3	7
Coris variegata	1	7
Ecsenius mandibulatus	3	12
Asteropteryx semipunctatus	9	37
Chaetodon plebius	1	3
Canthigaster valentini	1	2
Total fish present	100	100

Similarity between census A and B based on:
species presence/absence: Jaccard Index = 0.90
species relative abundance: Czekanowski Index = 0.50

[a] For details see text.

Bimini, as did Talbot *et al.* (1978) and Ogden and Ebersole (1981) for data they obtained.

Conversely, one can note that while nearly all species are represented in both censuses, the relative numbers of individuals per species are very different. This finer-resolution analysis suggests that there has been substantial change between censuses. One of the simpler similarity indices which takes account of species abundances, Czekanowski's proportional similarity index (Schoener, 1968; Feinsinger *et al.,* 1981), yields a value far from the 1.0 of perfect identity. This index was used by Sale and Douglas (1984) for analyzing patch reef assemblages. Gladfelter *et al.* (1980) and Sale and Dybdahl (1975, 1978) used comparable indices.

Even if different authors agree to use the same index, and use one including information on species abundances, there is still the possibility of putting substantially different glosses on the same results. This occurs because such indices rarely yield the unambiguous 1.0 or 0.0 extreme values of total constancy or total change. Instead they yield values in between. If the calculation of a proportional similarity index yields a value of 0.56 (e.g., Sale and Douglas, 1984), this can be written of as 56% similar, over 50% the same, or as 44% dissimilar, nearly 50% different in composition. Different authors, with different perspectives and expectations, will in this way write about the same numerical results in quite different ways. And what they write is interpreted by others from their own particular perspective. In the early 1970s, I was surprised at the degree of temporal variation which we had detected in assemblages of fish in isolated coral heads (Sale and Dybdahl, 1975), and I stated that "the chaos view of community structure comes closer to reality . . . than does the order view." Many others read this to mean that I believed these assemblages were totally chaotic in organization (C. L. Smith, 1978; Bohnsack, 1983b). I did not believe this, but I wrote in a way suggesting to others that I did.

An example of how methods and expectations interact can be seen in the interesting case of Randall's reef—a single artificial concrete-block reef built at Little Lamesure Bay, St. Thomas, Virgin Islands, by Jack Randall in 1960. Randall (1965) used his observations of this reef to form one of the first quantitative and experimental confirmations of the impact of herbivores on the standing crop of algae on coral reefs, although subsequent workers showed that echinoids may have played a larger role than the fishes Randall identified.

In the course of this work, Randall had netted and used rotenone to collect all fishes aggregated at the reef in 1962—largely disassembling the reef in the process. Ogden and Ebersole (1981) made four censuses at the reef over a 4-year period from 1975 to 1979. They concluded, on the basis of careful quantitative analysis of their data that the structure of the fish assemblage was remarkably consistent over that 16-year span. They also suggested that the difference between their results and those of Sale and Dybdahl (1975, 1978) or Talbot *et al.* (1978) on the Great Barrier Reef was due to the fact that Randall's reef was several times larger and supported several times the number of fish than the sites used in Australia. That is, they attributed these differences among studies to an inferred effect of spatial scale. Because their study was done at the larger scale, they concluded their results were the more relevant ones.

This conclusion was based on the fact that Jaccard Coefficient values among their censuses were about twice as large as those among the censuses of Talbot *et al.* (1978). Ogden and Ebersole (1981) also published their raw census data, and using these, I later showed (Sale and Douglas, 1984) that the degree of persistence of structure was quite similar in their data and in my own on small patch reefs, when Czekanowski's proportional similarity index was calculated. That is, in my analysis, there was no scale effect, and assemblage structure was variable. Two questions, of course, remain: Is my impression that these studies all document quite variable assemblage structure, or the view of Ogden and Ebersole that they do not, the correct one? And is it even correct to expect that there will be only one correct pattern found in all coral reef regions?

III. Survey of the Data: Patterns

A. Introduction

In the following section, I summarize the data available at present on the structure of reef fish assemblages, with particular emphasis on Great Barrier Reef and Caribbean studies. The twin aims are to assess the comparability, quality, and extent of the data available in different places, and to determine any general patterns or regional differences suggested by the available data.

B. Descriptions of Community Structure

1. Local Scale

a. Snapshot Descriptions The greatest amount of data defining the structure of reef fish assemblages comes from local-scale studies, particularly of small patch reef or artificial reef sites. Examples of descriptions of natural patch reef assemblages are those by D. M. Williams (1980) and Sale and Douglas (1981, 1984) for the Great Barrier Reef; Smith and Tyler (1972, 1975), Gladfelter and Gladfelter (1978), Gladfelter *et al.* (1980), and Clarke (1988) for the Caribbean; Brock *et al.* (1979), Gladfelter *et al.* (1980), and Brock (1982) for the mid-Pacific; and Molles (1978) and Gilligan (1980) for the Gulf of

California. Comparisons among patch reef studies are hampered somewhat by the variety of sampling methods and by differences in the size and location of the patch reefs.

Surveys of contiguous reef surfaces are still less easy to compare because the sampling methods vary more substantially, and because there is a greater diversity of habitat among studies. Williams (see Chapter 16) has been able to make significant progress in doing this, because he has sampled using constant methods in a variety of geographic locations. Other examples giving information on community structure of fishes on contiguous reef habitat are Anderson *et al.* (1981), Walsh (1985), Galzin (1987a,b), Galzin and Legendre (1988), Goldman and Talbot (1976), and Harmelin-Vivien (1989) for Pacific reefs; Talbot (1965) for east Africa; and Clarke (1977), Jones and Thompson (1978), and Bohnsack and Bannerot (1986) for Caribbean reefs.

Figure 2 (from Sale, 1980a) shows that for small patch reefs from a wide geographic range, there is a uniform relationship between the number of species of fish present and the number of individuals present. While this observation can be dismissed as just another example of a species–area curve, at the time it was intriguing to me that the relationship appears to be the same both in places of very high and in places of much lower total diversity of

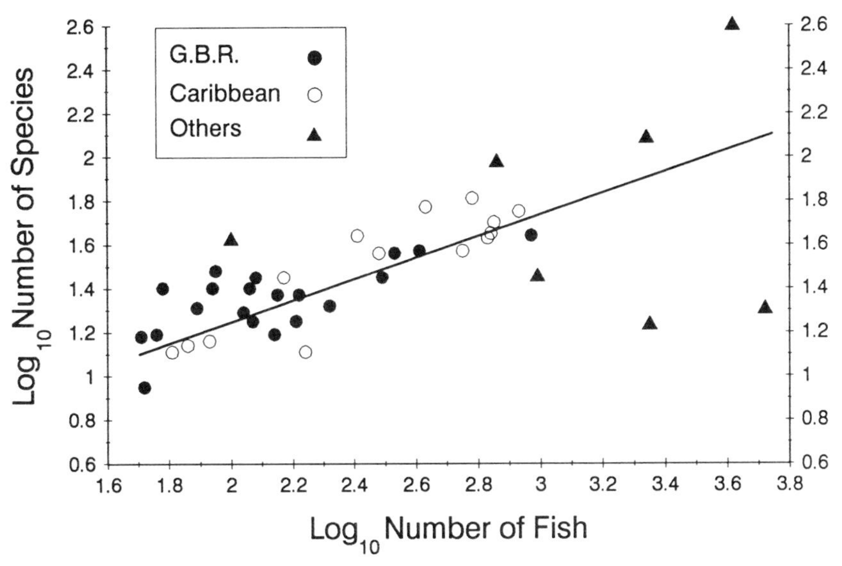

Figure 2 The relationship between the number of species represented and the number of individual reef fish present on a patch reef is remarkably consistent among geographic regions despite marked geographic variation in overall fish diversity. Each point represents a single censused patch reef. (Redrawn from Sale, 1980a, which also identifies the data sources used.)

species. It suggests that within-habitat diversity may be relatively constant among locations despite substantial differences in total regional diversity.

The taxonomic composition of local patch reef assemblages has received little attention. Yet taxonomic or trophic consistency among similar sites might be expected if community structure is strongly determined. In fact, there appear to be some substantial differences in proportional abundance at the family level when Caribbean and Great Barrier Reef assemblages from the same kind of habitat are compared. The differences are greater than one might expect, given the near equivalence of the two faunas at the family level. Figure 3 shows the distribution by family of fishes on 20 lagoonal patch reefs at One Tree Reef, and on 20 similar patch reefs from the back reef at Teague Bay, St. Croix, USVI, which I censused in 1981. The strong representation (at least in terms of density of individuals) of Apogonidae, Eleotridae, and Pomacentridae at the southern Great Barrier Reef site is neither matched at St. Croix nor replaced by species of other families having similar ecological roles. Figure 3 also shows that the fish fauna at these two sites is moderately different in trophic composition. There does not appear to be an overriding pattern of ecological organization among patch reef fish assemblages in these two geographic regions.

Artificial reef data also provide information on density, diversity, and taxonomic and trophic composition of fish assembled on patch reef-like units. Great Barrier Reef sources are Talbot and co-workers (Russell *et al.,* 1977; Talbot *et al.,* 1978). In the Caribbean, Bohnsack (1983b; Bohnsack and Talbot, 1980) constructed patch reefs identical in size and material (concrete blocks) to Talbot's, while Molles (1978) used a very similar design in the Gulf of California. Other artificial reef studies, using reefs of different size or construction material, include Ogden and Ebersole's (1981) re-survey of Randall's large concrete-block reef, Shulman's (1984, 1985a,b) work with small conch shell reefs or small concrete-block reefs (Shulman *et al.,* 1983), Hixon's concrete-block reefs (Hixon and Beets, 1989; Hixon, Chapter 17), all in the Virgin Islands, and work by Walsh (1985) in Hawaii, among numerous others. A comparison of the structure of fish assemblages reported for these artificial structures would be interesting because artificial reefs in different locations are presumably much more similar habitats than are natural reefs. While assemblages on artificial reefs need not be "natural" (the artificial structure might lack resources needed by certain species), we might expect them to be more similar in trophic structure from place to place than is the case for natural reefs. That comparison waits to be made.

b. Temporal Monitoring and Experiments The discovery that local assemblages varied in structure through time was made slowly and accidentally. In the 1970s this was not an expected result. In the Caribbean, Smith and Tyler

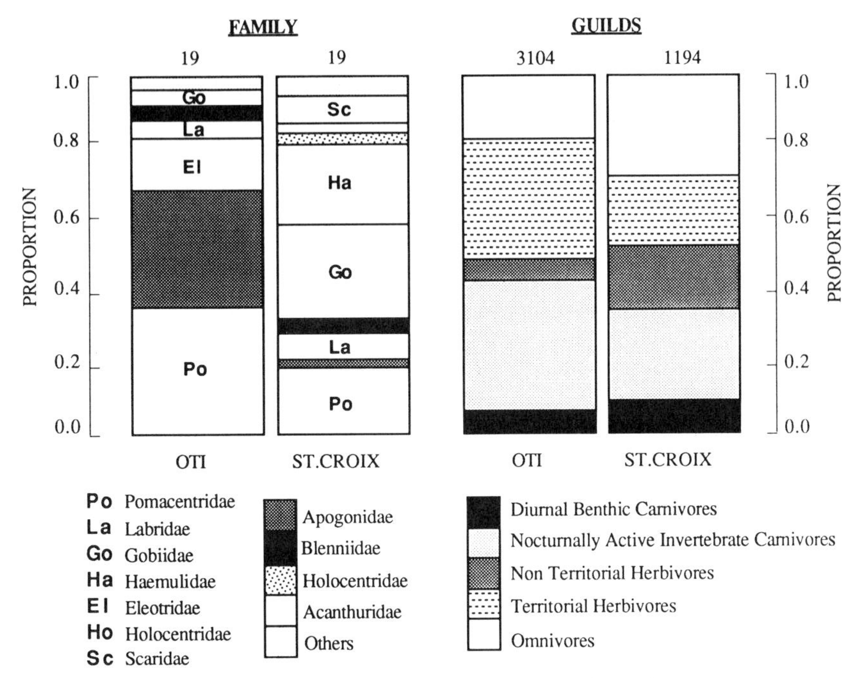

Figure 3 Distribution among families and among trophic guilds of fishes in similar patch reef habitat at One Tree Reef, Great Barrier Reef, and at Teague Bay, St. Croix, USVI. One Tree data are from Sale *et al.* (1991); St. Croix data were collected in 1981 (previously unpublished information).

(1975), by sampling a series of *Montastrea* coral domes in various stages of decay, had shown how fish assemblage structure might vary over relatively long periods of time, in effect tracking a changing habitat. But although they paid some attention to re-censusing patch reefs, they generally concluded that in the absence of habitat change, fish assemblages remained constant in structure. A primary reason for drawing this conclusion may have been that they measured change between successive censuses primarily by comparing species presence and absence rather than relative abundances.

Also in the early 1970s, Talbot and co-workers began their artificial reef studies on the Great Barrier Reef. They constructed identical replicate concrete-block reefs and fully expected them to fill with species to a more-or-less constant final assemblage. Yet, as they subsequently reported (Talbot *et al.*, 1978), they found substantial variation among reefs in the rates of settlement of different species. Different assemblages developed as a consequence.

In 1975, I commenced what was to have been a standard perturbation experiment using assemblages of fishes on natural patch reefs (Sale, 1980b). I denuded three similar patch reefs of fishes and monitored three additional reefs as controls. Since these reefs were located in a scientific reserve, I was confident that the control reefs were truly undisturbed by human interference over the following two years. The purpose of the experiment was to determine the rate and extent of "recovery" of community structure.

The results of this experiment (Fig. 4) were surprising to me. I had not really expected complete recovery of structure, because, by that time, those of us working in Australia had come to expect some temporal variation in assemblage structure due to the vagaries of recruitment and survivorship. But what I did not anticipate was the extent and rapidity of change in structure of the undisturbed assemblages (Fig. 4). Using Czekanowski's proportional similarity index to measure degree of similarity between successive censuses and the initial census of each assemblage, I found that, with the fairly rapid recolonization, the assemblages on denuded reefs began to return to their starting condition within three months. The rate of "recovery" slacked off and had virtually ceased with the assemblages being only about 50% similar to their starting conditions at the end of the experiment 29 months later. But during this same time, the assemblages on the control patch reefs deviated in structure from their starting condition, until they too retained only about 50% similarity to their starting condition. Since there was no indication that

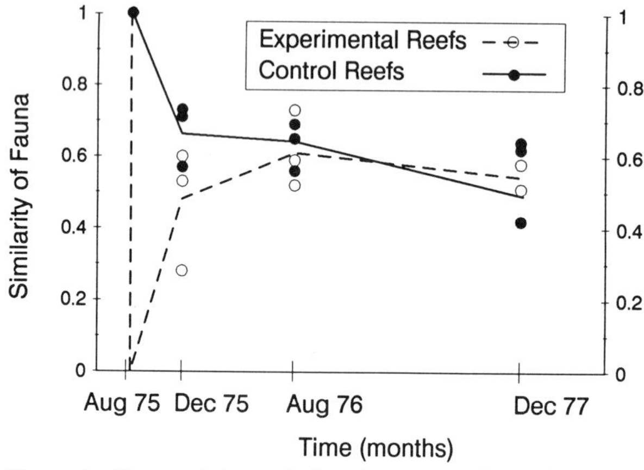

Figure 4 Temporal change in faunal structure of assemblages on experimentally denuded reefs and on undisturbed control reefs. Points at each date express faunal similarity of the assemblage present to that present on the same reef at the start of the experiment. Similarity is measured using Czekanowski's proportional similarity index which ranges from 0.0 to 1.0. Experimental reefs had all fish removed on day 0. For details see text. (Redrawn from Sale, 1980b.)

the patch reefs were changing in physical structure during this time, I concluded that this was not a case of the fish assemblage tracking a changing habitat. At the scale of small natural patch reefs, assemblages vary naturally in composition such that successive censuses show substantial differences to one another.

In viewing Fig. 4, it is important to remember the nature of similarity indices. Although the two curves suggest a process which changes in rate until reaching an assymptote at about 50% similarity, the curves measure the degree of faunal difference from starting conditions. In fact the processes responsible, the additions and losses of individual fishes, continue to occur at a more or less constant rate, and the later censuses of particular assemblages were about as different from one another as they were from the starting structure. Bohnsack (1983b) has discussed the sensitivity of similarity measures to the frequency of censusing in reef fish assemblages. My point here is that censuses of an assemblage can all differ to about the same degree from a single earlier (starting) census while still differing substantially from each other!

Subsequent to this experimental study, I commenced a nonmanipulative monitoring study of patch reef assemblages which was destined to continue for 10 years. The results of this study (Sale and Douglas, 1984; Sale and Steel, 1986, 1989; Sale and Guy, 1991) provide abundant evidence for the patterns of temporal change which can occur in patch reef fish assemblages. The study included 20 patch reefs ranging in size from 2.54 to 31.73 m^2 arrayed as 10 reefs in the southern (windward) quadrant and 10 in the northeastern quadrant of the enclosed lagoon of One Tree Reef. All stood on a sandy lagoonal floor at 3 to 4 m depth at low tide and were sufficiently isolated (5 m or more of open sand) from each other and other hard substrata so that migration of fish was not a serious problem. A number of physical attributes of the reefs, including substratum type and diversity, topographic rugosity, and size, were recorded early in the study (Sale and Douglas, 1984) and again at its end (Sale *et al.*, 1991). Change in reef structure during the 10 years was negligible.

Fish present on each patch reef were censused using a "multiple-visit" approach, which had been tested for precision using similar patch reef communities (Sale and Douglas, 1981), and compared to complete ichthyocide collections (Sale, 1980b). This census method generates as extensive a list of fishes present as does a total collection, although it records more of the active species and misses some of the cryptic species present in the deep interstices of the reefs (Sale, 1980b). The censuses are sufficiently precise that successive censuses a few days apart show Czekanowski similarity = 0.85 in faunal composition (Sale and Douglas, 1981). The technique produces a reliable record of all species which are diurnally active, or which are nocturnal but shelter in open caves and overhangs.

A primary result of this study was the thorough documentation of type and extent of variation in assemblage structure. Using Czekanowski's proportional similarity index, the 20 censuses of assemblages at the same patch reef showed average similarity to one another of 0.510 ± 0.013 S.E., slightly lower than the 0.568 ± 0.009 S.E. obtained for comparisons among the first 8 censuses (Sale and Douglas, 1984). On every reef, the total number of individuals recorded at a census varied at least twofold during the course of the study. On most reefs, the variation was greater than this, and episodes of high abundance were not coincident across patch reefs (Fig. 5). Among the commoner species, abundances fluctuated across censuses, and fluctuations were differently patterned among patch reefs (Fig. 6).

Using the Coefficient of Variation among censuses of total numbers of a species (pooled across 20 patch reefs), we classified the 40 commonest species in the data (Sale *et al.,* 1991). A low CV implies a tendency to vary little in overall abundance from one census to another. Concordance among censuses with which each species ranked patch reefs (as measured by abundance of that species present) was a second classifying dimension—a high concordance indicates a tendency to be similarly distributed among patch reefs from one census to another. These 40 species accounted for over 95% of all sightings during the 10 years. They showed substantial differences to one another, but many showed relatively low concordance and relatively high coefficients of variation (Fig. 7).

A large proportion of the structure evident in these data could be generated by a relatively simple model of random placement of recruits into patch reefs (Sale and Steel, 1986). This model assumed that patch reefs differed in attractiveness to recruits—perhaps simply because patch reefs differed in size–and used total abundance of fish present as a measure of attractiveness. Otherwise, it assumed that fishes colonized patch reefs independently and randomly. We found that once a proportion of habitat specialists or otherwise nonrandomly distributed species were eliminated from consideration, the model gave a good fit to observations. Subsequently we closely examined the possibility of associations among species of fish by means of a cluster analysis of distributions of fish among patch reefs at each of the 20 censuses (Sale and Steel, 1989). There were pronounced changes from one census to another in the pattern of association of species (Fig. 8), adding further support to the developing picture of these assemblages as highly plastic associations of fish.

This long-term study does not appear to be contradicted by other shorter-term studies on Australian reefs. D. M. Williams (1980) had earlier shown very similar patterns of erratic variation in numbers for the pomacentrid component of a series of natural patch reef assemblages, and Talbot *et al.* (1978) had documented considerable variation in composition among replicate small artificial reefs and within reefs during the first two years following their construction. There is also little evidence that the situation is fundamentally different in the Caribbean, apart from the fact that the lower overall

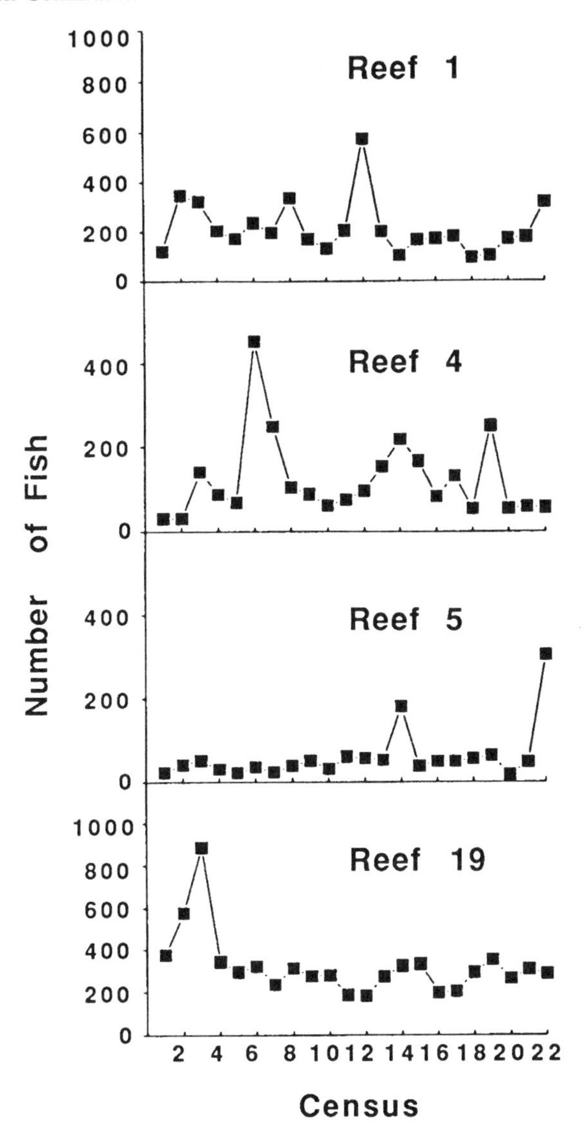

Figure 5 Total number of fish present on each of four typical patch reefs monitored over 10 years at One Tree Reef. Note extent and asynchrony of fluctuations.

diversity of the fish fauna there might be expected to reduce the extent of differences detected either among replicate units of habitat or across successive censuses. Apart from the study by Ogden and Ebersole (1981), already discussed, monitoring of assemblage structure over several years has been

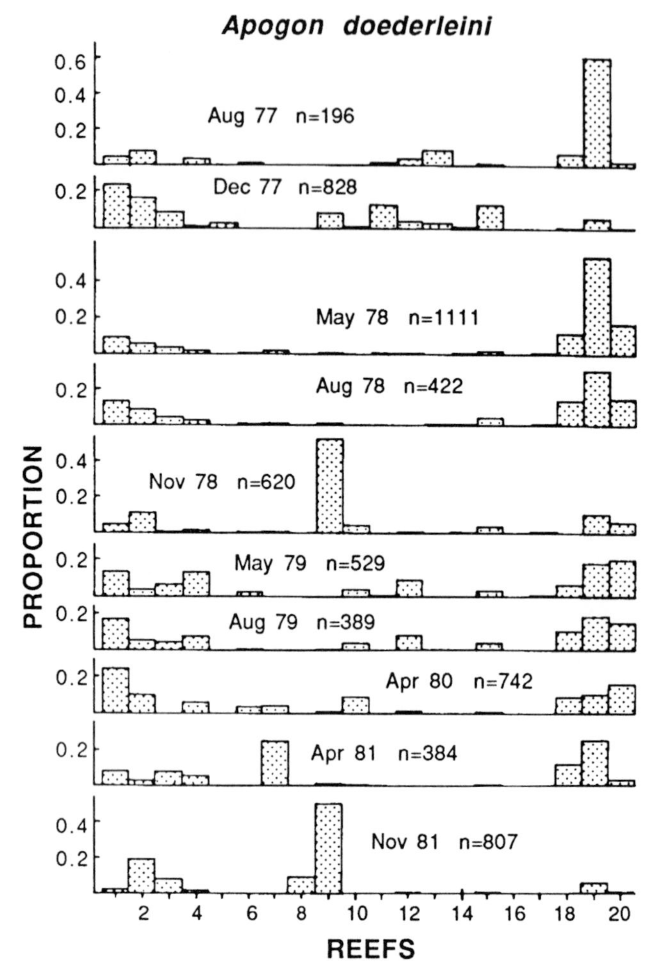

Figure 6 Fluctuations in abundance of *Apogon doederleini* (the most abundant species) at each of 20 patch reefs monitored at One Tree Reef. (Redrawn from Sale and Steel, 1989.)

done by Smith (1973), Smith and Tyler (1972, 1975), Bohnsack and Talbot (1980), Bohnsack (1983b), Clarke (1988), and Hixon and Beets (1989). Robertson (1988a,c) has recently provided patch reef data for particular species.

All of these studies of temporal persistence have used small patch reefs or artificial reefs as the habitat to be examined. Patch reefs, by their nature, provide a degree of local isolation to the resident fishes which is not present in

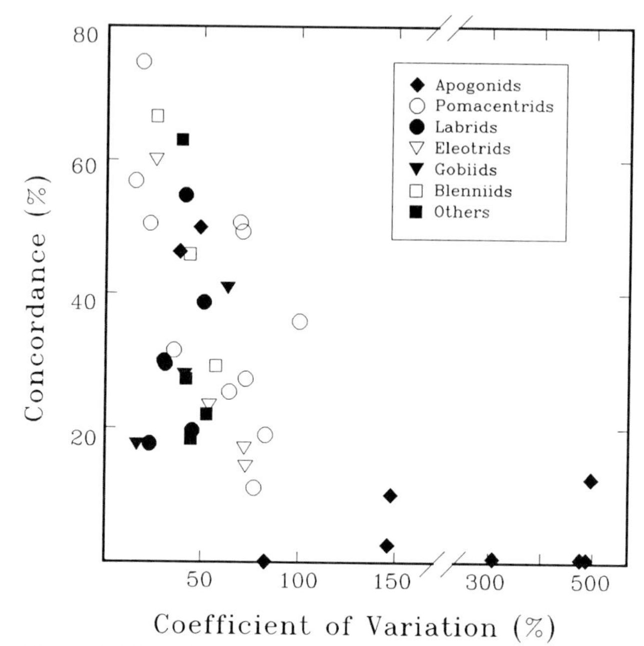

Figure 7 Dispersion of the 40 species most common on a set of 20 lagoonal patch reefs in a space defined by their variation in abundance among censuses (measured by the Coefficient of Variation) and by the consistency with which they distributed themselves among patch reefs at successive censuses (measured by Kendall's *W*). The species differ considerably in the ways in which they respond in space and time. For details see text.

other types of coral reef habitat. Whether local assemblage structure is as temporally variable over continuous reef habitat, where fish may be freer to move about, is not known.

2. *Reef and Regional Scales*

a. Snapshots On the Great Barrier Reef, Williams (1982, Chapter 16; Williams and Hatcher 1983) and Russ (1984a,b) provided data at the scale of a physiographic zone, using comparable techniques. Both authors documented broad-scale changes in faunal composition as one proceeds from coastal toward more offshore reefs over a span of some 200 km. The patterns they describe are complex, including change in species composition and change in trophic organization. Anderson *et al.* (1981) provided more limited data for chaetodontid fishes on this scale.

Elsewhere there are data on the scale of a physiographic zone, but system-

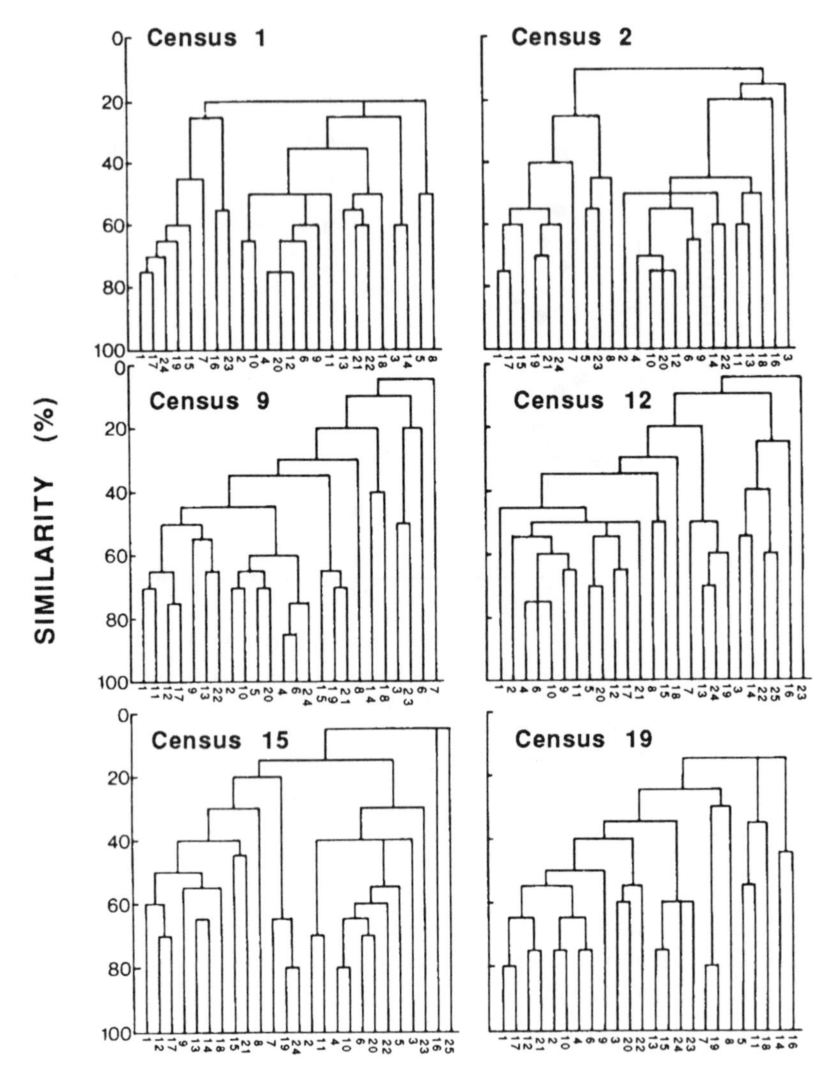

Figure 8 Dendrograms showing pattern of association among the 20 commonest species on lagoonal patch reefs at 6 of 20 censuses. Note the differences among dendrograms. (Redrawn from Sale and Steel, 1989.)

atic regional comparisons have been rare. Comparisons among the published accounts are hampered by the differing sampling techniques used. Gladfelter *et al.* (1980) compared the fish assembled on large patch reefs in the Virgin Islands, Caribbean, and in Enewetak Atoll, Marshall Islands, and Harmelin-

Vivien (1989) has recently made a detailed comparison for fish on reefs at Tulear, Madagascar, and at Moorea, Society Islands, at opposite edges of the Indo-Pacific region. The reef sampled in Tulear supported 552 species in 71 families, while that at Moorea yielded only 280 species in 48 families. While 43 families were common to the two sites, only 136 species were shared. Despite these strong differences in richness and taxonomic composition, Harmelin-Vivien reported considerable similarity between the two fish faunas in functional characteristics, such as patterns of distribution of reproductive type, size distribution, distribution of trophic types, and so on.

Community structure of Caribbean fishes has not been examined at a regional scale. It would be interesting to use published data on assemblage structure within particular reef zones to make some preliminary regional comparisons for further comparison with the Pacific, particularly Australian, results. At present, it is clear that assemblages of fish in different geographic regions or physiographic zones can sometimes differ substantially in what we might call ecological structure. These differences reflect differences in the availability of types of fish in the different regions or habitat zones. While taxonomic differences, such as those which exist between the Caribbean and the Great Barrier Reef, are an expected product of separate evolutionary histories, the trophic (= ecological ?) differences are not. These suggest fundamentally different trophic webs in different places, with fish playing roles of differing importance in the industry of transferring organic production through different reef communities. Why this should be so is a tantalizing question for the future, and one which should call into question assumptions about the grand design of nature.

b. Temporal Monitoring I know of no data available for any region in which community structure has been monitored over time on a regional scale. This is both methodologically and logistically a demanding task. That recruitment has been monitored at these scales (Sale *et al.*, 1984a; Doherty and Williams, 1988a; Victor, 1984, 1986b; Robertson *et al.*, 1988) suggests the task is feasible. It would be very interesting to see if the regional-scale patterns being found are persistent through time.

IV. Survey of the Data: Processes

A. Interactions among Resident Fishes

1. *Interspecific Interactions*

a. General There has been a strong tendency, until recently, to explain observed patterns in assemblage structure by recourse to interspecific interactions among resident reef fishes. Most emphasis has been placed on competitive interactions, although predatory and commensal interactions have also been invoked. This bias in favor of competitive processes and a near blindness to other possibilities were shared with ecologists working in other systems (see discussions by Wiens, 1977; Strong *et al.*, 1984a; Sale, 1988b).

Such interpretations occur in studies done in the Great Barrier Reef, the Caribbean, and other locations. Earlier studies have been well reviewed by Ehrlich (1975), Goldman and Talbot (1976), and Sale (1980a). Patterns interpreted have been ones expressed at a wide range of spatial scales, from the very local scale of microhabitat (e.g., Ebersole, 1985) to much broader geographic scales (e.g., Anderson *et al.*, 1981). In particular, the restriction of fish species to particular physiographic zones on a reef has frequently been ascribed to resource partitioning due to past or present competition with other species. Despite the popularity of such explanations, simply invoking them does not establish that these processes are actually operating!

For the Great Barrier Reef, there is now sufficient information on patterns of settlement of recruiting juveniles that we can be certain that these broader-scale patterns are not due to present-day competition. The juveniles settle into the physiographic zones also occupied by adults of their species (Sale, unpublished observations; Doherty and Williams, 1988a; Williams, Chapter 16) so there is little opportunity for competitive interactions with ecologically similar species which occupy other zones. The zonal separation may be a result of "ghostly" past competition (*sensu* Connell, 1980), but as I argue in the following sections, there are other explanations. The situation may be different in the Caribbean, where use of shallow sea grass nursery habitats by young juveniles is reportedly widespread (McFarland *et al.*, 1985; Shulman, 1984, 1985a). Little attention has yet been paid to the interactions among fish as they move to adult habitats.

Williams (Chapter 16) has surveyed the patterns in species distribution and recruitment. My intention here is to review studies in which the biotic interactions have been explicitly examined, particularly those in which manipulative methods have been used.

b. Interspecific Competition Despite the enthusiasm for competition-based explanations of species patterns, rather few experimental studies have been done. The best studies have been ones where some or all of the interacting species were territorial and the competition was for space. In particular, studies of territorial damselfishes of the genera *Pomacentrus*, and *Stegastes* (=*Eupomacentrus*) have provided a wealth of examples in which competitive interactions among species have been documented. One class of these studies has emphasized the interactions between territory-holders and nonterritorial herbivores or egg predators of other species (Low, 1971; Myrberg and Thresher, 1974; Robertson *et al.*, 1976; Thresher, 1976; Itzkowitz, 1977b; Shulman *et al.*,

1983). Other studies have emphasized interactions among co-occurring species which are all territorial (Robertson *et al.*, 1979; Robertson and Polunin, 1981; Robertson, 1984; Ebersole, 1985; see below). In these studies, the prevailing approach has been experimental, and the usual result has been to demonstrate that territoriality modifies the access by other fish to territorially claimed space. The one experiment which might have shown that this sole occupancy of defended sites enhances fitness gave a negative result: Robertson *et al.* (1981) reduced available space for a group of *Stegastes planifrons* by about 50% with no effects over one year on survivorship, growth, or reproductive condition of the fishes.

Recently, Sweatman (1983, 1985a, 1988) has reported on the interactions of residents of two pomacentrids which occur in small groups occupying heads of branching corals on the Great Barrier Reef. *Dascyllus aruanus* and *D. reticulatus* collectively defend their home corals from approaches by other fish. Sweatman concentrated primarily on the possible effect of this active defense on recruitment of newly arrived juvenile fish and has confirmed that settlers recruit preferentially to corals already occupied by conspecifics, while avoiding corals occupied by the congener. He also demonstrated that the congeners were more strongly separated in their distribution among neighboring corals than would occur if they were not interacting negatively with each other, and that certain other species were less likely to recruit to corals occupied by either fish than to empty coral heads. However, since the process operating appears to be habitat choice by the settling juveniles, rather than aggression by adults (they recruit at night while the adults are sleeping), the segregation of these two species is technically not a direct consequence of competition.

Interspecific competitive interactions among nonterritorial species are less commonly investigated. In the Caribbean, Robertson and Sheldon (1979) examined the apparent competition for sleeping sites between the wrasse *Thalassoma bifasciatum* and its own and other species. They used a manipulative approach and were able to show that although fishes fought over or defended their sleeping sites, there was not a shortage of supply of such sites in the area. Technically, the fighting that was observed is not a consequence of ongoing competition for shelter, although it may persist as an adaptive trait because of past periods of intermittent shortage of shelter (the same old competitive ghost).

Most studies of competitive interactions have not directly addressed the possible impact on community structure. Ebersole (1985) stressed the way in which each of two species of *Stegastes* was favored in each of two microhabitats at St. Croix, thus providing an explanation of their distribution in nature. It is possible that this kind of differential competitive success is important in determining microhabitat segregation of many groups of species which occupy the same general reef habitats. As noted earlier, documenting patterns does not establish processes, and experimental studies such as Ebersole's to address this question have been rare. Neither Talbot *et al.* (1978) nor Sale and co-workers (Sale and Dybdahl, 1975, 1978; Sale and Douglas, 1984; Sale and Steel, 1989) were able to establish temporally consistent patterns of association among the species on their patch reefs. This result suggests that competitive interactions were unimportant for determining species distributions in their systems.

c. Lottery Competition for Territories In the early 1970s, my study of three species of permanently territorial pomacentrids (*Stegastes apicalis, Plectroglyphidodon lacrymatus,* and *Pomacentrus wardi*) convinced me that they were in continuous competition for territory space, but that the competition among them was not likely to lead to either niche diversification or competitive exclusion! Individuals of these species defend small territories throughout the year, which supply food, shelter, and, for males, nest sites. The sites for this work were located on a sheltered shallow reef slope at Heron Island, Great Barrier Reef, and consisted of patches of dead coral plates and coral rubble substrata surrounded by dense stands of staghorn acroporid corals. These fish required dead, algae-covered rocky surfaces on which to establish their territories, and all available space in each site was occupied by contiguous but nonoverlapping territories belonging to members of the three species. Over a period of 40 months as resident fish died and new individuals recruited, the total area of habitat occupied remained nearly constant. Most of the recruitment was by juveniles from the plankton, although a sizeable fraction of immigrants of *P. wardi* were older juveniles which had been apparently living in suboptimal sites within the staghorn thickets.

My studies, which included some (weak) experiments (e.g., Sale, 1976) as well as long-term monitoring of natural changes (Sale, 1974, 1975, 1978a), indicated that in this location, all three species were capable of obtaining and successfully defending sites, and that sites which became available were allocated by chance among the three species. This "lottery" process was able to operate because the life histories of these species decoupled present competitive success from future recruitment by individuals of that species (Sale, 1977, 1979a). Subsequently, Chesson and Warner (1981; Warner and Chesson, 1985) pointed out that in addition to the decoupling produced by the bipartite life history, it was essential that species in a lottery were iteroparous (as most reef fishes are), so that the effects of periods of strong recruitment would be stored over a number of successive reproductive periods. In a lottery, each individual gambles against genetic extinction by producing and dispersing numerous larvae or equivalent propagules. By chance, some of its offspring will gain priority of access to needed space resources and retain their use despite the (unsuccessful) later arrival of other individuals of its own or another species.

There are other possible processes which might have permitted the coexistence I observed on these rubble patches (Sale, 1979a). Natural disturbances or predation might remove resident fishes so that free space is kept available. New recruits to these sites might come from other places where each species is competitively dominant. Competitively mediated resource partitioning during larval life might determine the coexistence of the three species. But I still consider, as I did at the end of my study, that lottery competition is the simplest explanation of their coexistence in those sites, and the one most compatible with known information (Sale, 1979a). Importantly, lottery competition and other essentially nonequilibrial processes (cf. Connell, 1978, who called lottery competition an equilibrial mechanism) which obviate competitive exclusion are not incompatible with each other. At present, I see no way in which organisms with the type of life history typical in reef fishes could engage in interspecific competition according to the tenets of conventional theory (e.g., MacArthur, 1972; Diamond, 1978). Reef fishes live in open systems. They are iteroparous and highly fecund, and have dispersive larvae and high early mortality. There are thus lots of ways in which they may find themselves below "carrying capacity." Yet when their numbers rise and competition takes place, they are predestined to enter into lotteries.

d. Predation The possible role of piscivory in determining the structure of reef fish assemblages has received far less attention than has that of competition. Hixon (Chapter 17) provides an excellent review, so my treatment here is kept brief.

The presence of piscivores must influence the chance of survival of newly recruited reef fishes. Smith (1977) spoke of a "predator screen" facing recruiting juvenile fishes, and Talbot *et al.* (1978) suggested that piscivory might be the reason for the variability they observed in assemblage structure on their artificial reefs. Yet, piscivory is very difficult to study directly (Sweatman, 1984). Doherty and Sale (1986) experimentally attacked the question of whether piscivores had a significant impact on survival of recruits during their first few days following settlement, and Hixon and Beets (1989) have looked more generally at whether the presence of larger, piscivorous fishes modifies the structure of assemblages which form on small artificial reefs.

Doherty and Sale (1986) worked on the reef slope of One Tree Reef and, despite experimental artifacts (which resulted in a strong plea to avoid caging experiments if at all possible!), were able to establish for certain species of fish that predation played a major role in causing mortality during the first two weeks in the reef environment. The data suggested that mortality declined greatly by the end of this time (see Hixon, Chapter 17, Fig. 7), a result confirmed by subsequent studies of early survivorship at the southern Great Barrier Reef study site (Eckert, 1987; Fowler, 1988; Mapstone, 1988; Sale and Ferrell, 1988). Hixon and Beets (1989) used artificial reefs differing in number and sizes of shelter holes to demonstrate that the presence of larger predaceous species had predictable effects on assemblage structure. It is clear that far more attention needs to be directed to this process.

e. Mutualism I include mutualism in this discussion for completeness rather than because there is anything much to report about its possible role in influencing community structure. The major mutualistic interaction involving coral reef fishes is "cleaning symbiosis," a relationship which may well be parasitic (Gorlick *et al.,* 1987). The early suggestion by Slobodkin and Fishelson (1974) that cleaning stations set up by *Labroides* spp. might serve to heighten local fish diversity on Indo-Pacific reefs has not been taken further, and Gorlick *et al.* (1987) showed that removal of *L. dimidiatus* had no effect on fish diversity.

Other possible examples of mutualism between fish species include the trophic associations which sometimes develop around larger benthic-feeding carnivores (e.g., Itzkowitz, 1977b; Diamant and Shpigel, 1985; Aronson and Sanderson, 1987) and the mixed-species schools of nonterritorial herbivores (e.g., Robertson *et al.,* 1976; Reinthal and Lewis, 1986; Wolf, 1987). These associations may be too transitory and nonspecific to play a significant role in determining community structure, but they have not been investigated from that perspective.

Associations of fish with particular invertebrates can be more readily explored as examples of microhabitat selection. Available evidence concerning *Amphiprion* spp. and anemones (Allen, 1972; Hanlon and Kaufman, 1976), gobiids and burrow-building alphaeid shrimps (Karplus *et al.,* 1972), or pomacentrids, labrids, and gobiids and particular species of branching corals or sponges (Tyler, 1971; Sale, 1972b; Greenfield and Greenfield, 1982; Eckert, 1985a; Sale *et al.,* 1984b) suggests that the associations are usually between groups of species of fish and groups of species of invertebrate. (Some authors have written as if their data showed narrower associations than they do.) A tighter bonding, pairing each fish species with a particular invertebrate species, would be necessary for such associations to play a major role in determining the structure of fish assemblages.

2. *Intraspecific Interactions*

a. Intraspecific Competition Curiously, little attention has been paid until recently to intraspecific competition in reef fishes. Yet conspecifics are likely to be the neighbors most similar ecologically to a reef fish, and the reef fish life history appears to make it unlikely that neighbors will be close relatives (Avise and Shapiro, 1986). Without altruism derived through kin selection, intraspecific competition should be especially prevalent.

Myrberg and Thresher (1974) established for a Caribbean territorial pomacentrid that territorial defense was more intense when the intruder was a conspecific. Sale *et al.* (1980) demonstrated this for two Great Barrier Reef pomacentrids. Both studies suggest a differential (more negative) response to conspecifics.

By experimentally removing adult *Amphiprion*, Allen (1972) suggested that adults limited the growth of juveniles resident in the same anemone. Such an experimental approach to directly examine possible effects of intraspecific competition has not been pursued further until quite recently. Doherty (1983a) pioneered and Jones (1987a) perfected a much more complex (and more powerful) type of controlled field experiment to directly test the effects of competition from members of the cohort, and from resident adults, on the growth and survivorship of newly recruited juvenile reef fishes. In these experiments, small reefs are built of locally available natural dead and living coral substrata to allow designs with sufficient replicates, but with natural substrata. Young, newly recruited fish are collected elsewhere and added to the experimental reefs in desired numbers with and without adults also collected and transferred from elsewhere. Doherty worked with two territorial pomacentrids (*Pomacentrus wardi* and *P. flavicauda*) and Jones used the nonterritorial but site-attached *P. amboinensis*. Recently, Mapstone (1988) and Forrester (1990) have done similar work with *P. molluccensis* and *Dascyllus aruanus,* respectively, both site-attached planktivorous pomacentrids.

Each of these studies has demonstrated that intraspecific competitive effects occur, but that they are usually effects on growth and maturation rather than on survivorship. The effects can be quite complex, differing with life stage of the young fish. G. J. Eckert (personal communication), G. P. Jones (1986), and Forrester (1990) have all demonstrated that the competition observed is at least partly for limited food. (All three authors studied species which were planktivorous at least when juvenile. Eckert worked with the labrid *Thalassoma lunare.*) At present there is a need for far more work of this type, using different families of reef fish. The work to date has all been done at One Tree Reef. This topic is well reviewed by Jones (Chapter 11).

b. Other Intraspecific Interactions Sweatman's (1983, 1985a,b, 1988) work on nonrandom recruitment by juvenile *Dascyllus aruanus* or *D. reticulatus* (discussed earlier) stands as the only detailed study of the attraction of new recruits to groups of conspecifics. This phenomenon may be more prevalent, especially considering the frequency with which (at least on the Great Barrier Reef) relatively rare species are found in small groups rather than as randomly dispersed individuals (personal observations). Too little attention has been paid to it so far.

3. Conclusions

Despite the emphasis which theory has placed on interspecific interactions in determining community structure, remarkably little empirical work has directly addressed these ideas. What has been done is tantalizing, because there are suggestions of species which in some ways attract conspecific recruits while repelling others, and other (territorial) species which act in ways to reduce the likelihood of new recruits of their own species settling nearby. Recent work on intraspecific competitive interactions demonstrates complex changes in competitive effects depending on age of the (juvenile) fish studied, but shows most effects to be on rates of growth and maturation rather than on survivorship.

There is also limited evidence for the importance of piscivory in determining survivorship, particularly of newly settled recruits, and evidence that the intensity of such predation varies among species of recruit in the same area. Mutualistic relationships between fish species have been much less closely examined, apart from the influence of "cleaning stations" on Pacific reefs acting to concentrate other fishes in a local area.

Available data do not suggest any regional differences among the processes likely to be operations. However, very little work has yet been done, and almost all of it has been done on the Great Barrier Reef. The work so far has also strongly emphasized pomacentrid species, although the Pacific pomacentrids come from a greater diversity of ecological types than do Caribbean species. Caribbean shallow-water pomacentrids are mostly aggressively territorial herbivores.

B. Patterns of Recruitment: Impact of Recruitment Variation on Assemblage Structure

It is now recognized that the recruitment of young reef fishes is a highly variable process, in both time and space and on a number of scales. The patterns in recruitment have been ably reviewed by Doherty (Chapter 10; Doherty and Williams, 1988a) and will not be discussed here. The temporally variable structure of local assemblages discussed earlier can be readily understood when one realizes that each assemblage is subject to a discontinuous, episodic pattern of addition of new juvenile fish of various species. What is not clear is how large a role recruitment variation plays in determining assemblage structure.

In speaking of "recruitment," reef fish ecologists usually refer to the time at which fish settle from the plankton to the reef and become juveniles. Given that this process is highly variable in time and space, its role in determining assemblage structure may vary in importance. At one extreme, recruitment variability alone might determine the structure of local reef fish assemblages. At the other extreme, this variability might have no impact on the structure of assemblages. Importantly, the temporal constancy, or the spatial uniformity of assemblage structure, cannot be used to decide the importance of the role of recruitment variation. This is because its importance depends on the strength of the patterns (spatial or temporal) it sets up, relative to the patterns set up by other postrecruitment processes: associations between fish species and particular microhabitats; competitive, predatory, and mutualistic interactions among resident fishes; and physical disturbances which act to disrupt or to remove patterns established by other processes. I agree with the authors of several of the earlier chapters (in particular, Jones, Chapter 11; Hixon, Chapter 17) that our present challenge in understanding the factors responsible for determining community structure is to move beyond attempts to identify a single organizing process to assessment of relative roles of processes, being aware always that relative importance may itself vary from place to place or through time.

The treatment of assemblage data discussed earlier (Fig. 7) provides one approach to assessing relative roles of processes. Those species which exhibit relatively high concordance in distribution among patch reefs from census to census, and which also exhibit relatively low coefficients of variation in total numbers present, make a contribution to assemblage structure which is not strongly influenced by recruitment variation. This may arise for several different reasons. Recruitment may be strongly determined by microhabitat responses so that spatial variation is uniform through time. Recruitment may generate abundance patterns which are rapidly eliminated by high early mortality which is site dependent because they are highly succeptible to particular predators or competitors. Or the species may be long-lived with low rates of juvenile mortality, and therefore they recruit in proportionately small numbers at any one time. On our patch reefs (Sale *et al.*, 1991), such species appear to be either long-lived ones which recruit only in low numbers or habitat specialists. Recruitment variation is swamped by a large standing crop of older fishes, or is strongly determined by habitat requirements.

Less can be said about the species arrayed in other regions of Fig. 7. They represent species which:

1. have varied little in overall numbers, but have been distributed differently across reefs at each census;
2. have varied both in overall abundance and in distribution among patch reefs; and
3. have varied in overall abundance, but have tended to retain the same pattern of distribution among patch reefs (these are least well represented in Fig. 7).

All three patterns are compatible with recruitment variation playing a major role in determining structure. Pattern (1) requires a temporally uniform but spatially heterogeneous pattern of recruitment. Pattern (2) might be generated entirely by patterns of spatial and temporal variation in recruitment.

Pattern (3) is compatible with a temporally variable recruitment acting in conjuction with other factors determining a consistent differential response to patch reefs. Possible factors might be preference by the settling fishes for some but not other of the patch reefs, or consistent differential survival among patch reefs, perhaps because of the preference for some reefs by piscivores or superior competitors.

All three patterns can also be generated by factors other than recruitment variability. Disturbances, predation, and competitive interactions, acting at the appropriate scales of space or time, could generate variations in survivorship which can act in exactly the same way as variations in recruitment to determine assemblage structure. Perhaps the chief value of Fig. 7 is to reemphasize that the species making up an assemblage can be expected to have differing patterns of recruitment, early survivorship, and longevity, differing degrees of specialization with respect to microhabitat, and differing susceptibility to competition, predation, and physical disturbances. The structure of an assemblage, and the relative importance of factors in determining that structure, will depend on the particular mix of species available in the pool from which that assemblage is drawn.

C. Physical and Weather Disturbances

Coral reefs are biotically derived habitats, and in recent years it has become apparent that they change in structure, sometimes quite quickly. Storms can cause sudden and dramatic change in the characteristics of a reef, although the modifications can also be quite limited in areal extent. A limited amount of information has been obtained on the effects of disturbances to the reef habitat on the resident fishes. Severe storms can cause the disappearance (presumed death) of reef fishes (Lassig, 1983), and they certainly can cause pronounced local change in habitat features (Connell, 1978; Woodley *et al.*, 1981; Walsh, 1983). In subtropical regions like Florida, cooler than usual winter temperatures can cause extensive mortality of reef fishes (Bohnsack, 1983a), leaving vacant habitat. A disease-caused die-off of the herbivorous urchin *Diadema antillarum* throughout the Caribbean during the early 1980s led to measurable increases in the abundance of turf algae (Hay and Taylor, 1985).

Reef fishes seem to be slow to take advantage of increases in availability of habitat or food following storms or other disturbances. Extensive coral death, for whatever cause, is usually followed by development of turf algae on the resulting rubble. Neither Williams (1986a), who monitored the effects of crown-of-thorns starfish predation, nor Wellington and Victor (1985), who followed El Niño-induced coral mortality due to high water temperature or heightened insolation, could find any effects on the densities of herbivorous species, although the loss of obligate corallivores and other close associates of living coral was quick. (This loss was due to both a departure or death of residents and a failure of subsequent recruitment.) Hay and Taylor (1985) reported that the increase in turf algae following loss of urchins correlated with a small increase in the local abundance of herbivorous fishes. Reciprocal disturbances, which eliminate rubble banks or algal beds and generate living coral, are unfortunately not available for study.

The relative importance of physical processes in structuring assemblages may be slight, although we still lack sufficient information on the demography of reef fishes to correctly assess this. El Niños of the severity of the 1982–1983 episode are infrequent events but affect coral reefs on broad geographic scales. Storms are more frequent (every three to four years on the southern Great Barrier Reef) but have more localized effects. On the central Great Barrier Reef, crown-of-thorns outbreaks have impacted the majority of midshelf reefs during the past 15 years. Since many reef fishes live 10 or more years, it is possible that part of the pattern in assemblages is set by such disturbances in much the same way that part of the pattern in forests is set by fire or storms (e.g., Romme, 1982; Whelan, 1985; Runkle and Yetter, 1987).

D. Interactions of Structuring Processes

The data available at present suggest that the structuring of local reef fish assemblages is determined by several interacting processes. Variation in recruitment, on quite small spatial and temporal scales, seems to be characteristic of many species (see Victor, Chapter 9; Doherty, Chapter 10) and provides a starting distribution of the youngest cohort of fish. This initial pattern is modified to a varying degree, or even disrupted totally, by several other processes: biotic interactions and physical disturbances. How conspicuous the initial pattern is depends on how large a proportion new recruits are of the standing stock and how strong postrecruitment processes are relative to recruitment variation. Local assemblages can be expected to vary markedly through time and to differ in neighboring locations.

On larger spatial scales (i.e., physiographic zones or larger) the situation is less clear. Differences in species composition between physiographic zones or over longer geographic transects are expected to be consistent through time, but data do not yet exist to confirm that this is true. Furthermore, the censusing methods which have been used on larger scale studies may be insufficiently precise to measure any but the most extreme temporal changes. At least some of these broader-scale differences may be attributed to differential availability of larvae to settle, although strong annual variation in recruitment success has been documented at these broad spatial scales (Doherty and Williams, 1988a). This suggests that the observed broad spatial patterns may not be due entirely to differential larval availability (cf. Williams, Chapter 16) and may not be uniform over time.

V. Conclusions

A. Patterns at Several Spatial Scales

In the preceding pages, I have attempted to fairly evaluate the data available concerning the organization of reef fish assemblages. In my assessment, the view which I outlined at the start of this chapter is well supported by available data, at least in its broader aspects. On all spatial scales up to the biogeographic scale, reef fish occupy a spatially patchy environment. The patches are ephemeral: large-scale ones on geological time scales, and smaller-scale ones on ecological time scales.

On a small spatial scale of meters or tens of meters, reef fish assemblages are nonequilibrial associations of species which vary through time and space because of the action of a series of processes which themselves vary through time and space. These processes include recruitment from the planktonic larval phase, competitive and predatory interactions both within and among species of fish, and physical, biotic, or weather disturbances which alter habitat and which may also kill resident fish. These processes may differ in relative importance from one geographic location to another.

Patterns expressed at the larger spatial scale characteristic of physiographic zones on reefs—a scale of tens to hundreds of meters—are likely to be more persistent than those apparent at smaller scales. Evidence that this is the case is lacking, as is information on the causes of such patterns. Great Barrier Reef data indicate that these larger scale spatial patterns arise because of consistent habitat preferences by settling larvae. Such preferences may be the consequence of evolutionarily earlier periods of interspecific competition, but they also may have arisen for other reasons having to do with the evolving adaptation of a genotype to its environment quite independent of the presence of other species. Given the pronounced differences—physical, chemical, biotic—as one moves from one zone to another on a reef, we should expect that most species will have evolved adaptations fitting them to particular zones and adaptations ensuring that they recruited to those zones.

Patterns expressed at the still larger scale of tens to hundreds of kilometers have only been sought within the Great Barrier Reef Province. Evidence indicates that these patterns of species presence and absence are due at least partly to patterns of availability of those species as pelagic larvae at this same large scale. At present it is unclear whether these large-scale patterns are permanent features or themselves subject to temporal change. Finally, there are patterns available at the biogeographic scale of thousands to tens of thousands of kilometers, which are primarily the result of availability of species in the particular biogeographic region. At this largest scale, it is perhaps most interesting to look at ecological patterns, such as guild structure of assemblages, rather than at presence and relative abundances of species. The very limited data presently available suggest that there is biogeographic variation in these.

B. New Data Needs

The data available at present are not only a geographically biased sample but are also strongly biased toward small species in the patch reef habitat. We have relatively good information on pomacentrids and on the smaller labrids, and lesser information on acanthurids, scarids, chaetodontids, haemuliids, gobiids, and blenniids. We have next to no information on apogonids, the larger scarids and labrids, and the larger species of a wide range of families which make up a reef fish fauna. Pomacentrids are sedentary, relatively long-lived species which may be distributed in quite different patterns because of the action of quite different processes than are large lutjanids or serranids. The patch reef habitat is a habitat of particularly clearly demarcated patches. Species occupying it may be more strongly influenced by patchiness than is the case in less subdivided habitats. There is thus a real need for new studies which explore the patterns and processes which are most important for other kinds of species in other kinds of habitats.

There is also a need for more studies in regions outside the Great Barrier Reef to test some of the ideas developed there. Particular attention needs to be paid to using comparable methods and comparable spatial and temporal scales. It will be particularly interesting to explore larger-scale patterns in regions such as the Caribbean, the Southeast Asian region, or some of the central Pacific archipelagos to see if there are generalizations possible at larger scales as there now appear to be at smaller scales. A close comparison of continental reefs with oceanic reefs will also be worthwhile. While the present tendency is to emphasize the degree to which larval fish are advected away from their natal reefs, it is possible that we will discover that fish are capable of some considerable control of this process. Such control may be better expressed in oceanic regions, where advection may well be tantamount to death at sea.

C. Benefits of Understanding Community Structure

A fuller understanding of the nature of reef fish assemblages, and of the processes responsible, will be of direct benefit to managers, both those who manage reef fisheries and those who manage reefs for tourism and other purposes. This understanding is needed now and, on present indications, is likely to dictate that management be planned on a broader spatial scale than is currently the case.

If my present view of assemblage organization is confirmed by further data, reef fish assemblages can be used as living laboratories for investigating processes in complex communities. They are eminently manipulable systems. They operate on time scales which are not too long for ecological experiments. And they could become experimental models for systems in which dispersal in a patchy habitat is a major structuring process.

If this is insufficient justification, there is also the joy of gaining greater insight into the workings of nature. The rich assemblages of fish on coral reefs doubtless have new lessons to teach us.

PART VI

Fisheries and Management

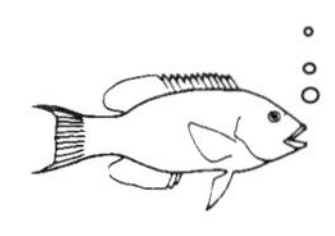

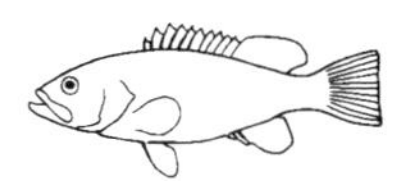

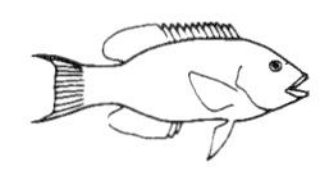

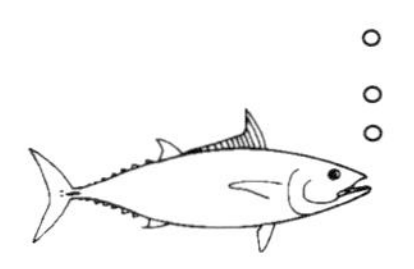

CHAPTER 20

Coral Reef Fisheries: Effects and Yields

Garry R. Russ
Department of Marine Biology
James Cook University
Townsville, Queensland
Australia

I. INTRODUCTION

Fishes are an important resource on coral reefs, particularly in the developing world. S. V. Smith (1978) estimated the fisheries potential of coral reef areas worldwide at 6 million metric tons per year, which represents about 7% of current world marine capture fisheries. The actual yield in 1983 [0.48 million metric tons (Longhurst and Pauly, 1987)] was well short of this "potential." The importance of coral reef fisheries may not be so much in terms of absolute magnitude of the catch as in their contribution to the catch of fishermen on low incomes and with few alternative opportunities for employment. Fishing is no doubt the most important human exploitative activity on coral reefs. What are the effects of fishing on populations and communities of coral reef fishes?

It is often suggested that coral reef fishes may be more vulnerable to overexploitation than fishes which are targeted commonly in northern high-latitude areas (e.g., clupeoids, gadoids, pleuronectids). The reasons for such suggestions relate to certain life history traits of reef fishes and to presumed trophodynamic characteristics of coral reefs. Coral reef fishes are usually restricted to the reef environment, are often territorial, or have small home ranges, with limited ranges of habitat and depth preferences (Munro and Williams, 1985). Furthermore, coral reefs have often been viewed as existing in "nutrient deserts" and relying upon rapid and tight nutrient cycling to retain high levels of production (Odum and Odum, 1955; Stevenson and Marshall, 1974; Marshall, 1980). It has been suggested that, despite the high primary productivity of reefs (Lewis, 1977) and the high standing stocks of reef fishes (Russ, 1984c), the harvestable component of reef fish stocks was probably quite limited and could be exhausted quite rapidly by intensive exploitation (e.g., Stevenson and Marshall, 1974). These life history traits and trophodynamic models suggest that fishermen may be able to concentrate their effort upon, and deplete stocks of, coral reef fishes to a far greater extent than preindustrial fishermen in higher latitudes who exploited more wide-ranging species (Koslow *et al.,* 1988).

Collection of detailed knowledge of the effects of fishing on coral reef fishes is made difficult by several factors. The fisheries are usually multispecific with fishing effort spread among a variety of gears. Effort is often unevenly distributed spatially and there is frequently a large number of artisanal fishermen landing their catch at a large number of sites spread over a wide area. Thus there are difficulties with collection of even the most basic information such as catch and effort. Furthermore, management agencies often view coral reef fisheries to be not important enough economically to warrant spending money on detailed research. Notable exceptions in this regard are the development of detailed research and management plans for the subtropical reef fisheries in the Gulf of Mexico and the U.S. South Atlantic (e.g., Huntsman *et al.,* 1982, 1983; Mahmoudi *et al.,* 1984; Huntsman and Waters, 1987) and for the Hawaiian deep-sea hand-line fishery (e.g., Ralston and Polovina, 1982; Polovina, 1987).

On the other hand, coral reefs are excellent microcosms in which to address questions of the effects of fishing on tropical, multispecies stocks. Effects of exploitation of marine communities and management of multispecies fisheries are the subjects of a large literature (e.g., May *et al.,* 1979; Beddington and May, 1982; Beddington, 1984; Gulland and Garcia, 1984; Larkin, 1982, 1984), with a great deal of interest directed to effects of fishing on and management of tropical multispecies stocks (e.g., Pauly and Murphy, 1982). Larkin (1982) drew attention to the need for more research to be directed toward understanding the mechanisms which determine the species composition of tropical, multispecies stocks subjected to varying levels of fishing effort. Several authors have advocated the need for an empirical, experimental approach to such questions (Walters and Hilborn, 1978; Larkin, 1982; Sainsbury, 1982, 1988; Walters, 1984).

This chapter reviews evidence for the effects of fishing on populations and communities of coral reef fishes. It is not an exhaustive review but has examined the literature to the extent of highlighting the major effects. The chapter begins with an introduction to the basic concepts of stock assessment and effects of fishing on stocks in general, before proceeding to review the effects of fishing on stocks of coral reef fishes. A brief section on yields of fishes from coral reefs is included.

II. Basic Concepts of Stock Assessment and Effects of Fishing on Stocks

Fisheries exploit stocks of wild animals which are living in their natural environment. In order to exploit these stocks in a manner which is both sustainable and in some manner (e.g., biologically or economically) optimal, it is essential to be able to assess:

1. The condition of the stocks at a given point in time.
2. The effects on these stocks of particular fishing strategies.

A. Basic Steps in Stock Assessment

The following are the basic steps of stock assessment. The steps are based on those presented in Gulland (1983). This reference should be consulted for an expanded discussion of each step.

1. Definition or description of the **UNIT STOCK.** In its simplest form a unit stock can be thought of as a homogeneous, reproductively independent unit. A unit stock can be maintained in the same geographical area by major current systems, for example.
2. Collection of **CATCH** data from the fishery.
3. Collection of data on **FISHING EFFORT** made by the fishery.
4. Collection of a time-series of data on **CATCH PER UNIT EFFORT.**
5. Collection of data on **BIOLOGICAL CHARACTERISTICS OF THE STOCK.** Principally, this involves determination of rates of (i) **GROWTH** (ii) **MORTALITY,** and (iii) **RECRUITMENT** (particularly stock-recruitment relationships).
6. Using 2 to 5 above, make **ESTIMATES OF EXPECTED YIELD UNDER DIFFERENT CONDITIONS OF FISHING PRESSURE.** The two approaches used most commonly in making such estimates are known as Production Models and Yield-Per-Recruit Models.
   (i) **Production or Surplus-Yield Models** require a time-series of catch and effort data only. These models assume that the growth in biomass through time of a stock is described by the logistic curve and that a level of effort exists which reduces the biomass of the stock to a level such that its growth rate (and thus production) is maximal (a point known as **Maximum Sustainable Yield**). Although simplistic, a large number of stocks of fishes, particularly in the tropics, are assessed in this manner.
   (ii) **Yield-Per-Recruit Models** require detailed information on the biological characteristics of the stock (see 5 above), namely, rates of growth and mortality. A yield-per-recruit analysis considers the fate of a brood or year class once they have been recruited to the fishery. Estimates of the yield (biomass) from a given year class of recruits can be calculated based simply on knowledge of their rates of growth and mortality. Yield per individual recruit can be calculated under different conditions of fishing mortality (F) or size (or age) of first capture.

7. Providing **ADVICE TO MANAGERS.** A manager will want to know if the stock is overexploited, underexploited, or being exploited optimally. Having performed 6 above, a fisheries biologist can offer advice on levels of fishing effort or fishing mortality to be applied to the stock, and/or make recommendations on the size (or age) at first capture to maximize yield from the stock. In addition to the preceding steps, two highly desirable steps in stock assessment are the following.
8. Carry out **SURVEYS** independent of the fishery itself using either standard or experimental fishing gears. Such information can provide estimates of stock abundance, for example.
9. Carry out **EXPERIMENTS.** Experiments of various forms can provide valuable information on the status of a stock and the possible effects of different fishing strategies on the stock. Large spatial-scale manipulations of fishing effort and gear types (e.g., Sainsbury, 1988) would be an example.

B. Effects of Fishing on Stocks

In the course of assessing the condition of a stock through time and making estimates of expected yield under different fishing strategies, the fisheries biologist is in a position to assess the effects of fishing on a stock. In fact, certain effects act as signposts which aid the fisheries biologist in providing advice to managers on fishing strategies. If a fishery were acting upon a single, uniform population in a uniform environment, then the condition of the stock could be assessed quite adequately by following the steps outlined above and in the course of the assessment the effects of fishing could be identified. Few fisheries operate on a single species or stock, particularly in the tropics, where multispecies fisheries are predominant. Thus, interactions between species need to be considered. Effects of fishing can thus act at the level of the **population** and the **community.** Furthermore, effects of fishing can operate **directly** on the population or community by removal of individuals or **indirectly** if the fishing techniques modify the habitat of fishes.

The direct effects of fishing at the population level are many and varied. One of the simplest to detect is reduction in catch per unit effort (CPUE) and eventually reduction in total catch. Detection of such an effect requires no specific information on the biological characteristics of the population. A common, nonbiological effect of fishing which can be detected with knowledge of CPUE and costs of fishing is **economic overfishing** (where fishing effort exceeds that required to obtain maximum economic yield, i.e., maximum profit). Fisheries often reach levels of economic overfishing rapidly.

Clearly, fishing should cause a detectable increase in rate of mortality. Since fishing mortality is often directed more effectively at larger, older individuals of a population (largely because many fishing gears act somewhat selectively on these fishes), fishing affects the size and age structure of populations. A reduction in the proportion of larger, older individuals may lead to an increase in the growth rates of individuals remaining in the population, so that fishing may result in a population in which fishes are larger for a given age. When fishing intensity reaches a point where fish are caught before they have time to grow (**growth overfishing**), there is a substantial reduction in the proportion of the large size classes. As fishing intensity increases further, there is likely to be a measurable reduction in both density and biomass of the population. If fishing reduces the size of the adult stock to a point where production of larvae and subsequent recruitment are impaired, the effect is termed **recruitment overfishing.** Other possible direct effects of fishing on populations include alteration of the sex ratio if individuals undergo sex reversal naturally at a certain size or age, changes in the behavior of individuals of the population, and small-scale changes in the distribution of fishes.

Direct effects of fishing at the community level involve effects of removal of predators, prey, or competitors from communities of fishes. When fishing is of such intensity that it results in changes in the relative abundance of species or the species composition of the community, the multispecies stock is said to have reached a point of **ecosystem overfishing** (Pauly, 1979, 1988). Such effects include decline in biomass of originally abundant species and increases in biomass of other species. There is usually an overall reduction in CPUE since species that increase in biomass do not "compensate" for declines in others (Pauly, 1987, 1988).

Finally, indirect effects of fishing, at the level of both populations and communities, involve modification of the habitat of either the exploited species or species in the community not targeted directly by the fishery.

III. Effects of Fishing on Coral Reef Fishes

A. Effects at the Population Level

1. *Direct Effects (Removal of Fishes)*

a. On Catch per Unit Effort and Total Catch Long time-series of catch per unit effort data are rare for coral reef fisheries. Some reef fisheries on small coral islands or atolls are ideal for collection of such information if there are a limited number of fishermen and landing sites for the catch. More often, the fishery supports a large number of artisanal or subsistence fishermen landing their catch at a multitude of sites. Furthermore, the value of (or importance placed upon) the fishery often is considered by fisheries agencies not to warrant the expense of collection of catch and effort data.

Even when catch and effort data are available, experience from two reef fisheries (Gulf of Mexico/U.S. South Atlantic and Hawaii) suggests that their value in the development of management plans may have substantial limitations. Huntsman and Waters (1987) provide an enlightening history of the development of management plans for reef fisheries in the Gulf of Mexico and the U.S. South Atlantic since the mid-1970s. Although commercial catch records were available, they were found to be of limited value. Catches were recorded by location of landing rather than location of fishing. Separation of species in the catch records were either lacking, improper, or not applied consistently. For example, "grouper" referred to over twelve species in two genera. Also, information on recreational catches was either totally lacking or "so fragmented that it was useless." Of even greater importance, little information on fishing effort was available and no information was available on the economic and social aspects of the reef fisheries in the two areas (Huntsman and Waters, 1987). These limitations on the usefulness of catch and effort data for the fisheries remained as the management plans were implemented, with the managers eventually opting for a management strategy based on yield-per-recruit rather than surplus-yield models (Huntsman and Waters, 1987).

Ralston and Polovina (1982) examined catch and effort statistics for the deep-water line fishery in the Hawaiian Islands for the period 1959–1978. They examined statistics for 13 species from 4 banks and attempted to fit Schaefer surplus-yield models to the data. No fits were statistically significant (Ralston and Polovina, 1982; Polovina, 1987). Fits were improved by combining species into three groups based on cluster analysis and a significant fit was obtained for all species combined for one of the banks. Overall, catches were maintained at high levels but catch per unit effort appears to have declined with increasing effort (Ralston and Polovina, 1982). An impressive series of catch information (composition, weight) has been collected at Tuamotu Archipelago (French Polynesia) over 25 years (Stein, 1988). Unfortunately no effort information for the fishery appears to be available.

The paucity of long time-series of information on catch and effort for most coral reef fisheries means that changes in catch per unit effort or total catch with changing effort are based either on measurements taken many years apart (e.g., Koslow *et al.*, 1988) or through comparisons of various sites at one time with different levels of fishing effort (e.g., Gulland, 1979; Munro, 1983; Wass, 1982; Marten and Polovina, 1982; Ferry and Kohler, 1987; Koslow, in press).

In a particularly informative paper, Koslow *et al.* (1988) compared changes in total catch and catch per unit effort of the reef fishery in Jamaica over a period of approximately 15 years (early 1970s to mid-1980s), a period of substantial increase in fishing effort. Total effort on the South Jamaica Shelf increased 30% from 1968 to 1986 but total catch did not change. During the same period, fishing effort by motorized canoes increased by 80%. On the offshore Pedro Bank south of Jamaica, fishing effort increased in some places by 108 to 178% and total catch for this bank more than doubled.

However, the picture at specific locations was more disturbing. Koslow *et al.* (1988) demonstrated that catch per unit effort of traps declined by 82% by weight over a 15-year period (1969–1973 to 1986) at SE Pedro Bank, associated with an increase in fishing effort of 108–178%. Catch per unit effort declined by 33% by weight over 15 years at Port Royal Cays, where effort changed from "heavy" to "very heavy." Catch per unit effort did not decline at SW Pedro Bank, where effort changed from virtually nil to "lightly-to-moderately exploited." Catch rates at this lightly fished site were 10 times higher by weight than at the two heavily fished sites in 1986. In 1969–1973, catch rates were three times higher at SE Pedro than at Port Royal Cays, but in 1986 catch rates were roughly equal at the sites. Koslow *et al.* (1988) recorded significant reductions in catch per unit effort of large and small serranids, lutjanids, haemulids, mullids, large scarids, and acanthurids at the two heavily fished sites and similar reductions in a further six families at SE Pedro Bank. Even at SW Pedro Bank, significant declines in catch rates were recorded for six families (mostly large predators). In addition, increases in catch per unit effort in the face of increasing effort were recorded for less "desirable" reef fishes such as chaetodontids, holocentrids, and tetraodontids. These latter changes, presumably related to changes in fishing intensity, catchability and interactions between species, are discussed further in Section III,B, on effects at the level of the community.

Catch per unit effort information is not commonly available for coral reef fishes at the level of species. Munro (1983) showed that catch rates (all species combined) in traps were generally higher on the more lightly exploited off-shore banks and cays than on the South Jamaica shelf (see Munro, 1983, Fig. 5.3). This applied to haemulids (Gaut and Munro, 1983, Table 10.36), mullids (Munro, 1983, Fig. 11.10), and *Balistes vetula* (Aiken, 1983b, Table 15.10). Munro demonstrated a significant decline also in catch per unit effort (all species) as effort increased in various Jamaican parishes, demonstrating that total catch was showing evidence of decline at high effort (see Munro, 1983, Fig. 2.4).

Koslow (in press) compared catch rates of traps in lightly to moderately fished sites in the Bahamas and Belize, in moderately to heavily fished sites on Pedro Bank off Jamaica, and in heavily fished reefs on the Jamaican coast. He showed that catch rates of traps were consistently lower in the areas of highest fishing effort.

Gulland (1979) was able to demonstrate a clear reduction in total catch at high effort in a trap fishery on coralline continental shelves off eastern Africa. However, both Marten and Polovina (1982), reviewing data from various sources, and Wass (1982), examining data from a number of fishing villages in American Samoa (see Munro and Williams, 1985, Fig. 4), were unable to demonstrate any clear pattern of decline in total catch over the range of fishing efforts applied to the respective fisheries. Ferry and Kohler (1987) reported that there was little difference in catch per unit effort of traps between two reefs with very different levels of fishing effort.

Alcala and Russ (1990) were able to document a significant decline in catch per unit effort of artisanal fishermen on the fringing coral reef at Sumilon Island, central Philippines, between 1983/1984 and 1985/1986 and related this to a significant decline in density of the major target groups (mostly caesionids) in an area which had been a reserve in 1983/1984 but had been fished heavily by 1985/1986. They also documented a 54% decline in total catch by weight of reef fishes from the island (1983/1984: 36.9 $mt/km^2/year$; 1985/1986: 19.9 $mt/km^2/year$; reef area approx. = 0.5 km^2).

On the Great Barrier Reef, Goeden (1982, 1986) suggested that densities of *Plectropomus* (a serranid) decreased with increasing proximity to a major center of human population (Cairns), and Craik (1986) showed that catch of *Plectropomus* per unit effort increased with increasing distance from Cairns.

b. On Mortality Rates Although it is intuitively obvious that fishing will impose an extra component of mortality on populations, there are surprisingly few direct measurements of total mortality (Z) and fishing mortality (F) of coral reef fishes at sites with different regimes of fishing intensity. This paucity of information relates to the great difficulties of estimating mortality rates of fishes, particularly rates of natural mortality (M) (Ralston, 1987). Munro (1983, Table 17.2), working in Jamaica, compared Z and F values of species on offshore banks which were either exploited lightly (Pedro Cays) or exploited at intermediate levels (California Bank) with those on the heavily exploited Port Royal Cays. Values of Z and F of five species were shown to be substantially higher at locations with higher fishing effort. Estimation of Z usually requires some knowledge of age structure or growth parameters of a stock (see Pauly, 1984). It is possible to estimate a closely related parameter, Z/K, without knowledge of age or growth data. K is the growth coefficient of the von Bertalanffy growth function, the formula used most commonly to describe growth in fisheries models. Munro (1983) showed that in six species the ratio of M/K on an unexploited portion of Pedro Bank ($M = Z$ where exploitation is zero) was substantially lower than the ratio of Z/K at exploited reefs. Russ and St. John (1988) made estimates of fishing mortality of scarids based on annual yields and estimated annual standing stock at three islands where fishing intensity differed in the Philippines. They estimated the highest values of F at the island of greatest fishing pressure. Wright *et al.* (1986)

reported a higher value of Z for *Lutjanus bohar* in shallow rather than deep water in Papua New Guinea, with the deep-water population being unfished.

Ralston (1987) summarized information on F/M ratios for tropical lutjanids and serranids. Gulland (1971) suggested that an F/M ratio of 1.0 represented an optimal level of F. The range of values of F/M reported by Ralston were 1.25–3.3 for four species of lutjanids and 0.27–3.00 for eight species of serranids. Of those species associated with coral reefs, the ratios for two species of lutjanid in deep waters off Hawaii were 1.33 and 1.92, for a serranid in the Virgin Islands 2.52, for a serranid in Florida 3.00, and for two species of serranid at lightly exploited banks of Jamaica 0.28 and 0.32.

c. On Size and Age Structure Most fishing techniques on coral reefs tend to target the larger (and presumably older) individuals in populations, and so fishing would thus be expected to reduce the proportion of large and old individuals in populations. Some degree of compensation for this effect may occur as density-dependent increases in growth rates occur among the individuals remaining in the population (Gulland, 1983). Such compensation would be detected as increased size for a given age. Little evidence for such an effect could be found in the literature on coral reef fishes, which is basically a reflection of the real or perceived difficulties of determining age of these fish reliably (Manooch, 1987). For the same reason there is little direct evidence of reduction due to fishing in the proportion of individuals in the higher age classes in any coral reef fish. Ralston and Kawamoto (1985) were able to demonstrate a substantial decrease in the age of entry to the fishery of the deep-water lutjanid *Pristipomoides filamentosus* in Hawaii associated with a substantial increase in fishing mortality. In the Gulf of Mexico and the U.S. South Atlantic, reef fisheries are assessed on a yield-per-recruit basis with emphasis on control of size at first capture (e.g., Huntsman *et al.*, 1982, 1983; Mahmoudi *et al.*, 1984). Currently, such fisheries, with detailed knowledge of size at age, are best placed to detect effects of fishing on the age structure of populations of reef fish.

There is ample evidence of **growth overfishing** in coral reef fishes, all of it from comparisons of mean size or size structure on reefs with differing fishing effort. Munro (1983) and colleagues compared heavily and lightly exploited areas in Jamaica and produced evidence of substantial shifts in size structure of holocentrids (Wyatt, 1983), serranids (Thompson and Munro, 1983a) (Fig. 1), carangids (Thompson and Munro, 1983b), lutjanids (Thompson and Munro, 1983c), pomadasyids (Gaut and Munro, 1983), mullids (Munro, 1983), chaetodontids (Aiken, 1983a), scarids (Reeson, 1983a), acanthurids (Reeson, 1983b), and balistids (Aiken, 1983b). Koslow *et al.* (1988) repeated fish trap surveys in Jamaica 15 years after the extensive studies of Munro

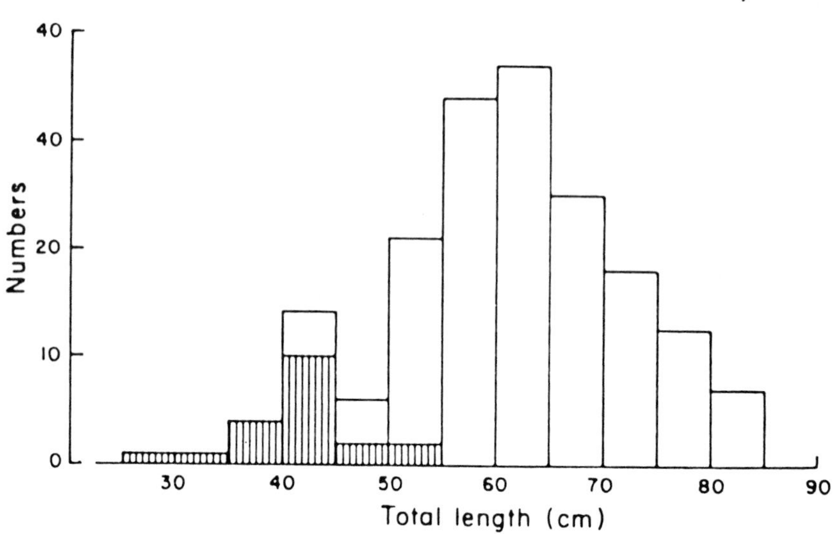

Figure 1 Length–frequency distributions of catches of *Epinephelus striatus* from unexploited oceanic banks off Jamaica (open bars) and from the heavily fished Port Royal reefs on the south Jamaica coast (hatched bars). [From R. Thompson and J. L. Munro (1978). Aspects of the biology and ecology of Caribbean reef fishes: Serranidae (hinds and groupers). *J. Fish Biol.* **12,** 115–146.]

(1983) and reported that the largest fishes caught regularly in traps in Jamaica (serranids, lutjanids, and scarids) virtually disappeared from catches at all study sites (Port Royal Cays, SE and SW Pedro Bank) over a 15-year period. Visual censuses confirmed the absence of large serranids at the Port Royal Cays and their virtual absence at Pedro Bank. Koslow (in press) concluded similarly that these fishes in Jamaica showed substantial evidence of growth overfishing when compared with the relatively lightly fished reefs of Belize and the Bahamas.

Ferry and Kohler (1987) compared two reefs in Haiti which had been subject to substantially different levels of exploitation by trap fisheries. They showed that the modal sizes of scarids and chaetodontids (the most abundant groups in the catch) were significantly lower on the more heavily exploited reef. Huntsman and Waters (1987) pointed out that a major impetus to the inception of a management plan for the reef fisheries of the Gulf of Mexico was the evidence that stocks of reef fish were depleted quickly after initial large catches of old, large fish, and that most subsequent catches consisted of much smaller fish. Ralston and Kawamoto (1988) presented evidence that three species of deep-water lutjanid in Hawaii were overfished to the point of growth overfishing.

Bohnsack (1982), working in Florida, used visual censuses to compare a reef subjected to heavy spearfishing pressure with two others which had been protected from spearfishing for approximately 20 years. The sizes of predatory fishes were generally smaller on the fished reef and the predatory species with individuals which attained large size (e.g., lutjanids and haemulids) displayed the greatest differences in density between fished and unfished reefs (Bohnsack, 1982, Figs. 1 and 2). Within a species, mean size of predator was generally smaller at the fished reef. In a similar type of study, Russ (1984d, 1985) used visual censuses at seven sites in the central Philippines to show that large serranids, lutjanids, and lethrinids were generally abundant only at a site which had been protected from fishing for almost 10 years. All other sites were either open to fishing or had been protected for a short period only. Samoilys (1988) used visual censuses to compare abundance and size structure of populations of fishes on coral reefs subjected to different levels of fishing pressure in Kenya. She suggested that the average size of fish is lowered by more intense fishing.

On the Great Barrier Reef (GBR), Craik (1981a) showed that the mean weight of individual fish caught by recreational fishermen in the area of Townsville declined from approximately 2.6 kg in 1961 to 1.3 kg in 1977 (Fig. 2). Ninety-eight percent of this catch consisted of *Lethrinus miniatus* (= *chrysostomus*) and the serranids *Plectropomus* spp. Using visual censuses, Craik (1981b) was able to show that the mean size of *Plectropomus leopardus* at

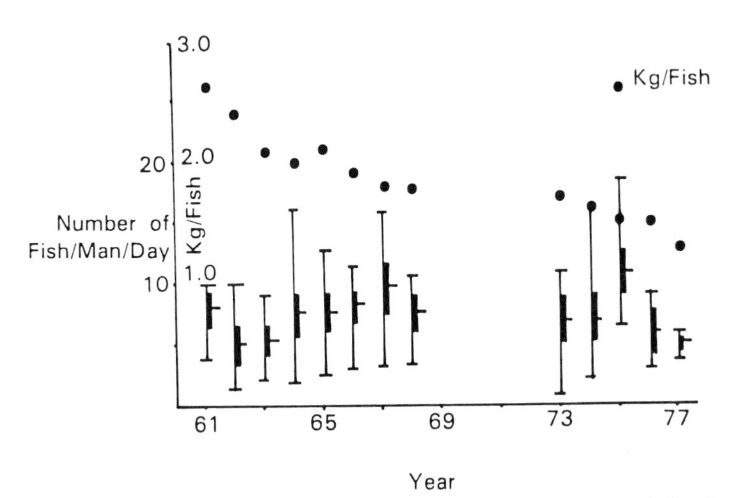

Figure 2 Catch per unit effort of recreational fishermen and mean weight of fish (mostly *Plectropomus leopardus* and *Lethrinus chrysostomus*) from coral reefs off Townsville, Australia, from 1961 to 1977. [From W. Craik (1981a). Recreational fishing on the Great Barrier Reef. *Proc. 4th Int. Coral Reef Symp.* **1**, 47–52.]

an unfished reef was considerably larger than that at four out of five fished reefs in the Capricornia region of the GBR. Size structure generally differed considerably between the fished reefs and the unfished reef and Craik commented on "the almost total absence of larger *P. leopardus* and the low densities of the populations at most of the unprotected reefs." The far more extensive visual surveys of *Plectropomus* in the Capricornia region of the GBR confirm these patterns (Ayling and Ayling, 1986). Beinssen (1988), working in the same Capricornia region, noted that the average size of *Plectropomus* on a protected reef (Boult Reef) was much greater than that on nearby unprotected ones.

d. On Total Density and Biomass Munro (1983) showed that the biomass of large, predatory fishes (mostly serranids, lutjanids, and lethrinids) was much higher on unexploited, offshore banks in Jamaica than on exploited offshore banks and that the biomass of predators on such banks was far greater than that on the heavily exploited inshore reefs. Bohnsack (1982) reported significantly lower densities and biomass of large fishes on a fished reef than on two reefs protected from spear fishing. This was particularly the case for large predatory species (at least five species of *Lutjanus* and eight species of haemulid showed substantial differences in density). One species each of scarid, acanthurid, and mullid had considerably reduced density on the fished reef.

Koslow (in press) used visual census and trapping studies to compare populations and communities of coral reef fishes at sites subjected to light, moderate, and heavy fishing pressure both within and between the Bahamas, Belize, the island of Jamaica, and Pedro Bank (a bank offshore from Jamaica). Both sampling techniques indicated that fishing reduced the density and biomass of many species, particularly the large serranids, balistids, acanthurids, and pomacanthids. Even within the Bahamas and Belize, where fishing was assessed as light to moderate, fishing reduced the abundance of many species. The large piscivores displayed disproportionately high reductions in biomass, comprising 5–10% of fish surveyed in areas subjected to light to moderate fishing (mainly Belize and the Bahamas) but only 1–2% of fish in areas subjected to heavy fishing pressure (Jamaican coast and offshore banks).

Russ (1984d, 1985) compared the abundance of coral reef fishes on the reef slope of a site protected from fishing for almost 10 years (the Sumilon Island reserve) with various reef slope sites throughout the central Philippines which had been open to fishing. The site which had been protected had the highest overall abundance of fishes, including significantly higher density and biomass of large piscivores (serranids, lutjanids, and lethrinids) and a significantly higher biomass of serranids. As with any correlative study, the differences in abundance of fishes between fished and unfished sites could not be attributed unequivocally to the effects of fishing. It was possible that the protected site had naturally higher densities than the others before protection. Russ (1985) did conclude that protection from fishing had been very important in *maintaining* the high density and biomass of many groups of fishes within the Sumilon Island reserve.

A direct test of this contention was reported by Russ and Alcala (1989). The Sumilon Island reserve was fished intensively for an 18-month period, beginning 3 months after the initial censuses made by Russ (1985). The dramatic increase in fishing pressure within the 750-m-long reserve involved fishing by up to 100 municipal fishermen using traps, hand-lines, gill nets, spears, and occasionally more destructive, nonselective fishing methods such as explosives and "muro-ami" drive nets (Carpenter and Alcala, 1977; Gomez *et al.*, 1987). Fishing in the reserve resulted in a significant decrease in the total abundance of coral reef fishes, including significant declines in density of pomacentrids, caesionids, chaetodontids, lutjanids, and lethrinids (Russ and Alcala, 1989). Lutjanids and lethrinids together declined by 94% in density. Substantial, but not statistically significant, declines in density of serranids, carangids, and scombrids occurred also.

This natural experiment in the Philippines was akin to a depletion experiment in which the stock is fished down so rapidly that the effects of recruitment, immigration, and natural mortality are negligible. In a well-designed experiment on a coral reef fishery, Beinssen (1988) performed such a depletion experiment at Boult Reef, in the Capricornia section of the Great Barrier Reef Marine Park. Beinssen tagged 375 *Plectropomus* in the last few months of a 3-year closure to fishing of the reef. The reef was reopened to line fishing in December 1986, and for the first 14 days the reef was subjected to a strong pulse of fishing by professional and recreational line fishermen. The aims of the experiment were to determine the rate at which fish stocks on a typical coral reef could be "fished out" (of *Plectropomus*), to measure the catchability coefficient (q) of *Plectropomus,* and to estimate standing stock. Beinssen (1988) showed that 25% of the individuals in the stock of *Plectropomus* were removed in 14 days, providing a graphic illustration of the direct effects of line fishing on the abundance of a large, predatory reef fish. In a similar type of depletion experiment on a deep-water (150–275 m) pinnacle in the Marianas, Polovina (1986) showed that 13 days of intensive line fishing reduced the stock of three species of lutjanid by an estimated 48%.

Ayling and Ayling (1985) used visual censuses to show that the serranid *Plectropomus* was more abundant on reefs with low fishing pressure than on reefs with high pressure in the Townsville region of the Great Barrier Reef Marine Park. Ferry and Kohler (1987) could not detect significant differences in total abundance of fishes between two reefs with very different fishing pressures in Haiti. Samoilys (1988) was unable to demonstrate significant differences in density of all fishes and fishes of commercial importance on reefs with different levels of fishing effort. She attributed this lack of significance to confounding effects of siltation by rivers and occasional use of dynamite "fishing" in sites that were protected as marine parks or reserves. Samoilys did, however, demonstrate a significant correlation between high biomass of both commercially important fish and serranids and reduced fishing pressure.

e. On Recruitment The term recruitment is used here as defined by Doherty and Williams (1988a): "increases in the abundance of young of fish produced by larval settlement." Fisheries biologists define recruitment as the number of individuals entering a fishery, which may involve fishes which have increased in size and age considerably from the 0+ stage and have attained or are near sexual maturity. Effects of fishing on the process converting 0+ fish to those considered by a fisheries biologist to be "recruited" (i.e., growth) were considered earlier (see Section III,A,1,c).

This section will deal strictly with evidence that fishing may affect the relationship between stock size and subsequent recruitment of coral reef fishes. If "stock" is considered to be the population on an individual coral reef (as often considered by many reef fish ecologists), then unless the reef is largely self-recruiting, a relationship between stock and recruitment will be weak to nonexistent. Studies on the genetic structure of populations of coral reef fishes suggest that only the most isolated oceanic islands and atolls will be totally independent of recruitment from other sources. Most reefs are not as isolated as such oceanic islands and atolls, being located instead on the shallow-water continental margins. In such situations, most recruitment to reefs is likely to come from beyond the reef itself, that is, most reefs will not be self-recruiting (Doherty and Williams, 1988a; Williams *et al.*, 1984).

If, on the other hand, "stock" is considered to be the genetic stock, whose size is determined by the degree of spatial dispersal of the pelagic larvae, stock-recruitment relationships should exist but be of an extremely variable nature, even in unfished stocks (see the review by Doherty and Williams, 1988a). At this scale Pauly (1987) suggests that fishing may cause a reduction in the number of age groups in the stock and a reduced "buffering" of recruitment fluctuations, leading to increased risk of fluctuations in the abundance of the stock. There will consequently be more frequent occurrences of extremely low catches and an increasing risk of occasional failure of recruitment, including a total collapse of the fishery. Such cases of **recruitment overfishing** are difficult to detect and it appears that most cases are of an anecdotal nature for coral reef fishes.

Munro (1983) pointed out that about half of 37 species studied in detail in the Jamaican trap fishery become catchable well before maturity (e.g., lutjanids, carangids, haemulids, and *Balistes*). Munro suggested that negligible numbers of fishes may survive long enough to spawn in the most heavily exploited areas (e.g., the north Jamaican shelf) and that recruitment to such areas may be dependent largely on spawning from elsewhere on the island or even elsewhere in the Caribbean. The relatively low catches at high levels of fishing effort on the north Jamaican shelf (e.g., Hanover, St. Ann, St. Mary) produced a declining right-hand arm of a surplus-yield curve for the Jamaican coast (Munro, 1983, Fig. 2.4). Munro argued that this provides evidence of high rates of exploitation (with small meshed traps) which adversely affect recruitment.

A further piece of evidence of a potential effect of fishing on recruitment comes from the oceanic Pedro Bank, offshore from Jamaica. Munro *et al.* (1973) observed major variation in stock size of target species among unfished oceanic banks, particularly SW Pedro Bank, and suggested that such variation may be due to chronic recruitment limitation. Current flows westward over Pedro Bank, so that the likely source of larvae for SW Pedro Bank will come from the much more heavily fished eastern side of the bank. Koslow *et al.* (1988) observed substantial reduction in catch rates of reef fishes at SE Pedro Bank between 1969–1973 and 1986. They also noted significant reductions in catch rates of large and small serranids, carangids, lutjanids, haemulids, sciaenids, and acanthurids at SW Pedro, despite only relatively moderate increases in fishing effort. They suggested that this depletion of large fishes at SW Pedro may be due to recruitment overfishing in the sense that overfishing of the stocks in the "upstream" portion of the bank (at SE Pedro) may have reduced recruitment downstream.

Ralston and Kawamoto (1985) estimated that the age of entry of the lutjanid *Pristipomoides filamentosus* to the deep-water line fishery in Hawaii changed from 4 to 1.8 years between 1980 and 1984 in the face of increasing fishing effort. Polovina (1987) estimated that such a change (with fishing mortality remaining close to 0.5/year) would reduce spawning stock biomass from 28% to 10% of its unexploited level. Beddington and Cooke (1983) have suggested that a substantial reduction in recruitment will occur if the spawning stock biomass is reduced below 20% of its unexploited level. This suggests, therefore, that the level of effort on this species may be high enough to lead to recruitment overfishing.

Evidence of recruitment overfishing at geographic scales is lacking, although it seems likely that larger predators such as sharks and large serranids have reached the stage of recruitment overfishing in the Philippines. This statement is based simply on the low densities of such large predators in many parts of the country (e.g., Carpenter, 1977; Carpenter *et al.*, 1981; Russ, 1984d, 1985; Russ and Alcala, 1989).

Brief mention should be made of one fishing strategy which may be likely to lead to recruitment overfishing very rapidly. If reef fisheries were specifically to target the well-known spawning aggregations of many reef fishes (e.g., Johannes, 1978b, 1981) on a large scale, and no controls over catch or effort could be imposed during the spawning periods, stock sizes could be reduced so rapidly as to potentially affect recruitment.

A considerable amount of research has been carried out on the effects of conspecifics and heterospecifics, both juvenile and adult, on the recruitment and subsequent survival of coral reef fishes (reviewed by Doherty and Williams, 1988a). Doherty and Williams (1988a) argued that there is little evidence to suggest that density of other fishes, con- or heterospecifics, affects rates of recruitment or survivorship of coral reef fishes (but see Shulman, 1985a; Shulman and Ogden, 1987). Under such circumstances, reduction in stock size by fishing would be unlikely to have substantial effects on rates of settlement and survivorship. Although the majority of the evidence supports the idea that recruitment rates are independent of densities of residents, some notable exceptions to this generality have been reported by Sweatman (1983, 1985a) and Jones (1987a). These two authors demonstrated clear cases of preferential settlement with conspecifics (for the pomacentrids *Dascyllus aruanus* and *Chromis caerulea*) and inhibition of settlement of heterospecifics (e.g., *D. aruanus* and *D. reticulatus*). If such patterns of dependence of recruitment on residents were to apply to any of the larger, exploited reef fishes, then fishing pressure could theoretically have direct effects on recruitment.

f. On Sex Reversal and Social Structure Many families of fishes have members which are sequential hermaphrodites (e.g., serranids, lethrinids, scarids, labrids, pomacanthids, pomacentrids, and sparids). The effect of fishing pressure on the sex ratio of such populations may be a function largely of the mechanism controlling sex reversal (Munro and Williams, 1985). If sex reversal is age or size dependent (and thus endogenously controlled), removal of the largest individuals could lead to a massive decline in the proportion of males in a protogynous, monandric population. Evidence in support of this scenario is provided by Thompson and Munro (1983a). In the Jamaican trap fishery, transformation of serranids to males occurs at sizes well above the mean length of recruitment to the fishery. The male/female ratio of the serranid *Epinephelus guttatus* on lightly fished, offshore oceanic banks was 1 : 2.81, but 1 : 5.60 on the heavily fished Port Royal Cays. Male/female ratios of three other species of serranid on oceanic banks were 1 : 0.72, 1:0.85, and 1:2.14, and for another species at the heavily fished Port Royal Cays it was 1:6.00 (Thompson and Munro, 1983a, Table 7.15). It should be noted that Reeson (1983a) could find no evidence of fishing pressure affecting sex ratio of scarids (see Reeson, 1983a, Table 13.8). If sex reversal is under social (exogenous) control, the effects of fishing pressure on the sex ratio may not be as severe. For example, in labrids, experimental removal of a male has been shown to cause sex transformation in one or more females (e.g., Robertson, 1972).

Bannerot *et al.* (1987) reviewed the topic of reproductive strategies and the management of tropical serranids and lutjanids and modeled the resiliency of protogynous and gonochoristic populations to exploitation. Generally, protogynous populations should be favored provided the availability of males (sperm) does not become limiting. However, as mortality reduces frequency of contact between males and females, the relative reproductive success of protogynous hermaphrodites declines.

g. On Distribution There is no direct evidence that fishing has modified patterns of distribution of coral reef fishes on scales of tens of kilometers or greater. It is likely that intense fishing pressure may lead to local extinctions of highly vulnerable species (e.g., sharks or large serranids) on individual coral reefs in places of such intensive fishing pressure as the Philippines and perhaps the north Jamaican shelf.

Intense fishing pressure in the Sumilon Island reserve in the central Philippines has been shown to alter the pattern of zonation of large herbivorous reef fishes on a scale of hundreds of meters (Russ and Alcala, 1989) (Fig. 3). A detailed study of the pattern of zonation of reef fishes in this reserve before fishing began (Russ, 1989) demonstrated that herbivores, in particular scarids, occurred in significantly higher densities on the shallow reef flat of the fringing reef than on the reef crest or reef slope (Fig. 3). An intense bout of fishing for an 18-month period (from May 1984 to December 1985), which involved use of the drive-net system known as "muro-ami" (Carpenter and Alcala, 1977; Gomez *et al.*, 1987), led to a significant reduction in the number of algal grazers and both the number of species and individuals of scarids on the reef flat (Fig. 3). This change in the pattern of zonation was partially due to direct removal. However, Russ and Alcala (1989) recorded a significant increase in density of adult scarids on the shallow reef slope between 1983 and 1985 (pre- and postfishing), just beneath the crest, suggesting a large migration of scarids (mostly *Scarus bleekeri* and *S. dimidiatus*) away from their "preferred" habitat on the flat. Note that this reef flat is never exposed at low tide and tidal influence on the abundance of herbivores is minimal (Russ, 1989).

h. On Behavior Behavior would appear to be one of the most obvious but least documented of the direct effects of fishing, at least on a geographic scale. Species of *Plectropomus* and *Cephalopholis* on the relatively lightly fished Great Barrier Reef show substantially less fear of divers than their counterparts on heavily fished Philippine reefs (unpublished observations). The same applies to scarids (unpublished observations; D. R. Bellwood, personal communication) and a wide variety of the larger reef species. Johannes (1981) cites a good example of the large scarid *Bolbometapon muricatum* changing the location of its sleeping sites away from the shallow reef flat to the deeper reef slope

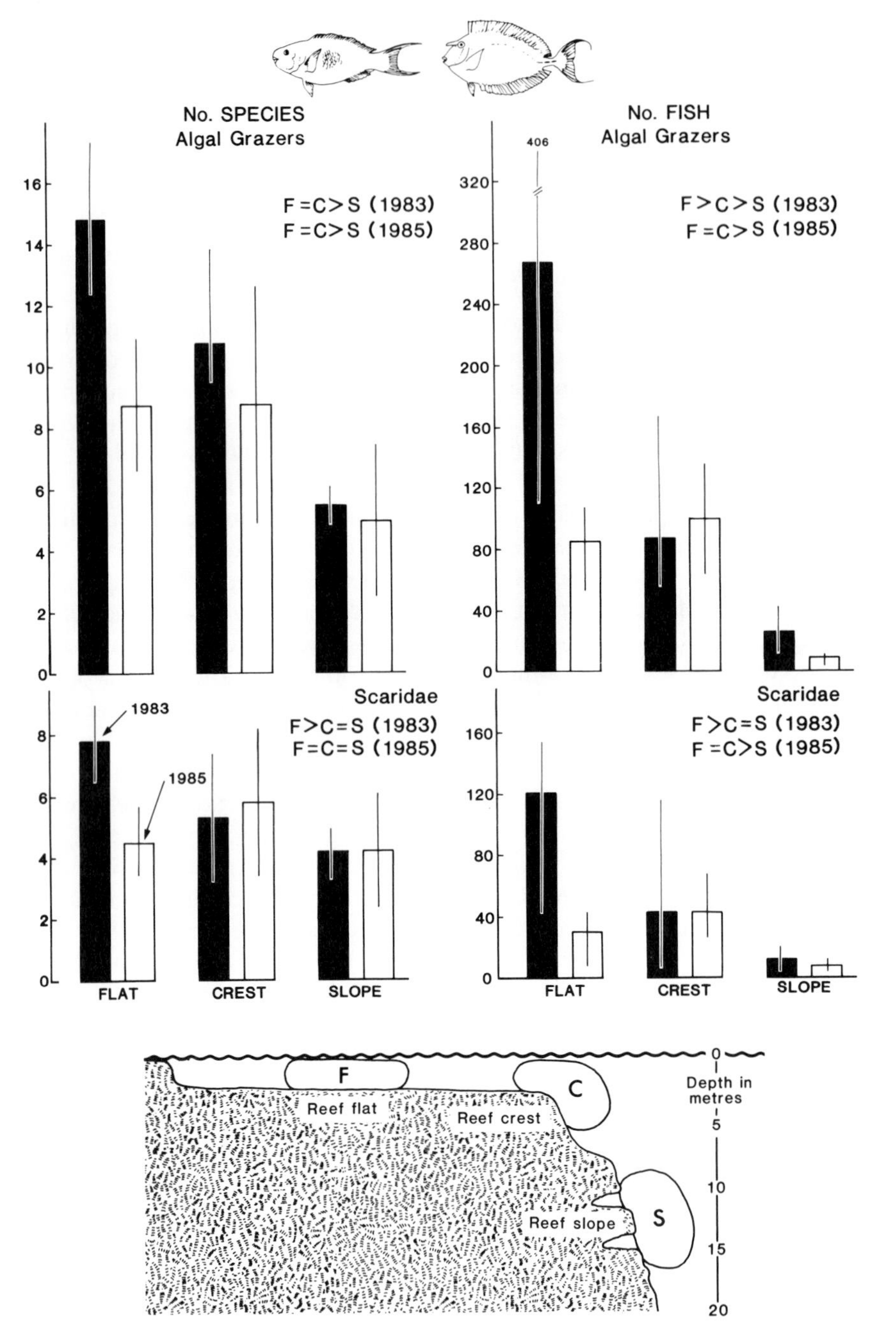

Figure 3 Numbers of species and individuals of all algal grazers and of scarids in three zones (reef flat, reef crest, and reef slope) of the Sumilon Island Reserve, Philippines, in 1983 before fishing (solid bars) and in 1985 after fishing (open bars) by artisanal fisherman. Patterns of zonation changed significantly, with reduced abundances and numbers of species of all algal grazers and scarids on the reef flat.

in Palau in response to increasing pressure from spearfishing at night. The shift in distribution of scarids off the reef flat at Sumilon Island in the face of intensive fishing (Fig. 3) could be explained partly by direct removal of fish, partly by the indirect effect of habitat destruction, or partly by a change in behavior of the fish.

2. *Indirect Effects (Habitat Modification)*

Several fishing techniques used on coral reefs are destructive to the benthic habitat, often causing reduction in the percentage cover of live coral or reducing structural heterogeneity of the substratum and thus reducing availability of hiding places for fishes. The most serious of these are explosives (reviewed by Alcala and Gomez, 1987) and drive nets [the "muro-ami" and "kayakas" techniques (see Carpenter and Alcala, 1977; Gomez *et al.*, 1987)]. Muro-ami and kayakas drive-net fishing involves a cordon of swimmers, each holding a vertical scareline, driving fish toward a large bag-net. Scarelines have short strips of plastic or coconut leaves tied to them and have a 2- to 4-kg stone weight attached to their base to offset their buoyancy. These weights are continually lifted up and dropped back on to the coral substratum to scare fish toward the net. The use of large traps can cause small-scale, incidental damage to corals as they are placed or dropped on the bottom. Some traps are set with the use of corals as weights or corals may be used to camouflage traps (Johannes, 1981; Gomez *et al.*, 1987).

a. Reduction in Live Coral Cover Russ and Alcala (1989) reported that muro-ami drive-net fishing and explosives led to a substantial reduction in cover of live corals on the shallow reef slope of the Sumilon Island reserve in the central Philippines. This resulted in a significant decrease in density (79% over 18 months) of the obligate coral-feeding chaetodontids. Significant reduction in cover of live corals due to *Acanthaster planci* outbreaks (Williams, 1986a) or channel construction (Bell and Galzin, 1984) has led to reductions in density of chaetodontids similar to those seen at Sumilon Island.

Reduction in cover of live coral can also lead to decreased recruitment of those species which require live coral as a substratum for settlement. Williams *et al.* (1986) reported failure in the recruitment of the pomacentrids *Pomacentrus moluccensis* and *Chromis atripectoralis* on midshelf reefs in the central Great Barrier Reef for 3 years following destruction of most of the live coral cover by *A. planci*. Long-term destruction of coral cover by intense fishing could presumably have similar effects.

It should be noted that long-term, large-scale destruction of live coral cover (e.g., by *A. planci* or by extremely destructive fishing methods) will result in coral cover being replaced by benthic, turfing algae (Moran, 1986). This increase in availability of algae does not necessarily imply that densities of herbivorous fishes will increase. Williams (1986a) failed to detect a positive numerical response of herbivorous fishes 3 years after live coral cover had been "converted" to turf-algal cover. Similarly, Wellington and Victor (1985) could not detect any increase in density of herbivorous pomacentrids on a coral reef following destruction of live coral cover by an El Niño event. Herbivorous fishes can display rapid functional responses to increases in algal cover in the form of increased grazing rates (Carpenter, 1988). This lack of a numerical response of herbivorous fishes in the face of a massive increase in algal biomass supports the notion that their abundance may be "recruitment limited" (Doherty and Williams, 1988a).

Fishing techniques which are destructive to the benthic substratum also tend to increase the percentage cover of coral rubble (Russ and Alcala, 1989). This could have the effect of attracting species which are specialists at feeding on and/or settling onto coral rubble. Certain species of labrids in the genera *Thalassoma* and *Cirrhilabrus* appear to fit these criteria. Russ and Alcala (1989) reported significant increases in density of *Cirrhilabrus* sp., *Thalassoma hardwicki,* and *T. lunare* following intense fishing and substantial destruction of live coral cover within the Sumilon Island reserve in the central Philippines. Although no effects of fishing on the benthic substratum were reported or implied by Bohnsack (1982), it is interesting to note that *Thalassoma bifasciatum* occurred in significantly higher densities on a fished reef than on two unfished reefs in Florida. Koslow (in press) reported significantly higher abundances of the labrids *Halichoeres* spp. at the more lightly fished reefs of Belize and the Bahamas but reported that *Thalassoma bifasciatum* was more abundant on lightly fished reefs.

b. Reduction in Structural Heterogeneity Munro *et al.* (1987), in reviewing the biological effects of intensive fishing upon coral reef communities, suggested that loss of shelter caused by destructive fishing techniques could be particularly serious for juvenile fishes which require places for concealment from predators. A number of studies have shown that abundance of reef fishes can be correlated with topographic relief (e.g., Carpenter *et al.,* 1981) and microscale heterogeneity (e.g., Kaufman and Ebersole, 1984). It should be noted, however, that Robertson and Sheldon (1979) demonstrated experimentally that holes in the reef (sleeping sites) were not limiting for labrids on a coral reef.

On a larger spatial scale, if the standing crop and productivity of reef fishes can be correlated to topographic relief of the substratum (Saila and Roedel, 1980; Marshall, 1985), then any fishing techniques which cause large-scale and long-term damage could affect populations of reef fishes and reef fisheries significantly.

B. Effects at the Community Level

1. Direct Effects (Removal of Fishes)

This section deals with the effects of removing individuals of species which may interact with individuals of other species via predation, competition, or mutualism. Munro and Williams (1985) stressed the critical need for knowledge of how biological interactions influence densities of coral reef fishes as fishing removes individuals, but pointed to the paucity of data available on the subject. They point out that changes in species composition or relative abundance of species in multispecies stocks depend largely on the level of intensity of fishing, the relative catchabilities of the species, and the magnitudes of the interactions between species. Many models of multispecies fisheries assume large mean interaction strengths between species, with competition and predation considered the most likely forms of interaction [e.g., (1) temperate systems: May *et al.,* 1979; Beddington and May, 1982; Beddington, 1984; (2) coral reefs: Munro, 1980; Polovina, 1984; Munro *et al.,* 1987]. Such models have led to suggestions that complex systems such as communities of tropical fishes may be more vulnerable to perturbations than simpler ones (May, 1976, 1984). Other models of multispecies fisheries stress the need for a more empirical approach, making use of large-scale natural or experimental manipulations (e.g., Walters and Hilborn, 1978; Larkin, 1982; Sainsbury, 1982, 1988; Walters, 1984).

A critical point in most of these models is the strength of the mean interaction terms between species. This section reviews the evidence of effects of fishing on communities of coral reef fishes, particularly those relating to interactions between species.

a. Removal of Predators Most coral reef fisheries tend to have a disproportionate negative effect on the abundance of larger piscivorous or predatory species. Much of the evidence for this has been reviewed in previous sections of this chapter. The evidence will be reiterated briefly before discussing the responses of communities of fishes to removal of these "top" predators.

(i) Predator Removal There are many reasons why large piscivorous and other predatory species of reef fishes are more vulnerable to reef fisheries than other groups. These species are generally favored food fishes and their aggressive, predatory nature, large size, and deeper bodies make them vulnerable to most fishing gears (Munro and Williams, 1985). Such fishes are the first to be attracted to baited hooks and traps, or to traps containing other fishes. Furthermore, most of these large predators have low rates of growth, attainment of sexual maturity, natural mortality, and recruitment, making them far more sensitive to overfishing than smaller species with higher rates of intrinsic

natural increase (Adams, 1980). Once depleted, recovery of stocks of large predators is likely to be slow.

Fishing has been demonstrated to reduce the proportion of large individuals in populations of predatory fishes particularly rapidly (see Section III,A,1,c, particularly references to Craik, 1981a,b; Bohnsack, 1982; Thompson and Munro, 1983a–c; Russ, 1985; Ayling and Ayling, 1986; Huntsman and Waters, 1987; Koslow *et al.*, 1988). There is a large body of evidence demonstrating that fishing has a disproportionately high, negative effect on the density and biomass of large predatory fishes on coral reefs (see Section III,A,1,d, particularly references to Bohnsack, 1982; Munro, 1983; Ayling and Ayling, 1985; Russ, 1985; Beinssen, 1988; Samoilys, 1988; Russ and Alcala, 1989; Koslow, in press) and on catch rates and total catch of large predatory fishes (see Section III,A,1,a, particularly references to Munro, 1983; Craik, 1986; Koslow *et al.*, 1988; Koslow, in press). Fishing has been shown also to have substantial effects on species richness and species composition of large predatory reef fishes (e.g., Bohnsack, 1982; Russ, 1985; Russ and Alcala, 1989).

Further evidence is available based on the increase in both species richness and density of large predatory fishes (serranids, lutjanids, and lethrinids) within a reserve at Apo Island in the central Philippines which has been protected from fishing since 1981 (Fig. 4). Large and consistent increases in species richness and density of these fishes have occurred over a 5-year period within the reserve but not at an adjacent fished site.

(ii) Release of Prey A common theme in the literature on responses of communities of fishes to removal of top predators is the likelihood of changes in the overall abundance and relative abundances of prey (e.g., Beddington and May, 1982; Beddington, 1984; Grigg *et al.*, 1984; Munro *et al.*, 1987; Koslow *et al.*, 1988). Depletion of top predators resulting in an increase in the abundance of prey has been well documented in closed systems such as tropical lakes (Pauly, 1979; Jones, 1982), but the evidence for such events on coral reefs remains equivocal, both at the level of individual species of prey and at the level of "total prey biomass" in box models (e.g., Polovina, 1984; Munro *et al.*, 1987).

Evidence of massive increases in density of a species of prey fish following decrease in density of one of its predators caused by fishing is scant on coral reefs. Bohnsack (1982) showed that the small labrid *Thalassoma bifasciatum* was twice as abundant on a reef subjected to spearfishing than on two other reefs protected from such fishing. However, he did concede that such evidence was suggestive but not conclusive concerning the role of large predators in determining abundance of their prey, and that evidence of a controlling role of

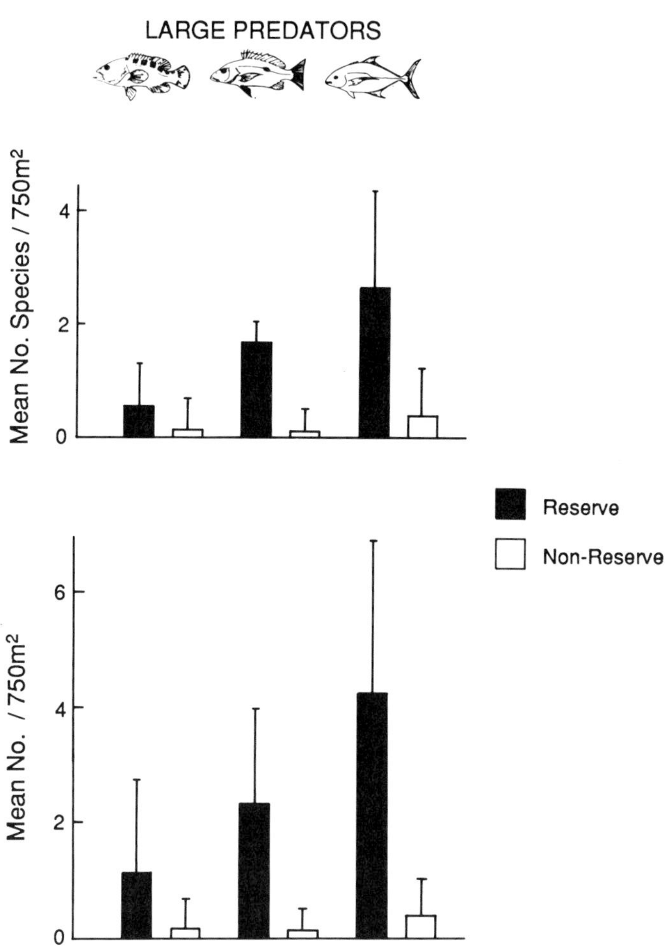

Figure 4 Increase in number of species and number of individuals of large predators (mostly serranids, lutjanids, and carangids) in the Apo Island Reserve, Philippines, over a 5-year period. The reserve was established in 1981. Note that species richness and abundance of large predators have not changed significantly in an adjacent area open to fishing.

predation by piscivores was premature. Thompson and Munro (1978) reported a similar pattern of increased abundance of prey on reefs where densities of large serranids had been reduced by trap fishing. Wyatt (1983) suggested that the relatively large numbers of large holocentrids at Pedro Cays in Jamaica may have been a result of a decrease in the rate of natural mortality, which itself was a result of selective exploitation of predatory species.

At the level of individual species, at least two attributes of assemblages of coral reef fishes suggest that removal by fishing of a species of top predator is unlikely to lead to a spectacular increase in abundance of a prey species. First, there are many species of carnivorous fishes on coral reefs, many of which are generalists and highly opportunistic in their feeding habits. Any increase in abundance of a species of prey following removal of one of its predators is likely to lead to functional responses of other predators which may switch their attention to the abundant prey item. Second, although postsettlement predation may have an important effect on population density (Shulman and Ogden, 1987; Doherty and Williams, 1988a) and secondary production (Grigg *et al.*, 1984) of coral reef fishes, current ideas on what determines abundance suggest that larval supply (recruitment) is more likely to be the ultimate control on abundance of populations than is predation (Doherty and Williams, 1988a). Since recruitment of coral reef fishes is highly variable in space and time, any response of a population of prey to depletion of one of its predators may be very difficult to detect. Depletion of a large number of carnivorous species may not necessarily lead to detectable increases in abundance of a species of prey. Russ (1985), in a correlative study, showed that a site with the highest density of carnivorous and piscivorous fishes was also the site of highest density of small prey (pomacentrids and anthiids), suggesting that some factor other than postsettlement predation was controlling abundance of prey. Similarly, Bohnsack (1982) noted that abundance patterns of most species in communities of reef fishes at one fished and two unfished reefs did not differ significantly (see his Table 4) despite significantly lower densities of piscivores on the fished reef. Koslow (in press) concluded that there was no evidence of a higher abundance of species preyed upon by piscivores at heavily fished sites.

Also, it is often suggested that removal of top predators may lead to large changes in relative abundance of prey species (e.g., Beddington and May, 1982; Beddington, 1984; Jones, 1982). The theme of such arguments involves predation holding the abundance of prey to levels below which competition for resources is important in regulating population density. In the presence of predation those species of prey which produce large numbers of offspring quickly will tend to be favored (Beddington and May, 1982). Relaxation of predator pressure leads to increase in abundance of prey to a point where competition for some limiting resource occurs and those species best able to compete for these resources tend to increase in abundance, resulting in a shift in the relative abundance of prey. Such scenarios rely upon the assumption that communities of coral reef fishes are structured by processes of predation and competition. Doherty and Williams (1988a) have argued convincingly that environmental factors affecting recruitment are more likely to control abundance of populations of coral reef fishes than either predation or competition, although renewed interest is being shown in the role of postrecruitment processes (Jones, 1987b), particularly predation (e.g., Shulman, 1985a; Shulman and Ogden, 1987; Warner and Hughes, 1988).

It is far more likely that large-scale removal of predators from communities of coral reef fishes by fishing will permit increases in total prey biomass but direct evidence from coral reefs is lacking. Munro (1980) formalized this concept in a theoretical relationship between rate of natural mortality (M), fishing mortality (F), and total mortality (Z) in a community in which predators are exploited most heavily (Fig. 5). Natural rates of mortality of prey are considered high initially and, as predators are removed, the natural and thus total mortality of prey decreases with increasing fishing effort. Eventually, as the intensity of exploitation becomes very high, fishing effort is switched to prey populations so that the total mortality rate of prey rises, despite natural

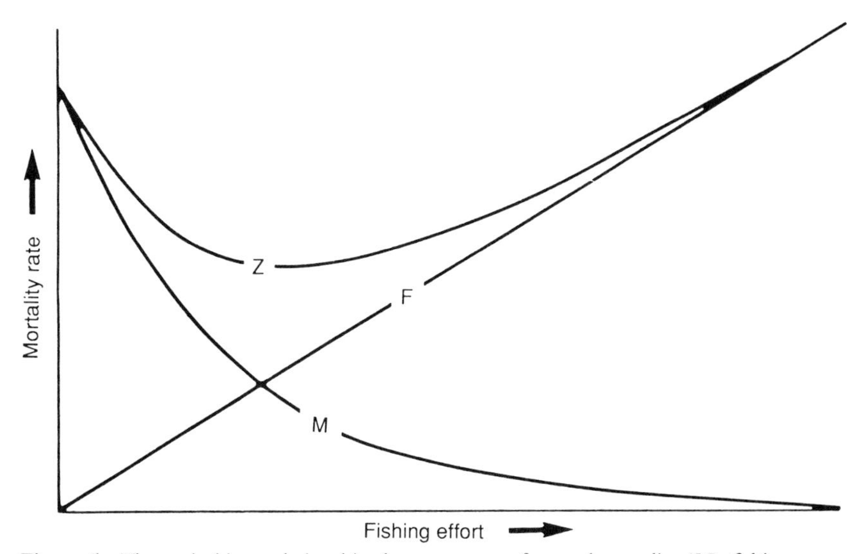

Figure 5 Theoretical interrelationships between rates of natural mortality (M), fishing mortality (F) and total mortality (Z) which should exist if M in an exploited community declines because of a disproportionately high rate of exploitation initially on predatory species. [From J. L. Munro (1983). "Caribbean Coral Reef Fisheries Resources," ICLARM Studies and Reviews, No. 7.]

mortality continuing to decrease. Munro and Williams (1985) and Grigg *et al.* (1984) both suggest that removal of predators will lead to increases in biomass and catch of prey, since humans are replacing the previous predators (see also Pauly, 1979; Munro *et al.*, 1987). In fact, both Munro and Williams (1985) and Grigg *et al.* (1984) advocate an active and sustaining fishing campaign for the predators, thus ensuring that potential or real catches of prey will be enhanced. Clearly, if quality rather than quantity of catch were a more important criterion for managers, a different fishing strategy would be necessary.

Notwithstanding the comments made earlier regarding effects of predator removal at the level of individual species, it is quite possible, particularly in light of estimates made by Grigg *et al.* (1984) that an average of 85% of the production of each species group in the French Frigate Shoals was consumed by the next higher trophic level, that decreases in biomass of predators may lead to substantial increases in biomass of prey. Nevertheless, although it is likely that predation does have a major regulatory role on secondary production of coral reef fishes once they have recruited, annual variability in recruitment is perhaps more likely to ultimately control yields. Such variation, although dismissed by Grigg *et al.* (1984) "from the equilibrium or long term average perspective that we view the coral reef ecosystem," is probably critical to the fisheries manager attempting to deal with year-to-year variability in catch.

b. Survivorship of New Recruits A potentially significant effect of fishing on coral reefs involves removal of predators and subsequent alterations in survivorship of newly settled fishes. Newly settled coral reef fishes may have mortality rates in the first year ranging from 25 to 92%, with a mean of around 50% (Shulman and Ogden, 1987; Doherty and Williams, 1988a). Much of this mortality, presumed to be due to predation, occurs within the first week of benthic life. Shulman (1985a) found that tethered prey disappeared faster from sites closer to areas of natural reef. Munro (1983) noted that some smaller species in the Jamaican trap fishery such as haemulids and acanthurids had relatively more recruits in exploited areas. He suggested that predation on juveniles might be an important factor determining recruitment rates and that if some of the more lightly exploited areas were to become more heavily exploited, patterns of recruitment (settlement and subsequent survival) may shift in favor of these smaller species.

c. Removal of Prey Many fisheries on coral reefs target fishes lower in the food chain, such as herbivores in Tikehau Lagoon, French Polynesia (Stein, 1988), Palau (Johannes, 1981), Papua New Guinea (Wright and Richards, 1985; Dalzell and Wright, 1986), and Palmerston Atoll, Cook Islands (A. Wright, personal communication). In other cases, herbivores or planktivores constitute a high proportion of the yield and catches of predatory or piscivorous fishes are low, either because such fishes are proportionally less abundant naturally and/or because their abundance has been reduced by fishing. Examples include high proportions of herbivores in the catches in Jamaica (Munro, 1983; Koslow *et al.,* 1988) and very high proportions of planktivores in catches from the central Philippines (e.g., Alcala, 1981, 1988; Alcala and Luchavez, 1981; Savina and White, 1986; Bellwood, 1988a; Alcala and Russ, 1990).

Again, little evidence exists that removal of fishes lower in the trophic chain of communities of coral reef fishes causes substantial and measurable effects at the level of individual species or total biomass of another trophic level. There appears to be no speculation that removal of prey will lead to reductions in predator populations, for example. Pauly (1979), based on a study of the Gulf of Thailand trawl fishery, hypothesized that prey species in tropical ecosystems are exploited by their natural predators near the point of their maximum sustainable yield before any fishing mortality by humans is imposed. He argued that this makes such prey more sensitive to additional, fishery-induced, mortality. Pauly (1979) noted that small prey species declined very rapidly relative to other major groups of fishes in the Gulf of Thailand following onset of intensive trawl fishing, with predators such as lutjanids being less rapidly affected. Trawl fisheries would appear to be far less selective than most reef fisheries and no cases analogous to that of the Gulf of Thailand appear to be available from coral reefs.

d. Removal of Competitors Doherty and Williams (1988a) reviewed the evidence for competition and resource limitation and found that much of it was unconvincing. Evidence that fishing may affect levels of competition between species of coral reef fishes is anecdotal at best. Bohnsack (1982) noted that high densities of the small serranid *Epinephelus cruentatus* on a reef open to spearfishing relative to on unfished reefs may have been due to reduced competition from other (presumably larger) species of predator. In the same study an abundant species of haemulid on the fished reef appeared to be "replaced" by a very similar and abundant species on the unfished reefs. Bohnsack (1982) noted that interspecific competition may have been an important factor in determining this pattern. Koslow *et al.* (1988) noted that *Balistes vetula*, the dominant species trapped on SE Pedro Bank in 1969–1973, was "replaced" by *Xanthichthys ringens* in 1986. They did not imply that some form of competitive replacement had taken place but later speculated that small, untrapped fish groups, such as labrids and pomacentrids, may increase as a result of reduced competition and predation from species commonly trapped. However, Koslow (in press) concluded, from a study over a larger geographical extent, that there was little evidence of "secondary effects" (increases in prey species or of nontrapped potential competitors) on the community of reef fishes due to fishing.

e. Changes in Community Structure There are several examples of intensive fishing on assemblages of coral reef fishes which conform to the definition of **ecosystem overfishing** given earlier in this chapter. Russ and Alcala (1989) demonstrated that an intensive pulse of fishing over an 18-month period in an area previously protected from fishing for 10 years in the Philippines caused a significant change in the community structure of reef fishes. Both the species richness and total abundance of fishes in the assemblage decreased significantly. Abundance of several major groups of fishes declined significantly, including caesionids, pomacentrids, chaetodontids, lutjanids, and lethrinids, with many other groups declining substantially in abundance and some (small labrids and scarids) increasing in abundance as a result of the intensive fishing pressure. The intensive fishing pressure had both direct and indirect effects and a far wider impact on the assemblage of fishes than the predicted significant decreases in abundance of target species. In addition to these massive changes in community structure within the reserve, Alcala and Russ (1990) reported a significantly lower catch per unit effort and total yield from the island after the intensive bout of fishing.

Koslow *et al.* (1988) reported significant decreases in catch rate of a wide range of families together with some increases in catch rates of commercially less desirable groups such as chaetodontids, holocentrids, and tetraodontids. Munro and Williams (1985) list several characteristics which may contribute to groups defined as "trash fish" eventually becoming a major component of the catch: low desirability as food fish, high ratios of natural mortality (M) to growth (K), high fecundity, and low catchability. Although catch rates declined and catch composition changed significantly at two heavily fished locations in Jamaica, it was unclear if total catch declined over the 15 years.

Koslow (in press) used principal component analysis to identify three distinct community types related to fishing pressure: (1) a community characterized by a high proportion of large piscivores in the lightly to moderately exploited reefs of Belize and the Bahamas; (2) a community occurring on the lightly to moderately exploited offshore reefs of Pedro Bank (Jamaica); and (3) a "depleted community" at heavily fished sites on both offshore Pedro Bank and reefs close to Kingston, Jamaica.

Hay (1984b) suggested that fishing on Caribbean coral reefs may have caused a substantial shift in the relative abundance of major herbivores (fishes and urchins). The two mechanisms hypothesized to account for such a shift were direct removal of herbivorous fishes (acanthurids and scarids) by fishermen and removal of a wide variety of predatory fishes which fed on urchins (e.g., balistids and pomadasyids). Hay showed that on overfished reefs, urchins were relatively more abundant and the magnitude of their grazing was comparable to that of herbivorous fishes. On reefs subject to little fishing pressure, densities of urchins were low and herbivorous fishes accounted for most grazing.

Goeden (1982, 1986) argued that intensive fishing on a 'keystone' predatory fish on the Great Barrier Reef—*Plectropomus*—led to increased instability (increased "community flux") in the assemblage of other predatory fishes in the community. His arguments were based on relationships between densities of *Plectropomus* and relative abundances of other predatory fishes (other serranids, lethrinids, and plectorhyncids) along a 1000-km section of the Great Barrier Reef. Goeden's data are confounded by changes in density of *Plectropomus* and other predatory species [e.g., *Lethrinus miniatus (=chrysotomus)*] which are completely independent of fishing pressure (see data of A. M. Ayling presented as Fig. 3 in Munro and Williams, 1985). Furthermore, it is difficult to define mechanisms by which reduction in density of a piscivore would lead to increased instability in an assemblage of fishes which feed largely on benthic invertebrates (lethrinids and plectorhyncids).

In cases where fishing is less intensive than many of the examples cited, one would not expect substantial shifts in the relative abundance of species or species composition due solely to the effects of fishing. Ferry and Kohler (1987) reported that there was little difference in the species composition of, or catch per unit effort from, two reefs subjected to very different levels of fishing pressure in Haiti. Bohnsack (1982) noted that communities of reef fishes were broadly similar on reefs of differing fishing pressure in Florida. Russ (1985) suggested that variates such as overall species richness and species richness of major groups may not be particularly good indicators of fishing pressure on coral reefs.

Assemblages of reef fishes traditionally have been viewed as equilibrial, resource-limited systems with strong mean interaction strengths between species, the most important of which being competition and predation (see the review by Sale, 1980a). These views form the basis of the development of models for management of multispecies fisheries in other ecosystems (e.g., May *et al.,* 1979; Beddington, 1984), although they have been questioned in the context of development of multispecies models for management purposes (e.g., Larkin, 1982; Sainsbury, 1982, 1988; Walters, 1984). Development of more realistic models of the effects of fishing on assemblages of coral reef fishes will occur if such assemblages are viewed as more open, non-resource-limited, nonequilibrial systems not necessarily characterized by strong interactions between species (Sale, 1982b; postrecruitment predation perhaps may be an important exception—see Shulman and Ogden, 1987; Doherty and Williams, 1988a), but influenced strongly by variability in recruitment (Doherty and Williams, 1988a).

2. *Indirect Effects (Habitat Modification)*

Any large-scale reductions in cover of live coral or structural heterogeneity of the benthic substratum are likely to have substantial effects on communities of coral reef fishes. Much of the evidence in support of this contention comes from detailed studies of communities of coral reef fishes affected by destructive storms (e.g., Lassig, 1983; Walsh, 1983) or destructive fishing techniques (e.g., Russ and Alcala, 1989). Carpenter *et al.* (1981) showed that strong positive correlations existed between live coral cover and diversity and abundance of fishes, and that biomass of fishes increased with structural heterogeneity of the substratum.

On a whole-reef scale, gross physiographic features may influence yields of reef fishes substantially (Saila and Roedel, 1980; Marshall, 1985). Fishing methods which are extremely destructive to the benthic habitat and which are used extensively may have a far greater and long-term impact than many of the direct effects listed earlier. For example, Samoilys (1988) found that some reefs in Kenya, where the benthic habitat had been damaged substantially by use of dynamite, generally had low standing stocks of fishes. Alcala and Gomez (1985) reported that four coral reefs they studied in the Philippines had been damaged by dynamite and drive-net fishing and the reefs with greatest damage to the benthic habitat were those with the lowest total fish yields.

C. Levels of Overfishing on Coral Reefs

The effects of fishing on coral reef fishes reviewed here provide examples of four different levels of overfishing: growth, recruitment, economic, and ecosystem overfishing. Several areas, notably Jamaica and the Philippines, have reached or are rapidly approaching a fifth level of overfishing aptly defined by Pauly (1988) as **Malthusian overfishing**—the situation where there are too many fishermen and not enough fish and where the fishermen initiate wholesale resource destruction in an effort to maintain their incomes. Small-scale fishermen fishing coral reefs in these areas are usually poor and currently have no prospects of alternative employment. Their numbers are increasing rapidly as a result of high birth rates (the doubling time of the human population in the Philippines is approximately 25 to 30 years) and because unemployed or displaced workers often turn to fishing as a last resort. The result is that such fishermen, faced with declining catches (and hence incomes), fish more intensively, often with gears destructive to the habitat, until the resource collapses. Koslow *et al.* (1988) have shown that fishing effort on the South Jamaica Shelf has increased dramatically in the past 15 years with no increase in total catch. Dalzell (1988), using catch-effort data dating from the early 1940s for small pelagic fishes in the Philippines, was able to show a massive decline in catch per unit effort associated with massive increases in fishing effort over a 42-year period. The status of Philippine coral reef fisheries is likely to be worse because reef fish are far more accessible and sedentary than small pelagics. The only solutions to Malthusian overfishing are creation of alternative employment opportunities for fishermen and stabilization of human population growth. As Pauly (1988) points out, if such solutions are not attempted we can forget about management options for the coral reef fisheries and indeed the coral reefs themselves in these areas.

IV. Some Comments on Yields of Fishes from Coral Reefs

Yields of finfishes from coral reefs have been reviewed by Stevenson and Marshall (1974), Marshall (1980, 1985), Marten and Polovina (1982), Munro (1983), Russ (1984c), and Munro and Williams (1985). These references should be consulted for a detailed account of the subject. Early reviews by Stevenson and Marshall (1974) and Marshall (1980) found that yields ranged from 0.8 to 5 metric tons/km^2/year. Much of the data on which early reviews were based came from studies of fish production from large areas of coralline shelf [e.g., 11760 km^2 in Jamaica (Munro, 1969); see Russ (1984c, Table 2) for a more complete list] rather than small areas of actively growing coral reef. Most of the reviews cited here noted studies in American Samoa (Hill, 1978; Wass, 1982) and the Philippines (Alcala, 1981; Alcala and Luchavez, 1981) which documented yields in the range 8.0–23.7 mt/km^2/year for intensively exploited, small areas of coral reef. These figures were treated as somewhat anomalous in earlier reviews.

Since 1985 a number of studies of intensively fished, small areas of coral reef in the Pacific have documented yields generally well above 5 mt/km^2/year, suggesting that earlier results from American Samoa and the Philippines may not have been anomalous. Results of these studies, together with earlier ones by Hill (1978), Wass (1982), Alcala (1981), and Alcala and Luchavez (1981), are summarized in Table 1. Yields reported from the more recent studies range from 0.42 to 36.9 mt/km^2/year. Eight of ten studies since 1985 have reported yields in excess of 5 mt/km^2/year and six of ten studies have reported yields in excess of 10 mt/km^2/year.

Several points of caution should be made about the data in Table 1. First, yield is clearly a function of the amount of fishing effort in a location. Although effort information is not presented in Table 1, it is presumed that effort is high at all sites but the Tigak Islands in Papua New Guinea. Second, yield figures can change quite substantially depending on the perception of the investigator as to the maximum depth of fishing and thus "area of reef fished." Bellwood (1988a) made the point that his yield estimate of 24.9 mt/km^2/year would have been 48.79 mt/km^2/year if he had considered the yield of fish from Apo Island to be derived from reef above 20 m in depth. Zann *et al.* (in prep.) pointed out that, had they have adopted a maximum depth of reef

Table 1 Yield of Fishes from Small Areas of Actively Growing Coral Reef

Location	Area of reef (km^2)	Depth used in area estimate (m)	Yield ($mt/km^2/year$)	Approximate % planktivores in catch	Reference
American Samoa	5.4	"Fringing reef"	8.0	—	Hill (1978)
American Samoa	3.6	8	18.3	13	Wass (1982)
Philippines (Sumilon Island)	0.5	40	9.7 (1976)	59 (5-year average)	Alcala (1981)
Philippines (Sumilon Island)	0.5	40	14.0 (1977)	59	Alcala (1981)
Philippines (Sumilon Island)	0.5	40	15.0 (1978)	59	Alcala (1981)
Philippines (Sumilon Island)	0.5	40	23.7 (1979)	59	Alcala (1981)
Philippines (Sumilon Island)	0.5	40	19.9 (1980)	59	Alcala (1981)
Philippines (Apo Island)	1.5	60	11.4	29	Alcala and Luchavez (1981)
Philippines (Selinog Island)	1.26	30	6.0	—	Alcala and Gomez (1985)
Philippines (Hulao-hulao Island)	0.5	15	5.2	—	Alcala and Gomez (1985)
Papua New Guinea (Tigak Islands)	208.0	30	0.42	5	Wright and Richards (1985)
Papua New Guinea (Port Moresby)	—	—	5.0	14	Lock (1986)
Philippines (Apo Island)	0.7	20	22.1	43	Savina and White (1986)
Philippines (Pamilacan Island)	1.8	20	10.7	15	Savina and White (1986)
Philippines (Apo Island)	1.06	60	24.9	23	Bellwood (1988b)
Philippines (Sumilon Island)	0.5	40	36.9 (1983/1984)	65	Alcala and Russ (1990)
Philippines (Sumilon Island)	0.5	40	19.9 (1985/1986)	>50	Alcala and Russ (1990)
Western Samoa (Upolu Island)	300.0	40	11.4	—	Zann *et al.* (in prep.)

from which yield was derived of 8 m as Wass (1982) had done, their yield estimate would have been 22.5 mt/km^2/year. Clearly, very close attention will have to be paid to defining the exact area of reef from which a yield is derived. Zann *et al.* (in prep.) suggested 40 m depth as a possible standard, since this is the approximate maximum depth limit of both coral growth and of most artisanal fishing techniques. Third, making a clear definition of whether fish are truly "reef-associated" or not will have a substantial effect on estimated "reef yield." Bellwood (1988a), for example, calculated that 79% of the biomass of fish caught at Apo Island, Philippines, consisted of "reef species," "reef-associated" species, or "reef planktivores." The remaining catch consisted of species characteristic of deep or "off-reef" habitats (e.g., scombrids). Fourth, the sample of studies in Table 1 is biased heavily toward the Philippines, and in particular small islands in this country. Thus, generalizations from Table 1 must be treated with caution.

Nevertheless, several points of interest derive from data in Table 1. First, it is clear that sustainable yields from intensively fished, small areas of actively growing coral reef on the order of 10–20 mt/km^2/year seem feasible. Estimates of this order (or greater) have been made for seven separate years at Sumilon Island and three separate years at Apo Island, Philippines. Munro (1984) suggested a **maximum** sustainable harvest of finfishes from intensively fished, small areas of coral reef on the order of 10–20 mt/km^2/year. Information in Table 1 indicates that this suggestion may have been correct. Second, all the yields in excess of 20 mt/km^2/year have high or very high proportions of small, schooling, planktivorous fishes in the catch. It must be conceded that all of these derive from two small islands in the Philippines. On the other hand, it is likely that few reef fisheries in the Pacific target schooling planktivores as effectively as the fisheries on small islands in the Philippines. Very high yields of finfishes may be possible in reef fisheries which target small, schooling planktivores effectively.

Traditionally, coral reefs have been viewed as closed ecosystems in nutrient-poor waters, with little or no external input of nutrients and materials. Such a view held that reefs rely heavily upon tight nutrient recycling to maintain productivity, with benthic production being the basis for secondary production of fishes (Odum and Odum, 1955). In such systems, secondary production of fishes may well be quite limited and removal of a significant biomass of fish quite disruptive to the ecosystem. A large proportion of coral reefs worldwide are in fact close to landmasses and are likely to be subject to large input of organic matter and nutrients. This contention is supported by circumstantial evidence that many communities of coral reef fishes are dominated both numerically and in terms of biomass by planktivores (e.g., Williams and Hatcher, 1983; Russ, 1984c; Samoilys, 1988). Many of these planktivores are most abundant on windward sides of reefs and evidence now suggests that this substantial biomass of fishes may be heavily dependent on inputs of plankton impinging onto reefs from outside (Hamner *et al.,* 1988). The "classic" food chain thought to be the major basis of secondary production of coral reef fishes (benthic algae→herbivores→primary and secondary carnivores) may not apply strictly to a large number of coral reefs worldwide. Many of the high yields in Table 1, for example, may well not have been derived through the "classic" food chain based on production of benthic algae.

V. Conclusions and Recommendations for Future Research

Some long-term studies of the effects of fishing on catch and catch per unit effort now exist (e.g., Ralston and Polovina, 1982; Alcala, 1988; Koslow *et al.,* 1988). There is, however, a desperate need for collection of basic catch and effort information over long time periods from coral reef fisheries.

Few studies have made reliable estimates of the rate of total mortality of species of coral reef fishes and partitioned this into natural and fishing mortality (some notable exceptions are Huntsman *et al.,* 1982; Mahmoudi *et al.,* 1984; see also the review by Ralston, 1987). The evidence for effects of fishing on size structure of populations is good, with cases of growth overfishing from reefs with such different fishing intensities as Jamaica and the Great Barrier Reef. However, evidence of effects of fishing on age structure, particularly reduction in the proportion of older age classes and alteration of patterns of size at age, is rare. A strong need exists for study of age structure of coral reef fishes, particularly the larger, commercially, and recreationally important species. Research should involve not only determination of age by examination of daily rings in otoliths (e.g., Brothers, 1980; Ralston and Williams, 1989) but a more concerted effort to search for seasonal or annual banding patterns such as those described by Manooch (1987) and Ferrell (1988). Age structures open up the possibilities of determining age-specific growth and mortality rates. Far more use could be made also of mark–recapture techniques to estimate growth and mortality rates.

The evidence that fishing on coral reefs has affected recruitment is generally not good, except perhaps at geographic scales, for example, the possible recruitment overfishing of sharks and large predatory fishes in northern Jamaica and the Philippines. Although recruitment overfishing in its early stages is difficult to detect unequivocally, far more study of stock-recruitment relationships of exploited coral reef fishes is needed. Details of the spatial extent of larval dispersal and the degree of self-recruitment of coral reefs in different areas are particularly important subjects from the point of view of fisheries management.

The small amount of direct evidence of effects of fishing on the sex ratios of sequential hermaphrodites is surprising, given that this aspect of the vulnerability of certain types of coral reef fishes to fishing pressure was recognized a long time ago (e.g., Moe, 1969; see also the review by Bannerot *et al.*, 1987). More research needs to be directed specifically at the effects of fishing on sex ratios of sequential hermaphrodites.

A voluminous literature has documented selective removal of piscivorous and other large predatory fishes from communities of coral reef fishes. Evidence that such a fishing strategy leads to "prey release" is limited and equivocal, at the level of both species and total prey biomass. Far more research is required on these topics of selective removal of predators and prey release. A great deal of scope exists for experiments involving replicated removal of predators from patch reefs or even whole reefs.

Much of our knowledge of the effects of fishing on populations or communities of coral reef fishes comes from comparisons of reefs with differing levels of fishing pressure. Evidence of such effects on density and biomass from this type of study are reasonable. Future comparative studies should pay particular attention to detailed sampling designs and use rigorous methods of sampling to ensure that effects of fishing of a specified magnitude can be detected unequivocally by the sampling effort applied.

Evidence now exists that intensive fishing can alter community structure of coral reef fishes over short [e.g., 18 months (Russ and Alcala, 1989)] and long [e.g., 15 years (Koslow *et al.,* 1988)] time periods. There is a great need for development of realistic models which explain or predict effects of fishing on the multispecies stocks of coral reef fishes. The approach of using large-scale manipulations (e.g., Walters and Hilborn, 1978; Larkin, 1982; Sainsbury, 1982, 1988; Walters, 1984) as the source of empirical information for models seems a fruitful one for the future. Coral reefs are ideal units for experimental manipulation and the testing of various management schemes. There is a great need for experimental investigations of the effects of fishing on coral reef fishes, using reefs or parts of reefs as replicate experimental units.

Acknowledgments

This chapter was improved substantially by comments from A. S. Cabanban, W. J. S. Craik, J. C. Ogden, P. F. Sale, and D. McB. Williams. Thanks to W. J. S. Craik and J. L. Munro for permission to use Figs. 1, 2, and 5. A. Sharp and E. Howlett typed the manuscript.

REFERENCES

Abel, E. F. (1972). Problem der okologischen Definition des "Korrallenfisches." *Proc. Symp. Corals Coral Reefs* **1,** 449–456.

Abel, D. J., Williams, W. T., and Williams, D. McB. (1985). A fast classificatory algorithm for large problems and the Bray–Curtis measure. *J. Exp. Mar. Biol. Ecol.* **89,** 237–245.

Aboussouan, A. (1965). Oeufs et larves de teleosteens de l'Ouest africain—II. Distribution verticale. *Bull. Inst. Fondam. Afr. Noire, Ser. A* **27,** 1504–1521.

Abrams, P. A. (1984). Recruitment, lotteries, and coexistence in coral reef fish. *Am. Nat.* **123,** 44–55.

Adams, P. B. (1980). Life history patterns in marine fishes and their consequences for fisheries management. *Fish. Bull.* **78,** 1–12.

Adey, W. H., and Steneck, R. S. (1985). Highly productive eastern Caribbean reefs: Synergistic effects of biological, chemical, physical and geological factors. *In* "The Ecology of Coral Reefs" (M. L. Reaka, ed.), NOAA Symp. Ser. Undersea Res., Vol. 3, pp. 163–188. Natl. Oceanic Atmos. Adm., Rockville, Maryland.

Ahlstrom, E. H. (1971). Kinds and abundance of fish larvae in the eastern tropical Pacific, based on collections made on EASTROPAC I. *Fish. Bull.* **69,** 3–77.

Ahlstrom, E. H. (1972). Kinds and abundance of fish larvae in the eastern tropical Pacific on the second multi-vessel Eastropac survey, and observations on the annual cycle of larval abundance. *Fish. Bull.* **70,** 1153–1242.

Aiken, K. (1983a). The biology, ecology and bionomics of the butterfly and angelfishes, Chaetodontidae. *ICLARM Stud. Rev.* **7,** 166–177.

Aiken, K. (1983b). The biology, ecology and bionomics of the triggerfishes, Balistidae. *ICLARM Stud. Rev.* **7,** 191–205.

Alcala, A. C. (1981). Fish yields of coral reefs of Sumilon Island, central Philippines. *Bull., Natl. Res. Counc. Philipp.* **36,** 1–7.

Alcala, A. (1987). Algunos aspectos de la reproduccion de la rabirubia, *Ocyurus chrysurus* (Pisces: Lutjanidae), en el Parque Nacional Archipelago Las Roques. Graduate thesis, Universidad Central de Venezuela, Caracas, Venezuela.

Alcala, A. C. (1988). Effects of marine reserves on coral fish abundances and yields of Philippine coral reefs. *Ambio* **17,** 194–199.

Alcala, A. C., and Gomez, E. D. (1985). Fish yields of coral reefs in the central Philippines. *Proc. Int. Coral Reef Congr., 5th* **5,** 521–524.

Alcala, A. C., and Gomez, E. D. (1987). Dynamiting coral reefs for fish: A resource-destructive fishing method. *In* "Human Impacts on Coral Reefs: Facts and Recommendations" (B. Salvat, ed.), pp. 51–60. Antenne Museum EPHE, Moorea, French Polynesia.

Alcala, A. C., and Luchavez, T. (1981). Fish yield of the coral reef surrounding Apo Island, central Visayas, Philippines. *Proc. Int. Coral Reef Symp., 4th* **1,** 69–73.

Alcala, A. C., and Russ, G. R. (1990). A direct test of the effects of protective management on abundance and yield of tropical marine resources. *J. Cons., Cons. Int. Explor. Mer* **46,** 40–47.

Aldenhoven, J. M. (1986a). Different reproductive strategies in a sex-changing coral reef fish *Centropyge bicolor* (Pomacanthidae). *Aust. J. Mar. Freshwater Res.* **37,** 353–360.

Aldenhoven, J. M. (1986b). Local variation in mortality rates and life-expectancy estimates of the coral-reef fish *Centropyge bicolor* (Pisces: Pomacanthidae). *Mar. Biol.* **92,** 237–244.

Alevizon, W. S. (1975). Spatial overlap and competition in congeneric surfperches (Embiotocidae) off Santa Barbara, California. *Copeia* **1975,** 352–356.

Alevizon, W. S., and Brooks, M. G. (1975). The comparative structure of two western Atlantic reef-fish assemblages. *Bull. Mar. Sci.* **25,** 482–490.

Alevizon, W. S., Richardson, R., Pitts, P., and Serviss, G. (1985). Coral zonation and patterns of community structure in Bahamian reef fishes. *Bull. Mar. Sci.* **36,** 304–318.

Alexander, R. M. (1967). The functions and mechanisms of the protrusible upper jaws of some acanthopterygian fish. *J. Zool.* **151,** 43–64.

Alheit, J. (1981). Feeding interactions between coral reef fishes and the zoobenthos. *Proc. Int. Coral Reef Symp., 4th* **2,** 545–552.

Alheit, J., and Scheibel, W. (1982). Benthic harpacticoids as a food source for fish. *Mar. Biol.* **70,** 141–147.

Alldredge, A. L., and King, J. M. (1977). Distribution, abundance, and substrate preference of demersal reef zooplankton at Lizard Island Lagoon, Great Barrier Reef. *Mar. Biol.* **41,** 317–333.

Alldredge, A. L., and King, J. M. (1980). Effects of moonlight on the vertical migration patterns of demersal zooplankton. *J. Exp. Mar. Biol. Ecol.* **44,** 133–156.

Allen, G. R. (1972). "Anemonefishes." TFH Publ., Neptune City, New Jersey.

Allen, G. R. (1975). "Damselfishes of the South Seas." TFH Publ., Neptune City, New Jersey.

Allen, G. R. (1981). "Butterfly and Angelfishes of the World," Vol. 2. Mergus, Melle, Germany.

Allen, G. R., and Hoese, D. F. (1975). A review of the pomacentrid fish genus *Parma*, with descriptions of two new species. *Rec. West. Aust. Mus.* **3,** 261–293.

Allen, G. R., and Steene, R. C. (1979). "The Fishes of Christmas Island, Indian Ocean." Aust. Gov. Publ. Serv., Canberra, Australia.

Allen, G. R., and Swainston, R. (1988). "The Marine Fishes of Northwestern Australia." West. Aust. Mus., Perth, Western Australia.

Amesbury, S. S., and Myers, R. F. (1982). "Guide to the Coastal Resources of Guam," Vol. 1. Univ. of Guam., Agana, Guam.

Anderson, T. A. (1987). Utilization of algal cell fractions by the marine herbivore the luderick, *Girella tricuspidata* (Quoy and Gaimard). *J. Fish Biol.* **31,** 221–228.

Anderson, G. R. V., Ehrlich, A. H., Ehrlich, P. R., Roughgarden, J. D., Russell, B. C.,

and Talbot, F. H. (1981). The community structure of coral reef fishes. *Am. Nat.* **117,** 476–495.
Andrew, N. L. (1989). Contrasting ecological implications of food limitation in sea urchins and herbivorous gastropods. *Mar. Ecol. Prog. Ser.* **51,** 189–193.
Andrew, N. L., and Mapstone, B. D. (1987). Sampling and the description of spatial pattern in marine ecology. *Oceanogr. Mar. Biol.* **25,** 39–90.
Andrewartha, H. G., and Birch, L. C. (1954). "The Distribution and Abundance of Animals." Univ. of Chicago Press, Chicago.
Arnold, G. P. (1975). The measurement of irradiance with particular reference to marine biology. *In* "Light as an Ecological Factor" (G. C. Evans, R. Bainbridge, and O. Rackham, eds.), Vol. II, pp. 1–25. Blackwell, Oxford, England.
Aronson, R. B. (1983). Foraging behavior of the west Atlantic trumpetfish, *Aulostomus maculatus*: Use of large, herbivorous reef fishes as camouflage. *Bull. Mar. Sci.* **33,** 166–171.
Aronson, R. B., and Sanderson, S. L. (1987). Benefits of heterospecific foraging by the Caribbean wrasse, *Halichoeres garnoti* (Pisces: Labridae). *Environ. Biol. Fishes* **18,** 303–308.
Atema, J., Fay, R. R., Fay, A. N., and Tavolga, W. N. (eds.) (1988). "Sensory Biology of Aquatic Animals." Springer-Verlag, New York.
Atkinson, M., Smith, S. V., and Stroup, E. D. (1981). Circulation in Enewetak Atoll Lagoon. *Limnol. Oceanogr.* **26,** 1074–1083.
Avise, J. C., and Shapiro, D. Y. (1986). Evaluating kinship of newly-settled juveniles within social groups of the coral reef fish *Anthias squamipinnis*. *Evolution (Lawrence, Kans.)* **40,** 1051–1059.
Ayling, A. M. (1981). The role of biological disturbance in temperate subtidal encrusting communities. *Ecology* **62,** 830–847.
Ayling, A. M., and Ayling, A. L. (1985). "A Biological Survey of Selected Reefs in the Great Barrier Reef Marine Park," Rep. Great Barrier Reef Mar. Park Authority, pp. 1–53. Great Barrier Reef Mar. Park Authority. Townsville, Australia.
Ayling, A. M., and Ayling, A. L. (1986). "A Biological Survey of Selected Reefs in the Capricorn Section of the Great Barrier Reef Marine Park," Rep. Great Barrier Reef Mar. Park Authority, pp. 1–61. Great Barrier Reef Mar. Park Authority. Townsville, Australia.
Babcock, R. C., Bull, G., Harrison, P. L., Heyward, A. J., Oliver, J. K., Wallace, C. C., and Willis, B. L. (1986). Synchronous multispecific spawnings of 107 scleractinian coral species on the Great Barrier Reef. *Mar. Biol.* **90,** 379–394.
Bailey, K. M. (1981). Larval transport and recruitment of Pacific hake, *Merluccius productus*. *Mar. Ecol. Prog. Ser.* **6,** 1–9.
Bailey, T. G., and Robertson, D. R. (1982). Organic and caloric levels of fish feces relative to its consumption by coprophagous reef fishes. *Mar. Biol.* **69,** 45–50.
Bailey, K. M., and Houde, E. D. (1989). Predation on eggs and larvae of marine fishes and the recruitment problem. *Adv. Mar. Biol.* **25,** 1–83.
Baird, T. A. (1988). Female and male territoriality and mating system of the sand tilefish, *Malacanthus plumieri*. *Environ. Biol. Fishes* **22,** 101–116.
Bakun, A. (1986). Local retention of planktonic early life stages in tropical reef/bank demersal systems: The role of vertically-structured hydrodynamic processes. *In* "IOC/FAO Workshop on Recruitment in Tropical Coastal Demersal Commu-

nities," (D. Pauly and A. Yanez-Arancibia, eds.), IOC Workshop Rep. 44 (suppl.), pp. 15–32. UNESCO, Paris.

Bakun, A. (1989). L'ocean et la variabilité des populations marines. *In* "L'homme et les Écosystemes Halieutiques" (J. P. Troadec, ed.). Gauthier Villard, Paris.

Bakun, A., Beyer, J., Pauly, D., Pope, J. G., and Sharp, G. D. (1982). Oceanic sciences in relation to living resources. *Can. J. Fish. Aquat. Sci.* **39,** 1059–1070.

Bakus, G. J. (1964). The effects of fish-grazing on invertebrate evolution in shallow tropical waters. *Allan Hancock Found. Publ.* **27,** 1–29.

Bakus, G. J. (1966). Some relationships of fishes to benthic organisms on coral reefs. *Nature (London)* **210,** 280–284.

Bakus, G. J. (1967). The feeding habits of fishes and primary production at Eniwetok, Marshall Islands. *Micronesica* **3,** 135–149.

Bakus, G. J. (1969). Energetics and feeding in shallow marine waters. Int. Rev. Gen. Exp. Zool. **4,** 273–369.

Bakus, G. J. (1981). Chemical defense mechanisms on the Great Barrier Reef, Australia. *Science* **211,** 497–499.

Bakus, G. J. (1983). The role of fishes in the structuring of coral reef communities. Proceedings of the International Conference on Marine Science in the Red Sea. *Bull. Egypt. Inst. Oceanogr. Fish.* **9,** 186–192.

Bakus, G. J., and Green, G. (1974). Toxicity in sponges and holothurians: A geographic pattern. *Science* **185,** 951–953.

Baltz, D. M. (1984). Life history variation among female surfperches (Perciformes: Embiotocidae). *Environ. Biol. Fishes* **10,** 159–171.

Bannerot, S. P., Fox, W. W., Jr., and Powers, J. E. (1987). Reproductive strategies and the management of snappers and groupers in the Gulf of Mexico and Caribbean. *In* "Tropical Snappers and Groupers: Biology and Fisheries Management" (J. J. Polovina and S. Ralston, eds.), pp. 561–603. Westview, Boulder, Colorado.

Bardach, J. E. (1958). On the movements of certain Bermuda reef fishes. *Ecology* **39,** 139–146.

Bardach, J. E. (1959). The summer standing crop of fish on a shallow Bermuda reef. *Limnol. Oceanogr.* **4,** 77–85.

Bardach, J. E., Smith, C. L., and Menzel, D. W. (1958). "Bermuda Fisheries Research Program. Final Report." Bermuda Trade Dev. Board, Hamilton, Bermuda.

Bardach, J. E., Winn, H. E., and Menzel, D. W. (1959). The role of senses in the feeding of the nocturnal reef predators *Gymnothorax moringa* and *G. vivinus*. *Copeia* **1959,** 133–139.

Barlow, G. W. (1975). On the sociobiology of four Puerto Rican parrotfishes (Scaridae). *Mar. Biol.* **33,** 281–293.

Barlow, G. W. (1981). Patterns of parental investment, dispersal and size among coral-reef fishes. *Environ. Biol. Fishes* **6,** 65–85.

Barlow, H. B. (1957). Purkinje shift and retinal noise. *Nature (London)* **179,** 255–256.

Barlow, H. B. (1972). Dark and light adaptation: Psychophysics. *In* "Handbook of Sensory Physiology, Vol. VII/4." (D. Jameson and L. M. Hurvich, eds.), pp. 1–28. Springer-Verlag, Berlin.

Barton, M. G. (1982). Intertidal vertical distribution and diets of five species of central California stichaeoid fishes. *Calif. Fish Game* **68,** 174–182.

Beddington, J. R. (1984). The response of multispecies systems to perturbations. *In* "Exploitation of Marine Communities" (R. M. May, ed.), pp. 209–225. Springer-Verlag, Berlin.

Beddington, J. R., and Cooke, J. G. (1983). The potential yield of fish stocks. *FAO Fish. Tech. Pap.* **242,** 47 pp.

Beddington, J. R., and May, R. M. (1982). The harvesting of interacting species in a natural ecosystem. *Sci. Am.* **247,** 42–49.

Begon, M. (1984). Density and individual fitness: Asymmetric competition. *In* "Evolutionary Ecology" (B. Shorrocks, ed.), pp. 175–194. Blackwell, London.

Behrents, K. C. (1987). The influence of shelter availability on recruitment and early juvenile survivorship of *Lythrypnus dalli* Gilbert (Pisces: Gobiidae). *J. Exp. Mar. Biol. Ecol.* **107,** 45–59.

Beinssen, K. (1988). Boult reef revisited. *Reeflections* **21,** 8–9.

Belk, M. S. (1975). Habitat partitioning in two tropical reef fishes, *Pomacentrus lividus* and *P. albofasciatus*. *Copeia* **1975,** 603–607.

Bell, G. (1989). A comparative method. *Am. Nat.* **133,** 553–571.

Bell, J. D., and Galzin, R. (1984). Influence of live coral cover on coral-reef fish communities. *Mar. Ecol. Prog. Ser.* **15,** 265–274.

Bell, L. J., and Colin, P. L. (1986). Mass spawning of *Caesio teres* (Pisces: Caesionidae) at Enewetak Atoll, Marshall Islands. *Environ. Biol. Fishes* **15,** 69–74.

Bell, J. D., Harmelin-Vivien, M. L., and Galzin, R. (1985). Large scale spatial variation in abundance of butterflyfishes (Chaetodontidae) on Polynesian reefs. *Proc. Int. Coral Reef Congr., 5th* **5,** 421–426.

Bell, J. D., Westoby, M., and Steffe, A. S. (1987). Fish larvae settling in seagrass: Do they discriminate between beds of different leaf density? *J. Exp. Mar. Biol. Ecol.* **111,** 133–144.

Bellwood, D. R. (1988a). Seasonal changes in the size and composition of the fish yield from reefs around Apo Island, central Philippines, with notes on methods of yield estimation. *J. Fish Biol.* **32,** 881–893.

Bellwood, D. R. (1988b). Ontogenetic changes in the diet of early post-settlement *Scarus* species (Pisces: Scaridae). *J. Fish Biol.* **33,** 213–219.

Bellwood, D. R. (1991). A new fossil fish *Phyllopharyngodon longipinnis gen. et. sp. nov.* (family Labridae) from the Eocene, Monte Bolca, Italy. *Studi e Ricerce sui Giacimenti Terziari di Bolca* **6,** In press, Museo Civico die Storia Naturale di Verona.

Bellwood, D. R., and Choat, J. H. (1989). A description of the juvenile phase colour patterns of 24 parrotfish species (family Scaridae) from the Great Barrier Reef, Australia. *Rec. Aust. Mus.* **41,** 1–41.

Bellwood, D. R., and Choat, J. H. (1990). A functional analysis of grazing in parrotfishes (family Scaridae): The ecological implications. *Environ. Biol. Fishes* **28,** 189–214.

Bellwood, D. R., and Schultz, O. (1991). A review of the fossil record of the parrotfishes (Labroidei: Scaridae) with a description of a new *Calotomus* species from the middle Miocene (Badenian) of Austria. *Naturhist. Mus. Wein* **92,** 55–71.

Belyanina, T. N. (1975). Preliminary results of the study of ichthyoplankton of the Caribbean Sea and the Gulf of Mexico. *Proc. P. P. Shirshov Inst. Oceanol.* **100,** 127–146 (Engl. transl. 1981 by Al Akram Cent. Sci. Transl.).

Belyanina, T. N. (1987). Ichthyoplankton of the Gulf of Tonkin (composition, distribution and seasonal changes in populations). *J. Ichthyol. (Engl. Transl.)* **27,** 51–57.

Berger, W. H., Smetacek, V. S., and Wefer, G. (eds.) (1989). "Productivity of the Ocean: Present and Past." Wiley, New York.

Bernays, E. A., Driver, G. C., and Bilgener, M. (1989). Herbivores and plant tannins. *Adv. Ecol. Res.* **19,** 263–302.

Bernstein, B. B., and Jung, N. (1979). Selective pressures and coevolution in a kelp canopy community in southern California. *Ecol. Monogr.* **49,** 335–355.

Bertness, M. D. (1981). Predation, physical stress, and the organization of a tropical rocky intertidal hermit crab community. *Ecology* **62,** 411–425.

Bertness, M. D., Garrity, S. D., and Levings, S. C. (1981). Predation pressure and gastropod foraging: A tropical–temperate comparison. *Evolution (Lawrence, Kans.)* **35,** 995–1007.

Bertram, B. C. R. (1978). Living in groups: Predator and prey. *In* "Behavioural Ecology: An Evolutionary Approach" (J. R. Krebs and N. B. Davies, eds.), pp. 64–96. Blackwell, Oxford, England.

Billheimer, L. E., and Coull, B. C. (1988). Bioturbation and recolonization of meiobenthos in juvenile spot (Pisces) feeding pits. *Estuarine, Coastal Shelf Sci.* **27,** 335–340.

Birkeland, C. (1977). The importance of rate of biomass accumulation in early successional stages of benthic communities to the survival of coral recruits. *Proc. Int. Coral Reef Symp., 3rd* **1,** 15–21.

Birkeland, C. (ed.) (1987). Comparison between Atlantic and Pacific tropical marine coastal ecosystems: Community structure, ecological processes, and productivity. *UNESCO Rep. Mar. Sci.* **46,** 1–262.

Birkeland, C. (1988). Geographic comparisons of coral reef community processes. *Proc. Int. Coral Reef Symp., 6th* **1,** 211–220.

Bjorndal, K. A. (1987). Digestive efficiency in a temperate herbivorous reptile, *Gopherus polyphemus. Copeia* **1987,** 714–720.

Black, K. P., and Gay, S. L. (1987). Hydrodynamic control of the dispersal of Crown-of-Thorns starfish larvae. 1. Small scale hydrodynamics on and around schematized and actual reefs. *Victorian Inst. Mar. Sci. Tech. Rep.* **8,** 1–67 (plus unnumbered appendices).

Black, K. P., Gay, S. L., and Andrews, J. C. (1990). Residence times of neutrally buoyant matter such as larvae, sewage or nutrients on coral reefs. *Coral Reefs* **9,** 105–114.

Blaxter, J. H. S. (1975). Reared and wild fish—How do they compare? *Eur. Symp. Mar. Biol., 10th* pp. 11–26.

Blaxter, J. H. S. (1986). Development of sense organs and behaviour of teleost larvae with special reference to feeding and predator avoidance. *Trans. Am. Fish. Soc.* **115,** 98–114.

Blaxter, J. H. S. (1988). Sensory performance, behavior, and ecology of fish. *In* "Sensory Biology of Aquatic Animals" (J. Atema, R. R. Fay, A. N. Popper, and W. N. Tavolga, eds.), pp. 203–232. Springer-Verlag, New York.

Blot, J. (1980). La fauna ichthyologique des gisements du Monte Bolca (Province de Vérone, Italy). Catalogue systematique présentant l'état actual des recherches concernant cette fauna. *Bull. Mus. Nat. Hist.* **2**(C;4), 339–396.

Blot, J., and Voruz, C. (1975). La famille de Zanclidae. *In* "Studi e Ricerche sui Giacimenti Terziari di Bolca," Vol. II, pp. 233–271. Mus. Civ. Storia Nat. Verona, Verona, Italy.

Blueweiss, L., Fox, H., Kudzma, V., Nakashima, D., Peters, R., and Sams, S. (1978). Relationships between body size and some life history parameters. *Oecologia* **37,** 257–272.

Bodkin, J. L., Van Blaricom, G. R., and Jameson, R. J. (1987). Mortalities of kelp forest fishes associated with large oceanic waves off central California, 1982–1983. *Environ. Biol. Fishes* **18,** 73–76.

Boehlert, G. W. (1977). Timing of the surface-to-benthic migration in juvenile rockfish, *Sebastes diploproa*, off southern California. *Fish. Bull.* **75,** 887–890.

Boehlert, G. W. (1986). An approach to recruitment research in insular ecosystems. *In* "IOC/FAO Workshop on Recruitment in Tropical Coastal Demersal Communities" (D. Pauly and A. Yanez-Arancibia, eds.), I0C Workshop Rep. 40 (suppl.), pp. 15–32. UNESCO, Paris.

Bohlke, J. E., and Chaplin, C. C. G. (1968). "Fishes of the Bahamas and Adjacent Territorial Waters." Livingston, Wynnewood, Pennsylvania.

Bohnsack, J. A. (1982). Effects of piscivorous predator removal on coral reef fish community structure. *In* "Gutshop '81: Fish Food Habits Studies" (G. M. Cailliet and C. A. Simenstad, eds.), pp. 258–267. Wash. Sea Grant Publ., Seattle, Washington.

Bohnsack, J. A. (1983a). Resiliency of reef fish communities in the Florida Keys following a January 1977 hypothermal fish kill. *Environ. Biol. Fishes* **9,** 41–53.

Bohnsack, J. A. (1983b). Species turnover and the order versus chaos controversy concerning reef fish community structure. *Coral Reefs* **1,** 223–228.

Bohnsack, J. A. (1989). Are high densities of fishes at artificial reefs the result of habitat limitation or behavioral preference? *Bull. Mar. Sci.* **44,** 631–645.

Bohnsack, J. A. (1991) Habitat structure and the design of artificial reefs. *In* "Habitat Structure: The Physical Arrangement of Objects in Space" (S. S. Bell, E. D. McCoy, and H. R. Mushinsky, eds.), pp. 412–426. Chapman & Hall, New York.

Bohnsack, J. A., and Bannerot, S. P. (1986). A stationary visual technique for quantitatively assessing community structure of coral reef fishes. *NOAA Tech. Rep. NMFS* **41,** 1–15.

Bohnsack, J. A., and Talbot, F. H. (1980). Species packing by reef fishes on Australian and Caribbean reefs: An experimental approach. *Bull. Mar. Sci.* **30,** 710–723.

Bortone, S. A. (1976). Effects of a hurricane on the fish fauna at Destin, Florida. *Fl. Sci.* **39,** 245–248.

Bouchet, P. (1981). Evolution of larval development in eastern Atlantic Terebridae (Gastropoda), Neogene to Recent. *Malacologia* **21,** 363–369.

Bouchon-Navaro, Y. (1980). Quantitative distribution of the Chaetodontidae on a fringing reef of the Jordanian coast (Gulf of Aqaba, Red Sea). *Tethys* **9,** 247–251.

Bouchon-Navaro, Y. (1981). Quantitative distribution of the Chaetodontiadae on a reef of Moorea Island (French Polynesia). *J. Exp. Mar. Biol. Ecol.* **55,** 145–157.

Bouchon-Navaro, Y. (1986). Partitioning of food and space resources by chaetodontid fishes on coral reefs. *J. Exp. Mar. Biol. Ecol.* **103,** 21–40.

Bouchon-Navaro, Y., and Bouchon, C. (1989). Correlations between chaetodontid fishes and coral communities of the Gulf of Aqaba (Red Sea). *Environ. Biol. Fishes* **25,** 1–3.

Bouchon-Navaro, Y., and Harmelin-Vivien, M. L. (1981). Quantitative distribution of herbivorous fishes in the Gulf of Aqaba (Red Sea). *Mar. Biol.* **63,** 79–86.

Bouchon-Navaro, Y., Bouchon, C., and Harmelin-Vivien, M. L. (1985). Impact of coral degradation on a chaetodontid fish assemblage (Moorea, French Polynesia). *Proc. Int. Coral Reef Congr., 5th* **5,** 427–432.

Bourret, P., Binet, D., Hoffschir, C., Rivaton, J., and Velayoudon, H. (1979). "Evaluation de 'l' effet d'île' d'un atoll: Plancton et micronecton au large de Mururoa (Tuamotus)." Cent. ORSTOM Noumea, New Caledonia, French Polynesia.

Bowen, H. M. (1968). Diver performance and effects of cold. *Hum. Factors* **10,** 445–463.

Bowen, S. H. (1979). A nutritional constraint in detritivory by fishes: The stunted population of *Sarotherodon mossambicus* in Lake Sibaya, South Africa. *Ecol. Monogr.* **47,** 17–31.

Bowen, S. H. (1984). Detrital amino acids and the growth of *Sarotherodon mossambicus*— A reply to Dabrowski. *Acta Hydrochim. Hydrobiol.* **12,** 55–59.

Boyce, M. S. (1979). Seasonality and patterns of natural selection for life histories. *Am. Nat.* **114,** 569–583.

Bradbury, R. H., and Goeden, G. B. (1974). The partitioning of the reef slope environment by resident fishes. *Proc. Int. Coral Reef Symp., 2nd* **1,** 167–178.

Bradbury, R. H., and Reichelt, R. E. (1983). Fractal dimension of a coral reef at ecological scales. *Mar. Ecol. Prog. Ser.* **10,** 169–171.

Bradshaw, A. D. (1965). Evolutionary significance of phenotypic plasticity in plants. *Adv. Genet.* **13,** 115–155.

Bray, R. N. (1981). Influence of water currents and zooplankton densities on daily foraging movements of blacksmith, *Chromis punctipinnis*, a planktivorous reef fish. *Fish. Bull.* **78,** 829–841.

Bray, R. N., and Ebeling, A. W. (1975). Food, activity, and habitat of three "picker-type" microcarnivorous fishes in the kelp forests off Santa Barbara, California. *Fish. Bull.* **73,** 815–829.

Bray, R. N., and Hixon, M. A. (1978). Night-shocker: Predatory behavior of the Pacific electric ray (*Torpedo californica*). *Science* **200,** 333–334.

Bray, R. N., Miller, A. C., and Geesey, G. G. (1981). The fish connection: A trophic link between planktonic and rocky reef communities. *Science* **214,** 204–205.

Bray, R. N., Purcell, L. J., and Miller, A. C. (1986). Ammonium excretion in a temperate-reef community by a planktivorous fish, *Chromis punctipinnis* (Pomacentridae), and potential uptake by young giant kelp, *Macrocystis pyrifera* (Laminariales). *Mar. Biol.* **90,** 327–334.

Breder, C. M. (1949). On the taxonomy and the postlarval stages of the surgeon-fish, *Acanthurus hepatus*. *Copeia* **1949,** 296.

Breen, P. A., and Mann, K. H. (1976). Destructive grazing of kelp by sea urchins in eastern Canada. *J. Fish. Res. Board Can.* **33,** 1278–1283.

Breitburg, D. L. (1984). Residual effects of grazing: Inhibition of competitor recruitment by encrusting coralline algae. *Ecology* **65,** 1136–1143.

Breitburg, D. L. (1987a). Interspecific competition and the abundance of nest sites: Factors affecting sexual selection. *Ecology* **68,** 1844–1855.

Breitburg, D. L. (1987b). Larval schooling behavior of a non-schooling benthic fish species. *Eos* **68,** 1750.

Breitburg, D. L. (1989). Demersal schooling prior to settlement by larvae of the naked goby. *Environ. Biol. Fishes* **26,** 97–103.

Briggs, J. C. (1974). "Marine Zoogeography." McGraw-Hill, New York.

Briggs, J. C. (1985). [Species richness among the tropical shelf regions.] *Mar. Biol. (U.S.S.R.)* **6,** 3–10.

Brock, R. E. (1979). An experimental study on the effects of grazing by parrotfishes and role of refuges in benthic community structure. *Mar. Biol.* **51,** 381–388.

Brock, R. E. (1982). A critique of the visual census method for assessing coral reef fish populations. *Bull. Mar. Sci.* **32,** 269–276.

Brock, V. E. (1954). A method of estimating reef fish populations. *J. Wildl. Manage.* **18,** 297–308.

Brock, V. E., and Chamberlain, T. C. (1968). A geological and ecological reconnaissance off western Oahu, Hawaii, principally by means of the research submarine "Asherah." *Pac. Sci.* **22,** 373–401.

Brock, R. E., Lewis, C., and Wass, R. C. (1979). Stability and structure of a fish community on a coral patch reef in Hawaii. *Mar. Biol.* **54,** 281–292.

Brooks, J. L., and Dodson, S. I. (1965). Predation, body size, and composition of plankton. *Science* **150,** 28–35.

Brothers, E. B. (1980). Age and growth studies in tropical fishes. *In* "Stock Assessment for Tropical Small-Scale Fisheries" (S. B. Saila and P. M. Roedel, eds.), pp. 119–136. Univ. of Rhode Island, Kingston, Rhode Island.

Brothers, E. B. (1984). Otolith studies. *In* "Ontogeny and Systematics of Fishes" (H. G. Moser, ed.), pp. 50–57. Allen, Lawrence, Kansas.

Brothers, E. B., and McFarland, W. N. (1981). Correlations between otolith microstructure, growth and life history transitions in newly recruited French grunts [*Haemulon flavolineatum* (Desmarest), Haemulidae]. *Rapp. P.-V. Reun., Cons. Int. Explor. Mer* **178,** 369–374.

Brothers, E. B., and Thresher, R. E. (1985). Pelagic duration, dispersal, and the distribution of Indo-Pacific coral-reef fishes. *In* "The Ecology of Coral Reefs" (M. L. Reaka, ed.), NOAA Symp. Ser. Undersea Res., Vol. 3, pp. 53–69. Natl. Oceanic Atmos. Adm., Rockville, Maryland.

Brothers, E. B., Williams, D. M., and Sale, P. F. (1983). Length of larval life in twelve families of fishes at "One Tree Lagoon", Great Barrier Reef, Australia. *Mar. Biol.* **76,** 319–324.

Brouard, F., and Grandperrin, R. (1985). Deep-bottom fishes of the outer reef slope of Vanuatu. *Working Pap. South Pac. Comm., 17th Region. Tech. Meet. Fish., SPC/Fish.* **17/WP.12,** 1–127.

Brown, J. H. (1984). On the relationship between abundance and distributions of species. *Am. Nat.* **124,** 255–279.

Brown, J. L. (1964). The evolution of diversity in avian territorial systems. *Wilson Bull.* **76,** 293–329.

Brown, J. L., and Orians, G. H. (1970). Spacing patterns in mobile animals. *Annu. Rev. Ecol. Syst.* **1,** 239–262.

Bryan, P. G., and Madraisau, B. B. (1977). Larval rearing and development of *Siganus lineatus* (Pisces: Siganidae) from hatching through metamorphosis. *Aquaculture* **10,** 243–252.

Buckman, N. S., and Ogden, J. C. (1973). Territorial behavior of the striped parrotfish *Scarus croicensis* Bloch (Scaridae). *Ecology* **54,** 1377–1382.

Burge, R. T., and Schultz, S. A. (1973). The marine environment in the vicinity of Diable Cove with special reference to abalones and bony fishes. *Tech. Rep.—Calif., Dep. Fish Game Mar. Resour.* **19,** 1–433.

Burgess, W. E. (1965). Larvae of the surgeonfish genus *Acanthurus* of the tropical western Atlantic. Unpublished M.S. thesis, University of Miami, Miami, Florida.

Burgess, W. E. (1978). "Butterflyfishes of the World." TFH Publ., Neptune City, New Jersey.

Buri, P., and Kawamura, G. (1983). The mechanics of mass occurrence and recruitment strategy of milkfish *Chanos chanos* (Forsskal) fry in the Philippines. *Mem. Kagoshima Univ. Res. Cent. South Pac.* **3,** 33–55.

Burnett-Herkes, J. N. (1975). Contribution to the biology of the red hind, *Epinephelus guttatus*, a commercially important serranid fish from the tropical Western Atlantic. Unpublished Ph.D. thesis, University of Miami, Miami, Florida.

Burt, A., Kramer, D. L., Nakatsuru, K., and Spry, C. (1988). The tempo of reproduction in *Hyphessobrycon pulchripinnis* (Characidae), with a discussion of the biology of "multiple spawning" in fishes. *Environ. Biol. Fishes* **22,** 15–27.

Buss, L. W., and Jackson, J. B. C. (1979). Competitive networks: Nontransitive competitive relationships in cryptic coral reef environments. *Am. Nat.* **113,** 223–234.

Butman, C. A. (1987). Larval settlement of soft sediment invertebrates: The spatial scales of pattern explained by active habitat selection and the emerging role of hydrodynamical processes. *Oceanogr. Mar. Biol.* **25,** 113–165.

Caffey, H. M. (1985). Spatial and temporal variation in settlement and recruitment of intertidal barnacles. *Ecol. Monogr.* **55,** 313–332.

Cahill, M. M. (1990). Bacterial flora of fishes: A review. *Microb. Ecol.* **19,** 21–41.

Caldwell, D. K. (1962). Development and distribution of the short bigeye *Pseudopriacanthus altus* (Gill), in the western North Atlantic. *Fish. Bull.* **62,** 103–150.

Caldwell, M. C. (1962). Development and distribution of larval and juvenile fishes of the family Mullidae of the western North Atlantic. *Fish. Bull.* **62,** 403–457.

Campana, S. E., and Neilson, J. D. (1985). Microstructure of fish otoliths. *Can. J. Fish. Aquat. Sci.* **42,** 1014–1032.

Campana, S. E., Gagne, J. A., and Munro, J. (1987). Otolith microstructure of larval herring (*Clupea harengus*): Image or reality? *Can. J. Fish. Aquat. Sci.* **44,** 1922–1929.

Cardellina, J. H., II, Marner, F. J., and Moore, R. E. (1979). Seaweed dermatitis: Structure of lyngbyatoxin A. *Science* **204,** 193–195.

Carlson, B. A. (1975). A preliminary checklist of the fishes of Fiji. Unpublished manuscript (plus 1985 update).

Carlton, J. T. (1985). Transoceanic and interoceanic dispersal of coastal marine organisms: The biology of ballast water. *Oceanogr. Mar. Biol.* **23,** 313–371.

Carlton, J. T. (1987). Patterns of transoceanic marine biological invasions in the Pacific Ocean. *Bull. Mar. Sci.* **41,** 452–465.

Carpenter, K. E. (1977). Philippine coral reef fisheries resources. *Philipp. J. Fish.* **15,** 95–126.

Carpenter, R. C. (1985). Sea urchin mass mortality: Effects on reef algal abundance, species composition, and metabolism and other coral reef herbivores. *Proc. Int. Coral Reef Congr., 5th* **4,** 53–60.

Carpenter, R. C. (1986). Partitioning herbivory and its effects on coral reef algal communities. *Ecol. Monogr.* **56,** 345–363.

Carpenter, R. C. (1988). Mass mortality of a Caribbean sea urchin: Immediate effects on community metabolism and other herbivores. *Proc. Natl. Acad. Sci. U.S.A.* **85,** 511–514.

Carpenter, R. C.(1990). Mass mortality of *Diadema antillarum*. II. Effects on population densities and grazing intensity of parrotfishes and surgeonfishes. *Mar. Biol.* **104,** 79–86.

Carpenter, K. E., and Alcala, A. C. (1977). Philippine coral reef fisheries resources. Part II. Muro-ami and kayakas reef fisheries, benefit or bane? *Philipp. J. Fish.* **15,** 217–235.

Carpenter, K. E., Miclat, R. I., Albaladejo, V. D., and Corpuz, V. T. (1981). The influence of substrate structure on the local abundance and diversity of Philippine reef fishes. *Proc. Int. Coral Reef Symp., 4th* **2,** 497–502.

Carr, M. H. (1989). Effects of macroalgal assemblages on the recruitment of temperate zone reef fishes. *J. Exp. Mar. Biol. Ecol.* **126,** 59–76.

Carroll, R. L. (1987). "Vertebrate Paleontology and Evolution." Freeman, New York.

Carter, J. (1989). Grouper sex in Belize. *Nat. Hist.* **10,** 61–68.

Carter, J. W., Carpenter, A. L., Foster, M. S., and Jessee, W. N. (1985). Benthic succession on an artificial reef designed to support a kelp-reef community. *Bull. Mar. Sci.* **37,** 86–113.

Castilla, J. C. (1988). Earthquake-caused coastal uplift and its effects on rocky intertidal kelp communities. *Science* **242,** 440–443.

Caswell, H. (1978). Predator-mediated coexistence: A nonequilibrium model. *Am. Nat.* **112,** 127–154.

Cerri, R. D. (1983). The effects of light intensity on predator and prey behavior in cyprinid fish: Factors that influence prey risk. *Anim. Behav.* **31,** 736–742.

Chambers, R. C., and Leggett, W. C. (1987). Size and age at metamorphosis in marine fishes: An analysis of laboratory-reared winter flounder (*Pseudopleuronectes americanus*) with a review of variation in other species. *Can. J. Fish. Aquat. Sci.* **44,** 1936–1947.

Chapman, A. R. O. (1981). Stability of sea-urchin dominated barren grounds following destructive grazing of kelp in St. Margaret's Bay, eastern Canada. *Mar. Biol.* **62,** 307–311.

Chapman, V. J. (1970). "Seaweeds and Their Uses." Methuen, London.

Charnov, E. L. (1982). "The Theory of Sex Allocation." Princeton Univ. Press, Princeton, New Jersey.

Chase, I. D. (1982). Dynamics of hierarchy formation: The sequential development of dominance relationships. *Behaviour* **80,** 218–240.

Chase, I. D. (1985). The sequential analysis of aggressive acts during hierarchy formation: An application of the "jigsaw puzzle" approach. *Anim. Behav.* **33,** 86–100.

Chave, E. H. (1978). General ecology of six species of Hawaiian cardinalfishes. *Pac. Sci.* **32,** 245–270.

Chave, E. H., and Eckert, D. B. (1974). Ecological aspects of the distribution of fishes at Fanning Island. *Pac. Sci.* **28,** 297–317.

Checkley, D. M., Jr., Raman, S., Maillet, G. L., and Mason, K. M. (1988). Winter storm effects on the spawning and larval drift of a pelagic fish. *Nature (London)* **335,** 346–348.

Chen, Z., and Wei, S. (1978). A preliminary investigation on pelagic fish eggs and larvae from the waters around the Xisha Islands and Zhongsha Islands in the South China Sea. *In* "Investigation Report of Chinese Studies of the Biology of the Ocean near the Xisha and Zhongsha Islands," pp. 295–320. Science Press, Beijing, China (in Chinese; Engl. abstr.).

Chen, Z., and Wei, S. (1982). An investigation on pelagic fish eggs and larvae of the central area of South China Sea. *In* "Symposium on Research Reports on the Sea Area of South China Sea," pp. 251–268. Science Press, Beijing, China (in Chinese; Engl. abstr.).

Chess, J. R., Smith, S. E., and Fisher, P. C. (1988). Trophic relationships of the shortbelly rockfish, *Sebastes jordani*, off central California. *CalCOFI Rep.* **24,** 129–136.

Chesson, P. L. (1985). Coexistence of competitors in spatially and temporally varying environments: A look at the combined effects of different sorts of variability. *Theor. Pop. Biol.* **28,** 263–287.

Chesson, P. L. (1986). Environmental variation and the coexistence of species. *In* "Community Ecology" (J. Diamond and T. J. Case, eds.), pp. 240–256. Harper & Row, New York.

Chesson, P. L., and Warner, R. R. (1981). Environmental variability promotes coexistence in lottery competitive systems. *Am. Nat.* **117,** 923–943.

Choat, J. H. (1968). Feeding habits and distribution of *Plectropomus maculatus* (Serranidae) at Heron Island. *Proc. R. Soc. Queensl.* **80,** 13–18.

Choat, J. H. (1969). Studies on Labroid fishes. Part 2. A comparative study of the ecology of the Labridae and Scaridae. PhD dissertation. University of Queensland.

Choat, J. H. (1982). Fish feeding and the structure of benthic communities in temperate waters. *Annu. Rev. Ecol. Syst.* **13,** 423–449.

Choat, J. H., and Ayling, A. M. (1987). The relationship between habitat structure and fish faunas on New Zealand reefs. *J. Exp. Mar. Biol. Ecol.* **110,** 257–284.

Choat, J. H., and Bellwood, D. R. (1985). Interactions amongst herbivorous fishes on a coral reef: Influence of spatial variation. *Mar. Biol.* **89,** 221–234.

Choat, J. H., and Kingett, P. D. (1982). The influence of fish predation on the abundance cycles of an algal turf invertebrate fauna. *Oecologia* **54,** 88–95.

Choat, J. H., Ayling, A. M., and Scheil, D. R. (1988). Temporal and spatial variation in an island fish fauna. *J. Exp. Mar. Biol. Ecol.* **121,** 91–111.

Choat, J. H., Doherty, P. J., Kerrigan, B. A., and Leis, J. M. (1991). Larvae and pelagic young of coral reef fishes: Comparison of three towed nets, a purse seine and two light aggregation devices. *Fish. Bull.* In press.

Christy, J. H. (1978). Adaptive significance of reproductive cycles in the fiddler crab *Uca pugilator*: A hypothesis. *Science* **199,** 453–455.

Clarke, R. D. (1977). Habitat distribution and species diversity of chaetodontid and pomacentrid fishes near Bimini, Bahamas. *Mar. Biol.* **40,** 277–289.

Clarke, R. D. (1988). Chance and order in determining fish species composition on small coral patches. *J. Exp. Mar. Biol. Ecol.* **115,** 197–212.

Clarke, R. D. (1989). Population fluctuation, competition and microhabitat distribution of two species of tube blennies, *Acanthemblemaria* (Teleostei: Chaenopsidae). *Bull. Mar. Sci.* **44,** 1174–1185.

Clarke, T. A. (1970). Territorial behavior and population dynamics of a pomacentrid fish, the garibaldi, *Hypsypops rubicunda. Ecol. Monogr.* **40,** 189–212.

Clavijo, I. (1982). Aspects of the reproductive biology of the redband parrotfish *Sparisoma aurofrenatum.* Unpublished Ph.D. thesis, University of Puerto Rico, Mayaguez, Puerto Rico.

Clements, K. D., and Bellwood, D. R. (1988). A comparison of the feeding mechanisms of two herbivorous labroid fishes, the temperate *Odax pullus* and the tropical *Scarus rubroviolaceus. Aust. J. Mar. Freshwater Res.* **39,** 87–107.

Clements, K. D., Sutton, D. C., and Choat, J. H. (1989). Occurrence and characteristics of unusual protistan symboints from surgeon fishes (Acanthuridae) of the Great Barrier Reef, Australia. *Mar. Biol.* **102,** 403–412.

Clepper, H. E. (ed.) (1979). "Predator–Prey Systems in Fisheries Management." Sport Fishing Inst., Washington, D.C.

Clifton, K. E. (1989). Territory sharing by the Caribbean striped parrotfish, *Scarus iserti*: Patterns of resource abundance, group size and behaviour. *Anim. Behav.* **37,** 90–103.

Clutton-Brock, T., and Harvey, P. (1984). Comparative approaches to investigating adaptation. *In* "Behavioural Ecology: An Evolutionary Approach", 2nd edition (J. R. Krebs and N. B. Davies, eds.), pp. 7–29. Sinauer, Sunderland, Massachusetts.

Coates, D. (1980). Prey-size intake in humbug damselfish, *Dascyllus aruanus* (Pisces, Pomacentridae) living within social groups. *J. Anim. Ecol.* **49,** 335–340.

Coates, D. (1982). Some observations on the sexuality of humbug damselfish, *Dascyllus aruanus* (Pisces, Pomacentridae) in the field. *Z. Tierpsychol.* **59,** 7–18.

Cobb, J. S., and Rooney, R. (1987). Swimming by postlarval lobsters: Significance to patterns of dispersal. *Eos* **68,** 1750–1751.

Coe, W. R. (1953). Resurgent populations of littoral marine invertebrates and their dependence on ocean currents and tidal currents. *Ecology* **34,** 225–229.

Cole, K. S., and Robertson, D. R. (1988). Protogyny in the Caribbean reef goby, *Coryphopterus personatus*: Gonad ontogeny and social influences on sex-change. *Bull. Mar. Sci.* **42,** 317–333.

Coley, P. D., Bryant, J. P., and Chapin, F. S., III (1985). Resource availability and plant antiherbivore defense. *Science* **230,** 895–899.

Colin, P. L. (1975). "The Neon Gobies." TFH Publ., Neptune City, New Jersey.
Colin, P. L. (1976). Filter feeding and predation on the eggs of *Thalassoma* sp. by the scombrid fish *Rastrelliger kanagurta*. *Copeia* **1976,** 596–597.
Colin, P. L. (1978). Daily and summer–winter variation in mass spawning of the striped parrotfish, *Scarus croicensis*. *Fish. Bull.* **76,** 117–124.
Colin, P. L. (1982). Spawning and larval development of the hogfish, *Lachnolaimus maximus* (Pisces: Labridae). *Fish. Bull.* **80,** 853–862.
Colin, P. L. (1989). Aspects of the spawning of western Atlantic butterflyfishes (Pisces: Chaetodontidae). *Environ. Biol. Fishes* **25,** 131–141.
Colin, P. L., and Bell, L. J. (1991). Aspects of the spawning of labrid and scarid fishes (Pisces: Labroidei) at Enewetak Atoll, Marshall Islands with notes on other families. *Environ. Biol. Fishes* **31,** 229–260.
Colin, P. L., and Clavijo, I. E. (1978). Mass spawning by the spotted goatfish, *Pseudupeneus maculatus* (Bloch) (Pisces: Mullidae). *Bull. Mar. Sci.* **28,** 780–782.
Colin, P. L., and Clavijo, I. E. (1988). Spawning activity of fishes producing pelagic eggs on a shelf edge coral reef, southwestern Puerto Rico. *Bull. Mar. Sci.* **43,** 249–279.
Colin, P. L., Shapiro, D. Y., and Weiler, D. (1987). Aspects of the reproduction of two groupers, *Epinephelus guttatus* and *E. striatus,* in the West Indies. *Bull. Mar. Sci.* **40,** 220–230.
Collette, B. B., and Talbot, F. H. (1972). Activity patterns of coral reef fishes with emphasis on nocturnal–diurnal changeover. *Bull. Nat. Hist. Mus. Los Angeles County* **14,** 98–124.
Collins, S. P., and Pettigrew, J. D. (1988a). Retinal topography in reef fishes. I. Some species with well-developed areae but poorly developed streaks. *Brain Behav. Evol.* **31,** 269–282.
Collins, S. P., and Pettigrew, J. D. (1988b). Retinal topography in reef fishes. II. Some species with prominent horizontal streaks and high density areae. *Brain Behav. Evol.* **31,** 282–295.
Colwell, R. K., and Fuentes, E. R. (1975). Experimental studies of the niche. *Annu. Rev. Ecol. Syst.* **6,** 281–310.
Conacher, M. J., Lanzing, W. J. R., and Larkum, A. W. D. (1979). The ecology of Botany Bay. II. Aspects of the feeding ecology of the fan bellied leatherjacket *Monacanthus chinensis* in *Posidonia australis* seagrass beds in Quibray, Botany Bay, N.S.W. *Aust. J. Mar. Freshwater Res.* **30,** 387–400.
Confer, J. L., and Blades, P. I. (1975). Omnivorous zooplankters and planktivorous fish. *Limnol. Oceanogr.* **20,** 571–579.
Connell, J. H. (1974). Ecology: Field experiments in marine ecology. *In* "Experimental Marine Biology" (R. N. Mariscal, ed.), pp. 21–54. Academic Press, New York.
Connell, J. H. (1975). Some mechanisms producing structure in natural communities: A model and evidence from field experiments. *In* "Ecology and Evolution of Communities" (M. L. Cody and J. M. Diamond, eds.), pp. 460–490. Belknap–Harvard Univ. Press, Cambridge, Massachusetts.
Connell, J. H. (1978). Diversity in tropical rain forests and coral reefs. *Science* **199,** 1302–1310.

Connell, J. H. (1980). Diversity and coevolution of competitors, or the ghost of competition past. *Oikos* **35,** 131–138.
Connell, J. H. (1983). On the prevalence and relative importance of interspecific competition: Evidence from field experiments. *Am. Nat.* **122,** 661–696.
Connell, J. H. (1988). Maintenance of species diversity in biotic communities. *In* "Evolution and Coadaptation in Biotic Communities" (S. Kawano, J. H. Connell, and T. Hidaka, eds.), pp. 201–218. Univ. of Tokyo Press, Tokyo.
Conover, D. O., and Kynard, B. E. (1984). Field and laboratory observations of spawning periodicity and behavior of a northern population of the Atlantic silverside, *Menidia menidia* (Pisces: Atherinidae). *Environ. Biol. Fishes* **11,** 161–171.
Cook, S. B. (1980). Fish predation on pulmonate limpets. *Veliger* **22,** 380–381.
Corbin, J. S. (1977). Laboratory derived nitrogen and energy budgets for a juvenile goatfish, *Parapeneus porphyreus*, fed brine shrimp, *Artemia salina*, with a description of the pattern of growth in wild fish. Unpublished M.S. thesis, University of Hawaii, Honolulu, Hawaii.
Cott, H. B. (1940). "Adaptive Colouration in Animals." Methuen, London.
Coughlin, D. J., and Strickler, J. R. (1990). Zooplankton capture by a coral reef fish: An adaptive response to evasive prey. *Environ. Biol. Fishes* **29,** 35–42.
Cowen, R. K. (1985). Large scale pattern of recruitment by the labrid, *Semicossyphus pulcher:* Causes and implications. *J. Mar. Res.* **43,** 719–742.
Cox, E. F. (1986). The effects of a selective corallivore on growth rates and competition for space between two species of Hawaiian corals. *J. Exp. Mar. Biol. Ecol.* **101,** 161–174.
Coyer, J. A. (1979). The invertebrate assemblage associated with *Macrocystis pyrifera* and its utilization as a food resource by kelp forest fishes. Unpublished Ph.D. dissertation, University of Southern California, Los Angeles, California.
Craik W. J. S. (1981a). Recreational fishing on the Great Barrier Reef. *Proc. Int. Coral Reef Symp., 4th* **1,** 47–52.
Craik, W. J. S. (1981b). Underwater survey of coral trout *Plectropomus leopardus* (Serranidae) populations in the Capricornia section of the Great Barrier Reef marine park. *Proc. Int. Coral Reef Symp., 4th* **1,** 53–58.
Craik, W. J. S. (1986). Recreational fishing on the Great Barrier Reef: Research findings *In* "Fisheries Management. Theory and Practice in Queensland" (T. J. A. Hundloe, ed.), pp. 178–198. Griffith Univ. Press, Brisbane, Australia.
Crawford, T. J. (1984). What is a population? *In* "Evolutionary Ecology" (B. Shorrocks, ed.), pp. 135–173. Blackwell, London.
Crescitelli, F. (1991). Adaptations of visual pigments to the photic environment of the deep sea. *J. Exp. Zool.* (in press).
Cross, J. N. (1982). Resource partitioning in three rocky intertidal fish assemblages. *In* "Gutshop '81: Fish Food Habits Studies" (G. M. Cailliet and C. A. Simenstad, eds.), pp. 142–150. Wash. Sea Grant Publ., Seattle, Washington.
Cummings, W. C. (1968). Reproductive habits of the sergeant major, *Abudefduf saxatilis* (Pisces: Pomacentridae) with comparative notes on four other damselfishes in the Bahama Islands. Unpublished Ph.D. thesis, University of Miami, Miami, Florida.

Curio, E. (1976). "The Ethology of Predation." Springer-Verlag, Berlin.
Cushing, D. H. (1977). The problems of stock and recruitment. *In* "Fish Population Dynamics" (J. A. Gulland, ed.), pp. 116–135. Wiley, Toronto, Ontario, Canada.
Cushing, D. H. (1982). "Climate and Fisheries." Academic Press, New York.
Dale, G. (1978). Money-in-the-bank: A model for coral reef fish coexistence. *Environ. Biol. Fishes* **3,** 103–108.
Dalzell, P. (1988). Small pelagic fisheries investigations in the Philippines. Part 1: History of the fishery. *Fishbyte* **6,** 2–4.
Dalzell, P., and Wright, A. (1986). An assessment of the exploitation of coral reef fishery resources of Papua New Guinea. *In* "First Asian Fisheries Forum" (J. L. Maclean, C. B. Dizon, and L. V. Hosillos, eds.), pp. 477–481. Asian Fish. Soc., Manila, Philippines.
Dartnall, H. J. A., and Lythgoe, J. N. (1965). The spectral clustering of visual pigments. *Vision Res.* **5,** 81–100.
Davies, P. J., Marshall, J. F., and Hopley, D. (1985). Relationships between reef growth and sea level in the Great Barrier Reef. *Proc. Int. Coral Reef Congr., 5th* **3,** 95–103.
Davis, G. E. (1986). Kelp forest dynamics in Channel Islands National Park, California 1982–85. *Channel Islands Nat. Park Nat. Mar. Sanct. Nat. Sci. Study Rep.* **CHIS-86-001,** 1–11.
Davis, G. E., and Anderson, T. W. (1989). Population estimates of four kelp forest fishes and an evaluation of three *in situ* assessment techniques. *Bull. Mar. Sci.* **44,** 1138–1151.
Davis, R. E., and Northcutt, R. G. (eds.) (1983). "Fish Neurobiology," Vols. 1 and 2. Univ. of Michigan Press, Ann Arbor, Michigan.
Davis, W. P., and Birdsong, R. S. (1973). Coral reef fishes which forage in the water column. *Helgol. Wiss. Meeresunters.* **24,** 292–306.
Davis, B. J., DeMartini, E. E., and McGee, K. (1981). Gene flow among populations of a teleost (painted greenling, *Oxylebius pictis*) from Puget Sound to southern California. *Mar. Biol.* **65,** 17–23.
Day, R. W. (1977). Two contrasting effects of predation on species richness in coral reef habitats. *Mar. Biol.* **44,** 1–5.
Day, R. W. (1985). The effects of refuges from predators and competitors on sessile communities on a coral reef. *Proc. Int. Coral Reef Congr., 5th* **4,** 41–45.
Dayton, P. K. (1979). Ecology: A science and a religion. *In* "Ecological Processes in Coastal and Marine Systems" (R. J. Livingstone, ed.), pp. 3–18. Plenum, New York.
Dayton, P. K. (1984). Processes structuring some marine communities: Are they general? *In* "Ecological Communities: Conceptual Issues and the Evidence" (D. R. Strong, Jr., D. Simberloff, L. G. Abele, and A. B. Thistle, eds.), pp. 181–197. Princeton Univ. Press, Princeton, New Jersey.
Dayton, P. K. (1985). Ecology of kelp communities. *Annu. Rev. Ecol. Syst.* **16,** 215–245.
Dayton, P. K., and Tegner, M. J. (1984a). Catastrophic storms, El Niño, and patch stability in a southern California kelp community. *Science* **224,** 283–285.
Dayton, P. K., and Tegner, M. J. (1984b). The importance of scale in community

ecology: A kelp forest with terrestrial analogs. *In* "A New Ecology: Novel Approaches to Interactive Systems" (P. W. Price, C. N. Slobodchikoff, and W. S. Gaud, eds.), pp. 457–481. Wiley, New York.

Dayton, P. K., Currie, V., Gerrodette, T., Keller, B. D., Rosenthal, R., and Ven Tresca, D. (1984). Patch dynamics and stability of some California kelp communities. *Ecol. Monogr.* **54,** 253–289.

de Boer, B. A. (1978). Factors influencing the distribution of the damselfish *Chromis cyanea* (Poey), Pomacentridae, on a reef at Curaçao, Netherlands Antilles. *Bull. Mar. Sci.* **28,** 550–565.

de Ciechomski, J. D., and Sanchez, R. P. (1984). Field estimates of embryonic mortality of Southwest Atlantic anchovy (*Engraulis anchoita*). *Meeresforschung* **30,** 172–187.

Deevey, E. S. (1947). Life tables for natural populations of animals. *Q. Rev. Biol.* **22,** 283–314.

De Jong, G. (1979). The influence of the distribution of juveniles over patches of food on the dynamics of a population. *Neth. J. Zool.* **29,** 33–51.

Dekhnik, T. V., Haures, M., and Salabariya, D. (1966). Distribution of pelagic eggs and larvae of fishes in the coastal waters of Cuba. *Akad. Nauk SSSR, Inst. Biol. Yuzhnikh Morei Im. Akad. A. O. Kovalevskogo* pp. 189–241 (Engl. transl. U.S. Dep. Commer. TT70-57762).

DeMartini, E. E., and Anderson, M. E. (1978). Comparative survivorship and life-history of painted greenling (*Oxylebius pictus*) in Puget Sound, Washington and Monterey Bay, California. *Environ. Biol. Fishes* **5,** 33–47.

DeMartini, E. E., and Fountain, R. K. (1981). Ovarian cycling frequency and batch fecundity in the queenfish, *Seriphus politus*: Attributes representative of serial spawning fishes. *Fish. Bull.* **79,** 547–560.

DeMartini, E. E., and Roberts, D. (1982). An empirical test of biases in the rapid visual technique for species–time censuses of reef fish assemblages. *Mar. Biol.* **70,** 129–134.

Denton, E. J., and Warren, F. J. (1956). Visual pigments of deep-sea fish. *Nature (London)* **178,** 1059.

Denton, E. J., and Warren, F. J. (1957). The photosensitive pigments in the retinae of deep-sea fish. *J. Mar. Biol. Assoc. U.K.* **36,** 651–662.

Diamant, A., and Shpigel, M. (1985). Interspecific feeding associations of groupers (Teleostei: Serranidae) with octopuses and moray eels in the Gulf of Eilat (Aqaba). *Environ. Biol. Fishes* **13,** 153–159.

Diamond, J. M. (1978). Niche shifts and the rediscovery of interspecific competition. *Am. Sci.* **66,** 322–331.

Diamond, J., and Case, T. J. (eds.) (1986). "Community Ecology." Harper & Row, New York.

Dight, I. J., James, M. K., and Bode, L. (1988). Models of larval dispersal within the Great Barrier Reef: Patterns of connectivity and their implications for species distributions. *Proc. Int. Coral Reef Symp., 6th* **3,** 217–224.

Doherty, P. J. (1980). Biological and physical constraints on the populations of two sympatric territorial damselfishes on the southern Great Barrier Reef. Unpublished Ph.D. dissertation, University of Sydney, Sydney, Australia.

Doherty, P. J. (1981). Coral reef fishes: Recruitment-limited assemblages? *Proc. Int. Coral Reef Symp., 4th* **2,** 465–470.

Doherty, P. J. (1982). Some effects of density on the juveniles of two species of tropical, territorial damselfishes. *J. Exp. Mar. Biol. Ecol.* **65,** 249–261.

Doherty, P. J. (1983a). Tropical territorial damselfishes: Is density limited by aggression or recruitment? *Ecology* **64,** 176–190.

Doherty, P. J. (1983b). Recruitment surveys of coral reef fishes as tools for science and management. *In* "Proceedings: Inaugural Great Barrier Reef Conference" (J. T. Baker, R. M. Carter, P. W. Sammarco, and K. P. Stark, eds.), pp. 191–196, James Cook Univ. Press, Townsville, Australia.

Doherty, P. J. (1983c). Diel, lunar and seasonal rhythms in the reproduction of two tropical damselfishes: *Pomacentrus flavicauda* and *P. wardi*. *Mar. Biol.* **75,** 215–224.

Doherty, P. J. (1987a). The replenishment of populations of coral reef fishes, recruitment surveys, and the problems of variability manifest on multiple scales. *Bull. Mar. Sci.* **41,** 411–422.

Doherty, P. J. (1987b). Light traps: Selective but useful devices for quantifying the distributions and abundances of larval fishes. *Bull. Mar. Sci.* **41,** 423–431.

Doherty, P. J. (1988). Large-scale variability in the recruitment of a coral reef fish. *Proc. Int. Coral Reef Symp., 6th,* **2,** 667–672.

Doherty, P. J., and Sale, P. F. (1985). Predation on juvenile coral reef fishes: An exclusion experiment. *Coral Reefs* **4,** 225–234.

Doherty, P. J., and Williams, D. McB. (1988a). The replenishment of coral reef fish populations. *Oceanogr. Mar. Biol.* **26,** 487–551.

Doherty, P. J., and Williams, D. McB. (1988b). Are local populations of coral reef fishes equilibrial assemblages? The empirical database. *Proc. Int. Coral Reef Symp., 6th* **1,** 131–139.

Doherty, P. J., Williams, D. McB., and Sale, P. F. (1985). The adaptive significance of larval dispersal in coral reef fishes. *Environ. Biol. Fishes* **12,** 81–90.

Dominey, W. J., and Blumer, L. S. (1984). Cannibalism of early life stages in fishes. *In* "Infanticide" (G. Hausfater and S. B. Hrdy, eds.), pp. 43–64. Aldine, New York.

Domm, S. B., and Domm, A. J. (1973). The sequence of appearance at dawn and disappearance at dusk of some coral reef fishes. *Pac. Sci.* **27,** 128–135.

Donaldson, T. J. (1984). Mobbing behavior by *Stegastes albifasciatus* (Pomacentridae), a territorial mosaic damselfish. *Japan. J. Ichthyol.* **31,** 345–348.

Donaldson, T. J. (1989). Facultative monogamy in obligate coral-dwelling hawkfishes (Cirrhitidae). *Environ. Biol. Fishes* **26,** 295–302.

Donaldson, T. J. (1990). Reproductive behavior and social organization of some Pacific hawkfishes (Cirrhitidae). *Japan J. Ichthyol.* **36,** 439–458.

Done, T. J. (1982). Patterns in the distribution of coral communities across the central Great Barrier Reef. *Coral Reefs* **1,** 95–107.

Done, T. J. (1983). Coral zonation: Its nature and significance. *In* "Perspectives on Coral Reefs" (D. J. Barnes, ed.), pp. 107–147. Aust. Inst. Mar. Sci., Townsville, Australia.

Douglas, R., and Djamgoz, M. B. A. (eds.) (1990) "Vision in Fishes." Chapman & Hall, London.

Downing, J. A., Perasse, M., and Frenette, Y. (1987). Effect of interreplicate variance on zooplankton sampling design and data analysis. *Limnol. Oceanogr.* **32,** 673–680.

Dragesund, O., and Nakken, O. (1971). Mortality of herring during the early larval stage in 1967. *Rapp. P.-V. Reun., Cons. Int. Explor. Mer* **160,** 142–146.

Dragovich, A. (1970). The food of skipjack and yellowfin tunas in the Atlantic Ocean. *Fish. Bull.* **68,** 445–460.

Dubin, R. E. (1981). Social behavior and ecology in some Caribbean parrotfish (Scaridae). Unpublished Ph.D thesis, University of Alberta, Edmonton, Alberta, Canada.

Dubin, R. E. (1982). Behavioral interactions between Caribbean reef fish and eels (Muraenidae and Ophichthidae). *Copeia* **1982,** 229–232.

Duffy, J. E., and Hay, M. E. (1990). Seaweed adaptations to herbivory. *BioScience* **40,** 368–375.

Dufour, V. (1988). L'ichthyoplancton en milieu corallien. Unpublished Ph.D. thesis, Université Pierre et Marie Curie, Paris.

Duggins, D. O. (1988). The effects of kelp forests on nearshore environments: Biomass, detritus, and altered flow. *Ecol. Stud.* **65,** 192–201.

Duggins, D. O., Simenstad, C. A., and Estes, J. A. (1989). Magnification of secondary production by kelp detritus in coastal marine ecosystems. *Science* **245,** 170–173.

Dunn, D. F. (1981). The clownfish sea anemones: Stichodactylidae (Coelenterata: Actiniaria) and other sea anemones symbiotic with pomacentrid fishes. *Trans. Am. Philos. Soc.* **71,** 1–115.

Durbin, A. G. (1979). Food selection by plankton feeding fishes. *In* "Predator–Prey Systems in Fisheries Management" (H. E. Clepper, ed.), pp. 203–218. Sport Fishing Inst., Washington, D.C.

Earle, S. A. (1972). The influence of herbivores on the marine plants of Great Lameshur Bay, with an annotated list of plants. *Bull. Nat. Hist. Mus. Los Angeles County* **14,** 17–44.

Ebeling, A. W., and Bray, R. N. (1976). Day versus night activity of reef fishes in a kelp forest off Santa Barbara, California. *Fish. Bull.* **74,** 703–717.

Ebeling, A. W., and Laur, D. R. (1985). The influence of plant cover on surfperch abundance at an offshore temperate reef. *Environ. Biol. Fishes* **16,** 123–133.

Ebeling, A. W., and Laur, D. R. (1986). Foraging in surfperches: Resource partitioning or individualistic responses? *Environ. Biol. Fishes* **16,** 123–133.

Ebeling, A. W., and Laur, D. R. (1988). Fish populations in kelp forests without sea otters: Effects of severe storm damage and destructive sea urchin grazing. *Ecol. Stud.* **65,** 169–191.

Ebeling, A. W., Larson, R. J., and Alevizon, W. S. (1980a). Habitat groups and island–mainland distribution of kelp-bed fishes off Santa Barbara, California. *In* "The California Islands: Proceedings of a Multidisciplinary Symposium" (D. M. Power, ed.), pp. 403–431. Santa Barbara Mus. Nat. Hist., Santa Barbara, California.

Ebeling, A. W., Larson, R. J., Alevizon, W. S., and Bray, R. N. (1980b). Annual variability of reef-fish assemblages in kelp forests off Santa Barbara, California. *Fish. Bull.* **78,** 361–377.

Ebeling, A. W., Laur, D. R., and Rowley, R. J. (1985). Severe storm disturbances and reversal of community structure in a southern California kelp forest. *Mar. Biol.* **84,** 287–294.

Ebeling, A. W., Holbrook, S. J., and Schmitt, R. J. (1990). Temporally concordant structure of a fish assemblage: Bound or determined? *Am. Nat.* **135,** 63–73.

Ebersole, J. P. (1977). The adaptive significance of interspecific territoriality in the reef fish *Eupomacentrus leucostictus. Ecology* **58,** 914–920.

Ebersole, J. P. (1980). Food density and territory size: An alternative model and a test on the reef fish *Eupomacentrus leucostictus. Am. Nat.* **115,** 492–509.

Ebersole, J. P. (1985). Niche separation of two damselfish species by aggression and differential microhabitat utilization. *Ecology* **66,** 14–20.

Eckert, G. J. (1984). Annual and spatial variation in recruitment of labroid fishes among seven reefs in the Capricorn/Bunker Group, Great Barrier Reef. *Mar. Biol.* **78,**123–127.

Eckert, G. J. (1985a). Settlement of coral reef fishes to different natural substrata and at different depths. *Proc. Int. Coral Reef Congr., 5th* **5,** 385–390.

Eckert, G. J. (1985b). Population studies of labrid fishes on the southern Great Barrier Reef. Unpublished Ph.D. dissertation, University of Sydney, Sydney, Australia.

Eckert, G. J. (1987). Estimates of adult and juvenile mortality for labrid fishes at One Tree Reef, Great Barrier Reef. *Mar. Biol.* **95,** 167–171.

Edmunds, M. (1974). "Defense in Aminals: A Survey of Anti-predator Defenses." Longmans, Essex, England.

Ehrlich, P. R. (1975). The population biology of coral reef fishes. *Annu. Rev. Ecol. Syst.* **6,** 211–247.

Ehrlich, P. R., and Ehrlich, A. H. (1973). Coevolution: Heterotypic schooling in Caribbean reef fishes. *Am. Nat.* **107,** 157–160.

Eibl-Eibesfeldt, I. (1962). Freiwasserbeobachtungen zur Deutung des Schwarmverhaltens verschiedener Fische. *Z. Tierpsychol.* **19,** 165–182.

Ekman, S. (1953). "Zoogeography of the Sea." Sidgwick & Jackson, London.

Ellison, J. P., Terry, C., and Stephens, J. S. (1979). Food resource utilization among five species of embiotocids at King Harbor, California, with preliminary estimates of caloric intake. *Mar. Biol.* **52,** 161–169.

Emery, A. R. (1968). Preliminary observations on coral reef plankton. *Limnol. Oceanogr.* **13,** 293–303.

Emery, A. R. (1972). Eddy formation from an oceanic island: Ecological effects. *Caribb. J. Sci.* **12,** 121–128.

Emery, A. R. (1973). Comparative ecology and functional osteology of fourteen species of damselfish (Pisces: Pomacentridae) at Alligator Reef, Florida Keys. *Bull. Mar. Sci.* **23,** 649–770.

Emery, A. R. (1978). The basis of fish community structure: Marine and freshwater comparisons. *Environ. Biol. Fishes,* **3,** 33–47.

Emery, A. R. and Thresher, R. E. (eds) (1980). Biology of the damselfishes. *Bull. Mar. Sci.* **30,** 145–328.

Emlen, S. T., and Oring, L. W. (1977). Ecology, sexual selection, and the evolution of mating systems. *Science* **197,** 215–223.

Endean, R. (1976). Destruction and recovery of coral reef communities. *In* "Biology and Geology of Coral Reefs" (O. A. Jones and R. Endean, eds.), Vol. 3, pp. 215–254. Academic Press, New York.

Endler, J. A. (1986). Defense against predators. *In* "Predator–Prey Relationships: Perspectives and Approaches from the Study of Lower Vertebrates" (M. E. Feder and G. V. Lauder, eds.), pp. 109–134, Univ. of Chicago Press, Chicago.

Erdman, D. S. (1977). Spawning patterns of fish from the northeastern Caribbean. *FAO Fish. Rep.* **200,** 145–169.

Erickson, K. L. (1983). Constituents of *Laurencia. In* "Marine Natural Products: Chemical and Biological Perspectives" (P. J. Scheuer, ed.), Vol. 5, pp. 131–257. Academic Press, New York.

Estes, J. A., and Harrold, C. (1988). Sea otters, sea urchins, and kelp beds: some questions of scale. *Ecol. Stud.* **65,** 116–150.

Estes, J. A., Smith, N. S., and Palmisano, J. F. (1978). Sea otter predation and community organization in the western Aleutian Islands, Alaska. *Ecology* **59,** 822–833.

Estes, J. A., Duggins, D. O., and Rathbun, G. B. (1989). The ecology of extinctions in kelp forest communities. *Conserv. Biol.* **3,** 252–263.

Eyberg, I. (1984). The biology of *Parablennius cornutus* (L.) and *Scartella emarginata* (Gunther) (Teleostei: Blenniidae) on a Natal reef. *Invest. Rep. Oceanogr. Res. Inst. S. Afr.* **54,** 1–16.

Fagen, R. (1987). Phenotypic plasticity and social environment. *Evol. Ecol.* **1,** 263–271.

Fager, E. W. (1971). Pattern in the development of a marine community. *Limnol. Oceanogr.* **16,** 241–253.

Fagerstrom, J. A. (1987). "The Evolution of Reef Communities." Wiley, New York.

Fasham, M. J. R. (1978). The statistical and mathematical analysis of plankton patchiness. *Oceanogr. Mar. Biol.* **16,** 43–79.

Faulkner, D. J. (1984). Marine natural products: Metabolites of marine algae and herbivorous marine molluscs. *Nat. Prod. Rep.* **1,** 251–280.

Faulkner, D. J. (1986). Marine natural products. *Nat. Prod. Rep.* **3,** 2–33.

Faulkner, D. J. (1987). Marine natural products. *Nat. Prod. Rep.* **4,** 539–576.

Feder, M. E., and Lauder, G. V. (eds.) (1986). "Predator–Prey Relationships: Perspectives and Approaches from the Study of Lower Vertebrates." Univ. of Chicago Press, Chicago.

Feder, H. M., Turner, C. H., and Limbaugh, C. (1974). Observations on fishes associated with kelp beds in southern California. *Fish. Bull.—Calif. Dep. Fish Game* **160,** 1–144.

Feeney, P. (1976). Plant apparency and chemical defense. *Recent Adv. Phytochem.* **10,** 1–40.

Fei, X. G., and Neushul, M. (1984). The effects of light on the growth and development of giant kelp. *Hydrobiologia* **116,** 456–462.

Feinsinger, P., Spears, E. E., and Poole, R. W. (1981). A simple measure of niche breadth. *Ecology* **62,** 27–32.

Fenical, W. (1975). Halogenation in the Rhodophyta: A review. *J. Phycol.* **11,** 245–259.

Fenical, W. (1982). Natural products chemistry in the marine environment. *Science* **215,** 923–928.

Fernald, R. (1988). Aquatic adaptations in fish eyes. *In* "Sensory Biology of Aquatic Animals" (J. Atema, R. R. Fay, A. N. Popper, and W. N. Tavolga, eds.), pp. 435–466. Springer-Verlag, New York.

Ferrell, D. J. (1988). The biology of *Pseudochromis queenslandica.* Unpublished M.S. thesis, University of Sydney, Sydney, Australia.

Ferry, R. E., and Kohler, C. C. (1987). Effects of trap fishing on fish populations inhabiting a fringing coral reef. *North Am. J. Fish. Manage.* **7,** 580–588.

Findley, J. S., and Findley, M. T. (1985). A search for pattern in butterfly fish communities. *Am. Nat.* **126,** 800–816.

Findley, J. S., and Findley, M. T. (1989). Circumtropical patterns in butterflyfish communities. *Environ. Biol. Fishes,* **25,** 1–3.

Fine, M. L. (1970). Pelagic Sargassum fauna. Mar. Biol. **7,** 112–122.

Finucane, J. H., Grimes, C. B., and Naughton, S. P. (1991). Diets of young king and Spanish mackerel off the southeast United States. *Northeast Gulf Science* **11,** 145–153.

Fischer, E. A. (1984). Local mate competition and sex allocation in simultaneous hermaphrodites. *Am. Nat.* **124,** 590–596.

Fischer, A. G., and Arthur, M. A. (1977). Secular variations in the pelagic realm. *Spec. Publ.–Soc. Econ. Paleontol. Mineral.* **25,** 19–50.

Fischer, E. A., and Petersen, C. W. (1987). The evolution of sexual patterns in the seabasses. *BioScience* **37,** 482–489.

Fischer, W., and Bianchi, G. (1984). "FAO Species Identification Sheets for Fishery Purposes. Western Indian Ocean." Food Agric. Organ., Rome.

Fishelson, L. (1970). Protogynous sex reversal in the fish *Anthias squamipinnis* (Teleostei, Anthiidae) regulated by the presence or absence of a male fish. *Nature (London)* **227,** 90–91.

Fishelson, L., Popper, D., and Avidor, A. (1974). Biosociology and ecology of pomacentrid fishes around the Sinai Peninsula (northern Red Sea). *J. Fish. Biol.* **6,** 119–133.

Fishelson, L., Montgomery, W. L., and Myrberg, A. A. (1985a). A new fat body associated with the gonads of surgeon fishes (Acanthuridae: Teleostei). *Mar. Biol.* **86,** 109–112.

Fishelson, L., Montgomery, W. L., and Myrberg, A. A. (1985b). A unique symbiosis in the gut of tropical herbivorous surgeon fish (Acanthuridae: Teleostei) from the Red Sea. *Science* **229,** 49–51.

Fishelson, L., Montgomery, W. L., and Myrberg, A. A., Jr. (1987). Biology of surgeonfish *Acanthurus nigrofuscus* with emphasis on changeover in diet and annual gonadal cycles. *Mar. Ecol. Prog. Ser.* **39,** 37–47.

Fitch, W. T. S., and Shapiro, D. Y. (1990). Spatial dispersion and nonmigratory spawning in the bluehead wrasse (*Thalassoma bifasciatum*). *Ethology* **85,** 199–211.

Fitz, C. H., Reaka, M. L., Bermingham, E., and Wolf, N. G. (1983). Coral recruitment at moderate depths: The influence of grazing. *In* "The Ecology of Deep and Shallow Coral Reefs" (M. L. Reaka, ed.), NOAA Symp. Ser. Undersea Res., Vol. 1, pp. 89–96. Natl. Oceanic Atmos. Adm., Rockville, Maryland.

Fitzhugh, G. R., and Fleeger, J. W. (1985). Goby (Pisces: Gobiidae) interactions with meiofauna and small macrofauna. *Bull. Mar. Sci.* **36,** 436–444.

Forrester, G. E. (1990). Factors influencing the juvenile demography of a coral reef fish population. *Ecology* **71,** 1666–1681.

Foster, S. A. (1987a). The relative impacts of grazing by Caribbean coral reef fishes and *Diadema*: Effects of habitat and surge. *J. Exp. Mar. Biol. Ecol.* **105,** 1–20.

Foster, S. A. (1987b). Diel and lunar patterns of reproduction in the Caribbean and Pacific sergeant major damselfishes *Abudefduf saxatilis* and *A. troschelii. Mar. Biol.* **95,** 333–343.

Foster, S. A. (1989). The implications of divergence in spatial nesting patterns in the geminate Caribbean and Pacific sergeant major damselfishes. *Anim. Behav.* **37,** 465–476.

Foster, M. S., and Schiel, D. R. (1985). The ecology of giant kelp forests in California: a community profile. *U.S. Fish Wildl. Serv. Biol. Rep.* **85,** 1–152.

Foster, M. S., and Schiel, D. R. (1988). Kelp communities and sea otters: Keystone species or just another brick in the wall? *Ecol. Stud.* **65,** 92–115.

Fourmanoir, P. (1980). Deep bottom fishing in New Caledonia. *SPC Fish. Newsl.* **20,** 15–20.

Fowler, A. J. (1987). The development of sampling strategies for population studies of coral reef fishes. A case study. *Coral Reefs* **6,** 49–58.

Fowler, A. J. (1988). Aspects of the population biology of three species of chaetodont at One Tree Reef, southern Great Barrier Reef. Unpublished Ph.D. dissertation, University of Sydney, Sydney, Australia.

Fowler, A. J. (1989). Description, interpretation and use of the microstructure of otoliths from juvenile butterflyfishes (family Chaetodontidae). *Mar. Biol.* **102,** 167–181.

Fowler, A. J. (1990a). Spatial and temporal patterns of distribution and abundance of chaetodontid fishes at One Tree Reef, southern GBR. *Mar. Ecol. Prog. Ser.* **64,** 39–53.

Fowler, A. J. (1990b). Validation of annual growth increments in the otoliths of a small, tropical coral reef fish. *Mar. Ecol. Prog. Ser.* **64,** 25–38.

Frank, K. T., and Leggett, W. C. (1985). Reciprocal oscillations in densities of larval fish and potential predators: a reflection of present or past predation? *Can. J. Fish. Aquat. Sci.,* **42,** 1841–1849.

Fricke, H. W. (1973a). Individual partner recognition in fish: Field studies on *Amphiprion bicinctus. Naturwissenschaften* **60,** 204–205.

Fricke, H. W. (1973b). Okologie und sozialverhalten des Korallenbarches *Dascyllus trimaculatus* (Pisces, Pomacentridae). *Z. Tierpsychol.* **32,** 225–256.

Fricke, H. W. (1974). Oko-ethologie des monogamen Anemonenfisches *Amphiprion bicinctus* (Freiwasseruntersuchung aus dem Roten Meer). *Z. Tierpsychol.* **36,** 429–512.

Fricke, H. W. (1975). Evolution of social systems through site attachment in fish. *Z. Tierpsychol.* **39,** 206–210.

Fricke, H. W. (1979). Mating system, resource defence and sex change in the anemonefish *Amphiprion akallopisos. Z. Tierpsychol.* **50,** 313–326.

Fricke, H. W. (1980). Control of different mating systems in a coral reef fish by one environmental factor. *Anim. Behav.* **28,** 561–569.

Fricke, H. W. (1986). Pair swimming and mutual partner guarding in monogamous butterflyfishes (Pisces, Chaetodontidae): A joint advertisement for territory. *Ethology* **73,** 307–333.

Fricke, H. W., and Fricke, S. (1977). Monogamy and sex change by aggressive dominance in coral reef fish. *Nature (London)* **266,** 830–832.

Fricke, H. W., and Holzberg, S. (1974). Social units and hermaphroditism in a pomacentrid fish. *Naturwissenschaften* **61,** 367–368.

Fricke, H. W., and Kacher, H. (1982). A mound-building deep water sand tilefish of the Red Sea: *Haplolatilus geo* n.sp. (Perciformes: Branchiostegidae). Observations from a research submersible. *Senckenbergiana Marit.* **14,** 245–259.

Frith, C. A., Leis, J. M., and Goldman, B. (1986). Currents in the Lizard Island region of the Great Barrier Reef Lagoon and their relevance to potential movements of larvae. *Coral Reefs* **5,** 81–92.

Froelich, A. S. (1983). Functional aspects of nutrient cycling on coral reefs. *In* "The Ecology of Deep and Shallow Coral Reefs" (M. L. Reaka, ed.), NOAA Symp. Ser. Undersea Res., Vol. 1, pp. 133–139. Natl. Oceanic Atmos. Adm., Rockville, Maryland.

Frydl, P., and Stearn, C. W. (1978). Rate of erosion by parrotfish in Barbados reef environments. *J. Sediment. Petrol.* **48,** 1149–1168.

Fryer, G. (1959). The trophic interrelationships and ecology of some littoral communities of Lake Nyasa with especial reference to the fishes, and a discussion of the evolution of a group of rock-frequenting Cichlidae. *Proc. Zool. Soc. London* **132,** 153–281.

Fryer, G., and Iles, T. D. (1972). "The Cichlid Fishes of the Great Lakes of Africa. Their Biology and Evolution." TFH Publ., Hong Kong.

Furlani, D., Bruce, B. D., Thresher, R. E., and Gunn, J. S. (1991). Seasonal and spatial patterns of the distribution and abundance of larval fishes in Tasmanian coastal waters. Manuscript in preparation.

Furnas, M. G., and Mitchell, A. W. (1987). Phytoplankton dynamics in the central Great Barrier Reef. II. Primary production. *Continent. Shelf Res.* **7,** 1049–1062.

Fursa, T. I. (1969). Quantitative and qualitative characterization of the ichthyoplankton off the western shore of Hindustan. *Prob. Ichthyol.* **9,** 394–403.

Gaines, S. D., and Lubchenco, J. (1982). A unified approach to marine plant–herbivore interactions: II. Biogeography. *Annu. Rev. Ecol. Syst.* **13,** 111–138.

Gaines, S. D., and Roughgarden, J. (1987). Fish in offshore kelp forests affect recruitment to intertidal barnacle populations. *Science* **235,** 479–481.

Galzin, R. (1985). Ecologie des poissons recifaux de Polynesie francaise. Unpublished Ph.D. thesis, Université des Sciences et Techniques de Languedoc, Montpellier, France.

Galzin, R. (1987a). Structure of fish communities of French Polynesian coral reefs. I. Spatial scales. *Mar. Ecol. Prog. Ser.* **41,** 129–136.

Galzin, R. (1987b). Structure of fish communities of French Polynesian coral reefs. II. Temporal scales. *Mar. Ecol. Prog. Ser.* **41,** 137–145.

Galzin, R., and Legendre, P. (1988). The fish communities of a coral reef transect. *Pac. Sci.* **41,** 158–165.

Garrity, S. D., and Levings, S. C. (1981). A predator–prey interaction between two physically and biologically constrained tropical rocky shore gastropods: Direct, indirect and community effects. *Ecol. Monogr.* **51,** 267–286.

Gascon, D., and Miller, R. A. (1981). Colonization by nearshore fish on small artificial reefs in Barkley Sound, British Columbia. *Can. J. Zool.* **59,** 1635–1646.

Gascon, D., and Miller, R. A. (1982). Space utilization in a community of temperate reef fishes inhabiting small experimental artificial reefs. *Can. J. Zool.* **60,** 798–806.

Gaston, K. J., and Lawton, J. H. (1988). Patterns in body size, population dynamics, and regional distribution of bracken herbivores. *Am. Nat.* **132,** 662–680.

Gause, G. F. (1942). The relation of adaptability to adaptation. *Q. Rev. Biol.* **17,** 99–114.

Gaut, V. C., and Munro, J. L. (1983). The biology, ecology and bionomics of the grunts, Pomadasyidae. *ICLARM Stud. Rev.* **7,** 110–141.

Geffen, A. J. (1982). Otolith ring deposition in relation to growth rate in herring (*Clupea harengus*) and turbot (*Scophthalmus maximus*) larvae. *Mar. Biol.* **71,** 317–326.

Gerard, V. A. (1976). Some aspects of material dynamics and energy flow in a kelp forest in Monterey Bay, California. Unpublished Ph.D. dissertation, University of California, Santa Cruz, California.

Gerber, R. P. (1981). Species composition and abundance of lagoon zooplankton at Enewetak Atoll, Marshall Islands. *Atoll Res. Bull.* **247,** 1–22.

Gerber, R. P., and Marshall, N. (1974). Ingestion of detritus by the lagoon pelagic community at Eniwetok Atoll. *Limnol. Oceanogr.* **19,** 815–824.

Germann, I. (1988). Effects of the 1983 El Niño on growth and carbon and nitrogen metabolism of *Pleurophycus gardneri* (Phaeopyceae: Laminariales) in the northeastern Pacific. *Mar. Biol.* **99,** 445–455.

Getty, T. (1981). Competitive collusion: The preemption of competition during the sequential establishment of territories. *Am. Nat.* **118,** 426–431.

Getty, T. (1987). Dear enemies and the prisoner's dilemma: Why should territorial neighbors form defensive coalitions? *Am. Zool.* **27,** 327–336.

Ghiselin, M. T. (1969). The evolution of hermaphroditism among animals. *Q. Rev. Biol.* **44,** 189–208.

Gilbert, C. R. (1972). Characteristics of the western Atlantic reef-fish fauna. *Q. J. Fla. Acad. Sci.* **35,** 130–144.

Gilinsky, E. (1984). The role of fish predation and spatial heterogeneity in determining benthic community structure. *Ecology* **65,** 455–468.

Gill, F. B., and Wolf, L. L. (1975). Economics of feeding territoriality in the golden-winged sunbird. *Ecology* **56,** 333–345.

Gilligan, M. R. (1980). Beta diversity of a Gulf of California rocky-shore fish community. *Environ. Biol. Fishes* **5,** 109–116.

Gladfelter, W. B. (1979). Twilight migrations and foraging activities of the copper sweeper *Pempheris schomburgki* (Teleostei: Pempheridae). *Mar. Biol.* **50,** 109–119.

Gladfelter, W. B., and Gladfelter, E. H. (1978). Fish community structure as a function of habitat structure on West Indian patch reefs. *Rev. Biol. Trop.* **26,** 65–84.

Gladfelter, W. B., and Johnson, W. S. (1983). Feeding niche separation in a guild of tropical reef fishes (Holocentridae). *Ecology* **64,** 552–563.

Gladfelter, W. B., Ogden, J. C., and Gladfelter, E. H. (1980). Similarity and diversity among coral reef fish communities: A comparison between tropical western Atlantic (Virgin Islands) and tropical central Pacific (Marshall Islands) patch reefs. *Ecology* **61,** 1156–1168.

Gladstone, W. (1985). Behavioral ecology of the sharpnose pufferfish, *Canthigaster valentini* (Bleeker), at Lizard Island, Great Barrier Reef. Unpublished Ph.D. thesis, Macquarie University, North Ryde, Australia.

Gladstone, W. (1987). The eggs and larvae of the sharpnose pufferfish *Canthigaster valentini* (Pisces: Tetraodontidae) are unpalatable to other reef fishes. *Copeia* **1987,** 227–230.

Gladstone, W. (1991). Early reproduction and reduction in survivorship and lifetime reproductive success in a coral reef fish. *Environ. Biol. Fishes* (in press).

Gladstone, W., and Westoby, M. (1988). Growth and reproduction in *Canthigaster valentini* (Pisces, Tetraodontidae): A comparison of a toxic reef fish with other reef fishes. *Environ. Biol. Fishes* **21,** 207–221.

Glynn, P. W. (1973). Ecology of a Caribbean coral reef. The Porites reef-flat biotope: Part II. Plankton community with evidence for depletion. *Mar. Biol.* **22,** 1–21.

Glynn, P. W. (1985). Corallivore population sizes and feeding effects following El Niño (1982–1983) associated coral mortality in Panama. *Proc. Int. Coral Reef Congr., 5th* **4,** 183–188.

Glynn, P. W. (1988). Predation on coral reefs: Some key processes, concepts and research directions. *Proc. Int Coral Reef Symp., 6th* **1,** 51–62.

Glynn, P. W., Wellington, G. M., and Birkeland, C. (1979). Coral reef growth in the Galapagos: Limitation by sea urchins. *Science* **203,** 47–49.

Godwin, J. R., and Kosaki, R. K. (1989). Reef fish assemblages on submerged lava flows of three different ages. *Pac. Sci.* **43,** 289–301.

Goeden, G. B. (1989). Intensive fishing and a "keystone" predator species: Ingredients for community instability. *Biol. Conserv.* **22,** 273–281.

Goeden, G. B. (1986). The effects of selectively fishing reef fish stocks along the Great Barrier Reef. *In* "Fisheries Management. Theory and Practice in Queensland" (T. J. A. Hundloe, ed.), pp. 199–207. Griffith Univ. Press, Brisbane, Australia.

Goldman, B., and Talbot, F. H. (1976). Aspects of the ecology of coral reef fishes. *In* "Biology and Geology of Coral Reefs" (O. A. Jones and R. Endean, eds.), Vol. 3, pp. 125–154. Academic Press, New York.

Goldman, B., Stroud, G. J., and Talbot, F. (1983). Fish eggs and larvae over a coral reef: Abundance with habitat, time of day and moon phase. *In* "Proceedings: Inaugural Great Barrier Reef Conference" (J. T. Baker, R. M. Carter, P. W. Sammarco, and K. P. Stark, eds.), pp. 203–211. James Cook Univ. Press, Townsville, Australia.

Gomez, E. D., Alcala, A. C., and Yap, H. T. (1987). Other fishing methods destructive to coral. *In* "Human Impacts on Coral Reefs: Facts and Recommendations" (B. Salvat, ed.), pp. 67–75. Antenne Museum EPHE, Moorea, French Polynesia.

Gooding, R. M., and Magnuson, J. J. (1967). Ecological significance of a drifting object to pelagic fishes. *Pac. Sci.* **21,** 486–497.

Goodman, L. A., and Kruskal, W. H. (1954). Measures of association for cross-classification. *J. Am. Stat. Assoc.* **49,** 732–764.

Gordina, A. D., and Bladimirtsev, V. B. (1987). Distribution of ichthyoplankton in the regions of bottom elevations in the tropical part of the Indian Ocean. *J. Ichthyol.(Engl. Transl.)* **27,** 79–84.

Goreau, T. F., and Goreau, N. I. (1973). The ecology of Jamaican coral reefs. II. Geomorphology, zonation, and sedimentary phases. *Bull. Mar. Sci.* **23,** 399–464.

Goreau, T. F., Goreau, N. I., and Yonge, C. M. (1971). Reef corals: Autotrophs or heterotrophs? *Biol. Bull.* (*Woods Hole, Mass.*) **141,** 247–260.

Gorlick, D. L., Atkins, P. D., and Losey, G. S. (1987). Effect of cleaning by *Labroides dimidiatus* (Labridae) on an ectoparasite population infecting *Pomacentrus vaiuli* (Pomacentridae) at Enewetak Atoll. *Copeia* **1987,** 41–45.

Gosline, W. A. (1965). Vertical zonation of inshore fishes in the upper water layers of the Hawaiian Islands. *Ecology* **46,** 823–831.

Gosline, W. A. (1968). Considerations regarding the evolution of Hawaiian animals. *Pac. Sci.* **22,** 267–273.

Gosline, W. A. (1971). Functional morphology and classification of teleostean fishes. Press University Hawaii, Honolulu.

Gosline, W. A. (1981). The evolution of the premaxillary protrusion system in some teleostean fish groups. *J. Zool.* **193,** 11–23.

Govardovskii, V. I. (1976). Comments on the sensitivity hypothesis. *Vision Res.* **16,** 1363–1364.

Graham, N. E., and White, W. B. (1988). The El Niño cycle: A natural oscillator of the Pacific Ocean–atmospheric system. *Science* **240,** 1293–1302.

Green, G. (1977). Ecology of toxicity in marine sponges. *Mar. Biol.* **40,** 207–215.

Greene, L. E., and Alevizon, W. S. (1989). Comparative accuracies of visual assessment methods for coral reef fishes. *Bull. Mar. Sci.* **44,** 899–912.

Greenfield, D. W., and Greenfield, T. A. (1982). Habitat and resource partitioning between two species of *Acanthemblemaria* (Pisces: Chaenopsidae) with comments on the chaos hypothesis. *Smithson. Contrib. Mar. Sci.* **12,** 499–507.

Grieg-Smith, P. (1983). "Quantitative Plant Ecology." Butterworths, London.

Grigg, R. W. (1988). Paleoceanography of coral reefs in the Hawaiian–Emperor chain. *Science* **240,** 1737–1743.

Grigg, R. W., and Epp, D. (1989). Critical depth for the survival of coral islands: Effects on the Hawaiian archipelago. *Science* **243,** 638–641.

Grigg, R. W., Wells, J. W., and Wallace, C. (1981). *Acropora* in Hawaii. Part 1. History of the scientific record, systematics and ecology. *Pac. Sci.* **35,** 1–13.

Grigg, R. W., Polovina, J. J., and Atkinson, M. J. (1984). Model of a coral reef ecosystem. III. Resource limitation, community regulation, fisheries yield and resource management. *Coral Reefs* **3,** 23–27.

Grimes, G. B. (1987). Reproductive biology of the Lutjanidae: A review. *In* "Tropical Snappers and Groupers: Biology and Fisheries Management" (J. J. Polovina and S. Ralston, eds.), pp. 239–294. Westview, Boulder, Colorado.

Gronell, A. (1984). Courtship, spawning and social organization of the pipefish, *Corythoichthys intestinalis* (Pisces, Syngnathidae) with notes on two congeneric species. *Z. Tierpsychol.* **65,** 1–24.

Grossman, G. D. (1986). Food resource partitioning in a rocky intertidal fish assemblage. *J. Zool.* **1,** 317–355.

Gudger, E. W. (1941). Coelenterates as enemies of fishes. IV. Sea anemones and corals as fish eaters. *N. Engl. Nat.* **10,** 1–8.

Guillemot, P. J., Larson, R. J., and Lenarz, W. H. (1985). Seasonal cycles of fat and gonad volume in five species of northern California rockfish (Scorpaenidae: *Sebastes*). *Fish. Bull.* **83,** 299–311.

Gulland, J. A. (1971). "The Fish Resources of the Ocean." Fishing News Books, Surrey, England.

Gulland, J. A. (1979). Report of the FAO/IOD workshop on the fishery resources of the western Indian Ocean south of the equator. *IOFC/Dev/79/45.*

Gulland, J. A. (1982). Why do fish numbers vary? *J. Theor. Biol.* **97,** 69–75.

Gulland, J. A. (1983). "Fish Stock Assessment. A Manual of Basic Methods." FAO/Wiley Ser. Food Agric., Vol. 1. Wiley, Chichester, England.

Gulland, J. A., and Garcia, S. (1984). Observed patterns in multispecies fisheries. *In* "Exploitation of Marine Communities" (R. M. May, ed.), pp. 155–190. Springer-Verlag, Berlin.

Guzman, H. M., and Robertson, D. R. (1989). Population and feeding responses of the corallivorous pufferfish *Arothron meleagris* to coral mortality in the eastern Pacific. *Mar. Ecol. Prog. Ser.* **55,** 121–131.

Gygi, R. A. (1975). *Sparisoma viride* (Bonnaterre), the stoplight parrotfish, a major sediment producer on coral reefs of Bermuda? *Eclogae Geol. Helv.* **68,** 327–359.

Hailman, J. P. (1977). "Optical Signals." Indiana Univ. Press, Bloomington, Indiana.

Hairston, N. G., Li, K. T., and Easter, S. S. (1982). Fish vision and the detection of planktonic prey. *Science* **218,** 1240–1242.

Hallacher, L. E., and Roberts, D. A. (1985). Differential utilization of space and food by the inshore rockfishes (Scorpaenidae: *Sebastes*) of Carmel Bay, California. *Environ. Biol. Fishes* **12,** 91–110.

Halstead, B. W. (1978). "Poisonous and Venomous Marine Animals of the World." Darwin, Princeton, New Jersey.

Hamner, W. N., and Carleton, J. H. (1979). Copepod swarms: Attributes and role in coral reef ecosystems. *Limnol. Oceanogr.* **24,** 1–14.

Hamner, W. M., and Wolanski, E. (1988). Hydrodynamic forcing functions and biological processes on coral reefs: A status review. *Proc. Int. Coral Reef Symp., 6th* **1,** 103–113.

Hamner, W. M., Jones, M. S., Carleton, J. H., Hauri, I. R., and Williams, D. M. (1988). Zooplankton, planktivorous fish, and water currents on a windward reef face: Great Barrier Reef, Australia. *Bull. Mar. Sci.* **42,** 459–479.

Handbook of Sensory Physiology (1971–1981). Vol. I–IX. Springer-Verlag, Heidelberg.

Haney, J. F. (1988). Diel patterns of zooplankton behaviour. *Bull. Mar. Sci.* **43,** 583–603.

Hanlon, R. T., and Kaufman, L. (1976). Associations of seven West Indian reef fishes with sea anemones. *Bull. Mar. Sci.* **26,** 225–232.

Hansen, T. A. (1978). Larval dispersal and species longevity in Lower Tertiary gastropods. *Science* **199,** 885–886.

Harmelin-Vivien, M. L. (1977). Ecological distribution of fishes on the outer slope of Tulear reef (Madagascar). *Proc. Int. Coral Reef Symp., 3rd* **1,** 289–295.
Harmelin-Vivien, M. L. (1981). Trophic relationships of reef fishes in Tulear (Madagascar). *Oceanol. Acta* **3,** 365–374.
Harmelin-Vivien, M. L. (1989). Reef fish community structure: An Indo-Pacific comparison. *Ecol. Stud.* **69,** 21–60.
Harmelin-Vivien, M. L., and Bouchon, C. (1976). Feeding behavior of some carnivorous fishes (Serranidae and Scorpaenidae) from Tulear (Madagascar). *Mar. Biol.* **37,** 329–340.
Harmelin-Vivien, M. L., and Bouchon-Navaro, Y. (1983). Feeding diets and significance of coral feeding among chaetodontid fishes in Moorea (French Polynesia). *Coral Reefs* **2,** 119–127.
Harris, L. G., Ebeling, A. W., Laur, D., and Rowley, R. J. (1984). Community recovery after storm damage: A case of facilitation in primary succession. *Science* **224,** 1336–1338.
Harrold, C., and Reed, D. C. (1985). Food availability, sea urchin grazing, and kelp forest community structure. *Ecology* **66,** 1160–1169.
Hart, M. W., and Scheibling, R. E. (1988). Heat waves, baby booms, and the destruction of kelp beds by sea urchins. *Mar. Biol.* **99,** 167–176.
Hartline, A. C., Hartline, P. H., Szmant, A. M., and Flechsig, A. O. (1972). Escape response in a pomacentrid reef fish, *Chromis cyaneus*. *Bull. Nat. Hist. Mus. Los Angeles County* **14,** 93–97.
Harvey, P. H., and Greenwood, P. J. (1978). Anti-predator defense strategies: Some evolutionary problems. *In* "Behavioural Ecology: An Evolutionary Approach" (J. R. Krebs and N. B Davies, eds.), pp. 129–151. Blackwell, Oxford, England.
Harvey, P. H., and Mace, G. M. (1983). Foraging models and territory size. *Nature (London)* **305,** 14–15.
Hatcher, B. G. (1981). The interaction between grazing organisms and the epilithic algal community of a coral reef: A quantitative assessment. *Proc. Int. Coral Reef Symp., 4th* **2,** 515–524.
Hatcher, B. G. (1983). Grazing in coral reef ecosystems. *In* "Perspectives on Coral Reefs" (D. J. Barnes, ed.), pp. 164–179. Aust. Inst. Mar. Sci., Townsville, Australia.
Hatcher, A. I., and Hatcher, B. G. (1981). Seasonal spatial variation in dissolved inorganic material in One Tree Reef Lagoon. *Proc. Int. Coral Reef Symp., 4th* **1,** 419–424.
Hatcher, B. G., and Larkum, A. W. D. (1983). An experimental analysis of factors controlling the standing crop of the epilithic algal community on a coral reef. *J. Exp. Mar. Biol. Ecol.* **69,** 61–84.
Hatcher, B. G., and Rimmer, D. W. (1985). The role of grazing in controlling benthic community structure on a high latitude coral reef: Measurements of grazing intensity. *Proc. Int. Coral Reef Congr., 5th* **6,** 229–236.
Hatcher, B. G., Imberger, J. and Smith S. V. (1987). Scaling analysis of coral reef systems: An approach to problems of scale. *Coral Reefs* **5,** 171–182.
Haury, L. R., McGowan, J. A., and Wiebe, P. H. (1978). Patterns and processes in the

time–space scales of plankton distributions. *In* "Spatial Pattern in Plankton Communities" (J. H. Steele, ed.), pp. 277–327. Plenum, New York.
Hawkins, A. D., and Myrberg, A. A. (1983). Hearing and sound communication under water. *In* "Bioacoustics: A Comparative Approach" (B. Lewis, ed.), pp. 347–405. Academic Press, New York.
Hay, M. E. (1981a). Herbivory, algal distribution, and the maintenance of between-habitat diversity on a tropical fringing reef. *Am. Nat.* **118,** 520–540.
Hay, M. E. (1981b). The functional morphology of turf-forming seaweeds; persistence in stressful marine habitats. *Ecology* **62,** 739–750.
Hay, M. E. (1981c). Spatial patterns of grazing intensity on a Caribbean barrier reef: Herbivory and algal distribution. *Aquat. Bot.* **11,** 97–109.
Hay, M. E. (1984a). Predictable spatial escapes from herbivory: How do these affect the evolution of herbivore resistance in tropical marine communities? *Oecologia* **64,** 396–407.
Hay, M. E. (1984b). Patterns of fish and urchin grazing on Caribbean coral reefs: Are previous results typical? *Ecology* **65,** 446–454.
Hay, M. E. (1985). Spatial patterns of herbivore impact and their importance in maintaining algal species richness. *Proc. Int. Coral Reef Congr., 5th* **4,** 29–34.
Hay, M. E. (1986). Associational plant defenses and the maintenance of species diversity: Turning competitors into accomplices. *Am. Nat.* **128,** 617–641.
Hay, M. E. (1991). Seaweed chemical defenses: Their role in the evolution of feeding specialization and in mediating complex interactions. *In* "Ecological Roles for Marine Secondary Metabolites: Explorations in Chemical Ecology Series" (V. J. Paul, ed.). Comstock, Ithaca, New York. In press.
Hay, M. E., and Fenical, W. (1988). Marine plant–herbivore interactions: The ecology of chemical defense. *Annu. Rev. Ecol. Syst.* **19,** 111–145.
Hay, M. E., and Goertemiller, T. (1983). Between-habitat differences in herbivore impact on Caribbean coral reefs. *In* "The Ecology of Deep and Shallow Coral Reefs" (M. L. Reaka, ed.), NOAA Symp. Ser. Undersea Res. Vol. 1, pp. 97–102. Natl. Oceanic Atmos. Adm., Rockville, Maryland.
Hay, M. E., and Taylor, P. R. (1985). Competition between herbivorous fishes and urchins on Caribbean reefs. *Oecologia* **65,** 591–598.
Hay, M. E., Colburn, T., and Downing, D. (1983). Spatial and temporal patterns in herbivory on a Caribbean fringing reef: The effects on plant distribution. *Oecologia* **58,** 299–308.
Hay, M. E., Fenical, W., and Gustafson, K. (1987a). Chemical defense against diverse coral-reef herbivores. *Ecology* **68,** 1581–1591.
Hay, M. E., Duffy, J. E., Pfister, C. A., and Fenical, W. (1987b). Chemical defenses against different marine herbivores: Are amphipods insect equivalents? *Ecology* **68,** 1567–1580.
Hay, M. E., Duffy, J. E., and Fenical, W. (1988a). Seaweed chemical defenses: Among-compound and among-herbivore variance. *Proc. Int. Coral Reef Symp., 6th* **3,** 43–48.
Hay, M. E., Duffy, J. E., Fenical, W., and Gustafson, K. (1988b). Chemical defense in the seaweed *Dictyopteris delicatula*: Differential effects against reef fishes and amphipods. *Mar. Ecol. Prog. Ser.* **48,** 185–192.

Hay, M. E., Paul, V. J., Lewis, S. M., Gustafson, K., Tucker, J., and Trindell, R. N. (1988c). Can tropical seaweeds reduce herbivory by growing at night? Diel patterns of growth, nitrogen content, herbivory, and chemical versus morphological defenses. *Oecologia* **75,** 233–245.

Hay, M. E., Renaud, P. E., and Fenical, W. (1988d). Large mobile versus small sedentary herbivores and their resistance to seaweed chemical defenses. *Oecologia* **75,** 246–252.

Hay, M. E., Pawlik, J. R., Duffy, J. E., and Fenical, W. (1989). Seaweed–herbivore–predator interactions: Host–plant specialization reduces predation on small herbivores. *Oecologia* **81,** 418–427.

Hay, M. E., Duffy, J. E., and Fenical, W. (1990). Host–plant specialization decreases predation on a marine amphipod: An herbivore in plant's clothing. *Ecology* **71,** 733–743.

Heck, K. L., and McCoy, E. D. (1978). Long-distance dispersal and the reef-building corals of the eastern Pacific. *Mar. Biol.* **48,** 349–356.

Heck, K. L., Jr., and Orth, R. J. (1980). Seagrass habitats: The roles of habitat complexity, competition and predation in structuring associated fish and motile macroinvertebrate assemblages. *In* "Estuarine Perspectives" (V. S. Kennedy, ed.), pp. 449–464. Academic Press, New York.

Helfman, G. S. (1978). Patterns of community structure in fishes: Summary and overview. *Environ. Biol. Fishes* **3,** 129–148.

Helfman, G. S. (1981). Twilight activities and temporal structure in a freshwater fish community. *Can. J. Fish. Aquat. Sci.* **38,** 1405–1420.

Helfman, G. S. (1983). Underwater methods. *In* "Fisheries Techniques" (L. A. Nielsen and D. L. Johnson, eds.), pp. 349–369. Am. Fish. Soc., Bethesda, Maryland.

Helfman, G. S. (1986a). Behavioral responses of prey fishes during predator–prey interactions. *In* "Predator–Prey Relationships: Perspectives and Approaches from the Study of Lower Vertebrates" (M. E. Feder and G. V. Lauder, eds.), pp. 135–156. Univ. of Chicago Press, Chicago.

Helfman, G. S. (1986b). Fish behavior by day, night and twilight. *In* "The Behaviour of Teleost Fishes" (T. J. Pitcher, ed.), pp. 366–387. Croom-Helm, London.

Helfman, G. S. (1989). Threat-sensitive predator avoidance in damselfish–trumpetfish interactions. *Behav. Ecol. Sociobiol.* **24,** 47–58.

Helfman, G. S., and Schultz, E. T. (1984). Social transmission of behavioural traditions in a coral reef fish. *Anim. Behav.* **32,** 379–384.

Helfman, G. S., Meyer, J. L., and McFarland, W. N. (1982). The ontogeny of twilight migration patterns in grunts (Pisces: Haemulidae). *Anim. Behav.* **30,** 317–326.

Herman, S. S., and Beers, J. R. (1969). The ecology of inshore plankton populations in Bermuda. Part 2. Seasonal abundance and composition of the zooplankton. *Bull. Mar. Sci.* **19,** 483–503.

Heuter, R. E., and Cohen, J. (eds.) (1991). Vision in elasmobranches. Special Conference in April 1989 at the Mote Marine Laboratory, Sarasota, Florida. *J. Exp. Zool.* In press.

Hewitt, R. P., Theilacker, G. H., and Lo, N. C. H. (1985). Causes of mortality in young jack mackerel. *Mar. Ecol. Prog. Ser.* **26,** 1–10.

Hiatt, R. W., and Strasburg, D. W. (1960). Ecological relationships of the fish fauna on coral reefs of the Marshall Islands. *Ecol. Monogr.* **30,** 65–127.

Hillborn, R., and Stearns, S. C.(1982). On inference in ecology and evolutionary biology: The problem of multiple causes. *Acta Biotheor.* **31,** 145–164.

Hill, H. B. (1978). The use of nearshore marine life as a food resource by American Samoans. *Pac. Islands Program, Univ. Hawaii, Misc. Work Pap.* **1978:1,** 1–170.

Hinde, R. A. (ed.) (1983). "Primate Social Relationships: An Integrated Approach." Sinauer, Sunderland, Massachusetts.

Hirschfield, M. F. (1980). An experimental analysis of reproductive effort and cost in the Japanese medaka, *Oryzias latipes. Ecology* **61,** 282–292.

Hixon, M. A. (1980a). Competitive interactions between California reef fishes of the genus *Embiotoca. Ecology* **61,** 918–931.

Hixon, M. A. (1980b). Food production and competitor density as the determinants of feeding territory size. *Am. Nat.* **115,** 510–530.

Hixon, M. A. (1981). An experimental analysis of territoriality in the California reef fish *Embiotoca jacksoni* (Embiotocidae). *Copeia* **1981,** 653–665.

Hixon, M. A. (1983). Fish grazing and community structure of reef corals and algae: A synthesis of recent studies. *In* "The Ecology of Deep and Shallow Coral Reefs" (M. L. Reaka, ed.), NOAA Symp. Ser. Undersea Res., Vol. 1, pp. 79–87. Natl. Oceanic Atmos. Adm., Rockville, Maryland.

Hixon, M. A. (1986). Fish predation and local prey diversity. *In* "Contemporary Studies on Fish Feeding" (C. A. Simenstad and G. M. Cailliet, eds.), pp. 235–257. Junk, Dordrecht, The Netherlands.

Hixon, M. A. (1987). Territory area as a determinant of mating systems. *Am. Zool.* **27,** 229–247.

Hixon, M. A., and Beets, J. P. (1989). Shelter characteristics and Caribbean fish assemblages: Experiments with artificial reefs. *Bull. Mar. Sci.* **44,** 666–680.

Hixon, M. A., and Brostoff, W. N. (1983). Damselfish as keystone species in reverse: Intermediate disturbance and diversity of reef algae. *Science* **220,** 511–513.

Hixon, M. A., and Brostoff, W. N. (1985). Substrate characteristics, fish grazing, and epibenthic reef assemblages off Hawaii. *Bull. Mar. Sci.* **37,** 200–213.

Hixon, M. A., and Menge, B. A. (1991). Species diversity: Prey refuges modify the interactive effects of predation and competition. *Theor. Pop. Biol.* **39,** 178–200.

Hjort, J. (1914). Fluctuations in the great fisheries of northern Europe. *Rapp. P.-V. Reun., Cons. Int. Explor. Mer* **20,** 1–13.

Hobson, E. S. (1965). Diurnal–nocturnal activity of some inshore fishes in the Gulf of California. *Copeia* **1965,** 291–302.

Hobson, E. S. (1966). Visual orientation and feeding in seals and sealions. *Nature (London)* **210,** 326–327.

Hobson, E. S. (1968). Predatory behavior of some shore fishes in the Gulf of California. *Res. Rep.—U.S. Fish Wildl. Serv.* **73,** 1–92.

Hobson, E. S. (1971). Cleaning symbiosis among California inshore fishes. *Fish. Bull.* **69,** 491–523.

Hobson, E. S. (1972). Activity of Hawaiian reef fishes during the evening and morning transitions between daylight and darkness. *Fish. Bull.* **70,** 715–740.

Hobson, E. S. (1973). Diel feeding migrations in tropical reef fishes. *Helgol. Wiss. Meeresunters.* **24,** 361–370.

Hobson, E. S. (1974). Feeding relationships of teleostean fishes on coral reefs in Kona, Hawaii. *Fish. Bull.* **72,** 915–1031.

Hobson, E. S. (1975). Feeding patterns among tropical reef fishes. *Am. Sci.* **63,** 382–392.

Hobson, E. S. (1978). Aggregating as a defense against predators in aquatic and terrestrial environments. *In* "Contrasts in Behavior" (E. S. Reese and F. J. Lighter, eds.), pp. 219–234. Wiley, New York.

Hobson, E. S. (1979). Interactions between piscivorous fishes and their prey. *In* "Predator–Prey Systems in Fisheries Management" (H. E. Clepper, ed.), pp. 231–242. Sport Fishing Inst., Washington, D.C.

Hobson, E. S. (1982). The structure of fish communities on warm-temperate and tropical reefs. *NOAA Tech. Mem.,* **NMFS SEFC-80,** 160–166.

Hobson, E. S. (1984). The structure of reef fish communities in the Hawaiian archipelago. *Proc. Symp. Resour. Invest. Northwest. Hawaii. Islands, 2nd* **1,** 101–122.

Hobson, E. S. (1986). Predation on the Pacific sand lance, *Ammodytes hexapterus* (Pisces:Ammodytidae), during the transition between day and night in southeastern Alaska. *Copeia* **1986,** 223–226.

Hobson, E. S., and Chave, E. H. (1972). "Hawaiian Reef Animals." Univ. of Hawaii, Press, Honolulu, Hawaii.

Hobson, E. S., and Chess, J. R. (1973). Feeding oriented movements of the atherinid fish *Pranesus pinquis* at Majuro Atoll, Marshall Islands. *Fish. Bull.* **71,** 777–786.

Hobson, E. S., and Chess, J. R. (1976). Trophic interactions among fishes and zooplankters nearshore at Santa Catalina Island, California. *Fish. Bull.* **74,** 567–598.

Hobson, E. S., and Chess, J. R. (1978). Trophic relationships among fishes and plankton in the lagoon at Enewetak Atoll, Marshall Islands. *Fish. Bull.* **76,** 133–153.

Hobson, E. S., and Chess, J. R. (1979). Zooplankters that emerge from the lagoon floor at night at Kure and Midway atolls, Hawaii. *Fish. Bull.* **77,** 275–280.

Hobson, E. S., and Chess, J. R. (1986). Diel movements of resident and transient zooplankters above lagoon reefs at Enewetak Atoll, Marshall Islands. *Pac. Sci.* **40,** 7–26.

Hobson, E. S., McFarland, W. N., and Chess, J. R. (1981). Crepuscular and nocturnal activities of Californian nearshore fishes, with consideration of their scotopic visual pigments and the photic environment. *Fish. Bull.* **79,** 1–30.

Hodgson, E. S., and Mathewson, R. F. (eds.) (1978). "Sensory Biology of Sharks, Skates, and Rays." U.S. Off. Nav. Res., Dep. of the Navy, Arlington, Virginia.

Hoffman, K. S., and Grau, E. G. (1989). Daytime changes in oocyte development with relation to the tide for the Hawaiian saddleback wrasse *Thalassoma duperrey.* J. Fish Biol. **34,** 529–546.

Hoffman, S. G., and Robertson, D. R. (1983). Foraging and reproduction of two Caribbean reef toadfishes (Batrachoididae). *Bull. Mar. Sci.* **33,** 919–927.

Hoffman, S. G., Schildhauer, M. P., and Warner, R. R. (1985). The costs of changing sex and the ontogeny of males under contest competition for mates. *Evolution (Lawrence, Kans.)* **39,** 915–927.

Hoffman, A. J., and Ugarte, R. (1985). The arrival of propagules of marine macroalgae in the intertidal zone. *J. Exp. Mar. Biol. Ecol.* **92,** 83–95.

Hogan, M. E., Slaytor, M., and O'Brien, R. W. (1985). Transport of volatile fatty acids across the hindgut of the cockroach *Panesthia cribata* Saussure and the termite, *Mastotermes darwiniensis. J. Insect Physiol.* **31,** 587–591.

Hogeweg, P., and Hesper, B. (1983). The ontogeny of the interaction structure in bumble bee colonies: A mirror model. *Behav. Ecol. Sociobiol.* **12,** 271–283.

Holbrook, S. J., and Schmitt, R. J. (1984). Experimental analyses of patch selection by foraging surfperch (*Embiotoca jacksoni* Agazzi). *J. Exp. Mar. Biol. Ecol.* **79,** 39–64.

Holbrook, S. J., and Schmitt, R. J. (1986). Food acquisition by competing surfperch on a patchy environmental gradient. *Environ. Biol. Fishes,* **16,** 135–146.

Holbrook, S. J., and Schmitt, R. J. (1989). Resource overlap, prey dynamics, and the strength of competition. *Ecology* **70,** 1943–1953.

Holbrook, S. J., Carr, M. H., Schmitt, R. J., and Coyer, J. A. (1990). The effect of giant kelp on local abundance of demersal fishes: The importance of ontogenetic resource requirements. *Bull. Mar. Sci.* **47,** 104–114.

Hollowed, A. B., Bailey, K. M., and Wooster, W. S. (1987). Patterns in recruitment of marine fishes in the northeast Pacific Ocean. *Biol. Oceanogr.* **5,** 99–131.

Holt, R. D. (1984). Spatial heterogeneity, indirect interactions, and the coexistence of prey species. *Am. Nat.* **124,** 377–406.

Holt, R. D. (1987). Prey communities in patchy environments. *Oikos* **50,** 276–290.

Holzberg, S. (1973). Beobachtungen zur Okologie und zum Sozialverhalten des Korallenbarsches *Dascyllus marginatus* Ruppell (Pisces; Pomacentridae). *Z. Tierpsychol.* **33,** 492–513.

Horn, M. H. (1989). Biology of marine herbivorous fishes. *Oceanogr. Mar. Biol.* **27,** 167–272.

Horn, M. H., and Allen, L. G. (1978). A distributional analysis of California coastal marine fishes. *J. Biogeogr.* **5,** 23–42.

Horn, M. H., Murray, S. N., and Edwards, T. W. (1982). Dietary selectivity in the field and food preferences in the laboratory for two herbivouous fishes (*Cebidichthys violaceus* and *Xiphister mucosus*) from a temperate intertidal zone. *Mar. Biol.* **67,** 237–246.

Hoss, D. E., Peters, D. S., and Cummings, S. R. (1986). "Final Report to National Ocean Service, Division of Ocean Minerals and Energy, NOAA, on the Vertical Distribution and Abundance of Fish Larvae at Two Potential OTEC Sites in the Caribbean Sea." Natl. Oceanic Atmos. Adm., Rockville, Maryland.

Houde, E. D. (1987). Fish early life dynamics and recruitment variability. *Am. Fish. Soc. Symp. Ser.* **2,** 17–29.

Houde, E. D., and Lovdal, J. A. (1984). Seasonality of occurrence, foods and food preferences of ichthyoplankton in Biscayne Bay, Florida. *Estuarine, Coastal Shelf Sci.* **18,** 403–419.

Houde, E. D., and Lovdal, J. A. (1985). Patterns of variability in ichthyoplankton occurrence and abundance in Biscayne Bay, Florida. *Estuarine, Coastal Shelf Sci.* **20,** 79–103.

Houde, E. D., and Schekter, R. C. (1978). Simulated food patches and survival of larval bay anchovy, *Anchoa mitchilli,* and sea bream, *Archosargus rhomboidalis. Fish. Bull.* **76,** 483–487.

Houde, E. and Schekter, R. (1980). Feeding by marine fish larvae: Developmental and functional responses. *Environ. Biol. Fishes* **5,** 315–334.

Houde, E. D., and Schekter, R. C. (1981). Growth rates, rations, and cohort consumption of marine fish larvae in relation to prey concentrations. *Rapp. P.-V. Reun., Cons. Int. Explor. Mer* **178,** 441–453.

Houde, E. D., Leak, J. C., Dowd, C. E., and Berkely, S. A. (1979). Ichthyoplankton abundance and diversity in the eastern Gulf of Mexico. *Rep. Bur. Land Manage. Contr. AA550-CT7-28* **NTIS-PB-299839,** xxxii, 1–546.

Houde, E. D., Almatar, S., Leak, J. C., and Dowd, C. E. (1986). Ichthyoplankton abundance and diversity in the western Arabian Gulf. *Kuwait Bull. Mar. Sci.* **8,** 107–393.

Hourigan, T. F. (1989). Environmental determinants of butterflyfish social systems. *Environ. Biol. Fishes* **25,** 61–78.

Hourigan, T. F., and Reese, E. S. (1987). Mid-ocean isolation and the evolution of Hawaiian reef fishes. *Trends Ecol. Evol.* **2,** 187–191.

Hourigan, T. F., Tricas, T. C., and Reese, E. S. (1988). Coral reef fishes as indicators of environmental stress in coral reefs. *In* "Marine Organisms as Indicators" (D. F. Soule and G. S. Kleppel, eds.), pp. 107–136. Springer-Verlag, New York.

Hourigan, T. F., Stanton, F. G., Motta, P. J., Kelley, C. D., and Carlson, B. (1989). The feeding ecology of three species of Caribbean angelfishes (family Pomacanthidae). *Environ. Biol. Fishes* **24,** 105–116.

Hubbard, D. K. (1988). Controls of modern and fossil reef development. Common ground for biological and geological research. *Proc. Int. Coral Reef Symp., 6th* **1,** 243–252.

Hubbard, R. (1958). Bleaching of rhodopsin by light and by heat. *Nature (London)* **181,** 1126.

Hubbard, D. K., Miller, A. I., and Scaturo, D. (1990). Production and cycling of calcium carbonate in a shelf-edge reef system (St. Croix, U.S. Virgin Islands): Applications to the nature of reef systems in the fossil record. *J. Sediment. Petrol.* **60,** 335–360.

Hubbs, C. L. (1948). Changes in the fish fauna of western North America correlated with changes in ocean temperature. *J. Mar. Res.* **7,** 459–482.

Hubbs, C. L. (1960). The marine vertebrates of the outer coast. *Syst. Zool.* **9,** 134–147.

Hubbs, C. L. (1967). A discussion of the geochronology and archeology of the California islands. *In* "Proceedings of the Synposium on the Biology of the California Islands" (R. N. Philbrick, ed.), pp. 337–341. Santa Barbara Bot. Gard., Santa Barbara, California.

Hubbs, C. L. (1974). "Marine Zoogeography" by John C. Briggs. *Copeia* **1974,** 1002–1005.

Hubbs, C. (1976). The diel reproductive pattern and fecundity of *Menidia audens. Copeia* **1976,** 386–388.

Hughes, T. P. (1989). Community structure and diversity of coral reefs: The role of history. *Ecology* **70,** 275–279.

Hughes, T. P., Keller, B., Jackson, J. B. C., and Boyle, M. J. (1985). Mass mortality of the echinoid, *Diadema antillarum* Phillipi, in Jamaica. *Bull. Mar. Sci.* **35,** 377–384.

Hughes, T. P., Reed, D. C., and Boyle, M. J. (1987). Herbivory on coral reefs: Community structure following mass mortalities of sea urchins. *J. Exp. Mar. Biol. Ecol.* **113,** 39–59.

Hulsman, K. (1988). Seabird predation on coral reef fishes: Its magnitude and effects. *Proc. Int. Coral Reef Symp., 6th* **2,** 71–76.

Hunte, W., and Cote, I. M. (1989). Recruitment in the redlip blenny *Ophioblennius atlanticus*: Is space limiting? *Coral Reefs* **8,** 45–50.

Hunter, J. R. (1981). Feeding ecology and predation of marine fish larvae. *In* "Marine Fish Larvae: Morphology, Ecology and Relation to Fisheries" (R. Lasker, ed.), pp. 33–77. Univ. of Washington Press, Seattle, Washington.

Hunter, J. R. (1984). Inferences regarding predation on the early life stages of cod and other fishes. *Flodevigen Rapp.* **1,** 533–562.

Hunter, J. R., and Mitchell, C. T. (1967). Association of fishes with flotsam in the offshore waters of Central America. *Fish. Bull.* **66,** 13–29.

Hunter, J. R., and Mitchell, C. T. (1968). Field experiments on the attraction of pelagic fish to floating objects. *J. Cons., Cons. Int. Explor. Mer* **31,** 427–434.

Hunter, J. R., and Sanchez, C. (1976). Diel changes in swim bladder inflation of the larvae of the northern anchovy, *Engraulis mordax*. *Fish. Bull.* **74,** 847–855.

Huntsman, G. R. (1979). Predation's role in structuring reef fish communities. *In* "Predator–Prey Systems in Fisheries Management" (H. E. Clepper, ed.), pp. 103–108. Sport Fishing Inst., Washington, D.C..

Huntsman, G. R., Nicholson, W. R., and Fox, W. W. (1982). The biological bases for reef fishery management. *NOAA Tech. Memo* NMFS-SEFC-80.

Huntsman, G. R., and Waters, J. R. (1987). Development of management plans for reef fishes—Gulf of Mexico and U.S. South Atlantic. *In* "Tropical Snappers and Groupers: Biology and Fisheries Management" (J. J. Polovina and S. Ralston, eds.), pp. 533–560. Westview, Boulder, Colorado.

Huntsman, G. R., Manooch, C. S., and Grimes, C. B. (1983). Yield per recruit models of some reef fishes of the U.S. South Atlantic Bight. *Fish. Bull.* **81,** 679–695.

Hulberg, L. W., and Oliver, J. S. (1980). Caging manipulations in marine soft-bottom communities: Importance of animal interactions or sedimentary habitat modifications. *Can. J. Fish. Aquat. Sci.* **37,** 1130–1139.

Hutchings, P. A. (1986). Biological destruction of coral reefs. A review. *Coral Reefs* **4,** 239–252.

Hutchins, B., and Swainston, R. (1986). "Sea Fishes of Southern Australia." Swainston, Perth, Australia.

Hutchinson, G. E. (1961). The paradox of the plankton. *Am. Nat.* **95,** 137–145.

Ishida, K., and Kawamura, G. (1985). The early life history of marine fish: 7—Development of sense organs. *Aquabiology* **7,** 8–14 (in Japanese; Engl. abstr.).

Ishihara, M. (1987). Effect of mobbing toward predators by the damselfish *Pomacentrus coelestis* (Pisces: Pomacentridae). *J. Ethol.* **5,** 43–52.

Issacs, J. D., and Schwartzlose, R. A. (1965). Migrant sound scatterers: Interactions with the sea floor. *Science* **150,** 1810–1813.

Itzkowitz, M. (1977a). Spatial organization of the Jamaican damselfish community. *J. Exp. Mar. Biol. Ecol.* **28,** 217–242.

Itzkowitz, M. (1977b). Social dynamics of mixed-species groups of Jamaican reef fishes. *Behav. Ecol. Sociobiol.* **2,** 361–384.

Itkowitz, M. (1978). Group organization of a territorial damselfish, *Eupomacentrus planifrons*. *Behaviour* **65,** 125–137.

Itzkowitz, M. (1985). Aspects of the population dynamics and reproductive success of the permanently territorial beaugregory damselfish. *Mar. Behav. Physiol.* **12,** 57–69.

Ivlev, V. S. (1961). "Experimental Ecology of the Feeding of Fishes." Yale Univ. Press, New Haven, Connecticut.

Jablonski, D. (1986). Larval ecology and macroevolution in marine invertebrates. *Bull. Mar. Sci.* **39,** 565–587.

Jablonski, D., and Lutz, R. A. (1983). Larval ecology of marine benthic invertebrates: Paleobiological implications. *Biol. Rev. Cambridge Philos. Soc.* **58,** 21–89.

Jackson, G. A. (1986). Interaction of physical and biological processes in the settlement of planktonic larvae. *Bull. Mar. Sci.* **39,** 202–212.

Jackson, G. A., and Winant, C. D. (1983). Effect of a kelp forest on coastal currents. *Continent. Shelf Res.* **2,** 75–80.

Jackson, J. B. C., and Buss, L. W. (1975). Allelopathy and spatial competition among coral reef invertebrates. *Proc. Natl. Acad. Sci. U.S.A.* **72,** 5160–5163.

Jacoby, C. A., and Greenwood, J. G. (1988). Spatial, temporal, and behavioural patterns in emergence of zooplankton in the lagoon of Heron Reef, Great Barrier Reef, Australia. *Mar. Biol.* **97,** 309–328.

Jahn, A. E. (1987). Precision of estimates of abundance of coastal fish larvae. *Am. Fish. Soc. Symp.* **2,** 30–38.

Jamieson, G. S. (1986). Implications of fluctuations in recruitment in selected crab populations. *Can. J. Fish. Aquat. Sci.* **43,** 2085–2098.

Janekarn, V. (1988). Biogeography and environmental biology of fish larvae along the west coast of Thailand. Unpublished M.S. thesis, University of Newcastle-upon-Tyne, Newcastle-upon-Tyne, England.

Janekarn, V., and Boonruang, P. (1986). Composition and occurrence of fish larvae in mangrove areas along the east coast of Phuket Island, western peninsular Thailand. *Phuket Mar. Biol. Cent. Res. Bull.* **44,** 1–22.

Jeffries, M. J., and Lawton, J. H. (1984). Enemy free space and the structure of ecological communities. *Biol. J. Linn. Soc.* **23,** 269–286.

Jenkins, G. P., Milward, N. E., and Hartwick, R. F. (1984). Food of larvae of Spanish mackerels, genus *Scomberomorus* (Teleostei: Scombridae), in shelf waters of the Great Barrier Reef. *Aust. J. Mar. Freshwater Res.* **35,** 477–482.

Jerlov, N. G. (1968). "Optical Oceanography." Elsevier, London.

Jerlov, N. G. (1976). "Marine Optics." Elsevier, Amsterdam.

Johannes, R. E. (1978a). Reproductive strategies of coastal marine fishes in the tropics. *Environ. Biol. Fishes* **3,** 65–84.

Johannes, R. E. (1978b). Traditional marine conservation methods in Oceania and their demise. *Annu. Rev. Ecol. Syst.* **9,** 349–364.

Johannes, R. E. (1981). "Words of the Lagoon: Fishing and Marine Lore in the Palau District of Micronesia." Univ. of California Press, Berkeley, California.

Johannes, R. E., Coles, S. L., and Kuenzel, W. T. (1970). The role of zooplankton in the nutrition of some scleractinian corals. *Limnol. Oceanogr.* **15,** 579–586.

Johnson, G. D., and Washington, B. B. (1987). Larvae of the Moorish Idol, *Zanclus cornutus*, including a comparison with other larval acanthuroids. *Bull. Mar. Sci.* **40,** 494–511.

Jones, C. (1986). Determining age of larval fish with the otolith increment technique. *Fish. Bull.* **84,** 91–103.

Jones, G.P. (1984a). The influence of habitat and behavioural interactions on the local distribution of the wrasse, *Pseudolabrus celidotus. Environ. Biol. Fishes* **10,** 43–58.

Jones, G. P. (1984b). Population ecology of the temperate reef fish *Pseudolabrus celidotus* Bloch & Schneider (Pisces: Labridae). II. Factors influencing adult density. *J. Exp. Mar. Biol. Ecol.* **75,** 277–303.

Jones, G. P. (1984c). Population ecology of the temperate reef fish *Pseudolabrus celidotus* Bloch & Schneider (Pisces: Labridae). I. Factors influencing recruitment. *J. Exp. Mar. Biol. Ecol.* **75,** 257–276.

Jones, G. P. (1986). Food availability affects growth in a coral reef fish. *Oecologia* **70,** 136–139.

Jones, G. P. (1987a). Competitive interactions among adults and juveniles in a coral reef fish. *Ecology* **68,** 1534–1547.

Jones, G. P. (1987b). Some interactions between residents and recruits in two coral reef fishes. *J. Exp. Mar. Biol. Ecol.* **114,** 169–182.

Jones, G. P. (1988a). Ecology of rocky reef fish of north-eastern New Zealand: A review. N. Z. J. Mar. Reshwater Res. **22,** 445–462.

Jones, G. P. (1988b). Experimental evaluation of the effects of habitat structure and competitive interactions on the juveniles of two coral reef fishes. *J. Exp. Mar. Biol. Ecol.* **123,** 115–126.

Jones, G. P. (1990). The importance of recruitment to the dynamics of a coral reef fish population. *Ecology* **71,** 1691–1698.

Jones, G. P. (1991). Manuscript in preparation.

Jones, R. (1982). Ecosystems, food chains and fish yields. *ICLARM Conf. Proc.* **9,** 195–239.

Jones, R. S. (1968). Ecological relationships in Hawaiian and Johnston Island Acanthuridae (surgeonfishes). *Micronesica* **4,** 309–361.

Jones, G. P., and Andrew, N. L. (1990). Herbivory and patch dynamics on rocky reefs in temperate Australasia: The roles of fish and sea urchins. *Aust. J. Ecol.* **15,** 505–520.

Jones, G. P., and Norman, M. R. (1986). Feeding selectivity in relation to territory size in a herbivorous reef fish. *Oecologia* **68,** 549–556.

Jones, G. P., and Thompson, S. M. (1980). Social inhibition of maturation in females of the temperate wrasse *Pseudolabrus celidotus* and a comparison with the blennioid *Tripterygion varium. Mar. Biol.* **59,** 247–256.

Jones, R. S., and Chase, J. A. (1975). Community structure and distribution of fishes in an enclosed high island lagoon in Guam. *Micronesica* **11,** 127–148.

Jones, R. S., and Thompson, M. J. (1978). Comparison of Florida reef fish assemblages using a rapid visual technique. *Bull. Mar. Sci.* **28,** 159–172.

Jones, G. P., Sale, P. F., and Ferrell, D. J. (1988). Do large carnivorous fishes affect the ecology of macrofauna in shallow lagoonal sediments?: A pilot experiment. *Proc. Int. Coral Reef Symp., 6th* **2,** 77–82.

Jones, G. P., Ferrell, D. J., and Sale, P. F. (1990). Spatial pattern in the abundance and

structure of mollusc populations in the soft sediments of a coral reef lagoon. *Mar. Ecol. Prog. Ser.* **62,** 109–120.
Jones, G. P., Ferrell, D. J., and Sale, P. F. (1991). Fish feeding and the dynamics of soft sediment mollusc populations in a coral reef lagoon. (Submitted for publication.)
Jones, G. P., Ferrell, D. J., and Sale, P. F. Fish predation and the structure of soft-sediment communities within a tropical lagoon. *Mar. Ecol. Prog. Ser.* In press.
Jordan, D. S., and Evermann, B. W. (1905). The aquatic resources of the Hawaiian Islands. I. The shore fishes. *Bull. U. S. Fish Comm.* **23(1),** 1–574 (plus 73 plates).
Kami, H. T., and Ikehara, I. I. (1976). Notes on the annual juvenile siganid harvest in Guam. *Micronesica* **12,** 323–325.
Karowe, D. N. (1989). Differential effect of tannic acid on two tree-feeding Lepidoptera: Implications for theories of plant anti-herbivore chemistry. *Oecologia* **80,** 507–512.
Karplus, I., Tsurnamal, M., and Szlep, R. (1972). Analysis of the mutual attraction in the association of the fish *Cryptocentrus cryptocentrus* (Gobiidae) and the shrimp *Alpheus djiboutensis* (Alpheidae). *Mar. Biol.* **17,** 275–283.
Katzir, G. (1981). Aggresssion by the damselfish *Dascyllus aruanus* (L.) towards conspecifics and heterospecifics. *Anim. Behav.* **29,** 835–841.
Kaufman, L. S. (1976). Feeding behavior and functional coloration of the Atlantic trumpetfish, *Aulostomus maculatus. Copeia* **1976,** 377–378.
Kaufman, L. S. (1977). The threespot damselfish: Effects on benthic biota of Caribbean coral reefs. *Proc. Int. Coral Reef Symp., 3rd* **1,** 559–564.
Kaufman, L. S. (1983). Effects of hurricane Allen on reef fish assemblages near Discovery Bay, Jamaica. *Coral Reefs* **2,** 43–47.
Kaufman, L. S., and Ebersole, J. P. (1984). Microtopography and the organization of two assemblages of coral reef fishes in the West Indies. *J. Exp. Mar. Biol. Ecol.* **78,** 253–268.
Kawamura, G. (1984). The sense organs and behavior of milkfish fry in relation to collection techniques. *In* "Advances in Milkfish Biology and Culture" (J. V. Juario, R. P. Ferraris, and L. V. Benitez, eds.), pp. 69–84. Island Publ., Manila, Philippines.
Kawamura, G. and Hara, S. (1980). On the visual feeding of milkfish larvae and juveniles in captivity. *Bull. Jap. Soc. Sci. Fish* **46,** 1297–1300.
Kawamura, G., Tsuda, R., Kumai, H., and Ohiashi, S. (1984). The visual cell morphology of *Pagrus major* and its adaptive changes with shift from pelagic to benthic habitats. *Bull. Jap. Soc. Sci. Fish.* **50,** 1975–1980.
Kay, E. A. (1980). Little worlds of the Pacific, an essay of Pacific Basin biogeography. *Univ. Hawaii, Harold L. Lyon Arbor. Lect.* **9,** 1–40.
Kay, E. A., and Palumbi, S. R. (1987) Endemism and evolution in Hawaiian marine invertebrates. *Trends Ecol. Evol.* **2,** 183–186.
Keener, P., Johnson, G. D., Stender, B. W., Brothers, E. B., and Beatty, H. R. (1988). Ingress of postlarval gag *Mycteroperca microlepis* (Pisces: Serranidae), through a South Carolina barrier island inlet. *Bull Mar. Sci.* **42,** 376–396.
Keller, B. D. (1983). Coexistence of sea urchins in seagrass meadows: An experimental analysis of competition and predation. *Ecology* **64,** 1581–1598.
Kennelly, S. J. (1987a). Physical disturbances in an Australian kelp community. I. Temporal effects. *Mar. Ecol. Prog. Ser.* **40,** 145–153.

Kennelly, S. J. (1987b). Physical disturbances in an Australian kelp community. II. Effects on understorey species due to differences in kelp cover. *Mar. Ecol. Prog. Ser.* **40,** 155–165.

Kennelly, S. J. (1989). Effects of kelp canopies on understorey species due to shade and scour. *Mar. Ecol. Prog. Ser.* **50,** 215–224.

Keough, M. J. (1988). Benthic populations: Is recruitment limiting or just fashionable? *Proc. Int. Coral Reef Symp. 6th* **1,** 141–148.

Kerfoot, W. C., and Sih, A. (eds.) (1987). "Predation: Direct and Indirect Impacts on Aquatic Communities." Univ. Press of New England, Hanover, New Hampshire.

Kiene, W. E. (1988). A model of bioerosion on the Great Barrier Reef. *Proc. Int. Coral Reef Symp., 6th* **3,** 449–454.

Kingett, P. D., and Choat, J. H. (1981). Analysis of density and distribution patterns in *Chrysophrys auratus* (Pisces: Sparidae) within a reef environment: An experimental approach. *Mar. Ecol. Prog. Ser.* **5,** 283–290.

Kingsford, M. J. (1980). Interrelationships between spawning and recruitment of *Chromis dispilus* (Pisces: Pomacentridae). Unpublished M.S. thesis, University of Auckland, Auckland, New Zealand.

Kingsford, M. J. (1988). The early life history of fish in coastal waters of northern New Zealand: A review. *N. Z. J. Mar. Freshwater Res.* **22,** 463–479.

Kingsford, M. J. (1989). Distribution patterns of planktivorous reef fish along the coast of northeastern New Zealand. *Mar. Ecol. Prog. Ser.* **54,** 13–24.

Kingsford, M. J., and Choat, J. H. (1985). The fauna associated with drift algae captured with a plankton-mesh purse seine net. *Limnol. Oceanogr.* **30,** 618–630.

Kingsford, M. J., and Choat, J. H. (1986). Influence of surface slicks on the distribution and onshore movements of small fish. *Mar. Biol.* **91,** 161–171.

Kingsford, M. J., and Choat, J. H. (1989). Horizontal distribution patterns of presettlement reef fish: Are they influenced by the proximity of reefs: *Mar. Biol.* **101,** 285–297.

Kingsford, M. J., and MacDiarmid, A. B. (1988). Interrelations between planktivorous reef fishes and zooplankton in temperate waters. *Mar. Ecol. Prog. Ser.* **48,** 103–117.

Kingsford, M. J., and Milicich, M. J. (1987). Presettlement phase of *Parika scaber* (Pisces: Monacanthidae): A temperate reef fish. *Mar. Ecol. Prog. Ser.* **36,** 65–79.

Kingsford, M. J., Schiel, D. R., and Battershill, C. N. (1989). Distribution and abundance of fish in a rocky reef environment at the subantarctic Auckland Islands, New Zealand. *Polar Biol.* **9,** 179–186.

Kingsford, M. J., Wolanski, E., and Choat, J. H. (1991). Influence of tidally-induced fronts and langmuir circulations on the distribution and movements of presettlement fishes around a coral reef. *Mar. Biol.* **109,** 167–180.

Kirk, J. T. O. (1983). "Light and Photosynthesis in Aquatic Ecosystems." Cambridge Univ. Press, Cambridge, England.

Klumpp, D. W., and Nichols, P. D. (1983). Nutrition of the southern sea garfish *Hyporhamphus melanochir*: Gut passage rate and daily consumption of two food types and assimilation of seagrass components. *Mar. Ecol. Prog. Ser.* **12,** 207–216.

Klumpp, D. W., and Polunin, N. V. C. (1989). Partitioning among grazers of food resources within damselfish territories on a coral reef. *J. Exp. Mar. Biol. Ecol.* **125,** 145–169.

Klumpp, D. W., McKinnon, D., and Daniel, P. (1987). Damselfish territories: Zones of high productivity on coral reefs. *Mar. Ecol. Prog. Ser.* **40,** 41–51.

Klumpp, D. W., McKinnon, A. D., and Mundy, C. N. (1988). Motile cryptofauna of a coral reef: Abundance, distribution and trophic potential. *Mar. Ecol. Prog. Ser.* **45,** 95–108.

Knowlton, N., Lang, J. C., and Keller, B. D. (1988). Fates of staghorn coral isolates on hurricane damaged reefs in Jamaica: The roles of predators. *Proc. Int. Coral Reef Symp., 6th* **2,** 83–88.

Kobayashi, D. R. (1989). Fine-scale distribution of larval fishes: Patterns and processes adjacent to coral reefs in Kaneohe Bay, Hawaii. *Mar. Biol.* **100,** 285–293.

Kock, R. L. (1982). The pattern of abundance variation in reef fishes near an artificial reef in Guam. *Environ. Biol. Fishes* **7,** 121–136.

Kohda, M. (1988). Diurnal periodicity of spawning activity of permanently territorial damselfishes (Teleostei: Pomacentridae). *Environ. Biol. Fishes* **21,** 91–100.

Kohn, A. J. (1956). Piscivorous gastropods of the genus *Conus*. *Proc. Nat. Acad. Sci.* **42,** 168–171.

Kohn, A. J. (1985). Evolutionary ecology of *Conus* on Indo-Pacific coral reefs. *Proc. Int. Coral Reef Congr., 5th* **4,** 139–144.

Koslow, J. A. (1984). Recruitment patterns in northwest Atlantic fish stocks. *Can. J. Fish. Aquat. Sci.* **41,** 1722–1729.

Koslow, J. A. (1991). The influence of fishing on reef fish of the western Caribbean and Bahamas. *Mar. Biol.* (submitted).

Koslow, J. A., Hanley, F., and Wicklund, R. (1988). Effects of fishing on reef fish communities at Pedro Bank and Port Royal cays, Jamaica. *Mar. Ecol. Prog. Ser.* **43,** 201–212.

Kotrschal, K. (1988). Evolutionary patterns in tropical marine reef fish feeding. *Z. Zool. Syst. Evol.-forsch.* **26,** 51–64.

Kotrschal, K., and Thomson, D. A. (1986). Feeding patterns in eastern Pacific blennioid fishes (Teleostei: Tripterygiidae, Labrisomidae, Chaenopsidae, Bleniidae). *Oecologia* **70,** 367–378.

Krebs, C. R. (1978). "Ecology: The Experimental Analysis of Distribution and Abundance." Harper & Row, New York.

Kulbicki, M. (1988). Patterns in the trophic structure of fish populations across the SW lagoon of New Caledonia. *Proc. Int. Coral Reef Symp., 6th* **2,** 89–94.

Kulbicki, M. (1991). Manuscript in preparation.

Kulbicki, M., and Grandperrin, R. (1988). Survey of the soft bottom carnivorous fish population using bottom longline in the south-west lagoon of New Caledonia. *Backgr. Pap. S. Pac. Comm., Workshop Pac. Inshore Fish. Resour., SPC/Inshore Fish. Res.* **BP.15,** 1–25.

Kuwamura, T. (1984). Social structure of the protogynous fish *Labroides dimidiatus*. *Publ. Seto Mar. Biol. Lab.* **29,** 117–174.

Labelle, M., and Nursall, J. R. (1985). Some aspects of the early life history of the redlip blenny, *Ophioblennius atlanticus* (Teleostei: Blenniidae). *Copeia* **1985,** 35–59.

Lachner, E. A. (1953). Family Apogonidae: Cardinal fishes. *Bull.—U.S. Nat. Mus.* **202(1),** 412–498.

Lachner, E. A., and Karnella, S. J. (1980). Fishes of the Indo-pacific genus *Eviota* with descriptions of eight new species (Teleostei: Gobiidae). *Smithson. Contrib. Zool.* **315,** 1–127.

Lambert, T. C., and Ware, D. M. (1984). Reproductive strategies of demersal and pelagic spawning fish. *Can. J. Fish Aquat. Sci.* **41,** 1565–1569.

Larkin, P. A. (1982). Directions for future research in tropical multispecies fisheries. *ICLARM Conf. Proc.* **9,** 309–328.

Larkin, P. A. (1984). Strategies for multispecies management. *In* "Exploitation of Marine Communities" (R. M. May, ed.), pp. 287–301. Springer-Verlag, Berlin.

Larkum, A. W. D. (1983). The primary productivity of plant communities on coral reefs. *In* "Perspectives on Coral Reefs" (D. J. Barnes, ed.), pp. 221–230. Aust. Inst. Mar. Sci., Townsville, Australia.

Larson, R. J. (1980). Competition, habitat selection, and the bathymetric segregation of two rockfish (*Sebastes*) species. *Ecol. Monogr.* **50,** 221–239.

Larson, R. J., and DeMartini, E. E. (1984). Abundance and vertical distribution of fishes in a cobble-bottom kelp forest off San Onofre, California. *Fish. Bull.* **82,** 37–53.

Lasker, R. (1975). Field criteria for survival of anchovy larvae: The relation between inshore chlorophyll maximum layers and successful first feeding. *Fish. Bull.* **73,** 453–462.

Lassig, B. R. (1982). The minor role of large transient fishes in structuring full scale coral patch reef fish assemblages. Unpublished Ph.D. dissertation, Macquarie University, Sydney, Australia.

Lassig, B. R. (1983). The effects of a cyclonic storm on coral reef fish assemblages. *Environ. Biol. Fishes* **9,** 55–63.

Lassuy, D. R. (1980). Effects of "farming" behaviour by *Eupomacentrus lividus* and *Hemiglyphidodon plagiometapon* on algal community structure. *Bull. Mar. Sci.* **30,** 304–312.

Lauder, G. V., and Liem, K. F. (1981). Prey capture by *Luciocephalus pulcher*: Implications for models of jaw protrusion in teleost fishes. *Environ. Biol. Fishes* **6,** 257–268.

Lauder, G. V., and Liem, K. F. (1983). The evolution and interrelationships of the actinopterygian fishes. *Bull. Mus. Comp. Zool.* **150,** 95–197.

Laur, D. R., and Ebeling, A. W. (1983). Predator–prey relationships in surfperches. *Environ. Biol. Fishes* **8,** 217–229.

Lawrence, J. M. (1975). On the relationships between marine plants and sea urchins. *Oceanogr. Mar. Biol.* **13,** 213–286.

Lawton, J. (1987). Problems of scale in ecology. *Nature* (*London*) **325,** 206.

Leaman, B. M. (1976). The association between the black rockfish (*Sebastes melanops* Girard) and beds of the giant kelp (*Macrocystis intergrifolia* Bory) in Barkley Sound, British Columbia. Unpublished M.S. thesis, University of British Columbia, Vancouver, British Columbia, Canada.

Leaman, B. M. (1980). "The Ecology of Fishes in British Columbia Kelp Beds. I. Barkley Sound *Nereocystis* Beds," Fish. Dev. Rep. 22. Mar. Resour. Branch, Minist. Environ., Vancouver, British Columbia, Canada.

Leigh, E. G., Jr., Charnov, E. L., and Warner, R. R. (1976). Sex ratio, sex change, and natural selection. *Proc. Natl. Acad. Sci. U.S.A.* **73,** 3656–3660.

Leis, J. M. (1978a). Distributional ecology of ichthyoplankton and invertebrate macrozooplankton in the vicinity of a Hawaiian coastal power station. Unpublished Ph.D. thesis, University of Hawaii, Honolulu, Hawaii.

Leis, J. M. (1978b). Systematics and zoogeography of the porcupinefishes (*Diodon*, Diodontidae, Tetraodontiformes), with comments on egg and larval development. *Fish. Bull.* **76,** 535–567.

Leis, J. M. (1981). Distribution of fish larvae around Lizard Island, Great Barrier Reef: Coral reef lagoon as refuge. *Proc. Int. Coral Reef Symp., 4th* **2,** 471–477.

Leis, J. M. (1982a). Nearshore distributional gradients of larval fish (15 taxa) and planktonic crustaceans (6 taxa) in Hawaii. *Mar. Biol.* **72,** 89–97.

Leis, J. M. (1982b). Hawaiian creediid fishes (*Crystallodytes cookei* and *Limnichthys donaldsoni*): Development of eggs and larvae and use of pelagic eggs to trace coastal water movement. *Bull. Mar. Sci.* **32,** 166–180.

Leis, J. M. (1983). Coral reef fish larvae (Labridae) in the East Pacific Barrier. *Copeia* **1983,** 826–828.

Leis, J. M. (1984). Larval fish dispersal and the East Pacific Barrier. *Oceanogr. Trop.* **19,** 181–192.

Leis, J. M. (1986a). Vertical and horizontal distribution of fish larvae near coral reefs at Lizard Island, Great Barrier Reef. *Mar. Biol.* **90,** 505–516.

Leis, J. M. (1986b). Ecological requirements of Indo-Pacific larval fishes: A neglected zoogeographic factor. *In* "Indo-Pacific Fish Biology: Proceedings of the Second International Conference on Indo-Pacific Fishes, Tokyo" (T. Uyeno, R. Arai, T. Taniuchi, and K. Matsuura, eds.), pp. 759–766. Ichthyol. Soc. Jpn., Tokyo.

Leis, J. M. (1986c). Epibenthic schooling by larvae of the clupeid fish *Spratelloides gracilis*. Japan J. Ichthyol. **33,** 67–69.

Leis, J. M. (1987). Review of the early life history of tropical groupers (Serranidae) and snappers (Lutjanidae). *In* "Tropical Snappers and Groupers: Biology and Fisheries management" (J. J. Polovina and S. Ralston, eds.), pp. 189–237. Westview, Boulder, Colorado.

Leis, J. M. (1989). Larval biology of butterflyfishes (Pisces, Chaetodontidae): What do we really know? *Environ. Biol. Fishes* **25,** 87–100.

Leis, J. M. (1991). Vertical distribution of fish larvae in the Great Barrier Reef Lagoon, Australia. *Mar. Biol.* **109,** 157–166.

Leis, J. M., and Goldman, B. (1984). A preliminary distributional study of fish larvae near a ribbon coral reef in the Great Barrier Reef. *Coral Reefs* **2,** 197–203.

Leis, J. M., and Goldman, B. (1987). Composition and distribution of larval fish assemblages in the Great Barrier Reef Lagoon, near Lizard Island, Australia. *Aust. J. Mar. Freshwater Res.* **38,** 211–223.

Leis, J. M., and Miller, J. M. (1976). Offshore distributional patterns of Hawaiian fish larvae. *Mar. Biol.* **36,** 359–367.

Leis, J. M., and Moyer, J. T. (1985). Development of eggs, larvae and pelagic juveniles of three Indo-Pacific ostraciid fishes (Tetraodontiformes): *Ostracion meleagris, Lactoria fornasini* and *L. diaphana*. *Japan. J. Ichthyol.* **32,** 189–202.

Leis, J. M., and Reader, S. E. (1991). Distributional ecology of larval milkfish, *Chanos*

chanos (Pisces, Chanidae), in the Lizard Island region. *Environ. Biol. Fishes* **30(4),** 395–405.
Leis, J. M., and Rennis, D. S. (1983). "The Larvae of Indo-Pacific Coral Reef Fishes." New South Wales Univ. Press, Sydney, Australia, and Univ. of Hawaii Press, Honolulu, Hawaii.
Leis, J. M., and Trnski, T. (1989). "The Larvae of Indo-Pacific Shorefishes." New South Wales Univ. Press, Sydney, Australia, and Univ. of Hawaii Press, Honolulu, Hawaii.
Leis, J. M., Goldman, B., and Ueyanagi, S. (1987). Distribution and abundance of billfish larvae (Pisces: Istiophoridae) in the Great Barrier Reef Lagoon and Coral Sea near Lizard Island, Australia. *Fish. Bull.* **85,** 757–766.
Leis, J. M., Goldman, B., and Reader, S. E. (1989). Epibenthic fish larvae in the Great Barrier Reef Lagoon near Lizard Island, Australia. Japan. J. Ichthyol. **35,** 423–433.
Lenanton, R. C. L. (1982). Nearshore accumulations of detached macrophytes as nursery areas for fish. *Mar. Ecol. Prog. Ser.* **9,** 51–58.
Lessios, H. A. (1988). Mass mortality of *Diadema antillarum* in the Caribbean: What have we learned? *Annu. Rev. Ecol. Syst.* **19,** 371–393.
Leum, L. L., and Choat, J. H. (1980). Density and distribution patterns of the temperate marine fish *Cheilodactylus spectabilis* (Cheilodactylidae) in a reef environment. *Mar. Biol.* **57,** 327–337.
Levin, S. A. (1976). Population dynamic models in heterogeneous environments. *Annu. Rev. Ecol. Syst.* **7,** 287–310.
Levine, J. S., and MacNichol, E. F. (1979). Visual pigments in teleost fishes: Effects of habitat, microhabitat, and behavior on visual system evolution. *Sens. Processes* **3,** 95–131.
Levine, J. S., and MacNichol, E. F. (1982). Color vision in fishes. *Sci. Am.* **246,** 140–149.
Lewis, J. B. (1977). Processes of organic production on coral reefs. *Biol. Rev. Cambridge Philos. Soc.* **52,** 305–347.
Lewis, S. M. (1985). Herbivory on coral reefs: Algal susceptibility to herbivorous fishes. *Oecologia* **65,** 370–375.
Lewis, S. M. (1986). The role of herbivorous fishes in the organization of a Caribbean reef community. *Ecol. Monogr.* **56,** 183–200.
Lewis, S. M., and Wainwright, P. C. (1985). Herbivore abundance and grazing intensity on a Caribbean coral reef. *J. Exp. Mar. Biol. Ecol.* **87,** 215–228.
Lewis, S. M., Norris, J. N., and Searles, R. B. (1987). The regulation of morphological plasticity in tropical reef algae by herbivory. *Ecology* **68,** 636–641.
Liem, K. F., and Sanderson, S. L. (1986). The pharyngeal jaw apparatus of labrid fishes: A functional morphological perspective. *J. Morphol.* **187,** 143–158.
Liew, H.-C. (1983). Studies on flatfish larvae (Fam. Psettodidae, and Bothidae, Pleuronectiformes) in the shelf waters of the central Great Barrier Reef, Australia. Unpublished M.S. thesis, James Cook University, Townsville, Australia.
Limbaugh, C. (1955). "Fish Life in the Kelp Beds and the Effects of Kelp Harvesting," IMR Ref. 55-9. Univ. Calif. Inst. Mar. Resour., La Jolla, California.
Limbaugh, C. (1961). Cleaning symbiosis. *Sci. Am.* **205,** 42–49.

Lincoln-Smith, M. P. (1988). Effects of observer swimming speed on sample counts of temperate rocky reef fish assemblages. *Mar. Ecol. Prog. Ser.* **43,** 223–231.

Lindeman, K. C. (1989). Coastal construction, larval settlement and early juvenile habitat use in grunts, snappers and other coastal fishes of southeast Florida. *Bull. Mar. Sci.* **44,** 1068..

Lindsey, C. C. (1966). Body size of poikilotherm vertebrates at different latitudes. *Evolution (Lawrence, Kans.)* **20,** 456–465.

Little, M. C., Reay, P. J., and Grove, S. J. (1988). Distribution gradients of ichthyoplankton in an East African mangrove creek. *Estuarine, Coastal Shelf Sci.* **26,** 669–677.

Littler, M. M., and Littler, D. S. (1980). The evolution of thallus form and survival strategies in benthic marine macroalgae: Field and laboratory tests of a functional form model. *Am. Nat.* **116,** 25–44.

Littler, M. M., and Littler, D. S. (1984). Models of tropical reef biogenesis. *Prog. Phycol. Res.* **3,** 323–364.

Littler, M. M., Littler, D. S., and Taylor, P. R. (1983a). Evolutionary strategies in a tropical barrier reef system: Functional-form groups of marine macroalgae. *J. Phycol* **19,** 229–237.

Littler, M. M., Taylor, P. R., and Littler, D. S. (1983b). Algal resistance to herbivory on a Caribbean barrier reef. *Coral Reefs* **2,** 111–118.

Littler, M. M., Taylor, P. R., and Littler, D. S. (1986). Plant defense associations in the marine environment. *Coral Reefs* **5,** 63–71.

Littler, M. M., Littler, D. S., and Taylor, P. R. (1987). Animal–plant defense associations: Effects on the distribution and abundance of tropical reef macrophytes. *J. Exp. Mar. Biol. Ecol.* **105,** 107–121.

Lloyd, D. G., and Bawa, K. S. (1984). Modification of the gender of seed plants in varying conditions. *Evol. Biol.* **17,** 255–338.

Lobel, P. S. (1978). Diel, lunar, and seasonal periodicity in the reproductive behavior of the pomacanthid *Centropyge potteri,* and some other reef fishes in Hawaii. *Pac. Sci.* **32,** 193–207.

Lobel, P. S (1980). Herbivory by damselfishes and their role in coral reef community ecology. *Bull. Mar. Sci.* **30,** 273–289.

Lobel, P. S. (1981). Trophic biology of herbivorous reef fishes: Alimentary pH and digestive capabilities. *J. Fish Biol.* **19,** 365–397.

Lobel, P. S. (1989). Ocean current variability and the spawning season of Hawaiian reef fishes. *Environ. Biol. Fishes* **24,** 161–171.

Lobel, P. S., and Robinson, A. R. (1983). Reef fishes at sea: Ocean currents and the advecton of larvae. *In* "The Ecology of Deep and Shallow Coral Reefs" (M. L. Reaka, ed.), NOAA Symp. Ser. Undersea Res., Vol. 1, pp. 29–38. Natl. Oceanic Atmos. Adm., Rockville, Maryland.

Lobel, P. S., and Robinson, A. R. (1985). The potential role of ocean eddies in the life histories of Hawaiian fishes. *Spec. Publ.—Hawaii. Inst. Geophys.* **1985,** 61–85.

Lobel, P. S., and Robinson, A. R. (1986). Transport and entrapment of fish larvae by ocean mesoscale eddies and currents in Hawaiian waters. *Deep-Sea Res.* **33,** 483–500.

Lobel, P. S., and Robinson, A. R. (1988). Larval fishes and zooplankton in a cyclonic eddy in Hawaiian waters. *J. Plankton Res.* **10,** 1209–1223.

Lock, J. M. (1986). Fish yields of the Port Moresby barrier and fringing reefs. *Tech. Rep. Dep. Primary Ind.* (*Papua New Guinea*) **86/2,** 1–17.

Locket, N. A. (1977). Adaptations to the deep-sea environment. *Handb. Sens. Physiol.* **7,** 67–192.

Loew, E. R., and Lythgoe, J. N. (1978). The ecology of cone pigments in teleost fishes. *Vision Res.* **18,** 715–722.

Loew, E. R., and McFarland, W. N. (1990). The underwater visual environment. *In* "The Visual System of Fish" (N. R. Douglas and M. Djamgoz, eds.), pp. 1–43. Chapman & Hall, London.

Longhurst, A. R. (1984). Heterogeneity in the ocean—Implications for fisheries. *Rapp. P.-V. Reun., Cons. Int. Explor. Mer.* **185,** 268–282.

Longhurst, A. R., and Pauly, D. (1987). "Ecology of Tropical Oceans," Academic Press, Orlando, Florida.

Longley, W. H. (1917). Studies upon the biological significance of animal coloration. I. The colors and color changes of West Indian reef fishes. *J. Exp. Zool.* **23,** 533–601.

Loosanoff, V. L. (1964). Variations in time and intensity of settling of the starfish, *Asterias forbesi,* in Long Island Sound during a twenty-five year period. *Biol. Bull.* (Woods Hole, Mass.) **126,** 423–439.

Losey, G. S. (1972a). The ecological importance of cleaning symbiosis. *Copeia* **1972,** 820–833.

Losey, G. S (1972b). Predation protection in the poison-fang blenny, *Meiacanthus atrodorsalis,* and its mimics, *Escenius bicolor* and *Runula laudadus* (Blenniidae). *Pac. Sci.* **26,** 129–139.

Losey, G. S., Jr. (1974). Cleaning symbiosis in Puerto Rico with comparison to the tropical Pacific. *Copeia* **1974,** 960–970.

Love, M. S., and Ebeling, A. W. (1978). Food and habitat of three switch-feeding fishes in the kelp forests off Santa Barbara, California. *Fish. Bull.* **76,** 257–271.

Low, R. M. (1971). Interspecific territoriality in a pomacentrid reef fish, *Pomacentrus flavicauda* Whitley. *Ecology* **52,** 648–654.

Lowe-McConnell, R. H. (1979). Ecological aspects of seasonality in fishes of tropical waters. *Symp. Zool. Soc. London* **44,** 219–241.

Lubbock, R. (1980). The shore fishes of Ascension Island. *J. Fish Biol.* **17,** 283–303.

Lubchenco, J. (1978). Plant species diversity in a marine intertidal community: Importance of herbivore food preference and algal competitive abilities. *Am. Nat.* **112,** 23–39.

Lubchenco, J., and Gaines, S. D. (1981). A unified approach to marine plant–herbivore interactions. I. Populations and communities. *Annu. Rev. Ecol. Syst.* **12,** 405–437.

Lubchenco, J., Menge, B. A., Garrity, S. D., Lubchenco, P. J., Ashkensas, L. R., Gaines, S. D., Emlet, R., Lucas, J., and Strauss, S. (1984). Structure, persistence, and role of consumers in a tropical rocky intertidal community (Taboguilla Island, Bay of Panama). *J. Exp. Mar. Biol. Ecol.* **78,** 23–73.

Luckhurst, B. E., and Luckhurst, K. (1978a). Nocturnal observations of coral reef fishes along depth gradients. *Can. J. Zool.* **56,** 155–158.

Luckhurst, B. E., and Luckhurst, K. (1978b). Diurnal space utilization in coral reef fish communities. *Mar. Biol.* **49,** 325–332.

Luckhurst, B. E., and Luckhurst, K. (1978c). Analysis of the influence of the substrate variables on coral reef fish communities. *Mar. Biol.* **49,** 317–323.

Lundberg, B., and Lipkin, Y. (1979). Natural food of the herbivorous rabbitfish (*Siganus* spp.) in the northern Red Sea. *Bot. Mar.* **22,** 173–181.

Lythgoe, J. N. (1966). Underwater vision. *In* "British Sub-Aqua Club Diving Manual," 2nd ed. Eaton, London, England.

Lythgoe, J. N. (1968). Visual pigments and visual range underwater. *Vision Res.* **8,** 997–1012.

Lythgoe, J. N. (1979). "The Ecology of Vision." Oxford Sci. Publ., Oxford, England.

Lythgoe, J. N. (1984). Visual pigments and environmental light. *Vision Res.* **24,** 1539–1550.

Lythgoe, J. N. (1988). Light and vision in the aquatic environment. *In* "Sensory Biology of Aquatic Animals" (J. Atema, R. R. Fay, A. N. Popper, and W. N. Tavolga, eds.), pp. 75–82. Springer-Verlag, New York.

Lythgoe, J. N., and Northmore, D. P. M. (1973). Colours underwater. *In* "Colour '73," pp. 77–98. Hilger, London.

MacArthur, R. H. (1972). "Geographical Ecology." Harper & Row, New York.

MacDonald, C. D. (1976). Nesting rhythmicity in the damselfish *Plectroglyphidodon johnstonianus* (Perciformes, Pomacentridae) in Hawaii. *Pac. Sci.* **30,** 216.

MacDonald, C. D. (1981). "Reproductive strategies and social organization in damselfishes." Unpublished Ph.D thesis, University of Hawaii, Honolulu, Hawaii.

MacDonald, C. D. (1985). Oceanographic climate and Hawaiian spiny lobster larval recruitment. *Spec. Publ.—Hawaii. Inst. Geophys.* **1985,** 127–136.

Mahmoudi, B., Powers, J. E., and Huntsman, G. R. (1984). Assessment of vermilion snapper of the northern South Atlantic Bight. *NOAA Southeast Fish. Cent., U.S. Natl. Mar. Fish. Serv.,* **SAW/84/RFR/2,** 1–36.

Major, P. F. (1977). Predator–prey interactions in schooling fishes during periods of twilight: A study of the silverside *Pranesus insularum* in Hawaii. *Fish. Bull.* **75,** 415-426.

Major, P. F. (1978). Predator–prey interactions in two schooling fishes, *Caranx ignobilis* and *Stolephorus purpureus. Anim. Behav.* **26,** 760–777.

Mann, K. H. (1982). Ecology of coastal waters: A system approach. *Stud. Ecol.* **8,** 1–322.

Manooch, C. S., III (1987). Age and growth of snappers and groupers. *In* "Tropical Snappers and Groupers: Biology and Fisheries Management" (J. J. Polovina and S. Ralston, eds.), pp. 329–373. Westview, Boulder, Colorado.

Mapstone, B. D. (1988). Patterns in the abundance of *Pomacentrus molluccensis* Bleeker. Unpublished Ph.D. thesis, University of Sydney, Sydney, Australia.

Mapstone, B. D., and Fowler, A. J. (1988). Recruitment and the structure of assemblages of fish on coral reefs. *Trends Ecol. Evol.* **3,** 72–77.

Marliave, J. B. (1977). Development of behavior in marine fish. *Tech. Rep.—Mem. Univ. Mar. Sci. Res. Lab.* **20,** 240–267.

Marliave, J. B. (1981). Vertical migrations and larval settlement in *Gilbertidia sigalutes,* F. Cottidae. *Rapp. P.-V. Reun., Cons. Int. Explor. Mer* **178,** 349–351.

Marliave, J. B. (1986). Lack of planktonic dispersal of rocky intertidal fish larvae. *Trans. Am. Fish. Soc.* **115,** 149–154.

Marraro, C. H., and Nursall, J. R. (1983). The reproductive periodicity and behaviour of *Ophioblennius atlanticus* (Pisces: Blenniidae) at Barbados. *Can. J. Zool.* **61,** 317–325.

Marshall, N. (1980). Fishery yields of coral reefs and adjacent shallow-water environments. *In* "Stock Assessment for Tropical Small-Scale Fisheries" (S. B. Saila and P. M. Roedel, eds.), pp. 103–109. Univ. of Rhode Island, Kingston, Rhode Island.

Marshall, N. (1985). Ecological sustainable yield (fisheries potential) of coral reefs, as related to physiographic features of coral reef environments. *Proc. Int. Coral Reef Congr., 5th* **5,** 525–530.

Marten, G. G., and Polovina, J. J. (1982). A comparative study of fish yields from various tropical ecosystems. *ICLARM Conf. Proc.* **9,** 255–285.

Masuda, H., Amaoka, K., Araga, C., Uyeno, T., and Yoshino, T. (1984). "The Fishes of the Japanese Archipelago." Tokai Univ. Press, Tokyo.

Mattson, W. J. (1980). Herbivory in relation to plant nitrogen content. *Annu. Rev. Ecol. Syst.* **11,** 119–161.

May, R. C. (1967). Larval survival in the maomao, *Abudefduf abdominalis* (Quoy and Gaimard). Unpublished M.S. thesis, University of Hawaii, Honolulu, Hawaii.

May, R. M. (1976). Patterns in multispecies communities. *In* "Theoretical Ecology" (R. M. May, ed.), pp. 142–162. Saunders, Philadelphia, Pennsylvania.

May, R. M. (ed.) (1984). "Exploitation of Marine Communities." Springer-Verlag, Berlin.

May, R. M., Beddington, J. R., Clark, C. W., Holt, S. J., and Laws, R. M. (1979). Management of multispecies fisheries. *Science* **205,** 267–277.

Maynard Smith, J. (1974). "Models in Ecology." Cambridge Univ. Press, Cambridge, England.

McAllister, D. E., and Rubee, P. J. (1986). Aquino government approves program to end cyanide use in the Philippines for the collection of tropical marine fishes. *Environ. Biol. Fishes* **17,** 315–317.

McArdle, B. H., and Blackwell, R. G. (1989). Measurement of density variability in the bivalve *Chione stutchburyi* using spatial autocorrelation. *Mar. Ecol. Prog. Ser.* **52,** 245–252.

McClanahan, T. R. (1989). Kenyan coral reef-associated gastropod fauna: a comparison between protected and unprotected reefs. *Mar. Ecol. Prog. Ser.* **53,** 11–20.

McClanahan, T. R., and Muthiga, N. A. (1988). Changes in Kenyan coral reef community structure and function due to exploitation. *Hydrobiologia* **166,** 269–276.

McClanahan, T. R., and Muthiga, N. A. (1989). Patterns of predation on a sea urchin, *Echinometra mathaei* (de Blainville), on Kenyan coral reefs. *J. Exp. Mar. Biol. Ecol.* **126,** 77–94.

McClanahan, T. R., Shafir, S. H. (1990). Causes and consequences of sea urchin abundance and diversity in Kenyan coral reef lagoons. *Oecologia* **83,** 362–370.

McCormick, M. I. (1989). Spatio-temporal patterns in the abundance and population structure of a large temperate reef fish. *Mar. Ecol. Prog. Ser.* **53,** 215–225.

McCormick, M. I., and Choat, J. H. (1987). Estimating total abundance of a large temperate-reef fish using visual strip-transects. *Mar. Biol.* **96,** 469–478.

McCosker, J. E. (1977). Fright posture of the plesiopid fish *Calloplesiops altivelis*: An example of Batesian mimicry. *Science* **197,** 400–401.

McCosker, J. E., and Dawson, C. E. (1975). Biotic passage through the Panama Canal, with particular reference to fishes. *Mar. Biol.* **30,** 343–351.

McEnroe, F. J., Robertson, K. J., and Fenical, W. (1977). Diterpenoid synthesis in brown seaweeds of the family Dictyotaceae. *In* "Marine Natural Products Chemistry" (D. J. Faulkner and W. Fenical, eds.), pp. 179–189. Plenum, New York.

McFarland, W. N. (1985). Overview: The dynamics of recruitment in coral reef organisms. *In* "The Ecology of Coral Reefs" (M. L. Reaka, ed.), NOAA Symp. Ser. Undersea Res., Vol. 3, pp. 9–15. Natl. Oceanic Atmos. Adm., Rockville, Maryland.

McFarland, W. N. (1986). Light in the sea—Correlations with behaviors of fishes and invertebrates. *Am. Zool.* **26,** 389–401.

McFarland, W. N. (1990). Light in the sea—The optical world of elasmobranches. *J. Exp. Zool.* (in press).

McFarland, W. N. (1991). Light in the Sea: The optical world of Elasmobranchs. *J. Exp. Zoo.* Suppl. **5,** 3–12.

McFarland, W. N., and Hillis, Z. M. (1982). Observations on agonistic behavior between members of juvenile French and white grunts—family Haemulidae. *Bull. Mar. Sci.* **32,** 255–268.

McFarland, W. N., and Kotchian, N. M. (1982). Interaction between schools of fish and mysids. *Behav. Ecol. Sociobiol.* **11,** 71–76.

McFarland, W. N., and Munz, F. W. (1975a). Part II. The photic environment of clear tropical seas during the day. *Vision Res.* **15,** 1063–1070.

McFarland, W. N., and Munz, F. W. (1975b). Part III. The evolution of photopic visual pigments in fishes. *Vision Res.* **15,** 1071–1080.

McFarland, W. N., and Munz, F. W. (1975c). The visible spectrum during twilight and its implications to vision. *In* "Light as an Ecological Factor" (G. C. Evans, R. Bainbridge, and O. Rackham, eds.), Vol. II. Blackwell, Oxford, England.

McFarland, W. N., and Ogden, J. C. (1985). Recruitment of young coral reef fishes from the plankton. *In* "The Ecology of Coral Reefs" (M. L. Reaka, ed.), NOAA Symp. Ser. Undersea Res., Vol. 3, pp. 37–51. Natl. Oceanic Atmos. Adm., Rockville, Maryland.

McFarland, W. N., and Prescott, J. (1959). Standing crop, chlorophyll content and *in situ* metabolism of a giant kelp community in southern California. *Bull. Inst. Mar. Sci. Univ. Tex.* **6,** 109–132.

McFarland, W. N., Ogden, J. C., and Lythgoe, J. N. (1979). The influence of light on the twilight migrations of grunts. *Environ. Biol. Fishes* **4,** 9–22.

McFarland, W. N., Brothers, E. B., Ogden, J. C., Shulman, M. J., Bermingham, E. L., and Kotchian-Prentiss, N. M. (1985). Recruitment patterns in young french grunts, *Haemulon flavolineatum* (family Haemulidae) at St. Croix, U.S.V.I. *Fish. Bull.* **83,** 413–426.

McIntosh, R. P. (1986). "The Background of Ecology: Concept and Theory." Cambridge Univ. Press, Cambridge, England.

McManus, J. W. (1985). Marine speciation, tectonics and sea-level changes in southeast Asia. *Proc. Int. Coral Reef Congr., 5th* **4,** 133–138.

McWilliam, P. S., Sale, P. F., and Anderson, D. T. (1981). Seasonal changes in resident zooplankton sampled by emergence traps in One Tree Lagoon, Great Barrier Reef. *J. Exp. Mar. Biol. Ecol.* **52,** 185–203.

Mead, G. W. (1970). A history of South Pacific fishes. *In* "Scientific Exploration of the South Pacific" (W. S. Wooster, ed.), pp. 236–251. Natl. Acad. Sci., Washington, D.C.

Mead, P. (1979). Common bottom fishes caught by SPC fishing projects. *SPC Fish. Newsl.* **18,** 1–4.

Meekan, M. G. (1988). Settlement and mortality patterns of juvenile reef fishes at Lizard Island, northern Great Barrier Reef. *Proc. Int. Coral Reef Symp., 6th* **2,** 779–784.

Mendez, F. (1989). Contribución al estudio de la biologia y pesqueria del pargo guanapo, *Lutjanus synagris,* en el Parque Nacional Archipelago de Las Roques. Graduate thesis, Universidad Central de Venezuela, Caracas, Venezuela.

Menge, B. (1975). Brood or broadcast? The adaptive significance of different reproductive strategies in the two intertidal sea stars *Leptasterias hexactis* and *Pisaster ochraeceus. Mar. Biol.* **31,** 87–100.

Menge, B. A., and Lubchenco, J. (1981). Community organization in temperate and tropical rocky intertidal habitats: Prey refuges in relation to consumer pressure gradients. *Ecol. Monogr.* **51,** 429–450.

Menge, B. A., and Sutherland, J. P. (1987). Community regulation: Variation in disturbance, competition, and predation in relation to environmental stress and recruitment. *Am. Nat.* **130,** 730–757.

Menge, B. A., Lubchenco, J., and Ashkenas, L. R. (1985). Diversity, heterogeneity and consumer pressure in a tropical rocky intertidal community. *Oecologia* **65,** 394–405.

Meyer, J. L., and Schultz, E. T. (1985a). Migrating haemulid fishes as a source of nutrients and organic matter on coral reefs. *Limnol. Oceanogr.* **30,** 146–156.

Meyer, J. L., and Schultz, E. T. (1985b). Tissue condition and growth rate of corals associated with schooling fish. *Limnol. Oceanogr.* **30,** 157–166.

Meyer, J. L., Schultz, E. T., and Helfman, G. S. (1983). Fish schools: An asset to corals. *Science* **220,** 1047–1049.

Michanek, G. (1975). "Seaweed Resources of the Ocean." Food Agric. Organ., Rome.

Milicich, M. J. (1988). The distribution and abundance of presettlement fish in the nearshore waters of Lizard Island. *Proc. Int. Coral Reef Symp., 6th* **2,** 785–790.

Miller, J. M. (1974). Nearshore distribution of Hawaiian marine fish larvae: Effects of water quality, turbidity and currents. *In* "The Early-Life History of Fish" (J. H. S. Blaxter, ed.), pp. 217–231. Springer-Verlag, New York.

Miller, J. M. (1979a). Larval fish distributions. *In* "An Atlas of Common Nearshore Marine Fish Larvae of the Hawaiian Islands" (J. M. Miller, W. Watson, and J. M. Leis, eds.), Misc. Rep. 80-02, pp. 105–152. Univ. of Hawaii Sea Grant Program, Honolulu, Hawaii.

Miller, J. M. (1979b). Nearshore abundance of tuna (Pisces: Scombridae) larvae in the Hawaiian Islands. *Bull. Mar. Sci.* **29,** 19–26.

Miller, P. J. (1979). Adaptiveness and implications of small size in teleosts. *Symp. Zool. Soc. London* **44,** 263–306.

Miller, D. J., and Geibel, J. J. (1973). Summary of blue rockfish and lingcod life histories: A reef ecology study; and giant kelp, *Macrocystis pyrifera,* experiments in Monterey Bay, California. *Fish. Bull.—Calif. Dep. Fish Game* **158,** 1–173.

Miller, D., Colton, J. B. Jr., and Marak, R. R. (1963). A study of the vertical distribution of larval haddock. *J. Cons., Cons. Int. Explor. Mer* **28,** 37–49.

Milward, N. E., and Hartwick, R. F. (1986). Temporal and spatial distribution of fish larvae across the continental shelf lagoon of the central Great Barrier Reef. *In* "Indo-Pacific Fish Biology: Proceedings of the Second International Conference on Indo-Pacific Fishes, Tokyo" (T. Uyeno, R. Arai, T. Taniuchi, and K. Matsuura, eds.), pp. 748–758. Ichthyol. Soc. Jpn., Tokyo.

Mitchell, C. T., and Hunter, J. R. (1970). Fishes associated with drifting kelp, *Macrocystis pyrifera,* off the coast of southern California and northern Baja California. *Calif. Fish Game* **56,** 288–297.

Moe, M. A. (1969). Biology of the red grouper *Epinephelus morio* (Valenciennes) from the eastern Gulf of Mexico. *Prof. Pap. Ser.—Fla. Dep. Nat. Resour., Mar. Res. Lab.* **10,** 1–95.

Molles, M. C. Jr. (1978). Fish species diversity on model and natural reef patches: Experimental insular biogeography. *Ecol. Monogr.* **48,** 289–305.

Monkolprasit, S. (1981). Investigations of coral reef fishes in Thai waters. *Proc. Int. Coral Reef Symp., 4th* **2,** 491–496.

Montgomery, W. L. (1980). The impact of non-selective grazing by the giant blue damselfish, *Microspathodon dorsalis,* on algal communities in the Gulf of California, Mexico. *Bull. Mar. Sci.* **30,** 290–303.

Montgomery, W. L., and Gerking, S. D. (1979). Marine macroalgae as foods for fishes: An evaluation of potential food quality. *Environ. Biol. Fishes* **5,** 143–153.

Montgomery, W. L., Gerrodette, T., and Marshall, L. D. (1979). Effects of grazing by the yellowtail surgeon fish, *Prionurus punctatus,* on algal communities in the Gulf of California, Mexico. *Bull. Mar. Sci.* **30,** 901–908.

Montgomery, W. L., Myrberg, A. A., and Fishelson, L. (1989). Feeding ecology of surgeonfishes (Acanthuridae) in the northern Red Sea with particular reference to *Acanthurus nigrofuscus* (Forsskal). *J. Exp. Mar. Biol. Ecol.* **132,** 179–207.

Moore, R. E. (1977). Toxins from blue-green algae. *BioScience* **27,** 797–802.

Moore, R. E. (1981). Constituents of blue–green algae. *In* "Marine Natural Products: Chemical and Biological Perspectives" (P. J. Scheuer, ed.), Vol. 4, pp. 1–49. Academic Press, New York.

Moran, P. J. (1986). The *Acanthaster* phenomenon. *Oceanogr. Mar. Biol.* **24,** 379–480.

Moran, M. J., and Sale, P. F. (1977). Seasonal variation in territorial response, and other aspects of the ecology of the Australian temperate pomacentrid fish *Parma microlepis*. *Mar. Biol.* **39,** 121–128.

Moreno, C. A., and Jara, H. F. (1984). Ecological studies on fish fauna associated with

Macrocystis pyrifera belts in the south of Fueguian Islands, Chile. *Mar. Ecol. Prog. Ser.* **15,** 99–107.

Morin, J. G. (1988). Piscivorous behavior and activity patterns in the tropical ophiuroid *Ophiarachna incrassata* (Ophiuroidea: Ophiodermatidae). *In* "Echinoderm Biology" (R. D. Burke, P. V. Mladenov, P. Lambert, and R. L. Parsley, eds.), pp. 401–407. Balkema, Rotterdam, The Netherlands.

Morrison, D. (1988). Comparing fish and urchin grazing in shallow and deeper coral reef algal communities. *Ecology* **69,** 1367–1382.

Moser, H. G. (1981). Morphological and functional aspects of marine fish larvae. *In* "Marine Fish Larvae: Morphology, Ecology and Relation to Fisheries" (R. Lasker, ed.), pp. 89–131. Univ. of Washington Press, Seattle, Washington.

Moser, H. G., Ahlstrom, E. H., Kramer, D., and Stevens, E. G. (1974). Distribution and abundance of fish eggs and larvae in the Gulf of California. *CalCOFI Rep.* **17,** 112–128.

Moser, H. G., Richards, W. J., Cohen, D. M., Fahay, M. P., Kendall, A. W., Jr., and Richardson, S. L. (1984). Ontogeny and systematics of fishes. *Spec. Publ.—Am. Soc. Ichthyol. Herpetol.* **1,** 1–760.

Motta, P. J. (1983). Response by potential prey to coral reef fish predators. *Anim. Behav.* **31,** 1257–1259.

Motta, P. J. (1984). Mechanics and function of jaw protrusion in teleost fishes: A review. *Copeia* **1984,** 1–18.

Motta, P. J. (1988). Functional morphology of the feeding apparatus of ten species of Pacific butterflyfishes (Perciformes, Chaetodontidae): An ecomorphological approach. *Environ. Biol. Fishes* **22,** 39–67.

Moulton, L. L. (1977). An ecological analysis of fishes inhabiting the rocky nearshore regions of northern Puget Sound, Washington. Unpublished Ph.D. dissertation, University of Washington, Seattle, Washington.

Moyer, J. T. (1975). Reproductive behavior of the damselfish, *Pomacentrus nagasakiensis,* at Miyake-jima, Japan. *Japan. J. Ichthyol.* **22,** 151–163.

Moyer, J. T. (1976). Geographical variation and social dominance in Japanese populations of the anemonefish *Amphiprion clarkii. Japan J. Ichthyol.* **23,** 12–22.

Moyer, J. T. (1980). Influence of temperate waters on the behavior of the tropical anemonefish *Amphiprion clarkii* at Miyake-jima, Japan. *Bull. Mar. Sci.* **30,** 261–272.

Moyer, J. T. (1987). Quantitative observations of predation during spawning rushes of the labrid fish *Thalassoma cupido* at Miyake-jima, Japan. *Japan J. Ichthyol.* **34,** 76–81.

Moyer, J. T. (1990). Social and reproductive behavior of *Chaetodontoplus mesoleucus* (Pomacanthidae) at Bantayan Island, Philippines, with notes on pomacanthid relationships. *Japan J. Ichthyol.* **36,** 459–467.

Moyer, J. T., and Bell, L. J. (1976). Reproductive behavior of the anemonefish *Amphiprion clarkii* at Miyake-jima, Japan. *Japan J. Ichthyol.* **23,** 23–32.

Moyer, J. T., and Nakazono, A. (1978). Protandrous hermaphroditism in six species of the anemonefish genus *Amphiprion* in Japan. *Japan. J. Ichthyol.* **25,** 101–106.

Moy-Thomas, J. A., and Miles, R. S. (1971). "Palaeozoic Fishes." Chapman & Hall, London.

Munro, J. L. (1969). The sea fisheries of Jamaica: Past, present and future. *Jam. J.* **3,** 16–22.

Munro, J. L. (1980). Stock assessment models: Applicability and utility in tropical small-scale fisheries. *In* "Stock Assessment for Tropical Small-Scale Fisheries" (S. B. Saila and P. M. Roedel, eds.), pp. 35–47. Univ. of Rhode Island, Kingston, Rhode Island.

Munro, J. L. (ed.) (1983). Caribbean Coral Reef Fishery Resources. *ICLARM Stud. Rev.* **7,** 1–276.

Munro, J. L. (1984). Yields from coral reef fisheries. *Fishbyte* **2,** 13–15.

Munro, J. L., and Williams, D. M. (1985). Assessment and management of coral reef fisheries: Biological, environmental and socioeconomic aspects. *Proc. Int. Coral Reef Congr., 5th* **4,** 545–581.

Munro, J. L., Gaut, V. C., Thompson, R., and Reeson, P. H. (1973). The spawning seasons of Caribbean reef fishes. *J. Fish Biol.* **5,** 69–84.

Munro, J. L., Parrish, J. D., and Talbot, F. H. (1987). The biological effects of intensive fishing upon coral reef communities. *In* "Human Impacts on Coral Reefs: Facts and Recommendations" (B. Salvat, ed.), pp. 41–49. Antenne Mus. E.P.H.E., Moorea, French Polynesia.

Muntz, W. R. A., and Northmore, D. P. M. (1973). Scotopic spectral sensitivity in a teleost fish (*Scardinius erythrophthalmus*) adapted to different daylengths. *Vision Res.* **13,** 245–252.

Munz, F. W. (1957). Photosensitive pigments from the retinas of deep-sea fishes. *Science* **125,** 1142–1143.

Munz, F. W. (1958). Photosensitive pigments from the retinae of certain deep sea fishes. *J. Physiol (London)* **140,** 220–225.

Munz, F. W. (1964). The visual pigments of epipelagic and rocky-shore fishes. *Vision Res.* **4,** 441–454.

Munz, F. W., and McFarland, W. N. (1973). The significance of spectral position in the rhodopsins of tropical marine fishes. *Vision Res.* **13,** 1829–1874.

Munz, F. W., and McFarland, W. N. (1975). Part I: Presumptive cone pigments extracted from tropical marine fishes. *Vision Res.* **15,** 1045–1062.

Munz, F. W., and McFarland, W. N. (1977). Evolutionary adaptations of fishes to the photic environment. *Handb. Sens. Physiol.* **7,** 194–274.

Murdoch, W. W., and Oaten, A. (1975). Predation and population stability. *Adv. Ecol. Res.* **9,** 1–132.

Murdy, E. O., Ferraris, C. J., Jr., Hoese, D. F., and Steene, R. C. (1981). Preliminary list of fishes from Sombrero Island, Philippines, with fifteen new records. *Proc. Biol. Soc. Wash.* **94,** 1163–1173.

Myers, R. F. (1989). "Micronesian Reef Fishes." Coral Graphics, Agana Guam.

Mynderse, J. S., Moore, R. E., Kashiwaga, M., and Norton, T. R. (1977). Antileukemia activity in the Oscillatoriaceae: Isolation of debromoaplysiatoxin from *Lyngbya*. *Science* **196,** 538–540.

Myrberg, A. A. (1972). Social dominance and territoriality in the bicolor damselfish, *Eupomacentrus partitus* (Poey) (Pisces: Pomacentridae). *Behavior* **41,** 207–231.

Myrberg, A. A., and Thresher, R. E. (1974). Interspecific aggression and its relevance to the concept of territoriality in reef fishes. *Am. Zool.* **14,** 81–96.

Myrberg, A. A., Jr., Montgomery, W. L., and Fishelson, L. (1989). The reproductive behavior of *Acanthurus nigrofuscus* (Forskal) and other surgeonfishes (Fam. Acanthuridae) off Eilat, Israel (Gulf of Aqaba, Red Sea). *Ethology* **79,** 31–61.

Nagelkerken, W. (1977). The distribution of the graysby *Petrometopon cruentatum* (Lacepede) on the coral reef at the southwest coast of Curaçao. *Proc. Int. Coral Reef Symp., 3rd* **1,** 311–315.

Nagelkerken, W. P. (1979). Biology of the graysby, *Epinephelus cruentatus*, of the coral reef of Curaçao. *Stud. Fauna Curacao* **60,** 1–118.

Nellen, W. (1973a). Kinds and abundance of fish larvae in the Arabian Sea and the Persian Gulf. *In* "Ecological Studies: Analysis and Synthesis" (B. Zeitzschell, ed.), Vol. 3, pp. 415–530. Springer-Verlag, New York.

Nellen, W. (1973b). Fischlarven des Indischen Ozeans. *"Meteor" Forschungsergeb., Reihe D* **14,** 1–66.

Nelson, J. S. (1984). "Fishes of the World," 2nd ed. Wiley, New York.

Nelson, S. G., and Wilkins, S. D. (1988). Sediment processing by the surgeonfish *Ctenochaetus striatus* at Moorea, French Polynesia. *J. Fish Biol.* **32,** 817–824.

Neudecker, S. (1977). Transplant experiments to test the effect of fish grazing on coral distribution. *Proc. Int. Coral Reef Symp., 3rd* **1,** 317–323.

Neudecker, S. (1979). Effects of grazing and browsing fishes on the zonation of corals in Guam. *Ecology* **60,** 666–672.

Neudecker, S. (1985). Foraging patterns of chaetodontid and pomacanthid fishes at St. Croix (U.S. Virgin Islands). *Proc. Int. Coral Reef Congr., 5th* **5,** 415–420.

Neudecker, S. (1989). Eye camouflage and false eyespots: Chaetodontid responses to predators. *Environ. Biol. Fishes* **25,** 143–157.

Neudecker, S., and Lobel, P. S. (1982). Mating systems of chaetodontid and pomacanthid fishes at St. Croix. *Z. Tierpsychol.* **59,** 299–318.

Newell, N. D. (1971). An outline history of tropical organic reefs. *Am. Mus. Novit.* **265,** 1–37.

Nicol, J. A. C. (1988). "The Eyes of Fishes." Oxford Univ. Press (Clarendon), Oxford, England.

Nolan, R. S. (1975). The ecology of patch reef fishes. Unpublished Ph.D. dissertation, University of California, San Diego, California.

Norman, J. R. (1947). "A History of Fishes." Benn, London.

Norman, M. D., and Jones, G. P. (1984). Determinants of territory size in the pomacentrid reef fish, *Parma victoriae*. *Oecologia* **61,** 60–69.

Norris, J. E., and Parrish, J. D. (1988). Predator–prey relationships among fishes in pristine coral reef communities. *Proc. Int. Coral Reef Symp., 6th* **2,** 107–113.

Norris, J. N., and Fenical, W. (1982). Chemical defense in tropical marine algae. *Smithson. Contrib. Mar. Sci.* **12,** 417–431.

Norton, S. F. (1989). Constraints on the foraging ecology of subtidal cottid fishes. Unpublished Ph.D. dissertation, University of California–Santa Barbara, Santa Barbara, California.

Nursall, J. R. (1977). Territoriality in redlip blennies (*Ophioblennius atlanticus*—Pisces: Blenniidae). *J. Zool.* **182,** 205–223.

Nyberg, D. W. (1971). Prey capture in the largemouth bass. *Am. Midl. Nat.* **86,** 128–144.

Obrien, W. J. (1979). The predator–prey interaction of planktivorous fish and zooplankton. *Am. Sci.* **67,** 572–581.

Ochi, H. (1985). Temporal patterns of breeding and larval settlement in a temperate population of the tropical anemonefish, *Amphiprion clarkii. Japan J. Ichthiol.* **32,** 248–257.

Ochi, H. (1986a). Growth of the anemonefish *Amphiprion clarkii* in temperate waters, with special reference to the influence of settling time in the growth of 0-year olds. *Mar. Biol.* **92,** 223–229.

Ochi, H. (1986b). Breeding synchrony and spawning intervals in the temperate damselfish *Chromis notata. Environ. Biol. Fishes* **17,** 117–123.

O'Day, W. T., and Fernandez, H. R. (1974). *Aristomias scintillans* (Malacosteidae): A deep-sea fish with visual pigments apparently adapted to its own bioluminescence. *Vision Res.* **14,** 545–550.

Odum, H. T., and Odum, E. P. (1955). Trophic structure and productivity of a windward coral reef community of Eniwetok Atoll. *Ecol. Monogr.* **25,** 291–320.

Ogden, J. C. (1977). Carbonate sediment production by parrot fish and sea urchins on Caribbean reefs. *Stud. Geol.* (*Tulsa, Okla.*) **4,** 281–288.

Ogden, J. C. (1980). Faunal relationships in Caribbean seagrass beds. *In* "Handbook of Seagrass Biology: An Ecosystem Perspective" (R. C. Phillips and C. P. McRoy, eds.), pp. 173–198. Garland STPM Press, New York.

Ogden, J. C., and Buckman, N. S. (1973). Movements, foraging groups, and diurnal migrations of the striped parrotfish *Scarus croicensis* Bloch (Scaridae). *Ecology* **54,** 589–596.

Ogden, J. C., and Ebersole, J. P. (1981). Scale and community structure of coral reef fishes: A long-term study of a large artificial reef. *Mar. Ecol. Prog. Ser.* **4,** 97–103.

Ogden, J. C., and Ehrlich, P. R. (1977). The behavior of heterotypic resting schools of juvenile grunts (Pomadasyidae). *Mar. Biol.* **42,** 273–280.

Ogden, J. C., and Gladfelter, E. H. (1983). Coral reefs, seagrass beds and mangroves: Their interaction in the coastal zones of the Caribbean. *UNESCO Rep. Mar. Sci.* **23,** 1–130.

Ogden, J. C., and Lobel, P. S. (1978). The role of herbivorous fishes and urchins in coral reef communities. *Environ. Biol. Fishes* **3,** 49–63.

Ogden, J. C., and Zieman, J. C. (1977). Ecological aspects of coral reef–seagrass bed contacts in the Caribbean. *Proc. Int. Coral Reef Symp., 3rd* **1,** 378–382.

Ogden, J. C., Brown, R. A., and Salesky, N. (1973). Grazing by the echinoid *Diadema antillarum philippi*: Formation of halos around West Indian patch reefs. *Science* **182,** 715–717.

Okiyama, M. (ed.) (1988). "An Atlas of the Early Stage Fishes in Japan." Tokai Univ. Press, Tokyo (in Japanese).

Olla, B. L., Bejda, A. J., and Martin, A. D. (1974). Daily activity, movements, feeding, and seasonal occurrence in the tautog, *Tautoga onitis. Fish. Bull.* **72,** 27–35.

Olla, B. L., Bejda, A. J., and Martin, A. D. (1975). Activity, movements, and feeding behavior of the cunner, *Tautogolabrus adspersus*, and comparison of food habits with young tautog, *Tautoga onitis*, off Long Island, New York. *Fish. Bull.* **73,** 895–900.

Olla, B. L., Bejda, A. J., and Martin, A. D. (1979). Seasonal dispersal and habitat selection of cunner, *Tautogolabrus adspersus*, and young tautog, *Tautoga onitis*, in Fire Island Inlet, Long Island, New York. *Fish. Bull.* 77, 255–261.

Ormond, R. F. G. (1980). Aggressive mimicry and other interspecific feeding associations among Red Sea coral reef predators. *J. Zool.* **191,** 247–262.

Ortega, S. (1986). Fish predation on gastropods on the Pacific coast of Costa Rica. *J. Exp. Mar. Biol. Ecol.* **97,** 181–191.

Overpeck, J. T., Peterson, L. C., Kipp, N., Imbrie, J., and Rind, D. (1989). Climate change in the circum-North Atlantic region during the last deglaciation. *Nature (London)* **338,** 553–557.

Padilla, D. K. (1985). Structural resistance of algae to herbivores: A biomechanical approach. *Mar. Biol.* **90,** 103–109.

Padilla, D. K. (1989). Algal structural defenses: Form and calcification in resistance to tropical limpets. *Ecology* **70,** 835–842.

Paine, R. T. (1984). Ecological determinism in the competition for space. *Ecology* **65,** 1339–1348.

Paine, R. T., and Levin, S. A. (1981). Intertidal landscapes: Disturbance and the dynamics of pattern. *Ecol. Monogr.* **51,** 145–198.

Panella, G. (1971). Fish otoliths: Daily growth layers and periodical patterns. *Science* **173,** 1124–1127.

Panella, G. (1980). Growth patterns in fish sagittae. *In* "Skeletal Growth of Aquatic Organisms: Biological Records of Environmental Change" (D. C. Rhoads and R. A. Lutz, eds.), pp. 519–560. Plenum, New York.

Pankhurst, N. W. (1989). The relationship of ocular morphology to feeding modes and activity periods in shallow marine teleosts from New Zealand. *Environ. Biol. Fishes* **26,** 201–211.

Parin, N. V. (1967). Diurnal variations in the larval occurrence of some oceanic fishes near the ocean surface. *Oceanology* **7,** 115–121.

Parrish, J. D. (1987a). Characteristics of fish communities on coral reefs and in potentially interacting shallow habitats in tropical oceans of the world. *UNESCO Rep. Mar. Sci.* **46,** 171–218.

Parrish, J. D. (1987b). The trophic biology of snappers and groupers. *In* "Tropical Snappers and Groupers: Biology and Fisheries Management" (J. J. Polovina and S. Ralston, eds.), pp. 405–463. Westview, Boulder, Colorado.

Parrish, J. D. (1989). Characteristics of fish communities on coral reefs and in potentially interacting shallow habitats in tropical oceans of the world. *Mar. Ecol. Prog. Ser.* **58,** 143–160.

Parrish, J. D., and Zimmerman, R. J. (1977). Utilization by fishes of space and food resources on an off-shore Puerto Rican coral reef and its surroundings. *Proc. Int. Coral Reef Symp., 3rd* **1,** 297–303.

Parrish, J. D., Callahan, M. W., and Norris, J. E. (1985). Fish trophic relationships that structure reef communities. *Proc. Int. Coral Reef Congr., 5th* **4,** 73–78.

Parrish, J. D., Norris, J. E., Callahan, M. W., Callahan, J. K., Magarifuji, E. J., and Schroeder, R. E. (1986). Piscivory in a coral reef fish community. *In* "Contemporary Studies on Fish Feeding" (C. A. Simenstad and G. M. Cailliet, eds.), pp. 285–298. Junk, Dordrecht, The Netherlands.

Parrish, R. H., Nelson, C. S., and Bakun, A. (1981). Transport mechanisms and reproductive success of fishes in the California Current. *Biol. Oceanogr.* **1,** 175–203.

Parslow, J., and Grubic, A. (1989). Advection, dispersal and plankton patchiness on the Great Barrier Reef. *Aust. J. Mar. Freshwater Res.* **40,** 403–419.

Partridge, B. L. (1982). The structure and function of fish schools. *Sci. Am.* **246,** 90–99.

Patton, M. L. (1985). Changes in fish abundance in the Southern California Bight during a warm-water episode. *In* "Marine Environmental Analysis and Interpretation: San Onofre Nuclear Generating Station, Rep. on 1984 Data" (J. E. Yuge and J. L. Elliott, eds.), 85-RD-37, pp. 8-1–8-34. South. Calif. Edison Co., Pasadena, California.

Patton, M. L., Grove, R. S., and Harman, R. F. (1985). What do natural reefs tell us about designing artificial reefs in southern California? *Bull. Mar. Sci.* **37,** 279–298.

Paul, V. J. (1987). Feeding deterrent effects of algal natural products. *Bull. Mar. Sci.* **41,** 514–522.

Paul, V. J. (1991). "Ecological Roles for Marine Secondary Metabolites." Comstock, Ithaca, New York. In press.

Paul, V. J., and Fenical, W. (1986). Chemical defense in tropical green algae, order Caulerpales. *Mar. Ecol. Prog. Ser.* **34,** 157–169.

Paul, V. J., and Fenical, W. (1987). Natural products chemistry and chemical defense in tropical marine algae of the phylum Chlorophyta. *In* "Bioorganic Marine Chemistry" (P. J. Scheuer, ed.), Vol. 1, pp. 1–29. Springer-Verlag, Berlin.

Paul, V. J., and Hay, M. E. (1986). Seaweed susceptibility to herbivory: Chemical and morphological correlates. *Mar. Ecol. Prog. Ser.* **33,** 255–264.

Paul, V. J., and Van Alstyne, K. L. (1988a). Chemical defense and chemical variation in some tropical Pacific species of *Halimeda* (Halimedaceae; Chlorophyta). *Coral Reefs* **6,** 263–270.

Paul, V. J., and Van Alstyne, K. L. (1988b). Use of ingested algal diterpenoids by *Elysia halimedae* Macnae (Opisthobranchia: Ascoglossa) as antipredator defenses. *J. Exp. Mar. Biol. Ecol.* **119,** 15–29.

Paul, V. J., Hay, M. E., Duffy, J. E., Fenical, W., and Gustafson, K. (1987). Chemical defense in the seaweed *Ochtodes secundiramea* (Montagne) Howe (Rhodophyta): Effect of its monoterpenoid components upon diverse coral reef herbivores. *J. Exp. Mar. Biol. Ecol.* **114,** 249–260.

Paul, V. J., Ciminiello, P., and Fenical, W. (1988a). Diterpenoid feeding deterrents from the Pacific green alga *Pseudochlorodesmis furcellata. Phytochemistry* **27,** 1011–1014.

Paul, V. J., Wylie, C. R., and Sanger, H. R. (1988b). Effects of algal chemical defenses toward different coral-reef herbivorous fishes. *Proc. Int. Coral Reef Symp., 6th* **3,** 73–78.

Paul, V. J., Nelson, S. G., and Sanger, H. R. (1990). Feeding preferences of adult and juvenile rabbitfish *Siganus argenteus* in relation to chemical defences of tropical seaweeds. *Mar. Ecol. Prog. Ser.* **60,** 23–34.

Pauly, D. (1979). Theory and management of tropical multispecies stocks: a review with emphasis on the South East Asian demersal fisheries. ICLARM Studies and Reviews **1,** Manila, Philippines, 35pp.

Pauly, D. (1984). "Fish Population Dynamics in Tropical Waters. A Manual for Use with Programmable Calculators." ICLARM, Manila, Philippines.

Pauly, D. (1986). A simple method for estimating the food consumption of fish populations from growth data and food conversion experiments. *Fish. Bull.* **84,** 827–839.

Pauly, D. (1987). Theory and practice of overfishing: A South-East Asian perspective. *FAO RAPA Rep.* **1987/10,** 146–163.

Pauly, D. (1988). Some definitions of overfishing relevant to coastal zone management in Southeast Asia. *Trop. Coastal Area Manage.* **3,** 14–15.

Pauly, D., and Murphy, G. I. (eds.) (1982). Theory and management of tropical fisheries. *ICLARM Conf. Proc.* **9.** ICLARM, Manila, Philippines.

Paya, I., and Santelices, B. (1989). Macroalgae survive digestion by fishes. *J. Phycol.* **25,** 186–188.

Pearl, R. (1928). "The Rate of Living." Knopf, New York.

Pearl, R., and Miner, J. R. (1935). Experimental studies on the duration of life. XIV. The comparative mortality of certain lower organisms. *Q. Rev. Biol.* **10,** 60–79.

Pearre, S., Jr. (1979). Problems of detection and interpretation of vertical migration. *J. Plankton Res.* **1,** 29–44.

Pennington, M. (1983). Efficient estimators of abundance for fish and plankton surveys. *Biometrics* **39,** 281–286.

Pereyra, W. T., Pearcy, W. G., and Carvey, F. E., Jr. (1969). *Sebastodes flavidus*, a shelf rockfish feeding on mesopelagic fauna, with consideration of the ecological implications. *J. Fish. Res. Board Can.* **26,** 2211–2215.

Perez-Villanoel, A. (1982). Desarrollo gonadal en el mero Tofia, *Epinephelus guttatus* (Serranidae). Graduate thesis, Universidad Central de Venezuela, Caracas, Venezuela.

Peters, R. H. (1976). Tautology in evolution and ecology. *Am. Nat.* **110,** 1–12.

Petersen, C. W. (1990a). The relationships among population density, individual size, mating tactics, and reproductive success in a hermaphroditic fish, *Serranus fasciatus*. *Behaviour* **113,** 57–80.

Petersen, C. W. (1990b). Variation in reproductive success and gonadal allocation in the simultaneous hermaphrodite *Serranus fasciatus*. *Oecologia* **83,** 62–67.

Petersen, C. W. (1991b). Variation in reproductive success and gonadal allocation in the simultaneous hermaphrodite *Serranus fasciatus*. *Oecologia* (in press).

Peterson, C. H. (1979). Predation, competitive exclusion, and diversity in the soft-sediment benthic communities of estuaries and lagoons. *In* "Ecological Processes in Coastal Marine Systems" (R. J. Livingston, ed.), pp. 233–264. Plenum, New York.

Pfeffer, R. A., and Tribble, G. W. (1985). Hurricane effects on an aquarium fish fishery in the Hawaiian Islands. *Proc. Int. Coral Reef Congr., 5th* **3,** 331–336.

Pfister, C. A., and Hay, M. E. (1988). Associational plant refuges: Convergent patterns in marine and terrestrial communities result from differing mechanisms. *Oecologia* **77,** 118–129.

Philander, S. G. (1989). "El Niño, La Niña, and the Southern Oscillation." Academic Press, San Diego, California.

Pianka, E. R. (1973). The structure of lizard communities. *Annu. Rev. Ecol. Syst.* **4,** 53–74.

Pianka, E. R., and Parker, W. S. (1975). Age-specific reproductive tactics. *Am. Nat.* **109,** 453–464.

Pietsch, T. W., and Grobecker, D. B. (1978). The compleat angler: Aggressive mimicry in an antennariid anglerfish. *Science* **201,** 369–370.

Pietsch, T. W., and Grobecker, D. B. (1987). "Frogfishes of the World: Systematics, Zoogeography, and Behavioral Ecology." Stanford Univ. Press, Stanford, California.

Pillai, C. S. G., Mohan, M., and Kunhikoya, K. K. (1983). On an unusual massive recruitment of the reef fish *Ctenochaetus strigosus* (Bennet) (Perciformes: Acanthuridae) to the Minicoy Atoll and its significance. *Indian J. Fish.* **30,** 261–268.

Pimm, S. L., Jones, H. L., and Diamond, J. (1988). On the risk of extinction. *Am. Nat.* **132,** 757–785.

Pirenne, M. H. (1962a). Absolute thresholds and quantum effects. *In* "The Eye" (H. Davson, ed.), Vol. 2, pp. 123–140. Academic Press, New York.

Pirenne, M. H. (1962b). Quantum fluctuations at the absolute threshold. *In* "The Eye" (H. Davson, ed.), Vol. 2, pp. 141–158. Academic Press, New York.

Pitcher, C. R. (1987). Validation and application of otolith ageing techniques to some problems in the ecology of coral reef fishes. Unpublished Ph.D. thesis, Griffith University, Brisbane, Australia.

Pitcher, C. R. (1988a). Validation of a technique for reconstructing daily patterns in the recruitment of coral reef damselfish. *Coral Reefs* **7,** 105–111.

Pitcher, C. R. (1988b). Spatial variation in the temporal pattern of recruitment of a coral reef damselfish. *Proc. Int. Coral Reef Symp., 6th* **2,** 811–816.

Pitcher, T. J. (1986). Functions of shoaling behaviour in teleosts. *In* "The Behavior of Teleost Fishes" (T. J. Pitcher, ed.), pp. 294–337. Johns Hopkins Univ. Press, Baltimore, Maryland.

Policansky, D. (1987). Evolution, sex and sex allocation. *BioScience* **37,** 466–468.

Polis, G. A., Myers, C. A., and Holt, R. D. (1989). The ecology and evolution of intraguild predation: Potential competitors that eat each other. *Annu. Rev. Ecol. Syst.* **20,** 297–330.

Polovina, J. J. (1984). Model of a coral reef ecosystem. I. The ECOPATH model and its application to French Frigate Shoals. *Coral Reefs* **3,** 1–11.

Polovina, J. J. (1986). A variable catchability version of the Leslie model with application to an intensive fishing experiment on a multispecies stock. *Fish. Bull.* **84,** 423–428.

Polovina, J. J. (1987). Assessment and management of deepwater bottom fishes in Hawaii and the Marianas. *In* "Tropical Snappers and Groupers: Biology and Fisheries Management" (J. J. Polovina and S. Ralston, eds.), pp. 375–404. Westview, Boulder, Colorado.

Polunin, N. V. C. (1988). Efficient uptake of algal production by a single resident herbivorous fish on a reef. *J. Exp. Mar. Biol. Ecol.* **123,** 61–76.

Polunin, N. V. C., and Klumpp, D. W. (1989). Ecological correlates of foraging periodicity in herbivorous reef-fishes of the Coral Sea. *J. Exp. Mar. Biol. Ecol.* **126,** 1–20.

Polunin, N. V. C., and Koike, I. (1987). Temporal focusing of nitrogen release by a periodically feeding reef fish. *J. Exp. Mar. Biol. Ecol.* **111,** 285–296.

Popper, A. N., and Coombs, S. (1980). Auditory mechanisms in teleost fishes. *Am. Sci.* **68,** 429–440.

Popper, D., and Fishelson, L. (1973). Ecology and behaviour of *Anthias squamipinnis* (Peters, 1855) (Anthiidae, Teleostei) in the coral habitat of Eilat (Red Sea). *J. Exp. Zool.* **184,** 409–424.

Porter, J. W., and Porter, K. G. (1977). Quantitative sampling of demersal plankton migrating from different coral reef substrates. *Limnol. Oceanogr.* **22,** 553–556.

Potts, D. C. (1977). Suppression of coral populations by filamentous algae within damselfish territories. *J. Exp. Mar. Biol. Ecol.* **28,** 207–216.

Potts, D. C. (1983). Evolutionary disequilibrium among Indo-Pacific corals. *Bull. Mar. Sci.* **33,** 619–632.

Potts, D. C. (1984). Generation times and the quarternary evolution of reef-building corals. *Paleobiology* **10,** 48–58.

Potts, D. C. (1985). Sea-level fluctuations and speciation in Scleractinea. *Proc. Int. Coral Reef Congr., 5th* **4,** 127–132.

Potts, G. W. (1973). The ethology of *Labroides dimidiatus* (Cuv. and Val.) (Labridae, Pisces) on Aldabra. *Anim. Behav.* **21,** 250–291.

Potts, G. W. (1981). Behavioural interactions between Carangidae (Pisces) and their prey on the fore-reef slope of Aldabra, with notes on other predators. *J. Zool.* **195,** 385–404.

Powell, T. M. (1989). Physical and biological scales of variability in lakes, estuaries, and the coastal ocean. *In* "Perspectives in Ecological Theory" (J. Roughgarden, R. M. May, and S. A. Levin, eds.), pp. 157–176. Princeton Univ. Press, Princeton, New Jersey.

Power, J. H. (1984). Advection, diffusion, and drift migrations of larval fish. *In* "Mechanisms of Migration in Fishes" (J. D. McCleave, G. P. Arnold, J. J. Dodson, and W. H. Neill, eds.), pp. 27–37. Plenum, New York.

Power, M. E. (1990). Resource enhancement by indirect effects of grazers: Armored catfish, algae and sediment. *Ecology* **71,** 897–904.

Powles, H. (1977a). Island mass effects on the distribution of larvae of two pelagic fish species off Barbados. *FAO Fish. Rep.* **200,** 333–346.

Powles, H. (1977b). Description of larval *Jenkinsia lamprotaenia* (Clupeidae, Dussumieriinae) and their distribution off Barbados, West Indies. *Bull. Mar. Sci.* **27,** 788–801.

Powles, H. P., and Burgess, W. E. (1978). Observations on benthic larvae of *Pareques* (Pisces: Sciaenidae) from Florida and Colombia. *Copeia* **1978,** 169–172.

Preisendorfer, R. W. (1976). "Hydrologic Optics," Vol. 1. U.S. Dep. Commer., Washington, D.C.

Pressley, P. H. (1980). Lunar periodicity in the spawning of yellowtail damselfish, *Microspathodon chrysurus. Environ. Biol. Fishes* **5,** 153–159.

Pulliam, H. R. (1988). Sources, sinks, and population regulation. *Am. Nat.* **132,** 652–661.

Purcell, J. E. (1981). Dietary composition and diel feeding patterns of epipelagic siphonophores. *Mar. Biol.* **65,** 83–90.

Purcell, J. E. (1984). Predation on fish larvae by *Physalia physalis*, the Portuguese man-of-war. *Mar. Ecol. Prog. Ser.* **19,** 189–191.

Purcell, J. E. (1985). Predation on fish eggs and larvae by pelagic cnidarians and ctenophores. *Bull. Mar. Sci.* **37,** 739–755.

Qasim, S. (1956). Time and duration of the spawning season in some marine teleosts in relation to their distribution. *J. Cons. Cons. Int. Explor. Mer* **21,** 144–155.

Quast, J. C. (1968a). Some physical aspects of the inshore environment, particularly as it affects kelp-bed fishes. *Fish. Bull.—Calif., Dep. Fish Game* **139,** 25–34.

Quast, J. C. (1968b). Fish fauna of the rocky inshore zone. *Fish Bull.—Calif., Dep. Fish Game* **139,** 35–55.

Quast, J. C. (1968c). Estimates of the populations and the standing crop of fishes. *Fish. Bull.—Calif., Dep. Fish Game* **139,** 57–79.

Quast, J. C. (1968d). Observations on the food of the kelp-bed fishes. *Fish. Bull.—Calif., Dep. Fish Game* **139,** 109–142.

Quinn, N. J., and Kojis, B. J. (1985). Does the presence of coral reefs in proximity to a tropical estuary affect the estuarine fish assemblage? *Proc. Int. Coral Reef Congr., 5th,* **5,** 445–450.

Radakov, D. V. (1973). "Schooling in the Ecology of Fish." Halsted, New York.

Radovich, J. (1961). Relationships of some marine organisms of the northeast Pacific to water temperature, particularly during 1957 through 1959. *Fish. Bull.—Calif., Dep. Fish Game* **112,** 1–62.

Radtke, R. L. (1985). Life history characteristics of the Hawaiian damselfish, *Abudefduf abdominalis* defined from otoliths. *Proc. Int. Coral Reef Congr., 5th* **5,** 397–401.

Radtke, R. L. (1988). Recruitment parameters resolved from structural and chemical components of juvenile *Dascyllus albisella* otoliths. *Proc. Int. Coral Reef Symp., 6th* **2,** 821–826.

Radtke, R. L., Williams, D. F., and Hurley, P. C. F. (1987). The stable isotopic composition of bluefin tuna (*Thunnus thynnus*) otoliths: Evidence for physiological regulation. *Comp. Biochem. Physiol.* **87,** 797–801.

Radtke, R. L., Kinzie, R. A., and Folsom, S. D. (1988). Age at recruitment of Hawaiian freshwater gobies. *Environ. Biol. Fishes* **23,** 205–214.

Ragan, M. A., and Glombitza, K. W. (1986). Phlorotannins, brown algal polyphenolics. *Prog. Phycol. Res.* **4,** 129–241.

Ralston, S. (1976a). Age determination of a tropical reef butterflyfish utilizing daily growth rings of otoliths. *Fish. Bull.* **74,** 990–994.

Ralston, S. (1976b). Anomalous growth and reproductive patterns in populations of *Chaetodon miliaris* (Pisces, Chaetodontidae) from Kaneohe Bay, Oahu, Hawaiian Islands. *Pac. Sci.* **30,** 395–403.

Ralston, S. (1987). Mortality rates of snappers and groupers. *In* "Tropical Snappers and Groupers: Biology and Fisheries Management" (J. J. Polovina and S. Ralston, eds.), pp. 375–404. Westview, Boulder, Colorado.

Ralston, S. L., and Horn, M. H. (1986). High tide movements of the temperate-zone herbivorous fish *Cebidichthys violaceus* (Girard) as determined by ultrasonic telemetry. *J. Exp. Mar. Biol. Ecol.* **98,** 35–50.

Ralston, S., and Kawamoto, K. E. (1985). A preliminary analysis of the 1984 size structure of Hawaii's commercial opakapaka landings and a consideration of age at entry and yield per recruit. *Southwest Fish Cent., Honolulu Lab., Natl. Mar. Fish. Serv., NOAA, Adm. Rep.* **H-85-1,** 1–9.

Ralston, S., and Kawamoto, K. E. (1988). A biological assessment of Hawaiian bottom fish stocks, 1984–87. *Southwest Fish Cent., Honolulu Lab., Natl. Mar. Fish Serv., NOAA, Adm. Rep.* **H-88-8,** 1–60.

Ralston, S., and Polovina, J. J. (1982). A multispecies analysis of the commercial deep-sea handline fishery in Hawaii. *Fish. Bull.* **80,** 435–448.

Ralston, S., and Williams, H. A. (1989). Numerical integration of daily growth increments: An efficient means of ageing tropical fishes for stock assessment. *Fish. Bull.* **87,** 1–16.

Randall, J. E. (1961a). A contribution to the biology of the convict surgeonfish of the Hawaiian Islands, *Acanthurus triostegus sandvicensis. Pac. Sci.* **15,** 215–272.

Randall, J. E. (1961b). Overgrazing of algae by herbivorous marine fishes. *Ecology* **42,** 812.

Randall, J. E. (1962). Tagging reef fishes in the Virgin Islands. *Proc. Annu. Gulf Caribb. Fish. Inst.* **14,** 201–241.

Randall, J. E. (1963). An analysis of the fish populations of artificial and natural reefs in the Virgin Islands. *Caribb. J. Sci.* **3,** 31–47.

Randall, J. E. (1965). Grazing effects on seagrasses by herbivorous reef fishes in the West Indies. *Ecology* **46,** 255–260.

Randall, J. E. (1967). Food habits of reef fishes of the West Indies. *Stud. Trop. Oceanogr.* **5,** 665–847.

Randall, J. E. (1974). The effects of fishes on coral reefs. *Proc. Int. Coral Reef Symp., 2nd* **1,** 159–166.

Randall, J. E. (1983). "Red Sea Reef Fishes." IMMEL Press, London.

Randall, J. E. (1985a). Fishes. *Proc. Int. Coral Reef Congr., 5th* **1,** 462–481.

Randall, J. E. (1985b). "Guide to Hawaiian Reef Fishes." Harrowood, Newtown Square, Pennsylvania.

Randall, J. E. (1987). Introductions of marine fishes to the Hawaiian Islands. *Bull. Mar. Sci.* **41,** 490–502.

Randall, J. E., and Egana, A. C. (1984). Native fish names of Easter Island fishes, with comments on the origin of the Rapanui people. *Occas. Pap.—Bernice P. Bishop Mus.* **25**(12), 1–16.

Randall, J. E., and Kanayama, R. K. (1972). Hawaiian fish immigrants. *Sea Front.* **18,** 144–153.

Randall, J. E., and Kanayama, R. K. (1973). Marine organisms–introduction of serranid and lutjanid fishes from French Polynesia to the Hawaiian Islands. *In* "Nature Conservation in the Pacific" (A. B. Costin and R. H. Groves, eds.), pp. 197–200. Aust. Natl. Univ. Press, Canberra, Australia.

Randall, J. E., and Randall, H. A. (1960). Examples of mimicry and protective resemblance in tropical marine fishes. *Bull. Mar. Sci. Gulf Caribb.* **10,** 444–480.

Randall, J. E., and Randall, H. A. (1987). Annotated checklist of the fishes of Enewetak Atoll and other Marshall Islands. *In* "The Natural History of Enewetak Atoll" (D. M. Devaney, E. S. Reese, B. L. Burch, and P. Helfrich, eds.), Vol. 2, pp. 289–324. U.S. Dep. Energy, Washington, D.C.

Randall, J. E., Aida, K., Oshima, Y., Hori, K., and Hashimoto, Y. (1981). Occurrence of a crinotoxin and hemagglutinin in the skin mucus of the moray eel *Lycodontis nudivomer. Mar. Biol.* **62,** 179–184.

Randall, J. E., Bauchot, M. L., and Descoulte, M. (1985a). *Chromis viridis* (Cuvier,

1830), the correct name for the Indo-Pacific damselfish previously known as *C. caerulea* Cuvier, 1830 (Pisces:Pomacentridae). *Cybium* **9,** 411–413.

Randall, J. E., Lobel, P. S., and Chave, E. H. (1985b). Annotated checklist of the fishes of Johnston Island. *Pac. Sci.* **39,** 24–80.

Reaka, M. L. (1980). Geographic range, life history patterns, and body size in a guild of coral-dwelling mantis shrimps. *Evolution (Lawrence, Kans.)* **34,** 1019–1030.

Reaka, M. L. (1985). Interactions between fishes and the motile benthic invertebrates on reefs: The significance of motility vs. defensive adaptations. *Proc. Int. Coral Reef Congr., 5th* **5,** 439–444.

Reaka, M. L., and Manning, R. B. (1981). The behavior of stomatopod Crustacea, and its relationship to rates of evolution. *J. Crustacean Biol.* **1,** 309–327.

Reed, D. C., and Foster, M. S. (1984). The effects of canopy shading on algal recruitment and growth in a giant kelp forest. *Ecology* **65,** 937–948.

Reese, E. S. (1975). A comparative field study of the social behavior and related ecology of reef fishes of the family Chaetodontidae. *Z. Tierpsychol.* **37,** 37–61.

Reese, E. S. (1977). Coevolution of corals and coral feeding fishes of the family Chaetodontidae. *Proc. Int. Coral Reef Symp., 3rd* **2,** 268–274.

Reeson, P. H. (1983a). The biology, ecology and bionomics of the parrotfishes, *Scaridae. In* "Caribbean Coral Reef Fishery Resources" (J. L. Munro, ed.) pp. 166–177. *ICLARM Stud. Rev.* 7, Manila, Philippines.

Reeson, P. H. (1983b). The biology, ecology and bionomics of the surgeonfishes, Acanthuridae. *ICLARM Stud. Rev.* 7, 178–190.

Reidenauer, J. A., and Thistle, D. (1981). Response of a soft-bottom harpacticoid community to stingray (*Dasyatis sabina*) disturbance. *Mar. Biol.* **65,** 261–267.

Reinthal, P. N., and Lewis, S. M. (1986). Social behavior, foraging efficiency and habitat utilization in a group of tropical herbivorous fish. *Anim. Behav.* **34,** 1687–1693.

Reinthal, P. N., Kensley, B., and Lewis, S. M. (1984). Dietary shifts in the queen triggerfish, *Balistes vetula*, in the absence of its primary food item, *Diadema antillarum. P.S.Z.N.I. Mar. Ecol.* **5,** 191–195.

Reintjes, J. W., and King, J. W. (1953). Food of yellowfin tuna in the central Pacific. *Fish. Bull.* **54,** 92–110.

Richards, W. J. (1982). Planktonic processes affecting establishment and maintenance of reef fish stocks. The biological bases for reef fishery management. *NOAA Tech. Mem. NMFS SEFC* **NMFS SEFC-80,** 92–102.

Richards, W. J. (1984). Kinds and abundances of fish larvae in the Caribbean Sea and adjacent waters. *NOAA Tech. Rep.,* **NMFS SSRF-776,** 1–54.

Richards, W. J., and Lindeman, K. C. (1987). Recruitment dynamics of reef fishes: Planktonic processes, settlement and demersal ecologies, and fishery analysis. *Bull. Mar. Sci.* **41,** 392–410.

Richards, W. J., Potthoff, T., Kelley, S., McGowan, M. F., Ejsymont, L., Power, J. H., and Olvera, L, R. M. (1984). SEAMAP 1982—Ichthyoplankton. Larval distribution of Engraulidae, Carangidae, Clupeidae, Lutjanidae, Serranidae, Coryphaenidae, Istiophoridae, Xiphiidae and Scombridae in the Gulf of Mexico. *NOAA Tech. Mem.,* **NMFS SEFC-144,** 1–4.

Richardson, S. L., and Pearcy, W. G. (1977). Coastal and oceanic fish larvae in an area of upwelling off Yaquina Bay, Oregon. *Fish. Bull.* **75,** 125–145.

Richmond, R. H. (1987). Energetics, competency, and long-distance dispersal of planula larvae of the coral *Pocillopora damicornis*. *Mar. Biol.* **93,** 527–533.
Rimmer, D. W. (1986). Changes in diet and the development of microbial digestion in juvenile buffalo bream *Kyphosus cornellii*. *Mar. Biol.* **93,** 443–448.
Rimmer, D. W., and Weibe, W. J. (1987). Fermentative microbial digestion in herbivorous fishes. *J. Fish Biol.* **31,** 229–236.
Risk, M. J. (1972). Fish diversity on a coral reef in the Virgin Islands. *Atoll Res. Bull.* **153,** 1–6.
Risk, M. J., and Sammarco, P. W. (1982). Bioerosion of corals and the influence of damselfish territoriality: A preliminary study. *Oecologia* **52,** 376–380.
Roberts, C. M. (1985). Resource sharing in territorial herbivorous reef fishes. *Proc. Int. Coral Reef Congr., 5th* **4,** 17–22.
Roberts, C. M. (1987). Experimental analysis of resource sharing between herbivorous damselfish and blennies on the Great Barrier Reef. *J. Exp. Mar. Biol. Ecol.* **111,** 61–75.
Roberts, C. M., and Ormond, R. F. G. (1987). Habitat complexity and coral reef fish diversity and abundance on Red Sea fringing reefs. *Mar. Ecol. Prog. Ser.* **41,** 1–8.
Robertson, D. R. (1972). Social control of sex-reversal in a coral-reef fish. *Science* **177,** 1007–1009.
Robertson, D. R. (1973a). Sex change under the waves. *New Sci.* **58,** 538–540.
Robertson, D. R. (1973b). Field observations on the reproductive behavior of a pomacentrid fish, *Acanthochromis polyacanthus*. *Z. Tierpsychol.* **32,** 319–324.
Robertson, D. R. (1974). A study of the ethology and reproductive biology of the labrid fish, *Labroides dimidiatus*, at Heron Island, Great Barrier Reef. Unpublished Ph.D thesis, University of Queensland, Brisbane, Australia.
Robertson, D. R. (1981). The social and mating systems of two labrid fishes, *Halichoeres maculipinna* and *H. garnoti*, off the Caribbean of Panama. *Mar. Biol.* **64,** 327–340.
Robertson, D. R. (1982). Fish feces as fish food on a Pacific coral reef. *Mar. Ecol. Prog. Ser.* **7,** 253–265.
Robertson, D. R. (1983). On the spawning behavior and spawning cycles of eight surgeonfishes (Acanthuridae) from the Indo-Pacific. *Environ. Biol. Fishes* **9,** 192–223.
Robertson, D. R. (1984). Cohabitation of competing territorial damselfishes on a Caribbean coral reef. *Ecology* **65,** 1121–1135.
Robertson, D. R. (1987). Responses of two coral reef toadfishes (Batrachoididae) to the demise of their primary prey, the sea urchin *Diadema antillarum*. *Copeia* **1987,** 637–642.
Robertson, D. R. (1988a). Abundances of surgeonfishes on patch-reefs in Caribbean Panama: Due to settlement, or post-settlement events? *Mar. Biol.* **97,** 495–501.
Robertson, D. R. (1988b). Extreme variation in settlement of the Caribbean triggerfish *Balistes vetula* in Panama. *Copeia* **1988,** 698–703.
Robertson, D. R. (1988c). Settlement and population dynamics of *Abudefduf saxatilis* on patch reefs in Caribbean Panama. *Proc. Int. Coral Reef Symp., 6th* **2,** 839–843.
Robertson, D. R. (1990). Differences in the seasonalities of spawning and recruitment of some small neotropical reef fishes. *J. Exp. Mar. Biol. Ecol.* **144,** 49–62.
Robertson, A. I., and Duke, N. C. (1987). Mangroves as nursery sites: Comparisons of

the abundance and species composition of fish and crustaceans in mangroves and other nearshore habitats in tropical Australia. *Mar. Biol.* **96,** 193–205.

Robertson, D. R., and Foster, S. A. (1982). Off-reef emigration of young adults of the labrid fish *Epibulus insidiator. Copeia* **1982,** 227–229.

Robertson, D. R., and Gaines, S. D. (1986). Interference competition structures habitat use in a local assemblage of coral reef surgeonfishes. *Ecology* **67,** 1372–1383.

Robertson, D. R., and Hoffman, S. G. (1977). The roles of female mate choice and predation in the mating systems of some tropical labroid fishes. *Z. Tierpsychol.* **45,** 298–320.

Robertson, D. R., and Lassig, B. (1980). Spatial distribution patterns and coexistence of a group of territorial damselfishes from the Great Barrier Reef. *Bull. Mar. Sci.* **30,** 187–203.

Robertson, D. R., and Polunin, N. V. C. (1981). Coexistence: Symbiotic sharing of feeding territories and algal food by some coral reef fishes from the western Indian Ocean. *Mar. Biol.* **62,** 185–195.

Robertson, D. R., and Sheldon, J. M. (1979). Competitive interactions and the availability of sleeping sites for a diurnal coral reef fish. *J. Exp. Mar. Biol. Ecol.* **40,** 285–298.

Robertson, D. R., Sweatman, H. P. A., Fletcher, E. A., and Cleland, M. G. (1976). Schooling as a mechanism for circumventing the territoriality of competitors. *Ecology* **57,** 1208–1220.

Robertson, D. R., Polunin, N. V. C., and Leighton, K. (1979). The behavioral ecology of three Indian Ocean surgeonfishes (*Acanthurus lineatus, A. leucosternum,* and *Zebrasoma scopas*): Their feeding strategies, and social and mating systems. *Environ. Biol. Fishes* **4,** 125–170.

Robertson, D. R., Hoffman, S. G., and Sheldon, J. M. (1981). Availability of space for the territorial Caribbean damselfish *Eupomacentrus planifrons. Ecology* **62,** 1162–1169.

Robertson, D. R., Green, D. G., and Victor, B. C. (1988). Temporal coupling of reproduction and recruitment of larvae of a Caribbean reef fish. *Ecology* **69,** 370–381.

Robertson, D. R., Petersen, C. W., and Brawn, J. D. (1990). Lunar reproductive cycles of benthic-brooding reef fishes: Reflections of larval-biology or adult-biology? *Ecol. Monogr.* **60,** 311–329.

Robichaux, D. M., Cohen, A. C., Reaka, M. L., and Allen, D. (1981). Experiments with zooplankton on coral reefs, or will the real demersal plankton please come up. *Mar. Biol.* **2,** 77–94.

Robilliard, G. A. (1971). Feeding behavior and prey capture in an asteroid, *Stylasterias forreri. Syesis* **4,** 191–195.

Robins, C. R. (1971). Distributional patterns of fishes from coastal and shelf waters of the tropical western Atlantic. *In* "Symposium on Investigations and Resources of the Caribbean Sea and Adjacent Regions," Pap. Fish. Resour. pp. 249–255, Food Agric. Organ., Rome.

Robison, D. E. (1985). Variability in the vertical distribution of ichthyoplankton in lower Tampa Bay, Florida. *In* "Proceedings of the Tampa Bay Area Scientific Information Symposium (1982)" (S. A. F. Treat, J. L. Simon, R. R. Lewis, and R. L. Whitman, eds.), Rep. 65, Fla. Sea Grant Coll., pp. 359–383. Bellwether, Bronxville, New York.

Romme, W. H. (1982). Fire and landscape diversity in subalpine forests of Yellowstone National Park. *Ecol. Monogr.* **52,** 199–221.

Root, R. B. (1967). The niche exploitation pattern of the blue-gray gnatcatcher. *Ecol. Monogr.* **37,** 317–350.

Rosen, B. R. (1984). Reef coral biogeography and climate through the late Cainozoic: Just islands in the sun or a critical pattern of islands? *In* "Fossils and Climate" (P. J. Brenchley, ed.), pp. 201–264. Wiley, New York.

Rosen, B. R. (1988). Progress, problems and patterns in the biogeography of reef corals and other tropical marine organisms. *Helgol. Wiss. Meeresunters.* **42,** 269–301.

Rosen, B. R., and Smith, A. B. (1988). Tectonics from fossils? Analysis of reef-coral and sea-urchin distributions from late Cretaceous to Recent, using a new method. *In* "Gondwana and Tethys" (M. G. Audley-Charles and A. Hallam, eds.), Geol. Soc. Spec. Pub. 37, pp. 275–306. Oxford Univ. Press, Oxford, England.

Rosenblatt, R. H. (1963). Some aspects of speciation in marine shore fishes. *Systematics Association Publ.* **5,** 171–180.

Rosenblatt, R. H. (1967a). The osteology of the congrid eel *Gorgasia punctata* and the relationships of the Heterocongrinae. *Pac. Sci.* **21,** 91–97.

Rosenblatt, R. H. (1967b). The zoogeographic relationships of the marine shore fishes of tropical America. *Stud. Trop. Oceanogr.* **5,** 579–592.

Rosenblatt, R. H., and Walker, B. W. (1963). The marine shorefishes of the Galápagos Islands. *Occas. Pap. Calif. Acad. Sci.* **44,** 97–106.

Rosenblatt, R. H., and Waples, R. S. (1986). A genetic comparison of allopatric populations of shore fish species from the eastern and central Pacific Ocean: Dispersal or vicariance? *Copeia* **1986,** 275–284.

Ross, R. M. (1978a). Territorial behavior and ecology of the anemonefish *Amphiprion melanopus* on Guam. *Z. Tierpsychol.* **46,** 71–83.

Ross, R. M. (1978b). Reproductive behavior of the anemonefish *Amphiprion melanopus* on Guam. *Copeia* **1978,** 103–107.

Ross, S. T. (1986). Resource partitioning in fish assemblages: A review of field studies. *Copeia* **1986,** 352–388.

Ross, R. M., Losey, G. S., and Diamond, M. (1983). Sex change in a coral-reef fish: Dependence of stimulation and inhibition on relative size. *Science* **221,** 574–575.

Rothschild, B. J. (1986). "Dynamics of Marine Fish Populations." Harvard Univ. Press, Cambridge, Massachusetts.

Roughgarden, J. (1974). Species packing and the competition function with illustrations from coral reef fish. *Theor. Pop. Biol.* **5,** 163–186.

Roughgarden, J., and Iwasa, Y. (1986). Dynamics of a metapopulation with space-limited subpopulations. *Theor. Pop. Biol.* **29,** 235–261.

Roughgarden, J., Gaines, S. D., and Pacala, S. W. (1987). Supply side ecology: The role of physical transport processes. *In* "Organization of Communities: Past and Present" (J. H. R. Gee and P. S. Giller, eds.), pp. 491–518. Blackwell, London.

Roughgarden, J., Gaines, S., and Possingham, H. (1988). Recruitment dynamics in complex life cycles. *Science* **241,** 1460–1466.

Roughgarden, J., May, R. M., and Levin, S. A. (eds.) (1989). "Perspectives in Ecological Theory." Princeton Univ. Press, Princeton, New Jersey.

Rubenstein, D. I. (1986). Ecology and sociality in horses and zebras. *In* "Ecological Aspects of Social Evolution" (D. I. Rubenstein and R. W. Wrangham, eds.), pp. 282–302. Princeton Univ. Press, Princeton, New Jersey.

Rubenstein, D. I., and Wrangham, R. W. (eds.) (1986). "Ecological Aspects of Social Evolution." Princeton Univ. Press, Princeton, New Jersey.

Runkle, J. R., and Yetter, T. C. (1987). Treefalls revisited: Gap dynamics in the southern Appalachians. *Ecology* **68,** 417–424.

Russ, G. R. (1984a). Distribution and abundance of herbivorous grazing fishes in the central Great Barrier Reef. I. Levels of variability across the entire continental shelf. *Mar. Ecol. Prog. Ser.* **20,** 23–34.

Russ, G. R. (1984b). Distribution and abundance of herbivorous grazing fishes in the central Great Barrier Reef. II. Patterns of zonation of mid-shelf and outershelf reefs. *Mar. Ecol. Prog. Ser.* **20,** 35–44.

Russ, G. R. (1984c). A review of coral reef fisheries. *UNESCO Rep. Mar. Sci.* **27,** 74–92.

Russ, G. R. (1984d). Effects of fishing and protective management on coral reefs at four locations in the Visayas, Philippines (phase II). *UNEP–NRMC Coral Reef Monitor. Project, Nat. Resour. Manage. Cent., Philipp.* pp. 1–54.

Russ, G. R. (1985). Effects of protective management on coral reef fishes in the central Philippines. *Proc. Int. Coral Reef Congr., 5th* **4,** 219–224.

Russ, G. R. (1987). Is the rate of removal of algae by grazers reduced inside territories of tropical damselfishes? *J. Exp. Mar. Biol. Ecol.* **110,** 1–17.

Russ, G. R. (1989). Distribution and abundance of coral reef fishes in the Sumilon Island reserve, central Philippines, after nine years of protection from fishing. *Asian Mar. Biol.* **6,** 59–71.

Russ, G. R., and Alcala, A. C. (1989). Effects of intense fishing pressure on an assemblage of coral reef fishes. *Mar. Ecol. Prog. Ser.* **56,** 13–27.

Russ, G. R., and St. John, J. (1988). Diets, growth rates and secondary production of herbivorous coral reef fishes. *Proc. Int. Coral Reef Symp., 6th* **2,** 37–43.

Russell, B. C. (1977). Population and standing crop estimates for rocky reef fishes of north-eastern New Zealand. *N.Z. J. Mar. Freshwater Res.* **11,** 23–36.

Russell, B. C. (1983a). The food and feeding habits of rocky reef fish of north-eastern New Zealand. *N.Z. J. Mar. Freshwater Res.* **17,** 121–145.

Russell, B. C. (1983b). Annotated checklist of the coral reef fishes in the Capricorn–Bunker group, Great Barrier Reef, Australia. *Spec. Publ.—Great Barrier Reef Mar. Park Authority* **1,** 1–184.

Russell, B. C. (1987). New Australian fishes. Part 20. A new species of *Aplodactylus* (Aplodactylidae). *Mem. Mus. Victoria* **48,** 85–87.

Russell, B. C., Talbot, F. H., and Domm, S. (1974). Patterns of colonisation of artificial reefs by coral reef fishes. *Proc. Int. Coral Reef Symp., 2nd* **1,** 207–215.

Russell, B. C., Allen, G. R., and Lubbock, R. H. (1976). New cases of mimicry in marine fishes. *J. Zool.* **180,** 407–423.

Russell, B. C., Anderson, G. R. V., and Talbot, F. H. (1977). Seasonality and recruitment of coral reef fishes. *Aust. J. Mar. Freshwater Res.* **28,** 521–528.

Sadovy, Y. (1986). The sexual pattern and social organisation of the bicolor damselfish, *Stegastes partitus* (Poey) (Pisces: Pomacentridae). Unpublished Ph.D. thesis, University of Manchester, Manchester, England.

Sadovy, Y., and Shapiro, D. Y. (1987). Criteria for the diagnosis of hermaphroditism in fishes. *Copeia* **1987,** 136–156.

Saether, B.-E. (1988). Pattern of co-variation between life-history traits of European birds. *Nature* (*London*) **331,** 616–617.

Saila, S. B., and Roedel, P. B. (eds.) (1980). "Stock Assessment for Tropical Small-Scale Fisheries." Univ. of Rhode Island, Kingston, Rhode Island.

Sainsbury, K. J. (1982). The ecological basis of tropical fisheries management. *ICLARM Conf. Proc.* **9,** 167–194.

Sainsbury, K. J. (1988). The ecological basis of multispecies fisheries, and management of a demersal fishery in tropical Australia. *In* "Fish Population Dynamics" (J. A. Gulland, ed.), pp. 349–382. Wiley, London.

Sale, P. F. (1969). Pertinent stimuli for habitat selection by the juvenile manini. *Ecology* **50,** 616–623.

Sale, P. F. (1970). Distribution of larval Acanthuridae off Hawaii. *Copeia* **1970,** 765–766.

Sale, P. F. (1971). Extremely limited home range in a coral reef fish, *Dascyllus aruanus* (Pisces: Pomacentridae). *Copeia* **1971,** 324–327.

Sale, P. F. (1972a). Effect of cover on agonistic behavior of a reef fish: A possible spacing mechanism. *Ecology* **53,** 753–758.

Sale, P. F. (1972b). Influence of corals in the dispersion of the pomacentrid fish, *Dascyllus aruanus*. *Ecology* **53,** 741–744.

Sale, P. F. (1974). Mechanisms of co-existence in a guild of territorial fishes at Heron Island. *Proc. Int. Coral Reef Symp., 2nd* **1,** 193–206.

Sale, P. F. (1975). Patterns of use of space in a guild of territorial reef fishes. *Mar. Biol.* **29,** 89–97.

Sale, P. F. (1976). The effect of territorial adult pomacentrid fishes on the recruitment and survival of juveniles on patches of coral rubble. *J. Exp. Mar. Biol. Ecol.* **24,** 297–306.

Sale, P. F. (1977). Maintenance of high diversity in coral reef fish communities. *Am. Nat.* **111,** 337–359.

Sale, P. F. (1978a). Coexistence of coral reef fishes—A lottery for living space. *Environ. Biol. Fishes* **3,** 85–102.

Sale, P. F. (1978b). Reef fishes and other vertebrates: A comparison of social structures. *In* "Contrasts in Behavior. Adaptations in the Aquatic and Terrestrial Environments" (E. S. Reese and F. J. Lighter, eds.), pp. 313–346. Wiley (Interscience), New York.

Sale, P. F. (1979a). Recruitment, loss and coexistence in a guild of territorial coral reef fishes. *Oecologia* **42,** 159–177.

Sale, P. F. (1979b). Habitat partitioning and competition in fish communities. *In* "Predator–Prey Systems in Fisheries Management" (H. E. Clepper, ed.), pp. 323–331. Sport Fishing Inst., Washington, D.C.

Sale, P. F. (1980a). The ecology of fishes on coral reefs. *Oceanogr. Mar. Biol.* **18,** 367–421.

Sale, P. F. (1980b). Assemblages of fish on patch reefs—predictable or unpredictable? *Environ. Biol. Fishes* **5,** 243–249.

Sale, P. F. (1982a). Stock–recruit relationships and regional coexistence in a lottery competitive system: A simulation study. *Am. Nat.* **120,** 139–159.

Sale, P. F. (1982b). The structure and dynamics of coral reef fish communities. *ICLARM Conf. Proc.* **9,** 241–253.

Sale, P. F. (1984). The structure of communities of fish on coral reefs, and the merit of an hypothesis-testing manipulative approach to ecology. *In* "Ecological Communities: Conceptual Issues and the Evidence" (D. R. Strong, Jr., D. Simberloff, L. G. Abele, and A. B. Thistle, eds.), pp. 478–490. Princeton Univ. Press, Princeton, New Jersey.

Sale, P. F. (1985). Patterns of recruitment in coral reef fishes. *Proc. Int. Coral Reef Congr., 5th* **5,** 391–396.

Sale, P. F. (1988a). Perception, pattern, chance and the structure of reef fish communities. *Environ. Biol. Fishes* **21,** 3–15.

Sale, P. F. (1988b). What coral reefs can teach us about ecology. *Proc. Int. Coral Reef Symp., 6th* **1,** 19–27.

Sale, P. F. (1990). Recruitment of marine species: Is the bandwagon rolling in the right direction? *Trends Ecol. Evol.* **5,** 25–27.

Sale, P. F., and Douglas, W. A. (1981). Precision and accuracy of visual census technique for fish assemblages on coral patch reefs. *Environ. Biol. Fishes* **6,** 333–339.

Sale, P. F., and Douglas, W. A. (1984). Temporal variability in the community structure of fish on coral patch reefs and the relation of community structure to reef structure. *Ecology* **65,** 409–422.

Sale, P. F., and Dybdahl, R. (1975). Determinants of community structure for coral reef fishes in an experimental habitat. *Ecology* **56,** 1343–1355.

Sale, P. F., and Dybdahl, R. (1978). Determinants of community structure for coral reef fishes in isolated coral heads at lagoonal and reef slope sites. *Oecologia* **34,** 57–74.

Sale, P. F., and Ferrell, D. J. (1988). Early survivorship of juvenile coral reef fishes. *Coral Reefs* **7,** 117–124.

Sale, P. F., and Guy, J. A. (1991). Persistence of community structure: What happens when you change taxonomic scale? *Coral Reefs*. In press.

Sale, P. F., and Sharp, B. J. (1983). Correction for bias in visual transect censuses of coral reef fishes. *Coral Reefs* **2,** 37–42.

Sale, P. F., and Steel, W. J. (1986). Random placement and the structure of reef fish communities. *Mar. Ecol. Prog. Ser.* **28,** 165–174.

Sale, P. F., and Steel, W. J. (1989). Temporal variability in patterns of association among fish species on coral patch reefs. *Mar. Ecol. Prog. Ser.* **51,** 35–47.

Sale, P. F., and Williams, D. M. (1982). Community structure of coral reef fishes: Are the patterns more than those expected by chance? *Am. Nat.* **120,** 121–127.

Sale, P. F., Doherty, P. J., and Douglas, W. A. (1980). Juvenile recruitment strategies and the coexistence of territorial pomacentrid fishes. *Bull. Mar. Sci.* **30,** 147–158.

Sale, P. F., Douglas, W. A., and Doherty, P. J. (1984a). Choice of microhabitats by coral reef fishes at settlement. *Coral Reefs* **3,** 91–99.

Sale, P. F., Doherty, P. J., Eckert, G. J., Douglas, W. A., and Ferrell, D. J. (1984b). Large scale spatial and temporal variation in recruitment to fish populations on coral reefs. *Oecologia* **64,** 191–198.

Sale, P. F., Jones, G. P., Choat, J. H., Leis, J. M., Thresher, R. E., and Williams, D. M. (1985). Current priorities in ecology of coral reef fishes. *Search* **16,** 270–274.

Sale, P. F., Guy, J. A., and Steel, W. J. (1991). Manuscript in preparation.

Salt, G. W. (ed.) (1984). "Ecology and Evolutionary Biology." Univ. of Chicago Press, Chicago.

Sammarco, P. W., and Carleton, J. H. (1981). Damselfish territoriality and coral community structure: Reduced grazing, coral recruitment, and effects on coral spat. *Proc. Int. Coral Reef Symp., 4th* **2,** 525–535.

Sammarco, P. W., and Crenshaw, H. (1984). Plankton community dynamics of the central Great Barrier Reef Lagoon: Analysis of data from Ikeda *et al. Mar. Biol.* **82,** 167–180.

Sammarco, P. W., and Williams, A. H. (1982). Damselfish territoriality: Influences on *Diadema* distribution and implications for coral community structure. *Mar. Ecol. Prog. Ser.* **8,** 53–59.

Samoilys, M. A. (1988). Abundance and species richness of coral reef fish on the Kenyan coast: The effects of protective management and fishing. *Proc. Int. Coral Reef Symp., 6th* **2,** 261–266.

Samuel, M., Mathews, C. P., and Bawazeer, A. S. (1987). Age and validation of age from otoliths for warm water fishes for the Arabian Gulf. *In* "Age and Growth of Fish" (R. C. Summerfelt and G. E. Hall, eds.), pp. 253–265. Iowa State Univ. Press, Ames, Iowa.

Sano, M., Shimizu, M., and Nose, Y. (1984a). Changes in structure of coral reef fish communities by destruction of hermatypic corals: Observational and experimental views. *Pac. Sci.* **38,** 51–79.

Sano, M., Shimizu, M., and Nose, Y. (1984b). Food habits of teleostean reef fishes in Okinawa Island, southern Japan. *Jpn. Univ. Mus., Univ. Tokyo Bull.* **25,** 128 pp.

Sano, M., Shimizu, M., and Nose, Y. (1987). Long-term effects of destruction of hermatypic corals by *Acanthaster planci* infestation on reef fish communities at Iriomote Island, Japan. *Mar. Ecol. Prog. Ser.* **37,** 191–199.

Santelices, B., and Ugarte, R. (1987). Algal life-history strategies and resistance to digestion. *Mar. Ecol. Prog. Ser.* **35,** 267–275.

Savina, G. C., and White, A. T. (1986). Reef fish yields and non-reef catch of Pamilacan Island, Bohol, Philippines. *In* "The First Asian Fisheries Forum" (J. L. Maclean, L. B. Dizon, and L. V. Hosillos, eds.), pp. 497–500. Asian Fish. Soc., Manila, Philippines.

Schaeffer, B., and Rosen, D. E. (1961). Major adaptive levels in the evolution of the actinopterygian feeding mechanism. *Am. Zool.* **1,** 187–204.

Schaeffer, B., and Williams, M. (1977). Relationships of fossil and living elasmobranch. *Am. Zool.* **17,** 293–302.

Scheltema, R. S. (1986). Long-distance dispersal by planktonic larvae of shoal-water benthic invertebrates among central Pacific islands. *Bull. Mar. Sci.* **39,** 241–256.

Scheltema, R. S. (1988). Initial evidence for the transport of teleplanic larvae of benthic invertebrates across the East Pacific Barrier. *Biol. Bull.* (*Woods Hole, Mass.*) **174,** 145–152.

Schiel, D. R., and Foster, M. S. (1986). The structure of subtidal algal stands in temperate waters. *Oceanogr. Mar. Biol.* **24,** 265–307.

Schiel, D. R., Kingsford, M. J., and Choat, J. H. (1986). Depth distribution and abundance of benthic organisms and fishes at the sub-tropical Kermadec Islands. *N.Z. J. Mar. Freshwater Res.* **20,** 521–535.

Schmale, M. C. (1981). Sexual selection and reproductive success in males of the bicolor damselfish *Eupomacentrus partitus* (Pisces: Pomacentridae). *Anim. Behav.* **29,** 1172–1184.

Schmitt, P. D. (1984a). Aspects of the larval ecology of *Hypoatherina tropicalis* (Pisces: Atherinidae) at One Tree Reef, Great Barrier Reef, Australia. Unpublished Ph.D. thesis, University of Sydney, Sydney, Australia.

Schmitt, P. D. (1984b). Marking growth increments in otoliths of larval and juvenile fish by immersion in tetracycline to examine the rate of increment formation. *Fish. Bull.* **82,** 237–242.

Schmitt, P. D. (1986). Feeding by larvae of *Hypoatherina tropicalis* (Pisces: Atherinidae) and its relation to prey availability in One Tree Lagoon, Great Barrier Reef, Australia. *Environ. Biol. Fishes* **16,** 79–94.

Schmitt, R. J., and Coyer, J. A. (1983). Variation in surfperch diets between allopatry and sympatry: Circumstantial evidence for competition. *Oecologia* **58,** 402–410.

Schmitt, R. J., and Holbrook, S. J. (1986). Seasonally fluctuating resources and temporal variability of interspecific competition. *Oecologia* **69,** 1–11.

Schmitt, R. J., and Strand, S. W. (1982). Cooperative foraging by yellowtail (Carangidae) on two species of prey fish. *Copeia* **1982,** 714–717.

Schoener, T. W. (1968). Sizes of feeding territories among birds. *Ecology* **49,** 123–141.

Schoener, T. W. (1974). Resource partitioning in ecological communities. *Science* **185,** 27–39.

Schoener, T. W. (1983a). Simple models of optimal feeding–territory size: A reconciliation. *Am. Nat.* **121,** 608–629.

Schoener, T. W. (1983b). Field experiments in interspecific competition. *Am. Nat.* **122,** 240–285.

Schroeder, R. E. (1985). Recruitment rate patterns of coral-reef fishes at Midway Lagoon (northwestern Hawaiian Islands). *Proc. Int. Coral Reef Congr., 5th* **5,** 379–384.

Schroeder, R. E. (1987). Effects of patch reef size and isolation on coral reef fish recruitment. *Bull. Mar. Sci.* **41,** 441–451.

Schroeder, R. E., and Starck, W. A. (1964). Photographing the night creatures of Alligator Reef. *Nat. Geogr.* **125,** 128–154.

Schultz, E. T., and Warner, R. R. (1989). Phenotypic plasticity in life-history traits of female *Thalassoma bifasciatum* (Pisces: Labridae). I. Manipulations of social structure in tests for adaptive shifts of life-history allocations. *Evolution* (*Lawrence, Kans.*) **43,** 1497–1506.

Scott, F. J., and Russ, G. R. (1987). Effects of grazing on species composition of the epilithic algal community on coral reefs in the central Great Barrier Reef. *Mar. Ecol. Prog. Ser.* **39,** 293–304.

Seigel, J. A., and Adamson, T. A. (1983). Batesian mimicry between a cardinalfish (Apogonidae) and a venomous scorpionfish (Scorpaenidae) from the Philippine Islands. *Pac. Sci.* **37,** 75–79.

Sette, O. E. (1960). The long term historical record of meteorological, oceanographic and biological data. *Calif. Coop. Oceanic Fish. Invest. Rep.* **7,** 181–194.

Shaklee, J. B. (1984). Genetic variation and population structure in the damselfish, *Stegastes fasciolatus,* throughout the Hawaiian Archipelago. *Copeia* 1984, 629–640.

Shallenberger, R. J., and Madden, W. D. (1973). Luring behavior in the scorpionfish, *Iracundus signifer. Behaviour* **47,** 33–47.

Shanks, A. L. (1983). Surface slicks associated with tidally forced internal waves may transport pelagic larvae of benthic invertebrates and fishes shoreward. *Mar. Ecol. Prog. Ser.* **13,** 311–315.

Shanks, A. L. (1985). Behavioral basis of internal-wave-induced shoreward transport of megalopae of the crab *Pachygrapsus crassipes. Mar. Ecol. Prog. Ser.* **24,** 289–295.

Shanks, A. L., and Wright, W. G. (1987). Internal-wave-mediated shoreward transport of cyprids, megalopae, and gammarids and correlated longshore differences in the settling rate of intertidal barnacles. *J. Exp. Mar. Biol. Ecol.* **114,** 1–13.

Shapiro, D. Y. (1977). The structure and growth of social groups of the hermaphroditic fish *Anthias squamipinnis* (Peters). *Proc. Int. Coral Reef Symp., 3rd* **1,** 571–577.

Shapiro, D. Y. (1979). Social behavior, group structure, and the control of sex reversal in hermaphroditic fish. *Adv. Study Behav.* **10,** 43–102.

Shapiro, D. Y. (1980). Serial female sex changes after simultaneous removal of males from social groups of a coral reef fish. *Science* **209,** 1136–1137.

Shapiro, D. Y. (1981). Size, maturation and the social control of sex reversal in the coral reef fish *Anthias squamipinnis* (Peters). *J. Zool.* **193,** 105–128.

Shapiro, D. Y. (1983). On the possibility of kin groups in coral reef fishes. *In* "The Ecology of Deep and Shallow Coral Reefs" (M. L. Reaka, ed.), NOAA Symp. Ser. Undersea Res., Vol. 1, pp. 39–46. Natl. Oceanic Atmos. Adm., Rockville, Maryland.

Shapiro, D. Y. (1984). Sex reversal and sociodemographic processes in coral reef fishes. *In* "Fish Reproduction: Strategies and Tactics" (G. W. Potts and R. J. Wootton, eds.), pp. 103–118. Academic Press, Orlando, Florida.

Shapiro, D. Y. (1986a). Intragroup home ranges in a female-biased group of a sex-changing fish. *Anim. Behav.* **34,** 865–870.

Shapiro, D. Y. (1986b). Subgroup independence and group development in a sex-changing fish. *Anim. Behav.* **34,** 716–726.

Shapiro, D. Y. (1987a). Differentiation and evolution of sex change in fishes. *BioScience* **37,** 490–497.

Shapiro, D. Y. (1987b). Inferring larval recruitment strategies from the distributional ecology of settled individuals of a coral reef fish. *Bull. Mar. Sci.* **41,** 325–343.

Shapiro, D. Y. (1987c). Patterns of space use common to widely different types of social groupings of a coral reef fish. *Environ. Biol. Fishes* **18,** 183–194.

Shapiro, D. Y. (1987d). Reproduction in groupers. *In* "Tropical Snappers and

Groupers: Biology and Fisheries Management" (J. J. Polovina and S. Ralston, eds.), pp. 295–328. Westview, Boulder, Colorado.

Shapiro, D. Y. (1988a). Behavioural influences on gene structure and other new ideas concerning sex change in fishes. *Environ. Biol. Fishes* **23,** 283–297.

Shapiro, D. Y. (1988b). Variation of group composition and spatial structure with group size in a sex-changing fish. *Anim. Behav.* **36,** 140–149.

Shapiro, D. Y. (1989a). Inapplicability of the size-advantage model to coral reef fishes. *Trends Ecol. Evol.* **4,** 272.

Shapiro, D. Y. (1989b). Sex change as an alternative life-history style. *In* "Alternative Life-History Styles of Animals" (M. N. Bruton, ed.), pp. 177–195. Kluwer Academic, Dordrecht, The Netherlands.

Shapiro, D. Y., and Boulon, R. H., Jr. (1982). The influence of females on the initiation of female-to-male sex change in a coral reef fish. *Horm. Behav.* **16,** 66–75.

Shapiro, D. Y., and Boulon, R. H., Jr. (1987). Evenly dispersed social groups and intergroup competition for juveniles in a coral reef fish. *Behav. Ecol. Sociobiol.* **21,** 343–350.

Shapiro, D. Y., and Lubbock, R. (1980). Group sex ratio and sex reversal. *J. Theor. Biol.* **82,** 411–426.

Shapiro, D. Y., Hensley, D. A., and Appeldoorn, R. S. (1988). Pelagic spawning and egg transport in coral-reef fishes: A skeptical overview. *Environ. Biol. Fishes* **22,** 3–14.

Shapley, R., and Gordon, J. (1980). The visual sensitivity of the conger eel. *Proc. R. Soc. London, B* **209,** 317–330.

Shaw, E. (1978). Schooling fishes. *Am. Sci.* **66,** 166–175.

Shepard, J. W., and Myers, R. F. (1981). A preliminary checklist of the fishes of Guam and the southern Mariana Islands. *Tech. Rep.—Univ. Guam Mar. Lab.* **70,** 60–88.

Sherman, K. M., Reidenaur, J. A., Thistle, D., and Meeter, D. (1983). Role of a natural disturbance in an assemblage of marine free-living nematodes. *Mar. Ecol. Prog. Ser.* **11,** 23–30.

Sherman, K., Smith, W., Morse, W., Benua, M., Green, J., and Ejsymant, J. (1984). Spawning strategies of fishes in relation to circulation, phytoplankton products, and pulses in zoo-plankton off the northeastern United States. *Mar. Ecol. Prog. Ser.* **18,** 1–19.

Shpigel, M., and Fishelson, L. (1986). Behavior and physiology of coexistence in two species of *Dascyllus* (Pomacentridae, Teleostei). *Environ. Biol. Fishes* **17,** 253–265.

Shpigel, M., and Fishelson, L. (1989). Food habits and prey selection of three species of groupers from the genus *Cephalopholis* (Serranidae: Teleostei). *Environ. Biol. Fishes* **24,** 67–73.

Shulman, M. J. (1984). Resource limitation and recruitment patterns in a coral reef fish assemblage. *J. Exp. Mar. Biol. Ecol.* **74,** 85–109.

Shulman, M. J. (1985a). Recruitment of coral reef fishes: Effects of distribution of predators and shelter. *Ecology* **66,** 1056–1066.

Shulman, M. J. (1985b). Variability in recruitment of coral reef fishes. *J. Exp. Mar. Biol. Ecol.* **89,** 205–219.

Shulman, M. J., and Ogden, J. C. (1987). What controls tropical reef fish populations:

Recruitment or benthic mortality? An example in the Caribbean reef fish *Haemulon flavolineatum*. *Mar. Ecol. Prog. Ser.* **39,** 233–242.

Shulman, M. J., Ogden, J. C., Ebersole, J. P., McFarland, W. N., Miller, S. L., and Wolf, N. G. (1983). Priority effects in the recruitment of juvenile coral reef fishes. *Ecology* **64,** 1508–1513.

Sih, A., Crowley, P., McPeek, M., Petranka, J., and Strohmeier, K. (1985). Predation, competition, and prey communities: A review of field experiments. *Annu. Rev. Ecol. Syst.* **16,** 269–311.

Sikkel, P. C. (1990). Social organization and spawning in the Atlantic sharpnose puffer, *Canthigaster rostrata* (Tetraodontidae). *Environ. Biol. Fishes* **27,** 243–254.

Simenstad, C. A., and Cailliet, G. M. (eds.) (1986). "Contemporary Studies on Fish Feeding." Junk, Dordrecht, The Netherlands.

Sinclair, M., and Tremblay, M. J. (1984). Timing of spawning of Atlantic herring (*Clupea harengus harengus*) populations and the match–mismatch hypothesis. *Can. J. Fish. Aquat. Sci.* **41,** 1055–1065.

Singer, M. M. (1985). Food habits of juvenile rockfishes (*Sebastes*) in a central California kelp forest. *Fish. Bull.* **83,** 531–541.

Sissenwine, M. P. (1984). Why do fish populations vary? *In* "Exploitation of Marine Communities" (R. M. May, ed.), pp. 59–94. Springer-Verlag, Berlin.

Sivak, J. G. (1973). Interrelation of feeding behaviour and accommodative lens movements in some species of North American freshwater fishes. *J. Fish. Res. Board Can.* **30,** 1141–1146.

Slobodchikoff, C. N. (ed.) (1988). "The Ecology of Social Behavior." Academic Press, San Diego, California.

Slobodkin, L. B., and Fishelson, L. (1974). The effect of the cleaner-fish *Laboides dimidiatus* on the point diversity of fishes on the reef front at Eilat. *Am. Nat.* **108,** 369–376.

Smith, C. L. (1973). Small rotenone stations: A tool for studying coral reef fish communities. *Am. Mus. Novit.* **2512,** 1–21.

Smith, C. L. (1977). Coral reef fish communities—Order and chaos. *Proc. Int. Coral Reef Symp., 3rd* **1,** xxi–xxii.

Smith, C. L. (1978). Coral reef fish communities: A compromise view. *Environ. Biol. Fishes* **3,** 109–128.

Smith, G. B. (1979). Relationship of eastern Gulf of Mexico reef-fish communities to the species equilibrium theory of insular biogeography. *J. Biogeogr.* **6,** 49–61.

Smith, S. V. (1978). Coral-reef area and the contribution of reefs to processes and resources of the world's oceans. *Nature (London)* **273,** 225–226.

Smith, S. V. (1983). Net production of coral reef ecosystems. *In* "The Ecology of Deep and Shallow Coral Reefs" (M. L. Reaka, ed.), NOAA Symp. Ser. Undersea Res., Vol. 1, pp. 127–131. Natl. Oceanic Atmos. Adm., Rockville, Maryland.

Smith, C. L., and Tyler, J. C. (1972). Space resource sharing in a coral reef fish community. *Bull. Nat. Hist. Mus. Los Angeles County* **14,** 125–170.

Smith, C. L., and Tyler, J. C. (1973a). Direct observations of resource sharing in coral reef fish. *Helgol. Wiss. Meeresunters.* **24,** 264–275.

Smith, C. L., and Tyler, J. C. (1973b). Population ecology of a Bahamian suprabenthic fish assemblage. *Am. Mus. Novit.* **2528,** 1–38.

Smith, C. L., and Tyler, J. C. (1975). Succession and stability in fish communities of the dome-shaped patch reefs in the West Indies. *Am. Mus. Novit.* **2572,** 1–18.

Smith, C. L., Tyler, J. C., and Stillman, L. (1987). Inshore ichthyoplankton: A distinctive assemblage? *Bull. Mar. Sci.* **41,** 432–440.

Smith, S. V., Chave, K. E., and Kam, D. T. O. (1973). Atlas of Kaneohe Bay: A reef ecosystem under stress. *Univ. Hawaii Sea Grant Program* **UNIHI-Seagrant-TR-72-01,** 1–128.

Smith-Vaniz, W. F., and Allen, G. R. (1991). Manuscript in preparation.

Snodgrass, R. E., and Heller, E. (1905). Shore fishes from the Hopkins–Stanford Galápagos Expedition. *Proc. Wash. Acad. Sci.* **35,** 333–427.

Sorbini, L. (1979). Resultats de la révision des béryciformes et des perciformes generalisés du Monte Bolca. *Studi Ric. Graciment: Terziari Bolca* **4,** 41–48.

Sorbini, L. (1983a). "La Collezione Baja di Pesci e Piante Fossili di Bolca." Mus. Civ. Storia Nat., Verona, Italy.

Sorbini, L. (1983b). Littiofauna fossile di Bolca e le sue relazioni biogeografiche con i pesci attuali: Vicarianza o dispersione? *Bull. Soc. Paleontol. Ital.* **22,** 109–118.

Springer, V. G. (1982). Pacific Plate Biogeography, with Special Reference to Shore-fishes. *Smithson. Contr. Zool.* **367,** 1–182.

Springer, V. G., and Gomon, M. (1975). Revision of the blenniid fish genus *Omobranchus* with descriptions of three new species and notes on other species of the tribe Omobranchini. *Smithson. Contrib. Zool.* **177,** 135 pp.

Springer, V. G., and McErlean, A. J. (1962). A study of the behavior of some tagged south Florida reef fishes. *Am. Midl. Nat.* **67,** 386–397.

Springer, V. G., and Williams, J. T. (1990). Widely distributed Pacific Plate endemics and lowered sea-level. *Bull. Mar. Sci.* **49,** 631–640.

Stamps, J. A. (1988). Conspecific attractions and aggregation in territorial species. *Am. Nat.* **131,** 329–347.

Stamps, J. A., and Krishnan, V. V. (1990). How to acquire a larger territory by settling next to a neighbor. *Am. Nat.* **135,** 527–546.

Stanton, F. G. (1985). Temporal patterns of spawning in the demersal brooding blackspot sergeant *Abudefduf sordidus* (Pisces: Pomacentridae) from Kaneohe Bay (Hawaii). *Proc. Int. Coral Reef Congr., 5th* **5,** 361–366.

Starck, W. A. (1968). A list of fishes of Alligator Reef, Florida with comments on the nature of the Florida reef fish fauna. *Undersea Biol.* **1,** 1–39.

Starck, W. (1971). Biology of the gray snapper, *Lutjanus griseus* (Linnaeus), in the Florida Keys. *Stud. Trop. Oceanogr.* **10,** 12–150.

Starck, W. A., and Davis, W. P. (1966). Night habits of fishes of Alligator Reef, Florida. *Ichthyologica* **38,** 313–356.

Stearns, S. C. (1982). The role of development in the evolution of life histories. *In* "Evolution and Development" (J. T. Bonner, ed.), pp. 237–258. Springer-Verlag, Berlin.

Stearns, S. C., and Koella, J. C. (1986). The evolution of phenotypic plasticity in life-history traits: Predictions of reaction norms for age and size at maturity. *Evolution (Lawrence, Kans.)* **40,** 893–913.

Steele, J. H. (1984). Kinds of variability and uncertainty affecting fisheries. *In*

"Exploitation of Marine Communities" (R. M. May, ed.), pp. 245–262. Springer-Verlag, Berlin.
Steele, J. H. (1985). A comparison of terrestrial and marine ecological systems. *Nature (London)* **313,** 355–358.
Steger, R., and Benis-Steger, B. (1988). Abundance and distribution of piscivorous mantis shrimps around Moorea, French Polynesia. *Proc. Int. Coral Reef Symp., 6th* **2,** 115–118.
Stein, A. (1988). La peche lagonaire dans l'archipel des Tuamotu. *South Pac. Comm., Inshore Fish. Workshop, CPS/Inshore Fish. Res.* **BP81.** 10 pp.
Steinberg, P. D. (1985). Feeding preferences of *Tegula funebralis* and chemical defenses of marine brown algae. *Ecol. Monogr.* **55,** 333–349.
Steinberg, P. D. (1986). Chemical defense and susceptibility of tropical brown algae to herbivores. *Oecologia* **69,** 628–630.
Steinberg, P. D. (1988). Effects of quantitative and qualitative variation in phenolic compounds on feeding in three species of marine invertebrate herbivores. *J. Exp. Mar. Biol. Ecol.* **120,** 221–237.
Steinberg, P. D., and Paul, V. J. (1990). Fish feeding and chemical defenses of tropical brown algae in Western Australia. *Mar. Ecol. Prog. Ser.* **58,** 253–259.
Steneck, R. S. (1982). A limpet–coralline algal association: Adaptations and defenses between a selective herbivore and its prey. *Ecology* **63,** 502–522.
Steneck, R. S. (1983). Escalating herbivory and resulting adaptive trends in calcareous algal crusts. *Paleobiology* **9,** 44–61.
Steneck, R. S. (1985). Adaptations of crutose coralline algae to herbivory: Patterns in space and time. *In* "Paleobiology: Contemporary Research and Applications" (D. F. Toomey and M. H. Nitecki, eds.), pp. 352–366. Springer-Verlag, Berlin.
Steneck, R. S. (1986). The ecology of coralline algal crusts: Convergent patterns and adaptive strategies. *Annu. Rev. Ecol. Syst.* **17,** 273–303.
Steneck, R. S. (1988). Herbivory on coral reefs: A synthesis. *Proc. Int. Coral Reef Symp., 6th* **1,** 37–49.
Steneck, R. S., and Adey, W. H. (1976). The role of environment in control of morphology in *Lithophyllum congestum*, a Caribbean algal ridge builder. *Bot. Mar.* **19,** 197–215.
Steneck, R. S., and Watling, L. (1982). Feeding capabilities and limitations of herbivorous molluscs: A functional group approach. *Mar. Biol.* **68,** 299–319.
Stephens, J. S., and Zerba, K. E. (1981). Factors affecting fish diversity on a temperate reef. *Environ. Biol. Fishes* **6,** 111–121.
Stephens, J. S., Johnson, R. K., Key, G. S., and McCosker, J. E. (1970). The comparative ecology of three sympatric species of California blennies of the genus *Hypsoblennius* Gill (Teleostomi, Blenniidae). *Ecol. Monogr.* **40,** 213–233.
Stephens, J. S., Morris, P. A., and Zerba, K. E. (1984). Factors affecting fish diversity on a temperate reef: The fish assemblage of Palos Verdes Point, 1974–1981. *Environ. Biol. Fishes* **11,** 259–275.
Stephens, J. S., Jordan, G. A., Morris, P. A., Singer, M. M., and McGowen, G. E. (1986). Can we relate larval fish abundance to recruitment or population stability?

A preliminary analysis of recruitment to a temperate rocky reef. *Calif. Coop. Oceanic Fish. Invest. Rep.* **27,** 65–83.

Stevens, C. E. (1988). "Comparative Physiology of the Vertebrate Digestive System." Cambridge Univ. Press, Cambridge, England.

Stevens, G. C. (1989). The latitudinal gradient in geographical range: How so many species coexist in the tropics. *Am. Nat.* **133,** 240–256.

Stevenson, R. A. (1972). Regulation of feeding behavior of the bicolor damselfish (*Eupomacentrus partitus* Poey) by environmental factors. *In* "Behavior of Marine Animals" (H. E. Winn and B. L. Olla, eds.), Vol. 2. Plenum, New York.

Stevenson, D. K., and Marshall, N. (1974). Generalizations on the fisheries potential of coral reefs and adjacent shallow water environments. *Proc. Int. Coral Reef Symp., 2nd* **1,** 147–156.

Stiles, W. S. (1948). The physical interpretation of the spectral sensitivity curve of the eye. *In* "Transactions of the Optical Convention of the Worshipful Company of Spectacle Makers." Spectacle Makers, London.

Stimson, J., Blum, S., and Brock, R. (1982). An experimental study of the influence of muraenid eels on reef fish sizes and abundance. *Hawaii Sea Grant Q.* **4,** 1–6.

St. John, J., Jones, G. P., and Sale, P. F. (1989). The distribution and abundance of soft-sediment meiofauna and a predatory goby in a coral reef lagoon. *Coral Reefs* **8,** 51–57.

Stouder, D. J. (1987). Effects of a severe-weather disturbance on foraging patterns within a California surfperch guild. *J. Exp. Mar. Biol. Ecol.* **114,** 73–84.

Strasburg, D. W. (1953). The comparative ecology of two salariin blennies. Unpublished Ph.D. thesis, University of Hawaii, Honolulu, Hawaii.

Strong, D. R., Jr., Simberloff, D., Abele, L. G., and Thistle, A. B. (eds.) (1984a). "Ecological Communities: Conceptual Issues and the Evidence." Princeton Univ. Press, Princeton, New Jersey.

Strong, D. R., Lawton, J. H., and Southwood, T. R. E. (1984b). "Insects on Plants: Community Patterns and Mechanisms." Blackwell, Oxford, England.

Stroud, G. J., Goldman, B., and Gladstone, W. (1989). Larval development, growth and age determination in the sharpnose puffer *Canthigaster valentini* (Teleostei: Tetraodontidae). *Japan. J. Ichthyol.* **36,** 327–337.

Suchanek, T. H., and Colin, P. L. (1986). Rates and effects of bioturbation by invertebrates and fishes at Enewetak and Bikini atolls. *Bull. Mar. Sci.* **38,** 25–34.

Sutton, D. C., and Clements, K. D. (1988). Aerobic, heterotrophic gastrointestinal microflora of tropical marine fishes. *Proc. Int. Coral Reef Symp., 6th* **3,** 185–190.

Svaetichin, G. (1956). Spectral response curves from single cones. *Acta Physiol. Scand.* **39,** 17–46.

Svejkovsky, J. (1988). Remotely sensed ocean features and their relation to fish distributions. *In* "Marine Organisms as Indicators" (D. F. Soule and G. S. Keppel, eds.), pp. 177–197. Springer-Verlag, New York.

Sverdrup, H. U., Johnson, M. W., and Fleming, R. H. (1942). "The Oceans: Their Physics, Chemistry and General Biology." Prentice Hall, Englewood Cliffs, New Jersey.

Sweatman, H. P. A. (1983). Influence of conspecifics on choice of settlement sites by larvae of two pomacentrid fishes (*Dascyllus aruanus* and *D. reticulatus*) on coral reefs. *Mar. Biol.* **75,** 225–229.

Sweatman, H. P. A. (1984). A field study of the predatory behaviour and feeding rate of a piscivorous coral reef fish, the lizardfish *Synodus englemani. Copeia* **1984,** 187–193.

Sweatman, H. P. A. (1985a). The influence of adults of some coral reef fishes on larval recruitment. *Ecol. Monogr.* **55,** 469–485.

Sweatman, H. P. A. (1985b). The timing of settlement by larval *Dascyllus aruanus*: Some consequences for larval selection. *Proc. Int. Coral Reef Congr., 5th* **5,** 367–371.

Sweatman, H. P. A. (1988). Field evidence that settling coral reef fish larvae detect resident fishes using dissolved chemical cues. *J. Exp. Mar. Biol. Ecol.* **124,** 163–174.

Tachibana, K., Sakaitanai, M., and Nakanishi, K. (1984). Pavoninins: Shark-repelling ichthyotoxins from the defense secretion of the Pacific sole. *Science* **226,** 703–705.

Talbot, F. H. (1965). A description of the coral structure of Tutia Reef (Tanganyika Territory, East Africa), and its fish fauna. *Proc. Zool. Soc. London* **145,** 431–470.

Talbot, F. H., and Gilbert, A. J. (1981). A comparison of quantitative samples of coral reef fishes latitudinally and longitudinally in the Indo-West Pacific. *Proc. Int. Coral Reef Symp., 4th* **2,** 485–490.

Talbot, F. H., and Goldman, G. (1972). A preliminary report on the diversity and feeding relationships of the reef fishes on One Tree Island, Great Barrier Reef system. *Proc. Symp. Corals Coral Reefs* **1,** 425–442.

Talbot, F. H., Russell, B. C., and Anderson, G. R. V. (1978). Coral reef fish communities: Unstable, high-diversity systems? *Ecol. Monogr.* **48,** 425–440.

Tamura, T., and Wisby, W. J. (1963). The visual sense of pelagic fishes, especially the visual axis and accommodation. *Bull. Mar. Sci. Gulf Caribb.* **13,** 433–448.

Tanaka, M. (1985). Factors affecting the inshore migration of pelagic larval and demersel juvenile red sea bream *Pagrus major* to a nursery ground. *Trans. Am. Fish. Soc.* **114,** 471–477.

Targett, N. M., Targett, T. E., Vrolijk, N. H., and Ogden, J. C. (1986). Effects of macrophyte secondary metabolites on feeding preferences of the herbivorous parrotfish *Sparisoma radians. Mar. Biol.* **92,** 141–148.

Tarzwell, C. M. (1970). Thermal requirements to protect aquatic life. J.—Water Pollut. Control Fed. **42,** 824–828.

Tavolga, W. N., Popper, A. N., and Fay, R. R. (eds.) (1981). "Hearing and sound communication in fishes." Springer-Verlag, New York.

Taylor, R. J. (1984). "Predation." Chapman & Hall, New York.

Tegner, M. J., and Dayton, P. K. (1987). El Niño effects on southern California kelp forest communities. *Adv. Ecol. Res.* **17,** 243–279.

Tertschnig, W. P. (1989). Diel activity patterns and foraging dynamics of the sea urchin *Tripneustes ventricosus* in a tropical seagrass community and a reef environment (Virgin Islands). *P.S.Z.N.I. Mar. Ecol.* **10,** 3–21.

Theilacker, G. H. (1986). Starvation-induced mortality of young sea-caught jack mackerel, *Trachurus symmetricus*, determined with histological and morphological methods. *Fish. Bull.* **84,** 1–17.

Theilacker, G., and Dorsey, K. (1980). Larval fish diversity, a summary of laboratory and field research. *UNESCO Intergov. Oceanogr. Comm., Workshop Rep.* **28,** 105–142.

Thistle, D. (1981). Natural physical disturbances and communities of marine soft bottoms. *Mar. Ecol. Prog. Ser.* **6,** 223–228.

Thompson, J. M. (1976). Prey strategies of fishes in evolution and ecology—Or how to stay alive long enough to fertilize some eggs. *Environ. Biol. Fishes* **1,** 93–100.

Thompson, R., and Munro, J. L. (1978). Aspects of the biology of Caribbean reef fishes: Serranidae (hinds and groupers). *J. Fish Biol.* **12,** 115–146.

Thompson, R., and Munro, J. L. (1983a). The biology, ecology and bionomics of the hinds and groupers, Serranidae. *ICLARM Stud. Rev.* **7,** 59–81.

Thompson, R., and Munro, J. L. (1983b). The biology, ecology and bionomics of the jacks, Carangidae. *ICLARM Stud. Rev.* **7,** 82–93.

Thompson, R., and Munro, J. L. (1983c). The biology, ecology and bionomics of the snappers, Lutjanidae. *ICLARM Stud. Rev.* **7,** 94–109.

Thompson, S. M., and Jones, G. P. (1983). Interspecific territoriality and competition for food between the reef fishes *Forsterygion varium* and *Pseudolabrus celidotus*. *Mar. Biol.* **76,** 95–104.

Thomson, D. A., and Lehner, C. E. (1976). Resilience of a rocky intertidal fish community in a physically unstable environment. *J. Exp. Mar. Biol. Ecol.* **22,** 1–29.

Thomson, D. A., Findley, L. T., and Kerstitch, A. N. (1979). "Reef Fishes of the Sea of Cortez" Wiley, New York.

Thorson, G. (1950). Reproductive and larval ecology of marine bottom invertebrates. *Biol. Rev. Cambridge Philos. Soc.* **25,** 1–45.

Thorson, G. (1961). Length of larval life in marine bottom invertebrates as related to larval transport by ocean currents. *Am. Assoc. Adv. Sci. Publ.* **67,** 455–474.

Thresher, R. E. (1976). Field analysis of the territoriality of the threespot damselfish, *Eupomacentrus planifrons* (Pomacentridae). *Copeia* **1976,** 266–276.

Thresher, R. E. (1979). Social behavior and ecology of two sympatric wrasses (Labridae: *Halichoeres* spp.) off the coast of Florida. *Mar. Biol. 53,* 161–172.

Thresher, R. E. (1982). Interoceanic differences in the reproduction of coral-reef fishes. *Science* **218,** 70–72.

Thresher, R. E. (1983a). Environmental correlates of the distribution of planktivorous fishes in the One Tree Reef lagoon. *Mar. Ecol. Prog. Ser.* **10,** 137–145.

Thresher, R. E. (1983b). Habitat effects on reproductive success in the coral reef fish, *Acanthochromis polyacanthus* (Pomacentridae). *Ecology* **64,** 1184–1199.

Thresher, R. E. (1984). "Reproduction in Reef Fishes." TFH Publ., Neptune City, New Jersey.

Thresher, R. E. (1985a). Distribution, abundance, and reproductive success in the coral reef fish *Acanthochromis polyacanthus*. *Ecology* **66,** 1139–1150.

Thresher, R. E. (1985b). Interoceanic differences in the early life history of coral reef fishes: Australia versus the Western Atlantic. *Proc. Ecol. Soc. Aust.* **14,** 1–5.

Thresher, R. E. (1988). Latitudinal variation in egg sizes of tropical and sub-tropical North Atlantic shore fishes. *Environ. Biol. Fishes* **21,** 17–25.

Thresher, R. E., and Brothers, E. B. (1985). Reproductive ecology and biogeography of Indo-West Pacific angelfishes (Pisces: Pomacanthidae). *Evolution* (*Lawrence, Kans.*) **23,** 878–887.

Thresher, R. E., and Brothers, E. B. (1989). Evidence of intra- and inter-oceanic regional differences in the early life history of reef-associated fishes. *Mar. Ecol. Prog. Ser.* **57,** 187–205.

Thresher, R. E., and Colin, P. L. (1986). Trophic structure, diversity and abundance of fishes of the deep reef (30–300 m) at Enewetak, Marshall Islands. *Bull. Mar. Sci.* **38,** 253–272.

Thresher, R. E., and Moyer, J. T. (1983). Male success, courtship complexity and patterns of sexual selection in three congeneric species of sexually monochromatic and dichromatic damselfishes (Pisces: Pomacentridae). *Anim. Behav.* **31,** 113–127.

Thresher, R. E., Sainsbury, K. J., Gunn, J. S., and Whitelaw, A. W. (1986). Life history strategies and recent changes in population structure in the lizard fish genus, *Saurida*, on the Australian Northwest Shelf. *Copeia* **1986,** 876–885.

Thresher, R. E., Colin, P. L., and Bell, L. J. (1989). Planktonic duration, distribution and population structure of western and central Pacific damselfishes (Pomacentridae). *Copeia* **1989,** 420–434.

Tranter, D. J., and George, J. (1969). Nocturnal abundance of zooplankton at Kavaratti and Kalpeni, two atolls in the Laccadive Archipelago. *Proc. Symp. Corals Coral Reefs* **1,** 45–67.

Tribble, G. W., and Nishikawa, H. (1982). An analysis of the diets for four spatially overlapping damselfishes of the genus *Chromis*. *Japan J. Ichthyol.* **29,** 261–272.

Tribble, G. W., Bell, L. J., and Moyer, J. T. (1982). Subtidal effects of a large typhoon on Miyake-jima, Japan. *Publ. Seto Mar. Biol. Lab.* **27,** 1–10.

Tricas, T. C. (1985). The economics of foraging in coral-feeding butterflyfishes of Hawaii. *Proc. Int. Coral Reef Congr., 5th* **5,** 409–414.

Troyer, K. (1984). Microbes, herbivory and the evolution of social behaviour. *J. Theor. Biol.* **106,** 157–169.

Tsuda, R. T., and Bryan, P. G. (1973). Food preferences of juvenile *Siganus rostrata* and *S. spinus* in Guam. *Copeia* **1973,** 604–606.

Tyler, J. C. (1971). Habitat preferences of the fishes that dwell in shrub corals on the Great Barrier Reef. *Proc. Natl. Acad. Sci. U.S.A.* **123,** 1–26.

Tyler, J. C., Johnson, G. D., Nakamura, I., and Collette, B. B. (1989). Morphology of *Luvarus imperialis* (Luvaridae), with a phylogenetic analysis of the Acanthuroidei (Pisces). *Smithson. Contrib. Zool.* **485,** 1–78.

Underwood, A. J., and Denley, E. (1984). Paradigms, explanations and generalizations in models for the structure of intertidal communities on rocky shores. *In* "Ecological Communities: Conceptual Issues and the Evidence" (D. R. Strong, Jr., D. Simberloff, L. G. Abele, and A. B. Thistle, eds.), pp. 151–180. Princeton Univ. Press, Princeton, New Jersey.

Underwood, A. J.,and Fairweather, P. G. (1985). Intertidal communities: Do they have different ecologies or different ecologists. *Proc. Ecol. Soc. Aust.* **14,** 7–16.

Underwood, A. J., and Fairweather, P. G. (1989). Supply-side ecology and benthic marine assemblages. *Trends Ecol. Evol.* **4,** 16–19.

U.S. Department of Commerce (1982a). "Atlas of Pilot Charts for Central American Waters." Environ. Data Inf. Serv., Natl. Oceanic Atmos. Adm., Washington, D.C.

U.S. Department of Commerce (1982b). "World Weather Records 1961–1970," Vol. 3. Nat. Climat. Cent. Natl. Oceanic Atmos. Adm., Washington, D.C.

U.S. Department of Commerce (1989). "Atlas of Pilot Charts of the North Pacific Ocean." Environ. Data Inf. Serv., Natl. Oceanic Atmos. Adm., Washington, D.C.

Vaissiere, R., and Seguin, G. (1984). Initial observations of the zooplankton microdistribution on the fringing coral reef at Aqaba (Jordan). *Mar. Biol.* **83,** 1–11.

Valiela, I. (1984). "Marine Ecological Processes." Springer-Verlag, Berlin.

Van Alstyne, K. L. (1988). Herbivore grazing increases polyphenolic defenses in the intertidal brown alga *Fucus distichus. Ecology* **69,** 655–663.

Van Alstyne, K. L., and Paul, V. J. (1988). The role of secondary metabolites in marine ecological interactions. *Proc. Int. Coral Reef Symp., 6th* **1,** 175–186.

Van Alstyne, K. L., and Paul, V. J. (1990). The biogeography of polyphenolic compounds in marine macroalgae: Temperate brown algal defenses deter feeding by tropical herbivorous fishes. *Oecologia* **84,** 158–163.

Van Blaricom, G. R. (1982). Experimental analyses of structural regulation in a marine sand community exposed to oceanic swell. *Ecol. Monogr.* **52,** 283–305.

Van den Hoek, C., Breeman, A. M., Bak, R. P. M., and van Buurt, G. (1978). The distribution of algae, corals, and gorgonians in relation to depth, light attenuation, water movement and grazing pressure in the fringing coral reef of Curaçao, Netherlands Antilles. *Aquat. Bot.* **5,** 1–46.

Van Valen, L. (1974). Predation and species diversity. *J. Theor. Biol.* **44,** 19–21.

Vermeij, G. J. (1978). "Biogeography and Adaptation: Patterns of Marine Life." Harvard Univ. Press, Cambridge, Massachusetts.

Vermeij, G. J. (1986). Survival during biotic crises: The properties and evolutionary significance of refuges. *In* "Dynamics of Extinction" (D. K. Elliott, ed.), pp. 231–246. Wiley (Interscience), New York.

Vermeij, G. J. (1987). The dispersal barrier in the tropical Pacific: Implications for molluscan speciation and extinction. *Evolution* (*Lawrence, Kans.*) **41,** 1046–1058.

Vermeij, G. J., and Veil, J. A. (1978). A latitudinal pattern in bivalve shell gaping. *Malacologia* **17,** 57–61.

Via, S., and Lande, R. (1985). Genotype–environment interaction and the evolution of phenotypic plasticity. *Evolution* (*Lawrence, Kans.*) **39,** 505–522.

Victor, B. C. (1982). Daily otolith increments and recruitment in two coral-reef wrasses, *Thalassoma bifasciatum* and *Halichoeres bivittatus. Mar. Biol.* **71,** 203–208.

Victor, B. C. (1983a). Recruitment and population dynamics of a coral reef fish. *Science* **219,** 419–420.

Victor, B. C. (1983b). Settlement and larval metamorphosis produce distinct marks on the otoliths of the slippery dick, *Halichoeres bivittatus. In* "The Ecology of Deep and Shallow Coral Reefs" (M. L. Reaka, ed.), Symp. Ser. Undersea Res., Vol. 1, pp. 47–51. Natl. Oceanic Atmos. Adm., Rockville, Maryland.

Victor, B. C. (1984). Coral reef fish larvae: Patch size estimation and mixing in the plankton. *Limnol. Oceanogr.* **29,** 1116–1119.

Victor, B. C. (1986a). Duration of the planktonic larval stage of one hundred species of Pacific and Atlantic wrasses (family Labridae). *Mar. Biol.* **90,** 317–326.

Victor, B. C. (1986b). Larval settlement and juvenile mortality in a recruitment-limited coral reef fish population. *Ecol. Monogr.* **56,** 145–160.

Victor, B. C. (1986c). Delayed metamorphosis with reduced larval growth in a coral reef fish (*Thalassoma bifasciatum*). *Can. J. Fish Aquat. Sci.* **43,** 1208–1213.

Victor, B. C. (1987a). Growth, dispersal, and identification of planktonic labrid and pomacentrid reef-fish larvae in the eastern Pacific Ocean. *Mar. Biol.* **95,** 145–152.

Victor, B. C. (1987b). The mating system of the Caribbean rosy razorfish, *Xyrichtys martinicensis. Bull. Mar. Sci.* **40,** 152–160.

Victor, B. C. (1991). Comparative early life-history and settlement patterns in a pair of congeneric coral-reef fishes. *Mar. Biol.* In press.

Vine, P. J. (1974). Effects of algal grazing and aggressive behaviour of the fishes *Pomacentrus lividus* and *Acanthurus sohal* on coral-reef ecology. *Mar. Biol.* **24,** 131–136.

Virnstein, R. W. (1978). Predator caging experiments in soft sediments: Caution advised. *In* "Estuarine Interactions" (M. L. Wiley, ed.), pp. 261–273. Academic Press, New York.

Vivien, M. L. (1973). Contribution a la connaissance de l'ethologie alimentaire de l'ichthyofaune du platier interne des recifs coralliens de Tulear (Madagascar). *Tethys, Suppl.* **5,** 221–308.

Vivien, M. L., and Peyrot-Clausade, M. (1974). A comparative study of the feeding behavior of three reef fishes (Holocentridae) with special reference to reef cryptofauna as prey. *Proc. Int. Coral Reef Symp., 2nd* **2,** 179–192.

Von Arx, W. W. (1948). The circulation systems of Bikini and Rongelap lagoons. *Trans. Am. Geophys. Union* **29,** 861–870.

von Herbing, I. H. (1988). Reproduction and recruitment in the bluehead wrasse, *Thalassoma bifasciatum,* in Barbados. Unpublished M.S. thesis, McGill University, Montreal, Quebec, Canada.

Voris, H. K., and Voris, H. H. (1983). Feeding strategies in marine snakes: An analysis of evolutionary, morphological, behavioral and ecological relationships. *Am. Zool.* **23,** 411–425.

Walbran, P. D., Henderson, R. A., Jull, A. J. T., and Head, M. J. (1989). Evidence from sediments of long-term *Acanthaster planci* predation on corals of the Great Barrier Reef. *Science* **245,** 847–850.

Wald, G., Brown, P. K., and Brown, P. S. (1957). Visual pigments and depths of habitat of marine fishes. *Nature* (*London*) **180,** 969–971.

Waldner, R. E., and Robertson, D. R. (1980). Patterns of habitat partitioning by eight species of territorial Caribbean damselfishes (Pisces: Pomacentridae). *Bull. Mar. Sci.* **30,** 171–186.

Walls, G. L. (1967). "The Vertebrate Eye and Its Adaptive Radiation." (Reprinted in 1967 by Hafner, New York.)

Walsh, W. J. (1983). Stability of a coral reef fish community following a catastrophic storm. *Coral Reefs* **2,** 49–63.

Walsh, W. J. (1985). Reef fish community dynamics on small artificial reefs: The influence of isolation, habitat structure, and biogeography. *Bull. Mar. Sci.* **36,** 357–376.

Walsh, W. J. (1987). Patterns of recruitment and spawning in Hawaiian reef fishes. *Environ. Biol. Fishes* **18,** 257–276.

Walters, C. J. (1984). Managing fisheries under biological uncertainty. *In* "Exploitation of Marine Communities" (R. M. May, ed.), pp. 263–274. Springer-Verlag, Berlin.

Walters, C. J., and Hilborn, R. (1978). Ecological optimization and adaptive management. *Annu. Rev. Ecol. Syst.* **9,** 157–188.

Warburton, K. (1989). Ecological and phylogenetic contraints on body size in Indo-Pacific fishes. *Environ. Biol. Fishes* **24,** 13–22.

Warner, R. R. (1975). The adaptive significance of sequential hermaphroditism in animals. *Am. Nat.* **109,** 61–82.

Warner, R. R. (1980). The coevolution of behavioral and life history characteristics. *In* "Sociobiology: Beyond Nature–Nurture?" (G. W. Barlow and J. Silverberg, eds.), pp. 151–188. Westview, Boulder, Colorado.

Warner, R. R. (1984). Deferred reproduction as a response to sexual selection in a coral reef fish: A test of the life historical consequences. *Evolution (Lawrence, Kans.)* **38,** 148–162.

Warner, R. R. (1985). Alternative mating behaviors in a coral reef fish: A life-history analysis. *Proc. Int. Coral Reef Congr., 5th* **4,** 145–150.

Warner, R. R. (1987). Female choice of sites versus mates in a coral reef fish, *Thalassoma bifasciatum. Anim. Behav.* **35,** 1470–1478.

Warner, R. R. (1988a). Sex change and the size-advantage model, *Trends Ecol. Evol.* **3,** 133–136.

Warner, R. R. (1988b). Sex change in fishes: Hypotheses, evidence, and objections. *Environ. Biol. Fishes* **22,** 81–90.

Warner, R. R. (1988c). Traditionality of mating-site preferences in a coral reef fish. *Nature (London)* **335,** 719–721.

Warner, R. R. (1990). Resource assessment versus traditionality in mating site determination. *Am. Nat.* **135,** 205–217.

Warner, R. R., and Chesson, P. L. (1985). Coexistence mediated by recruitment fluctuations: A field guide to the storage effect. *Am. Nat.* **125,** 769–787.

Warner, R. R., and Downs, I. F. (1977). Comparative life histories: Growth vs. reproduction in normal males and sex-changing hermaphrodites of the striped parrotfish, *Scarus croicensis. Proc. Int. Coral Reef Symp., 3rd* **1,** 275–281.

Warner, R. R., and Hoffman, S. G. (1980a). Local population size as a determinant of mating system and sexual composition in two tropical marine fishes (*Thalassoma* spp.) *Evolution (Lawrence, Kans.)* **34,** 508–518.

Warner, R. R., and Hoffman, S. G. (1980b). Population density and the economics of territorial defense in a coral reef fish. *Ecology* **61,** 772–780.

Warner, R. R., and Hughes, T. P. (1988). The population dynamics of reef fishes. *Proc. Int. Coral Reef Symp., 6th* **1,** 149–155.

Warner, R. R., and Robertson, D. R. (1978). Sexual patterns in the labroid fishes of the western Carribbean. I. The wrasses (Labridae). *Smithson. Contrib. Zool.* **254,** 1–27.

Warner, R. R., Robertson, D. R., and Leigh, E. G. (1975). Sex change and sexual selection. *Science* **190,** 633–638.

Waser, P. M., and Wiley, R. H. (1979). Mechanisms and evolution of spacing in animals. *Handb. Behav. Neurobiol.* **3,** 159–223.

Wass, R. C. (1982). The shoreline fishery of American Samoa—Past and present. *In* "Ecological Aspects of Coastal Zone Management" (J. L. Munro, ed.), Proc. Semin. Mar. Coastal Processes Pac. pp. 51–83. UNESCO, Jakarta, Indonesia.

Wass, R. C. (1984). An annotated checklist of the fishes of Samoa. *NOAA Tech. Rep.* **NMFS SSRF-781,** 1–43.

Watabe, N., Tanaka, L., Yamada, J., and Dean, J. M. (1982). Scanning electron microscope observations of the organic matrix in the otolith of the teleost fish *Fundulus heteroclitus* (Linnaeus) and *Tilapia nilotica* (Linnaeus). *J. Exp. Mar. Biol. Ecol.* **58,** 127–134.

Watson, W. (1974). Diel changes in the vertical distributions of some common fish larvae in southern Kaneohe Bay, Oahu, Hawaii. Unpublished M.S. thesis, University of Hawaii, Honolulu, Hawaii.

Watson, W., and Leis, J. M. (1974). Ichthyoplankton of Kaneohe Bay, Hawaii: A one-year study of fish eggs and larvae. *Tech. Rep.—Univ. Hawaii Sea Grant Program* **TR-75-01,** 1–178.

Welden, C. W., and Slauson, W. L. (1986). The intensity of competition versus its importance: An overlooked distinction and some implications. *Q. Rev. Biol.* **61,** 23–44.

Wellington, G. M. (1982). Depth zonation of corals in the Gulf of Panama: Control and facilitation by resident reef fish. *Ecol. Monogr.* **52,** 223–241.

Wellington, G. M., and Victor, B. C. (1985). El Niño mass coral mortality: A test of resource limitation in a coral reef damselfish population. *Oecologia* **68,** 15–19.

Wellington, G. M., and Victor, B. C. (1988). Variation in components of reproductive success in an undersaturated population of coral-reef damselfish: A field perspective. *Am. Nat.* **131,** 588–601.

Wellington, G. M., and Victor, B. C. (1989). Planktonic larval duration of one hundred species of Pacific and Atlantic damselfishes (Pomacentridae). *Mar. Biol.* **101,** 557–567.

Welty, J. C. (1934). Experiments in group behavior of fishes. *Physiol. Zool.* **7,** 85–128.

Werner, E. E., and Gilliam, J. F. (1984). The ontogenetic niche and species interactions in size-structured populations. *Annu. Rev. Ecol. Syst.* **15,** 393–425.

West-Eberhard, M. J. (1988). Phenotypic plasticity and "genetic" theories of insect sociality. *In* "Evolution of Social Behavior and Integrative Levels" (G. Greenberg and E. Tobach, eds.), T. C. Schneirla Conf. Ser. Vol. 3, pp. 123–133. Erlbaum, Hillsdale, New Jersey.

Westneat, M. W., and Resing, J. M. (1988). Predation on coral spawn by planktivorous fish. *Coral Reefs* **7,** 89–92.

Whelan, R. J. (1985). Patterns of recruitment to plant populations after fire in Western Australia and Florida. *Proc. Ecol. Soc. Aust.* **14,** 169–178.

White, T. C. R. (1985). When is a herbivore not a herbivore? *Oecologia* **67,** 596–597.

Whittaker, R. H. (1970). "Communities and Ecosystems." Macmillan, New York.

Wickler, W. (1965). Zur Biologie und Ethologie von *Ecsenius bicolor* (Pisces, Teleostei, Blenniidae). *Z. Tierpsychol.* **22,** 36–49.

Wickler, W. (1968). "Mimicry in Plants and Animals." McGraw-Hill, New York.

Wiens, J. A. (1977). On competition and variable environments. *Am. Sci.* **65,** 590–597.

Wiens, J. A. (1989). Spatial scaling in ecology. *Funct. Ecol.* **3,** 385–397.

Wiley, R. H. (1981). Social structure and individual ontogenies: Problems of description, mechanism, and evolution. *Perspect. Ethol.* **4,** 105–133.

Wilkinson, C. R. (1987). Interocean differences in size and nutrition of coral reef sponge populations. *Science* **236,** 1654–1657.

Williams, A. H. (1978). Ecology of threespot damselfish: Social organization, age structure, and population stability. *J. Exp. Mar. Biol. Ecol.* **34,** 197–213.

Williams, A. H. (1979). Interference behavior and ecology of threespot damselfish (*Eupomacentrus planifrons*). *Oecologia* **38,** 223–230.

Williams, A. H. (1980). The threespot damselfish: A noncarnivorous keystone species. *Am. Nat.* **116,** 138–142.

Williams, A. H. (1981). An analysis of competitive interactions in a patchy back-reef environment. *Ecology* **62,** 1107–1120.

Williams, D. McB. (1979). Factors influencing the distribution and abundance of pomacentrids (Pisces: Pomacentridae) on small patch reefs in the One Tree Lagoon (Great Barrier Reef). Unpublished Ph.D. dissertation, University of Sydney, Sydney, Australia.

Williams, D. McB. (1980). Dynamics of the pomacentrid community on small patch reefs in One Tree Lagoon (Great Barrier Reef). *Bull. Mar. Sci.* **30,** 159–170.

Williams, D. McB. (1982). Patterns in the distribution of fish communities across the central Great Barrier Reef. *Coral Reefs* **1,** 35–43.

Williams, D. McB. (1983a). Daily, monthly and yearly variability in recruitment of a guild of coral reef fishes. *Mar. Ecol. Prog. Ser.* **10,** 231–237.

Williams, D. McB. (1983b). Longitudinal and latitudinal variation in the structure of reef fish communities. *In* "Proceedings, Inaugural Great Barrier Reef Conference" (J. T. Baker, R. M. Carter, P. W. Sammarco, and K. P. Stark, eds.), pp. 265–270. James Cook Univ. Press, Townsville, Australia.

Williams, D. McB. (1985). Temporal variation in the structure of reef slope fish communities (central Great Barrier Reef): Short-term effects of *Acanthaster planci*. *Mar. Ecol. Prog. Ser.* **28,** 157–164.

Williams, D. McB. (1986a). Temporal variation in the structure of reef slope fish communities (central Great Barrier Reef): Short-term effects of *Acanthaster planci* infestation. *Mar. Ecol. Prog. Ser.* **28,** 157–164.

Williams, D. McB. (1986b). Spatial and temporal scales of processes determining inter-annual variation in recruitment of fishes of the Great Barrier Reef: Some preliminary data. *Int. Oceanogr. Comm. Workshop Rep., Suppl.* **44,** 229–239.

Williams, D. McB. (1989). Geographical and environmental variation in shallow-water reef fishes of Vanuatu. *In* "Vanuatu Marine Resource Survey, March–April 1988. Final Report" (T. J. Done and K. F. Navin, eds.), Consult. Rep. Aust. Inst. Mar. Sci., Townsville, Australia.

Williams, G. C. (1966). "Adaptation and Natural Selection." Princeton Univ. Press, Princeton, New Jersey.

Williams, D. McB., and English, S. (1991). Geographical ecology of fishes of the Great Barrier Reef. Manuscript in preparation.

Williams, D. McB., and Hatcher, A. I. (1983). Structure of fish communities on outer slopes of inshore, mid-shelf and outer shelf reefs of the Great Barrier Reef. *Mar. Ecol. Prog. Ser.* **10,** 239–250.

Williams, D. McB., and Sale, P. F. (1981). Spatial and temporal patterns of recruitment of juvenile coral reef fishes to coral habitats within "One Tree Lagoon," Great Barrier Reef. *Mar. Biol.* **65,** 245–253.

Williams, V. R., and Clarke, T. A. (1983). Reproduction, growth, and other aspects of the biology of the gold spot herring, *Herklotsichthys quadrimaculatus* (Clupeidae), a recent introduction to Hawaii. *Fish. Bull.* **81,** 587–597.

Williams, D. McB., Wolanski, E., and Andrews, J. C. (1984). Transport mechanisms and the potential movement of planktonic larvae in the central region of the Great Barrier Reef. *Coral Reefs* **3,** 229–236.

Williams, D. McB., Russ, G., and Doherty, P. J. (1986). Reef fishes: Large-scale distributions, trophic interactions and life cycles. *Oceanus* **29,** 76–82.

Williams, D. McB., Dixon, P., and English, S. (1988). Cross-shelf distribution of copepods and fish larvae across the central Great Barrier Reef. *Mar. Biol.* **99,** 577–589.

Winn, H. E., and Bardach, J. E. (1959). Differential food selection by moray eels and a possible role of the mucous envelope of parrotfishes in reduction of predation. *Ecology* **40,** 296–298.

Winterbottom, R., Emery, A. R., and Holm, E. (1989). An annotated checklist of the fishes of the Chagos Archipelago, central Indian Ocean. *Life Sci. Contrib. R. Ontario Mus.* **145,** 1–226.

Wolanski, E., and Hamner, W. M. (1988). Topographically controlled fronts in the ocean and their biological influence. *Science* **241,** 177–181.

Wolf, N. G. (1987). Schooling tendency and foraging benefit in the ocean surgeonfish. *Behav. Ecol. Sociobiol.* **21,** 59–63.

Wolf, N. G., Bermingham, E. B., and Reaka, M. L. (1983). Relationships between fishes and mobile benthic invertebrates on coral reefs. *In* "The Ecology of Deep and Shallow Coral Reefs" (M. L. Reaka, ed.), Symp. Ser. Undersea Res., Vol. 1, pp. 69–78. Natl. Oceanic Atmos. Adm., Rockville, Maryland.

Womersley, H. B. S (1954). The species of *Macrocystis* with special reference to those on southern Australian coasts. *Univ. Calif. Publ. Bot.* **27,** 109–132.

Woodland, D. J. (1983). Zoogeography of the Siganidae (Pisces): An interpretation of distribution and richness patterns. *Bull. Mar. Sci.* **33,** 713–717.

Woodley, J. D., Chornesky, E. A., Clifford, P. A., Jackson, J. B. C., Kaufman, L. S., Knowlton, N., Lang, J. C., Pearson, M. P., Porter, J. W., Rooney, M. C., Rylaardam, K. W., Tunnicliffe, V. J., Wahle, C. M., Wulff, J. L., Curtis, A. S. G., Dallmeyer, M. D., Jupp, B. P., Koehl, M. A. R., Neigel, J., and Sides, E. M. (1981). Hurricane Allen's impact on Jamaican coral reefs. *Science* **214,** 749–755.

Woodroffe, C., and McLean, R. (1990). Micro atolls and recent sea level change on coral atolls. *Nature* (*London*) **344,** 531–534.

Wootton, R. J. (1979). Energy costs of egg production and environmental determinants of fecundity in teleost fishes. *Symp. Zool. Soc. London* **44,** 133–160.

Wootton, R. J. (1984). Introduction: Strategies and tactics in fish reproduction. *In* "Fish Reproduction" (G. W. Potts, and R. J. Wootton, eds.), pp. 1–12. Academic Press, Orlando, Florida.

Wrangham, R. W. (1986). Evolution of social structure. *In* "Primate Societies" (B. B. Smuts, D. L. Cheney, R. M. Seyfarth, R. W. Wrangham, and T. T. Struhsaker, eds.), pp. 282–298. Univ. of Chicago Press, Chicago.

Wright, A., and Richards, A. H. (1985). A multispecies fishery associated with coral reefs in the Tigak Islands, Papua New Guinea. *Asian Mar. Biol.* **2,** 69–84.

Wright, A., Dalzell, P. J., and Richards, A. H. (1986). Some aspects of the biology of the red bass, *Lutjanus bohar* (Forsskal), from the Tigak Islands, Papua New Guinea. *J. Fish Biol.* **28,** 533–544.

Wyatt, J. R. (1982). The distribution, abundance and development of young Jamaican reef fishes. *Res. Rep. Zool. Dep. Univ. West Indies* **6,** 1–111.

Wyatt, J. R. (1983). The biology, ecology and bionomics of the squirrelfishes, Holocentridae. *ICLARM Stud. Rev.* **7,** 50–58.

Wylie, C. R., and Paul, V. J. (1988). Feeding preferences of the surgeonfish *Zebrasoma flavescens* in relation to chemical defenses of tropical algae. *Mar. Ecol. Prog. Ser.* **45,** 23–32.

Wyllie-Echeverria, T. (1987). Thirty-four species of California rockfishes: Maturity and seasonality of reproduction. *Fish. Bull.* **85,** 229–250.

Ydenberg, R. C., and Dill, L. M. (1986). The economics of fleeing from predators. *Adv. Study Behav.* **16,** 229–249.

Yogo, Y. (1986). Protogyny, reproductive behavior and social structure of the Anthiine fish *Anthias* (*Franzia*) *squamipinnis*. *In* "Indo-Pacific Fish Biology: Proceedings of the Second International Conference on Indo-Pacific Fishes, Tokyo" (T. Uyeno, R. Arai, T. Taniuchi, and K. Matsuura, eds.). Ichthyol. Soc. Jpn., Tokyo.

Yoshino, T., and Nishijima, S. (1981). A list of fishes found around Sesoko Island, Okinawa. *Tech. Rep. Sesoko Mar. Sci. Lab.* **8,** 20–87.

Yoshioka, P. M., and Yoshioka, B. B. (1989). A multispecies, multiscale analysis of spatial pattern and its application to a shallow water gorgonian community. *Mar. Ecol. Prog. Ser.* **54,** 257–264.

Yoshiyama, R. M. (1980). Food habits of three species of rocky intertidal sculpins (Cottidae) in central California. *Copeia* **1980,** 515–525.

Yoshiyama, R. M. (1981). Distribution and abundance patterns of rocky intertidal fishes in central California. *Environ. Biol. Fishes* **6,** 315–332.

Young, P. C., Leis, J. M., and Hausfeld, H. F. (1986). Seasonal and spatial distribution of fish larvae in waters over the North West continental shelf of Western Australia. *Mar. Ecol. Prog. Ser.* **31,** 209–222.

Youngbluth, M. J. (1968). Aspects of the ecology and ethology of the cleaning fish, *Labriodes phthirophagus* Randall. *Z. Tierpsychol.* **25,** 915–932.

Zann, L. P., Bell, L., and Sua, T. (1991). A preliminary survey of the inshore fisheries of Upolu Island, Western Samoa. Manuscript in preparation.

Zaret, T. M. (1972). Predators, invisible prey, and the nature of polymorphism in the Cladocera (class Crustacea). *Limnol. Oceanogr.* **17,** 171–184.

Zaret, T. M. (1980). "Predation and Freshwater Communities." Yale Univ. Press, New Haven, Connecticut.

Zaret, T. M., and Kerfoot, W. C. (1975). Fish predation on *Bosmina longirostris*: Body size selection versus visibility selection. *Ecology* **56,** 232–237.

Zeller, D. C. (1988). Short-term effects of territoriality of a tropical damsel fish and experimental exclusion of large fishes on invertebrates in algal turfs. *Mar. Ecol. Prog. Ser.* **44,** 85–93.

Zinmeister, W. J., and Emerson, W. K. (1979). The role of passive dispersal in the distribution of hemipelagic invertebrates, with examples from the tropical Pacific Ocean. *Veliger* **22,** 32–40.

Zucker, W. V. (1983). Tannins: Does structure determine function? An ecological perspective. *Am. Nat.* **121,** 335–365.

INDEX